WITHDRAWN

OCEAN ENGINEERING

**UNIVERSITY OF CALIFORNIA
ENGINEERING AND PHYSICAL
SCIENCES EXTENSION SERIES**

Howard Seifert, Editor · Space Technology
Robert L. Pecsok, Editor · Principles and Practice of Gas Chromatography
Howard Seifert and Kenneth Brown, Editors · Ballistic Missile and Space Vehicle Systems
George R. Pitman, Jr., Editor · Inertial Guidance
Kenneth Brown and Lawrence D. Ely, Editors · Space Logistics Engineering
Robert W. Vance and W. M. Duke, Editors · Applied Cryogenic Engineering
Donald P. LeGalley, Editor · Space Science
Robert W. Vance, Editor · Cryogenic Technology
Donald P. LeGalley and Alan Rosen, Editors · Space Physics
Edwin F. Beckenbach, Editor · Applied Combinatorial Mathematics
Alan S. Goldman and T. B. Slattery · Maintainability: A Major Element of System Effectiveness
C. T. Leondes and Robert W. Vance, Editors · Lunar Missions and Exploration
J. E. Hove and W. C. Riley, Editors · Modern Ceramics: Some Principles and Concepts
J. E. Hove and W. C. Riley, Editors · Ceramics for Advanced Technologies
Joseph A. Pask · An Atomistic Approach to the Nature and Properties of Materials
John F. Brahtz, Editor · Ocean Engineering

OCEAN ENGINEERING

Goals, Environment, Technology

JOHN F. BRAHTZ, Editor

JOHN WILEY & SONS, INC. NEW YORK, LONDON, SYDNEY

Copyright © 1968 by John Wiley & Sons, Inc.

All rights reserved.
No part of this book may be reproduced by any means,
nor transmitted, nor translated into a machine language
without the written permission of the publisher.

Library of Congress Catalog Card Number: 67-30912
GB 471 09580X

Printed in the United States of America

CONTRIBUTORS

JOHN F. BRAHTZ, U.S. Naval Civil Engineering Laboratory and University of California

MILNER B. SCHAEFER, The Department of the Interior

JOHN P. CRAVEN, Deep Submergence Systems Project, U.S. Navy

JAMES G. WENZEL, Lockheed Missiles and Space Company

WILBERT M. CHAPMAN, Van Camp Sea Food Company

JOHN D. ISAACS, Scripps Institution of Oceanography

MYRL C. HENDERSHOTT, Scripps Institution of Oceanography

FRANCIS A. RICHARDS, University of Washington

JOSEPH J. HROMADIK, U.S. Naval Civil Engineering Laboratory

OWEN H. OAKLEY, U.S. Naval Ship Engineering Center

JAMES M. SNODGRASS, Scripps Institution of Oceanography

CAPTAIN GEORGE F. BOND (MC) USN, Deep Submergence Systems Project, U.S. Navy

WILLIAM H. HUNLEY, U.S. Naval Ship Engineering Center

FRED N. SPIESS, Scripps Institution of Oceanography

FRANCIS L. LAQUE, The International Nickel Company

JERRY D. STACHIW, U.S. Naval Civil Engineering Laboratory

FOREWORD

Several of the contributors to this book and members of the Advisory Committee for the University of California State-Wide Lecture Series on Ocean Engineering were directly involved in a previous *Planning Study of Marine Resources* (California and Use of the Ocean), which had been prepared for the California State Office of Planning by the University of California Institute of Marine Resources. At the time, it appeared natural, constructive, and promising to capture as much as possible of the momentum that had been generated at the conclusion of the planning study. This was accomplished by inviting certain members of the study team as well as four members of the California Governor's Advisory Commission on Ocean Resources to participate in the University's Ocean Engineering Lectures Series. By further incorporating the talents of outstanding workers in the field of ocean engineering from industry and the United States Government this book has ultimately evolved.

The work of the authors included appears highly pertinent to the planning goals for the state of California as well as to our national aims and objectives in marine resources development.

David Potter, Chairman

Governor's Advisory Commission
on Ocean Resources

Santa Barbara, California
March 1968

PREFACE

This book, in effect, summarizes the University of California State-Wide Lecture Series on Ocean Engineering, which was presented during the spring of 1966. Responding to a burgeoning interest in the oceans along with demands from professional and industrial groups for corresponding educational opportunities, the University of California Extension was the sponsor.

Specific conduct of the lecture series was the joint responsibility of the Engineering Extension and the Physical Sciences Extension at the University of California, Los Angeles, and the Engineering Extension at the University of California, Berkeley.

Academic and technical policy, including program objectives, scope, and selection of invited speakers, was the primary concern of an advisory committee which had been especially established for the ocean engineering lecture series.

Members of the Ocean Engineering Advisory Committee were appointed from the University staff, industry, and government in order to obtain a balanced consensus of continuing educational needs of the community of postgraduate engineers.

The contributing authors for this book generally have adhered to the same guidelines for purpose, scope, and depth of subject matter as they had observed earlier in presenting their lectures for the state-wide series. Consequently, the subjects for the lectures became the chapter topics for the book. The lecture series and the book were designed to satisfy first-generation informational needs of postgraduate engineers who find themselves operating as long-range planners, managers, technologists, or system designers in the newly developing fields of ocean engineering.

In organizing the book material as a composite work I have sought to exemplify the engineering systems approach, which appears to be an

Preface

indisputable characteristic for this new area of technology. Accordingly, Part 1 treats the *goals* and the *environment* for planning ocean systems and Part 2 treats the *technology* for designing them. It is my sincere hope that this framework for presenting the subject matter will prove to be in accord with the needs of the graduate seminar student or objective reference reader while still serving the requirements of the exploratory reader. Additionally, in writing Chapters 1 and 9, I have sought to relate the planning and design aspects of systems engineering to the subject matter of Parts 1 and 2, respectively.

As State-Wide Coordinator for the Ocean Engineering Lecture Series, I gratefully acknowledge the invaluable suggestions and guidelines provided by fellow members of the Advisory Committee in the conduct of the lecture program. Because these same suggestions and guidelines have been extended to the development of this book, I hope that the written presentation will be as well received as the lectures. The members of the Advisory Committee for the 1966 State-Wide Lecture Series were

Professor Clifford Bell, University of California, Los Angeles
Dr. John F. Brahtz, University of California, Los Angeles, and U.S. Naval Civil Engineering Laboratory, Port Hueneme, California
Captain Wayne J. Christensen (CEC) USN, Commanding Officer and Director, U.S. Naval Civil Engineering Laboratory, Port Hueneme, California
Mr. John C. Dillon, Head, Engineering Extension, University of California, Los Angeles
Professor Leonard Farbar, Head, Engineering Extension, University of California, Berkeley
Captain Robert E. Garrels, USN (Ret.), Physical Sciences Extension, University of California, Los Angeles
Professor John D. Isaacs, Scripps Institution of Oceanography, University of California at San Diego
Professor Joe W. Johnson, University of California, Berkeley
Dr. S. Russell Keim, Institute of Marine Resources, University of California at San Diego
Mr. John W. North, Lockheed Missiles and Space Company, Sunnyvale, California
Mrs. Bernice W. Park, Assistant Head, Engineering Extension, University of California, Los Angeles
Dr. David S. Potter, AC Electronics—Defense Research Laboratories, General Motors Corporation, Goleta, California
Captain Harold L. Tallman, USN (Ret.), Head, Physical Sciences Extension, University of California, Los Angeles

Preface

Mr. James G. Wenzel, Lockheed Missiles and Space Company, Sunnyvale, California

Professor Robert L. Wiegel, University of California, Berkeley,

Finally, I wish to express my appreciation for the patience and continued helpfulness of my fellow authors during the extensive and sometimes arduous period of manuscript preparation and assemblage. I hope that each of these outstanding contributors to this project will share my final satisfaction in our work.

John F. Brahtz

La Jolla, California
March 1968

CONTENTS

PART I SYSTEM PLANNING—THE GOALS AND THE ENVIRONMENT

1	**Introduction (John F. Brahtz)**	1
	1.1 Marine Science and National Policy	1
	1.2 Ocean Engineering	3
	1.3 System Planning	4
2	**Economic and Social Needs for Marine Resources (Milner B. Schaefer)**	6
	2.1 Kinds of Resources and Their Potentials	7
	2.2 Strategic Importance of the Ocean	32
	2.3 Some Problems in the Utilization of Marine Resources	33
3	**Sea Power and the Sea Bed (John P. Craven)**	38
4	**Systems-Development Planning (James G. Wenzel)**	58
	4.1 Background	60
	4.2 Needs	66
	4.3 Systems-Development Planning Methodology	73
	4.4 Application Examples	91
	4.5 Summary	109
5	**The Law of the Sea and Public Policy (Wilbert M. Chapman)**	112
	5.1 The New Technologies	118
	5.2 The Political Situation between 1945 and 1958	126
	5.3 The Law of the Sea Conference of 1958	134
	5.4 The Law of the Sea Conference of 1960	137
	5.5 The Law of the Sea in 1966	138

Contents

6	**General Features of the Oceans (John D. Isaacs)**		157
	6.1	The Natural Large-Scale Systems	157
	6.2	Opportunities for Man's Intervention	179
	6.3	Requirements for Man's Intervention	194
7	**Physical and Hydrodynamic Factors (Myrl C. Hendershott)**		202
	7.1	Salinity and Temperature Distribution; Stratification; Thermohaline Circulation	204
	7.2	The Wind-Driven Circulation	213
	7.3	Waves; Wind Waves; Tsunami; Storm Surges; Tides; Internal Waves; Inertial Motions	226
	7.4	Waves on a Beach	251
	7.5	Acoustic Phenomena	253
8	**Chemical and Biological Factors in the Marine Environment (Francis A. Richards)**		259
	8.1	The Chemical Environment of the Sea: the Nature of Seawater	259
	8.2	The Dissolved Gases in Seawater	267
	8.3	The pH, Alkalinity, and Buffer Capacity of Seawater	274
	8.4	Micronutrients in Seawater	275
	8.5	Other Minor Constituents of Seawater	279
	8.6	Processes Altering the Distribution of Chemical Properties in the Sea in Time and Space	281
	8.7	Some Biological and Chemical Problems of Special Interest to Engineers	295

PART 2 SYSTEMS DESIGN—THE TECHNOLOGY

9	**Introduction (John F. Brahtz)**		307
	9.1	Design of Ocean Systems	307
	9.2	A Reader's Viewpoint	308
10	**Deep-Ocean Installations and Fixed Structures (Joseph J. Hromadik)**		310
	10.1	Planning the Structure Installation	311
	10.2	Sea-Floor Soil Mechanics	317
	10.3	Anchorage Systems	327
	10.4	Submersible Test Units (STU's): Their Installation and Retrieval	328
	10.5	Future Applications	338
	10.6	Long-Range Forecast	346

Contents

11 Vehicles and Mobile Structures (Owen H. Oakley) — 350
  11.1 Surface Ships — 352
  11.2 Floating Platforms — 368
  11.3 Subsurface Craft — 374

12 Instrumentation and Communications (James M. Snodgrass) — 393
  12.1 Instrument Classification — 398
  12.2 Power Sources — 403
  12.3 Electrical Noise in the Sea — 408
  12.4 Expendable Instruments — 410
  12.5 The Ocean as an Electrical Filter — 426
  12.6 The Ball Breaker: A Deep-Water Bottom Signaling Device — 427
  12.7 Bottom Sediment Temperature-Gradient Recorder — 430
  12.8 Bureau of Ships Telemetering Current Meter — 439
  12.9 Acoustic Background Noise — 449
  12.10 Integrating Radiant-Energy Recorder — 451
  12.11 Panel Integrating Radiant-Energy Recorder — 452
  12.12 Integrating Irradiance Meter — 452
  12.13 Dual Radiance Meters: In Situ Recording — 454
  12.14 Communications — 460
  12.15 The "Giant" Buoy — 471

13 Undersea Ambient Environmental Habitation and Manned Operations (George F. Bond) — 478
  13.1 The External Environment — 481
  13.2 The Internal Environment — 481
  13.3 The Habitat — 484
  13.4 The Aquanaut — 485
  13.5 Aquanaut Team Selection and Training — 486
  13.6 Aquanaut Equipment — 488
  13.7 The Future — 490
  13.8 Conclusion — 491

14 Deep-Ocean Work Systems (William H. Hunley) — 493
  14.1 Undersea Mining — 493
  14.2 Offshore Oil-Well Drilling and Completion — 505
  14.3 Diving Work Systems, Salvage, and Recovery — 512
  14.4 Vehicle-Manipulator Systems — 528

15 Oceanographic and Experimental Platforms (Fred N. Spiess) — 553
  15.1 Environmental Considerations — 558
  15.2 General Considerations — 562
  15.3 Craft in Being — 570
  15.4 FLIP — 578

Contents

16 Materials Selection for Ocean Engineering (Francis L. LaQue) — 588
- 16.1 Steels — 588
- 16.2 Guides to Selection of Materials Other than Steel — 602
- 16.3 General Wasting — 603
- 16.4 Crevice Corrosion — 605
- 16.5 Fouling — 608
- 16.6 Velocity Effects — 610
- 16.7 Cavitation Erosion — 611
- 16.8 Galvanic Effects — 613
- 16.9 Selective Corrosion — 616
- 16.10 Stress Corrosion Cracking — 619
- 16.11 Wires and Ropes — 622
- 16.12 Hot Seawater — 622
- 16.13 Atmospheric Corrosion of Nonferrous Metals and Alloys — 624
- 16.14 Biological Deterioration — 625
- 16.15 Fiber-Glass Re-enforced Plastics — 627
- 16.16 Massive Glasses — 627
- 16.17 Concrete — 628
- 16.18 Conclusion — 628
- 16.19 Appendix: Identification of Alloys Referred to in Text — 628

17 Hydrospace—Environment Simulation (Jerry D. Stachiw) — 633
- 17.1 History of Hydrospace—Environment Simulation — 634
- 17.2 Hydrospace—Environment Simulation — 638
- 17.3 Pressure Vessels for the Simulation of Hydrospace — 645
- 17.4 Hoisting Equipment — 685
- 17.5 Hydraulic System for Simulated Hydrospace Pressurization — 687
- 17.6 Instrumentation of Simulated Hydrospace — 693
- 17.7 Testing in Simulated Hydrospace — 698
- 17.8 Summary — 709

PART 1

*SYSTEMS PLANNING—
THE GOALS AND THE
ENVIRONMENT*

CHAPTER 1

Introduction

JOHN F. BRAHTZ

In this book the individual authors have been concerned with the needs, opportunities, environment, and technology for ocean-centered operations. Their combined purpose is to indicate a rationale by which the problem-oriented activities and disciplines of professional engineers, scientists, and program planners can be integrated and polarized for support of the various national and state objectives for use of the oceans.

In Part 1 the specific aim has been to indicate planning opportunities for matching socioeconomic, civil, and military needs with resources of the ocean. Toward this goal, consideration has been given to environmental constraints and significant conflicts that might arise in the planning process.

In introducing Part 1 the writer seeks to indentify both the objectivity and connectivity of the authors' combined works. By doing so it is expected that a systems approach to ocean-centered program and project planning will be unmistakably implied as the underlying purpose. By this approach, a rationale emerges for integrating and polarizing the functions of all participants in a given program or project, including engineers, scientists, planners, managers, and educators. Also, such a systems-planning rationale will be found generally applicable at the several operating levels of the public sector and within individual organizations of the private sector, all serving the national program in marine science.

1.1 MARINE SCIENCES AND NATIONAL POLICY

The Eighty-Ninth Congress in 1966 passed the *Marine Resources and Engineering Development Act* (Public Law 89-454), which has been approved by the President as a new national policy for intensified study of the sea and economic utilization of marine resources for the benefit of mankind.

The new law has provided for the establishment of a National Council on Marine Resources and Engineering Development at the Cabinet level and an advisory Commission on Marine Science, Engineering, and Resources. The Council is chaired by the Vice President and it is additionally composed of five Cabinet members and three heads of other Federal agencies. The Council has statutory responsibility to advise and assist the President in the development of a comprehensive long-range, and coordinated national program in marine science. This responsibility includes the supporting policy planning and coordination of the marine-science activities of eleven Federal agencies. The Council employs a full-time professional staff composed of specialists in ocean sciences, engineering, national security affairs, economics, foreign affairs, and public administration. The activities of this staff are directed by the Council's Executive Secretary, who is appointed by the President.

The *Marine Resources and Engineering Development Act* of 1966 outlines national goals by declaring it to be the policy of the United States to develop, encourage, and maintain a coordinated, comprehensive, and long-range national program in marine science for the benefit of mankind to assist in (1.) protection of health and property; (2.) enhancement of commerce, transportation, and national security; (3.) rehabilitation of our commercial fisheries; (4.) increased utilization of these and other resources.

Related to the above national goals are eight specific objectives which have been identified under the Marine Sciences Act and are cited here:

1. The accelerated development of the resources of the marine environment.

2. The expansion of human knowledge of the marine environment.

3. The encouragement of private investment enterprise in exploration, technological development, marine commerce, and economic utilization of the resources of the marine environment.

4. The preservation of the role of the United States as a leader in marine science and resource development.

5. The advancement of education and training in marine science.

6. The development and improvement of the capabilities, performance, use, and efficiency of vehicles, equipment, and instruments for use in exploration, research, surveys, and the recovery of resources, and the transmission of energy in the marine environment.

7. The effective utilization of the scientific and engineering resources of the United States, with close cooperation among all interested agencies, public and private, in order to avoid unnecessary duplication of effort, facilities, and equipment, or waste.

8. The cooperation by the United States with other nations and groups of nations and international organizations in marine-science activities when such cooperation is in the national interest.

1.2 OCEAN ENGINEERING

In considering the eight objectives cited above, it is important to recognize that an expanded spectrum of ocean technologies is a dominant requirement for fulfilling these objectives. Attending the development of new and unique areas of ocean technology, the need appears for a custodial profession in ocean engineering, which can be expected to emerge under circumstances being fostered through the *Marine Resources and Engineering Development Act* of 1966.

It will be generally incumbent upon the ocean engineering fraternity to assume primary responsibility for planning and managing the design and development of ocean systems. Because of the extremeties and complexities of the environment within which ocean systems must operate, in addition to the multiplicity of disciplines involved, ocean engineers will, as a general practice, resort to advanced and sophisticated methods for the resolution of both planning and design problems.

Because of the timeliness and relative concurrency with which the several government agencies, industrial groups, institutions, and the public have responded to opportunities for developing marine resources, a highly favorable situation prevails for managing marine-science affairs in an effective and efficient manner. Moreover, this situation obtains at practically all levels of organization where marine sciences are of concern in both the public and private sectors of the national economy. For this reason, ocean engineers can hope to realize the full potential of systems engineering methods for efficiency in planning, designing, and managing ocean-centered operations involving men, equipment, and facilities.

Notwithstanding the gradual emergence of new and unique areas of technology for longer-range marine-science objectives, it would appear that existing technologies are in a prime state for being extended to meet the present needs. Similarly, the professional community of civil, mechanical, electronic, mining, metallurgical, and industrial engineers is effectively extending the respective engineering disciplines in their present formulation for purposes of managing the complexities of the oceans and meeting present needs. Transition from the current state of the professional arts to that required for fully satisfying long-range national objectives will be greatly facilitated by educational and research institutions acting under provisions of the *National Sea Grant College and Program Act* of 1966.

1.3 SYSTEM PLANNING

Systems engineering emphasizes the planning and design of systems for implementing operations and functions that will serve the needs of mankind. Ocean-centered operations and the included systems concern the marine environment, which at the present time is in the earliest stages of both scientific and geographical exploration. Therefore, systems engineering for the oceans at the present time is being applied primarily to a wide range of exploratory developments. Further, the exploratory nature of early stage ocean operations suggests the need for special emphasis on carefully and thoughtfully essayed long-range planning, particularly where large commitments of the national resource are involved.

The somewhat sensitive relationship between present planning decisions and the circumstances under which future decisions and operations will be performed is basically due to lack of social and economic infrastructure supporting ocean-centered activities. Hence the systems planner is now as much concerned with the impact of his ocean systems on the socio-economic environment as he is with the problems of matching needs with resources within the environment. Again for this reason, the importance of applying the most advanced professional planning techniques is indicated, rather than sole reliance on the blunt intuition of past experience.

A systems-engineering approach to program and project planning may involve a sequence of decision-making steps by which basic public needs are resolved into corporate or organizational goals and subsequently into specific program objectives. First steps in the planning sequence usually include the identification of basic needs, comparison of needs with corresponding opportunities, and evaluation of potential conflicts which might arise in the ensuing activity. The corporate or organizational planner will necessarily consider the many environmental factors affecting the class of systems within the scope of his marketing interest and accordingly evaluate the over-all performance capabilities of his organization.

In arriving at a set of program objectives that are compatible with corporate goals and recognized operating constraints, the systems planner will have thoroughly reviewed and evaluated pertinent environmental parameters that would be included in the technology, social and economic structure, and the physical characteristics of the ocean itself. A planning operation culminating in a set of viable program objectives would have then set the stage for subsequent systems-design studies which involve specific application of technology to design requirements.

In support of requirements for systems planning as outlined above, Part One of this book has been structured to include (a) treatments of social, economic and military needs; (b) a methodology for resolving needs and objectives into a rational systems development plan; and (c) a comprehensive review of the most significant aspects of the environment for planning purposes.

CHAPTER 2

Economic and Social Needs for Marine Resources

MILNER B. SCHAEFER

The current explosive growth of the world's human population is well known and widely discussed. We now number something over 3 billion, and the population will almost certainly rise to 6 billion by the end of this century. It is to be hoped, and indeed expected, that the human population will be brought into comfortable balance with the world's resources, but before that is accomplished there will be a great many more people to provide for than there are now. In order for the burgeoning population to live healthful and satisfying lives, there is demanded a great quantity of materials, services, and amenities. Indeed, the requirements grow much faster than the population because the world's people demand better standards of living. This pressure is especially evident in the less privileged nations, where a great many people are now living on the margin of existence, with less than minimal means for a healthy life. For example, at least a third of the world's population is poorly nourished, and some 500 million persons are suffering from protein-deficiency diseases that impair both their physical and mental growth, and handicap their productive capacity, which gravely impedes economic and social development.

Rapid population growth, crowding, and urbanization present critical problems not only in the lesser-developed parts of the world, but also in important sections of the advanced countries. For example, in California the population, is expected, because of both reproduction and immigration, to grow from its present level of 18 million to about 30 million by 1980. California's population is, and will continue to be, heavily concentrated in the coastal zone; more than 80% of the State's population is located near the southern and central coastline. This concentration near the margin of the sea is expected to continue, and to increase. The concentration of people in the coastal zone is also generally evident in the United States as a whole. Fifty-two million people live within a 50-mile belt along the coastline of the continental United States. This zone, representing about 8% of the total land area of the United States, is occupied by 29% of our population and contains a vast industrial complex. The population and the industrial development in this coastal strip have been increasing at about 2.5% a year, and the trend is expected to continue.

With the increased crowding of the land, and increased demands for resources to support the population, men are turning increasingly to the sea to satisfy a portion of their needs, not only for extractive materials, such as food, minerals, and water, but also for such "services" as transportation and waste disposal. In addition, particularly for the people living in urban concentrations near the shore, the sea is becoming of increasing importance for recreation and respite from the rigors of life in an urban, industrial society.

We also should not overlook the influence of the sea on our lives through its role in determining weather and climate. This is most evident, of course, near the coast. However, the interaction between the atmosphere and the ocean, operating as a single great heat engine, is a critically important part of the mechanism determining variations in weather and climate of all parts of our planet. Thus extended weather forecasting, which is of considerable economic importance, not to mention the possibility of future weather control, depends very much on the sea.

I will attempt herein to consider in a general fashion the potential of the ocean for supplying some of the important requisites for a human world population of 6 billion people, and to examine the relationships of the sea and its resources to the present and future economic and social needs of the United States. I will consider opportunities and problems of development of some specific resources. Finally, I will touch upon some economic and social problems, both domestic and international, that result from the fuller utilization of the resources of the sea.

2.1 KINDS OF RESOURCE AND THEIR POTENTIALS

Definition of marine resources

The resources of the sea, as one category of natural resource, consist of those properties of the ocean and its contents to which man may apply his activity to increase his welfare. It is important to remember that the term *resource* implies an economic context; that is, the ocean must not only have the physical capability of satisfying certain human needs, but it must be possible for man to attain the benefits while not using up more goods and services than are gained in the process, or than would be expended in securing the benefits in alternative ways. Thus whether a given aspect of the ocean is a resource or not depends both on its physical nature and on the state of our science and technology. For example, for extraction of food from the sea, some of the populations of organisms are economically harvestable now, but others are not, although they will foreseeably become so through improved technology. Resources in this latter category (ones that are not economically exploitable now but foreseeably will be so) may be regarded as *latent* resources. Although the ocean presently provides much to mankind, many of its resources are in the category of latent resources, which, through advances in science and engineering, can satisfy our growing needs.

It is important to note that the resources of the sea include not only the useful things we take out of it, the extractive resources, such as fish, minerals, petroleum, and water, but also the many ways in which we use it to our benefit, without taking anything out of it, such as its use for transportation, for recreation, and for defense. In the case of waste disposal the ocean is a resource because of what we put *into* it. That is, the capacity of the ocean, because of its large volume and its ceaseless mixing, to absorb vast quantities of domestic and industrial wastes is an extremely important resource, especially for coastal communities.

Another way of categorizing the resources of the sea, which is of particular importance in relation to the strategy of their utilization, is based on considering the rate of their renewal. The strategy of use, which often goes under the term "conservation," depends very much upon whether the resources are nonrenewable or renewable (flow) resources. In the latter case there is also a dichotomy of opinion as to whether or not man's use of the resource significantly affects the rate of renewal. Among the renewable resources affected by man's activities, the living resources are in a unique category, because their rate of renewal depends upon the amount of the resource left to perpetuate itself, which in turn depends on the rate of harvesting. These matters have been discussed more fully by Ciriacy-Wantrup (1963) and by Schaefer and Revelle (1959).

Food

The continuing explosive growth of the human population is placing enormous pressures on the food resources of this hungry planet. This is a critical problem that many people and agencies, both national and international, are wrestling with. There has been much discussion of whether the problem will be met by improvements in agriculture or will, perhaps, be met by feeding people from the sea. Careful examination of the problem reveals that the sea is a poor place to look to for the total food requirements of any large sector of the human population, but that it is an excellent source for satisfying the need for a most critical element of the human diet, animal protein. This is because of the fact that, although the total fixation of organic carbon is roughly the same in the sea as on land, the plants of the sea are almost entirely microscopic organisms, with very fast growth rates and short longevity, that are not amenable to economical harvesting, or to culture. The food harvest of the sea consists, and for the foreseeable future will undoubtedly continue to consist, of animals one or more steps above the plants in the food chain. Fortunately, there is really no general lack of food energy (i.e., calories), which is mostly supplied by the carbohydrates of plants, although there is maldistribution of some of the grains and other plant foods which are produced by agriculture in superabundance in some parts of the world. The production of cereals and other plant foods on the land can apparently be greatly increased in many locations. The really critical shortage promises to be for proteins, especially for the animal proteins that are essential for people's health and well being. The sea is a good place to obtain these. Already some 15% of the world's supply of animal proteins comes from the fisheries, the supply is rapidly increasing, and the potential of the sea is large.

A comparison of the potential harvest of the living resources of the sea at the second trophic level above the plants, with the requirements of 6 billion people, is presented in Table 2–1. The potential harvest is calculated from the net primary productivity, assuming transfer coefficients between trophic levels of 10 and 20%, which almost certainly bracket the true value (Schaefer, 1965). The population requirement assumes a per-capita diet of 2500 kcal per day, 80 g per day of total protein, and 15 g per

TABLE 2-1
Productivity of the Sea in Relation to Food Requirements of 6 Billion People

	Calories	Total Protein	Animal Protein
Population Requirement	5.4×10^{18} g-cal/yr	1.8×10^{14} g/yr	3.2×10^{13} g/yr
Total Net Primary Productivity of the Sea	1.9×10^{20} g-cal/yr	2.8×10^{16} g/yr	—
Probable Potential Yield at Second Trophic Level Above Plants	1.9×10^{18} to 7.6×10^{18} g-cal/yr	2.8×10^{14} to 1.1×10^{15} g/yr	2.8×10^{14} to 1.1×10^{15} g/yr
Portion of Need	0.35 to 1.4	1.6 to 6.1	8.7 to 34

day of animal protein. Although a large part of the present sea-fishery harvest is taken at higher trophic levels, a large and increasing share, currently about 40%, consists of herringlike fishes which feed at the second trophic level, and many of which feed at the first trophic level, above the plants (see Fig. 2–3). At this level, as may be seen from Table 2–1, the total potential yield would be inadequate to satisfy the human caloric requirement for 6 billion people. Besides, there are much better sources on land. With respect to total proteins, the potential yield is easily adequate to satisfy the total human requirement, and for animal proteins the potential yield at this trophic level is between 8 and 34 times the estimated requirement. Thus, even though the full potential yield may not be economically attainable, there is reason to believe that the sea can satisfy all or a large portion of the requirement for animal protein. We will see below that the current harvest of the sea fisheries, about 45 million metric tons per year, may, at a conservative estimate, be increased to 200 million metric tons per year, which, with no wastage, would correspond to the full animal-protein requirement estimated in Table 2–1.

In another paper (1965), I have considered the potential harvest of the world's sea fisheries, not only on the basis of similar theoretical calculations, based on rates of formation and transfer of organic material through the marine food chain, but also on what we know about the fishery resources, harvestable by present technology, in various parts of the sea, that are now unutilized or much underutilized. There are very large areas of the ocean, such as parts of the Southern hemisphere, and much of the Indian Ocean, which are scarcely being harvested, where we know from oceanographic studies, as well as direct observations, that there are large potential fishery resources. Even in the areas of the ocean that have been exploited for a long time, there are many fishery resources with large potential which are lying unused. For example, off the Pacific Coast of the United States there are populations of anchovies, hake, jack mackeral, saury, and squid that almost certainly can provide a sustainable harvest equal to the present total fishery production of the United States. Considering this kind of information, I have concluded that, at a conservative estimate, the world fishery production may be increased to 200 million metric tons per year, with no radical developments, such as fish farming or far-out new kinds of fishing gear.

Actually, the world fishery harvest is growing at a very satisfactory pace; it is growing much more rapidly than that of the world's human population. In Fig. 2–1, adapted from my paper cited above, the world harvest of the fisheries (total harvest, both marine and fresh water, being shown by the solid dots, and the marine harvest alone by the open circles) is shown. The sea fishery harvest has almost doubled during the last 10

Economic and Social Needs 11

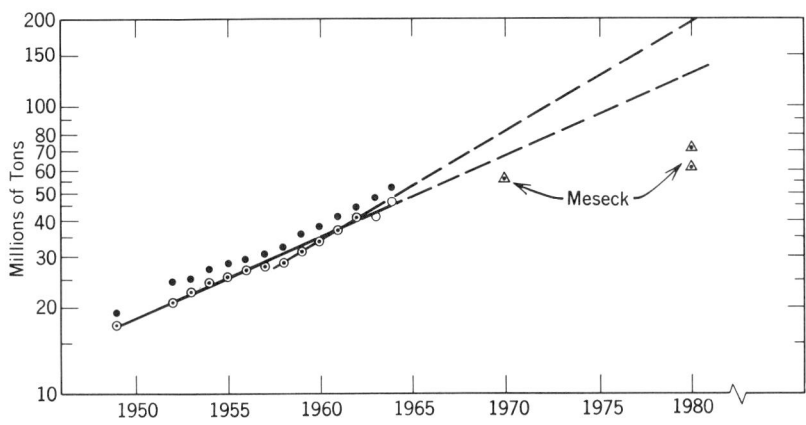

FIG. 2-1. Growth of the world fishery harvest.

years, and the recent growth rate is a bit faster. On Fig. 2–1 are drawn trend lines fitted to the long-term series since 1948, and fitted to the more recent points from 1958 through 1962. (The points for 1963 and 1964 have been added since the graph was originally drawn.) Although extrapolation of trends is a risky business, it does appear that the attainment of an eventual total production of 200 million tons, with foreseeable extension of present technology, is quite reasonable. This production might be attained as early as 1980. Also shown on this figure are some estimates made by Gerhard Meseck as recently as 1962. It looks as if his forecasts were perhaps conservative.

The recent growth of the fisheries, and comparison with a prewar year, are also shown in Fig. 2–2, which indicates the accelerated growth of the sea fisheries. Some of the details regarding the kinds of fish, the geographical regions, and the producers responsible for the recent growth of the fisheries are illuminating. Figure 2–3 shows the production in several broad categories of kinds of fish. Although there is a general upward trend in all categories, the really spectacular increase is in the catching of herrings, sardines, anchovies, and their relatives. As noted above, these are species low in the food chain, which on theoretical grounds should be capable of yielding a higher sustained production than the fishes, higher in the food chain, that feed upon them and their competitors. A good part of this harvest of herringlike fishes is due to the anchovy fishery off Peru which attained a production of 8 million tons in 1964, from almost nothing in 1956. However, there have also been large increases

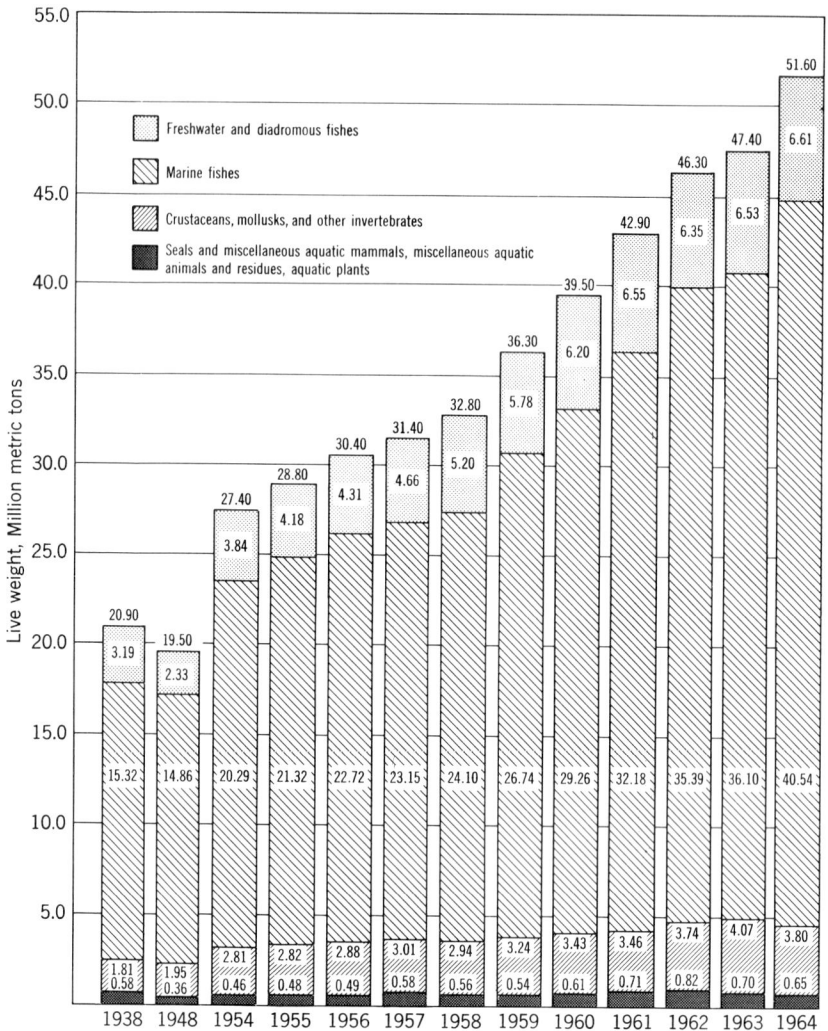

FIG. 2-2. World fisheries production. (From *FAO Yearbook of Fishery Statistics*, Vol. 18, for 1964.)

in the catch of this type of fish in other areas of the ocean, and we know where there are additional large stocks not being utilized.

The major share of the production of the herringlike fishes does not presently go directly to human consumption, but is used for the manufacture of fish meal which is employed as protein food supplements for

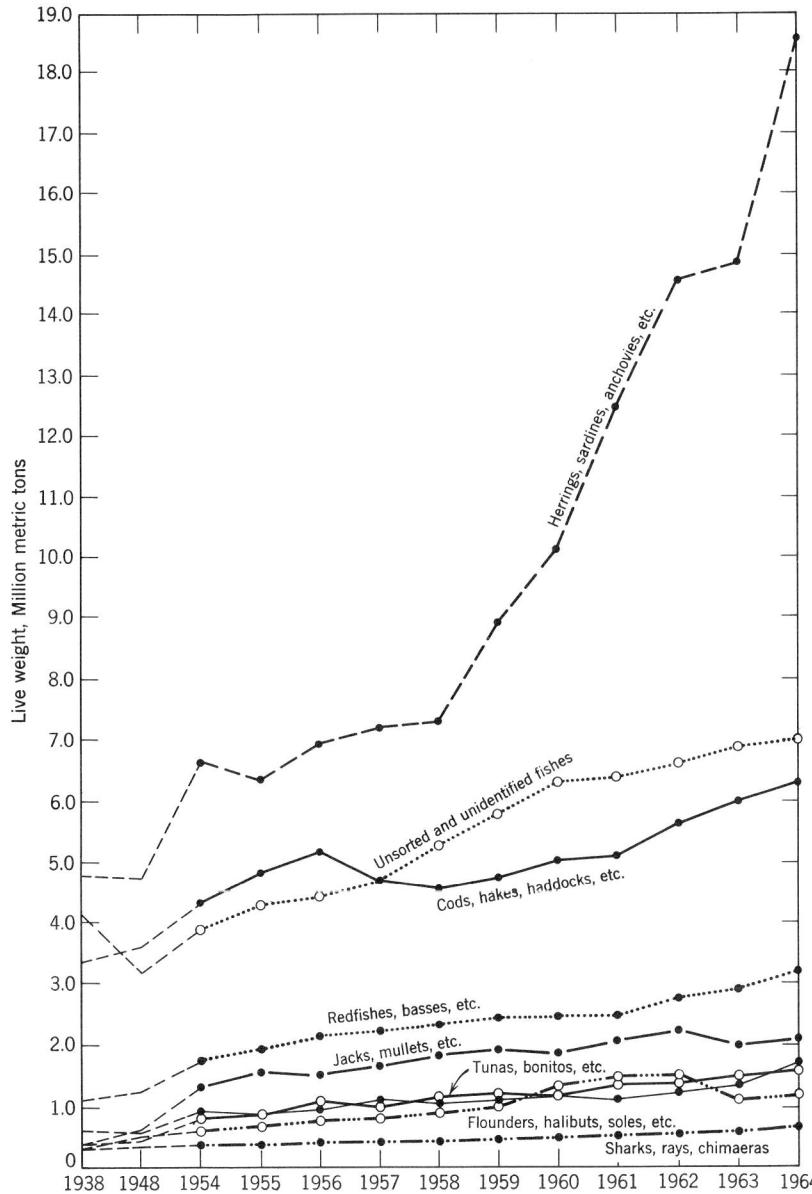

FIG. 2-3. World fisheries production by species groups. (From *FAO Yearbook of Fishery Statistics*, Vol. 18, for 1964.)

poultry and livestock, and thus reaches people indirectly. Since the protein conversion through these animals is quite high, of the order of 25 to 50%, the conversion of fish protein into other forms of animal protein that some people like better is not a terribly inefficient process. However, the direct consumption of these fish proteins by people would be more efficient. Much developmental work is being done toward commercial production of fish flour, and other products for human consumption, in acceptable form, and at low price. This promises to be an extremely important basis both for the expansion of the world fisheries and for the feeding of the undernourished millions in developing countries.

Figure 2–4 shows the fishery production by continents, and from it we may see that the most rapid rates of increase are in the fisheries based on Asia, South America, and the USSR. The increase in the Asian production is partly due to the increase in the Chinese production (shown in Fig. 2–5) through 1960, the data of which are suspect. Another considerable element is the expanding fishery of Japan, which fishes throughout nearly all the world ocean, although there have also been notable increases in the local fisheries of India, Thailand, and some other countries of Southeast Asia. The spectacular increase in the fisheries of South America are largely due to the burgeoning fishery of Peru, but there have also been notable increases elsewhere, for example in Chile and in Argentina. These fisheries, except for the fishery off South Africa, and some of the distant-water fisheries of Japan, are the beginning of the exploitation of the resources of the Southern Hemisphere, which until recent years were scarcely utilized at all. For example, as recently as 1954 the temperate zone of the Southern Hemisphere was producing only 6% of the total world production, whereas it is now producing about 30%. Who is taking the harvest is revealed in Fig. 2–5, which shows the recent trends of landings by the leading fishing nations. It is to be seen that Peru, Japan, and the USSR have rapidly increased their production, whereas production in the United States has run along about level for the last decade. This is not, as is commonly supposed, due to lack of growth of the market in the United States. From 1948 to 1964 the use of fish has increased from 2.8 million metric tons, round weight, to 6.0 million metric tons, round weight. Present use is about 63.5 pounds per capita per year, or nearly double what it was in 1948. A large portion of this increase is due, however, not to direct consumption of the fish and other marine organisms but to the conversion of fish into poultry and livestock.

So far I have not mentioned fish farming, which is a very popular subject nowadays. The culture and rearing of marine organisms is, as a matter of fact, a promising area of development in certain localities. The rearing of sessile organisms, such as oysters, clams, and mussels, has been carried

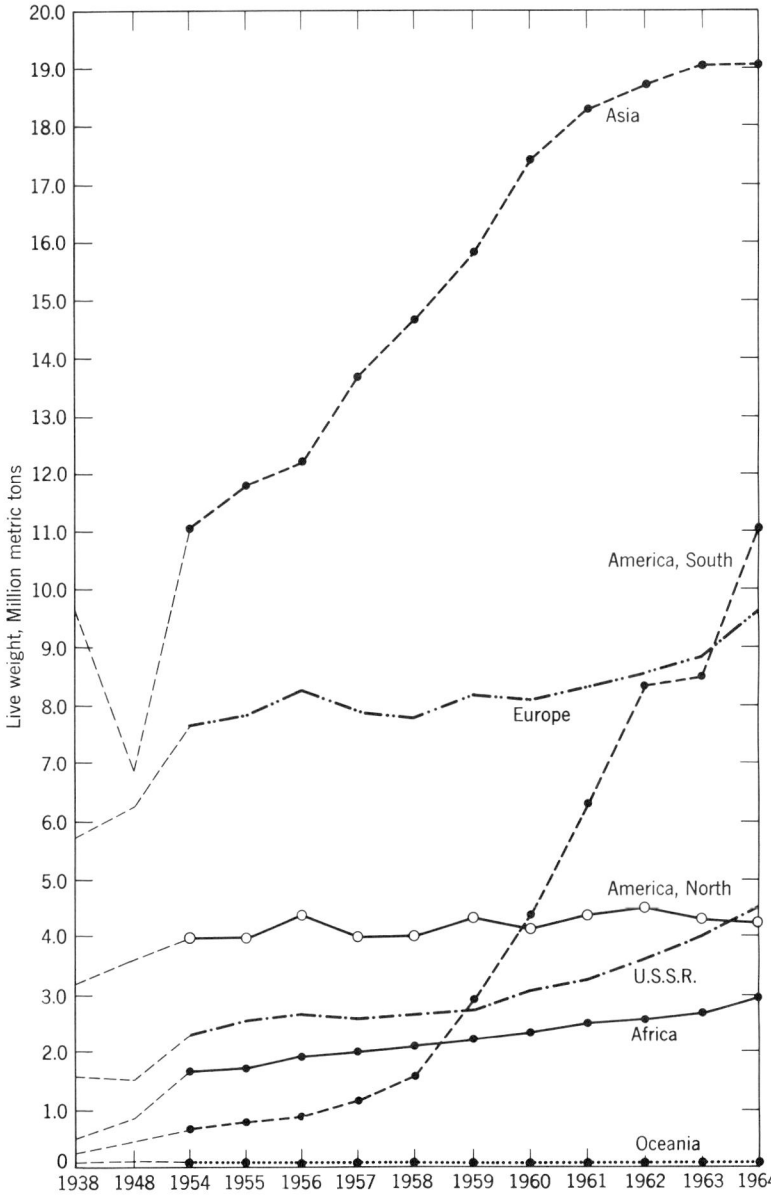

FIG. 2-4. World fisheries production by continents. (From *FAO Yearbook of Fishery Statistics*, Vol. 18, for 1964.)

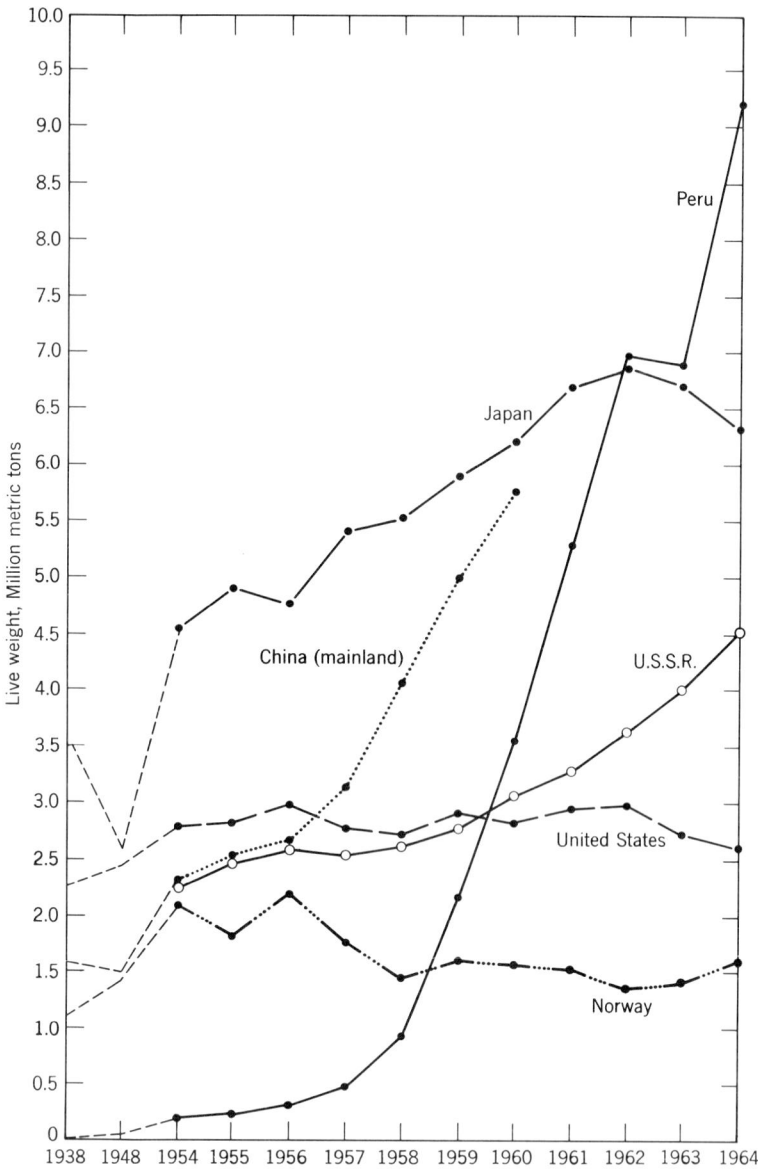

FIG. 2-5. World fisheries production by countries. (From *FAO Yearbook of Fishery Statistics*, Vol. 18, for 1964.)

on, both in the United States and elsewhere in the world, for centuries. Rearing of certain kinds of fish, and shrimp, in enclosed ponds or embayments is also an ancient industry. Certainly the development of fish culture in inshore waters, employing organisms which, at least in their adult stages, are sessile, or are amenable to impoundment, offers great promise for further development. However, in the vast reaches of the open sea, aquaculture is not likely to prove feasible for yet a very long time. On the other hand, it is perfectly feasible to manage the fisheries of the open sea, by selective harvesting, to encourage the maximum production of the kinds of fish we want, and to suppress the "weeds." The management of the high-seas fisheries is therefore much more nearly analogous to range management than to agriculture. I believe that this form of management has larger payoffs, in terms of total production, than inshore fish culture, although the latter will certainly continue to be of growing importance for the production of particularly desirable and high-priced marine products.

Minerals

The sea water and the sea bottom constitute a vast storehouse of many minerals. However, only some of these may be economically recovered now or in the foreseeable future. Some of the minerals both from the sea water and sea bottom are now being extracted in large quantities, and others can certainly become economical with improved technology.

Minerals in solution. Some of the minerals dissolved or suspended in sea water, in terms of the per capita share for a human population of

TABLE 2-2
Some Minerals in Seawater,
Per Capita, for 6 Billion People

Mineral	Amount
Water	2×10^8 tons
Salt (sodium chloride)	6×10^6 tons
Magnesium	2×10^5 tons
Calcium	9×10^4 tons
Potassium	9×10^4 tons
Bromine	1×10^4 tons
Aluminium	200 tons
Manganese	2 tons
Copper	460 pounds
Silver	140 pounds
Gold	3 pounds

6 billion, are shown in Table 2–2. The sea water will obviously continue to be an excellent source of salt, potash, magnesium, and bromine, which are being extracted now on a large scale. It is also obviously a remarkably good source of salt water, and, as we will see below, of fresh water under some circumstances. For such things as calcium and aluminum, although present in sea water in large quantities, there are much better sources on the land or on the sea floor. There is a possibility that future scientific discoveries may make possible the economical extraction of some of the rarer elements, as discussed by McIlhenny and Ballard (1963), but the significant mineral resources would seem to continue to be much as they are at present.

Deposits of the continental shelf. In a general but somewhat oversimplified sense the material of the earths crust may be divided into two kinds, the oceanic (heavy basaltic) and continential (lighter granitic). Beneath this is the solid "mantle" 5 or 6 km below the oceans and 10 to 12 km below the continents. The principal underwater margins of the continental masses are the continental slopes, the upper edges of which are 600 or more feet below the sea surface. From here to the dry continental shores are gentle slopes, the continental shelves. Their width varies from virtually nothing to several hundred miles, and they differ widely in their configuration and features. Large areas of the continental shelf are covered with sediments from various sources, but there are also outcrops of bare rock. Similarly, in the deep sea, much of the basement rock is covered with thick sediments both of geological and biological origin.

The continental shelf is simply the drowned margin of the continent, and much of it has been dry land, in past ages when the sea level was lower because much water was locked up in glacial ice sheets. The continental shelf, from a geological point of view, is essentially similar to the continent, and the types of mineral deposits found in the continental block may be expected in the shelf. The types of mineral deposits found or predicted for the shelf are (a) drowned land-formed placers, such as the cassiterite in drowned river bars off Indonesia, the diamond deposits off West Africa, and gold deposits off Alaska; (b) placers formed under water; (c) chemical rocks, such as the phosphorite nodules formed on the continental shelf off Southern California; (d) undersea extensions of land vein deposits, such as the coal mines in the shelf off Chile. Among the sedimentary deposits on the shelf, the economically most important at the present time is, of course, petroleum and natural gas. Whether petroleum will be found in the sediments in the deep sea is somewhat of an open question, although many geologists believe that it is unlikely for several reasons. Many of the organic source materials are oxidized before they can be incorporated into the bottom sediments, due to the slow rate of sedimentation in deep water; the thick sedimentary section necessary to provide source bed and entrapment for extensive deposits is not present; and widespread vulcanism in the deep sea would volatilize hydrocarbon in the surrounding sediments. However, as yet we know too little about the origin and mode of formation of petroleum to be certain of this.

Since the continental shelf is, as pointed out above, simply a drowned part of the continent, its mineral and petroleum reserves should perhaps be regarded as simply an extension of the reserves on land. The continental shelf, to a depth of 200 m, under the sea, has an area of 27.4×10^6 km^2, or approximately one-fifth of the area of the land. Thus, the mineral deposits of the continental shelf may be expected to increase the reserves on land by about 20%.

The rapid exploration and development of suboceanic petroleum deposits is well known. It has been reported that within the next decade the world petroleum industry plans to spend something like 2.5 billion dollars in offshore development, not including actual production. It is estimated that the U.S. petroleum industry is currently sponsoring private geophysical exploration at sea to the tune of nearly a million dollars a day, and that it is also paying the government about 250 million dollars a year for offshore leases and royalties. Petroleum fields in waters as deep as 300 or 400 feet, 50 miles from land, are today producing in the Persian Gulf, off Saudi Arabia, off Venezuela, Trinidad, and off the coast of the Gulf States and California. Fields are now under development off the coast of Alaska, Oregon, and Washington, and under the North Sea. The capability now exists to drill and operate anywhere on the continential shelf, and probably much deeper if necessary.

Productive operations for the mining of submarine deposits of tin, diamonds, gold, and iron in various parts of the world have been well advertized. It is less widely known that nearshore submarine deposits of sand and gravel are becoming increasingly important sources of these construction materials at a number of places in the United States and elsewhere.

Deep-sea mineral deposits. The sea bottom under the deep sea, beyond the continental slope, probably does not contain the common metaliferous deposits of copper, lead, and zinc, which are associated with granitic rocks, because the deep ocean basins are underlain by basaltic rocks. However, ultramafic (or ultrabasic) rocks on land produce much of the world supply of nickel, chrome, and platinum, so there may be deposits of these things in the basement rocks of the deep sea. In the ocean basins a cover of sediment blankets these basement rocks, except where volcanic seamounts emerge, so that exploration for primary deposits of these metals would be very difficult. The exploitable mineral deposits are therefore probably primarily the materials of the sediments, and chemically formed material at or near the surface, such as the ferro-manganese nodules.

The sedimentary deposits and precipitates on the deep-sea floor contain vast quantities of materials which may be useful to our present and future populations. Some of the more important of these are listed in Table 2–3, which is abstracted from Mero (1963). Red clay covers about 40 million square miles of the ocean floor. It is a very lean ore for aluminum, copper, and some other elements. It may become important in the distant future when the ores of higher grade on land run out, but that seems a long time hence. The calcareous oozes, consisting of the skeletons of Globigerina and some other organisms, cover some 50 million square miles of ocean floor. These oozes are nearly pure calcium carbonate, and compare favorably with ASTM Types I and II cement rock on land. They may become of importance in the future, just as the carbonate deposits of shallower waters, such as oyster shell, are commercially important in some locations now. The diatomaceous oozes, which cover some 11 million square miles of sea floor, are almost pure silica, and could serve the same purposes for which diatomaceous earth is now used. Again, this utilization seems somewhat remote because we still have large quantities on land.

The most promising deposits of the deep-sea floor appear to be the manganese nodules. These concretions were discovered nearly a hundred years ago by the famous Challenger Expedition, and surveys during the International Geophysical Year, and since, have revealed that they cover large areas of the sea bottom, and over large areas of the Central Pacific occur in concentrations of 50 thousand tons per square mile. Much interesting information concerning the occurrence of these nodules, their mineral content, and theories concerning their mode of formation is given by Menard (1964) and Mero (1965).

Pictures of some of these nodule deposits, giving some idea of what they look like, are shown in Figs. 2-6–2-8.

Mero (1963) estimates that on the floor of the Pacific there are about a trillion tons of these nodules, and he has estimated their average mineral composition from samples taken at widely scattered locations. From these data, I have prepared Table 2-4, which shows the quantity of some of the elements in these ferro-manganese nodules, on a per capita basis for 6

TABLE 2-3
Materials of Possible Economic Interest on the Deep-Sea Floor

Material	Tonnage Estimates	Elements of Interest
Red Clay	10^{16}	Cu, Al, Co, Ni
Calcareous ooze	10^{16}	$CaCo_3$
Diatomaceous ooze	10^{13}	SiO_2
Manganese Nodules	10^{12}	Mn, Cu, Co, Ni, Mo, V

FIG. 2-6. Ferro-manganese nodules, 5 to 10 cm diameter, on sea bottom in 3778 m depth near 20°N 114°W.

TABLE 2-4

Some Elements in Ferro-Manganese Nodules on the Floor of the Pacific Ocean, Per Capita, for 6 Billion People

Element	Amount
Manganese	60 tons
Iron	35 tons
Aluminium	7 tons
Nickel	2 tons
Copper	1 ton
Cobalt	1 ton

FIG. 2-7. Ferro-manganese nodules, 2 to 5 cm diameter, in sea bottom in 4104 m depth, near 20°N 120°W.

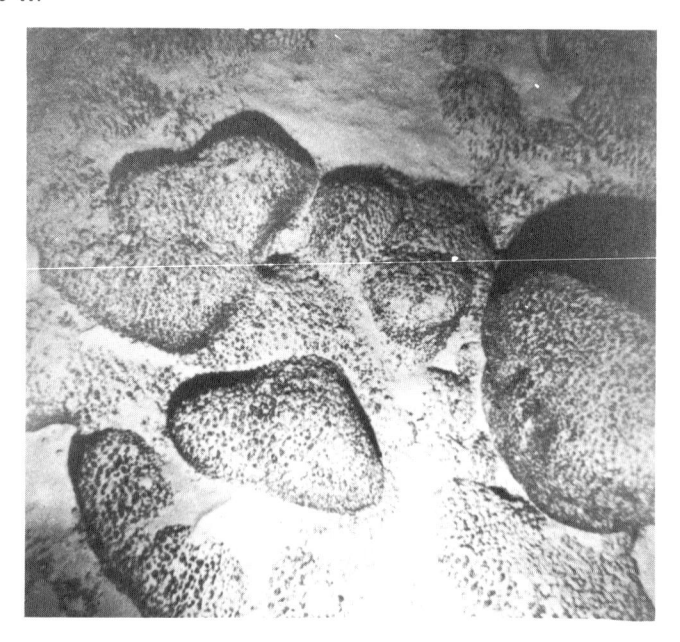

FIG. 2-8. Ferro-manganese deposits on Sylvania Guyot, 1350 m depth, near 12°N 165°E.

billion people. It is evident that, if they can be economically mined, they are important sources of manganese, nickel, copper, and cobalt. For iron and aluminum, of course, there are much better terrestrial sources for the foreseeable future. They are to be regarded as low-grade ores for nickel, copper, and cobalt, and perhaps for manganese. Interestingly enough, the relative abundance of the different metals in the nodules shows marked systematic geographical variations. For example, the nickel, cobalt, and copper content tends to be higher in certain parts of the Central Pacific than elsewhere, and in this region the cobalt content increases in the relatively shallower water on the topographic highs. Conversely, the manganese content is high and the content of cobalt, nickel, copper, and lead is low in certain locations in the Eastern Pacific, such as the mouth of the Gulf of California. With increased knowledge of the deposits in various parts of the ocean, and with advances in technology for their harvesting and processing, and in the face of the decreasing average grade of the ores extracted on land, it is highly probable that the ferro manganese nodules will become economically important.

One interesting aspect of the ferro-manganese nodules is that they are continually being formed in the deep sea. Although the rate of accumulation is probably, on the average, less than a millimeter per thousand years, the total area over which they are being precipitated is so large that the rate of accumulation of some of the metals, including manganese, cobalt, and nickel, is as large as, or larger than, the present total world consumption. Thus the ferro-manganese nodules consitute a renewable mine.

Fresh water

People in California are particularly aware of water as a critical resource. Several regions in the United States could become short of water in the next few decades, but none could be as short as Southern California. Within the borders of California precipitation will provide less than 20% of the water needed; the rest will have to be imported or produced from the sea. The most convenient source, and the most politically popular with other regions for Southern California, is the sea. The question is, of course, can the water be extracted economically? California is only one example of many parts of the world facing a growing water shortage. As we have already seen, the quantity of water available in the ocean is fantastically large, so the question of its utilization comes down simply to technology and economics.

The economics of water is an extremely complex field into which I scarcely dare to venture. However, it is useful to look at per capita cost for various uses of several sources of fresh water, as set forth in Table

TABLE 2-5

Approximate Prices of Water for Different Uses, from Various Sources

Source	Price of [a] Irrigation Water Per Capita Year	Price of [b] Drinking Water Per Capita Year	Price of [c] Domestic and Industrial Use Per Capita Year
Shallow pumped ground water	$ 6.00	0.0005	$ 0.30
Forage crops irrigation water	15.00	0.01	0.75
Wholesale irrigation water, Southern California	36.00	0.03	1.80
City domestic water	400.00	0.30	20.00
Desalinated seawater (Navy stills)	1,500.00	1.10	75.00
Desalinated seawater (attainable low)	300.00	0.25	15.00

[a] Required to raise food for one person, 3 acre feet/year.
[b] Drinking and cooking, one ton per capita per year.
[c] All municipal uses 1/6 acre foot per capita per year.

2-5. From this, it may be seen that the feasibility of using desalinated seawater depends very much on the purpose for which it is used. Looking at the second column, it may be seen that for water for drinking and cooking there is no real economic problem with any source. For domestic and industrial uses, in the third column, it may be seen that, using ordinary Navy stills, the added cost of producing required water, above the present cost of city domestic water, is only about $55.00 per person per year. Technological development to reduce the cost to the indicated attainable low is purely an industrial problem, on which great progress is being made. I have used an attainable low of 30 cents per thousand gallons, which is apparently easily within reach of present technology. However, this does not solve the agricultural problem, nor does it bring it anywhere near solution.

Development of large-scale desalination of seawater, using nuclear power, is progressing rapidly. Studies by an Interagency Subcommittee of the Office of Science and Technology (1964), indicated that large nuclear plants, producing both power and fresh water should be able to produce water at the plant for a cost in the neighborhood of 20 to 25 cents per thousand gallons, and delivered to nearby reservoirs for about 30 cents per thousand gallons, or less. More recently, a specific study made by a consulting engineering firm for the Metropolitan Water District (MWD) of Southern California, found that a large nuclear desalting – power plant, on an artificial island off the coast of Southern California, would be capable of producing fresh water at about 22 cents a thousand gallons, and delivering to the MWD distribution system would cost an additional 5 cents per thousand gallons.

It is to be emphasized that the production of fresh water from seawater appears to be economically feasible only along the coastal zone, because delivery to the interior, especially where lifting the water over hills or mountains is involved, would incur costs for power for pumping larger than the cost of the desalination. Even if the Pacific Ocean were fresh water, it would appear infeasible with the present technology to deliver the water to the Imperial Valley of California for economical use for agriculture. Until our science and technology develops extremely cheap sources of energy, such as energy from nuclear fusion, I do not believe that seawater desalination will have much impact on agriculture.

Power

Another of the very important requirements for modern man is power, cheap and abundant power being the key to the utilization of many other resources. At the present time the average power requirement in the United States is 5 kWh per capita per day. In California the rate of use is more than double the national average. The use of power is increasing in all parts of the United States, and in other parts of the world, at a rapid rate. If we assume that a population of 6 billion people at the end of this century will have per capita power requirements equal to the present average in the United States, the total power requirement will be 1.2×10^6 MW. The ocean does not seem to be a very practical source for the satisfaction of any significant share of this need.

Tidal power has been frequently considered. Yet the total dissipation of tidal energy, according to estimates by Isaacs and Schmitt (1963) is only about equal to the power requirement above estimated, and the quantity that is feasible to be developed is only a small fraction of the total, because a feasible situation involves an especially favorable configuration of the adjacent land, and a large tidal range. One such plant is under construction in the Rance Estuary in France. It will deliver, according to Duport et al. (1965) about 70 MW, and is therefore not a very large power plant. Plans to develop the tidal power potential at Passamaquoddy, between Maine and Canada, have been on-again off-again for a number of years. Favorable locations are, however, relatively few. Similarly, wind power and wave power (which is generated by the wind) amount in total to about the same order of magnitude as tidal power, and an extremely small fraction is feasible to use.

A large potential source of energy in the sea is the temperature difference between surface waters and deeper waters, and the temperature difference at the surface between areas of upwelling and adjacent non-upwelling zones. Theoretically, this would be capable of satisfying over a thousand times the needed power. Practically, however, utilization of this power source is not likely to be competitive with other sources except in very favorable locations, where thermal gradients are large and near to where power is needed. One power plant to operate on this principle, near Abidjan in West Africa, has been under development for a number of years, and is yet to get into productive operation. However, in special circumstances this may be a significant source of useful power, and would seem to be a fruitful area for engineering research.

The greatest potential sources in the sea for generation of energy are materials for atomic fission and fusion. Isaacs has pointed out that the thorium and uranium in the world ocean could, in principle, supply a power requirement an order of magnitude larger than I have estimated above for some 700,000 years, and the deterium and hydrogen could supply fusion power for times greater than the age of the solar system. However, for the foreseeable future there are better sources of these elements on land than in the sea.

It is probable that the principal use of the sea in relation to the generation of power will, at least during the next several decades, continue to be, as it is now, a source of cooling water for power plants using either fossil fuels or nuclear fission as power sources. In the typical power plant, nearly half of the energy of the fuel is lost as waste heat, which requires an efficient cooling system to remove it. For power plants located along the coast, the sea is an excellent source of cooling water for this purpose.

Transportation

The United States, like many other nations, depends on its foreign trade for many of the raw materials required in its commerce and industry, and also exports large quantities both of raw materials and manufactured goods. The United States is by no means self-sufficient; it requires the import of such raw materials as asbestos, tin, manganese, iron ore, bauxite, cobalt, nickel, chromite, and industrial diamonds. Wood and petroleum have shifted from net exports to net imports. Our imports of other essential raw materials and food products are rising steadily. Although visible exports still exceed imports, a continuing increase in exports is necessary for the maintenance and growth of our economy.

The sea is the major highway for the international transportation of heavy or bulky materials, and it will undoubtedly continue to be so for the indefinite future.

Ocean-borne transportation is even more important to many other countries, such as Japan, whose economy is almost totally dependent on importation of raw materials. It is particularly important to the less developed countries, because they are very largely dependent for their economic development on the overseas sales of raw materials and agricultural products, and on the importation of heavy machinery for industrialization, both of which are largely waterborne.

A recent study by an Inter-Agency Maritime Task Force (1965) indicates that the value of ocean-borne foreign commerce of the United States in 1966 will be 30 billion dollars, and is expected to rise by 1975 to 41 billion. The volume of foreign ocean-borne cargoes will be about 340 million long tons in 1966 and will rise to 471 million long tons by 1975. In addition, domestic ocean-borne commerce will amount to 167 million long tons in 1966, and will rise to 184 million long tons by 1975.

The volume of ocean-borne foreign commerce will require, by 1975, 24–35 million deadweight tons of cargo ships, partly under the United States flag and partly under foreign flags. With a 20-year life for these vessels, the world rate of ship construction for replacement of the tonnage carrying our trade will be about 1.5 million deadweight tons annually, and the rate of growth of shipping will require an additional million deadweight tons. The total world ship construction required for U.S. ocean trade for 1975 will thus be about 2.5 million deadweight tons annually. With half of this new tonnage in bulk carriers, at a construction cost of about $150 per deadweight ton, and half in smaller cargo ships at around $250 per deadweight ton, the annual cost of new construction will be in the neighborhood of $500 million per year. The freight bill for the U.S. ocean trade, with present technology, will be some 5 billion dollars a year by 1975.

Ocean transportation and trade consists of three elements: (a) waterborne trade, that is the commerce, imports and exports, using foreign or domestic ships, (b) operation of vessels; (c) ship building. Because of the international nature of the business, these industries are not necessarily mutually supporting in any given country. For example, Norway, which operates one of the world's largest fleets, has an insufficient import–export trade compared to her fleet capacity, and a relatively small, although growing, ship-building industry. The large foreign trade of the United States, on the contrary, is carried, for the most part, by ships flying foreign flags, and therefore supports relatively small shipping and ship building industries.

There is an obvious need for improved design of ships, to make them more efficient, and for improved and more efficient methods of ship building. Improved efficiency of ships can be obtained in several ways. Increasing the size of vessels itself improves efficiency, since the total energy required to carry a given tonnage a given distance at a given speed decreases as the cube root of the weight of the vessel. For tankers and some other bulk cargo carriers this appears to be the direction modern ship design is going. There are at present under construction vessels of 150,000 deadweight tons. However, a limitation on these very large ships is the depths of harbors, so that this in turn affects the design and operation of ports. Offshore cargo unloading facilities, such as pipe lines, is one way of solving that problem, and another is the construction of deeper and larger harbors, which might be accomplished in some localities with the use of atomic explosives. Another way of increasing vessel efficiency, on some routes, is the containerization of dry cargoes. This also affects the nature of the harbor and terminal facilities, because of the necessity for large container storage and sorting space, and other shore-based support. There would also appear to be considerable opportunity for automation in the operation of vessels, so that the manpower requirements are decreased, but this obviously involves difficult social problems.

The problem of optimizing ocean transportation involves, of course, more than the design and operation of ships. As we have already noted, the ships and the harbors from which they operate need to be made compatible. Decreasing the turn-around time can also be important, since about half of the time of a present-day ocean-going cargo ship is spent loading and unloading. It is further interesting to note that a study of the experience of three companies that operate U.S. flagships from the Pacific Coast of the United States to the Orient and return indicates that port costs, including cargo handling costs, accounted for 32% of the total operating costs of the vessels (including depreciation of the capital investment, insurance, maintenance, and administrative costs as well as the cost of running the vessels).

The movement of cargoes across the ocean, and their loading and unloading at the sea–land interface, is only a portion of the system, the objective of which is to move the goods from their point of origin to their point of consumption. Thus there should also be taken into account the transportation system in the hinterland from which the cargoes originate and to which they are destined. Thus it is desirable to consider the entire system of transportation from the shipper to the consignee, of which the ocean transportation sector is only one component, although a very large one. This is a formidable problem because the complexity of the transport system is enormous when everything is taken into account. Further, the situation is likely to deteriorate before it gets better, because in many parts of the United States much of the present investment is in long-lived equipment and facilities, in which large investments have already been sunk. There are also serious social problems resulting from rapid technological change. However, it is evident that this is an important area of marine resources requiring a systems approach.

Waste disposal

Disposal of domestic sewage and industrial wastes, dissolved or suspended in water, is conveniently accomplished by running them, either untreated or after partial treatment, directly into coastal embayments and estuaries, or into open coastal waters. Because of the increase in population density, causing a shift from individual septic systems, local collection networks, and sewage treatment plants to large-scale interceptor sewage networks and centralized treatment plants, the use of the marine environment as a dump for liquid waste is increasing at an even faster rate than that of population growth of the coastal regions.

The use of this resource of the sea—its capacity to assimilate such wastes—is due to its large volume and rapid mixing, which dilute the waste, and the micro-organisms in the sea which break down the organic constituents. Any sector of the marine environment can assimilate a certain amount of waste discharge without damage to its other uses; this is a valuable and legitimate use of this environment, provided the wastes are deposited in such amounts and in such a manner that the capacity of the environment is not exceeded. However, where large volumes are run into waters of nearly enclosed harbors, the rate of interchange with the open sea may be insufficient to provide rapid dilution, and high levels of waste products may be built up locally. Even on open coasts, it is necessary to take careful account of local oceanic conditions if large volumes of sewage and industrial waste are to be disposed of without harmful effects. It is possible in enclosed and nearshore waters to introduce even such fragile wastes as domestic sewage at a rate so great that dilution and decomposition are too slow to prevent the concentrations in the environment from reaching levels that are harmful to man and to his other uses of the environment.

Some of the more refractory materials, especially those that are toxic at very low concentrations, can challenge even the capacity of the vast high seas. The most notable of this class are the radioactive isotopes, the introduction of which into the sea has been approached with great caution. But other things, such as some of the pesticides, are building up in the open sea and its biota in easily measurable amounts, the ultimate effects of which cannot now be forecast.

The optimum use of the sea for disposal of wastes, including the near-shore disposal of liquid wastes, involves careful measurements of the factors of the local environment causing transport, diffusion, and biological degradation of the wastes, so that the cost of disposing of them, including the design of treatment plants, may be minimized while the other uses of the environment are still fully safeguarded. Because it is estimated that the provision of sewage systems and treatment plants for domestic sewage alone costs between 70 and 100 dollars per capita for capital costs, expenditures of the order of 2 billion dollars during the next 30 years will be required in the coastal belt, as well as annual operating costs of some 50 million dollars per year. The economic implications of efficient and effective design are obvious.

Another waterborne constituent that will be introduced into the water along the coastal margin of the sea in increasing quantities is waste heat from power plants. Here again we need to study the environment, and methods of introduction, in order to prevent damage to the living resources at a minimum cost. In this case there is the added possibility that, by introducing the waste heat into the sea in a properly judicious fashion, we can benefit rather than harm other resources; for example we could use the heat to make some of the very cold beach areas comfortable for swimming, or to enhance the local abundance of some of the finer fishes, or even to increase the biological productivity of the sea by using the heat to bring up nutrient-rich subsurface waters to fertilize the sunlit upper layer where the planktonic pasture nourishes the harvestable organisms.

The large volume of solid waste that a modern urban complex needs to dispose of is often overlooked. In a California municipality, for example, the quantity of refuse collected has been typically about two pounds per capita per day, of which about a half pound is garbage. To this are added paper salvaged in commercial districts, and commercial swill. Some of the waste from canneries, debris from the construction business, industrial solid waste, agricultural waste, sewage sludge, and animal manures must likewise be disposed of. A recent survey in the San Francisco Bay area reveals that it must dispose of a total of eight pounds of refuse per capita per day, which is probably typical of other urban areas also. There are a variety of ways in which these solid wastes may be disposed of, but the most economical method, so far, of disposing of unsegregated total solid waste of a community is the sanitary landfill. Its advantages are particularly attractive where there exist low-lying swampy areas not connected with fresh-water aquafers. Consequently, for coastal communities, the filling up of low-lying shallow areas on embayments is often very attractive. This has the difficulty, however, that it decreases the area of coast line, and destroys some of the habitat essential for certain of the living resources, both of which are important for recreational and commercial uses. This presents a formidable problem, as a consequence of which there is great need for developing improved methods of disposal of solid wastes, and especially for development of ways in which the solid waste can be used as construction materials for increasing the shore line rather than decreasing it. Certain kinds of solid wastes can also be used for creating offshore artificial reefs, which can enhance the abundance and concentration of the living resources, as has been shown by experiments, off both California and Florida, with this technique.

In the case of deep ocean, we are primarily concerned with those few kinds of waste which are toxic in such low concentrations that even the large capacity of the ocean is insufficient to accommodate them by dilution, and especially where organic degradation is nonexistant or slow. The most notable of these are the radioactive wastes and probably some of the pesticides. In the case of radioactive wastes, only very low-level, large-volume, wastes, are being introduced into the sea by the United States, or are likely to be in the foreseeable future. However, for some other countries, such as Japan (and perhaps our state of Hawaii), which have limited land area for accommodating high-level atomic wastes, it may be desirable to consider the possibility of sea disposal. Particularly attractive in these cases appears to be the possibility of disposing of high-level wastes in packages placed deep under the sediments in the deep sea, where they could remain undisturbed for thousands of years without fear of return to man.

Recreation

A visit to the crowded seashore of Southern California on a summer weekend will convince any observer that a majority of our citizens look to the sea as a source of relaxation and outdoor recreation. Sailing, swimming, sport fishing, surfing, skindiving, water skiing, and simple contemplation of the beauty and majesty of the sea, and of its inhabitants, provide rest and relaxation from the stress of living and working in a complex urban society. Throughout the United States the demand for outdoor recreation has been accelerating rapidly since the end of World War II, and much of this demand is related to the enjoyment of our water resources. In California alone nearly 100,000 boats are used for marine recreation, there are thousands of scuba divers, and over a million salt-water anglers. All of these, and other forms of marine recreation are increasing at a rate much more rapid than the population growth.

Although California is blessed with over 1200 miles of coastline, about half of which is beach, there is already a desperate shortage of shoreline for recreational activities, in addition to the other requirements that must be satisfied at the interface between the land and the sea. Other coastal states are in an even more desperate situation. The problem we face, and which will become more exacerbated in the future, is twofold. First, with the exception of the open ocean, the supply of the marine recreational resources is far less expandable than the demand. Second, coastal populations tend to be concentrated in a few areas, thus making great demands on some parts of the seashore and nearshore waters, whereas other suitable places are underused.

The accomodation of the need for marine recreation will require the most imaginative application and development of new technology in a systematic fashion, as well as the most careful planning for the various uses of the interface between the sea and the shore. We will need to create more shoreline, to expand the use of existing shoreline by means of effective multipurpose planning and use, and to provide greater and easier access to the underused portions of the shoreline. In some locations, such as Southern California, where beaches are disappearing due to the modification of the supply and transport of the sediments, new means of preserving and enhancing these features need to be sought.

2.2 STRATEGIC IMPORTANCE OF THE OCEAN

In Chapter Three, Dr. John Craven deals with military needs for ocean technology. I will not encroach on that topic. I would like, however, to point out that the strategic position of the United States can be profoundly affected by the coming intensive development of this new frontier, in other ways than through direct military challenges.

An obvious element of our strategy is to secure for our people a legitimate share of the resources of the high seas and of the underlying seabed. International law is proceeding toward the establishment of property rights of one sort or another in these resources. It may be that it will be of greatest advantage to the United States and other nations to vest the ownership of such resources in the United Nations, or some other international agency, as has been suggested. It may, however, be that this solution would be most unwise, and that other arrangements would be superior. In any case we may be certain that the nation which has the most knowledge of the ocean and capability to garner and use its resources will be in the best position to control its own destiny.

For purposes of defense, as well as for enhancement of the economy of our nation and that of our allies, we will have to maintain control of vast portions of the sea so that we may be able to use this broad highway to other lands, and to have access to its resources. Control of the sea by means other than military action is becoming increasingly important.

In addition to military force, the important element of the control of the sea is its use. The traditional ways in which nations have used the high seas in the past are for marine transportation and for fisheries. Our merchant fleet and our fishing fleets are by no means in vigorous condition, whereas the fleets of other nations are flourishing. It would seem important to reverse this tendency. Likewise, the exploitation of other resources of the high seas, and most particularly the minerals of the sea bottom, would be an important element of the use of the sea whereby we could maintain a modicum of control.

The ultimate element of peaceful control is, of course, occupation. The nation whose people actually continuously live on and draw their sustenance from a particular piece of this planet usually wind up controlling or owning it. We remember that the Spanish crown could not hold its New World claims because they were essentially based on the extraction and export of treasure, rather than full occupation and use. Likewise, the cattlemen could not hold the open range of our great plains against the settlers who occupied and farmed it. Who first learns to occupy and fruitfully use portions of the sea will have the highest probability of controlling them. It is important that we push vigorously toward this capability.

2.3 SOME PROBLEMS IN THE UTILIZATION OF MARINE RESOURCES

The foregoing discussion of the potentiality and limitations of some of the particular resources of the sea has indicated, by implication, important scientific, technical, and engineering problems that need to be solved if we are to be able fully to use these resources. We will not go into details of particular scientific and engineering problems here; many of these are covered in other chapters of this book. We may, however, briefly consider some of the problems which, for their solution, require not only improved science and technology, but the solution of important social and political problems, both domestic and international.

Many of these social and political problems arise from the fact that the ocean environment and its resources are, by their nature, rather different from the resources of the land, so that the institutional and legal arrangements that have developed on the land apply imperfectly to the sea. For example, many of the living resources are highly migratory, and not amenable to being fenced or domesticated. These resources, analogous to migratory birds on land, must be managed as the common property of many men and nations, rather than being reduced to individual property rights.

Another example is the fluid nature of the ocean, which is in constant motion, so that what is introduced in one place may rather quickly affect other locations. Waste materials introduced into the marine environment at a given location may influence people living at considerable distances away. Thus large areas of the sea need to be considered as a unit with respect to pollution problems. We have, of course, analogous, situations on land with respect to the pollution of the large water courses, and pollution of the atmosphere, where satisfactory institutional and jurisdictional frameworks have not yet been developed either.

Ownership and jurisdiction

The contents of the high seas, beyond a very narrow belt of territorial sea have been regarded internationally either as common property belonging to everyone or as property belonging to no one until removed and reduced to possession. This has applied to the resources of the sea bottom as well as to the resources of the water itself. Even within the territorial sea, individual nations or states have commonly handled the resources of the water, and often also the sea bottom, on this same basis. These concepts, which have served well in the past, are based on two assumptions: (a) that the resources of the sea are practically inexhaustible, and unlimited, so that one persons's use of them cannot seriously affect another's; and (b) that the contents of the sea are shifting about constantly so that ownership of a part of a marine resource is not feasible. The difficulty is that, with the vastly increased rate of exploitation of some of the sea's resources, such as the marine fisheries, it has been found that some of them are by no means inexhaustible, and that it is necessary, for the common good, to regulate their use.

For the resources of the seabed that are attached to or lie under the sea floor and therefore have a fixed location, the tendency both nationally and internationally is to grant property rights to units of the resource. At the international level, the *Convention on the Continental Shelf*, which came into force in June 1964, provides for the sovereign ownership by the coastal nations of the adjacent continental shelf and the continental terrace to as deep as their resources may be effectively exploited. It is possible that this concept will be progressively extended to the deep-sea bottom as the necessity for allocation and management of the resources thereof emerges.

This does not, however, affect the ownership, or jurisdiction, over the resources of the superjacent waters. This is for the very good reason that those resources do not have, in general, a fixed location, some of them migrating over thousands of miles, so that other means for their management need to be developed. For the living resources, a generally acceptable set of international rules and procedures has been codified and developed in the *Convention on Fishing and the Conservation of the Living Resources of the High Seas*, which came into force in March 1966. Likewise, with respect to the uses of the high seas for commerce, for communications, and for some kinds of waste disposal, a convention that has also come into force, the *Convention on the High Seas*, similarly codifies the preset generally accepted rules. In all of these areas there is room for, and probably will be need for, progressive future developments.

The fact that the rules among nations are becoming established, does not by any means solve all the problems. There still remain the problems flowing from the laws or other institutional arrangements within National or State jurisdictions, which concern which individuals may utilize the resources, and under what sort of a system—private ownership, licensing under state management, or some combination thereof. Major difficulties are that, as noted before, many of the resourses (such as stocks of fishes) come in very large units and that several different resources occupy the same space.

Conflicts

Conflicts arise both over different uses of the same resource, and over uses of different resources in the same location.

The most extensive, and historically oldest, disputes involving different uses of the same resource are those that have arisen over alternative uses of the living resources. For example, there are numerous disputes between sport fishermen and commercial fishermen, disputes among commercial fisherman using different types of fishing gear, and even disputes among different users, or between users and government agencies, concerning the products which are to be manufactured from a given kind of raw material. Many of these disputes involve social considerations where common agreement is very difficult. Take, for example, the question of whether recreational use or commercial use of a limited living resource should have highest priority. These disputes over alternative uses of a given resource should, I believe, be amenable to solution through elucidation of the facts concerning the natural nature of the resource and possible alternative uses, in the light of economic and social considerations, with full public discussion.

Equally difficult are the problems of the uses of different resources in the same region. These problems are most critical along the land interface with the sea, and the adjacent shore, and in the marginal sea some few miles from shore. It is in this zone, as we have seen, that the ocean is used for recreation in a variety of ways, where there are important commercial fisheries as well as sport fisheries, and where exist the greatest possibilities for fish farming. This zone also is where we most conveniently dispose of many of our domestic and industrial wastes. It is at the land–water interface that we must transfer ocean-borne cargoes, and where certain types of industrial plants for processing extractive resources from the sea must be located. Also, it is in the sea bottom underlying nearshore waters that the important, and most easily exploitable, deposits of minerals and petroleum are frequently encountered. Determination of the proper mixture of uses presents a most difficult and complex set of problems.

Fortunately, some of the conflicts that have arisen in the past have been more imaginary than real, and others have been capable of resolution by rather simple technical measures. There must remain, however, conflicts which are real, and where decisions will need to be made regarding alternatives. This category is bound to increase with the increasing utilization of the sea's resources. Choices that will have to be made, if they are to be made for our maximum benefit, must be based on adequate information, both in respect to the nature and extent of the resources, and in respect to the various economic and social benefits to be derived from them.

Unfortunately, some decisions once made and acted on are economically irreversible, as has already been pointed out by Ciriacy-Wantrup (1963). For example if we decide to fill an embayment to create additional land, such a decision is essentially economically irreversible, even though it is later determined that some other utilization would have been more beneficial. It is, I believe, therefore most urgent that we obtain as rapidly as possible a fairly complete inventory of the potential resources of the marginal sea, and of the lands lying along the interface between the land and the sea, as a basis of rational planning of the optimum uses thereof. In those cases where decisions are economically irreversible, we should be particularly cautious about making them, until really required to do so, so that we may preserve as many options as possible for future contingencies.

Conservation

Inherent in the problems of utilization of the marine resources, as for other resources, are important considerations concerning their conservation. For this purpose, conservation may be considered as the allocation of use of a resource over time, in relation to the expected flow of future benefits and costs. For some resources—the nonrenewable resources—there is a certain fixed quantity available for use, which may be used rapidly or may be used slowly. For the renewable, or flow, resources the problem is rather different. For some of these, where our use of the resource does not affect the quantity that will be available in the future, we have no reason to take heed for the future supply. For other flow resources, where our use *does* affect the future supply, we have carefully to consider the future supply. As inferred above, this is particularly important where the use of the resource is such that its diminution may become physically or economically irreversible.

I have only touched lightly upon several of the foregoing topics, and have not even begun to elucidate their complexities. It is hoped, however, that this brief discussion will provide at least a little insight into the economic and social constraints within which ocean technology must be developed.

REFERENCES

Ciriacy-Wantrup, S. V., 1963 *Resource Conservation, Economics and Policies* Revised Edition, U. of California Press, Berkeley.

Committee on Oceanography, 1964, "Economic Benefits from Oceanographic Research," Nat. Acad. Sci.—Nat. Res. Council, Publ. 1228.

Duport, J., C. LeMenestrel, C. Million, and E. Chapus, 1965, "*Power from the Tides,*" *Int. Sci. and Techn.*, May 1965, pp. 34–40.

Interagency Subcommittee on Large Nuclear Fired Sea Water Distillation Plants," 1964, "An Assessment of Large Nuclear Powered Sea Water Distillation Plants," Office of Science and Technology.

Interagency Maritime Task Force, 1965, The Merchant Marine in National Defense and Trade, A Policy and Program," U.S. Dept. of Commerce, Maritime Administration.

Institute of Marine Resources, 1965, "California and Use of the Ocean. A Planning Study of Marine resources, Prepared for the California State Office of Planning." U. of Calif., Inst. of Marine Resources, IMR Ref. 65–21.

Isaacs, J. D., and W. R. Schmitt, 1963, "Resources from the Sea", Int. Sci. and Techn., June 1963, pp. 39–45.

McIlhenny, W. F., and D. A. Ballard 1963 "The Sea as a Source of Dissolved Chemicals," Paper presented at Chemical Marketing and Economics, Am. Chem. Soc., Los Angeles, 2–4 April 1963, pp. 122–131.

Menard, H. W. 1964, *Marine Geology of the Pacific*, McGraw-Hill, New York.

Mero, John L., 1963, "The Sea as a Source of Insoluble Chemicals and Minerals", Papers presented at Chemical Marketing and Economics Division, Am. Chem. Soc., Los Angeles, 2–4 April 1953, pp. 139–159.

Mero, John L. 1965, *The Mineral Resources of the Sea* Elsevier, New York.

Schaefer, M. B., and R. R. Revelle, 1959, "Marine Resources", Ch. 4 in *Natural Resources*, M. R. Huberty and W. L. Flock, Eds. Mc Graw-Hill, New York, pp. 73–109.

Schaefer, Milner B. 1965, "The Potential Harvest of the Sea," Trans. Am. Fish. Soc., Vol. 94, No. 2, pp. 123–128.

Schaefer, Milner B. 1964, "California and the World Ocean", Proceedings of the Governor's Conference on California and the World Ocean, Los Angeles, 31 Jan.–1 Feb., 1964.

CHAPTER 3

Sea Power and the Sea Bed

JOHN P. CRAVEN

The impact of new technology on the evolutionary development of sea power and its influence in world affairs is not always clear or discernible at the time of its introduction. The bold innovators who introduced steam, aircraft, and nuclear power into naval development were well aware that some major effect on sea power was thus presaged, but could not discern at that time the battleship fleets of the first half of the century, nor the carrier squadrons and ballistic missile submarines of today. It may well be presumptuous, therefore, to attempt to forecast the future effects of a developing technology that permits occupation and manned operation of the sea bed. Unfortunately, the accelerated tempo of modern technology threatens to change the evolutionary development of international relationships into one that is more revolutionary in nature. It therefore behooves the naval strategist to anticipate, no matter how haltingly, the effect of each new technological development, and to suggest the time scale and scope of measures that must be taken to ameliorate disruptive changes. Fortunately, a number of major constraints on the use of the sea have remained unchanged for many centuries. We may therefore expect empirical relationships derived from observations of naval history to have validity today. There are also available for study many changes in the evolution of sea power—from raft, to oar, to sail, to steam, and to nuclear power—which permits some identification of the nature of the evolutionary process. In the hope of stimulating thought and insight, let us (a) restate in today's terminology those principles of sea power enunciated by Rear Admiral Alfred Thayer Mahan in 1890 which appear to have relevance today, (b) outline in brief an evolutionary pattern of the development of sea power, (c) summarize and project the state of art in the technology of naval capability at or near the sea bed, and (d) derive therefrom the probable effects of this technology on the future of sea power.

The foundation of this chapter is the assumption that three fundamental constraints on the use of the sea have not significantly changed since the dawn of sea power, viz., the law of the sea, the speed of transit of vehicles on or in the sea, and the geopolitical configuration of the sea–land interface. The first of these constraints, the law of the sea, has remained fundamentally unchanged since the ancient *Codes of Amalfi*, the *Rhodian Code* and the *Laws of Oleron*. It provides for the free use of the sea by all nations engaged in legitimate commerce and even for recognized belligerents engaged in the conduct of formally declared war. This particular provision, unique to the sea, is so ingrained in custom that it is rarely recognized as a prime constraint in the design of naval systems such as the Polaris submarine system, which must remain undetected in waters in which potential enemies may freely traffic and search.

The present embodiment of the law of the sea can be found in the partially ratified *United Nations Convention on the Law of the Sea* of 29 April 1958. Although agreement on this convention was hampered by unilateral attempts to extend the territorial waters from three miles to six, twelve, and even 200 miles, no significant effect on the basic rights and freedoms of the seas has yet been discernible.

The second constraint on the use of the sea, transit speed, results from the hydrodynamic power law which requires (to a first order of approximation) an increase in horsepower proportional to the cube of the velocity of vehicles immersed in the medium. Vehicle speeds significantly above those already attained will require more horsepower per unit volume than technology seems to indicate is possible. Thus, although remarkable gains have been made in the powering of ships, they have been evidenced primarily by the provision of assured and continuous speed rather than by major increases in absolute speed. Increases in speed will assuredly be realized, but even the most optimistic naval architect does not project for the future speeds in excess of 60 knots for vehicles which are fully wetted. Even at 60 knots, the nature of the speed constraint does not change, for this constraint has the curious effect of making credible national intent when a major task force is deployed to a particular conflict theater. So long as the redeployment time of such a force is long as compared with political action time, the original action cannot be construed as a feint or bluff. Therefore, for the foreseeable future, the sea will continue to be a mechanism for providing both political and military presence in a manner which transmits messages stronger than words.

The third constraint on sea power, the constancy of geopolitical or geologistical relationships, derives from the massive engineering under-

40   Systems Planning

takings required to alter them. Indeed, the relationships between sea and land masses have been changed in a major way only twice in history—with the construction of the Suez and Panama Canals—and in lesser ways with the construction of the Kiel Canal, the St. Lawrence Seaway, and other such interconnecting waterways.

The effect of these three constraints has been that throughout recorded

FIG. 3-1.   Main features of the sea bed, Eastern Hemisphere. (From *U. S. Naval Institute Proceedings*, April 1966. Relief map copyright Geographical Projects, Ltd., London.)

history, sea power has been deployed freely, credibly, surely, and with constrained, deliberate speed by those nations whose geologistics are favorably oriented with respect to the sea. The imaginative reader may quickly perceive how at least two of these constraints may be modified by an effective occupation of, or utilization of, the bottom of the sea. It is already clear that the law of the sea bed will be different from the law of the sea, and the interaction between these differing legal concepts will undoubtedly modify both. Less easily perceived is the effect of utilization of the sea bottom on the geologistics of the sea. The possibility of underwater or semisubmerged offshore loading platforms, for example, could introduce into competitive commerce those nations whose rugged coastline otherwise forbids such enjoyment of the sea. Such modifications have not as yet

FIG; 3-2. Main features of the sea bed, Western Hemisphere. (From *U. S. Naval Institute Proceedings*, April 1966. Relief map copyright Geographical Projects, Ltd., London.)

appeared even with the introduction of nuclear power and armament. We should therefore expect empirical laws derived while these constraints were still in force to be equally valid today.

These relationships were perhaps most effectively enunciated by Rear Admiral Mahan in his famous study of sea power in 1890. A restatement of these principles is essential to an analysis of the effect of occupation or operations at the sea bed.

The sea provides a domain in which national power and continuing political presence may be effectively projected to the territorial limits of other nations having boundaries on the sea. The ability of a nation to use this element of national powers depends (a) upon the conformation and topology of the land and water masses, (b) upon the nature of the coastline in terms of its capacity for harbors and access to the major sea routes, (c) upon the number of people in the vicinity of the sea having competence in and an understanding of the technology of the sea, and (d) upon the character of the people and their government.

The first of these determinants of sea power, the topology of the land and water masses, is crucial for the subsequent discussion of the effect of occupation of the sea bed. The topological relationship determines the ease with which a fleet can be deployed around the sea perimeters of a maritime nation as well as the size and disposition of land armies which are required to protect the land boundaries. Mahan accorded primacy in this element of sea power to nations for which the effective sea boundaries are singly connected, that is, the entire coastline can be patrolled by a single transit and without the necessity of passing or rounding the coastline of another nation. The island is the best example of this relationship. On this basis, Mahan correctly predicted the future importance of Malta in the Mediterranean as evidenced by its key role in holding the balance to Mussolini's Mare Nostrum. He predicted the future importance of Cuba in the Caribbean as now evidenced by its position as the single intrusion of Communist domination in the Western Hemisphere. His prime example, of course, was Great Britain and its effectiveness as a balance to the entire European continent. Had Rear Admiral Mahan looked to the Pacific, he would no doubt have identified the significance of the Japanese Islands and Formosa as balances to the Asiatic mainland. Had he looked toward Africa, he would have identified Madagascar as being similarly crucial.

Whereas islands and peninsulas such as the Italian Boot or the Portuguese Coast provide a positive basis for the use of sea power, nations having multiple coasts find themselves in a position of naval inferiority. In this instance, it is more than the necessity of maintaining two separate navies that creates the difficulty. Invariably, the competition between the several coasts for allocation of national resources for naval purposes is reinforced by the demands from the dwellers along the land borders for adequate armies. The political pressure is usually resolved by the inhabitants of the interior and there results a polarization along the land axis which favors the development of armies over the development of multiple competing navies. As Mahan pointed out, the United States would be in this position until a canal was built across the Isthmus of Panama. It was left to another great student of naval history, Theodore Roosevelt, to implement that foresight in 1910. In terms of modern commerce, however, this canal no longer has the capability or size to effectively maintain this geologistic relationship. In fact, we are again a doubly connected sea domain with competition for naval amd maritime power on both coasts and no longer the singly connected domain at the scale of commerce we enjoyed from 1914 through World War II.

The second element of the Mahan thesis derives from the value which accrues to a nation from a coastline that permits the development of harbor and inland transportation systems. Thus, Great Britain with its myriad of navigable waterways in the Thames, the Tyne, and the Clyde River basins is in a far superior position to use the sea than is its island neighbor, Ireland. The Eastern United States has been in the position of Britain with its extensive inland waterways, its network of great inland rivers such as the Mississippi, the Ohio, the Hudson, and the St. Lawrence which have made possible large cities as remote from the ocean as Minneapolis or Cincinnati. On the West Coast, however, the lack of navigable rivers has limited the development of large cities to relatively few good harbors, that is, San Diego, Los Angeles, San Francisco, and Seattle. Until the dredging of the Sacramento River, the West Coast had only one inland city, Portland, Oregon, connected to the ocean. At first, it would seem that only mighty engineering modifications to the coastlines could rectify such deficiencies, until it is realized that it is solely the free surface and its chaotic forces in open waters that restrict use of the majority of the seacoast. Were the creation of submerged offshore platforms technically feasible for the transport of goods along the sea bed to the Coast, the effect of this relationship would be minimized.

The third element of the Mahan restatement pertains to the number of people within the vicinity of the seacoast having a knowledge of maritime technology. An overly restrictive view of United States technology would lead to the false conclusion that the country is limited by its supply of oceanographers or by the total number of professionals classified as naval architects or in the purely maritime systems. Such a reckoning does not take into account the very large reservoir of maritme technology that has been acquired in military programs such as the Fleet Ballistic Missile Program, or the extensive antisubmarine warfare programs, or by the space programs in their recoveries at sea, or by the oil industry in its extensive offshore drilling operations. Indeed, it is not too difficult to identify at least 4 billion dollars worth of human resources in the United States which is annually expended in some phase of the technology of the sea.

The final element of a nation's ability to utilize sea power was identified as a function of its type of government. The free enterprise and initiative of the American economy should match this boundary condition of maritime pre-eminence should the trends of self-interest so dictate. The technology of the sea bed has already derived much from this aspect of the society, and current effort in industry points to many further contributions in this regard.

This brief review of the elements of sea power is indicative of the potential of the United States for its exploitation should the national economic interest demand, should international political or military developments force the issue, or should a significant evolutionary cycle in the use of the sea impend. This last possibility is inherent in any extensive or massive exploitation of the sea bottom. Its imminence may be assessed by a review of the basic evolutionary cycle in the expansion of sea power, and an assessment of the phase in the evolutionary process with which we are today confronted.

The briefest outline of history would not fail to encompass the era of the river society, the development of oared craft capable of navigating the Aegean Sea and the Nile Delta, the development of sail, the development of hull and structure of ocean transport, the introduction of steam, the introduction of the steel hull, and the introduction of nuclear propulsion. The student of history will not fail to recognize that each of these developments changed the basic logistics of the use of the sea and as a consequence, the scale of the land–water mass over which control could be exerted; nor will the student of history fail to recognize that each of these developments played a vital role in the final resolution of a cycle of national or international land conflict. Indeed, a broad evolutionary pattern of societal development having at least five distinct phases can be recognized. The identification of these phases is central to an estimate of the effect of the development of the sea bed.

The initial phase of an evolutionary cycle in a society is characterized by a level of maritime technology that provides for some land mass a natural means for defense, for commerce, for water supply, and for a means of waste disposal to the extent that a segment of the population can be freed from employment in the economic necessities of providing food, clothing, shelter, and military defense.

The second phase of the evolutionary cycle is characterized by the development of a priesthood or in more recent times, a scientific society interested in the acquisition of philosophy or knowledge for satisfaction of national cultural aspirations.

The third phase of the evolutionary cycle is characterized by the transition of the knowledge acquired by the scientific class into engineering arts that are applied to the construction of land-oriented logistic systems such as roads, aqueducts, warehouses, granaries, and sewage-systems.

The fourth phase of the evolutionary cycle is characterized by the invasion of the society along the land logistic systems by predators or barbarians who were otherwise denied access by means of the natural defenses provided by the land—water configuration. This phase continues until the land logistic systems are completely fractionated and the logistics of the conflict are forced to rely again on the sea.

The fifth phase which is essential to the evolutionary step (indeed, the first four phases can repeat themselves for a number of cycles) occurs when the engineering arts acquired in the construction of land systems and in the subsequent design of engines of defense are applied to the maritime and naval arts and a new level of maritime capability is achieved. At this juncture in history, the society will stabilize about some larger scale configuration of the land and water mass and a new evolutionary cycle will commence.

Historical evidence supports this pattern. The earliest known societies were lake communities such as the Neolithic villages uncovered in the Swiss lakes and similar remains in Scotland and Ireland. These lake people repaired to the lake islands for defense, while enjoying the collateral benefits of transportation, water supply, and waste disposal provided by the body of water. That these societies were not the forerunner of modern civilization is probably explained by the lack of geologic connectivity between the lakes and the world's major rivers; i.e., lakes are in general connected with the non-navigable headwaters of major rivers or constitute the major basin of a quite localized watershed. The honor of the birthplace of civilization, therefore, goes to the rivers.

It is of importance to note that the first river societies did not begin along the banks of rivers in arbitrary fashion, but began along rivers whose peculiar configuration provided a natural means of defense. Thus societies flourished between the protective rivers of the Tigris and the Euphrates in the Indus Valley, and the Yangtze and Huang Ho in China or between the protective sands of the deserts along the Nile. By virtue of their proximity to the Aegean Sea, the Egyptian, Babylonian, and Sumerian societies were destined to be the source from which sprang the next major societal development. In Babylon, Sumeria, and Egypt, societies waxed and waned in the cycle of priests, pyramids, hanging gardens, artisans, roads, canals, granaries, and the destruction thereof by barbarians until, as the evidence seems to indicate, the oared craft of the Egyptian delta were able to make their way to Crete and the technology of Babylon was transported across the mountains to Turkey, to the Western Mediterranean, and to the Aegean Sea. At this level of sea power, the great Hellenic chapter of society began.

The science which flourished in the sea protected Greek culture was transmuted to the engineering of the Roman Empire with the gentle transition of maritime technology from the Aegean Sea to the Mediterranean. The constraints of sea power dictated that Rome should dominate the Mediterranean and as long as the control of the sea was the deciding factor in attacks against this realm, as it was in the three Punic Wars, this empire dominated the civilized world. It was the great engineering feats which extended roads, aqueducts, forts, and logistic installations into Northern Europe that provided the means for the wagons of the Goths and Visigoths to destroy the Empire. There followed a period of almost 1000 years of land conflict resolved by the advent of the knight and the castle, but equally so by the development of a maritime technology which enabled commerce and defense to stabilize around the entire perimeter of Europe.

Three major societal elements contributed to this new scale in the use of the sea, the ships of Portugal and Spain, the rising influence of Britain as a maritime power and the establishment of the Hanseatic League. The stability provided to Europe by the Hanse permitted the onset of the new evolutionary cycle in the form of the Renaissance which characteristically was initially religious, artistic, and scientific in nature and was to precede the flowering of the engineering arts in the seventeenth and eighteenth centuries. So it was at the beginning of the twentieth century when Napoleon set out to conquer the European continent, the land logistical systems were at a technological peak. One can now easily see the pattern by which the land logistic system was destroyed with Napoleon's advance into Spain, and the resolution of that phase of the conflict by naval victory at Trafalgar, how it was further destroyed in the advance to and retreat from Moscow and finally resolved by sea-supported victory at Waterloo. Shortly thereafter, the same pattern was repeated in the United States when a great Civil War was fought whose resolution did not depend on the outcome of individual battles, but on the control by the North of the sea lanes. This fact became increasingly apparent to the South as its land logistics system was progressively destroyed and finally extinguished at Amelia Courthouse.

At this point in history, the introduction of steam marked the beginning of the next evolutionary cycle which changed the scale of sea power influence from European to Atlantic waters. The science of the late nineteenth century matured into the engineering of the early twentieth which flourished until World War I laid waste the resources of France and Germany in a shattering land conflict. This conflict was ultimately resolved by the United States and Great Britain through sea-borne logistics. In the armistice period between 1918 and 1939 the fruits of engineering were realized in the form of air power while contemporaneously, a new science and a new physics emerged from the works of Heisenberg and Planck. World War II produced the ultimate in the destruction of the industrial fabric of Britain, France, Germany, the Soviet Union, and Japan until the war was brought to a conclusion by the sea-borne invasion of the Normandy beaches and the isolation and destruction of the logistic support of the Japanese Islands by Pacific naval power.

The finale to that conflict provided the ominous warning that the next evolutionary cycle might be more compressed and that the engineering fruits of the new science, nuclear physics, could speed the development of land logistic systems and, concomitantly, the means for their destruction. At the same time, nuclear power provided the mechanism for a new scale in the use of sea power and thereby the basis for a new evolutionary cycle.

It is of central importance to note that of all the benefits that this new source of power confers on marine technology, the greatest is the ability to operate continuously within the sea, impervious to radioactivity and highly resistant to the overt pressures of nuclear attack.

To a larg extent this element of the new sea power has been realized in the Fleet Ballistic Missile System and in the nuclear attack submarines. The existence of nuclear power itself is a necessary, but not a sufficient condition to ensure that the sea bottom will play a significant part in this new episode of the sea. Other technologies must also be developed or have already been developed to make this domain more attractive or more competitive for control of the full dimension of the ocean. It is a part of this thesis that such technologies are already with us or are in the offing and their realization will not require a greater expenditure of resource than has been required in the development of military air power and certainly less than is required for the conquest of space.

Two major technologies that complement nuclear power must prove both feasible and economic before the sea bottom becomes of strategic or commercial importance. These are the technology of manned vehicles capable of operating at considerable depth and in close proximity to the bottom, and the technology of the physiology of man, which will permit him to operate as a free swimmer in the environment at depths at least as great as those encompassed by the world's continental shelves. The evidence now accumulating indicates that both of these technologies will be feasible at reasonable cost and with acceptable hazard.

The first self-powered vehicle to demonstrate the ability to carry man to the deepest part of the ocean was the bathyscaphe *Trieste*. This vehicle, designed by the Swiss scientist Auguste Piccard, was piloted by Jacques Piccard and Lieutenant Donald Walsh, U.S. Navy, to the deepest part of the Marianas Trench in 1960. In 1963 and 1964, the *Trieste* was called upon by the U. S. Navy to assist in the search for the *USS Thresher* (SSN-593). Piloted by Lieutenant Commander Donald Keach and Lieutenant George Martin during the first phase, and by Lieutenant Commander Bradford Mooney and Lieutenants Lawrence Shumaker and John Howland in the second phase, the *Trieste* was able to locate and to photograph substantial portions of the *Thresher's* hull. The difficulties experienced in this mission have created the impression that *Trieste*-type craft are inherently unwieldy for undersea engineering operations. This is an erroneous view. In her original concept the *Trieste* was designed for oceanographic observation and as such consisted basically of a pressure hull which was negatively buoyant and a gasoline-filled float of sufficient size to provide the necessary reserve buoyancy. The extremely limited budget and time scale associated with her original construction and her subsequent modification into the *Trieste II* precluded the introduction of anything but the barest elements of power and control. She was not, therefore, designed for the missions she was called upon to perform. It is a fact that the kindliness of the undersea for the mobility of large objects is such that a *Trieste*-type vehicle, if equipped with adequate power, appropriate sensors and integrated controls, could effectively perform many engineering missions in the sea. Indeed, the capability of this type of craft to support large payloads would appear to make them valuable for use in the commercial recovery of the contents of the sea bed.

There is a safety hazard, however, in the use of aviation gasoline for buoyancy, which has probably inhibited the commercial use of vehicles like *Trieste* and her French counterpart, the *Archimede*. This hazard severely limits the sea state in which these craft can successfully operate with surface support. This limitation is, however, only temporary, for the development of buoyancy materials such as the syntactic foams will permit the substitution of safer, rigid, and lighter materials. The syntactic foams are mixtures of small glass spheres in an epoxy resin. The glass spheres provide the necessary strength to resist the pressures of the deepest part of the ocan and at the same time the requisite buoyancy; the resin holds the spheres together. As will be further indicated, glass is an ideal material for deep-submergence application and in this first manifestation it holds the key to the continued use of the bathyscaphe concept. When these materials become commercially available, it will be possible to construct deep ocean vehicles at modest cost (less than that of aircraft).

With low cost, very deep-submergence vehicles of the bathyscaphe type may become commonplace, but their large size and limited power prevent them from being high-performance vehicles in the military sense. This deficiency is being alleviated by two other classes of vehicles which are now demonstrating a capability for operations on or near the ocean bottom. For each of these classes, however, the depths which have been obtained are, by bathyscaphe standards, relatively modest. These two classes are similar in that they both rely on reserve buoyancy within the manned pressure hull to provide the requisite safety, but are distinguishable in that they derive from different structural and fabrication concepts.

In the first class are those ships that are derivatives of the conventional ring-stiffened submarine hull. The depth capability is, in general, obtained by the use of stronger materials. Provision for near-bottom capability is made by some combination of ballast control and thrust devices. The first of such vehicles to be developed is the *Aluminaut*, whose conceptual design was made by Dr. Edward Wenk while he was at the David Taylor Model Basin. The *Aluminaut* consists of ring-stiffened cylinders of aluminum which are bolted together. Forged-aluminum hemispherical heads are employed for the end closures. Although the design depth of 15,000 ft has not been achieved at this writing, the capability to operate near the bottom in depths up to 6000 ft and for extended periods of time has been demonstrated. The second craft of this type is the *Auguste Piccard* designed by Jacques Piccard. This submersible is constructed of high-strength steel and has an operating depth of about 2000 ft. She has been operating in Lake Geneva, carrying 40 tourists to the maximum depth in the lake of about 1000 ft. Relatively few vehicles of this type have been built or are now planned, but progress must continue in this direction if high-performance vehicles with long staying power and modest crew or passenger capability are required.

Such a projection is being made by the Navy in its construction of the nuclear-powered, ocean-engineering submersible, the *NR-1*. This ship will have a near-bottom capability and will be built for the maximum practicable depth that can be safely assured within current technology. Initiated without fanfare, this submersible may be the most significant innovation in the technology of the sea bottom. With the ability to free herself from surface support, the *NR-1* should be the pioneering prototype of the sea-bottom vehicles which have all the requisites for revolutionizing our concepts of the utilization of the sea.

In the second class of vehicle are those that are derivatives of the precisely machined and precision-controlled welded pressure hull. In general, the vehicles have a positively buoyant spherical pressure capsule made of very high-strength material. Because they are limited in size, they are generally capable of supporting crews of not more than two or three. Notable vehicles in this configuration are the *Alvin* of the U.S. Navy and Woods Hole Oceanographic Institute; the *Deepstar* of the Jacques-Yves Cousteau and the Westinghouse Electric Corporation; the *Moray* of the Navy's Naval Ordnance Test Station; the *Deep Quest* of the Lockheed Corporation; and the *Asherah* of the Electric Boat Company.

The forerunner of these vehicles is the diving saucer, the *Soucoupe*, built by Captain Jacques-Yves Cousteau of France and employed by him in ocean exploration in the Red Sea, and subsequently off the West Coast of the United States in the Scripps Canyon. This vehicle is capable of depths of up to 1000 ft, and the pressure hull is built of fairly conventional steel. The *Alvin* hull is built of high-strength steel and has demonstrated a capability to operate at a depth of 6000 ft. Except for the *Moray*, which has a pressure hull of aluminum, all of the above vehicles are made of some form of high-strength steel either quenched and tempered, or of a relatively new type of maraging steel. All require surface support and use some form of battery power. Mission durations are therefore measured in hours. With the exception of the *Moray*, all are of relatively slow speed with maximums of about 5 knots. If, however, the speed were not limited by power considerations, it would be limited by visibility, the necessity for obstacle avoidance, or by the characteristics of other sensors. The *Moray*, which is not designed for near-bottom operation, has a considerably higher speed, albeit for shorter periods of time.

In the immediate future this type of vehicle will be expanded in capability in the U.S. Navy's Deep Submergence Rescue Vehicle (DSRV). In her preliminary design configuration the DSRV uses three high-strength steel spheres joined by cylindrical connections. The forward sphere is employed for the vehicle controls and the pilot and copilot; the after sphere will carry up to 14 passengers, and the center sphere is an access to a surface hatch and to a mating bell for attachment to a mother submarine. This latter feature makes the vehicle unique in her freedom from surface support and in her all-weather capability. When employed with a nuclear submarine as a mother craft, the combination permits continuous submerged operation for extended periods with a semicontinuous extension through the small vehicle to all of the sea bottom encompassed by the continental shelves. In a slightly longer time scale, the Navy is also developing a deep-ocean search submersible with a depth range of 20,000 ft, which will also be capable of mating with a conventional military submarine, providing a full ocean search and light work capability.

In the longer range, it can be anticipated that the two types of true submersibles will approach each other in concept. In order to attain greater depths, the ring-stiffened cylinder-type vehicles will have to employ increasingly sophisticated fabrication techniques and will have to employ more brittle materials such as the higher-strength steels, titanium, or aluminum. The small, spherical-hull-configured vehicles, in order to increase the payload, will be forced to evolve away from the structurally advantageous but volume-restraining spheres into ellipsoids or prolate spheroids, and will also require the use of higher-strength materials and increasingly sophisticated fabrication techniques.

If the trends herein outlined are projected forward, a variety of vehicle configurations will be possible for an extended range of ocean missions from search and oceanographic observation through numerous manipulative tasks, including large object recovery. However, the development path so far projected points to increasingly sophisticated and consequently increasingly costly vehicles, and the author's prognostication of an extensive commercial and military intrusion into the deep ocean would be rash indeed. Fortunately, a bright prospect in the offing is in the use of massive structural glass or ceramic materials. These materials augur a drastic change in the economics and technology of the deep ocean and the ocean bed. In a series of definitive studies, H. L. Perry of the Naval Ordnance Laboratory had demonstrated that sizable pressure hulls can be constructed of massive glass; that these structures are extremely light in weight; that they can resist the pressures of the deepest ocean without serious compromise of payload; that they can be adequately protected from shock on the surface; that their resistance to shock increases dramatically with depth; that the problems of creep and fatigue are minimal; that they can be inspected for flaws nondestructively during and after fabrication and during operation; that they can be protected against sympathetic implosion; and that in quantity production, the pressure hull will be a minor cost of the vehicle. These superior qualities of glass as the hull material for deep submergence have already been demonstrated in its use as the major constituent of the previously mentioned syntactic foams and as the essential element in deep-ocean unmanned oceanographic instruments such as a corer, which can collect samples in the deepest part of the ocean. A number of vehicles now on the drawing boards project the use of large glass spheres for additional buoyancy, and several of the glass companies are capable of producing glass spheres of sufficient size to carry a man. It is only a matter of time before a manned glass deep submersible becomes a reality.

The enthusiasm with which many naval architects approach the future of deep-ocean submersibles is conditioned by the fact that until this year all of the deep-ocean vehicles herein described, and all of the material developments which permit optimistic projection, were accomplished with extremely modest budgets and in the absence of any major focused national effort. With such a focus, a wide spectrum of vehicles could be obtained ranging from low-cost, one-man, glider-type vehicles capable of using the full ocean depth on each glide path, to large nuclear-powered, mother submersibles capable of tending and supporting numbers of fuel-cell or battery-powered vectorable craft analogous to the carrier aircraft systems of the surface Navy.

Vehicles are, however, only one extension of man into the sea. The second major technological development which will open the sea floor is the adaptation of the physiology of man to permit him to exist as a free swimmer in the ocean. The major innovation here was the introduction of the concept of saturation diving by Captains George Bond, Walter Mazzone, and Robert Workman while stationed at the New London Naval Medical Research Laboratory. The initial project, a series of chamber experiments appropriately entitled "Genesis," demonstrated the ability of animals and of man to exist for prolonged periods at high pressures while breathing appropriate mixtures of helium, nitrogen, and oxygen. Such adaptation to the environment imposes the limitation that man cannot make excursions into shallower water or return to atmospheric pressure without long and carefully controlled decompressions.

This limitation is more than compensated by the increased ability to make deep excursions and by the extensive work periods available at the saturation depth. The initial Genesis project matured into the Sea Lab I and Sea Lab II open-sea experiments which were conducted in depths of approximately 200 ft. The first was conducted off the Argus Island near Bermuda and was primarily to determine operational feasibility. Four men remained for a period of nine days in the Sea Lab and were carefully monitored during and after the experiment for evidence of significant physiological effects. The success of the experiment gave impetus to the Sea Lab II venture which was conducted at the mouth of the Scripps Canyon off La Jolla, California. This second, more extensive experiment used three teams of aquanauts, each of which spent approximately 15 days in the Laboratory. Commander Scott Carpenter was the team leader for the first and second teams, and lived under pressure for 30 days at a continuous stretch; Lieutenant Robert Sonnenberg was the medical doctor in the laboratory for the first and third teams and, as a consequence, was pressurized for 30 days in two discreet 15-day periods. This experiment included a great many measures of performance capabilities, work and salvage abilities, and evaluation of diver equipments as well as more extensive physiological and psychological measures than had been obtainable in Sea Lab I. The initial results of this experiment are highly optimistic in their indication of man's ability to operate effectively as a saturated diver at these modest depths and in their indication of the ability to project to considerably deeper operations in forthcoming experiments. These are designed to achieve, initially, an operational capability for man at 600 ft and, eventually, a capability to operate at 1000 ft. A parallel series of experiments is being conducted by Jacques-Yves Cousteau in France, and by Edwin Link of Ocean Systems Incorporated in the United States. Both of these investigators have conducted operations at deeper depths than those of the U.S. Navy Sea Lab, but with fewer subjects and with less extensive experimental measures. Commercial operations with saturated divers have already been undertaken by Marine Contracting, Inc., of Southport, Connecticut, at the Smith Mountain Dam in Virginia. The economic value of this operation is such that it is certain to extend to many commercial diving tasks now accomplished by more conventional diving techniques.

It may be too early to assess the optimism which may be legitimately employed in projecting this capability to the future, except to note three potential areas for extension of man's capability in the sea. The first results from the indication that it may be possible for man to make considerably deeper excursion dives from the depth for which he is saturated than he may make from the surface. That is, from the surface a short time excursion dive to a depth of 200 ft appears to be close to the maximum that can be safely tolerated; from a saturated depth of 200 ft, an excursion to 400 ft appears to be more easily tolerated and excursions to 500 ft may be possible. At a depth of 600 ft even greater excursions may be permitted. If this present indication (and it is only an indication) is correct, man's capability to perform useful tasks as a free swimmer on the continental shelf may be greatly accelerated. The second potential is covered in the gas mixtures which are employed by divers at deep submergence. At present, mixtures of helium, nitrogen, and oxygen are employed, such that the appropriate partial pressures of oxygen and nitrogen are maintained. Helium is chosen because it is both inert and light, but even helium is a heavy gas at ambient pressures of 600 ft or greater. The possibility of using a hydrogen gas mixture at these advanced depths appears initially attractive as a result of a very limited number of animal studies. The extent to which hydrogen enters into human metabolic processes raises certain notes of caution, but should this gas prove physiologically inert, man's extension into the sea will be again eased. The third possibility belongs to a distant, but foreseeable future and envisions the use of an appropriate fluid which fills the lungs, but contains and is resupplied with sufficient quantities of dissolved oxygen to sustain life. The ability of mammals to exist with such fluids in their lungs has already been proven, but the total set of physiological problems, to say nothing of the reacclimatization problems thereby raised, should keep investigators busy for at least a decade before any attempt to so condition man can be attempted. Should this possibility materialize, the depth potential for man as a free swimmer will extend to substantial portions of the ocean.

As in the case of vehicles, the enthusiasm with which the diving fraternity and the medical fraternity approach the future of man-in-the-sea, is conditioned by the fact that all of the accomplishments to date have been accomplished with modest or marginal application of development resource. The exponential effect of accomplishment, economics, and national interest should in short order eliminate this constraint. The pace of development should then be limited only by the time required to train the necessary specialists in the field.

The forecast of the technology of the sea, based on these modestly funded but impressive beginnings, can thus be summarized in terms of the technology of vehicles and of man as a function of the depth of operation.

For the greater portion of the continental shelf (0 to 3000 ft), the economic use of man and machines for extensive and prolonged engineering operations is virtually assured within the next decade. The magnitude of national effort devoted to this technology is still contingent upon the assessment of the military significance of the continental shelf, and the assessment of its resources and of its scenic and recreational potential. For the next geographic areas of interest, the ocean ridges and sea mounts (6000 to 8000 ft), the capability to deploy vehicles and to mount installations is virtually assured within the next decade. In the next two decades the commercial deployment of vehicles and machines in these areas is highly probable; and by the end of the century, the economic use of men and machines in these areas is not inconceivable. The magnitude of national effort in these areas is again contingent upon an as-yet-unassessed military, economic, and cultural potential. For the final areas of interest, the broad ocean basins (12,000 to 20,000 ft), within the next decade the capability to visit selected areas of the bottom and to perform scientific and light engineering missions is virtually assured; within the next two decades, the deployment of vehicles and machines in these areas is highly likely; and by the end of the century, the extensive economic deployment of a wide class of commercial and military vehicles is a distinct possibility.

The acceptance of this forecast permits an assessment of the effect of the technology on the bottom on the fundamental constraints of sea power and the laws of the sea that derive therefrom. Of the three fundamental constraints earlier enumerated (the law of the sea, the speed of transit, and the geopolitical, geologistical relationship between land and sea mass), two may be drastically altered. The first of these is the law of the sea. It is already established that emoluments of sovereignty or ownership already obtain in the sea bed. The relatively recently ratified *Treaty of the Continental Shelf* confers sovereign rights to the continental shelf out to a depth of 200m, or to the depth of practicable exploitation, to those nations having borders on the sea. If we grant the technology forecast here, and use the latter of these definitions, it is clear that the ability to exert sovereign rights in the entire sea bed has already received tacit approval.

The next question that will certainly be raised with respect to the law of the sea is the extent to which these sovereign rights in the sea bed confers additional rights in the waters immediately above the sea bed, the waters well above the sea bed, and the free surface above the sea bed. Certainly some rights akin to those which have delineated the legal rights on land from those in the air will develop. If an exact analogy were drawn (though it probably will not be), the right to transit over that portion of the sea bed whose sovereignty has been established would be the prerogative of the sovereign and the cherished freedom of the sea would thereby be abolished. While this extreme limit may not be reached, the right to regulate the jettison of material over occupied areas of the sea bed, the right to regulate traffic above and around occupied areas of the sea bed, and the right to discriminate between peaceable and belligerent transit over occupied areas of the sea bed may well result as emoluments of sovereignty. The establishment of boundary lines, particularly along elements of the bottom that define major geographic entities across which commercial and military passage will be regulated, is not a far-fetched extension of the trend of law and technology.

The second constraint on sea power, speed of transit, will not be markedly affected by the occupation and utilization of the sea bottom. In fact the limited capability of either optical or acoustic sensors will limit speeds of advance on or near the bottom to, at most, ten knots. It should be noted, however, that from a military–political standpoint, this constraint becomes an advantage particularly in the development of military systems having a potential for arms controllability. In such instance a near-bottom capability would permit deployment, covertly or overtly, at the option of the deployer and at the same time to permit credible deployment within or out of range of the threatened protagonist in accordance with the level of political or military crisis.

The third constraint on sea power—the geopolitical, geologistical relationship between sea and land mass—will, on the other hand, be completely altered. This drastic change is best appreciated if each of Mahan's basic laws of the sea is reviewed against the projected law of the sea bottom and future technology of the ocean. The first of these principles relates the ability to use sea power to the topological relationship between land and water masses. If, for example, sovereignty over the bottom is established to a depth of 3000 ft and the accompanying technology developed, then the British Isles, Cuba, and the other Caribbean Islands will lose their status as islands. The Soviet Union and the United States will have a common border no longer separated by the Bering Strait. The entire makeup of the complex of islands and continents which constitute the Southeast Asian island of Sumatra, Borneo, Australia and New Guinea will be completely altered in appearance forming a vast new domain of contiguous sovereign territories. If sovereignty is further established to a depth of 10,000 ft and the accompanying technology develops then the Atlantic Ocean will be cleaved by a great ridge in the Northern and Southern hemispheres, which divides it into an eastern and western basin. In the Pacific, the extensive chains of sea mounts divide the Pacific Ocean into a significant number of basins which are now identifiable by the sea mounts which constitute Wake, Guam, the New Hebrides, the Fijis, the Gilberts, the Marshalls, the Ryukyus, the Kuriles, etc. Even now, these islands are important elements in the strategic outer periphery of the Asian land mass. The occupation and utilization of the undersea portion of these strategic barriers will make even more effective the utilization of the outer islands as a commercial, political, and military balance to the mainland.

If, ultimately, sovereignty is established to the full depths of the ocean, effectively 20,000 ft, then there exists the complex and politically hazardous international task of dividing a territory more than three times as large as that of the present land mass and the establishment of appropriate international relationships which will still admit of the control, management, and utilization of the mineral, animal, and plant rights which are contained therein.

The second of Mahan's principles relates to the conformation of the coastline and its capacity for harbors. Here again the technology of the sea bottom will have a drastic effect. The purpose of a harbor is to provide calm waters to permit the transfer of cargoes across the land–sea interface. The provision of calm water is equally well obtained by going beneath the surface to a significant depth. Offshore platforms for the transfer of oil are already commonplace; offshore submerged transfer points for bulk cargoes have not as yet even been envisaged. Nevertheless their development would change the effective nature of the coastline, transforming California, for example, from a coastal state with limited harbors to a continuous quay for the transfer of goods and products in transpacific commerce.

The third element of the nation's ability to use the sea is found in the number of people in the vicinity of the seacoast having an understanding in the technology and competence of the sea. The technology of the sea bottom has a curious quality. It must employ not only the technology normally associated with naval architecture and marine engineering, but in addition must employ technologies of the type developed for space, for space medicine, and in medical physiology. The development of sea-bed capabilities will alter the balance of national capabilities with respect to technological capabilities. Thus, the paucity of commercial ship construction in the United States because of severe foreign competition is more than compensated for by the extensive aerospace industry, the defense industry, and the advanced capabilities in medicine. This spectrum of technology should prove fortunate for the United States and if properly employed could re-establish U.S. supremacy in the commercial exploitation of the seas.

The fourth element of a nation's ability to use the sea, the character of the people and its government, is probably unaltered by the development of a deep-ocean technology. The initiative of the U.S. system of free enterprise has already played a significant role in the developing capability. It is gratifying to note that the vast majority of the deep-ocean vehicles described herein were developed by private industry with its own funds and that the Navy's Man-In-The-Sea Program is paralleled by ambitious development under private initiative.

The ingredients for an evolutionary step in the use of the sea are present and at the outset are favorable to the United States. With vigor and skill and constant preparedness, there is every hope that a nuclear conflict, which would most assuredly fractionate civilization's most recent and proudest age of land logistics, can and will be averted. Should it occur, the final resolution of the conflict would most surely devolve on the ships which ply the nondestructible highways of the sea and on sea-based military elements which, by virtue of the ocean environment, are least vulnerable to the effects of nuclear weapons. To continue to prevent its occurrence, the manifestation of this sea-based capability must ever be present as even now it is present in the form of our tactical Fleets and Polaris deterrent. With vigor and skill and resolve, the technology of the ocean bottom can and will be advanced. In this advance, the United States must be in the forefront so that when the inevitable and Gordian problems of sovereignty on the ocean floor are raised at the international conference table, the ability to resolve them on terms favorable to international peace and stability is matched by the capability for enforcement.

The thesis is here complete. History, law, technology, and the principles of sea power have been parochially invoked to state the case for "Inner Space." The test of time, or even more quickly the test of analysis, may demonstrate the hypothesis faulty and the assumptions rashly made.

The challenge of the deep ocean may not be the most important international problem of the last half of this century—but it may.

REFERENCES

Caldwell, T. F., 1966, "Seapower: Key to New Ocean Markets," *Undersea Tech.* (Jan. 1966).

Kroll, D. W., 1965, "Oceanography and Naval Warfare," *Aeronautics* **3**, No. 7 (1965).

Parks, Larry G., 1966, "The Law of—and Under—the Sea," *U.S. Naval Inst. Proc.* (Feb. 1966).

Searle, Jr., W. F., "A History of Man's Deep Submergence," *U.S. Naval Inst. Proc.* (March 1966).

Society of American Military Engineers, 1967, "The Challenge of Inner Space: Advances in Ocean Engineering," Technical Proceedings: Northeast Regional Meeting (1967).

Sweitzer, H. W., 1966, "Sovereignty and the SLBM," *U.S. Naval Inst. Proc.* (Sept. 1966).

Terry, Richard D., 1966, *The Deep Submersible*, Western Periodical Co., North Hollywood. California.

CHAPTER 4

Systems-Development Planning

JAMES G. WENZEL

In Chapters 2 and 3 the social, economic, and military needs that have the potential of being met by a vigorous and systematic exploitation of the world's oceans have been outlined. I want to emphasize the terms *vigorous* and *systematic*, because man has to some degree satisfied his needs for survival since the beginning of history by the most simple forms of exploitation of the ocean environment. However, when considered in the aggregate of what the future holds for an ocean systems industry, we have barely penetrated the surface of this potential.

The exploitation of the oceans has all the earmarks of a potentially explosive growth, particularly technologically. In this country we have experienced explosive technological growth patterns several times, such as the national space program and nuclear weapon technology. Because of the breadth and complexity of the field of undersea technology and its system applications, however, the present Department of Defence (DOD) environment, and the very large and dominant commercial potential, the industrial planner is faced with a unique, difficult, and complex task in coping with this growth. It is the purpose of this chapter to present a general synopsis of the ocean systems development situation from the point of view of industry, defining the factors of major significance and outlining methods available to the system development planner to meet the military and commercial needs in this field. When I say *industry*, I refer to any industrial enterprise that is engaged in, or has the potential of being engaged in, business ventures directly related to the military and commercial hardware and service markets associated with the sea.

It perhaps is also important to recognize at the outset, in analyzing the fundamentals of system development planning, that it is not new. The basic principles involved are as old as free enterprise and have been applied by almost every successful businessman in history. To a certain extent, therein lies a danger: namely, overconfidence or neglect due to familiarity.

An industrial firm, to be successful in our dynamic economy, must be critically aware of the changing environments and markets that require product innovations, new technology, and managerial and organizational adaptations. Growth in most successful companies has occurred through frequent product innovations to meet changing demands, and, of equal importance, a willingness to diversify into other product lines. Investigation of the companies most successful in maintaining their position over extended periods indicates that their product mix and markets have changed significantly. These companies have seen many of their products move through a stage of maturity and decline, but they had the foresight to have new products in the innovational and developmental stages, drawing on the basic technological resources of the company to develop them. The recent interest of the aerospace industry in ocean system developments is certainly illustrative of this dynamic process. Industry's ability to predict the dynamic conditions of its environment and react to them through early investment of skills, capital, and financial resources is the objective and result of the systematic programming involved in development planning.

Koontz and O'Donnell (1959) present the following definition of planning:

"*Industrial Planning*—Planning is the function of selecting the enterprise objectives and the policies, programs, and procedures for achieving them. Planning is, of course, decision making, since it involves choosing among alternatives and is strongly influenced by profit/risk aspects of associated objectives."

In the past, key management personnel have carried out this decision-making process primarily with solid "horse sense" and an adventurous spirit—and with great success. Most of the formal planning, however, has been essentially on the back of an envelope, relegated to a minor role in the industrial programming process, with reliance placed more on emotional factors to establish business objectives. Such an approach is untenable in today's increasingly complex military and commercial environment, and our development planning must similarly become more sophisticated. Industry, as a whole, must now formalize application of the principles of system development planning in establishing its business objectives in order to most efficiently utilize our nation's resources in meeting the military and commercial market needs of the future.

A great deal has been said and written in this country on the subject of "program" planning on a systems basis. The need for such planning and the methods involved have not only been accepted by both government and industry, but they have become a way of life for the major system suppliers to DOD. However, the concern in this case has been primarily with the development of *implementation* plans for bringing an accepted system concept into operational status. Because of the complexities of the new business environment associated with the oceans, I would advocate the application of such methods much earlier in the business planning process, that is, to the development planning concerned with initial selection of new hardware and service business objectives. In short, I would suggest that systematic development planning must be a vital and vigorous element in the formation and sustenance of any successful national or industrial ocean systems program.

In approaching this broad subject, let us consider the outline shown in Fig. 4-1. First I would like to review a little of the background that has been covered in Chapters 2 and 3, particularly as affecting industry's present approach to this field. After a brief summary of the military and commercial needs, I will describe the specific methodology available to the planner, emphasizing the necessity for, and an approach to, focusing the required effort to minimize planning costs, to accelerate the availability of meaningful results and to provide maximum flexibility of application. Finally, I will give a few specific examples to add emphasis to the principles involved.

4.1 BACKGROUND

Man is increasing the over-all utility of the oceans for both military and nonmilitary purposes, but the rate of utility growth or the probability of satisfying any specific need is very dependent upon the primary factors shown in Fig. 4-2.

The need must be adequately defined; there must be adequate motivation to have the need satisfied (desire, fear, profits, etc.); and the credibility of satisfying the need must be established. These require sufficient evidence that the technology will be adequate and that the return on the investment will be attractive—that is, dollar profits, or in the military sense, a satisfactory increase in combat effectiveness. In reality these three factors are fundamental requirements for any oceanic operation or enterprise that would be conducted to satisfy any specific need. This criterion can be used to explain the past dominance of military ocean activities over nonmilitary ocean activities: (a) the dictates of military needs are easier to define, (b) military needs are usually accompanied by a strong sense of urgency (high motivation), and (c) operations credibility is easier to establish for military ocean operations.

The U.S. Navy has been the primary instigator of ocean technology advancement, and it does so with public funds for the public welfare and safety. Another very important champion of ocean capability development is found within the other U.S. Government agencies, again using public funds, of course (although certainly miniscule to date) for the ultimate public benefit. A few very significant commercial (nongovernment) ocean exploitation operations are being conducted today, such as shipping, oil and natural gas extraction from worldwide tidelands, fishing, recovery of gems from African offshore deposits, and extraction of minerals and chemical compounds from seawater. The nonmilitary and commercial exploitation potential of the oceans is enormous, and yet this potential remains virtually untapped. In general it can be said that, with few exceptions, the business opportunities attendant upon nonmilitary and commercial ocean utilization have not been sufficiently defined to induce widespread major capital investment on the part of industry. Operations credibility has not been sufficiently established.

From industry's point of view, commercial ocean exploitation suffers from the existence of a technological deficiency (at least from an engineering standpoint) and the lack of an adequate exploration and survey background. It is generally conceded that industry cannot afford to develop the technology and perform all the exploration and survey work necessary to commercially exploit the oceans on any appreciable scale without Government (public) assistance. It is also generally recognized that, at least in its broadest terms, ocean utilization is a vital public matter, and therefore the Government must lead in its development and control in these areas where the public's best interests are affected. A practical and efficient course or strategy, then, would be for Government and industry to join in a team assault on ocean technology development which will ultimately provide the capability needed for military and commercial exploitation of the oceans and their bottom lands.

- Background and industry approach
- Summary of the needs
- Development planning methodology
- Examples of application

FIG. 4-1. Outline.

- Need definition
- Adequate motivation
- Credibility of need satisfaction

FIG. 4-2. Factors affecting need satisfaction probability.

Figure 4-3 graphically presents the evolutionary pattern for the development of marine science and ocean-engineering technology. Various needs are recognized by the Government and the commercial operators who then divert the development of the necessary capabilities, utilizing appropriate in-house laboratories, universities, and industrial organizations. The university's role primarily relates to science—the foundation for development of an engineering technology. Industry carries out contracted R&D from Government and commercial sources, plus its own independent R&D programs. The role of industry relates primarily to engineering technology development and/or the design and supply of operational hardware and services. However, some excellent basic marine-science development is being conducted in a few industrial research laboratories. The Government needs or requirements have first call on technological developments sponsored by the Government, with military interests being served first. Commercial application of industry-sponsored research is obviously direct. However, there is also a major technology fallout from the military programs. It is this commonality of technology that has become so attractive and forms the logical basis for a major Government/industry team assault on the technology requirements for future ocean exploitation.

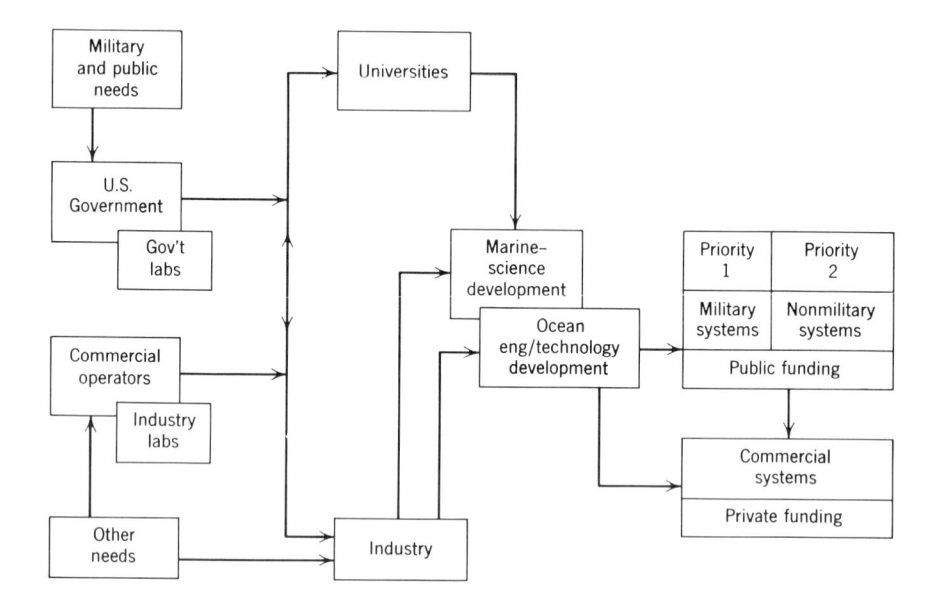

FIG. 4-3. Evolutionary pattern—ocean engineering technology.

The Industrial Approach

The business world is becoming increasingly sensitive to the possible emergence of large-scale ocean exploitation as a major new field for business endeavor; for the present, however, the business world in general has adopted a "proceed with caution" policy for the major reasons shown in Fig. 4-4. The specific degree to which the Government should be involved in the development of our capability to exploit the world's oceans is not understood by either Government or industry and is now the subject of much debate in high Government circles. There are many knotty problems relative to legalities concerning state governments and the cognizance and degree of control to be exercised by various federal government agencies. Because of these and many other unresolved and complex problems, large-scale ocean exploitation enterprise is, with but few exceptions at this time, considered to be a relatively high-risk endeavor for the business world.

In spite of the aforementioned "big-picture" situation from the general business-world point of view a certain segment of industrial suppliers *is* in a unique opportunity position and *can* realize a significant return on its investments in ocean exploitation, while at the same time establishing a good competitive position for future large-scale enterprise. This particular segment of industry is largely composed of firms that are the current major suppliers to the DOD, that is, shipbuilders, aerospace, electronics. Industry members are responding to scientific and military needs, and in doing so they are making major contributions to the development of ocean technology. Although industry's current major effort in ocean systems work is primarily in response to military needs, it is also seriously probing exploitation possibilities in certain specific areas of nonmilitary and commercial market needs, seeking means of applying both old and new ocean technology to future ocean-oriented systems and equipments.

Proceed With Caution
- The degree of government envolvment is not understood
- Problems in international law are not sufficiently resolved
- Business opportunities not sufficiently defined
- Diffused market
- Inadequate exploration and survey information
- Technological adequacy is questionable
- Evolutionary time phasing complicated by numerous unknowns

FIG. 4-4. Industry's approach to ocean exploitation.

Various industry analyses have established that future business opportunities in nonmilitary ocean exploitation will primarily relate to the categories of endeavor or operations and equipments shown in Fig. 4-5. A similar analysis prepared for military operations, particularly undersea, would show a marked similarity. It can easily be shown, therefore, that military and nonmilitary ocean operations have a common meeting ground in these same common denominators of equipment or technology. Weaponry is the only basic technological dissimilarity between military and nonmilitary ocean systems applications.

Major nonmilitary ocean operations	Techniques and equipment common denominators
• Shipping/transportation • Ocean and ocean bottom detailed survey/exploration • Extraction and processing of natural resources • Scientific operations • Public services operations	1. Vehicles/platforms 2. Navigation systems 3. Sensing systems 4. Instrumentation 5. Work manipulations/handling systems 6. Communication systems 7. Information systems/data management 8. Man in sea/man in loop 9. Bottom basing systems
Military ocean operations	Common denominators (1–9) not applicable to weaponry

FIG. 4-5. Military/nonmilitary ocean operations technological base similarity.

The most critical and basic dissimilarity between military and nonmilitary ocean exploitation, from an industry point of view, is in the area of marketing. Many major DOD suppliers (such as aerospace firms) have had little or no experience in dealing with nonmilitary markets, especially in the purely commercial market areas. Commercial ocean exploitation requires a general understanding, a method of approach, and types of operations that are somewhat foreign to DOD suppliers. In supplying systems, for the military (with the exception of fixed-price contracts), a substantial part of the business risk is removed from industry by the customer, whereas in supplying systems, equipments, and/or services for commercial enterprise, industry must assume a substantial share, if not all, of the total actual financial risk. In dealing with the military customer on sales of highly sophisticated systems and/or equipments that are pushing the state of the art industry is actually marketing highly specialized technical personnel services, unique facilities use, and product assurance. In commercial enterprise, industry is chiefly marketing operations efficiency in the furnishing of materials, goods, and services. In dealing with the military industry is a provider of systems and equipments for the user whereas in the case of nonmilitary ocean exploitation industry is both provider and user of systems and equipments which it must operate to earn a profit. In either case, industry must always evaluate its own enterprise effectiveness in terms of its return on investment.

The following, then, is a hypothesis that receives considerable industry support at this time: ocean exploitation *will* emerge as a major new field of business and industrial endeavor. It appears that the most viable approach for industry's participation in the ultimate superior business potential of large-scale nonmilitary ocean exploitation is through supporting present and future military needs. The technology required for the ultimate large-scale nonmilitary ocean exploitation will be developed mainly via military ocean programs. Industry should conduct nonmilitary ocean exploitation development activities in parallel with the development of military ocean programs, and it should seek to adapt the existing and the newly derived ocean technology as rapidly as is proven economically feasible.

Industry advanced planning

The foregoing industry hypothesis appears relatively straightforward and rational, but it presents a most difficult challenge for industry's advanced business planners. The job of the advanced planner in industry is to chart the most near-optimum path for his particular firm's maximum future business development. A firm's policies, structure, capabilities, operations, product lines, and so on, evolve under a compliant "grand strategy" or long-range plan of action, which may or may not appear in tangible form. A firm's long-range plans are usually most successful when all of its varied present and planned operations and actions are compatible or reinforce each other. In the planning for future participation in large-scale ocean exploitation, the industry planner is faced with an almost overwhelming number of unknowns and/or alternatives. Obviously, the advanced business planning problem is most critical for the larger, more diversified industrial firms. Under the hypothesis defined above, the crux of the industrial planner's difficulties lies in the *correlation of parallel* military and nonmilitary ocean exploitation programs for his company that maximize the firm's return on investment.

As indicated earlier, the basic trends in military ocean operations are quite well established for both present and future, and the major problem areas are known. The military establishment usually performs the necessary threat/missions/operations analyses, technical-feasibility analyses, and military-cost effectiveness studies which simultaneously justify and specify its needs requirements. Industry received the benefit of these military studies in that it can respond directly to specified requirements.

The situation in the commercial ocean systems area is quite different. Except in the grossest sense, industry's long-range market predictions and selection of product lines or services for nonmilitary ocean exploitation must be accomplished with far less confidence, particularly if the selection criterion is that of a near-optimum approach for maximizing future business development. The major reason for this state of affairs is that the nonmilitary ocean operations are immature or underdeveloped. This immaturity of nonmilitary ocean exploitation markets is primarily due to their being in competition with land-based enterprise opportunities, as pointed out so well by Dr. Schaefer in Chapter 2.

As previously indicated, most industrial firms are keying their advanced business planning to their support of present and future military operations. Instead of conducting a frontal analytical and experimental assault on the total spectrum of nonmilitary ocean exploitation possibilities to find the most near-optimum complementary enterprise, they are *selectively probing* the future business potential of *specific areas* in considerable depth. The probing of specific areas in detail involves the requirements shown in Fig. 4-6.

4.2 NEEDS

In chapters 2 and 3 we have presented some explicit and implicit present and future needs associated with operations in the ocean environment. Fundamentally, needs or requirements associated with the ocean can be classified by type as follows:

1. Scientific needs: those needs whose satisfaction serves to increase our basic understanding of the ocean environment (oceanography) and its interface with manmade systems. Satisfaction of scientific needs serves to benefit implementation of all other ocean-associated needs. Sponsorship—mostly public.

2. Military needs: those needs that relate to military support/enforcement of national goals and policy. The U.S. Navy has prime responsibility for U.S. military ocean systems development and operations. Sponsorship—public.

3. Nonmilitary needs: fall into two basic categories: (a) public—those needs whose satisfaction serves the public general welfare. (sponsorship—

- Market studies
- Technical feasibility studies
- Costs analysis/costs effectiveness studies
- Resources compatibility studies
- Statistical and fiscal data gathering and analysis
- Merger/acquisition studies
- Limited experimental systems design and fabrication
- Limited experimental field operations and surveys

FIG. 4-6. Commercial exploitation probing requirements.

public); (b) commercial—those needs whose satisfaction permits operations exploitation by an individual or company for a profit (sponsorship—private).

Scientific needs

Our capability to exploit the oceans is directly dependent upon the status of our basic understanding of the ocean environment, because ocean science or oceanography is the foundation for ocean engineering. Effective ocean operations of any type are dependent upon equipment and operating modes that are compatible with operations in the environment. The very broad field of ocean science is, at long last, receiving deserved attention and sponsorship, and considerable progress is being made. Practically all work in oceanography has been, and is being, underwritten by the various world governments, and the bulk of the ocean research is strongly military oriented.

The primary obstacle to rapid advancement in oceanology has been the relatively low level of governmental financial support. A second critical problem is the lack of adequately trained professionals and technicians. The third most critical problem is the lack of accurate and reliable tools and equipments necessary to make the kinds of measurements needed. A summary of the pricipal needs of the oceanologist is included in Fig. 4-7. In addition to the equipment needs in this listing, the oceanologist also requires various kinds of underwater photographic and television cameras, sampling devices, shipborne computers, shipborne modular laboratories, improved scuba gear, buoys, and various types of equipment and personnel deployment vehicles.

Military needs

When we speak of the rapidly growing worldwide interest in ocean exploitation, we must be careful to put this in the proper perspective. The U.S. Navy has been operating on the sea and under its surface for many

- High-resolution ocean-bottom topographical mapping.
- Ocean-bottom sediments and substrata physical structure and chemical charting and analysis.
- Ocean-bottom magnetic-field and gravimetric measurements and analysis.
- Ocean-bottom radioactivity measurements and analysis.
- Ocean-bottom temperatures, thermal characteristics, and mass heat-transfer analysis.
- Turbidity currents detection and measurements.
- Ocean-surface wave measurements and wave spectra analysis.
- Seismic wave detection and measurement, and analysis.
- Internal wave detection and measurement.
- Large-mass water transport measurements.
- Current velocity and eddy measurement.
- Water-temperature and temperature-profile measurement and analysis.
- Salinity, salinity variations, and general physical and chemical properties analysis.
- Water-density measurements.
- Water-pressure measurements.
- Direct reading velocity of sound measurements.
- Ocean ambient-noise measurements and charting including marine-life contributions.
- Acoustic energy propagation through seawater measurements and analysis with variable water depth, and temperature condition, bottom conditions, scutting conditions, ocean-surface roughness, and other ambient noise.
- Electromagnetic energy two-way propagation through the sea–air interface measurements and analysis.
- Visible-light transmission through seawater measurements under various conditions.

FIG. 4-7. Equipment needs.

years, and the Government's expenditures for antisubmarine and prosubmarine warfare, as well as for other sea-based system objectives, has not been exactly small. What we are really referring to is the potential utilization of the resources of the world's oceans for commercial purposes and an extension of the operational freedom of our military forces into the ocean depths. The importance and potential utilization of the ocean bottom for military purposes has been clearly defined by Dr. Craven in Chapter 3. Major extension of our military capabilities to the ocean bottom is some

Systems-Development Planning   69

years away, however, and many existing operational missions of the Navy, such as antisubmarine warfare and navy logistics, carry both great technological challenges and potential markets for the ocean systems industry. With the exception of weaponry (as previously noted) there is also a large potential technology fallout for commercial operations in these areas—for example, communication, search, and transportation.

A list of the basic tasks that must be implemented by the Navy is provided in Fig. 4-8. The list also indicates where responsibilities are shared by other services.

In addition, some of the needs for future military ocean operations are summarized in Fig. 4-9. Obviously a list of this nature is subject to many changes depending on the world political situation. Of particular interest, as outlined by Dr. Craven in Chapter 3, is the increased emphasis on deep submergence design and development, including man in sea.

Nonmilitary needs

Nonmilitary ocean operations are conducted primarily as parallels of extensions of land-based operations and with few exceptions they will probably

Offense missions/operations
- Strategic bombardment—sea based   USN–USAF
- Tactical bombardment—sea based   USN
- Enemy sea supply line and operations interdiction   USN
- Blockade   USN

Defense missions/operations
- Own shipping protection against all types of attack   USN
- Own fleet protection against all types of attack   USN
- Defense of CONUS against ocean-borne assault   USN
- Defense of CONUS against enemy SLBM   USN–USAF–ARMY

Support missions/operations
- Troop and equipment transport   USN–ARMY
- Assault forces shore landing   USN
- Assault forces logistics support   USN
- Salvage and rescue   USN–Coast Guard
- CONUS attack early warning   USN–USAF
- Fleet readiness and effectiveness   USN
- Reconnaissance and intelligence   USN–USAF–ARMY

FIG. 4-8. Military missions.

- Continued support of the deterrent mission
- Emphasis on underseas operations of all types including man-in-sea
- Strong ASW flavor in general
- Increased emphasis on deep-subergence design development
- Increased emphasis on ocean surface-ship traffic and enemy-submarine surveillance
- Emphasis on enemy submarine long-range detection
- Emphasis on limited-war techniques and equipment development
- High-speed underwater vehicle development (40–100 Knots)
- High-speed underwater weapon development (40–120 Knots)
- Hydrofoil and/or surface-effects vehicle development (40–160 Knots)
- Renewed interest in naval mine warfare
- Increased emphasis on communications and Navy Command and Control (C&C) development
- Submarine defensive measures development
- High-speed automatic integrated fire-control systems development emphasis
- Automatic oceanographic data gathering and processing
- Increased emphasis on oceanology personnel recruitment and training

FIG. 4-9. Military ocean systems needs.

always have nearshore continental land ties. The world continental-shelf areas and their waters appear to offer the most promise for nonmilitary ocean exploitation. The world continental-shelf total area is roughly equal to 8% of the world ocean total area, or an area about the size of Africa. For most practical purposes, a 600-ft depth capability would permit exploitation of about 10 million square miles. In the Americas, the shelves total about 3.7 million square miles.

The continental-shelf geological structure is an extension of the continental land mass structure and therefore probably contains the same basic minerals and other materials. However, as indicated by Dr. Schaefer in Chapter 2, there is strong evidence that the geological structure of the ocean basins is quite different from the continental-shelf structure and may be quite devoid of most concentrations of minerals. The sediment layers of the deep ocean basins are another matter, and in some areas have given evidence of containing valuable minerals such as manganese nodules. Other materials found in considerable concentration in the abyssal sea-floor sediments are cobalt, iron, nickel, copper, and radium.

The world continental-shelf areas constitute a new arena for the oil industry. Natural gas and petroleum are already being extracted from the tidelands in the United States and the tidelands of other areas in the world. Other continental-shelf areas may some day profitably yield basic materials for agricultural fertilizers, such as phosphorus pentoxide (P_2O_5).

The continental shelves and their shallow covering waters contain the essence of the marine-life food cycle. These areas provide an environmental zone suitable for growth of marine plants and marine organisms that are the food for higher orders of marine life (fish, and others). These shoal-like waters are also important as fish spawning grounds and as a habitat for young fish which attract the larger fish.

Obviously, the specific commercial ocean market is undefined. However, for planning purposes, present and future nonmilitary ocean exploitation needs will generally be related to the areas indicated in Fig.4-10.

Some of the ocean exploitation possibilities listed are suitable for enterprise only on the basis of serving the public welfare and/or public interests, whereas others would be suitable for enterprise on a private or commercial basis. Examination of this list in its entirety, however, dramatically points up to the need for Government concern for, and involvement in, all facets of ocean exploitation or utilization.

Summary

It is evident from my previous comments on the present status of ocean technology that a joint Government/industry approach is the best strategy, and that satisfaction of military needs and requirements constitutes the most obvious approach by industry for entry into the ocean systems business.

This situation is largely due to the inadequate definition of commercial needs and requirements, as well as the lack of sufficient technological background upon which to base design of hardware. It is generally conceded that, aside from weaponry, military hardware developments and associated technological advancements will be applicable for use on commercial system developments and that this commonality represents a minimum-risk approach for industry.

The extent to which industry realizes the benefits of this approach will be dependent on the ability of advanced planning functions to envision the direction of ocean science developments and fit them to the capabilities and desires of industry. The lack of general knowledge, the broad spectrum of possible ocean missions, and the high-risk aspects of the problem require the application of planning disciplines and methods that provide order and allow for successive decision points, and resulting focusing of the effort. It is also important that these methods provide flexibility that permits rapid determination of objectives, modification to meet the dynamic market

Transportation and shipping operations
   Surface craft
   Submersible vehicles
   Ground effects vehicles
   Water-based helicopters
   Hydrofoil/hydroski craft
   Amphibious craft
   Manned or unmanned bottom crawlers
   Cargo handling systems
   Underwater tubes and pipelines

Ocean Natural Resources Exploitation Operations
   Ocean-bottom mineral and chemical mining
   Oil and natural-gas prospecting and extraction
   Commercial fishing and fish farming
   Ocean vegetation harvest and processing
   Sea/fresh-water conversion
   Seawater materials extraction
   Sea motion electric-power generation
   Sea-based thermal-electric power generation
   Ocean-bottom exploration, survey/geological analysis
   Ocean physical and chemical state charting
   Oceanographic/meteorological analysis and forecasting
   Ocean-floor seismic stations
   Marine microbiological engineering

Miscellaneous Ocean Exploitation Operations
   Submerged material and goods storage facilities
   Ocean-bottom housing
   Harbor and waterway development
   Marine equipment anticorrosion/antifouling systems
   Navigation and communication aids
   Ocean and harbor traffic recording and control
   Marine data recording, processing, analysis, dissemination
   Quick reaction rescue operations
   Personnel and equipment safety gear
   Ocean sports and recreation
   Waste disposal systems
   Shoreline development

FIG. 4-10. Nonmilitary ocean exploitation needs.

condition, and continuous iteration for updating of previous management decisions.

4.3 SYSTEMS-DEVELOPMENT PLANNING METHODOLOGY

As previously discussed or implied, development planning is essential in new-business programming, since it represents a systematic approach to the creation and marketing of profitable industrial products. Its methods and disciplines create management capabilities geared for rapid assessment of complex alternatives, while giving objective considerations to profit/risk motivations inherent in any business venture. Essential features of system development planning as previously discussed are shown in Fig. 4-11.

The principal factors involved and the results to be achieved are exactly the same as for formulation of business plans in any other field. The differences are primarily in the difficulties encountered in carrying out the planning and in the details. The principal factors involved are expanded in Fig. 4-12 to show the raw material which the planner must have or must develop as he carries out his task. Emphasis here is placed on the complicating factors of military/commercial technology commonality and the interrelated risk and potential profits.

In carrying out a systematic planning process two approaches have been used in the past: selection by exhaustion and selection by exception, as depicted in Fig. 4-13.

Purpose:	Development of business objectives and plans for their management
Approach:	Systematic isolation of critical alternatives readily facilitating key business decisions
Principal factors:	• Market analysis • Technology assessment • Business resources
Results: established	• Long-range product plans • Business strategy • Marketing plans • Research and development plans • Investment Plans

FIG. 4-11. System development planning.

Marketing
- Customer needs and budgets
- Competition
- Market potential

Technical
- Cost effectiveness
- Technology state of art
- Technical advancement requirements
- Military/commercial commonality

Business
- Manpower and facility resources
- Funding requirements
- Risk/profit potential

FIG. 4-12. Planning Factors.

The decision on which approach to take has been a major pitfall for planners in the past. In the first approach, major emphasis is placed on the technical aspects of the problem early in the planning process, and the systems analyst and his scientific support staff have a "field day." In an ever-branching network of technological, market, and industrial alternatives, all business possibilities are enumerated. Every aspect of each field is exposed, since this information will ultimately be required by the decision makers in weighing the business potential of various courses of action. At first glance, the exhaustive nature of this approach appears to be most attractive, since the more information available to the planner the better the decision and assessment of risk; however, experience has shown that all too often this approach fails by virtue of its inherent complexity. The enormous number of alternatives begins to exceed the planner's ability to handle them, and the time required exceeds the patience of top management so that the efforts generally come to naught. It is this sort of exercise that has made many a business executive wary of the planner who speaks glibly about employing systems-analysis assistance in accomplishing his objectives.

Another factor of note is that the results of any analysis are only as good as the input data. Since the field of ocean engineering is in its infancy, marketing data and customer technical requirements are not well enough understood, and excessively detailed analysis involved in the exhaustive approach are not warranted.

The second approach, selection by exception, assumes a strong initial management input based on understanding of the field and the company's

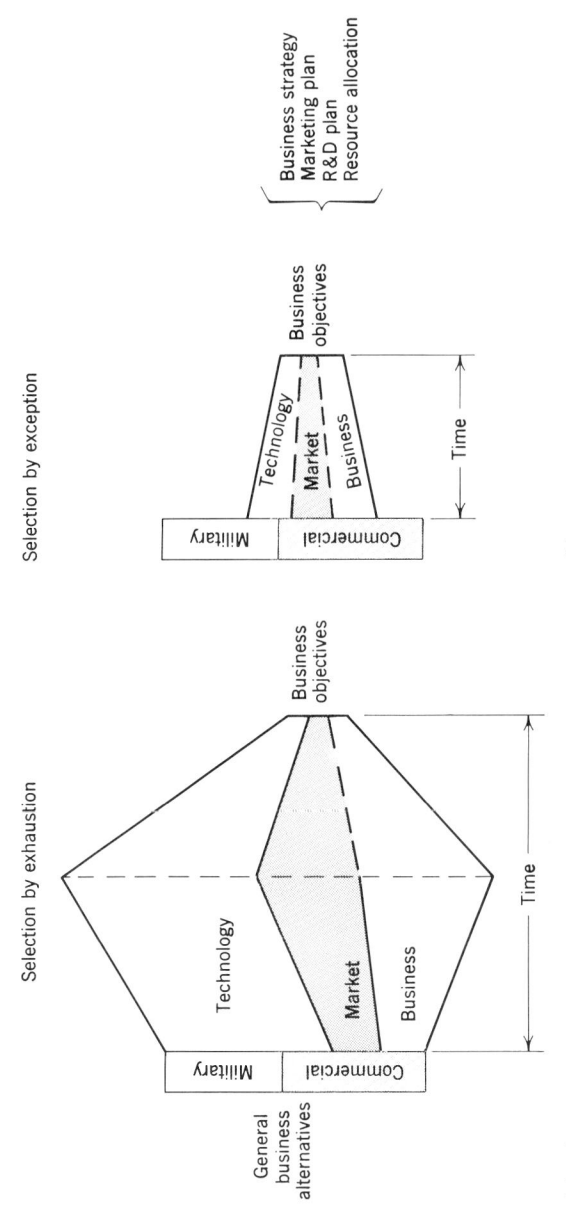

FIG. 4-13. Approaches to selection of business objectives.

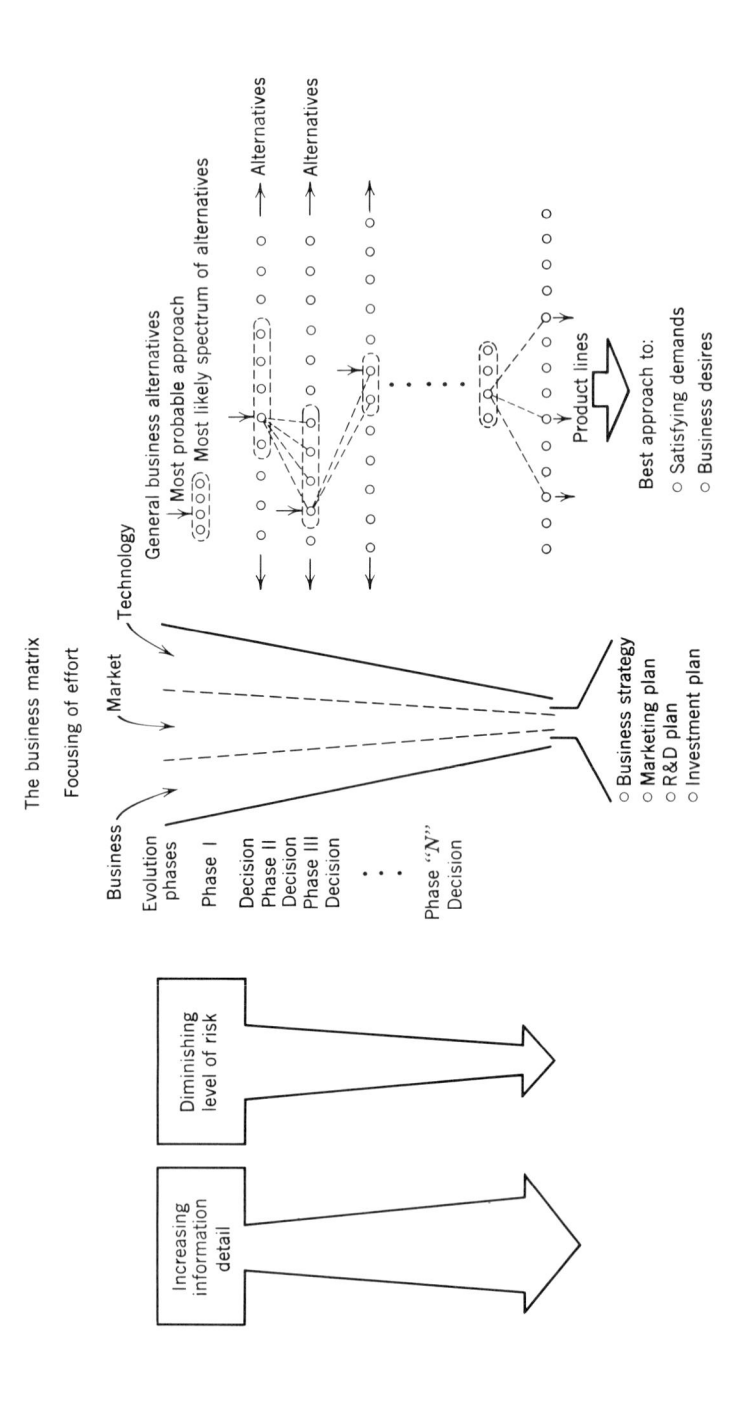

FIG. 4-14. Selection of business objectives by exception.

basic capabilities and business desires. There is a much better balance between the input of the major factors involved, and the time permitted for airing the first results is much shorter. It should be recognized that selection by exception carries with it an element of risk in that early elimination of alternatives can result in reduced business scope. This disadvantage is offset, however, by the earlier availability of useful results, which can always be refined as additional data become available.

In carrying out the selection by exception process, the planning operation is programmed using a "business matrix," as illustrated in Fig. 4-14.

The study is broken down into phases which carry along principal business, market, and technology factors in parallel. At the end of each phase, a series of alternatives is presented, and decisions are made as to the most profitable or logical approach, eliminating less attractive avenues. The planning is initiated considering broad mission, market, and business aspects, increasing the level of information detail during each successive phase, narrowing the field of interest, and decreasing the level of risk. Finally, the remaining elements are assembled into a business strategy and its associated marketing, research and development (R&D), and investment plans.

A typical matrix for a 5-phase study is shown in Fig. 4-15.

Mission analyses and requirements are utilized to define major system and subsystem hardware elements and to forecast trends in performance requirements as a function of time. An assessment of the state of the art (SOTA) in applicable technology permits determination of limiting problem areas and of advancements required in the SOTA to make achievement of predicted hardware performance requirements a reality.

The marketing requirements are straightforward but severe, as pointed out earlier, because of the immaturity of the entire field of ocean systems. Military and commercial customer budgets are analyzed and future business potential is forecast, first, by broad missions, and then by hardware products. Competition is evaluated in order to assess the market penetrability. The availability of R&D funding for technology advancement is predicted, and the degree of commonality assessed for application of the technology to both military and commercial products.

Any development planning effort must begin by determining the constraints under which the new business program must be developed. These constraints include existing product plans, commitments and investments, and existing facility and manpower resources, as well as the normal budget constraints. New-system hardware development decisions are preceded by a preliminary decision to either make or buy such components. If the hardware area is adopted or considered as a potential product line, the facility, manpower, and R&D funding requirements associated with technology advancement requirements must be determined. The resulting potential return on

	Phase I	Phase II	Phase III	Phase IV	Phase V
Technical	Mission analysis — Data acquisition and mission requirements	System definition and requirements forecast	Subsystem hardware definition and performance requirements	Technology SOTA assessment and limiting problem definition	Business strategy • Long-range product plan • Marketing plans • R & D budgets forecast • Facility investment plan • Acquisition plans
Marketing	Customer budget analysis — Market forecast by missions	Customer budget and programming analysis — Hardware market forecast	Competition assessment	Technology market analysis — Military/commercial commonality	
Business	Existing product plan review — Commitments and investments	Capability assessment • Facilities • Manpower • Budgets	Make or buy — Facility requirements	Skill requirements — R & D funding requirements — ROI	

FIG. 4-15. Typical business matrix.

investment (ROI) may cause a re-evaluation of the initial make-or-buy decision. The results of these various studies are then integrated to establish the elements of the new business strategy.

The business matrix represents an aggregate of technical, economic, and business information ordered to provide a rational basis for the selection of business objectives. It is characterized by phases of general description and detail becoming progressively more specific as selected mission functions and requirements are delineated. In order that this branching process does not get out of hand there must be an organized program plan that requires that decisions be made at appropriate points during the evaluation process. The necessary focusing of effort is illustrated in Fig. 4-16 as a rolling mill where missions of interests are reduced in number at the end of Phase I by consideration of basic product plans and commitments and the apparent market potential. At the completion of Phases II and III, the resulting hardware system and subsystem elements of interest are reduced in number by consideration of the resource requirements and the level of competition.

Following the definition in Phase IV of the advancements in technology required, the areas of interest are still further reduced by consideration of the R&D funding requirements, the probability of successful achievement of the required advancements within the available funding, and the degree of application of the results to both military and commercial hardware system objectives.

The risks of premature elimination of a marginal area by this selection-by-exception process, as mentioned earlier, are obvious. Such risks can be greatly reduced by simply making a second iteration, and in effect deferring rather than completely eliminating consideration of some marginal product areas.

The process described here permits a great deal of flexibility in establishing the length of time to be permitted for the effort and hence the level of detail to which each step can be carried. By far the greatest variability exists on the technical side, where the level of effort can vary from simple experience judgments to detailed operations and systems analyses. Anything worth doing is worth doing well, but this must be moderated by the depth of detail available in other areas. If the market analysis is but a series of educated guesses, it is senseless to waste time on extensive systems analysis and optimization.

Because of the large variability on the technical side of the planning effort, it is worthwhile to dwell on the major aspects of this process, indicating factors of importance and suggesting some specific methods of interest.

Mission Selection

Missions are a manifestation of economic demands and needs which, in a general manner, are descriptive of some technological, military,

80   Systems Planning

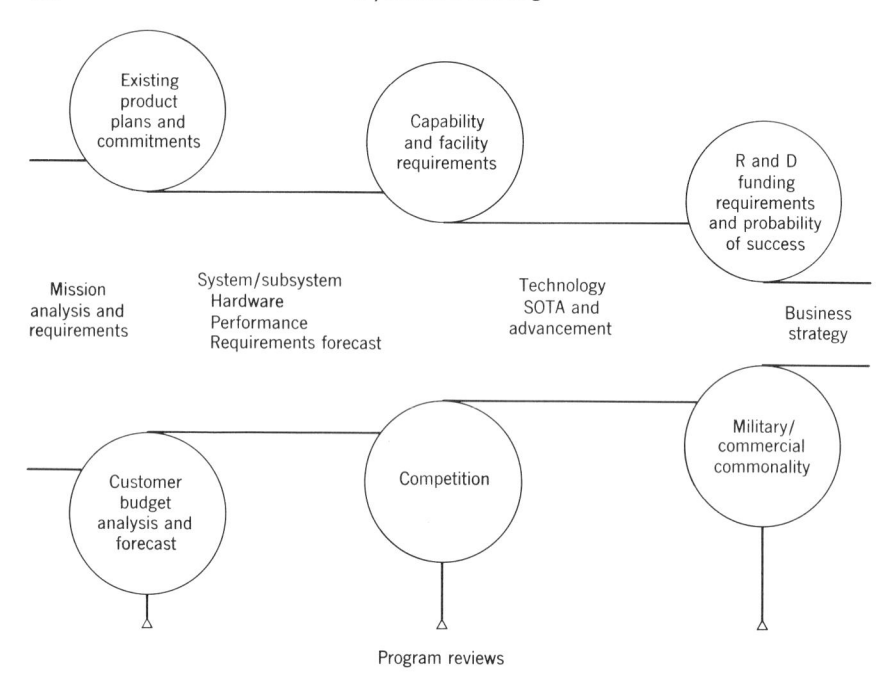

FIG. 4-16. Focusing of effort.

and/or industrial requirement. Military missions can be general in nature, as illustrated by the terms "strategic," "defensive," and "offensive," or they can be more specific, as indicated by terms such as antisubmarine warfare (ASW), ship defense, and sea-based bombardment. In a similar fashion, commercial missions can be described in general terms such as "sea logistics," "transportation," and "mining," or in more specific terms such as "surface shipping," "well-head completion," and "shallow-water mining." Hence the planner must identify and categorize the spectrum of missions in total so that management has ample opportunity to select those missions that satisfy its desires and fit the company's capabilities.

In selecting potential business areas management must use its industrial intelligence organizations to their fullest, Marketing, sales, and technical divisions must maintain pace with economic and technological demands, user capabilities and interests, competition, and their own abilities so that, when called on to assist management in selection of missions, they are fully capable of doing so.

Mission selection in the broad sense carries with a relatively small risk, since management can envision technological advancements and their probable impact quite easily. The principal problem lies in determination of which hardware alternatives will be most effective in meeting the customer's needs and the company's business objectives. Thus the prime risk in management decisions at this stage is in selecting the hardware approach to pursue and determining how well this approach satisfies the objective of profit and growth.

Hardware definition

As a result of the mission analysis, one or more mission alternatives will evolve, selected by management as best suiting its desires and capabilities. Various conceptual systems will also have been identified for purposes of defining functional hardware requirements and equipment. Hardware definition can be accomplished in a comprehensive and orderly manner using methods that expose the technical and operational ramifications of a given concept in increasing detail. This is done by means of an ordered, interrelated series of documents which break down the operational environment into functional elements which ultimately can be associated with hardware requirements. The basic approach to hardware definition is as follows: (1.) translate mission objectives into functional terms; (2.) develop preliminary design requirements and constraints; (3.) develop functional flow diagrams; (4.) perform engineering design and tradeoff studies; (5.) translate design requirements into integrated hardware specifications. The initial process in the basic approach is shown in Fig. 4-17.

The basic functional elements required to carrry out the missions are defined in ever-increasing detail at various levels until sufficient detail is established to permit writing preliminary hardware design specifications for hardware components of interest.

Mission requirements are obviously specified at the first level. Major system elements can be deduced from the second level, and preliminary definition of subsystem hardware components from the third level. Commonality of subsystem functional requirements, as determined from preliminary requirements allocation sheets (RAS), are then used to define preliminary hardware design specification.

However, this type of functional analysis is inadequate to complete the job for the following reasons: (1.) functions may be inadvertently omitted; (2.) tradeoff studies are not clearly indicated; (3.) interface requirements are not defined. All of these information gaps can be filled by extending the analysis of consideration of the actual operational sequence required to meet the mission objectives, through the use of functional flow diagrams (FFD). Typical functional flow diagrams are shown in Fig. 4-18. The basic functions are expanded and shown as specific operational events. Alternatives are immediately apparent at all levels, permitting tradeoff

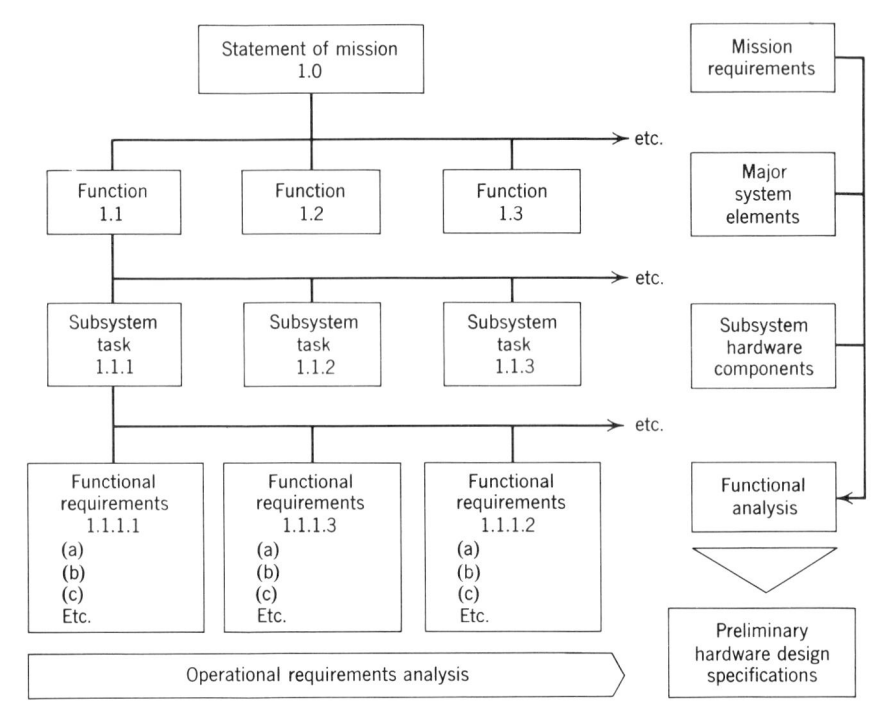

FIG. 4-17. Functional analysis.

studies to be made as necessary. The preliminary RAS's can now be modified to incorporate any changes or additions which resulted from preparation of the FFD's.

A typical form of RAS is shown in Fig. 4-19. Design requirements and constraints in carrying out the specific function are noted on the sheet. Integration of appropriate RAS's defines specific hardware-component specifications and also the interface between related requirements.

Technology

The unique aspects of hardware definition permit the application of technological disciplines in the determination of limiting problems. This phase is probably the most important consideration to the doing functions and management, since feasibility, and hence business success, is dependent on the exposure and early solution of critical limiting technology problems. In the exploding technological growth of any field, a company's competitive advantage depends heavily on technical excellence and innovation. Commonality of future requirements and existing equipment determines the

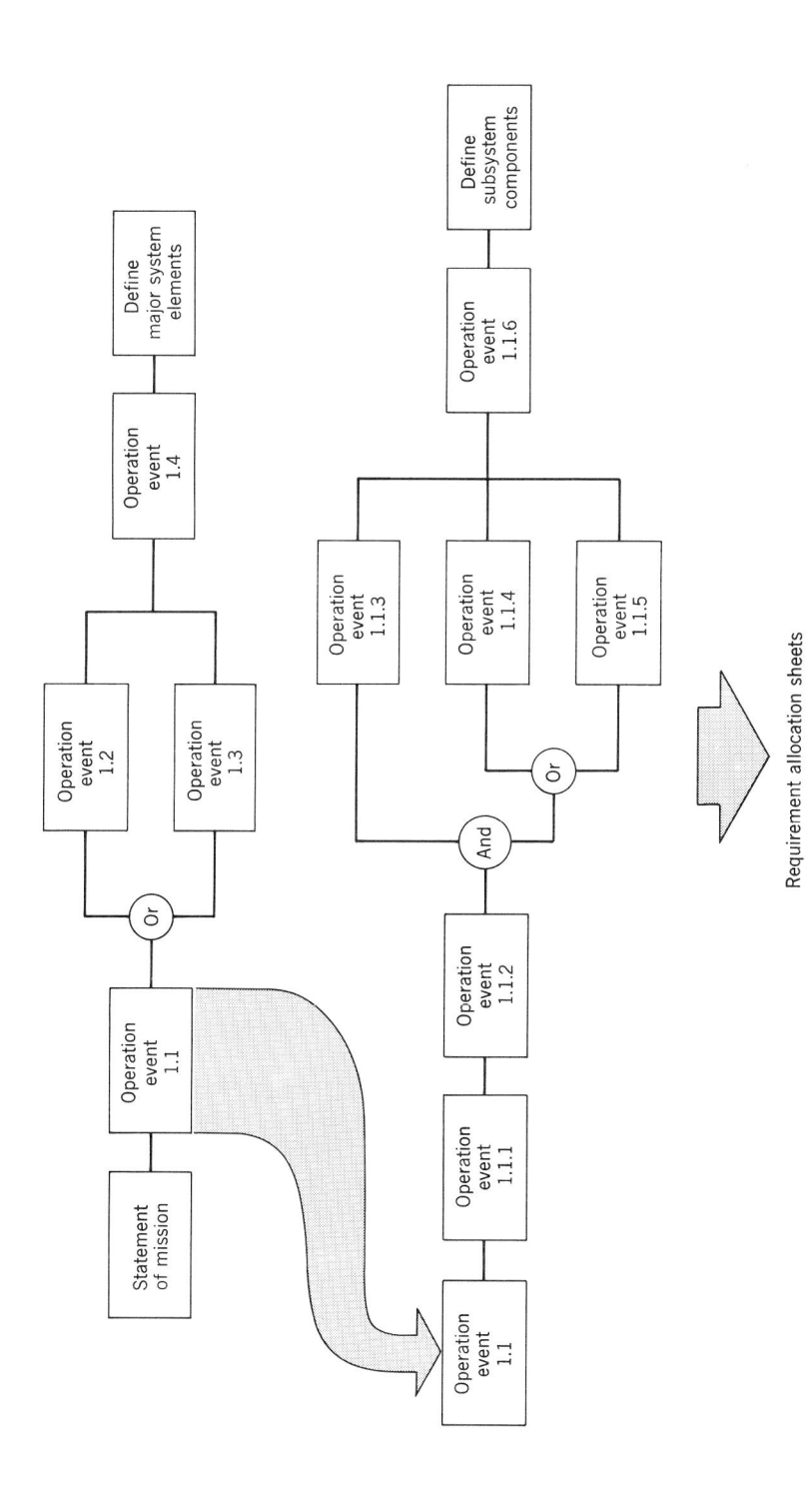

FIG. 4-18. Requirement allocation sheets.

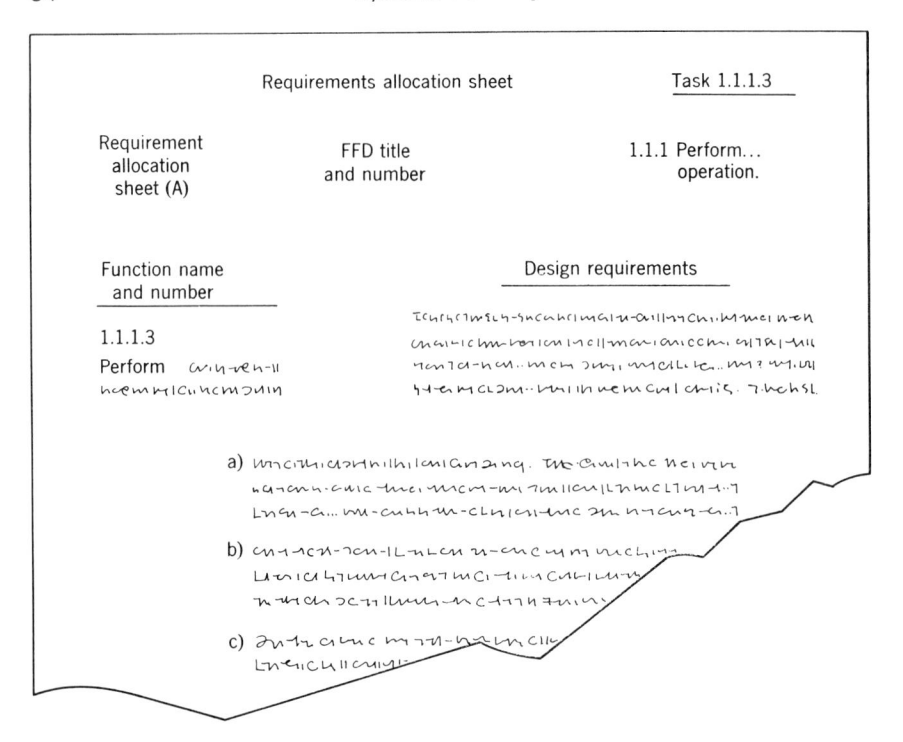

FIG. 4-19. Requirements analysis.

degree to which a company can expand its capabilities into broader mission satisfaction. State-of-the-art assessment determines the extent to which present capabilities are applicable. Differences between hardware performance requirements and limiting technological problems establish the R&D investments required for demonstration of feasibility so essential in the present DOD and commercial competitive environment.

Assessment of the technology advancement requirements is another area in the planning process which can easily be overdone. A simple method of minimizing the effort required and maximizing the output is the use of a system hardware/technical discipline matrix indicating limiting technological problem areas. Such a matrix is shown in Fig. 4-20. All hardware elements of the systems of interest can be listed with technical disciplines involved across the top. Areas where limiting technological problems must be resolved if basic performance requirements are to be met are designated by one symbol; problem areas which are not limiting but which require new developments are designated by another symbol.

Another useful output of this effort can be derived by simultaneously

Systems-Development Planning

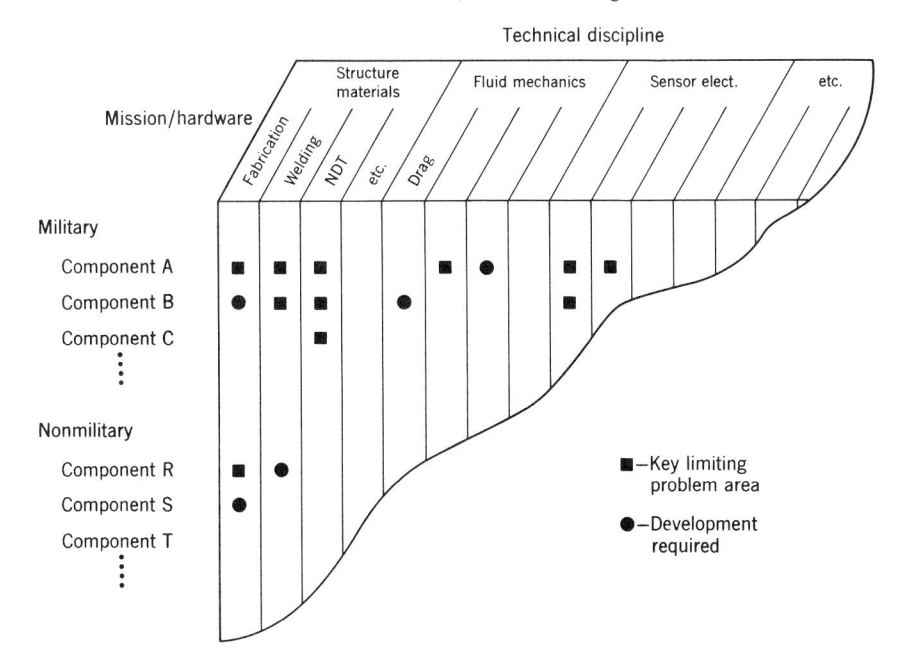

FIG. 4-20. System hardware/technology matrix.

[Matrix figure showing Technical discipline (Structure materials: Fabrication, Welding, NDT, etc.; Fluid mechanics: Drag, etc.; Sensor elect.; etc.) vs. Mission/hardware (Military: Components A, B, C, ...; Nonmilitary: Components R, S, T, ...). Symbols: ■ — Key limiting problem area; ● — Development required.]

evaluating the existence of limiting problems and the company's ability to achieve a solution. A simple designation in the matrix of company capability or knowledgeability will give an immediate qualitative evaluation for the best match of limiting problems and the ability to solve them—a vital factor in the development planning process.

The limiting problems must obviously be defined to permit a more detailed analysis. This can be done using a problem definition sheet, as depicted in Fig. 4-21. The amount of detail on this sheet can be expanded to provide, in essence, candidate work statements for the industrial research (IR) and industrial development (ID) required.

Business strategy

With the completion of the first four phases of the business matrix, all the groundwork has been laid for defining the business objectives and the corresponding strategy.

Profit and risk tradeoffs can be made against hardware alternatives, using the results of market, technology, and capability analyses. It is now feasible to evolve the business strategy. The what, where, when, how, who, and why are known, or at least the steps that must be taken to acquire this knowledge have been defined. Management is now in a position to make major product decisions with a full understanding of the risks involved.

Following the establishment of the basic system and product objectives, all the information necessary to lay out a long-range implementation plan is also available. Limiting problem areas requiring R&D action are planned, scheduled, and funded to meet the selected system schedules and requirements. Hardware and service development plans are initiated, facility and capability acquisitions are programmed, and suitable technology and hardware marketing plans are detailed.

- Statement of the problem
- Significance
- Recommended ID/IR effort and priority
- Minimum practical funding/time span
- Military/commercial hardware applications
- Related existing company programs
- Experienced personnel

FIG. 4-21. Limiting problem/capability definition.

In summary, system developement planning—beginning with management decisions on the initial course of action, and the planner's focusing of effort through assessment of technological, market, and company capabilities—has resulted in the development of potentially profitable, minimum-risk business objectives. The methods outlined are an outgrowth of the need for an orderly procedure in evaluating a broad spectrum of alternatives in a logical sequence. In an iterative manner, initial management mission hardware selections are modified to reflect refinement of knowledge. Industrial intelligence functions are brought to bear on the problem of minimizing risk, with the result that a business strategy is developed. Management plans to achieve the various business objectives are generated, and procurement, marketing, and research and engineering plans are executed.

The principle of selection by exception must be used if meaningful results are to be obtained in a reasonable time. The iterative process used in the evaluation of hardware serves to minimize risk and increase the ease with which this planning approach can be applied. It also makes possible a continued upgrading of the business strategy as the environmental dynamics shape the course of business.

Unique commercial aspects

With but a few relatively minor changes, the methodology just described is applicable to both military and commercial development planning. The primary differences are in the relative emphasis on the various aspects within the business matrix. The unique aspects of planning associated with commercial exploitation can be summarized as shown in Fig. 4-22.

In describing the technical approach to the business matrix heavy emphasis was placed on mission hardware. The functional flow diagrams and RAS's were described as primary tools in defining major system and subsystem hardware elements and their associated performance requirements.

Hardware is obviously of major interest to the commercial planner also. However, much of the profit potential and market interest lies in the supplying of operational services. This is unique to the commercial field. Immediately, then, another advantage of functional flow diagramming becomes evident, since, as was shown in Fig. 4–18, all operational support service requirements are defined simultaneously with system hardware requirements. These services can then be treated as an endproduct in exactly the same manner, considering investment requirements, capabilities, and profit potential.

> Technical
>   Services and/or hardware
> Marketing
>   Proprietary data
>   Market intelligence difficulties
> Business
>   Early emphasis on economics (ROI)
>   Investment risks/profits potential greater
>   Patent coverage

FIG. 4-22. Unique commercial aspects.

The field of commercial marketing is quite different from military marketing, and a completely different approach must be taken. This can be the subject of a special treatment in itself, and no attempt to go into it in detail will be made here. Of particular note, however, is the recognition that it is different, requires highly experienced people, and, in a new field such as ocean engineering, is frought with difficulties. Industrial secrets are much more jealously guarded, both on the part of a potential customer regarding his future plans and on the part of the contractor regarding industrial innovations. When this is added to the higher investment requirements, the increase in business risks is obvious.

Some of the differences in emphasis are illustrated in Fig. 4-23.

In discussing the business matrix, the importance of a balanced effort in technology, marketing, and business aspects of planning was emphasized. This is primarily true for development planning associated with military products. In the commercial area, much heavier emphasis is placed on the economics and marketing aspects early in the planning process, reflecting the risks involved and deferring technical emphasis until a considerable focusing of interest areas has been carried out. The effectiveness of the marketing efforts may be reduced, however, as a result of the difficulties described in commercial industrial intelligence.

A vital and overriding consideration in the planning associated with any business venture is the establishment of the break-even points. "When will I get my money back?" is a question of key importance to any top business executive. The differences that can be expected between military and commercial business ventures in ocean systems are illustrated in Fig. 4-24.

Obviously, this relationship is heavily influenced by the particular field of interest, such as deep submergence or fish farming, and the state of maturity of the particular area of interest. In general, however, there is a period of rather heavy investment prior to any significant penetration of the market. The rate of investment can, in most cases, be somewhat reduced after successful sales to a sustaining level required to maintain the competitive lead. The break-even point is defined by the crossover of the accumulated cost and income curves. Military-program investments can have an

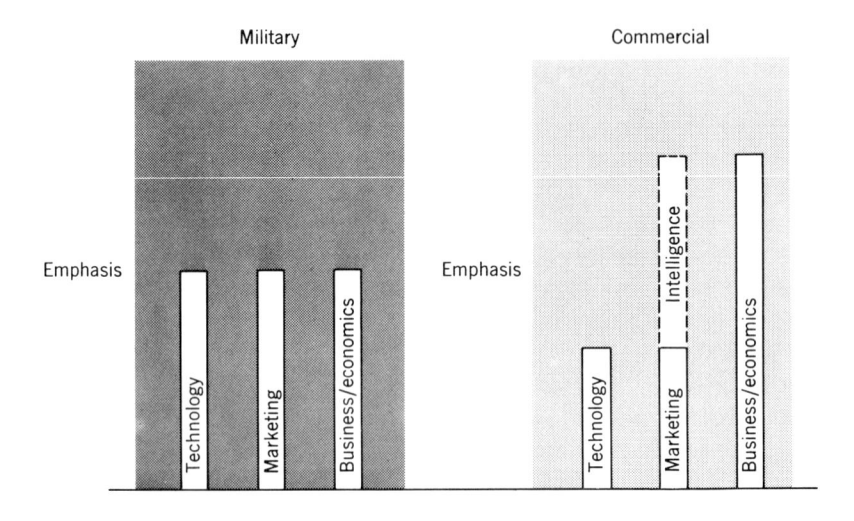

FIG. 4-23.   Differences in emphasis.

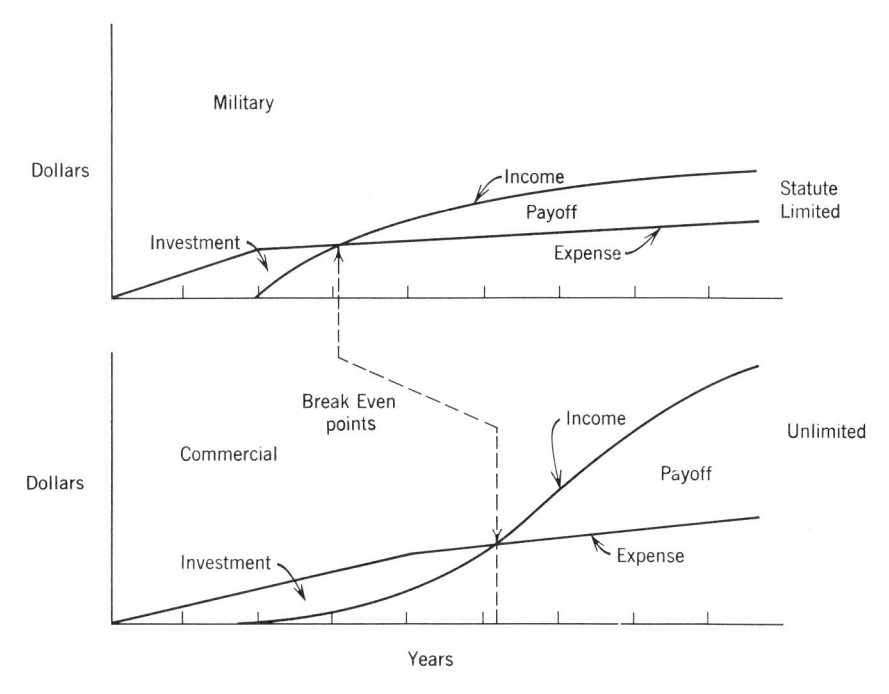

FIG. 4-24. Investment alternatives.

earlier payoff with less risk, but with profits definitely limited by statute. The commercial break-even points are farther downstream, in the 1970's and 1980's but the potential profits are enormous. The job of the development planner is to predict these break-even points and to move them as close together as possible as rapidly as possible.

Implementation: systems versus component approach

There is an interesting sidelight that is important to discuss at this time relative to the implementation of the plans defined for management approval, particularly in the commercial field. Once the complete system has been defined—the basic hardware requirements and the key technological problems—management is faced with the following questions:

Are we in the systems business or in the component business?

If in the systems business, should we proceed with development of (a) the complete system or (b) key components on a systems basis?

If in the component business, should we select a single key component and (a) optimize its application to several systems or (b) optimize its application to a single mission?

These are key questions. The decisions carry major funding and business potential implications.

This leads to the general subject of the systems approach versus the component approach to the development of future systems. Defining our terms: The *systems approach* is the planning of a system-development program in its entirety, including equipment, subsystems, and all the other elements which comprise the total system, including services. The *component, or nonsystems, approach* is the planning for the development of components, subsystems, equipment, and other elements which may or may not be carried as part of a specific system-development plan.

Obviously, the selection of which approach should be taken is very much a function of the specific system being considered, its operational availability, dates involved, and the business interests and capabilities of the specific company doing the planning. For many small companies, no option exists, and the component approach is mandatory. Larger companies do have this choice, and a brief comparison of key aspects of the two approaches is of interest.

The systems approach, the principal factors of which are shown in Fig. 4-25, is fairly obvious and permits advances in the state of the art of several areas to be carried out concurrently, minimizing total system development time, and permits "total package" cost forecasting and funding. It is obvious that system development planning, by definition, assumes that the systems approach is at least of potential interest to the company's management.

The principal considerations involved in selection of the component approach are shown in Fig. 4-26. In the very immature state of development of the commercial ocean systems field, the component approach is quite attractive, even for a company that is basically systems oriented.

There are cases where, despite the existence of a requirement, complete systems development (and, consequently, complete systems planning) may not be warranted. This may occur where the unknowns are great; the concept is complex; technology is to be achieved. Perhaps technical feasibility has been established through paper studies and analyses. But before investment and operating funds should be obligated, R & D would be funded to proceed only far enough to demonstrate physically a critical technique, component subsystem, or even a vehicle.

In many of the subsystem technologies, there are justifiable grounds for proceeding with state-of-the-art technological improvements without waiting for a specific system requiring the improvement. Predominantly, the component approach does result in components and subsystems of high technical excellence.

Of particular note is the attractiveness of the combined approach of carrying out development planning on a complete systems basis, and

Systems-Development Planning

- Optimizes total system performance and schedule
- Permits "concurrency" approach
- Ensures coordinated SOTA advances
- Facilities component development funding
- Permits forecasting of total system cost/funding
- Permits cost/mission effectiveness tradeoffs
- Responsive to mission requirements analysis

FIG. 4-25. Systems approach.

- Advances SOTA in specific areas
- Increases component schedule/performance accuracy
- Tends to base system performance on SOTA rather than mission requirements
- Minimize investment commitment
- Permits playing the "wait and see" game
- Risks potential lack of key technical development
- Jeopardizes timely system support planning

FIG. 4-26. Component approach.

then developing only those key components where technology limits the achievement of the complete systems objective. By this approach, a company can stay with the field, make a major contribution with a minimum investment, and play the "wait and see" game before getting in too deep. This approach characterizes the ocean system industry today—and the importance of system development planning to ensure that the right components are being worked on is obvious.

4.4 APPLICATION EXAMPLES

Military systems

The military needs summary given in Figs. 4-8 and 4-9 shows that many mission and system needs require and obtain major industrial attention. Perhaps few, however, will have as much direct and immediate impact on the development of ocean engineering in this country as that associated with the Deep Submergence Program of the U.S. Navy.

Figure 4-27 indicates some of the military missions and potential nonmilitary mission commonality associated with deep submergence. In 1963, following the tragic loss of the Thresher, the Navy's Deep Submergence Systems Review Group defined many of the military and commercial benefits of an aggressive deep-submergence technology program. The ocean systems planner did not have to be very astute to recognize that the Deep-Submergence Systems Project (DSSP) would give the entire ocean systems field the incentive it needed to get underway. Examination of Government budgets, and plans, technical problems, and the strong applicability of aerospace technology made it clear that the aerospace industry would play a major role in this new field and that the Navy's Deep Submergence Rescue Vehicle (DSRV). Program would be the first significant business potential. As indicated, these facts were all clearly defined essentially 2 years before the Navy actually solicited bids on the program.

The basic mission requirements postulated the desired system in broad terms, defining such elements as a submersible rescue vehicle which was air transportable, transported by submarine or surface vessel to the disaster site, etc. To get a "first cut" at major system and subsystem elements, a functional analysis can be carried out as shown in Fig. 4-28.

Military	Nonmilitary
• Bottom-based bombardment	• Mobile deep-ocean research
• Bottom-based ASW	• Homesteading/mining
• Bottom mapping	• Survey/prospecting
• Secure logistics	• Submerged cargo transportation
• Data acquisition	• Environmental research
• Rescue/recovery	• Rescue/recovery/disposal
• Salvage	• Salvage
• Cable laying	• Cable laying

FIG. 4-27. Deep submergence missions.

Note that at the first level only the major functions are listed, breaking each down into its functional elements, until there is sufficient definition to write the associated RAS's. Remember that these are preliminary, in order to get a first approach to the subsystem hardware performance requirements.

Extension of the analysis to functional flow diagrams is shown in the next three figures. Level 1 is shown in Fig. 4-29. Immediately, refinements in the system hardware requirements become evident:

1. If the ground transport is also air transportable, not only is its availability assured but only one needs to be procured.

2. The maintenance system should be air transportable and also modular, precluding the need for separate home base, dockside, and sea transport checkout and maintenance gear.

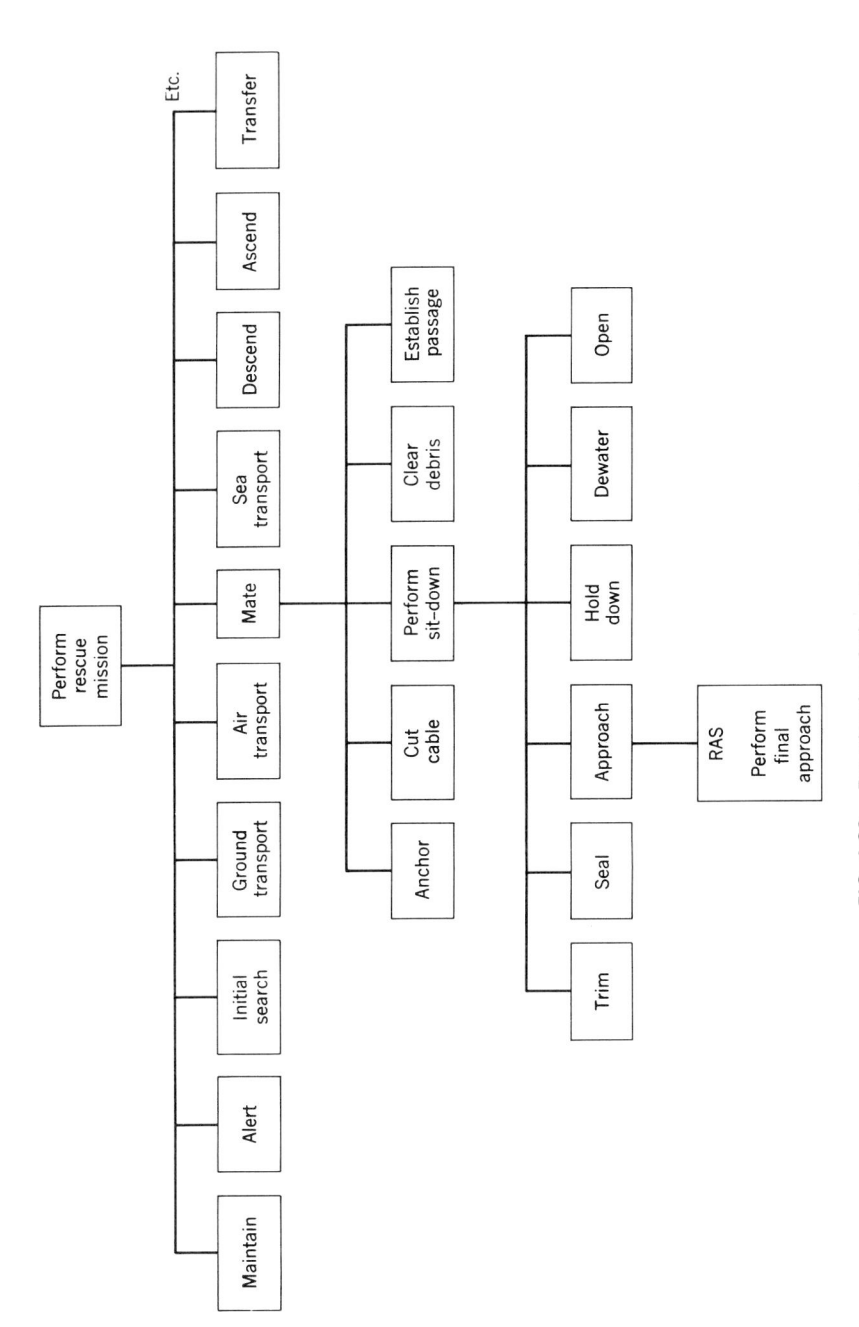

FIG. 4-28. Functional analysis—rescue mission.

94   Systems Planning

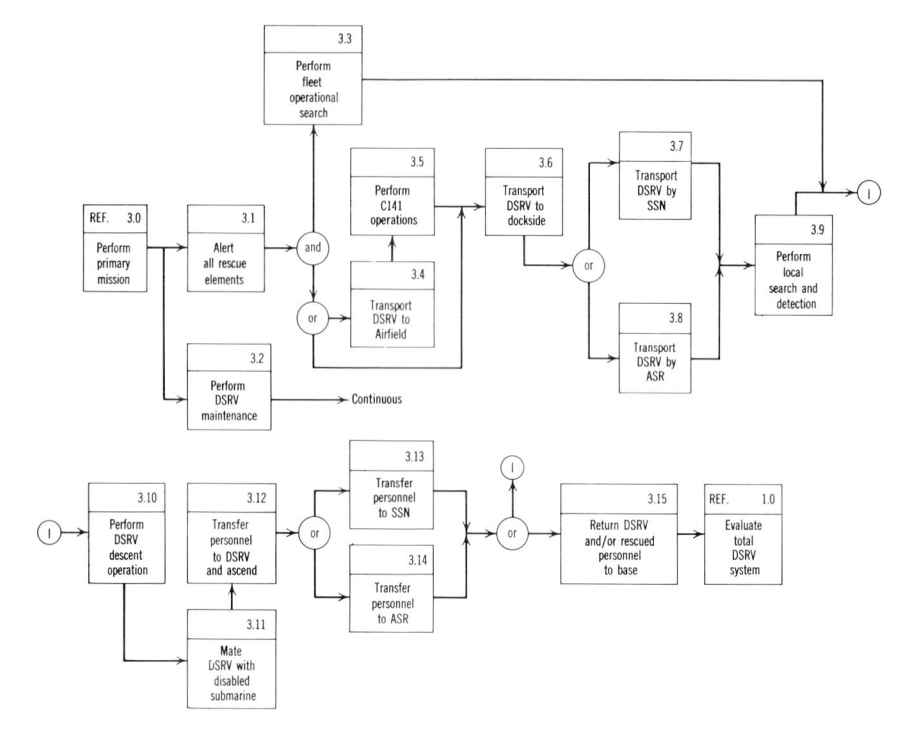

FIG. 4-29. Perform primary mission.

3. Two seat transport modes are possible—submarine and surface ship.

This last point is a prime example of a tradeoff study which can now be initiated to determine whether one or the other, or both, should be included in the planning.

Levels 2 and 3 are shown in Figs. 4-30 and 4-31. Breaking down the functions to this level begins to indicate subsystem elements. For example, the mating operation probably will require that the DSRV carry homing beacons, an anchoring system, a manipulator for cutting cable and simple debris clearance, a mating skirt, a skirt dewatering system, etc.

The third level provides sufficiently detailed definition so that an RAS such as that shown in Fig. 4-32 can be written for the block "perform final vehicle approach."

RAS—final approach

It is important to note that the RAS comprises the requirements associated with a specific *functional* element and not a given piece of hardware. That is the reason the sheets are termed "allocation sheets"; they

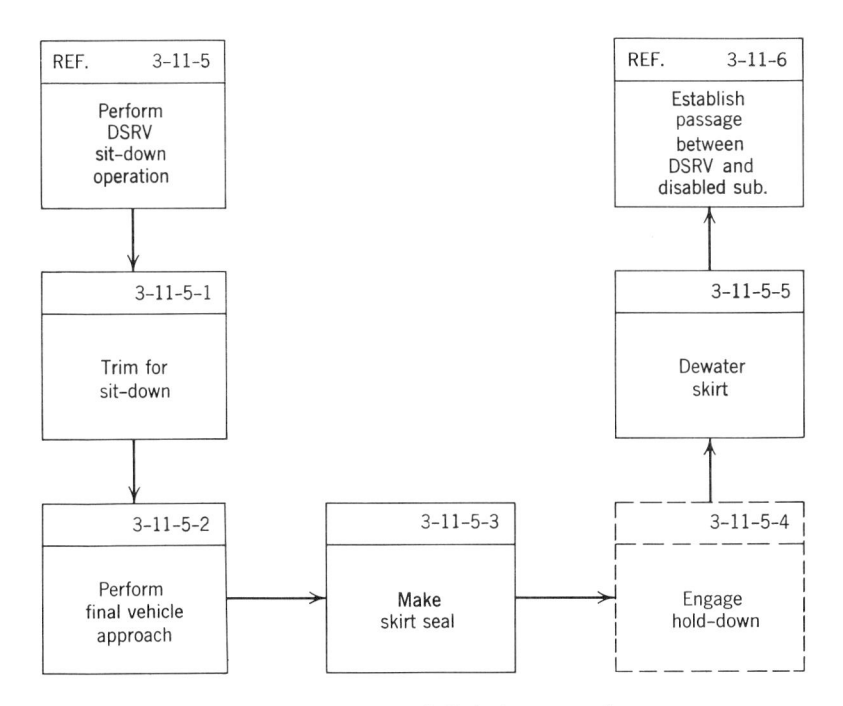

FIG. 4-30. Perform DSRV sit-down operation.

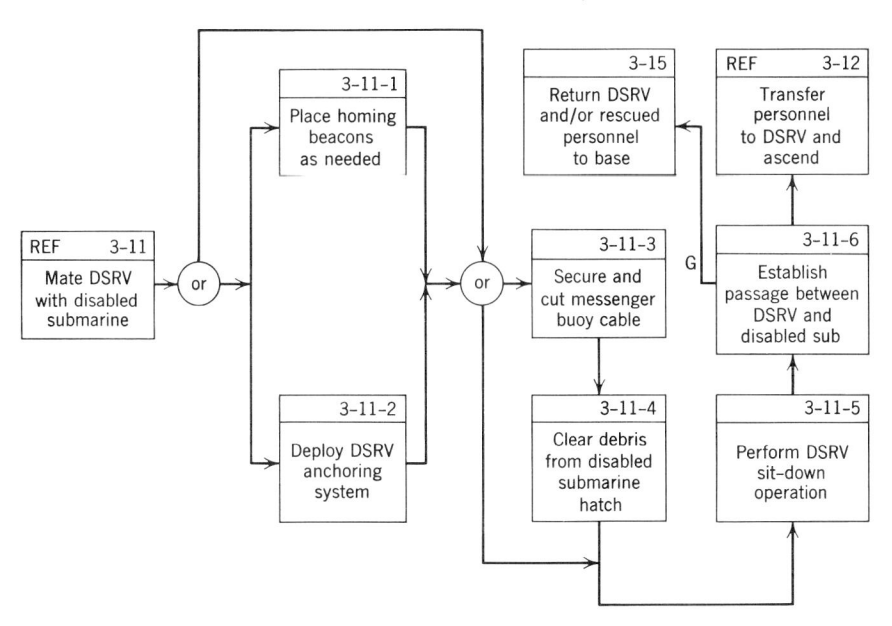

FIG. 4-31. Mate DSRV with distressed submarine.

95

define requirements that are allocated to, or associated with, a specific function. By examining all the RAS's, a set of design specifications can be prepared, both for the major system elements and for the sub-system hardware elements.

Another major aspect of the planning problem is to establish the appropriate system operational availability date, since working backward from this key milestone gives the time available for any subsystem research and development that may be required.

As indicated earlier, the DSRV requirement was clear in late 1963. With a desired operational availability date of 1968, and a two-year development period, only about 2 years were available for organization and development of major technology requirements. The rescue vehicle is therefore really an example of relatively short-range planning. The impact of the system-development planner can be much more significant when one can look farther downstream. For example, Dr. Craven defines the need for future bottom-based tactical and strategic weapon systems, the basic missions for which are not entirely clear, and the operational availability date for which is 10 to 15 years away. The process just defined is still completely valid, but now the level of confidence is lower (since mission requirements are much less clear), and therefore the amount of detail needed should become lower. But *some detail* is important, or the functional elements will not be clear enough to give real meaning to the state-of-the-art technology assessment task.

FIG. 4-32. Requirements allocation sheet for function 3.11.5.2.
(*Below*)

Requirements Allocation Sheet (A)	FFD Title & Number	3.11.5 Perform DSRV Sit-Down Operation
Function Name and Number		Design Requirements
3.11.5.2 Perform Final Vehicle Approach		The final vehicle approach shall bring the DSRV transfer skirt mating surface into contact with the rescue hatch of the disabled submarine without damage to the DSRV or submarine for the purpose of accomplishing a seal between the skirt and hatch.
		The vehicle propulsion and control system shall be the primary means of producing the forces and

moments required to perform the final vehicle approach. The approach shall be possible under the following conditions:

1. Ambient current—1 knot
2. Vehicle inclination from vertical—up to 45 deg in pitch and/or roll
3. Maximum visibility—3 ft
4. Maximum depth—design collapse depth of existing nuclear submarines

A supplemental approach system shall be provided which will augment the forces and moments produced by the propulsion system, in the event that a 1-knot ambient current is exceeded and/or the unsteady flow effects exceed those resulting from a 1-knot ambient current. The significant feature of the auxiliary approach system will be a physical link between the DSRV and the disabled submarine. For detail on a supplemental approach system see Design Study No. ———.

(a) Forces and Moments shall be provided by the propulsion and control system as follows:

	Magnitude	Rate
Forces	(lb)	(lb/min)
Longitudinal		
Lateral		
Normal		
Moments	(ft-lb)	(ft-lb/min)
Pitch		
Roll		
Yaw		

The propulsion and control system shall be capable of producing these forces and moments at any level up to the maximum magnitudes listed above for periods up to ——— min. The lateral and normal forces and the moments shall vary sinusoidally with amplitudes equal to the maximum rates listed above. Steady state and/or oscillating forces and moments shall be produced simultaneously only to the extent that the maximum values listed above are not exceeded.

(b) A shock absorbing system shall be provided which will limit the impact deceleration to 0.60 g. The maximum relative design velocity between the DSRV and the disabled submarine just prior to contact is 2ft/sec.

(c) Sensors shall be provided to obtain relative position and attitude of the mating skirt and disabled submarine hatch at nominal distances from 0 to 3 ft as follows:
  (1) Normal distance to $\pm$ 0.5 in.
  (2) Tangential distances to $\pm$ 0.25 in.
  (3) Relative attitude in DSRV pitch and roll planes to $\pm$ 0.5 deg
  (4) Relative attitude in DSRV yaw plane to $\pm$ 2.0 deg

Design Requirements

(d) A total of ——— kW-hr power shall be provided for the final vehicle approach maneuvers.

(e) Provisions shall be made to jettison any part or all of the auxiliary approach and mating systems to ensure a free departure of the DSRV from the disabled submarine in the event of system failure. Operation of the release shall be possible in the event of a main and/or auxiliary power system failure.

(f) Means shall be provided to automatically maintain a preselected relative position and attitude between the DSRV and the disabled submarine. The relative position and attitude shall be maintained within the following tolerances:———. An automatic hold capability is required during normal approach operation and in conjunction with the approach and mating systems.

(g) The reliability allocation for final vehicle approach shall be as follows for the various systems:———.

FIG. 4-32. Requirements allocation sheet for function 3.11.5.2.
(*Conclusion*)

A system hardware/technology matrix which can be used to define limiting technological problems has already been described. Such a matrix can be applied at almost any level of detail desired—for example, to the DSRV system, where the subsystem hardware elements would be the ordinate, or to a general field such as deep submergence. In the latter case the ordinates can be various systems to operate at or near the ocean bottom. The detailed technology matrix can be summarized for purposes of management presentation of the over-all business strategy.

As of 1963, technological problems in the field of deep submergence were concentrated on the areas indicated in Fig. 4-33.

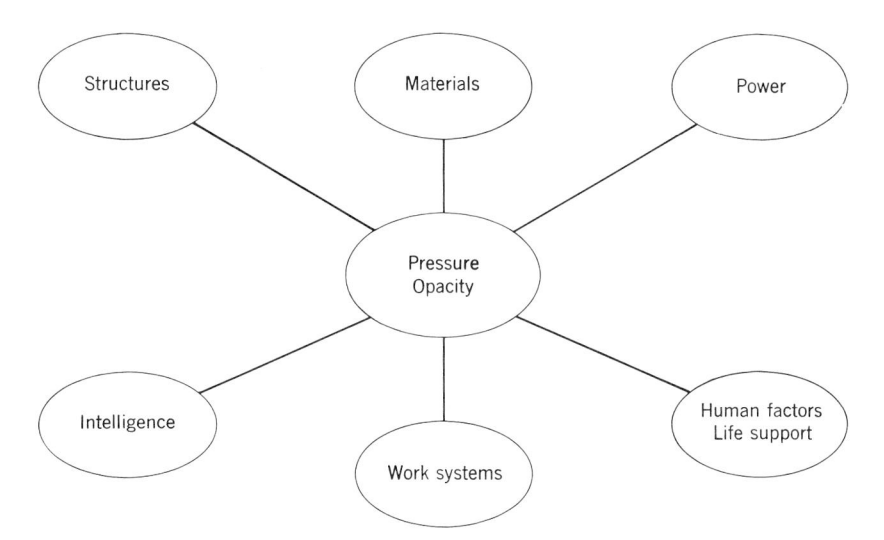

FIG. 4-33. Deep submergence —technology problem areas.

The two major differences in the environment affecting the design of system hardware are the severe pressure due to depth and the opacity of the medium. The high pressure requires that man be protected by a strong and lightweight, shell, which introduces limiting problems in the area of structure and materials—not limiting in regard to survival, since man has gone to the deepest known parts of the ocean, but limiting in terms of structural efficiency as influencing payload, hydrodynamic efficiency, etc. Relative to power systems, the pressure influence is direct in requiring pressure balancing and is indirect by presenting a weight and space problem. For example, a limiting problem for a bottom-based mobile bombardment system is the need for a lightweight, compact reactor power system. Similar problems can be defined in each of the areas shown.

The opacity of the medium introduces serious technical limitations in the area of intelligence subsystems, such as sensors, the ability to see, and the ability to communicate. The ability to perform useful work is affected both by the ambient pressure and its resultant pressure capsule and by the inability to see, particularly in the presence of any sediment in the water.

In the area of life support and human engineering development work is required—particularly in the areas of contaminants possible in extremely long missions and the life-support problems associated with extending man's depth capability as a free diver.

A typical summary of limiting technology areas and those in which development is required is given in Fig. 4-34.

It is easily seen that when deep-submergence systems are considered as a whole, the majority of the problems lie in two areas: (a) Structure and materials, and (b) intelligence subsystems. However, for recovery, salvage, or repair missions, there are very serious limitations in the technology associated with work systems. At shallow depths, it is a diver tool development problem; in the deeper depths, it is a manipulator development problem.

As mentioned earlier, when the system development planning process has been carried to the point of the missions, hardware, and services of interest, and the limiting technology or development problems have been assessed, the implementation plan requires that the decision on approach be made. In the case of deep submergence the systems approach could, for instance, include interest in fabrication of complete vehicles and their associated support system hardware. If the planner were not completely

Missions	S/M	P	I	H/L	W
Bottom-based mobile bombardment	■	■	■	□	
Bottom-based military/homesteading	■	□	□	□	
Bottom survey, exploration and mapping	■		■		
Secure personnel/cargo transport		□			
Data acquisition	■		■		
Rescue/search	■	□	■		
Recovery/salvage/disposal	■		■	■	■
Cable laying/inspection/repair	■	□			■

Limiting problem areas*

■ Limiting technical problems
□ Development problems

*Refer to Fig. 4-33.

FIG. 4-34. Technology status.

Systems-Development Planning 101

convinced of the market this new field would present (and there are many unknowns), he might reason that for these vehicles to be sold in large numbers, the fundamental problems of doing useful work—namely, visibility, controllability, and tools—must be solved; he might therefore elect to concentrate on one or more of these components, or he might recommend going both routes.

For some missions, the large systems firm has no choice and must go both routes. For example in antisubmarine warfare, the primary limiting problems lie in intelligence systems, detection, location, communications, etc. Therefore a firm interested in large vehicle platforms must also conduct major developments in sensors. Such a firm may not elect to produce the electronic hardware, but it must do development work in that area, since the future of the system depends on it.

A most significant fact for the planner to recognize in preparing his implementation recommendations is the level of investment required to support the chosen approach. Management must be shown the complete package, accompanied by the appropriate predictions of return on investment, and the alternatives available. As indicated at the beginning of this chapter, when we view the future potential and character of the ocean-systems field, the stakes and risks are high.

The Lockheed Deep Quest research submarine (Fig. 4-35), and the major commitment which it implies, is an almost classic example of implementation of the planning process just described.

FIG. 4-35. Deep Quest.

102   Systems Planning

As Dr. Craven defines in Chapter 3, the trends in operational depths of submarine vehicles is clear, with operation in the deep-ocean basins a long-range goal. There are many near-term missions concerned with continental-shelf operations, both in the development of concepts such as mobile man-in-sea and those associated with petroleum and mineral exploitation. Both basic research, such as in underwater acoustics or mineral exploration, and component development such as prototype power plants, etc., require large payload capabilities. In addition, the earliest potential market for Government-funded vehicle system hardware was associated with the objectives of the Navy's Deep Submergence Systems Project. These factors were given serious consideration, and so was born the concept of Deep Quest.

The general arrangements and physical data for Deep Quest are shown in Fig. 4-36. Relative to 20,000-ft technology, the pressure-hull configuration and material were aimed at developing and providing the potential of multiple intersecting-sphere pressure capsules and 200,000-psi-yield maraging steel. To develop the concepts and techniques of mating with

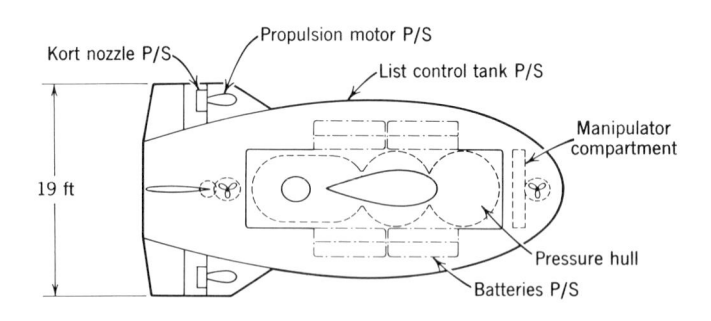

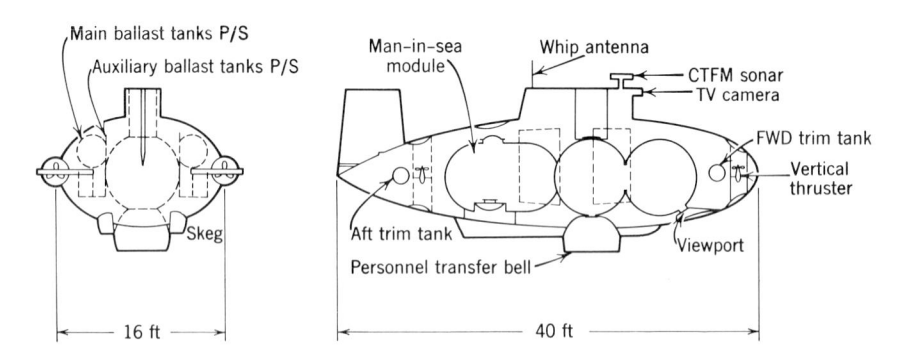

FIG. 4-36. General arrangements.

submerged submarines or habitats, the boat was equipped with a mating skirt and 6 deg of control (5 integrated). A large modular payload capability was provided to give maximum flexibility as a development hardware test bed. The design payload weights were set by the corers required for geophysical exploration, and the payload volumes set by provision for mobile diver support as shown. Provisions were made for inclusion of advanced concepts in search and work sensors when they become available. Backfit fuel-cell capability was included; and due to the development status of manipulators, a large well providing full y–z motion of the manipulator mount and maximum flexibility for incorporation of advanced manipulator concepts was provided.

In presenting the concept to management the return on investment was estimated on the basis of its near-term lease potential, defining the break-even points on this conservative basis, and also the much more significant long-range impact on future submarine and ocean-system hardware and operational services business.

Commercial systems

It is becoming increasingly evident to industry planners that *two areas* of nonmilitary ocean enterprise almost completely dominate the scene: (a) transportation and shipping, and (b) recovery of natural resources. Dr. M. B. Schaefer in Chapter 2 points out that as the world becomes more populous and closer knit, the sea as an over all resource must be increasingly and more effectively utilized. The world logistics problem of the future will be singularly crucial, and will most certainly require a complete systems-type approach for effective solutions. The sea transport and terminal elements will be of the most vital importance. The past 20 years experienced by our railroads during their program of modernization is only a mild analog of what is in store for the maritime shipping industry.

The oceans and their bottom lands will eventually be required to contribute their share of sustenance and materials needed by man. A mere glance at the world's fish-catch growth trend, shown in Fig. 4-37 and which is discussed in detail by Dr Schaefer in Chapter 2, vividly indicates the future demands to be placed upon industry and the ocean technology. Certain materials are already in short supply as demand growth rate has exceeded the growth rate of land source development. Oil, natural gas, certain metals, and agricultural chemical fertilizer ingredients are but a few examples of these materials. The ocean deposit of such materials must and will be developed to augment our land supplies.

As pointed out earlier, the same basic planning methodology can be utilized for both military and commercial systems, incorporating some changes in emphasis. The requirements were defined for earlier application

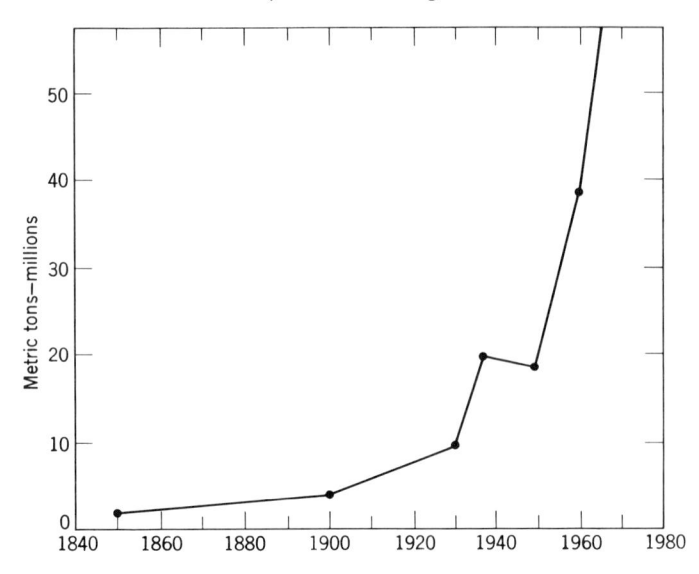

FIG. 4-37. World fish-catch growth. U.S. Dept. of the Interior—Bureau of Commercial fisheries data.

of economic analysis and tradeoff studies as a means of focusing the efforts, particularly within a given field. Considering ocean mining as an example, Fig. 4-38 depicts a typical evaluation procedure. Broad system definition and predesign efforts are initially carried only to the detail necessary to obtain first-cut costing of major functions. For ocean mining these are as shown—exploration and sites development, the actual ocean mining operation, benefication of the ore, and the transportation operations. A system-design synthesis model is then exercised considering such operating variables as the mining rate, value of the ore, and depth of the mining operations.

To such a parametric analysis must be added the basic technical, marketing, and business constraints. For example, the ore concentration must be above a minimum value to be given further consideration. Reasonable estimates of the probability of market penetration, in view of competitive companies, land sources, etc., must be made. The company may put some constraint on the acceptable profitability (for example, not less than 30%) and the total capital investment it is willing to make. The result is definition of a group of feasible products for management to consider. Eliminating the unattractive and marginal candidates permits initiating the development planning process previously described in more detail and with fewer candidate product areas.

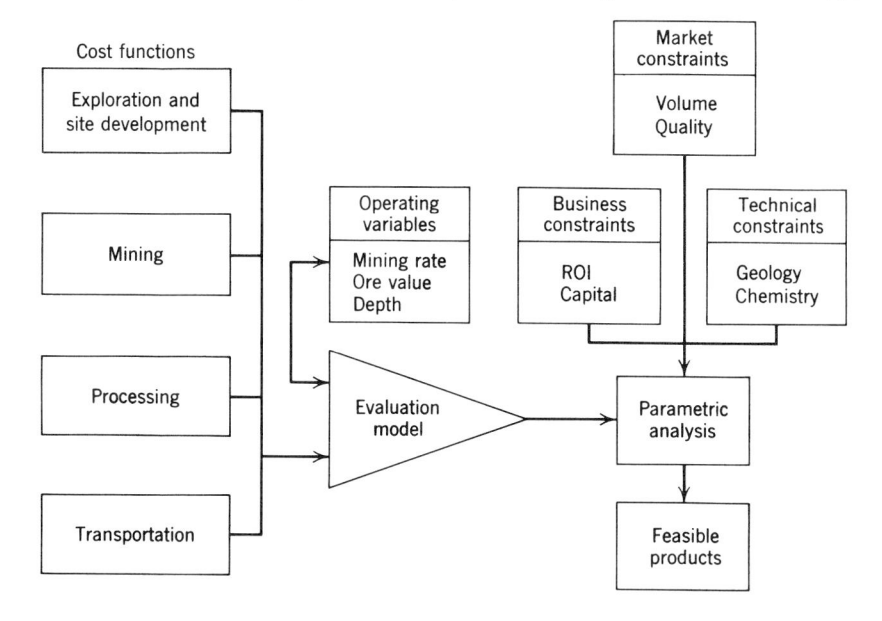

FIG. 4-38. Economic evaluation methodology.

A typical profitability analysis format is shown in Fig. 4-39, indicating the type of input information which is required. Expenses such as percentages of sales, tax levels, and working capital need to be ascertained to permit computation of the return on investment.

Sales = Average realization × product rate
Expenses
- Selling and administrative = $X\%$ of sales
- Royalties = $Y\%$ of sales
- Production = operating costs + % development capital

Taxes = T (Profit before taxes − depletion allowance)
Net Profit = profit before taxes − taxes

Total capital
- Facilities
- Development
- Working = $W\%$ of sales

Annual return on investment (ROI) = $\dfrac{\text{net profit} \times 100}{\text{total capital}}$

FIG. 4-39. Profitability analysis format.

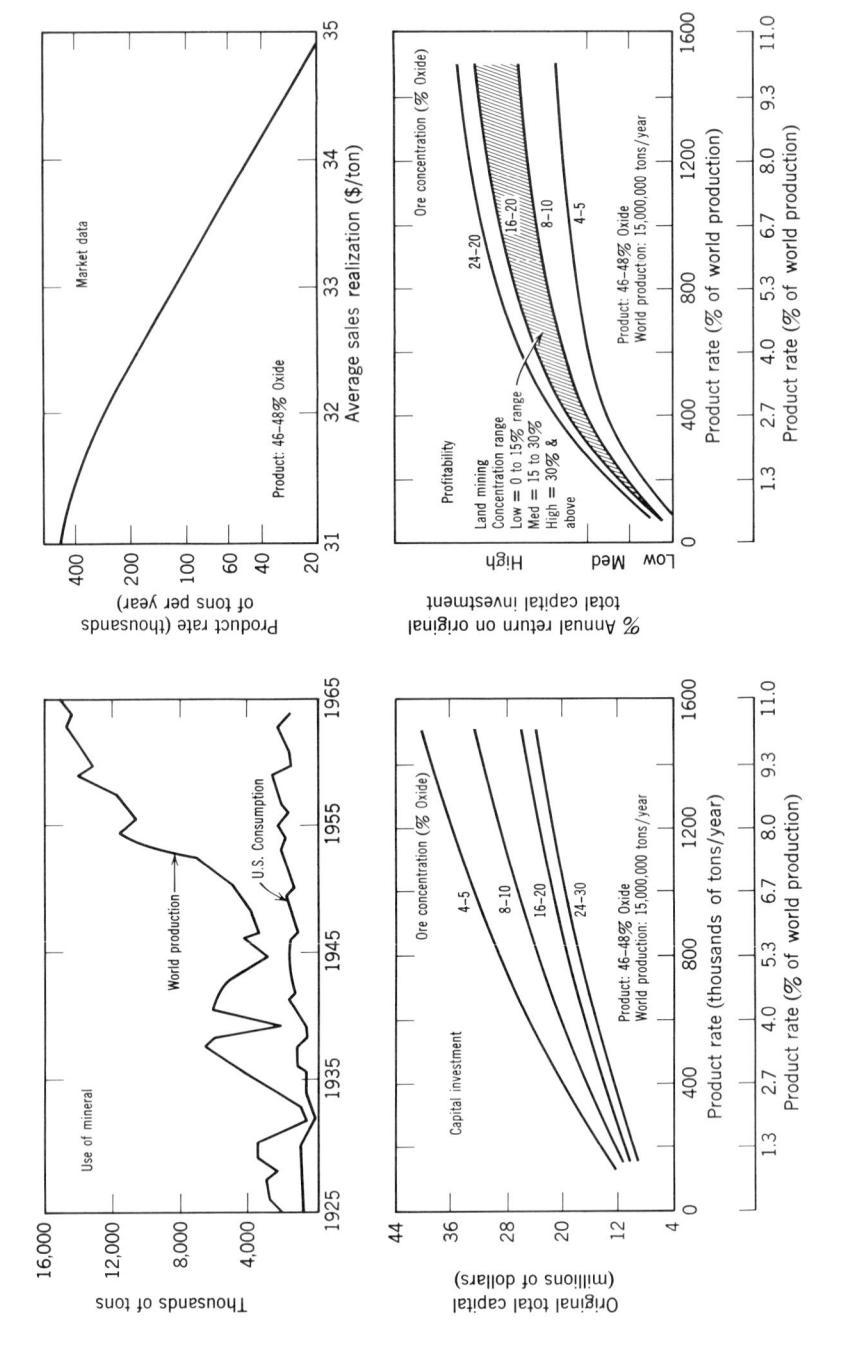

FIG. 4-40. Economic analysis mid-Pacific manganese mining.

Typical results of such an anlysis for mid-Pacific manganese are presented in Fig. 4-40. The expanding world production of manganese is shown, indicating the expanding non-U.S. market. Sales potential has been presented as a function of the production rate for ore of a given quality. Capital investment is also presented as a function of production rate and the amount of beneficiation required. Finally, the return that can be expected on the investment is given as a function of the same parameters.

In any ocean exploitation program involving the recovery of natural resources transportation and shipping costs play a major role. As indicated by Dr. Schaefer, this is a prime area for systems engineering. Obviously the development planning must also be done an a systems basis. Time does not permit any serious treatment of this subject here, but the principles outlined in this chapter are vital to such an analysis.

Figure 4-41 indicates several alternative ocean mining systems, emphasizing the various transportation modes involved. Depending on the actual location and depth of the ore, as well as its quality, we might either spend R&D on mining type devices or concentrate only on the transportation system. Within the transportation system, one might concentrate only on the undersea transport portion of the system and elect to build the plant close enough to eliminate problems of surface transport.

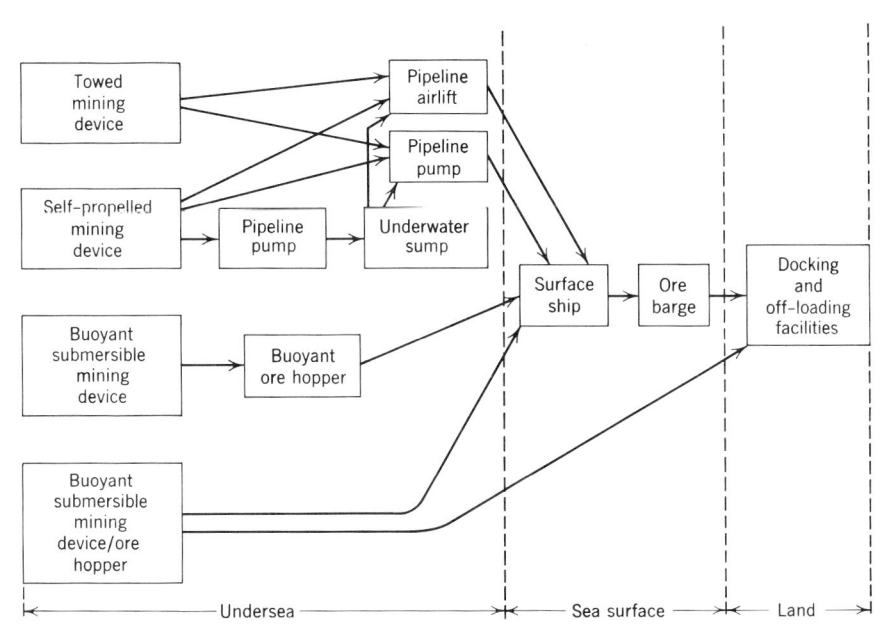

FIG. 4-41. Alternative ocean mining systems.

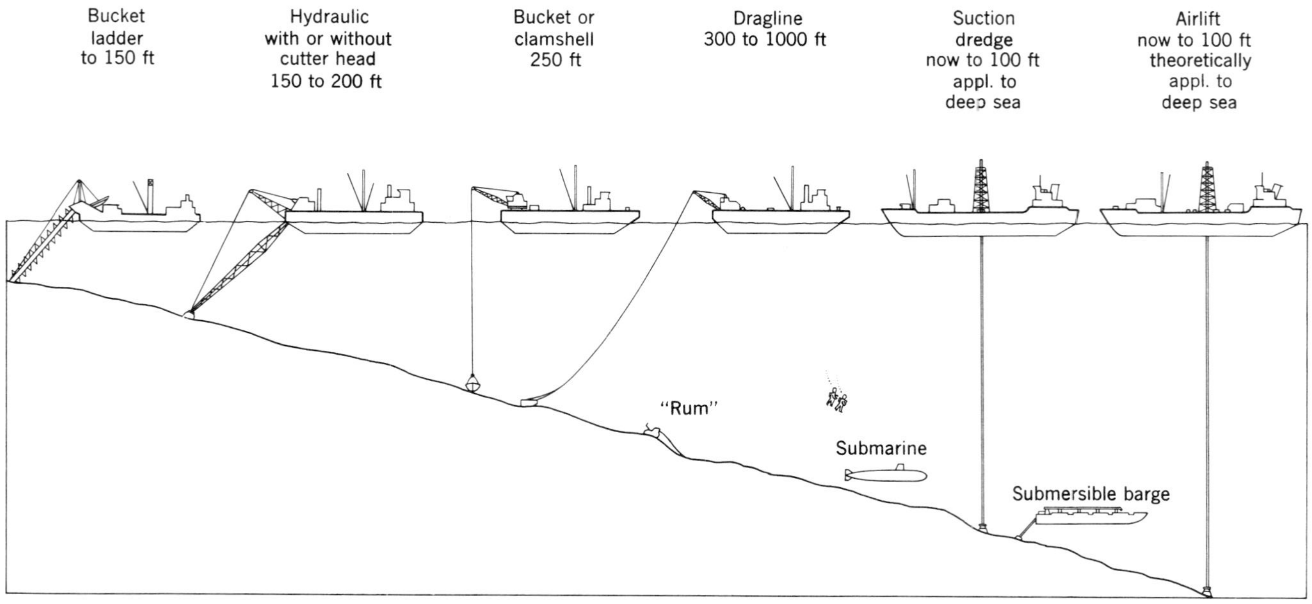

FIG. 4-42. Ocean exploration and exploitation types.

Systems-Development Planning

The effects of depth on the type of system required are shown in Fig. 4-42. The costs associated with shallow water mining are obviously very low when compared with the sophistication of deep systems. This is shown graphically in Fig. 4-43, where for 2 million tons per year the costs associated with operations at 20,000 ft exceed those at 1,000 ft by a factor of 3. The differences are primarily in the underwater transport costs and design, and economic tradeoffs are obviously necessary. For example, in the submersible barge approach, we can vary the degree of integration such as shown in Fig. 4-44.

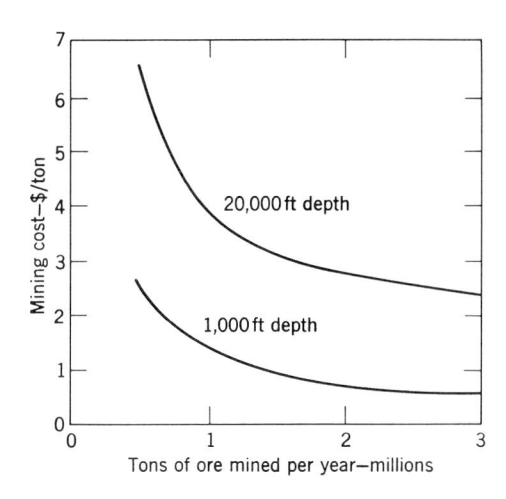

FIG. 4-43. Ocean mining cost versus mining depth.

4.5 SUMMARY

In summary, then, an outline has been presented of the industrial and market environment in which the ocean systems development planner must operate. Major emphasis has been placed on the vital nature of taking an orderly and a systematic approach to such planning, and some of the pitfalls that must be avoided have been defined.

There are many ways in which the planner can achieve his objectives, and one method, proved successful in many applications both within ocean systems and in other fields, has been suggested and defined in detail.

This is an exciting business, and we are on the threshold of recognizing, in this country, the tremendous potential which exists within the world's oceans. From almost every conceivable yardstick of value recognized by man—wealth, military security, humanitarianism, scientific knowledge,

Semi-integrated concept

Integrated concept

FIG. 4-44. Buoyant submersible mining devices.

curiosity, adventure—the oceans present challenges and opportunities the recognition of which is inevitable and only a function of time. It is the job of the systems-development planner to identify and define those opportunities, with alternatives, in time for corporate management to take action in a timely and profitable manner.

REFERENCES

Booda, L. L., 1966, "President Releases PSAC Panel Report," *Undersea Technol.*, **7**, 8, 23 (August 1966).

Covey, C. W., 1967, "Applying the Systems Approach to Military and Scientific Oceanic Problems," *Undersea Technol.*, **8**, 3, 14 (March 1967).

Gaber, N. H., and B. F. Reynolds, Jr., 1965, "Economic Opportunities in the Ocean," *Battelle Tech. Rev.*, **14**, 12, 5-11 (December 1965).

Interagency Committee on Oceanography, Federal Council for Science and Technology: Publication No. 10, "Long Range National Oceanographic Plan, Years 1963-1972"; Publication No. 15, "National Oceanographic Plan Fiscal Year 1965"; Publication No. 17, "National Oceanographic Plan Fiscal Year 1966"; Publication No. 24, "National Oceanographic Plan Fiscal Year 1967."

Koontz, H., and C. O'Donnell, 1959, *Principles of Management*, McGraw-Hill, New York.

Report of the Panel on Oceanography, President's Science Advisory Committee, 1966, "The Effective Use of the Sea," June 1966.

Staff Report, 1966, "Quest into the Deep," *Undersea Technol.* **7**, 5, 39-41 (March 1966).

Staff Report, 1966, "Bigger Benefits per Dollar Can Be Realized from Ocean Exploitation," *Undersea Technol.*, **7**, 8, 27 (August 1966).

Staff Report, 1967, "Hardware Development Paces Undersea Activity" *Undersea Technol.*, **8**, 1 (January 1967).

Tilson, S., 1966, "The Ocean," *Intern. Sci. Technol.*, **50** (February 1966), 26-37.

CHAPTER 5

The Law of the Sea and Public Policy

WILBERT McLEOD CHAPMAN

The ocean and its interconnecting seas form a continuous territory that covers 71% of the earth's surface. It is so large and diverse geographically that no one group of men organized as a sovereign nation, confederacy or otherwise, has yet been able to possess all of it. It is so important to those who dwell around it, from the standpoints of defense, transportation, and food production, that all efforts by one group to gain control over it have been successfully resisted by other groups who also needed to use it for their benefit or survival.

Attempts to control the use of the ocean have been a persistent and major cause of war from the earliest recorded history to the present day. Thucydides in the fourth century B.C. detailed in classic terms (*History of the Peloponnesian War*) the principles underlying the eternal contest between the land power (Sparta) and the sea power (Athens). Nothing essentially new has been added to the strategic concepts of the struggle.

It was a very old story in the time of Thucydides. Homer, writing about 400 years before him, had described, in the *Iliad* and *Odyssey*, aspects of the violent struggle for control of the Bosporus that resulted in the destruction of "Troy," which had been built on that site for the control of access between the Black Sea and the Aegean Sea and between the grain-rich Russian steppes and the food-poor Aegean populations, and destroyed because of this. The first "Troy" was built in the first half of the third millenium (Cottrell, 1960). The great Law of the Sea conference of 1958 swung on a contest involving rather subtle aspects of control of the ocean. In the course of the intervening five millenia, empires and civilizations by

112

the dozen repeatedly grew, flourished, and were destroyed by sea power, or its lack, in all parts of the world. The issue has now grown to be global in scope, and never before in this 5000-year period has the struggle over control of the sea involved more resources, affected more peoples, or been more tense than it is at the present time.

Yet war is neither a desirable state nor one that can be maintained at full force indefinitely. Great powers, like people, get tired and must rest and recover from their warlike efforts. In the periods of peace there must be public order, which is the reverse of war. The ocean is of such supervening importance to everyone, that it has been not only a cause of war, but also a cause of the creation of public order through treaties and agreements among the nations which could not uniquely control it but needed to use it, while keeping down piracy between the wars so that the ocean could be used for peaceful purposes.

Control by imperial power

It is probable that control of sea areas by imperial power was an ancient practice when first we learn of it, and we do know that in the sixth century B.C. (Polybius gives the date as the first year of the Roman Republic, which he put in 508–507 B.C.) there existed a written treaty between Rome and Carthage that gave Carthage preferential rights to the use of the western Mediterranean (Warmington, 1964).

Rome eventually reduced the entire Mediterranean to its control and there grew up in Roman law regulations under which the Mediterranean could be used by all. This was not, however, regulation under international law but under Roman law. Yet some concepts from that body of law concerning the use of the ocean carry over to the modern international Law of the Sea because the principles were the same. Piracy needed to be kept under as close control as possible, or eliminated entirely, so that private and public commerce could flow over the ocean. Public order needed to be maintained so that society could flourish and grow. The power supporting the public order had to be protected or there would be only strife and anarchy. The problems are about the same today as they were then.

With the breakup of the Roman power, strife and anarchy became the rule in the Mediterranean again for a long while. In this long period there were attempts by Venice, for instance, to establish ownership by itself over the Adriatic. Other powers tried similar ventures and these were successful for brief periods, when the sea power was strong enough to enforce control by might. Beginning in the fourteenth century the combined kingdom of Denmark–Norway attempted to maintain the whole Norwegian Sea as a *mare clausum* (Heinzen, 1959). Beginning toward the end of the fifteenth century, Portugal and Spain asserted claims to exclusive rights over most of the Atlantic and Pacific, supported by a papal bull of May 1493 and modified by the *Treaty of Tordesillas* between Spain and Portugal of 1494 (Heinzen, 1959).

Freedom of the seas

In the second half of the sixteenth century, however, a new trend in the Law of the Sea was established by both Poland and England, on new grounds that have persisted to this day. Poland formed a coalition in the eastern Baltic to oppose Danish practice in the Baltic and informed the Danes that the use of the sea was common to all (United Nations: Conference Document No. A/Conf. 13/40, 1958). Queen Elizabeth disputed the claims of Denmark, Portugal, and Spain from the beginning of her reign, employed armed force to support her contentions, and in 1580 boldly rejected Spanish complaints over Sir Francis Drake's expeditions in these words:

"The use of the sea and air is common to all; neither can any title to the ocean belong to any people or private man, for as much as neither nature nor regard of the public use permitteth any possession thereof."

England in 1580 was a "small" power. Spain was the ruler of the sea and, through its Hapsburg emperor, of a good deal of the rest of western Europe, including the Netherlands. England was not even a big power relative to Denmark–Norway, Poland, or Portugal, the other principal contenders in this struggle. It is important to realize that this statement of policy by Queen Elizabeth expressed the need of small nations to use freely the sea and air, and that the expression of it in this form was diametrically opposed to the old Roman enforcement of public order on the sea as a perquisite of Roman Imperial Power and Sovereignty. Elizabeth expressed British policy as an attribute of the common right of all humanity, with no special privilege under it for British citizens. These words marked a sharp break in philosophy underlying the public order of the ocean which has lasted to this day and forms the basis for existing Law of the Sea.

These words seemed to be an empty boast in 1580, but with the defeat of the Spanish Armada by British arms (and North Sea weather) eight years later they began to take on lasting significance. Shortly after the turn of the seventeenth century, England temporarily reversed its advocacy of the freedom of the sea and the concept of *res communis*. Under Elizabeth's successor, England joined forces with the Danes and suddenly claimed sovereignty over "Our coasts and seas" (Heinzen, 1959). This was an outgrowth of the Dutch fishing for herring off the British coast, and the Dutch were quite prepared to use naval force to retain their right to fish there. In the ensuing negotiations in 1610 the Dutch held firmly and successfully to the formerly ennunciated English doctrine that "The boundless and rowlinge Sea are as Common to all people as the ayre wch no prince can prohibite." The trend set in motion by Poland and England, reinforced by Dutch firmness, came to prevail. The French soon became strong advocates of the freedom of the seas and the English themselves rather quickly returned to the Elizabethan position (Heinzen, 1959).

The territorial sea

Out of this controversy emerged the concepts of the freedom of the seas and *res communis* which are still held by the nations of the world, but there also arose from it the controversy over the breadth of the territorial sea that also is still with us. Both the Dutch and Elizabethan positions noted above included the concept that the coastal state should have some limited jurisdiction over the sea directly off its coast. The Dutch, by the time of the 1610 negotiations, were prepared to recognize some supervisory jurisdiction by the coastal state within cannon range from its shore, and thus the cannon-shot rule for the breadth of territorial sea came into existence. In 1703 the Dutchman Bynkershoek declared flatly that a nation could exercise sovereignty over waters within cannon range of its shore, and that the range of cannon then was three miles. Thus the three-mile rule for the territorial sea came into existence (Heinzen, 1959).

Although Bynkershoek gets the credit for ennunciating the three-mile rule, a search of archives of the seventeenth century shows that nations were treating the cannon-shot rule as established law well before Bynkershoek's time, that the range of cannon in his time, and for a century thereafter, was substantially less than three miles, and that seventeenth and eighteenth century practice of the cannon-shot rule had nothing to do with territorial jurisdiction or sovereignty in the modern sense (Heinzen, 1959). It pertained only to area actually within cannon range, not to a band of water of uniform breadth along a coast.

The three-mile rule

So far as is known the United States was the first nation which adopted a zone of uniform breadth along its whole coast three marine miles in breadth in which it would take sovereignty. This was for purposes of neutrality, and to protect itself from the warring privateers of Britain and France which were then operating on commerce off the American coast. It was, again, the act of a small power wishing to protect itself against larger powers.

The decision by this small power was provisional and was enclosed in a note from Thomas Jefferson, then Secretary of State, to the British in 1793. It is not even clear why the United States claimed three miles, because Jefferson referred in his note to unspecified treaties as containing three-mile precedents which do not seem to have existed, and to the fact that the narrowest distance claimed by other nations was "The utmost range of a cannon ball, usually stated at one sea league." In any event a federal statute of the United States adopted in 1794 provided that "the district courts shall take cognizance of complaints of whomever instituted, in cases of capture made within the waters of the United States, or within a marine league of the coasts or shores thereof." [Act of June, 5, 1794, Chapter 50, Paragraph 6, 1 stat. 384 (1845)] (Heinzen, 1959).

In any event, during the first years of the nineteenth century there began to be practiced among nations an entirely new concept, that of the modern territorial sea of 1-league breadth in which, under international law, the coastal state can exercise all the rights that it can in those portions of its territory that are land, except for interference with innocent passage. Except in the Scandinavian countries, one marine league has, since the turn of the nineteenth century, generally been understood by nations to be three nautical miles. A nautical mile is taken to be 1 min of latitude at the equator, or approximately 1853.2 m, or 1.15 English statute miles. The league referred to in latter eighteenth century Scandivian ordinances equaled about 7420 m, or four nautical miles (Heinzen, 1959).

The United States continued to practice a three-mile zone for purposes of neutrality, and English courts began to follow the American practice. In 1805 an English court upheld the United States three-mile zone for purposes of neutrality in the case of "The Anna," and in 1807 England refused to ratify an agreement it had signed with the United States establishing a "special" neutrality zone five miles in breadth off the American coast. With the establishment of a three-mile fishing limit off certain British North American coasts by the treaty of 1818, England and the United States have been consistent in maintaining three miles for purposes of territorial (as contrasted with purely neutrality purposes) jurisdiction down to the present time (Heinzen, 1959).

The nineteenth century was a period of consolidation of the concepts of the territorial sea and the three-mile rule in the international law of the sea. One by one the principal users of the sea agreed to these concepts. This included not only European sea powers, but many non-European coastal states, and many states not then major sea powers as well. By 1900 the three-mile rule, or one-league limit, had been positively acknowledged by 20 of the 21 nations that claimed or acknowledged a territorial sea at that time and may therefore be said to have been generally accepted as a customary rule of international law at that time (Heinzen, 1959).

The customs zone

Whereas the growth of the concept will not be detailed here, it must be noted that growing side by side with the concepts of the freedom of the seas, the territorial sea, and the one-league breadth for the latter, there grew and consolidated by the same deliberate process of national practices, during the period from the fifteenth century, the concept of a customs zone under which a coastal state could search and seize vessels, suspected of intent to violate its fiscal laws, at a distance somewhat greater from its shores than the breadth of its territorial sea. This greater distance gradually settled to 12 marine miles, but the concept has remained distinct from that of the territorial sea and has been kept for its purely fiscal purpose.

The Hague Conference for the progressive codification of international law

Other countries adhered to the concepts of the freedom of the seas, the territorial sea, and the three-mile rule during the first quarter of the twentieth century and during World War I no coastal state was able to maintain more than one league for purposes of neutrality, although some tried. In 1930 an international conference was called for the progressive codification of international law. Thirty-eight coastal and two noncoastal nations attended. In spite of the fact that only a small proportion of the coastal states attending the conference claimed more than one league of territorial sea at the time, the conference adopted no provision on the breadth of the territorial sea and it was, for this reason, accounted a failure.

The prime reason for this was that Great Britain and a few other strong advocates of a three-mile limit refused to recognize either the right of a coastal state to exercise jurisdiction in a contiguous zone for fiscal purposes, or the historical claim of some Scandinavian nations to a league of four marine miles for the purpose of measuring the breadth of their territorial seas. As a result of the controversy over these issues the conference became deadlocked, and never took a formal vote on the breadth of the territorial sea.

Informal expressions of preference indicated that 20 of the 38 states would have accepted either the three-mile limit or the three-mile limit plus a contiguous zone; three Scandinavian countries favored a four-mile rule for themselves; 12 countries favored a six-mile rule; two coastal states and one landlocked state expressed no preference; the other landlocked state favored the greatest possible freedom of navigation. No more than eight states attending the conference actually claimed more than one league of territorial sea at the time, and the combined merchant-marine tonnage at that time of the nations who would have supported the three-mile limit together with a contiguous zone was nearly 80% of the world total (Heinzen, 1959).

5.1 THE NEW TECHNOLOGIES

Up to the date of the Hague Conference, the Law of the Sea was concerned almost exclusively with the liquid ocean and events on it. In the long period of its formulation, particularly following the breakup of the Spanish maritime empire, great changes transpired in world society and these had to be taken into account in the international law applying to the sea. At the same time the old problems persisted. Pressing most vigorously on the Law of the Sea was the impact of the new technologies arising out of the industrial revolution of the eighteenth and nineteenth centuries and the scientific revolution of the twentieth century. Since these factors are still at work, some special note requires to be taken of several of them in order to understand existing law of the sea.

Piracy

Piracy has come to be defined rather precisely as consisting of any of the following acts:

"(1) any illegal acts of violence, detention or any act of depredation, committed for private ends by the crew or the passengers of a private ship or a private aircraft, and directed:

(a) on the high seas, against another ship or aircraft, or against persons or property on board such ship or aircraft;

(b) against a ship, aircraft, persons or property in a place outside the jurisdiction of any state;

(2) any act of voluntary participation in the operation of a ship or aircraft with knowledge of facts making it a pirate ship or aircraft;

(3) any act of inciting or of intentially facilitating an act described in sub-paragraph 1 or sub-paragraph 2 of this article." (Article 15, 1958 Convention on the High Seas).

Piracy has always, in recorded history, flourished when and where public order on the ocean has broken down, and it will always do so. In this decade it has been practiced in southeast Asia. It is not a new problem, although the new technologies have aided the pirates of the Sulu archipelago, for instance, by providing them with fast speedboats and machine guns.

The line between piracy and national practice is not as clearly drawn as the above treaty article would indicate. Queen Elizabeth issued letters of marque to individual captains under which Spanish trade with the new world was harassed in a quasipublic fashion. The United States in its early days armed "privateers" to serve its public purposes. The Bey of Algiers gave harborage and protection to vessels raiding peaceful commerce. The practices of certain Latin American countries have verged toward this in limited aspects from time to time as will be noted under "Fisheries" below. The situation in southeast Asia, outside the actual war area, is at present somewhat confused in this respect.

It must always be expected that in the absence of public order on the ocean, piracy will erupt in the same manner as brigandage does ashore under similar circumstances, and that these private acts will merge indistinguishably into public acts blinked at, or condoned, by sovereigns as the power enforcing public order on the ocean decays. The fact that public order on the ocean has been maintained firmly over most of the world for nearly three centuries should not be permitted to push this universal truth aside.

It will be noted that the "small" power is a principal beneficiary of the "great" power, or powers, suppressing piracy of a private or quasipublic nature because the flow of its commerce is protected without it bearing the cost of the protection.

Freedom of navigation

Commerce among sovereign peoples was a matter pretty well arranged among sovereigns, not by private traders, in history until quite recent times. It was always the purpose of sovereigns during the greater period of history to lay taxes upon commerce as it passed through their area of control—either their sovereign territory or the zone outside it to which their effective control reached. The capture and control of trade routes was a principal preoccupation of sovereigns throughout history until recent time and the basic instinct is still there to be exercised at any time, and in any place, where public order of an international nature breaks down.

The enormous increase in commerce among nations that opened up with the exploration of searoutes in the period from the fifteenth through the seventeenth century, the need to bring raw materials to manufacturing centers and distribute manufactured products to markets that blossomed with the industrial revolution in the eighteenth and nineteenth centuries, and the close knitting together of the whole world that has developed with the scientific revolution of the twentieth century, have made the free flow of commerce among the nations no longer a luxury, but a necessity. A considerable part of the earth's population today could not exist in the absence of ocean-borne commerce. Should this be disrupted for a prolonged period, the human population of the world would shrink because of the famine, disease, and warfare that would ensue until new equilibria with the new conditions of commerce were created.

For this reason the international community has increasingly insisted upon the utmost freedom for peaceful commerce upon the ocean. Even the narrow territorial sea, which is agreed to be for all other purposes to be the same as the sovereign land territory of a nation, must freely bear commerce among the nations, and the right of innocent passage through it is not challenged in international law, or very often in national practice. The enormous power of this international need for absolutely free, peaceful navigation is not fully understood by all apologists for some of the new users of the sea. It goes so far as the requirement to nations not to obstruct innocent passage through their territorial sea unreasonably by the construction of artificial works in it, and to the requirement for the posting of warning lights and signals in it so that mariners strange to that coast will come to no harm from obstructions, natural or artificial.

Going hand in hand with this overpowering need of humanity as a whole for increased commerce among its parts, the old selfish desire to capture trade routes and industries mentioned above still exists. Whenever or wherever the public order of the ocean declines, this will emerge. To prevent its emergence nations will war upon one another because they no longer can exist as they are without the commerce.

Neutrality

The United States first moved formally to the three-mile limit, as noted above, in order to protect its neutrality in time of war among the great powers. A small power, in order to protect its neutrality in time of war among others, must ensure that its territorial sea or inland waters are not used by one belligerent, thus giving cause for the other belligerent to take warlike action against it, or the belligerent sheltering in its waters. This need still exists for nations that wish to stay out of a war and has been zealously guarded by them through all the wars of the past 200 years.

This vital necessity, particularly for "small" powers, is in contrast to the general, if submerged, desire to reach out and lay taxes upon offshore commerce. A 200-mile territorial sea, for instance, may be quite attractive for commercial purposes to a nation in peacetime, but in time of war, if a nation wishes to remain neutral, it wishes to have a quite narrow territorial sea within which it can certainly detect belligerent acts and defend against them occurring in its own territory.

This matter of neutrality was also to play a powerful role in the Geneva Conferences on the Law of the Sea as is noted below. It is not certain that the new modes of warfare or new weapon systems have much affected this powerful urgency for a narrow territorial sea. If anything, they seem to have firmed up the desire and need for a narrow territorial sea.

Fishing

There has been no more powerful force at work in the formulation of the Law of the Sea over the past five centuries than disputes among sovereigns over fishing rights. A majority of the treaties and arbitrations among nations that framed the precepts in existing Law of the Sea arose out of disputes over fishing rights. The individual instances are too numerous to mention. This has been a powerful force in shaping United States policy on these issues as well, and throughout its history to the present time.

Before the first decade of the twentieth century, these fishery disputes all arose from the desire of nations to gain preferential rights to the harvest of a particular fishery resource for its nationals or, in reverse, to prevent the nationals of another nation from obtaining preferential rights over their nationals in the fishery. This is still the most active and powerful motivation of nations in this aspect of the Law of the Sea field and it grows stronger as the worldwide need for animal protein for human nutrition becomes more pressing. This motivation was based initially on the belief that the living resources of the sea were each inexhaustible and, like the sea and air, were so abundant that they needed to be common to all.

In the first half of this century the systematic study of the ocean and its resources first got underway on a large scale off northwest Europe. One of the things that has emerged from this ocean research to affect this subject is an understanding of the *natural* laws under which such resources are provided by nature.

For every living resource of the ocean, these same biological rules apply. In a state of nature, before a fishery begins, there is a balance of nature in which one egg in one generation begets, on average, one egg in succeeding generations. The whole population of a particular fish species, or stock thereof, is in balance with the other populations of natural living things with which it lives. The rate of natural mortality equals the rate of increase of the population.

There is resilience in the population provided to safeguard it against fluctuations in the natural environment that would affect this balance in rates of increase and decrease adversely to the welfare of the population. When the population is subjected to a fishery (or adverse environmental conditions) the average age of the individual in the population decreases, the total number and weight of the individuals in the population decreases, and the total reproductive efficiency of the population increases because of this resiliency.

As the catch of the fishery from the population increases, a point comes when the total mortality (including the fishing mortality) not only equals the rate of growth plus the rate of recruitment, but also fully absorbs this resiliency of the population to produce. This point is called the "point of maximum sustainable yield." If we fish harder than necessary to reach this point, then the harder we fish beyond it, the less fish we get from that population. Not only do the number and average weight (and average age) of the individuals continue to decrease as it has done since the fishery began, but now so does the catch.

It needs to be kept in mind that this point of maximum sustainable yield is a result of balance between the rate of recruitment and the rate of growth (the increase) against the rate of fishing mortality and the rate of natural mortality (the decrease), and that the resiliency is really made up of those fish caught by the fishing which, in a state of nature, would have died from natural causes had they lived longer. The rate of recruitment, the rate of growth, and the rate of natural mortality are determined by natural factors in the ocean and on wild stocks and are not yet under the control of man in a consequential manner. As a matter of fact, the goal of fishery oceanography in the present state of science is to learn to predict, not control, these natural factors.

In some quite important populations of fish (particularly those having naturally a short life span) the rate of recruitment may be influenced sharply by changes in the physical nature of the ocean in succeeding years, or periodically. For instance, it appears that the point of maximum sustainable yield from the anchovy population off Peru may vary from 7 million to 9 million tons even in succeeding years because of the effect of changing ocean conditions (reflected by temperature of the water, but involving trade-wind strength and resultant upwelling), on the success of survival of larval anchovy in the first few weeks, or even days, of life. This does not affect the general thesis set out above; it merely requires more sophisticated and extensive research for its successful application to the management of the use of such resources.

There are three general principles that arise from this new understanding that have affected the Law of the Sea:

1. the only factor in this equation that can be affected by man in a consequential manner on major ocean living resources is the amount of fishing pressure expended on the population;

2. the use of more fishing pressure than is required to attain the maximum sustainable yield is wasteful both economically and biologically because the harder you fish beyond that point, the less weight of fish (and ordinarily dollars) you catch;

3. the use of less fishing pressure than is required to attain the maximum sustainable yield is wasteful biologically (and perhaps economically) because the fish produced beyond that point by the population, and not caught, die naturally, decay, and go back into the web of life in the ocean, unused by man.

Obviously the interest of the owner of these resources is to maximize the yield from each of them. If the owner were private, or perhaps even if the owner were a nation (although this has not been much in evidence where a nation does own a whole resource of this nature), he would wish to maximize the economic yield from the resource. It can be demonstrated that this point always requires less fishing effort to reach than the point of maximum sustainable yield (Schaefer, 1959). Since the owner under international law is all mankind, and mankind is broken up into sovereign units having different economic systems, a common definition of maximum economic yield has so far not been found. Accordingly, the general interest of mankind to this date has been to maximize the yield of food from the resource, and it is the sustainable physical yield that the nations have agreed to maximize.

Without at all doing away with the natural desire to have a preferential right in the fishery, or even mitigating that urge, this new understanding has introduced new usage into the management of the harvest of living resources of the sea and the law making that possible. Multiple users can fish harder and harder until all but one or a few is driven out of the fishery, or they can stop their combined fishing effort on the particular population at that point which will maximize the biological yield from the resource, and then decide how that yield is to be divided among themselves by diplomatic or other means.

The nations have not worked out any general formulas by which this new understanding can be uniformly used, but they have made some promising beginnings. In 1911 a treaty among Russia, England (for Canada), United States, and Japan stopped fur-seal hunting by their nationals on the high seas and left the management of the North Pacific fur-seal populations to the nations owning the rookeries. Each of these nations gave a share (established in the treaty) of the furs harvested each year to the others who refrained from the harvest. The fur-seal herds, which by 1911 had been almost wiped out, responded quickly to this treatment, returned to the point as near maximum sustainable yield as practical, and have been kept at that point to the present time.

A considerable number and variety of such international fisheries treaties have been initiated among different groupings of nations to cover different fisheries in different parts of the world. They provide a wide variety of regimes ranging from simply investigatory, through regulatory, to sharing of the maximum sustainable yield by formula. Those existing in 1955 are described in detail in Herrington and Kask (1956).

During the whole of the twentieth century that this new information and usage has been increasing, other new information and new technologies have been developing that have sharply increased the difficulty and complexity of the high-seas fishery problems with which the international community has had to deal.

It has been learned, for instance, that many, if not most, of the major living resources of the sea are widely migratory. Many of them migrate thousands of miles in their normal life cycles and may migrate completely across such large oceans as the North Pacific or the North Atlantic. Almost all major living resources aside from those that are sedentary in their harvestable stages, move sufficiently so that at some time they are available outside the territorial sea (measured at any breadth between three and twelve miles) to foreign fishermen.

The increased efficiency of motors, and the increased ability to carry out processing (freezing or other) on board ship economically, has made the fishermen, in a good many cases, even more migratory and mobile than the fish. Thus Roumanian vessels fish off New Zealand, Spanish vessels off Argentina, Californian vessels off Chile, Norwegian vessels off New York, and Russian and Japanese vessels everywhere, and on a large scale.

Additionally the use of electronic navigation systems, echo-ranging gear for locating fish, new sorts of catching equipment and methods, etc., has greatly intensified the total fishing effort on a worldwide basis rather rapidly. In 1900 the world catch of fish and shellfish (excluding whales) was about 4.0 million metric tons, and in 1950, 20.2 million tons (Moiseev, 1965). In 1964 it was 51.6 million metric tons (Food and Agriculture Organization, 1965). Known resources will permit an expansion of the sea fisheries to at least 200 million tons by methods now known (Schaefer, 1965) and the theoretical upper limit of fish and shellfish production lies closer to 2 billion tons per year (Chapman, 1966).

Fishery management problems were the prime element affecting changed position of nations toward territorial limits in the period between the Hague Conference in 1930 and the Geneva Conference in 1958 (Heinzen, 1959). They were of controlling importance at the 1958 and 1960 *Geneva Conferences on the Law of the Sea* (Chapman, 1960, and 1965). They are still of major importance in the pressures for restricting agreement.

Air transport

Adventitiously Queen Elizabeth covered the situation of both air and sea transportation in her dictum of 1580. The great development of air transportation in this century has not had a major disturbing effect on the Law of the Sea except in one respect. It has been settled in international law that aircraft have freedom to overfly the high seas, but they do not have the right of innocent passage in the air space over the territorial sea. This turned out to have some significance in the 1958 conference on the Law of the Sea, as is noted below.

New weapons systems

It might be thought that the coming into being, and practical use, of inter-continental ballistic missiles armed with nuclear or other warheads, the increasing flexibility of satellite use, and various other new technologies in the making of war, might have had a major effect on the Law of the Sea. In matter of fact, they have not. If anything, the posture of the United States toward a narrow territorial sea has strengthened as these new weapon systems have developed. The need to navigate freely the world ocean became, if anything, more important in the post-Hiroshima world, for both military and mercantile purposes than before, if the power that maintained the public order of the ocean (no matter among how many nations it was divided) was to be preserved.

In particular this took the form of need for wider sea space in which to navigate a naval task force because of dispersal patterns required in defense against the employment of nuclear weapons; the effect of a broader territorial sea in restricting the use of military aircraft over the sea in certain critical localities; and the possibility of closure to the passage of belligerent vessels of straits heretofore considered to be international. These factors had a major effect on the activities at the Geneva Law of the Sea Conferences of 1958 and 1960.

Petroleum and other minerals

Before the Hague Conference the Law of the Sea had been concerned almost entirely with the liquid ocean and its interface with the dry land. The advent of air transport, as noted above, brought little change to the concepts involved in it. The advancing technology of drilling for petroleum, however, brought a whole new dimension to the Law of the Sea—the regime of the sea bed. It was discovered that oil pools extended out under the sea in several places—Persian Gulf, Gulf of Mexico, coast of California. Technology was developed for drilling offshore in the relatively shallow waters of the continental shelf. Great sums of money became involved and it became necessary and important to determine who held title to these large and valuable resources and what governance there would be of their harvest.

This was of quite important international concern in the shallow Persian Gulf. It became of even greater importance within the United States as a contest developed between the Federal Government and the State Governments over which one owned the offshore oil reserves, could manage their harvesting, and collect the revenues from them. This purely domestic controversy inside the United States was to have very grave international consequences as its effect impinged upon the international community. The reason for this was that the United States did not have clear title internationally to the sea bed and its resources beyond the territorial sea where large petroleum resources were now known to exist. The squabble between the States and the Federal Government was difficult to settle internally until it was determined to what the United States itself had title, and this could be settled only internationally.

5.2 THE POLITICAL SITUATION BETWEEN 1945 AND 1958

Even such a cursory review of the history of the Law of the Sea as has been given above indicates that this subject has been near the root of agreement or disagreement among sovereigns for centuries. At one time, disputes under it have been settled by negotiation or arbitration; at other times, and not infrequently, the issues have been settled by war. Always the international political situation, the power balance, has been a major factor in the conclusions reached. Never has public order on the ocean been maintained exclusively out of human good will. Always there has been the power of a sovereign, or sovereigns, available to enforce it. The degree of enforcement has always been a result of an amalgam of agreement and the amount of force that was free at that time from other international call upon it and available for its enforcement where agreement did not exist. Never in this long history has the issue been a simple one. A fishery dispute in one isolated corner of the world always had the possibility of disturbing the navigational or neutrality situation in another part of the world, or the power balance among sovereigns.

At the end of World War II, the political situation in the world was incredibly complex and kept shifting with kaleidoscopic speed and lack of predictability. Germany and Japan (two normally important sea powers) were so defeated as to be temporarily out of the international community. The great overseas empires of several European powers, which had been such major factors throughout the period when the modern Law of the Sea was being developed, were in the process of dissolution, leaving a number of international power vacuums. Into these power vacuums had been sucked Russia and the United States as the two overwhelmingly most powerful sovereigns on the world scene, and no sooner was World War II concluded than these two allies, now the depositories of overwhelmingly the greatest power in the world, became contestants who were able to maintain peace between themselves, and in the world, only by major diplomacy, confrontations, brandishment of arms, and alarms and excursions of diverse sorts.

The United Nations was formed in 1945 just before this struggle between the two great remaining powers was fully formulated. This action created a powerful new force in the world because issues short of war could now be decided by votes among sovereigns, and the results of such votes would be applicable to sovereigns if they did not wish to fight over the issue. Neither of the two great remaining great powers could escape the effect of this new situation, although both tried on several occasions. The weight of world public opinion had become too great to contest against except in the most dire situation. It required to be listened to no matter what the military power of the particular sovereign, and supporting votes needed to be gathered from the smallest power, each of whose vote had the same weight in the final tally at the General Assembly of the United Nations as had that of the most puissant sovereign.

An impartial observer might well comment that these postwar times of excessively shifting political strength in the international arena was the worst possible time to precipitate conclusions on a subject so complex, delicate and basic as the Law of the Sea. On the other hand one must realize that all of these matters were not so clear in 1945 as they are by hindsight in 1968. In any event the technological changes noted above, and their effect internally in the United States, would not wait, and the United States opened the Pandora's box of the Law of the Sea almost before the ink was dry on the document concluding the Pacific phase of hostilities.

Two prime forces were at work in the United States:

1. The petroleum people wanted the issue of who owned the offshore oil resources resolved so that they could get on with the job of developing their harvest. The United States decided on a bold move to make a major change in the Law of the Sea to the end that the continental shelf adjacent to each maritime nation, with its contained resources, would effectively belong to that nation. It made diplomatic inquiries and found that this was generally agreeable among the nations so long as the nature of the superjacent waters as high seas was not affected by the change.

2. Just before the outbreak of war in the Pacific, Japanese salmon fishermen had entered Bristol Bay (in 1938) to fish experimentally. This had resulted in a terrific uproar in Alaska and the Pacific Northwest. The tensions that were to result in the Pacific War were already building and the Department of State (under Cordell Hull) made strong representations to

Japan. As a result Japan withdrew its fishermen from the area at that time. The Pacific War soon broke out.

The salmon people in the Bristol Bay area of Alaska kept well aware during the war of this sword of Damocles hanging over their heads, certain that upon the establishment of peace the Japanese would be back in the offshore waters of Bristol Bay and drive them out of business by intercepting the salmon runs before they got into the Alaskan territorial sea. Accordingly, they also wanted to change international law so that Japanese (or others) could be kept from fishing salmon that were destined for Alaskan rivers.

The Department of State explored this issue diplomatically also and found that although there was support for an apparatus to prevent overfishing on high-seas resources, there was no agreement whatever that this would give the fishermen of one nation preference in the fishery over those of another. Still, the Department of State had to deal with the internal political problem of the Pacific Northwest.

The compromise of these two situations was a clever one. The petroleum people won, but the salmon people lost. This was masked by language. In September 1945 President Truman issued two proclamations simultaneously. Under one the United States laid claim to exclusive jurisdiction over the continental shelf adjacent to its coast, and its contained resources, without attempting to affect the nature of the superjacent waters as high seas. Under the other the United States claimed the right to establish conservation zones off its coast but clearly without giving preferential fishing rights therein to its own citizens (Chapman, 1949).

The United States thus attempted to deal with its international problems but had not reckoned sufficiently with what would happen internationally when this act was done.

In October 1945 the President of Mexico also made a declaration and sent to the Mexican Congress proposed amendments to the Mexican Constitution to carry out the terms of the declaration. The effect would have been for Mexico to claim rights of jurisdiction over the sea bed and the adjacent seas to such distance as Mexico should decide from time to time in the future to be appropriate. The amendments to the Constitution were not adopted in this form and this concept apparently was not adopted into Mexican policy. At the time, however, it stirred vigorously activity in the United States shrimp fishery, which had recently begun fishing substantially off the coast of Mexico in the Gulf of Mexico. Mexico had seized four shrimp vessels fishing there in September 1945 (MacChesney, 1957). It stirred similar sentiments in California, where the tuna fishery (as well as others) operated extensively in the high seas off the west coast of Mexico. In this action Mexico led the way in Latin America for the tying together of sovereignty over the continental shelf and the superjacent water, which was precisely the contrary of what the United States had attempted by its Presidential Proclamation.

In October 1946 Argentina issued a Presidential Proclamation claiming sovereignty over the epicontinental sea and continental shelf (MacChesney, 1957). Since there were no foreign fishing vessels then operating in the area, and the Argentine fisheries themselves were not well developed, this action drew little, if any, derivation from fishery problems. It grew out of broader national objectives of a territorial nature, among which was the strengthening of claims of the Falkland Islands which rise from the continental shelf off Argentina. The great expanse of the continental shelf off Argentina gave dimension to the exaggerated claim for sovereignty offshore, which the Mexican decree had left undefined.

In June 1947 Chile issued a Presidential Proclamation which is ordinarily referred to as the first 200-mile zone (MacChesney, 1957), but this is not the correct description of the decree. The decree proclaimed *sovereignty* by Chile over the ocean and its resources *to a minimum distance* of 200 miles off its coast.

Although this later became enmeshed in fishery problems, when it was first promulgated it was almost solely in reaction to the Argentine claim. Chile had a very narrow continental shelf and could not use that as a reason for claiming a broad territorial sea. Accordingly it boldly proclaimed sovereignty over the sea itself without this underlying rationale. It incorporated the best feature (for it) of both the Mexican and Argentine claims, taking a large area of sea (200 miles for sure) as Argentina did, and leaving the outer boundary indeterminate for later appropriate action (as the Mexican Proclamation had done).

In August 1947 Peru followed Chile by issuing a Presidential decree substantially identical with that issued by Chile (MacChesney, 1957). As in the case with Chile, this latterly became involved in fishery disputes, particularly with the United States, but did not arise therefrom because in 1947 (or previously) United States vessels did not customarily fish off Peru.

Costa Rica followed the Chilean pattern of claiming sovereignty to at least 200 miles of ocean off its coast in July 1948, amended by another decree in November 1949. In this instance the California tuna fishery was directly affected because it not only customarily fished in the high seas of Costa Rica, but passed within 200 miles of Costa Rica on its transit to more southerly fishing grounds. Because of the ownership by Costa Rica of Cocos Island far off its coast, the Costa Rica decree, if valid and enforced, would have given Costa Rica taxing privileges on the whole American tuna fishery to the south by inserting a band of Costa Rica territory about 500 miles in breadth between the Californians and the southern tuna.

In September 1950 El Salvador included in its constitution the following paragraph:

"Article 7. The territory of the Republic within its present boundaries is irreducible. It includes the adjacent seas to a distance of two hundred sea miles from low water line and the corresponding air space, subsoil and continental shelf." (MacChesney, 1957).

This covered another large swath of ocean in which the California tuna fishermen customarily fished and passed through.

In August 1952 Chile, Ecuador, and Peru signed an agreement establishing a maritime zone off their countries and establishing a Commission for the Exploitation and Conservation of the Maritime Resources of the South Pacific. This agreement was subsequently ratified by the three countries and adhered to by Costa Rica. The declaration contains the following paragraphs (copied from MacChesney, 1957):

"II. The Governments of Chile, Ecuador and Peru therefore proclaim as a principle of their international maritime policy that each of them possess sole sovereignty and jurisdiction over the area of sea adjacent to the coast of its own country and extending not less than 200 nautical miles from the said coast."

"III. Their sole jurisdiction and sovereignty over the zone thus described including sovereignty and jurisdiction over the seabed and subsoil thereof."

"IV. The zone of 200 nautical miles shall extend in every direction from any island or group of islands forming part of the territory of a declarant country. The maritime zone of an island or group of islands belonging to one declarant country and situated less than 200 nautical miles from the general maritime zone of another declarant country shall be bounded by the parallel of latitude drawn from the point at which the land frontier between the two countries reaches the sea."

The United Nations quickly established as a continuing body the International Law Commission (ILC) composed of persons acting independently in their capacity as experts in international law. At its inaugural meeting in 1947 ILC instituted a comprehensive inquiry into the Law of the Sea which it continued through 1956. In the course of this it solicited views of governments. The United States was hard pressed to make reply to these inquiries because it had important interests on both sides of many of the issues involved and these were not easy to rationalize. Some of the difficulties are treated by Chapman (1965).

By the time of the General Assembly meeting in the fall of 1954, the United States had its internal position on these matters reasonably well rationalized and had informed the ILC of this. In the report of its 1953 session the ILC had made a preliminary report on the Law of the Sea noting that it was probably deficient on the fishing side of the problems because ILC had not got competence in this field and it would benefit from an international conference of fishery experts to consider these matters.

The United States had, in the intervening years, rationalized a large share of the petroleum aspect of its internal problems concerning the Law of the Sea by the passage of the *Submerged Lands Act* of 1953 and the *Outer Continental Shelf Act* of 1954. There were still actions interpreting these measures and related matters pending before the U. S. Supreme Court, but the main lines of division between the Federal Government and the several coastal State governments concerning title to any resources in the continental shelf over which the United States Government had jurisdiction, had been settled by these two Acts.

The United States was now prepared to move in the international field to secure a change in international law (for which it felt adequate diplomatic preparation had been made) whereby exclusive jurisdiction over the sea-bed resources of the adjacent continental shelf would pertain to the coastal nation. It took this issue to the General Assembly in 1954 with some confidence that it would go through without much trouble.

The tuna and shrimp people had been so alarmed by developments in Latin America early in 1954 that they agreed that the United States should take the fishery disputes issue to the General Assembly also in 1954. The way that this would be done would be to request the General Assembly to call an international conference on the conservation of the living resources of the ocean in order to fill the need expressed by the ILC for competent technical input from fishery experts on these Law of the Sea problems. In order to secure the agreement of the Pacific Northwest salmon people to this action the tuna and shrimp people agreed that the United States could espouse the abstention "principle" and try to get it adopted into international law.

The abstention principle, roughly described, held that if a nation or nations were fully utilizing a resource of the high seas and had it under management regulations, based on scientific findings, designed to produce the maximum sustainable yield from that resource, other nations should prevent their nationals from entering the fishery upon that resource. This principle arose from a treaty negotiated among United States, Canada and Japan toward the end of the postwar occupation of Japan, which was designed to keep the Japanese from fishing on salmon originating from American streams.

This "principle" was antithetical to the desires of the tuna and shrimp people, who fished off foreign coasts. What they wanted was multilateral regulation to secure the maximum sustainable yield from any resource of the high seas where that had been shown by scientific research to be necessary, but for entry in the fishery not to be limited by a general principle of international law when that point was reached.

The salmon and halibut people also wanted this but wanted it to be in conjunction with the "principle" of abstention. The tuna and shrimp people were so hard pressed to get these problems out of the arena of the Organization of American States, where they had few supporters, into the arena of the United Nations, where all of the other "fishing" states would insert their viewpoints, that they agreed not to oppose the United States espousing simultaneously a program for multilateral conservation for high-seas resources and the abstention "principle." They gambled that, given a free choice in a United Nations conference, the nations would be in favor of multilateral conservation regulations designed to secure the maximum sustainable yield from high-seas resources, and would turn down the abstention "principle." As is noted below, this gamble paid off.

Thus armed with internal agreement, the United States went to the General Assembly in 1954 with two proposals which it considered to be quite separate, and which it wished to have dealt with separately.

The first was a proposal that the resources of the sea bed of the continental shelf adjacent to a nation would be subject to its sole jurisdiction and use. It expected this to be adopted with little colloquoy in about the form it submitted. It was not felt, by this time, to be a very controversial issue internationally and, in fact, it was not.

The second was a proposal for the United Nations to convene a "Technical Conference on the Conservation of the Living Resources of the High Seas" prior to the 1955 meeting in Geneva of the International Law Commission. This was also not felt to be a very controversial matter because it responded to a need that had been requested by ILC in its 1953 interim report on the Law of the Sea.

The delegate of Iceland, however, had a quite different problem. His government wished to eliminate all fishing by nationals of other sovereigns on or over the Icelandic continental shelf, or come as close to that as possible. He conceived that this could best be done by keeping the continental shelf and the fisheries conservation issues both open and under simultaneous action in the international field so that he could best exercise such bargaining power as he had on the issue of interest to his country. In the outcome he was able, by exceedingly competent diplomacy, to control the swing votes that the United States needed on both of its proposals. The outcome was a resolution adopted by the General Assembly to the effect that the continental shelf and fishery conservation issues should be considered and solved simultaneously, in the United Nations apparatus, and a second resolution authorizing the convening of a "Technical Conference on the Conservation of the Living Resources of the High Seas" in Rome in the spring of 1955.

This conference was held. It became a contest not over the principles of conservation but between the "coastal" states, who wished to unilaterally regulate fisheries in the high seas off their coast and the "fishing" states, who wished to do such regulation multilaterally on the basis of scientific information indicating a need for such regulation to prevent overfishing. The two forces were almost equally represented at the conference. This issue was finally disposed of by voting on a proposal that the conference was not competent to deal with problems involving jurisdiction on the high seas. The proposal carried by one vote.

The conference then went on to adopt a "Report of the International Technical Conference on the Conservation of the Living Resources of the Sea" (United Nations, July 1955, A/Conf. 10/6), which set out the objectives of fishery conservation, the types of scientific information required for a fishery conservation program, listed the principal specific international fishery conservation problems of the world for the resolution of which international measures and procedures had been instituted, discussed the applicability of existing types of international conservation measures and procedures to other international fishery conservation problems, and set out a number of general conclusions.

Put briefly, the conference espoused the solution of international fishery conservation problems by multilateral conservation measures, based on scientifically demonstrated need, for the purpose of maintaining conditions so that any stock would produce its maximum sustainable yield, refused to deal with limiting entry into fisheries, and declared itself incompetent to express any opinion as to the appropriate extent of the territorial sea, the extent of the jurisdiction of the coastal state over fisheries, or the legal status of the superjacent waters of the continental shelf. The extreme views of the 200-mile states (Chile, Ecuador, and Peru) were rejected and those delegations filed reservations. The extreme views of the abstention states (United States and Canada) won so little support that those countries decided not to put the matter to a vote.

This proved to be a fundamentally important conference. The International Law Commission, meeting directly after it in Geneva, used its conclusions extensively in a basic revision of its report on the Law of the Sea, which it submitted that year to the member nations for comment.

In 1956 the International Law Commission completed its work on the Law of the Sea [Report of the International Law Commission covering the work of its eighth session 23 April – 4 July, 1956, General Assembly, Official Records, eleventh session, Supplement No. 9 (A/3159), New York,

1956] and submitted it to the General Assembly in the form of a treaty of 73 Articles, each with commentary, which might serve as the basis for a conference of plenipotentiaries to meet and reach agreement among the nations, on the Law of the Sea. It noted that, in its view, the various sections of the Law of the Sea hung together and were so closely interdependent that it was extremely difficult to deal with only one part and leave the others aside. It recommended that the General Assembly summon an international conference of plenipotentiaries to examine the Law of the Sea, taking account not only of the legal but also of the technical, biological, economic, and political aspects of the problem, and to embody the result of its work in one or more international conventions or such other instruments as it might deem to be appropriate.

The General Assembly followed this recommendation by adopting resolution 1105 (xi) at its 658th plenary meeting, 25 February 1957. Pursuant to this resolution there was convened an "International Conference of Plenipotentiaries to Examine the Law of the Sea," in Geneva, 24 February to 27 April, 1958.

5.3 THE LAW OF THE SEA CONFERENCE OF 1958

The above account of the political situation between 1945 and 1958 has been entirely too brief to give any realistic impression of the political atmosphere in which the 1958 conference was convened, and there is not space available to do that properly here. Some other factors than those already noted may be mentioned briefly as follows:

1. The settlement of the Anglo-Norwegian Fisheries Case by the International Court of Justice in 1951 had confirmed that under certain conditions a coastal state might draw straight baselines between headlands and islands off its coast, enclosing inside them internal waters over which it had complete sovereignty. The victory for the Scandanavian league of 4 marine miles had been substantial (stipulated by United Kingdom before adjudication) and the over-all effect of loosening up the measurement of the breadth of the territorial sea had been considerable. Iceland wished to extend the concepts of this judgment to its somewhat different problem, and this developed a schism over fishery limits among the NATO powers, who wanted a narrow territorial sea for military purposes.

2. The United States not only had the internal strife among its own tuna and shrimp people on one hand and the salmon and halibut people on the other hand to contend with, but among its vitally necessary (at that moment in history) allies were two whose economies were very dependent on fisheries and which had precisely opposite views. Japan wanted to fish everywhere with as little interference from other nations or international

law as was possible; Iceland wanted to have sole jurisdiction over all fishing on and over its broad continental shelf.

3. The Arab nations wished a 12-mile territorial sea because of their view that this would aid them in their attempts to limit access by Israel to the Gulf of Aqaba. They were not particularly interested in the fishery or many other aspects of the questions, but on this they were firmly united.

4. The Communist-bloc countries were intent on obtaining agreement to a 12-mile breadth for the territorial sea or at least preventing agreement on a narrower territorial sea. This was primarily on the grounds of military necessities noted previously, and for the same reasons the United States and its allies were absolutely intent on obtaining agreement to a narrow breadth of the territorial sea. Curiously enough the Russian and United States delegations were in quite close agreement on fishery matters and had worked in a complementary manner on them at the Rome 1955 conference and the 1955 meeting of ILC. But from that time on the delegations of the two countries engaged, each with all its allies, in vigorous struggle over the breadth of the territorial sea and substantially ignored their fishery interests in this vital struggle while developing compromises to win votes on that issue.

5. The 200-mile countries (Chile, Ecuador, Peru, with Costa Rica and El Salvador occasionally) had by no means given up their struggle to get jurisdiction of the fisheries in broad areas of the high seas off their coasts just because they had been defeated at the Rome conference and the International Law Commission. On entirely different grounds (the archipelago envelope principle), the Philippines and Indonesia could be counted among the voters for an extremely broad territorial sea and jurisdiction. (Russia did not favor their views because of its worldwide fishery interests.)

6. The United States and Canada were still pushing for the abstention principle, which was opposed by their European allies.

7. Afganistan and Bolivia, in particular, as landlocked countries wished to get freedom of access to the sea across the territories of countries which did not particularly wish to give it to them.

8. On top of these major issues were a variety of minor fishing problems in different parts of the world which were of only local importance but kept intruding into larger issues. An example was provided by the stake-net fisheries offshore of Bombay, India, and the chank fishery of the Gulf of Mannar.

What made all of these various political stresses and strains important was that the rules of the conference required substantive issues to receive a two-thirds majority to be adopted. On most days the casual observer in Geneva during the spring of 1958 would have despaired of obtaining a two-thirds vote on the date the next Sunday would fall on. It is all the more remarkable that this conference was so tremendously successful.

The conference divided its work among five main committees, each of which was assigned a section of the conference's work. The first committee was assigned the Territorial Sea and the Contiguous Zone. The second committee was assigned the High Seas and the General Regime. The third committee was assigned the High Seas Fishing and Conservation of Living Resources. The fourth Committee was assigned the Continental Shelf. The fifth Committee was assigned the question of free access to the sea of landlocked states.

The greatest controversy concentrated around the breadth of the territorial sea and the jurisdiction by the coastal state over fisheries lying in the adjacent high seas. These fell within the purview of the first committee and chiefs of delegation, as well as important diplomatic and military delegates, pretty well concentrated on that committee, leaving the technical members of delegations reasonably free to concentrate on the other committees. Those two issues were not resolved, but substantially all of the rest of the Law of the Sea was codified, established, and agreed to by suitable majorities in four conventions which were opened for ratification. These were

1. *The Convention on the Territorial Sea.*
2. *The Convention on the High Seas.*
3. *Convention on Fishing and Conservation on the Living Resources of the High Seas.*
4. *Convention on the Continental Shelf.*

The first, second, and fourth of these came into force some time ago and the third came into force on March 20, 1966.

Eighty-six nations were gathered together, the largest group of sovereigns that had ever gathered for any purpose to that time. Questions of the utmost gravity as well as technical complexity were under consideration with such intensity that every foreign office in the world was dealing with these issues on practically a daily basis and was in daily communication with its delegation. Each day the tension grew, but there were intervals of comic relief.

The Society for the Prevention of Cruelty to Animals had an observer present. He wished to press for a resolution that would result in fisheries being operated with the least pain and suffering to the animals being harvested, but he was without voice or vote as an observer. Finally one day when controversy was at its height in Committee Three, the delegate of Nepal arose and introduced a resolution to this effect: as a devout Buddist representing a Buddist Government, he would prefer no killing of any animal. Communists, monarchists, and democrats alike dropped their other arguments to deal in common with this threat to their fishing industries. They could not live with it if it passed. A subcommittee was appointed to deal with this matter with the greatest of difficulty because nobody wished to serve. There resulted a resolution on the humane killing of marine life adopted by the conference which "Requests States to prescribe by all means available to them, those methods for the capture and killing of marine life, especially of whales and seals, which will spare them suffering to the greatest possible extent." (A/Conf. 13/38, Vol. II, p. 144.)

The conference, aside from the four basic conventions, adopted nine resolutions and an optional protocol of signature concerning the compulsory settlement of disputes.

5.4 THE LAW OF THE SEA CONFERENCE OF 1960

By the eight of its resolutions the 1958 conference requested the General Assembly to study, at its thirteenth session, the advisability of convening a second international conference of plenipotentiaries for further consideration of the questions it left unsettled. The General Assembly did study this suggestion and on December 10, 1958 adopted a resolution calling for a second conference of plenipotentaries to consider just two questions: (a) the breadth of the territorial sea, and (b) the extent of fishery limits.

This conference convened in Geneva on March 17, 1960, and ran to April 27 of that year. Eighty-eight nations were represented.

The political problems of this second conference were more intense even than that of the first. All attention was concentrated on getting, or blocking, a favorable vote on the breadth of the territorial sea. Again Russia and her associates wanted a 12-mile territorial sea, or to block a vote on anything less than that while leaving history to take its course (1960 was the key year for new countries from Africa to arrive in the community of nations). The United States wanted in the worst way to get as narrow a territorial sea as possible but was willing to settle for a 6-mile breadth. I have treated the breath-taking events of the last days of this conference elsewhere (Chapman, 1960) and will not repeat them here save to say that at the very last the United States was prepared to sacrifice its high-seas fisheries to attain a 6-mile breadth for the territorial sea and that it lost in this attempt on the last vote by one *abstention*, not one full vote.

This was a most peculiar, as well as breath-taking, conference, in that every nation or group that was trying to put something across lost. Even though the Russians had dressed up their 12-mile limit proposal in the most palatable form possible and had added much to attract the odd vote, it got less votes than it did in 1958, and short of a simple majority. The radical wing of the Arab bloc, which was aimed simply at restricting Israel's economy, played its hand too strongly and the Arab bloc broke up in its votes rather badly toward the end of the conference. Those countries that wanted in the worst way to extend their jurisdiction far out to sea over foreign fishermen overplayed their position at the last moment and lost everything. The United States military and diplomatic people who were prepared to trade all fish away for a narrow territorial sea had barely lost and perhaps they gained more by losing so narrowly than they would have by winning through such a narrow margin. Iceland, which had led the fight in favor of extending fishery jurisdiction on the basis of its special case, lost its special case in the welter of other special cases that sprang up like weeds at this conference. Canada, which had labored so hard to obtain special rights vis-a-vis American fishermen, lost even that modest effort.

5.5 THE LAW OF THE SEA IN 1966

The 1960 conference neither added to, nor subtracted from the accomplishment of the 1958 conference. In the intervening years the four treaties that came from the 1958 conference have been accumulating ratifications. They are now all in force. Accordingly, they represent current Law of the Sea. Having in mind that the Law of the Sea is not static, but is continuously being modified by the practice of nations, new treaties, and so on, the following comments based on the 1958 treaties may nevertheless be useful.

The law of the superjacent atmosphere

The gradual growth of public order in space will affect the law of the superjacent atmosphere as time goes on, and already has. For the present purpose it is sufficient to say that the present Law of the Sea provides for freedom to overfly the high seas but does not provide the right to overfly the territorial sea in innocent passage or otherwise without the consent of the sovereign.

In 1958 and 1960 this had a great bearing on the votes at the conferences. For instance, under a 3-mile breadth for the territorial sea there is much of the high seas in the Aegean Sea and an observation plane from a carrier therein could have broad surveillance over adjacent territories without the consent of their sovereigns. Under a 12-mile breadth for the territorial sea, almost the whole of the high seas disappears in the Aegean Sea, so that a carrier plane could not get in close for photographic or other reconnaisance that the sovereign of adjacent territories did not want. With advancing technology of photography and other surveillance from satellites this particular aspect has now become of modest importance, but in 1958 it was very important.

In another example the claim of Indonesia and the Philippines to broad areas of high seas as territorial sea under the archipelago envelope principle would disturb commercial air travel routes from Australia to Asia in a major way.

The law of the land under the sea

Under internal waters. The land under the internal waters of a nation belong to it the same as its dry land and no sovereign may do things thereon without the consent of the sovereign of the internal waters.

The change from conditions before 1958 arises from the right of a nation, under particular conditions set out in Article 4 of the convention on the Territorial Sea and the Contiguous Zone, to draw baselines enclosing as internal waters some areas that had formerly been considered to be high seas.

The continental shelf. Prior to the 1958 conference, the continental shelf had been a concept of geographers and oceanographers and related to the nature of the earth where there is normally a shelf around the continents of greater or less breadth which plunges downward to the general depth of the ocean at about a depth of 200 m, or 100 fathoms. Most navigational charts note this feature by a line demarking the 200-m or 100-fathom depth.

The 1958 conference, with great deliberateness, and after intense and prolonged debate and voting, established a new legal concept for the term "continental shelf" and this cannot be described more accurately or succinctly than by quoting from Article 1—*Convention on the Continental Shelf*.

"For the purposes of these articles, the term 'continental shelf' is used as referring (a) to the seabed and subsoil of the submarine areas adjacent to the coast but outside the area of the territorial sea, to a depth of 200 meters, or, beyond that limit, to where the depth of the superjacent waters admits of the exploitation of the natural resources of the said areas; (b) to the seabed and subsoil of similar areas adjacent to the coasts of islands."

The word "adjacent" is a key and purposive word which writers favoring one aspect or another of the governance of harvesting of minerals from the sea bed like to overlook, but it is there, and it is there on purpose. The conference debated this issue extensively but in the last event was guided much by the report of the ILC, and the commentary of the ILC is thus part of the legislative history of the term. It reads, in part, (Article 1—*Convention on the Continental Shelf*, A/3159, p. 42) (8):

"In the special cases in which submerged areas to a depth less than 200 meters, situated fairly close to the coast, are separated from the part of the continental shelf adjacent to the coast by a narrow channel deeper than 200 meters, such shallow area could be considered as adjacent to the part of the shelf. It would be for the State relying on this exception to the general rule to establish its claim to an equitable modification of the rule. In case of dispute it must be a matter for arbitral determination whether a shallow submarine area falls within the rule as here formulated."

The terms "fairly close" and "narrowly" in this comment confine the term "adjacency" pretty closely to contiguousness and straight geological connection with the coast.

Leaving the outer edge of the newly defined "continental shelf" defined by technological ability to harvest resources despite the depth of the superjacent water was also the deliberate action of the Commission as well. The Conference, having in mind that technology was advancing more rapidly than codified international law was able to do, and not wishing to make the right to exploit resources subject to the prior necessity of securing change in law, provided the alternative of the flexible limit it adopted.

Technology has moved rapidly in the intervening years and drilling for oil can now be practically accomplished on what, prior to 1958, was known as the continental slope (and is still so known to oceanographers). Accordingly such areas where this can be done as are adjacent to the old continental shelf become automatically part of the new continental shelf.

The importance of this is that the Convention on the Continental Shelf gives to the adjacent nation exclusive right to explore and exploit the natural resources of the continental shelf as so defined. These natural resources are defined as "the mineral and other nonliving resources of the seabed and subsoil together with living organisms belonging to sedentary species, that is today, organisms which, at the harvestable stage, either are immobile on or under the seabed or are unable to move except in constant physical contact with the seabed or the subsoil."

It was not hard to deal with the mineral and other nonliving resources in this definition. On the other hand, it was impossible to deal in a clearcut way with the living resources because the animal kingdom is so various that it provides examples capable of defeating any definition that can be constructed to separate the resources of the sea bed from the resources of the water, which it was not desired to affect.

Even with this closely debated and voted definition, exceptions that are debatable exist and are in the process of being dealt with. A fishery for king crab has become important on the continental shelf of Alaska in recent years. Russia, Japan, and the United States have participated heavily in it. Recently the United States contended that these king crab was a resource of the continental shelf within the definition in the Convention on the Continental Shelf, and initiated discussions separately with Russia and Japan with a veiw to them withdrawing their fishermen from this fishery. Russia agreed that the king crab was a resource of the continental shelf and that its fishermen would fish there for them for two years at a reduced level and then cease to do so. The Japanese, although agreeing to reduce their catch of king crab there as a conservation measure, did not agree that these king crab were a resource of the continental shelf belonging to the United States.

There is no equivalent of the right of innocent passage on bottom on the continental shelf. It is necessary to obtain the consent of the coastal state before undertaking research on the continental shelf, although it is suggested in the Convention that the coastal State should not normally withhold its consent if the request is submitted by a qualified institution with a view to purely scientific research into the physical or biological characteristics of the continental shelf.

Under the high seas. The conventions do not specifically deal with the resources of the sea bed and subsoil under the high seas outside the continental shelf, as broadly defined, except to state that all states shall be entitled to lay submarine cables and pipelines on the bed of the high seas. Paragraph 2 of Article 26, Convention on the High Seas, also provides: "Subject to its right to take resonable measures for the exploration of the continental shelf and the exploitation of its natural resources, the coastal state may not impede the laying or maintenance of such cables or pipelines."

I can find nothing in these conventions that affects the status of the resources of the deep sea bed away from *res communis*, the legal condition under which they become the property of him who first reduces them to his possession.

It may be noted that occupation is still a valid tenet under international law through which to gain sovereignty over unoccupied land whether dry or under the sea, and that advancing technology is removing such possibilities from the realm of idle speculation. As a matter of fact this is dealt with in reverse by Paragraph 3, Article 2, Convention on the Continental Shelf, which says:

"The rights of the coastal State over the continental shelf do not depend on occupation effective or notional, or on any express proclamation."

The law of the waters

For most of history such Law of the Sea as existed applied only to the liquid ocean. Developing technology has required public order to be brought by international law to the atmosphere above the ocean and the seabed below it. It still remains the case that most of the Law of the Sea pertains to the liquid ocean and its resources. From a juridical viewpoint there are four sorts of waters of the ocean: the internal waters, the territorial sea, the contiguous zone and the high seas, and each has its separate legal connotation.

Internal waters. Before World War II, the concept of international ocean waters was not very prominent. By the end of the 1958 conference, it had become of considerable importance because of the Anglo-Norwegian Fishery case decision, the loosening up of the method of demarcating bays, roadsteads, and so on, in the *Convention on the Territorial Sea and the Contiguous Zone*, and the general desire of many coastal states to push out their territories into the ocean.

Internal waters are those that lie shoreward of the baseline from which the territorial sea is measured outward. They have the same relationship to the sovereignty of the adjacent land as does the land. There are no rights of innocent passage except in instances where the internal waters had formerly been considered to be territorial sea or high seas. They are the sovereign territory of the sovereign and everything in them are his property and for his exclusive use.

The normal baseline for measuring the breadth of the territorial sea is the low-water line along the coast as marked on large-scale charts officially recognized by the coastal state. Where the coast line is deeply indented or cut into, or where there is a fringe of islands along the coast in its immediate vicinity, straight baselines joining appropriate points may be employed in drawing the baseline from which the breadth of the territorial sea is measured. They must not depart to an appreciable extent from the general direction of the coast. They shall not be drawn from or to low-tide elevations unless lighthouses or similar installations of a permanent nature have been built on them. They may not be employed in such a manner as to cut off from the high seas, the territorial sea of another nation.

Also the outermost permanent harbor works which form an integral part of a harbor system can be regarded for this purpose as forming part of the coast. If the distance between low-water marks of natural entrance points of a bay do not exceed 24 miles, the waters inside a line joining those marks may be considered internal waters. If a river flows directly into the sea, a baseline shall be a straight line across the mouth of the river between points on the low-tide line of its banks. When a low-tide elevation is situated wholly or partially at a distance not exceeding the breadth of the territorial sea from the mainland or an island, the low-water line on that elevation may be used as the baseline for measuring the breadth of the territorial sea. "Historic" bays are internal waters.

Thus a considerable amount of ocean has now been withdrawn from the international common of the high seas when all of these pieces of internal waters are added together.

The territorial sea. The territorial sea is a juridically distinct body of water that everywhere separates the territory of the sovereign from the high seas. It is the sovereign territory of the state and all resources in it or under it are its possession. Its sovereignty extends to the air space over it and to its bed and subsoil.

Although the coastal state exercises sovereignty over the territorial sea it does so subject to the provisions of the articles of the Convention on the Territorial Sea and the Contiguous Zone and to other rules of international law. The international community thus retains specific rights in the territorial sea.

Chief among these rights is the right of innocent passage through the territorial sea. Such passage is innocent so long as it is not prejudicial to the peace, good order, or security of the coastal state. It includes stopping and anchoring but only when this is incidental to ordinary navigation or is rendered necessary by *force majeure* or by distress. Foreign fishing vessels must observe such laws and regulations as the coastal state may make and publish in order to prevent them from fishing in the territoral sea. Submarines must navigate on the surface and show their flag. No charge may be levied on foreign ships by reason only of their passage through the territorial sea, but charges may be levied (without descrimination) for specific services rendered.

The coastal State is required to give appropriate publicity to any dangers to navigation, of which it has knowledge, within its territorial sea.

The delimitation of the territorial sea is beautifully clear and distinct. Its inner limit is the baseline that marks the low-water line along the coast, or the outer limit of internal waters as noted above. Where two States are opposite or adjacent to each other the line separating their territorial sea is that line every point of which is equidistant from the nearest points on the baselines from which the breadth of the territorial sea of each of the two States is measured, except where historic title or other special circumstance is at variance with this method of measurement. The outer limit of the territorial sea is the line every point of which is at a distance from the nearest point of the baseline equal to the breadth of the territorial sea.

The nub of the matter is, of course, what is the breadth of the territorial sea, and this requires some explanation.

There is no agreed breadth of the territorial sea and there never has been. A nation has a perfect right to claim any breadth of territorial sea that it desires, and, in the absence of any internationally agreed breadth for the territorial sea, this is the only way in which the breadth of a nation's territorial sea can be established. For instance, the United States, in the *Outer Continental Shelf Act of 1954* established the breadth of its territorial sea at three marine miles, with certain minor exceptions.

On the other hand, a nation is under no obligation to recognize the breath of territorial sea claimed by another nation that is outside the usual norms. For instance, the United States has consistently refused to recognize a breadth for the territorial sea for any other nation of more than 3 marine miles. It regularly files protests with nations that claim more, stating that it considers their claims beyond the breadth of three miles to be without effect on its vessels.

Although no breadth for the territorial sea attained the two-thirds majority required for adoption at the 1958 or 1960 conferences there are norms which have some sanction under international law. In its 1956 report, which formed the bases for the 1958 conference, the ILC dealt with this matter as follows:

"The Commission noted that the right to fix the limit of the territorial sea at three miles was not disputed. It states that international law does not permit that limit to be extended beyond twelve miles as regards the right to fix the limit between three and twelve miles, the Commission was obliged to note that international practice was far from uniform. Since several States have established a breadth of between three and twelve miles, while others are not prepared to recognize such extensions, the Commission was unable to take a decision on the subject, and expressed the opinion that the question should be decided by an international conference of plenipotentiaries".

In the many votes that were taken on proposals respecting the breadth of the territorial sea at both the 1958 and 1960 conferences, it was clear that the Commission was correct in saying that the number lay between three and 12, and not beyond that, but there was no clear vote on particular limits in that rather narrow range. It was clear that proposals involving a 6-mile territorial sea were favored much more than those involving a 12-mile territorial sea, and that this was even more the case at the 1960 conference than at the 1958 conference. The Canadian–United States proposal at the 1960 conference came within one vote of winning. In fact had one of the negative voters only abstained the measure would have won. The proposal involving a 12-mile limit received not quite a simple majority.

On the other hand all of these proposals were so clouded with additional features designed to win votes by compromise on fishing rights that there was no real test of what number of nations favored any particular breadth for the territorial sea. It was a striking fact that the principal exponents of the six-mile limit did not favor it but favored a three-mile limit. They had moved to espousal of the proposal involving the six-mile limit in a futile attempt to win the few votes needed for a two-thirds majority beyond what they thought they could get for a three-mile limit. The three-mile limit had never been submitted for a vote and it was not clear that many of those who had voted for the proposal involving the six-mile vote would not have preferred to have voted for the same proposal had it involved a three-mile limit for the territorial sea.

Accordingly nations such as the United States, which stood solidly in favor of the three-mile limit, could defend its contentions by military power, by simply avoiding trouble on the issue, or by diplomatic measures when its vessels were molested on the basis of claims with which it was not in agreement.

The United States has used all three of these approaches to attending to these problems both before and after the 1958 and 1960 conferences. In the Matsu Island area it had freely displayed its military power inside the claimed 12-mile limit of Mainland China, but not within the 3-mile limit. On the other hand it had ordered its vessels to avoid penetrating the 12-mile territory claimed by Russia and had thus avoided a test of strength on that issue at those times and places.

Its fishing vessels had been subjected to harassment and seizure outside three miles off several Latin American countries from time to time over the past 20 years. It has dealt with these problems through diplomatic action with these countries. Since 1954 it has repaid to its vessels owners whose vessels had been so seized under claims it did not recognize, the fines they had found necessary to pay to the seizing country.

There is no clear evidence that any end is in sight to this situation or that there will ever be an agreed international breadth to the territorial sea between three and 12 miles. The stampede of the new countries to a 12-mile limit which had been anticipated after the 1960 conference has not developed, and there is no reason at the present to expect that it will.

The contiguous zone. As noted previously, the concept of the contiguous zone developed side by side with, but separately from, the concept of a territorial sea. There is no disagreement on what it is for or what its measurements are. It may not extend beyond 12 miles from the baseline from which the territorial sea is measured. It is demarked as between neighboring States in the same manner as the territorial sea. In it a State may exercise controls necessary to (a) prevent infringement of its customs, fiscal, immigration or sanitary regulations within its territory or territorial sea; (b) punish infringements of the above regulations committed within its territory or territorial sea.

That is all there is to it. The 1958 conference did not change the contiguous zone except as it incidentally altered the baseline from which it was measured.

The high seas. The high seas are the common property of all nations. It is particularly necessary to point out that they are the property of nations and not people. The high seas are governed under international law. Individual persons can be the objects of international law but only sovereigns are its subjects. Translated into more clear language a fishing vessel does not operate on the high seas under any right pertaining to it under international law, but only in exercising the rights that pertain to the sovereign whose flag the vessel wears.

Although most of the fussing in international circles over the Law of the Sea has been concentrated on the territorial sea and the continental shelf, most of the surface of the earth is covered by high seas, whether the territorial sea is three or 12 miles in breadth. It is difficult to draw or see a 12-mile zone on an ordinary globe or map of the world because it is so close to land and so little affects the extent of the high seas.

It is in the high seas that the vast food resources are found, and under it are the vast stores of minerals. It is the property of all nations; its resources are the property of the nation that first reduces them to its possession.

The 1958 Convention on the High Seas did little to alter the regime of the high seas. It mainly codified existing rules. The high seas being open to all nations, no nation may validly purport to subject any part of them to its sovereignty. There is freedom for all to navigate, fish, lay submarine cables and pipelines, and overfly the high seas. Freedom is not license. These and other freedoms recognized under the general principles of international law shall be exercised by all nations with reasonable regard for the interests of other nations in their exercise of the freedom of the high seas.

It is laid down in the Convention on the High Seas that all nations have a right to sail ships under their flags on the high seas whether they have a seacoast or not, and that coastal nations should give access to the sea to noncoastal nations.

There must exist a genuine link between a ship and the nation whose flag it flies, and a ship shall sail under only one flag except under specified particular situations. The ship is, in effect, a mobile part of the territory of the sovereign whose flag it carries, is responsible to that sovereign, and its sovereign is responsible for it and its actions. The slave trade is illegal. Piracy is defined and methods for its elimination set down.

The conditions under which a ship may be molested by other than its sovereign on the high seas are set down, and liability for damages resulting from molestations on inadequate grounds is established. The narrowly defined rights to board ships on the high seas are spelled out and the rights of hot pursuit by a sovereign from its territorial sea to the high seas are carefully defined.

The responsibility of nations not to pollute the high seas from its ships or its territory is made clear. The responsibilities of all respecting submarine cables and pipelines under the sea are defined, as well as the rights of all to engage in such action.

There is little about the conduct of ships on the high seas that is left unclear by this convention.

Fishing and the conservation of the living resources of the high seas

Whereas the Convention on the High Seas was primarily a codification of national practices, the *Convention on Fishing and the Conservation of the Living Resources of the High Seas* was primarily new legislation that broke new ground in international law. In view of the major nature of conflicts over fishing rights, their complexity and their variety, it is all the more remarkable that such foresighted legislation could be contrived in such tense conference atmosphere, and that sufficient nations would subsequently ratify it to put it into force.

The convention first reiterates the rights of all nations for their nationals to engage in fishing on the high seas subject to their treaty obligations and the terms of the Convention. It then states clearly the duty of each nation to see that its nationals, in exercising this right, do not overfish.

Conservation is clearly defined as the aggregate of those measures rendering possible the optimum sustainable yield from particular resources so as to secure a maximum supply of food and other marine products from them. The duty of nations to adopt measures to provide for such conservation, and to cooperate with other nations to do this, is clearly set out.

A series of conditions are dealt with separately and means provided for attending to them. These conditions are: (1.) when a nation does not fish a particular resource but has an interest in it, has reasons to believe overfishing is taking place, and wants it stopped; (2.) when only the nationals of one nation are fishing the resource; (3.) when the nationals of more than one nation fish the resource.

The genius of the Convention, aside from getting agreement on what conservation means, was the establishment of a balance of power between the coastal nation and other nations fishing on resources in the high seas off its coast that makes possible the attainment of conservation regulations when needed, and lack of interference with fishermen when not needed.

If the coastal nation has reason to believe that overfishing is taking place off its coast, it can take the matter up with the nation whose fishermen are doing it and, if it does not get satisfaction in a reasonably short time institute regulatory measures unilaterally needed to achieve conservation. It must, however, do this on the basis of sound scientific evidence and not on whim. Also it cannot discriminate in form or fact against foreign fishermen in such regulation.

The foreign nation has recourse. Criteria are set down in the convention under which the coastal nation can act in this fashion. If the coastal nation has acted outside those criteria the foreign nation, can seek redress diplomatically and, if satisfactory recourse cannot be had in that manner, it can call for a review of the regulations by a special arbitral commission established from among qualified experts for that particular case. The method for ensuring the impartiality of the arbitral commission are set out in the convention. It must review the matter in the light of criteria established in the convention. It can at once set aside the regulations while it investigates, if it thinks that course to be appropriate. On the other hand its findings are binding upon the disputants.

There are two prime problems involved in international fishery disputes: (a) a need to protect high-seas resources from overfishing, and (b) the selfish desire of fishermen and nations to secure special benefits from particular fisheries for themselves as against other nations.

The Convention on Fishing and the Conservation of the Living Resources of the High Seas provides a perfectly satisfactory mechanism for settling any sort of dispute that has so far arisen in the world over conservation, as defined in the convention. If nations will, in actuality, perform as they have committed themselves to do, there is no reason why the conservation half of this problem cannot be handled satisfactorily by this mechanism.

How is the second half of the problem, the selfish desires of the nations and their fishermen, is to be dealt with is not treated by the convention. What the convention does, in substance, is to provide a mechanism by which the resources can be kept in a condition so that they can provide their maximum sustainable yield. This attends to safeguarding the general human interest in these resources. It also preserves something for the nations to fight over instead of having the resources killed off one by one. The economic, political, and social problems of dividing up the profit resulting from conservation, however, remain in the field of diplomacy so far. It can be handled among the nations not by generalized action but among those nations appropriate to the specific problem. There is no general formula yet discovered adequate to deal peacefully with all of the selfish yearnings of all men.

Continuing problems

Most of the continuing problems in the Law of the Sea have been mentioned above but they can be noted again by way of brief review.

Overlapping and indistinct jurisdiction. There is a great deal of discussion about unsettled problems respecting the Law of the Sea currently in the United States, and particularly in industry. A prime reason for this is not that the Law of the Sea is either obscure or incomplete. Many new industries and firms have turned their attention to the ocean recently for the first time. They find the Law of the Sea different than the bodies of law with which they have been dealing ashore and because it does not always accord with industry practices that arose from land-oriented problems they think it to be improper and inadequate. A great deal of the puzzlement and dismay over the Law of the Sea arises from the different aspect given ocean-resource-development activities by the common-property nature of so many of the sea resources. They are not owned by any entity and therefore no entity can obtain their exclusive or preferential use.

Upon closer experience with the existing Law of the Sea, however, the mining and petroleum people find greater dissatisfaction with it as they approach the shore. The Law of the High Seas may be not agreeable to ordinary industry experience but it is quite clear. Also the problems related to it are not so immediate, because exploitation of resources beyond the continental shelf is not going to take place as rapidly as of those under shallower waters.

The shallow waters near shore in the United States come within the jurisdiction of a state government. The bodies of law in the states that apply to aquatic problems are complex, hard to sort out as to detail, and often have derived from extensions of land problems in such a manner that they very well may not fit aquatic problems arising from new outputs of science and technology. It is not infrequent that the jurisdiction of more than one state agency applies to the problem. Local bodies such as municipalities, harbor commissions, or county governments often have overlapping jurisdiction within that state. The several states have their own separate bodies of law, and regional approaches to regional aquatic problems in the state laws are infrequent.

The inner edge of the area, characterized by the name tidelands, is sometimes indistinct geographically, imperfectly surveyed, or has been materially altered by filling or dredging; thus there is often not a hard, finite edge from which to begin measuring outward.

Industrialists who are wishing to discover or exploit finite bodies of ore or oil wish to know with some precision the dimension of the claim or lease they have made or purchased and the navigational accuracy of being able to pinpoint a second time what was thought securely located at first often leaves much to be desired.

These problems that lie within the purview of state or local law must be attended to within those bodies of law, but do not entirely avoid Federal law. There are navigational, pollution, communication, military, and other aspects that are partially or wholly within the purview of Federal law while still within State jurisdiction geographically.

As we move out to the edge of the territorial sea (3 marine miles from land in most states), we encounter again the problem of the indistinct baseline from which the delineation began, and in not a few places around the United States a person does not know very precisely when he leaves the jurisdiction of the State and enters the Federal jurisdiction purely. Even on reaching certain Federal territory, all aspects of state regulations are not left behind. In the case of bottom resources, harvesting on Federal lands is not oblivious to the regulation of the same on the adjacent state lands. In the case of swimming resources, states often regulate the harvest by limiting rights to land catch, thus extending their regulatory authority to sea an indeterminate distance.

In deeper water and farther from shore the navigational problem of refinding the body or ore, for instance, and its geographical boundaries becomes more troublesome. The monuments from which precise triangulations can be made have not yet been placed, and the type of Loran available on most parts of the coast leave surveying of this sort of precision difficult.

The policies under which either the states or the Federal government will operate the harvesting of bottom resources either inshore or offshore are not in as clear a condition as they are on the dryland, and when they will be more precisely settled upon is problematical. We are reminded that it took more than 100 years for United States society to shape its rules and regulations to the effective use of the arid Great Plains area after emerging onto it from the humid, forested eastern part of the continent.

In actuality, for the United States the legal problems that exist out in the international domain of the high seas are clear and wholesome when compared with those of a domestic nature further inshore, and in a good many cases the entrepreneur can breathe a sigh of relief when he passes out over the edge (as indistinct as it may be) where the resource belongs to him who first reduces it to his possession, where piracy is defined and prohibited, and where simple treaty law rules the waves.

The Law of the Sea, which needs clarification, and codification, and rationalization to modern economic and social needs to permit the effective use of the sea by the United States begins most urgently with State law on the shoreside, then Federal law, and, lastly, international law.

Breadth of the territorial sea. This is not as large a practical problem as it appears to be in the literature and the popular press. Regardless of views to the contrary by a few countries, it is unlikely that international law will recognize a breadth of territorial sea greater than 12 marine miles in the forseeable future or that any pretensions to claims greater than 3 miles will be permitted to change existing navigational patterns on the ocean for either mercantile, military, or aircraft in any substantial manner for any great period of time. The use of the ocean and the air above it as a highway is so vital to the whole society that any serious effort to modify it substantially will be met by whatever diplomatic or military effort is required to correct the situation. There never has been an agreed breadth to the territorial sea during the period of the development of international trade, yet it has developed. The more it develops the more free the navigational use of the sea must be. Society will keep it so.

Chile, Ecuador, and Peru still maintain their claims to sovereignty for a minimum distance of 200 miles from land. The United States has requested each of the three individually and collectively to join it in taking a test case to the International Court of Justice (Department of State, 1955) but so far each has refused to accept the jurisdiction of the court in this matter, which is some indication of their faith in their broad claims.

The 12-mile fishing rule. In 1966, the United States, among other countries, adopted a 12-mile rule for exclusive fisheries jurisdiction (actually a zone nine miles broad beyond its territorial sea). For reasons set out in another connection (Chapman, 1963) the adoption of such a limit has little effect on the fishery problems it seeks to solve either one way or the other The principle reason is that the resources supporting the major fisheries of the United States, and of the world, either occur regularly beyond 12 miles from land, or move out there at some stage in their life history, where they are still liable to capture by foreign fishermen.

Thus a 12-mile fishing rule will not much impede the operations of foreign fishermen, give much competitive advantage to the coastal fisherman, or contribute much to the conservation of major resources.

What it will do is excite the coastal fisherman to further political adventures aimed at extending his protection from competition further to sea, lay the basis for further international discord from this cause, and to some degree further erode support for the three mile rule for the territorial sea.

Fishery disputes. Fishery disputes among nations can be expected with great certainty to increase in number and virulence. There is no nation, including the United States, that shows any inclination to support adequately the research required to husband the common property resources of the high seas either by their national agencies, through international fishery commissions, or through FAO. In no country is ocean research accorded the budgetary priority that agricultural research is.

Yet almost all countries are urging forward the development of their fisheries, the industrial countries in particular are actively subsidizing the growth of their fisheries, and FAO, the Special Fund, and many bilateral schemes are being pushed to develop the sea fisheries of the developing countries. The result is a rapid expansion of unregulated fishing effort throughout the world ocean.

The fishing effort grows rapidly, overfishing problems grow more rapidly than the means for their governance develop, and the research is not undertaken with which to detect the approach of overfishing or to devise the means for its treatment.

The result, inevitably, will be greater pressure for extending national fishery limits out further and further into the high seas. But this leads up a blind alley because the resources supporting major fisheries are so mobile that no set or arbitrary boundaries will fit the problems.

There is no way to handle the husbandry of the fishery resources of the ocean except by international cooperation. There is no way to husband the resources until research shows what needs to be done. Nobody will pay for the research except the owners of the resources, who can get the economic benefit from the husbandry. The sovereign nations, not the fishermen, own the resources and until they accept their responsibilities as owners fishery disputes will continue to mount.

It has always been fishery disputes that have kept the Law of the Sea unsettled. It was fishery disputes that have made it impossible to gain agreement to the breadth of the territorial sea at the 1958 Law of the Sea Conference, or to anything at the 1960 Conference. This situation will continue until the nations who use these sources are prepared to pay for the expenses of husbanding them. It will grow much worse before it begins to get better.

Indistinct nature of the outer edge of the continental shelf. Much is made of the geographically indistinct way in which the Convention on the Continental Shelf defines the outer edge of the shelf. This should in actuality, result in little interaction among the nations. The idea of Russia or China boring a hole in the bottom of the ocean a little further seaward of the United States than "adjacent" to its continental shelf for some malevolent purpose makes good material for the tabloid press, but is a little ridiculous for the real world.

The technology of drilling in the deep-sea bed is with us. The technology of dredging mineral deposits from the deep sea bed does not seem formidable. In both cases it is cost, not technology, that prevents the use of these deeper resources. So long as the term "adjacency" is fixed by the coastal country it can extend its sovereign jurisdiction and exclusive ownership down the continental slope to the adjacent deep sea bed under present law as rapidly as it can afford to do so. Certainly this should be of no concern to a nation so technically competent and so well supplied with capital as is the United States.

It should not be thought that present law permits the coastal state by improved technology to work its way down its adjacent continental shelf, across the sea bottom, and up the continental slope of its competitor to where its competitor's poorer technology reaches. The first rule of international law is the rule of reason. It will undoubtedly be the case that if the United States or Russia, or Ghana improves its technology so that the depth of the superjacent water does not prevent the utilization of bottom resources in 1500 fathoms of water off its coast, and therefore permits it to claim sovereignty to that depth, that the same rule will apply to all coastal states regardless of the state of their technology.

Indistinct nature of innocent passage for fishing vessels. Paragraph 5, Article 14 of the Convention on the Territorial Sea and the Contiguous Zone singles out the right of innocent passage for fishing vessels in this manner:

"Passage of foreign fishing vessels shall not be considered innocent if they do not observe such laws and regulations as the coastal State may make and publish in order to prevent these vessels from fishing in the territorial sea."

This has not as yet caused any considerable amount of interaction and is not likely to if it is confined within the concept of a narrow territorial sea. This is probably the case when it is confined to a 12-mile fishery zone also. When it will be troublesome, and already has been in a few cases, is when there is an attempt to enforce it in respect of a broader zone of exclusive fishery jurisdiction.

The organization of ocean governance. A basic problem at the root of public order on the ocean is that society generally has arranged itself into institutions that deal with land-oriented problems and for the most part has left ocean-oriented problems to be dealt with by these same institutions.

Some reference has been made above to problems this causes within the United States in governance of ocean activities within the near shore area where the laws of the states are effective and the regulations and activities of Federal agencies also intervene. Recommendations for the reorganization of ocean affairs in the United States Federal Government have recently been made, by a Panel on Oceanography of the President's Science Advisory Committee (PSACPOO, 1966), that would go a good way toward mitigating this situation if implemented. Recently enacted legislation has established "The National Council on Marine Resources and Engineering Development," composed of Cabinet officers and chaired by the Vice President, to examine the whole range of the United States' use of the sea, its organization to deal with ocean-oriented problems, a policy for this, and a program to carry out the policy. This should lead to some rationalization of the present rather inchoate attitude of the United States toward ocean problems.

The same sort of confusion exists in the governance of ocean affairs at the international level, and this is much more serious because of the international character of the high seas and most problems connected with it.

Here again a chief problem is that institutions erected for the management of land-oriented problems are used for dealing with ocean-oriented problems. In the same manner, responsibility for ocean problems is divided among many international agencies.

The fishery function of the United Nations is lodged in the Food and Agriculture Organization, which is dominated by Departments of Agriculture in the member countries. The oceanography function is lodged in UNESCO which is dominated by Ministries of Education in the member countries. The Weather and Meteorology function is lodged in the World Meteorological Organization which is dominated by the Weather Bureaus of the member countries who, having most of their observation points on dry land, have difficulty realizing that the ocean dominates the global climate pattern. The International Hydrographic Bureau is separate from all of these as is the Inter-Governmental Maritime Consultative Organization. Some aspects of ocean pollution fall within the purview of the Inter-Governmental Maritime Consultative Organization, others within the International Atomic Engery Agency, others within FAO, and it is uncertain where the main responsibility lies. Important other aspects of ocean activity such as telecommunications, public health, and the developing Law of the Sea fall within other specialized agencies of the United Nations family. Such international regulation of fisheries on the high seas as exists is accomplished by something over a dozen specialized intergovernmental fisheries commissions established under separate treaties among two or a few nations, quite outside the United Nations framework for the most part.

It is obvious that at the very least the whole subject of international agencies and institutions dealing with international ocean affairs needs the thorough going re-examination at high level in the United Nations structure that it is being given in the United States Government, with a view to making the attainment of public order on the ocean somewhat more rational and providing better for the governance that is necessary for the more effective use of the sea.

Futuristic problems. Other writers for this book have considered the sorts of legal problem that will arise when there are dwellings, groups of dwellings, or cities both upon and under the sea.

This field of speculation may still be characterized as fantastic but it is no longer entirely fanciful. The scientific problems concerned with man living and working in depths of 100 fathoms, that is on the continental shelf, are already worked out. The technology of doing this is well advanced. Not only is the Cousteau and Link adventurism extending our experience in this realm but the Deep-Sea Submergence Program of the United States Navy is moving forward solidly and vigorously despite being greatly under-financed. It is certain that within a short period of time funding will be adequate so that this field of human activity will spurt forward. It is not generally recognized that the continental shelf of the world ocean (that area covered by 100 fathoms of water or less) is greater in area than that of the moon (Smith, 1966), that it offers great material wealth for those who learn to take it, and the Convention on the Continental Shelf converted this to national ownership. By this step the United States alone gained more new sovereign territory than was contained in the original thirteen states, and almost as much as was gained from the Louisianna Purchase.

It is certain that the United States will move forward to the physical occupation and use of the Continental Shelf in good time and that when this happens an accomodation will require to be made between the national use of the bottom and the international use of the superjacent water. Speculation as to what form this accomodation will need to take is perhaps idle at this point except to repeat, as said above, that the governance of ocean affairs on both the national and international level will have to be tidied up materially before such problems can be dealt with effectively.

REFERENCES

Chapman, W. M., 1949, "United States High Seas Fishery Policy," Bull. U.S. Dept. State.

Chapman, W. M., 1960, "Effect of the 1960 Law of the Sea Conference on the High Seas Fisheries," Proc. Gulf and Carib. Fish. Inst., 13th Ann. Sess. pp. 38–53.

Chapman. W. M., 1963, "The Theory and Practice of the 12-Mile Fishery Limit," Proc. Gulf and Carib. Fish, Inst., 16th Ann. Sess. pp. 9–24.

Chapman, W. M., 1965, "The Law of the Sea and Public Policy," Serial No. 89-13, 89th Congress, 1st Sess. pp. 388–407.

Chapman, W. M., 1966, "Resources of the Ocean and Their Potentialities for Man," *Food Techn.* 20, No. 7, 45–51 (July 1966).

Cottrell, L., 1960, *The Concise Encyclopedia of Archaeology*, Hawthorn Books, Inc., New York.

Department of State, 1955, Santiago Conference on Problems of Fisheries Conservation.

Food and Agriculture Organization, 1965, *Yearbook of Fisheries Statistics, 1964*.

Heinzen, B. G., 1959, "The Three-Mile Limit: Preserving the Freedom of the Seas," *Stanford Law Rev.* 11, 597–664 (1959).

Herrington, W. C., and J. L. Kask, 1956, "International Conservation Problems, and Solutions in Existing Conventions," in *United Nations* A/Conf. 10/7, pp. 145–166.

MacChesney, B., 1957, *International Law Situations and Documents*, 1956, Navy War College, Navpers 15031, Vol. II.

Moiseev, P. A., 1965, The Present State and Perspective for the Development of the World Fisheries, FAO, EPTA, No. 1937-II, pp. 69–84.

President's Science Advisory Committee, 1966, "Effective Use of the Sea," Report of the Panel on Oceanography, The White House.

Schaefer, M. B., 1959, Biological and Economic Aspects of the Management of Commercial Fisheries. *Trans. Amer. Fish Soc.* 965–970 (1959).

Schaefer, M. B., 1965, "The Potential Harvest of the Sea," Trans. Amer. Fish Soc. (1965).

Smith, F. G. Walton, 1966, "What is there for Industry in Oceanography," *Res. Management* 9, No. 3, pp. 193–200 (1966).

United Nations, General Assembly, 1955, "Report of the International Technical Conference on the Conservation of the Living Resources of the Sea," 18 April to 10 May, 1955, Rome. A/Conf. 10/6, 17 pp.

United Nations, General Assembly, 1956, "Report of the International Law Commission," Covering the Work of Its Eighth Session, 23 April – 4 July, 1956; Gen. Ass. Off. Rec. Eleventh Session; Suppl. No. 9 (A/3159)

United Nations, General Assembly, 1958, "United Nations Conference on the Law of the Sea," Official Records A/Conf. 13/38, 7 volumes.

Warmington, B. H., 1964, *Carthage*, Penguin (Pelican Books, A 598), Harmonsworth, England.

CHAPTER 6

General Features of the Ocean

JOHN D. ISAACS

It is my purpose in this chapter to present the general features of the ocean in some over-all perspective, and at the same time to place engineering in this over-all context.

To get from one to the other of the extremes of my subject, I am going to have to sneak around and slither through the vast and inclusive blocks of the subject that are covered by Dr. Hendershott and Dr. Richards in Chapters 7 and 8, both of which include the physical, chemical, and biological factors. Please therefore forgive me if I encroach on the subject matter of Chapters 7 and 8 which follow. I also encroach on the subjects of at least two of the previous chapters, those authored by Dr. Schaefer and Dr. Craven, because I want to integrate the needs for ocean engineering with the subject of this chapter.

In the section 6-1 of this treatment, I attempt to put this planet and the oceans in some encompassing, hand-waving type of perspective. In the section 6-2 I deal with some examples of complex interaction that are of importance to man, and in section 6-3 present some ideas of how engineering and the engineering method can be expanded to a more significant compass of the oceans.

6.1 THE NATURAL LARGE-SCALE SYSTEMS

The striking fact about our universe is its astonishing complexity. We often hear repeated the words of some scientist (usually a physicist) of the last century who had the temerity to foretell that his science would soon be completed. Such remarks were made mainly prior to the discovery of radioactivity.

Since then there has been a proliferation of complexity—the discovery of relativity, a vast number of "fundamental" particles defined and described, superconductivity and superfluidity, supernovae, isotopes, the genetic messengers DNA and RNA, etc. These are the complexities of experimentation and analysis. The analysis of the earth and its oceans are no exception. Presumably among the galaxies of this universe there are all possible seas: deep spherical oceanic shells, equatorial rings of water; methane and ammonia oceans, magnetic seas, and oceans that boil on the morning side and freeze at sunset. But we are dealing with the specific oceans of Earth—unideal, ungeometric, and inhomogeneous. In the study of the ocean and, in fact, in all natural sciences, probability has only a mathematical significance. The most probable planet in the universe is undoubtedly also the most abundant, but Earth is what it is, and what *is* is more important to us than what *is probable!*

Hence, dealing with specific oceans of Earth, we are irrevocably bound to a high degree of empiricism, observation, and qualitative thinking about a specific object of study—the sea—and we are required to reassemble the analytical results of disciplinary science into a coherent understanding.

The very elements of which the earth is wrought are of complex origin. Complex processes, and apparently not a single astrophysical event, could give rise to the elemental and isotopic composition that we find on the earth's crust and in the meteorites. The isotopic composition of this crust must have been formed by slow processes, fast processes, from parts of giant stars, parts of space dust, gas, and exploding stars with the addition of protons and cosmic rays. All these were necessary, and other as-yet unknown and unelucidated processes must have contributed in the elaboration of the very elements of which the earth and its oceans are put together. Well-understood processes in the main sequences of stars account for the abundance of the stable isotopes of hydrogen, oxygen, carbon, nitrogen, silicon, iron, the main building stones of life. But it is apparently only in the brief factory of the cores of the collapsing-exploding suns that the neutron-rich heavy elements, which constitute our resources of fissionable materials, could have been formed.

These seas of Earth are several thousand times wider than they are deep. They generally have the proportions of a sheet of writing paper, and the oceans of this watery planet can be compared with the proportions of water on a wet basketball and with a human as a bacterium on a dry spot.

The ocean we trust to remain level within a fractional percentage of its depth. As a matter of fact, a majority of the earth's population lives on land no higher above sea level than 1% of the sea's average depth, and a 5% change is sea level, as in a tsunami, is a disaster. Though we conceive of the sea to be lively and vigorous where its tides and waves are large in comparison with manmade objects, in truth it is sluggish and placid. It is a sink with almost no internal excitation or forces, and, with the exception of the tide-producing forces, all excitations are applied at its boundaries. Yet its docility is really an expression of its massiveness, its stability, and of the relative minuteness of the forces that act upon it. The ocean is not unresponsive, however. The tiny tide-producing forces, capable of bulging the oceans only a meter, raise tides many times this height and produce important currents and mixing in coastal waters.

The principle that the oceans mainly are affected by things happening around them also applies to their form and distribution. The most logical earth would be wholly stratified—a heavy metal core, then heavy rock, light rock, water, and atmosphere, each in successive perfectly smooth shells. The water shell of this earth would be some 2,600 m thick. But the solid earth *does* possess an internal excitation, and powerful forces acting in the earth's mantle have deemed otherwise and have pushed up continents and archipelagoes that have herded the oceans into basins and, except in the Arctic and Antarctic zonal seas, have denied them zonal continuity. Some dynamic process deep in the ocean is at work to preserve this general situation.

I would like to examine further the processes that have given rise to this first gross curious aspect of the planet's surface. The earth is not arranged in these concentric spheres of the different materials. Rather, vast blocks of granite rear through the water and sustain themselves through a millenia of erosion and attrition. It can be shown that this truly is a surprising condition—that the continents can not be sustained, and that the oceans can not be confined in a purely static condition as we think of a building standing; but rather these entities are maintained by dynamic forces as real as those that support an aircraft, although with a much longer time scale, of course. Something must operate continuously to maintain this configuration, or the continents would have long ago eroded, spread, and disappeared beneath a continuous sheet of water. At present the most generally accepted theory states that the oceans and continental masses are maintained by giant convection cells slowly moving deep in the earth's mantle. Referring to Fig. 6-1, continents are seen as a sort of light froth of granite collecting at regions of downwelling, like the pattern of foam on a boiling jelly kettle. Recent heat-flow measurements through the ocean bottom support this theory. The heat flow along trenches is somewhat less than over swells. Present convection cells probably are not symmetrical. In the trenches that ring the Pacific the branch of the cell coming from the basin is apparently far more active than the opposed branch. The source of the energy to drive the convection is probably radiogenic, that is, heat resulting from the radioactive decay of materials in the mantle rocks. However, there are difficulties with attributing this heat to radioactivity. For

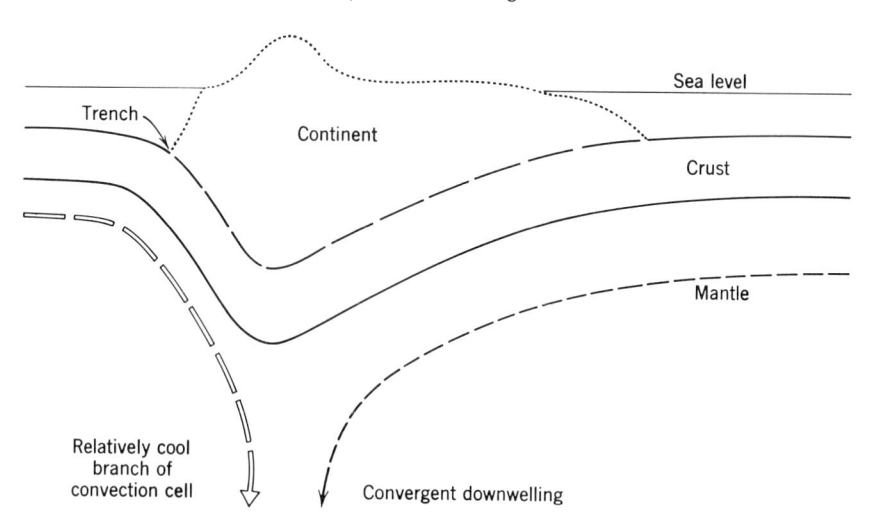

FIG. 6-1. Convection cell—downwelling component.

example, the heat flow under the continents and under the sea are very closely equal. This requires either a miracle of distribution of radioactivity or immense lateral transfers of heat. Also, to develop the instability necessary for large-scale convection, the heat must be generated deep in the earth's mantle. Recently, Dr. Hugh Bradner and I inquired into the possibilities that neutrinos penetrating the earth are the dominant source of geothermal energy (Isaas and Bradner, 1964). The flux of such neutrinos even from the sun is 8 or 10 thousand times that necessary to account for all geothermal heat, and the neutrino flux from supernovae may be very much greater over some short periods. Hence only a very small interaction of these extremely penetrating radiations would be needed to account for all geothermal heat. A cosmic flux of almost infinite magnitude of very weak neutrinos is also a possible generating condition. Generation of heat from such penetrating radiation would solve many present theoretical difficulties. The heat would be produced deep in the interior, and would be produced at the same rate under oceans and continents. Although present nuclear theory does not support this neutrino model, it cannot be ruled out.

The great dynamic processes transpiring deep within the earth and driven by heat generate the continents, trenches, islands, and probably the earth's magnetic field. Presently I will go into some of the profound effects of these great meridional barriers of the continents and oceans maintained by this process. The conditions that they impose upon the seas give rise to many of its most important qualities, and exercise a profound influence upon mixing, circulation, organic productivity, underwater sound, etc., and on the activities of man bent on bending the sea to his understanding and his

General Features of the Ocean 161

needs. For example, land and water in the high latitudes of this earth are very curiously arranged. The north polar region is in an odd circular sea and the south polar region is in an odd discoidal continent. Both of these now deny the oceans the heat exchange between tropics and arctic that may have existed in the past when the poles may have been over the deep ocean and the earth's climate apparently included neither steaming tropics nor frozen arctic.

For completeness of the portion of this treatment of the Oceans for engineers to which I am assigned, let me go back a bit into a field in which I am by no means an expert—the origin of the planet, of its interior substance and its oceans. Evidence on elemental abundances, particularly the relative absence within the earth of the light noble gases, and the relative abundance of volatiles such as chlorine, hydrogen, and oxygen in the form of water, argue that no great heating occurred in the early accumulation of the planet before a strong gravitational field was established. We can conceive then of some sort of slow accumulation of meteoritic material into a larger protoplanet, subject only to heating associated with local violent collisions

FIG. 6-2. Planetary nebula NGC 7293 (by courtesy of Mt. Wilson and Mt. Palomar Observatories).

(as evidenced on the moon), and to heat from radioactivity and the sun, which was than being born. Processes that result in ejection of material from suns can be observed, as in the planetary nebula, NGC 7293 (Fig. 6-2). In this curious object, located in our galaxy and not unlike the ring nebula in Lyra, we can see what must almost certainly be bodies and streaming material ejected from a central star.

As the planet grew larger, however, and gravitational contraction took place, a heating and eventual interior liquification occurred from the accumulation of radiogenic heat and the release of gravitational energy. This was associated with the separation of light and heavy material under the influence of gravity, producing a core, which is probably iron. This differentiation produced a slow development of a gradient of gravitational acceleration from what initially was essentially a linear gradient from the surface to the center, to an almost undiminished gravitational acceleration extending far down into the earth's interior. Figure 6-3 (Bascom, 1961)

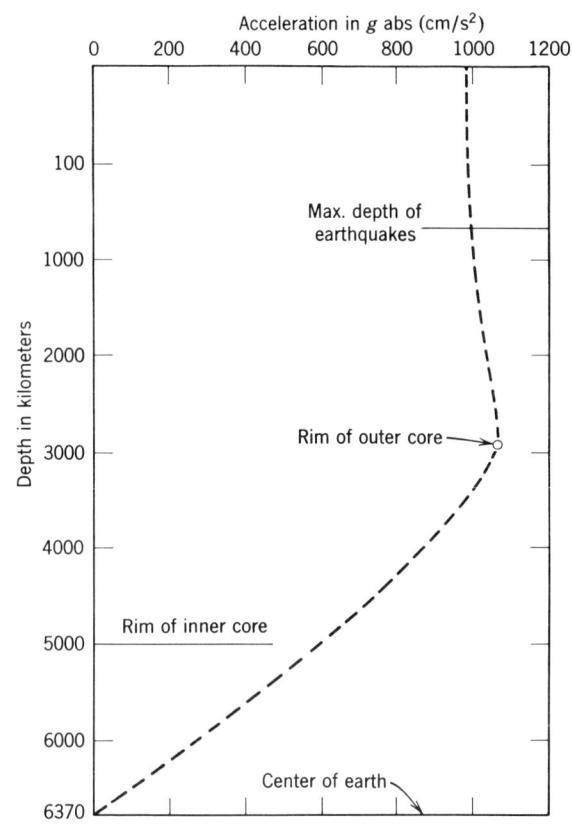

FIG. 6-3. Change of earth's gravity with depth (from Bascom, 1961).

indicates the nature of the present distribution of gravity below the earth's crust. Thus at some depths during differentiation, gravity increased by a factor of almost 3 over that of the undifferentiated earth at the depth. As gravity increased within the earth's mantle, the instabilities that gave rise to convection, and vulcanism became more readily attained. That is, the forces tending to set up convective processes in unstable material are directly related to the local gravitational attraction. It can be readily seen that if a hotter and less dense material underlies a cooler and more dense material, the force that tends to exchange the position of the two materials is directly related to the local gravitational force.

We can assume then that, even though radiogenic heat generation became less, the radiogenic heat accumulation became greater, temperatures increased, and, because of the differentiation, convection in the earth's mantle could be set up rather more easily. (Heat deposited by neutrinos may also have played an important role.) At pesent the transport of heat from the earth's mantle is about 40 cal per cm^2 per year, quite enough to drive the convective transport, which requires a velocity no more than perhaps 1 cm per year. It is probable that this convective transport is responsible for the distribution of the emerged continents, of ocean basins, of the oceanic swells, and of the trenches and troughs. Curiously, as we well know, the great deeps of the ocean are almost within sight of land, and surround the Pacific in a region of great seismic and volcanic activity often referred to as the "Pacific Ring of Fire." The large land masses of the Pacific are all girded by trenches.

It would be grossly inadequate if I did not also mention one of the old theories of the last century which is now rapidly receiving increased support and interest by the geophysicists. This is the theory of continental drift.

Wegener originally proposed this theory. He was excited, as had many others been before, by the manner in which the continents of the world appeared to be a vast and somewhat distorted jigsaw puzzle, which could be fitted together with quite surprising coherence. Ocean soundings later delineated the Mid-Atlantic Ridge and added another feature that fits surprisingly well. Thus a present theory of the continents' history is that they result from a breakup of an original proto-continent whose fragments migrate about the earth under the force of convective processes. Also quite possibly a general expansion of the crust due to internal heating occurred. Paleontological evidences of related fossil life forms in now distant areas seem to support this theory. In this view, oceanic swells are regions of a line of crustal upwelling resulting from plastic convergence deep within the mantle, diverging near the crust.

Figure 6-4 shows a sketch of how these swells must look in cross section. The high heat-flow measurements at the crest of these oceanic swells indicate

164   Systems Planning

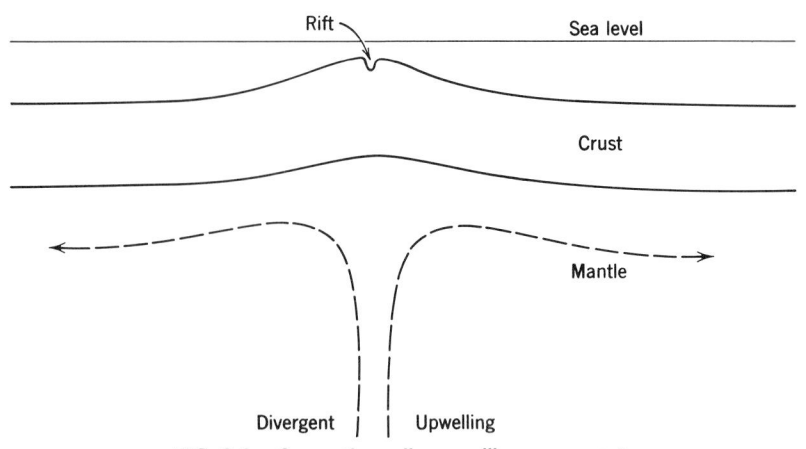

FIG. 6-4. Convection cell—upwelling component.

the increased transport of heat in these regions. Volcanic activity takes place on the surface of these, and the volcanoes then actually may be transported outward down the slopes of the swells and in time become the submerged seamounts like those that dot the Pacific. Simple calculations on the age of these seamounts and their depths seem again to support this motion of 1 cm per year (see Fig. 6-5). In this same view the trenches such

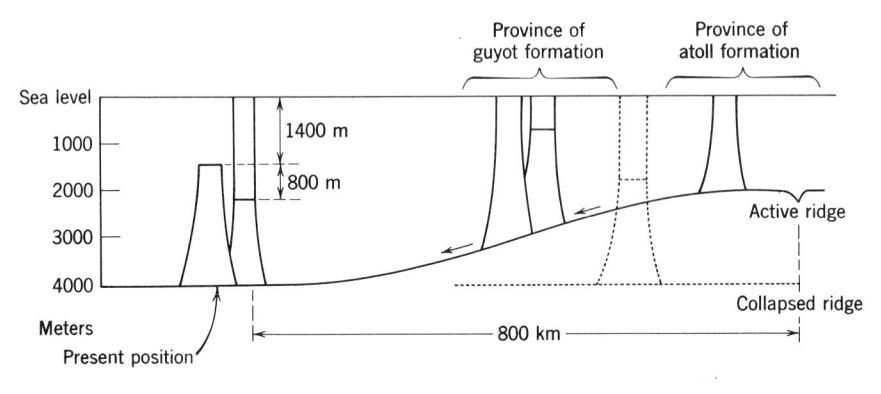

FIG. 6-5. Estimate of speed of crustal motion in Central Pacific (by W. R. Schmitt). *Hypothesis*: Guyots are younger volcanic cones than are the basements of nearby atolls. They were formed when their rate of subsidence—as consequence of downslope migration or collapse—was greater than the rate of *juvenile* coral buildup.

Age difference guyot – atoll basement	3×10^7 yr
Depth difference guyot – atoll basement	8×10^4 cm
Mean slope of ridge	1 in 400

$$\text{Speed} = \frac{8 \times 10^4 \times 400}{3 \times 10^7} \approx 1 \text{ cm/yr.}$$

as those surrounding the Pacific are youthful downwelling regions, which carry the floss and the foam on the jelly kettle into the surface convergence region and pile it up as great granitic protrusions, somewhat depressed because of the denser and cooler downwelling rock that underlies them. When a particular convection cell then becomes aged, enfeebled, and finally ceases, the downwelling stops, there is a reheating of the originally downwelled material, and the entire granitic mass is raised up in new mountain or island ranges. Other evidences to larger-scale continental motion come from paleomagnetism. In Fig 6-6 (Bascom, 1961), there is a comparison of the magnetic anomalies on two sides of one of the great rifts or fracture zones off the California coast. Comparing these will show that they have been displaced relative to one another by some 200 miles or so, directly into or away from the continents. New evidence shows a remarkable worldwide repeated succession of magnetic anomalies that gives great support to the continental-drift hypothesis (Vine, 1966).

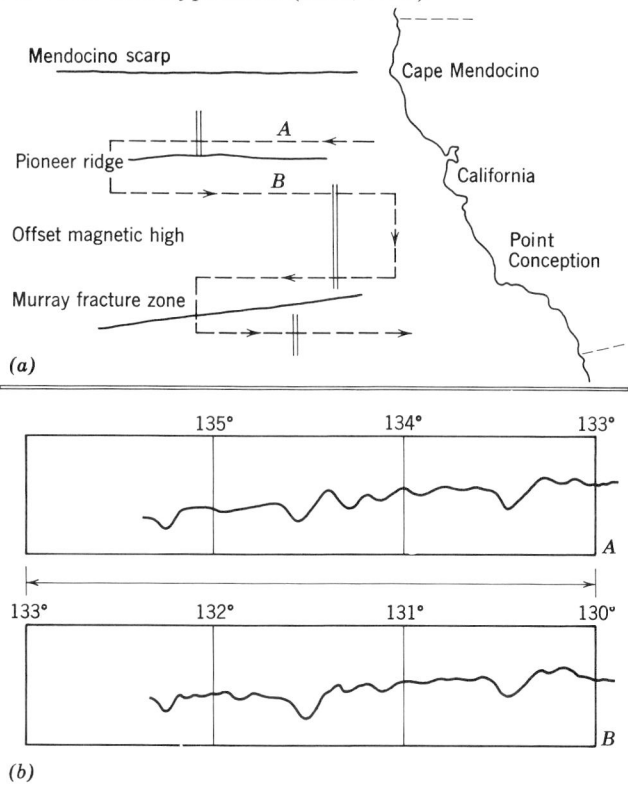

FIG. 6-6. Displacement of magnetic anomalies (from V. Vacquier, 1959). (a) Track of magnetic survey ship that discovered the movement on the Pioneer Ridge fault. (b) The two magnetic profiles made along A and B match up perfectly when B is offset 138 miles to the east.

Traces of polar wandering also show much motion. Sediments, sedimentary rocks, and lavas apparently record the direction of the earth's magnetic field of the past. There are many complexities to this that I cannot go into. For example, there is evidence that the magnetic poles reverse on occasions, but this evidence may be due to certain partial changes in the

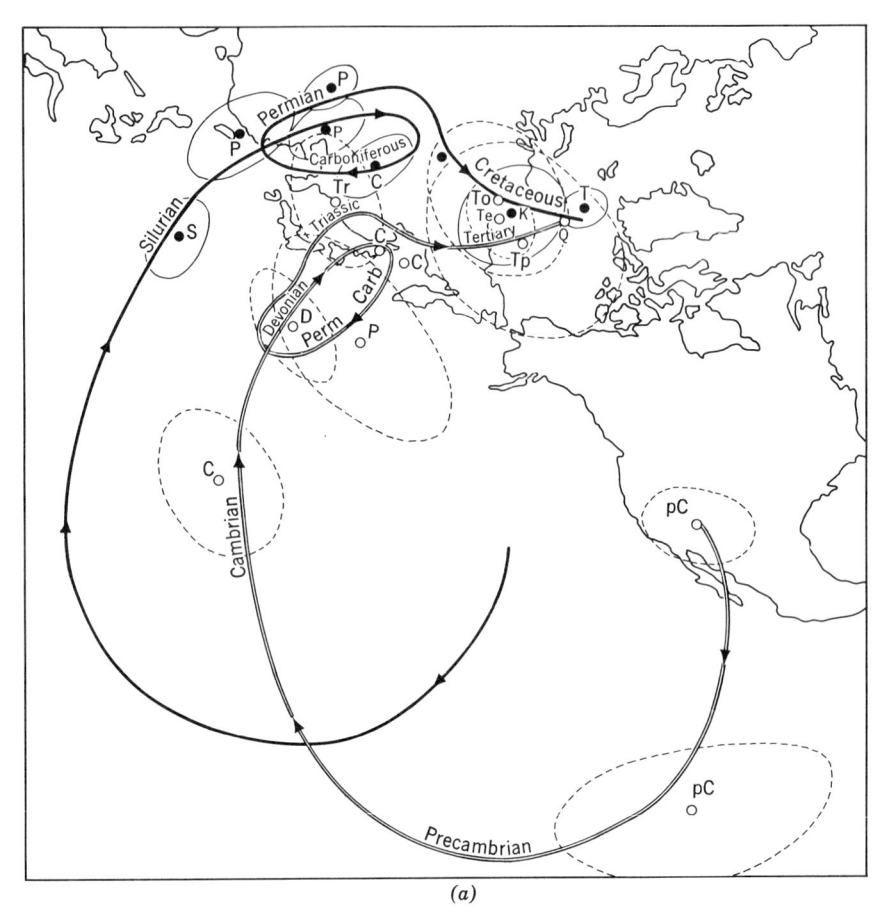

FIG. 6-7. Pole shift traced from Europe and North America (from Carey, 1958). (a) Pole positions measured from North America and Britain; (b) pole positions when Alaskan Orocline is restored.

magnetic components of the rocks. Assuming, however, that the position of the magnetic and geographical poles of Earth are always roughly correlated, we can take the traces of the magnetic poles of the past as recording somewhat the position of the geographic poles. Thus Fig 6-7 portrays two matters. It is a trace of the position of the magnetic poles, one taken from European rocks and one from North American rocks (Carey, 1958). When the Alaskan Orocline is eliminated by straightening out the Rocky Mountain–Aleutian Island ranges, not only does North America fit reasonably well into the jigsaw-puzzle piece of Europe but the paleomagnetic traces of the pole are brought into correspondence. This is evidence both for

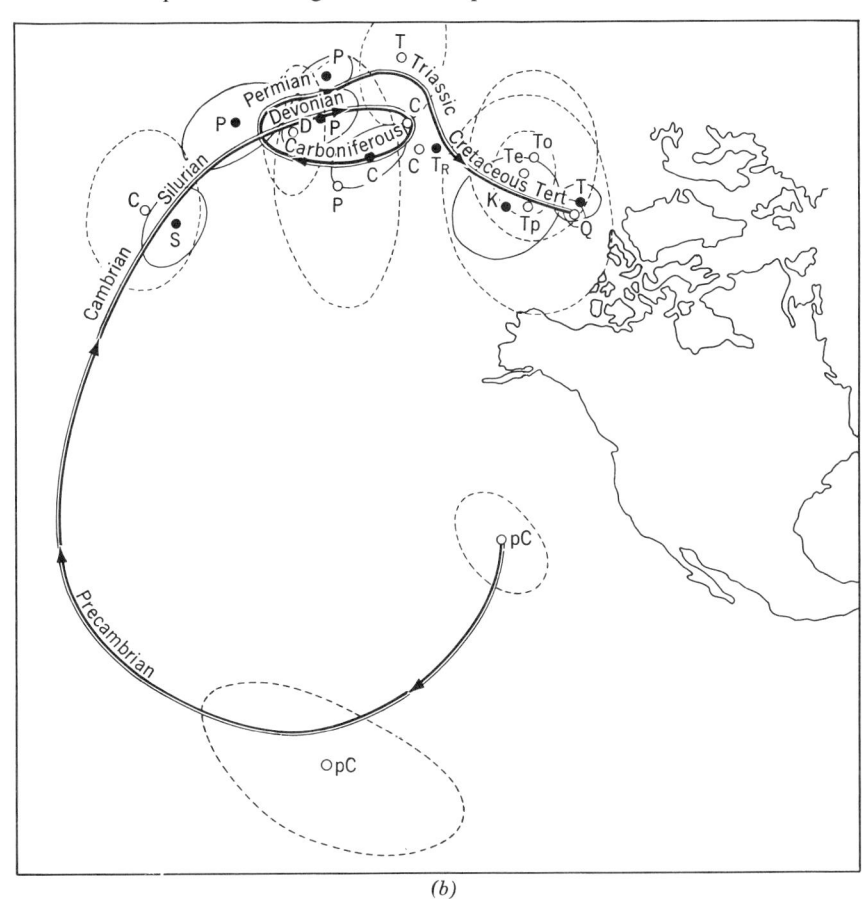

(b)
FIG. 6-7. *(Continued)*.

continental drift and, possibly, of a deep-sea location of the geographical poles in past geological ages.

These events, while controlling the gross structure, also are responsible for much of the detail. In some parts of the world, as volcanoes are moved below the ocean surface (see Fig. 6-5), conditions of temperature and organic productivity have been sufficient for the combined growth of corals and the calcareous algae to produce some of the most gigantic monuments that exist among the monuments of nature. The coral atolls are of the order of tens of miles in diameter and of a thickness of a mile or so, and are entirely of organic accumulation—skeletons of organisms. The history of atolls and the probable sequence of events that gave rise to these remarkable formations was deduced by Darwin in his early voyage through the South Seas. He noted these coral islands in various stages of development and re-elevation. In some cases the processes of wave erosion and coral growth take place rapidly. For example, a new volcanic island that was several hundred feet above the surface 40 years ago is now wholly submerged and bears a rich growth of coral.

In this rather broad presentation of the manner in which the oceans have come to be as they are, we will have to admit that very little is known about the growth of the oceans in so far as their depth and total water content is concerned. Despite the great number of sites where ancient marine sediments have been raised to the continental levels, essentially all of these are of shallow-water origin and nowhere in the world is anything that is conceived to be of the nature of deep-sea sediment that has been raised up into coastal or continental rocks. Indeed the present evidence is that the total thickness of the sediments of the deep sea is probably less than a kilometer. It is very difficult to see how the oceans have been on this globe for some 4 billion years in anything resembling their present state, if this very small thickness of sediments is the total record.

Now, it is quite true that sedimentation processes may have cemented the sediments that are buried below the level of 1000 m or so, so that to seismic exploration they have appeared to be another phase of rock, the second layer. It is also possible that the oceans have not existed in their present state for more than a few hundred million years; or that some process associated with continental drift (and perhaps trenches) continuously or periodically removes the sediments, cooking them back up to make more continents of them, as it were.

I should really say a bit, also, about the chemical content of the oceans and the nature of the distribution of the sediment. The dissolved material leached and swept out of the continents from the continental rocks (and to some extent from erosion and weathering of the exposed rocks of the ocean bottom) and lava, ash, cosmic materials and dust, and airborne desert sand and continental dust are all carried to the sea. The sediments of the deep sea are principally oxidized materials of continental origin and minerals elaborated in the seawater itself. These are the red clays. Many of the organisms of the overlying waters of the sea produce skeletons of silica or calcium carbonate. The silicious skeletons are fairly insoluble and are added to these sediments. The calcium carbonate skeletons, however, although surviving quite well in the relatively shallow waters down to perhaps 4000 m go into solution rather rapidly below this level. Carbonaceous sediments are thus restricted to the shallower waters of shelves, continental slopes, swells, and perhaps seamounts, and they occur in very deep water only where the deposition is so rapid that the carbonaceous component does not dissolve before it becomes buried. Later in this treatment I discuss some very special nearshore sediments that are highly stratified and have maintained essentially a calendar of a short period of events. These may give us some real insight into short-term climate changes and the changes of oceanographic conditions in brief periods of time, such as 25 years. Thus this record, unlike most stratigraphic evidence, is significant within the span of a human life.

Right now, however, I would like to return to the effects of the restriction of the ocean waters into the great basins of the world—these vast deeps typified by the Pacific Basin where the hydrosphere is of the order of three miles thick. I already have pointed out how thin they are in comparison with their lateral dimensions. Yet physical communication from top to bottom is very slow. The water that occupies the deepest portions of the basins is similar to the densest that occurs on the ocean surface. This deep water is of fairly high salinity and very cold, actually being below the freezing point of fresh water. In the Pacific, such water originates only from the Antarctic Ocean. In the Atlantic it is produced both in the North Atlantic and in the Antarctic Ocean. This very dense water moves through the deepest portion of the basins and slowly spreads, mixes, and—after some heating—eventually becomes part of the intermediate water, which is brought up to the surface mainly in the northwestern Pacific.

In Chapter 7 Dr. Hendershott discusses the circulation of the ocean; hence I go no further into this general matter here. However, it should be pointed out that this deep and intermediate circulation gives rise to one of the most profound conditions of the ocean, that is the condition of stability, in which it requires energy to bring water up from the depths and into the surface layers against gravity (see Fig. 6-8). This stability or density gradient is the characteristic that controls mixing, circulation, organic productivity, the condition that produces the remarkable underwater sound channels which conduct sound from one end of the Pacific Basin to another, etc. To a great extent, this stability is responsible for the relatively low productivity of the ocean. As we will see presently, the nutrients—the simple plant fertilizers: nitrate, phosphate, trace elements, that the oceanic plants need as well as do land plants—are trapped below this lid of stability. The oceanic plants naturally are restricted to the surface layer where there is sufficient light for their growth. Here they take up nutrients. They are then fed upon by the grazers, the small crustacea, arrowworms, fish. As dead organisms, moults, feces, and organic debris sink below the lighted layers of the surface, bacterial action releases the nutrients, down in the dark and

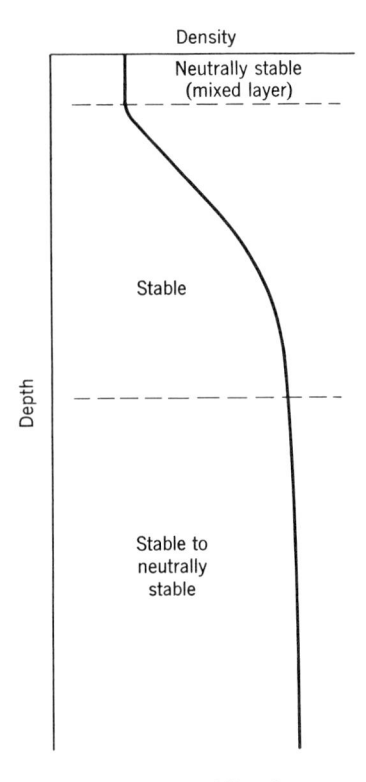

FIG. 6-8. Density gradient—stability of water masses.

cold and unlit layers where no plant can have access to them. As a consequence of this process the entire ocean, which is two and one half times the area of the land, is probably not quite as productive as the land in the total elaboration of organic material. This same stability or lid exists in almost all parts of the ocean, and we might think that it would put limitations upon the circulations of other chemicals such as radioactive material introduced into the deep ocean water. We will see presently that the lid is not very effective in this case.

I have mentioned the fact that the ocean has little internal excitation, but of course in this way it differs very little from the remainder of the planet. As everybody well knows the earth surface would be dead, lifeless, and inactive if it were not for the input of our great natural nuclear furnace—the sun. This great flux of radiant energy pours over the earth, and falls upon the land and upon the sea at the rate of about 10^{32} ergs per year. In Fig. 6-9 is shown the cascade of solar energy as it degrades down through the atmosphere into the organic realm. Several aspects of this diagram make

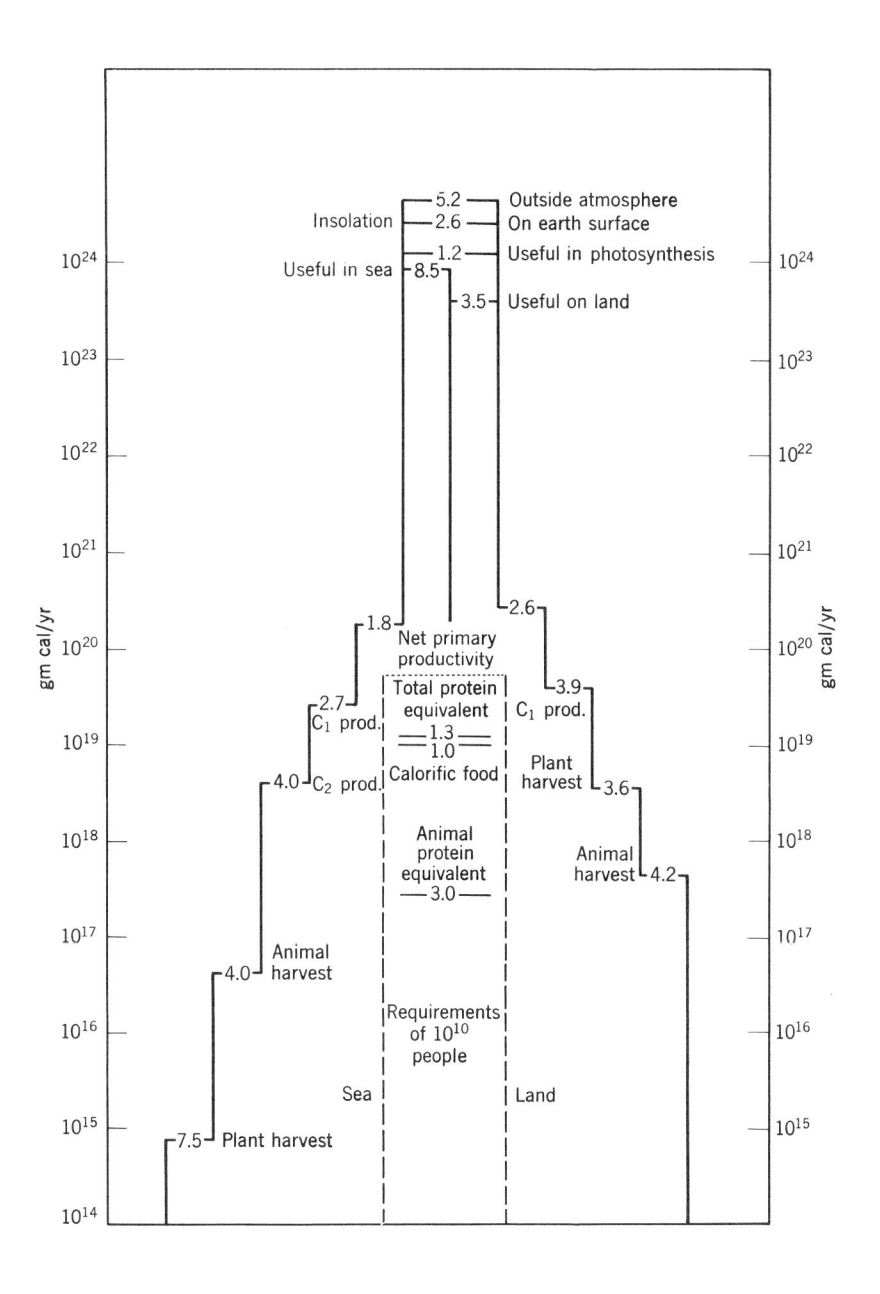

FIG. 6-9. Cascade of energy in organic productivity.

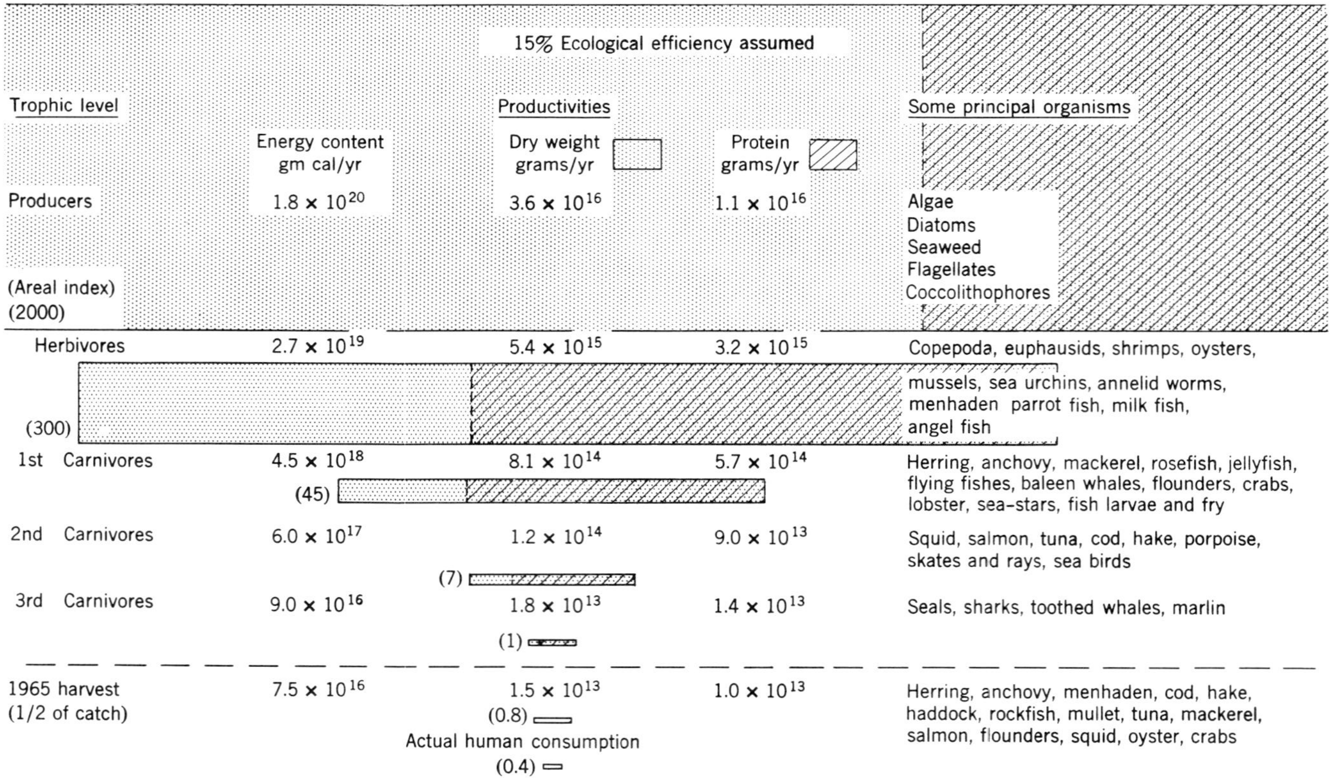

Trophic level	Energy content gm cal/yr	Productivities — 15% Ecological efficiency assumed		Some principal organisms
		Dry weight grams/yr	Protein grams/yr	
Producers	1.8×10^{20}	3.6×10^{16}	1.1×10^{16}	Algae, Diatoms, Seaweed, Flagellates, Coccolithophores
(Areal index) (2000)				
Herbivores (300)	2.7×10^{19}	5.4×10^{15}	3.2×10^{15}	Copepoda, euphausids, shrimps, oysters, mussels, sea urchins, annelid worms, menhaden parrot fish, milk fish, angel fish
1st Carnivores	4.5×10^{18} (45)	8.1×10^{14}	5.7×10^{14}	Herring, anchovy, mackerel, rosefish, jellyfish, flying fishes, baleen whales, flounders, crabs, lobster, sea-stars, fish larvae and fry
2nd Carnivores	6.0×10^{17}	1.2×10^{14} (7)	9.0×10^{13}	Squid, salmon, tuna, cod, hake, porpoise, skates and rays, sea birds
3rd Carnivores	9.0×10^{16}	1.8×10^{13} (1)	1.4×10^{13}	Seals, sharks, toothed whales, marlin
1965 harvest (1/2 of catch)	7.5×10^{16}	1.5×10^{13} (0.8) Actual human consumption (0.4)	1.0×10^{13}	Herring, anchovy, menhaden, cod, hake, haddock, rockfish, mullet, tuna, mackerel, salmon, flounders, squid, oyster, crabs

FIG. 6-10. Oceanic food chain with principal organisms. *Note:* The food chain should more accurately be thought of as a *food web*, in which most organisms feed on more than one trophic level, changing diet with their age (especially when young) and the availability of food.

interesting reading. Although a greater proportion of light energy falls on the ocean than on the land due to the greater area of the seas, the total productivity of the land and sea are probably about the same. However, if we look on the land side of this diagram, we notice that the food energy available to man on land is probably far greater than at sea because of the greater availability of the primary product—the land plants. In the sea the primary products are mainly the microscopic floating plants, the phytoplankton, which are passed in a number of steps through the food chain to increasingly larger organisms. Each of these steps is very inefficient, yielding only perhaps as much as 10 or 15% of the respective inputs (see Fig. 6-10). The fish, which are advanced in the food chain a step or two, have a potentiality of yielding only a small portion of the initial productivity of the sea. Thus it is clear that one of the greatest advances that can be made in increasing the productivity of the sea from the standpoint of man, is not to produce more plants but to eliminate steps in the food chain, for with each step removed there is about a sevenfold gain. Consequently, the greatest fisheries in the world are based on the more primitively feeding fishes—that is the menhaden, herrings, anchovies, that feed down in the food chain close to the plants where there is more available food. The higher-level predators such as shark, tuna, cod, and squid, can represent only a small part of the available protein.

Another way of increasing the productivity of the sea from man's standpoint (by eliminating some of the steps in the food chain) is by providing a substratum of solid materials on which large filter-feeding organisms—mussels, oysters, etc.—can attach themselves to feed directly on the microscopic phytoplankton and detritus of the sea. Such natural or artificial areas are spectacularly productive. Figure 6-11 is a picture of some debris (automobiles, streetcars, and, curiously, old movie sets) on the ocean bottom placed there purposely by the California Department of Fish and Game. This is attracting and, perhaps, developing fish. It is well fouled and provides hiding places and grazing and food material for a greater number of organisms than could have survived on the original flat bottom. In a similar way the mussel fisheries of Europe and particularly of the Bay of Taranto in Italy, where wands are placed for mussel attachment, are certainly the most productive areas under man's control on this planet. They produce some 60 tons of organic material—animal material—per acre annually. This is an order of magnitude greater than the best farmland.

To put this all in different terms, the basic problem of man utilizing the sea for food is that it has almost no "grass" or "large plants," hence no "cows." The sea does not support many large organisms that feed directly on plant life, as the land does. Many of the present fisheries such as the tuna fishery are capturing carnivores far up the food chain, equivalent to tigers

FIG. 6-11. Fish attracted to artificial environment (by J. Radovich).

or even to predators of tigers on land. No society in its right mind ever considered trying to feed itself on tigers. It requires entirely too many cows.

To get back to solar energy, there are two other interesting aspects of this. One is that the total of fossil fuels that people have estimated exists on this earth (2.2×10^{22} gm cal) is only a small part of the annual flux of thermal energy from the sun (2.6×10^{24} gm cal/year) and represents only a hundred or so years accumulation of the total organic productivity (4.4×10^{20} gm cal/year). We can put a prima facie limit on the total amount of fossil fuels through a sort of reverse estimate as to the total energy that would be released if all the oxygen of the atmosphere were burned with some hydrocarbon (4×10^{24} gm cal). This simple model is based on the assumption that the oxygen in the atmosphere represents that part of the oxygen released in the photosynthetic process of the past in which the carbonaceous portion of the reaction was locked into the earth. We see here then that the total fossil-fuel reserves of the earth by this model might be several hundred times those estimated. There is some evidence from

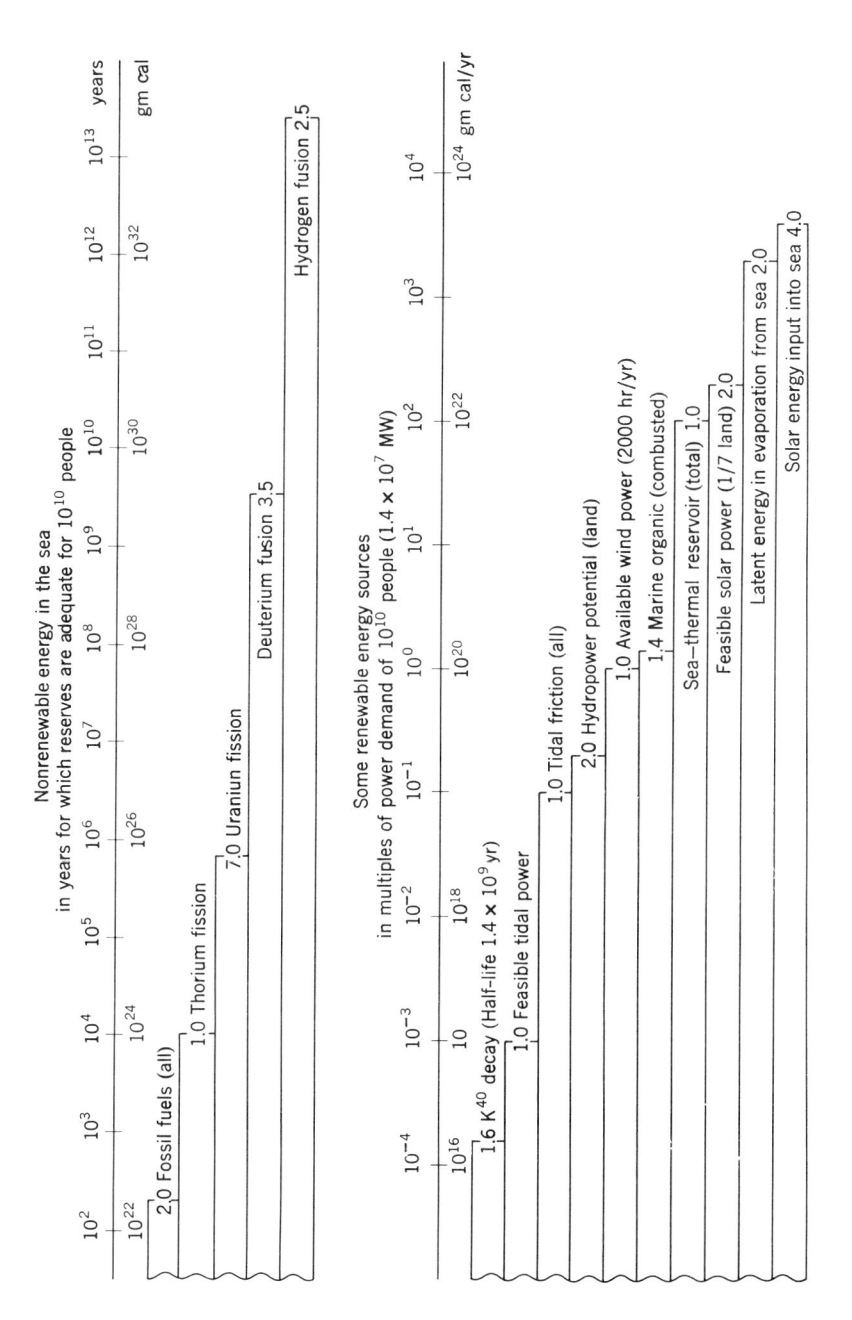

FIG. 6-12. Sources of energy in the sea.

planetary nebula (objects similar to the one in Fig. 6-2) that hydrocarbons of the aromatic type are produced by inorganic processes in these nebula. It is not impossible, therefore, that the ring-type hydrocarbons of petroleum deposits are not all produced by organic processes, but rather are a part of an original inventory. Thus the totals of fossil fuels might be even greater.

While we are on the matter of energy in the ocean, let us take a look at some of the other sources of energy in the Fig. 6-12. This is in terms of 1.4×10^7 MW, the power demand of 10 billion people at United Sates rates of consumption. In the renewable spectrum we see that even the total tidal dissipation of the earth is small—roughly 10% of the probable eventual needs for heat and power. As a matter of fact, aside from solar energy, the only great reservoirs are thermal power (represented by the thermal gradient of the sea) and wind power. In the nonrenewable spectrum we have the potential lifetimes of nuclear sources at the same power demand. They are gigantic.

At this point it is interesting to look at one of the anecdotes of the ocean relating to the power that is represented in the thermal structure of the sea. This is the so-called "salt fountain." Over the majority of the ocean the surface water, although somewhat more saline than submerged water, is lighter because of its higher temperature. It is clear that if it were possible to exchange the temperature between these two bodies of water but not to exchange the salt, it would then be possible to set up a continuous circulation with either the deep water pouring up to the surface or the surface water pouring down into the deep water. Thus if we were to place a long, open-ended copper tube in the sea and start the water in it circulating in either direction, it would continue to flow in that direction indefinitely. In this manner and without the input of any other energy we could induce the artificial upwelling of the deeper water or the downwelling of the surface waters. This is one of the possibilities of elevating deep waters to the surface, and so, in principle, increasing the productivity of the oceans.

I would now like to make a few more points regarding the ocean and its general conditions before proceeding onto the next portion of this Chapter, in which I will come back to these subjects through the medium of the various problems that engineering of the ocean faces and to a few examples on which some progress has been made.

Part of the effect of the cascade of energy onto the ocean is the evaporation of seawater, which is estimated to be of the order of a meter per year. Thus the water of the sea has a turnover time of something of the order of 4000 years—or 12,000 years, on the outside, by the great hydrocycle: coming from the sea, falling on the land, and returning via evaporation, rainfall, and river runoff. The tremendous river, the Amazon, some 400 times

TABLE 6-1
Discharge of the World's Great Rivers (Approximate Mean)
in million acre ft/year (by W. R. Schmitt)

Amazon	5,200	Lena	235
La Plata–Parana	2,000	Yenisei	235?
Congo	1,450	Ob	230?
Yangtze	565	Danube	230
Ganges–Brahmaputra	515	Orinoco	225?
Miss.–Missouri	450	Zambezi	220?
Mekong	435	Indus	215
Mackenzie	325	Amur	200?
Nile	305	Niger	190?
St. Lawrence	290	Columbia	170
Volga	255	Hwang Ho	160?

NOTE: Precipitation: on land, 80,000; over oceans, 270,000.

larger than the Colorado, alone could fill the ocean basins in 200,000 years. Table 6-1 describes the surprising data on rivers of the world. This one meter or so of rainfall is poorly distributed around the earth's surface. Nevertheless, the average rainfall over the land surface of the earth is just about the optimum for the growth of land plants. I could go into the curious reasons for the apparent water shortage on this earth, attributing it to a peculiar cultural drive to cultivate deserts rather than rainforests, but at this moment it will suffice for me to discuss the matter illustrated in Figure 6-13 (Putnam et al., 1960) and Table 6-2 on the general distribution of land types around the world. We see that the very productive, heavily watered regions of the world—the tropical and temperate rain forests—are as large as the *potentially* very productive deserts of heat, which we always attempt to cultivate.

I should discuss, however, one other aspect of this matter of evaporation. This is the matter of the changes that in the past have given rise to periods of glaciation. Many people have assumed that during the periods of the growth of the glaciers there were profound changes in the circulation of the atmosphere and oceans that resulted in greatly increased evaporation, giving rise to deep falls of snow and accumulation of ice in the great glaciers of the Arctic. A bit of arithmetic will show us that the changes in mean evaporation, precipitation, and runoff need be only very small to account for glaciation. During the great glacial ages the total sea level was reduced by perhaps a 100 or a 150m, which is represented by the total evaporation of perhaps 150 years. As this takes place over some 25 to 100,000 years, the imbalance evaporation–precipitation–runoff needs to have been only a few parts in a thousand to produce the great glacial ages. This argues for neither great change nor for highly unusual oceanic and meteorological circulation. Every few years we probably see the weather conditions similar to those

178   Systems Planning

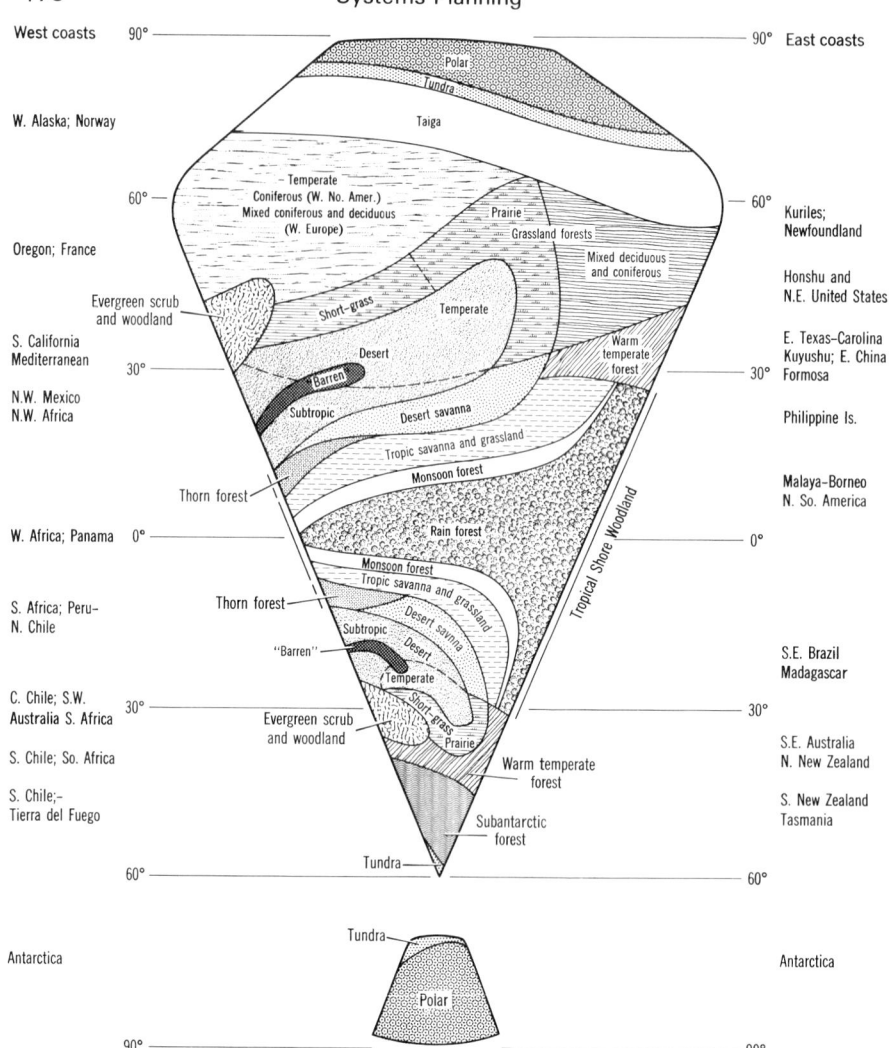

FIG. 6-13. Vegetational character of idealized continent (from W. C. Putnam et al., 1960).

that existed in glacial periods. These conditions needed only to occur more frequently during the glacial periods.

This brings me to climate intervention. Much has been argued about possible bold intervention into climate via the oceans. It is true that we probably now have at our command energy sources and materials that could permit us to make large changes in the circulation of the earth's atmosphere. However, we have only the tools. We do not have the understanding to make the results of intervention predictable. In some cases possible interventions

TABLE 6-2
Character of the Environment of
the World's Land Surface (36.7 × 10^9 Acres) (by W. R. Schmitt)

	Percent	Acres
Cultivated Land	11	4.1×10^9
Potentially Cultivable with Present Technology		
From forests	2	0.7×10^9
From grasslands	5	1.8×10^9
Uncultivable with Present Technology		
Lakes and rivers	2	0.7×10^9
Humid lands		
Wetlands	2	0.7×10^9
Tropical rain forest	13	4.8×10^9
Temperate rain forest	2	0.7×10^9
Forest, conifers and deciduous	12	4.4×10^9
Grasslands	4	1.5×10^9
Semiarid Grasslands	13	4.8×10^9
Deserts		
Heat	15	5.5×10^9
Frost	18	6.6×10^9
Altitude	1	0.4×10^9
	100	

might be able to take advantage of trigger mechanisms that would have an eventual effect far greater than the initial induced change. One of these possibilities has been explored by the Russians—the damming of the Bering Straits. It is clear that such an intervention may be of real substance in changing the Arctic Ocean's circulation and hence the nature of its ice cover, thus changing the reflectivity of this region, causing the eventual warming of the Arctic. The Arctic is fed mainly by relatively warm, highly saline Atlantic water. From the North Pacific, however, comes a colder, less saline and less dense flow through the Bering Straits. This latter then occupies the surface of half of the Arctic in a fairly thin sheet, and dominates the heat exchange far out over the Arctic Ocean. It is this water, and the stability associated with it and the screening of the warmer Atlantic water, that controls the icing of much of the Arctic Ocean.

6.2 OPPORTUNITIES FOR MAN'S INTERVENTION

In this part of the treatment of the oceans I would like to emphasize the nature of the interdisciplinary relationships in ocean engineering by a series of examples. In Section 6-1 I attempted to set forth a few of the conditions of the ocean that produce constraints or allow freedom in the understanding and employment of the ocean in mankind's need. I think that these will be clearer in this portion when I give you specific examples of such matters. My first example is the case of the determining of the amount of radioactivity that the ocean can bear, from the standpoint of man's use. Initially, looking at the problem of introducing radioactive material purposely as wastes or inadvertently through fallout or nuclear accident, we might approach it as a problem of physical diffusion and dilution; that is, as a physical problem. The nature of dilution and eddy diffusivity is reasonably well known; hence when we put some soluble or finely particulate material into the ocean, we can predict its dilution and eventual path with reasonable accuracy. However, in this case biology enters prominently. If humans utilized the sea by drinking seawater, the problem of the distribution of radioactivity in the sea might well be attacked on a purely physical basis. The majority of humans, however, avoid drinking seawater as vigorously as possible, and their principal intake from the sea is, of course, seafood. Thus the radioactive elements that *enter into the life* of the sea are the nuclides that limit introduction from man's viewpoint. Looking at the matter from this standpoint, radioactive elements that are limiting on land are not the limiting ones at sea. I like to see in this the interaction of chemical and biological evolution, which *a priori* could be used to predict that there were great and meaningful barriers set up between radioactive materials in the ocean and mankind. Let me see if I can make this interaction clear. The dissolved materials of the ocean are derived from those that are leached from the land, washed to the ocean and remain in solution, neither precipitating out nor being evaporated away. Hence the ocean has a particular spectrum of elements and other substances of great abundance, and these tend to be the ones that have been leached out of the land and so in many cases are relatively absent in the land. At the same time, the land has an abundance of materials that are present in only small quantities in the ocean. Now, as life began in the sea, the creatures of the sea had need to develop certain abilities to concentrate the trace elements needed in their metabolism and growth. This they did, and these creatures utilize and absorb and concentrate and retain certain trace elements with great efficacy. This evolution of creatures in the sea eventually reached the point at which animals and plants emerged from the sea and started growing on the land. In this new environment the spectrum of elements was quite different, and for the reasons of dissolving and leaching that I have just noted, most of the trace elements that they had found necessary to concentrate at the sea were now found in great abundance. Thus they lost these abilities, neither concentrating greatly nor long retaining these substances. At the same time, new and different needs for their concentrating of elements arose. These were elements in abundance in the sea but deficient on land because they had been leached out and carried to the sea. The land animals then developed the ability to greatly concentrate certain of the elements that they still required, but which were now deficient in this new environment.

We can see what will happen if we now place into the sea a spectrum of artificially radioactive elements participating with the nonradioactive natural isotopes. One of two things will occur, either of which will effectively erect barriers between the radioactive materials introduced into the sea and the human race.

First, elements that marine creatures greatly concentrate are at the same time elements that land animals do not concentrate and that have a short biological half-life in land animals, particularly man. The health hazard of these isotopes is thus greatly reduced. Examples of such elements are copper, zinc, iron, silicon, and manganese. What about elements that the land biota greatly concentrates? Iodine, strontium, calcium, sodium, and others are of this nature. When such radioactive substances are placed into the ocean, not only do marine creatures not greatly concentrate them, but the radioactive elements are immediately isotopically diluted by the large quantities of nonradioactive natural isotopes existing there and which are there by virtue of the fact that they have been leached out of the land and remained dissolved in seawater. We see the beauty and curious nature of the interaction of this dual chemical and biochemical evolution on the problems of disposal of radioactivity in the ocean. The results of this are that the elements that are most worrisome on the land are among the least worrisome in the ocean, and the concentration of radioactive strontium in the bones of the fish at Bikini, where many millions of curies of radioactive waste have been "disposed" in the last decades is not as high as that in the bones of the sheep of Wales, a highly leached country.

Second, viewing the problem of disposal of atomic waste from the standpoint of purely physical phenomenology, we would assume that when radioactive waste material is disposed in the deep ocean it is less likely to enter the human environment than that disposed in the surface, because of the stability trapping the materials below the surface. This is, of course, true. But, we can not assume that the stability constitutes an impermeable lid on radioactive materials put into the deep water. This is because that most important component of the ocean environment—the organisms—penetrates this lid very commonly. Throughout the entire ocean a large fraction of the zooplankton, the larger of the minute feebly swimming animals of the open sea, make a daily roundtrip migration from a depth of 200 fathoms or 300 fathoms to the surface, rising at dusk, descending at dawn. The total transport of these organisms, and their introduction into the food materials of the upper-surface-layer fishes, directly circumvents the supposed lid that would be assumed to exist when looking at the ocean on a strictly physical basis.

While we are on the subject of these diurnally migrating organisms, we might look at another case of interdisciplinary interaction between the various components of the ocean. As I have said, these migrating zooplankters are very commonly distributed about the ocean, and carry out these deep migrations daily. They are thus creatures with a clearly predictable type of migration, and we can explore the manner in which they interact with various other characteristics of the ocean. I have looked into two of these. One is the interaction of these diurnal migrants with the internal waves that are produced near a coast that is subject to day–night wind alternation, common on the temperate coasts of the world. In these regions, during the day, the strengthening sea breeze pushes surface waters towards the coast, depressing the deeper water, and causing it to move offshore. At night the process reverses. Breezes off the land shove the surface waters away from the coast and cause the deep water to upwell. This also is a daily alternation and it is interesting to explore the manner in which this motion interacts with the diurnal rhythmic migration of the zooplankton. The zooplankters are down during the day when that water is moving offshore, and at night they are near the surface when that water is now moving offshore also. Thus they "rectify" this oscillation of the current into a total over-all offshore motion that keeps them off the beach. In fact, it may concentrate them in certain bands along the coast. An interesting sideline to this is that in certain seasons some of these zooplankters, when gravid (that is when bearing eggs), reverse the procedure, being near the surface in the daytime and deep at night. At these times they are often thrown up on the beaches in windrows.

Another type of interaction that I have explored more thoroughly is a simpler one. These migrant creatures of the deep sea are sometimes swept into shallower water, where they encounter bottom for the first time in their lives. The sequence in Fig. 6-14 shows a case of this nature off the Mexican Coast where the California Current commonly sweeps over depths that are much shallower than the usual migration of these creatures. Presumably the migration in the deep sea, in part, is a tactic to minimize predation. When they are caught against the bottom in daylight (ridge at night) they are undoubtedly more available to the predators, both to the predators that accompany them and to resident predators. This is nicely exemplified in this sequence of precision depth records showing these organisms—which are "visible" to the echo sounder because they scatter sound—as they are carried over the shelf at night, and as they descend onto the ridge of Banco San Isidro and into a valley Fig. 14(a)–(d). Here are evidences of a mass of resident predators feeding on this daily breakfast when these innocent creatures of the deep, vertical migrant group descend onto the ridge. Darkness in the valley protects them and they survive and

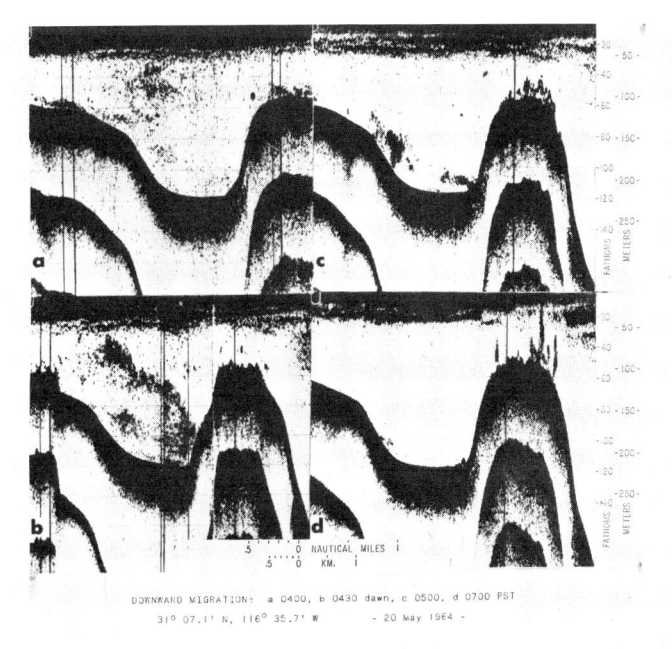

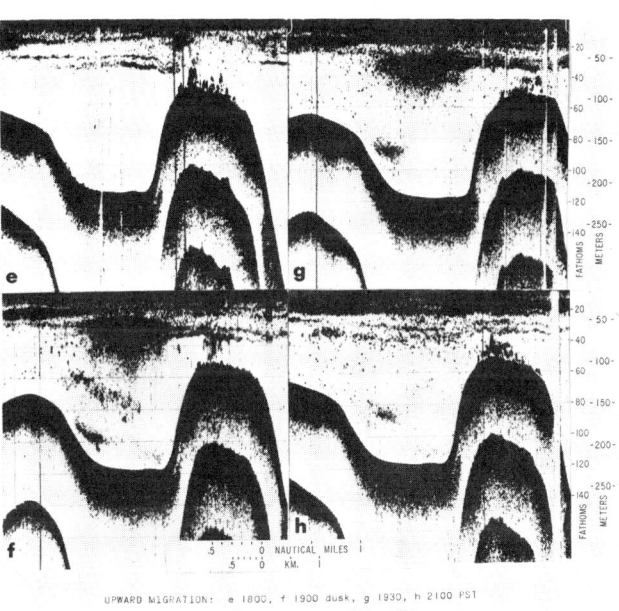

FIG. 6-14. Diurnal vertical migration of scattering organisms.

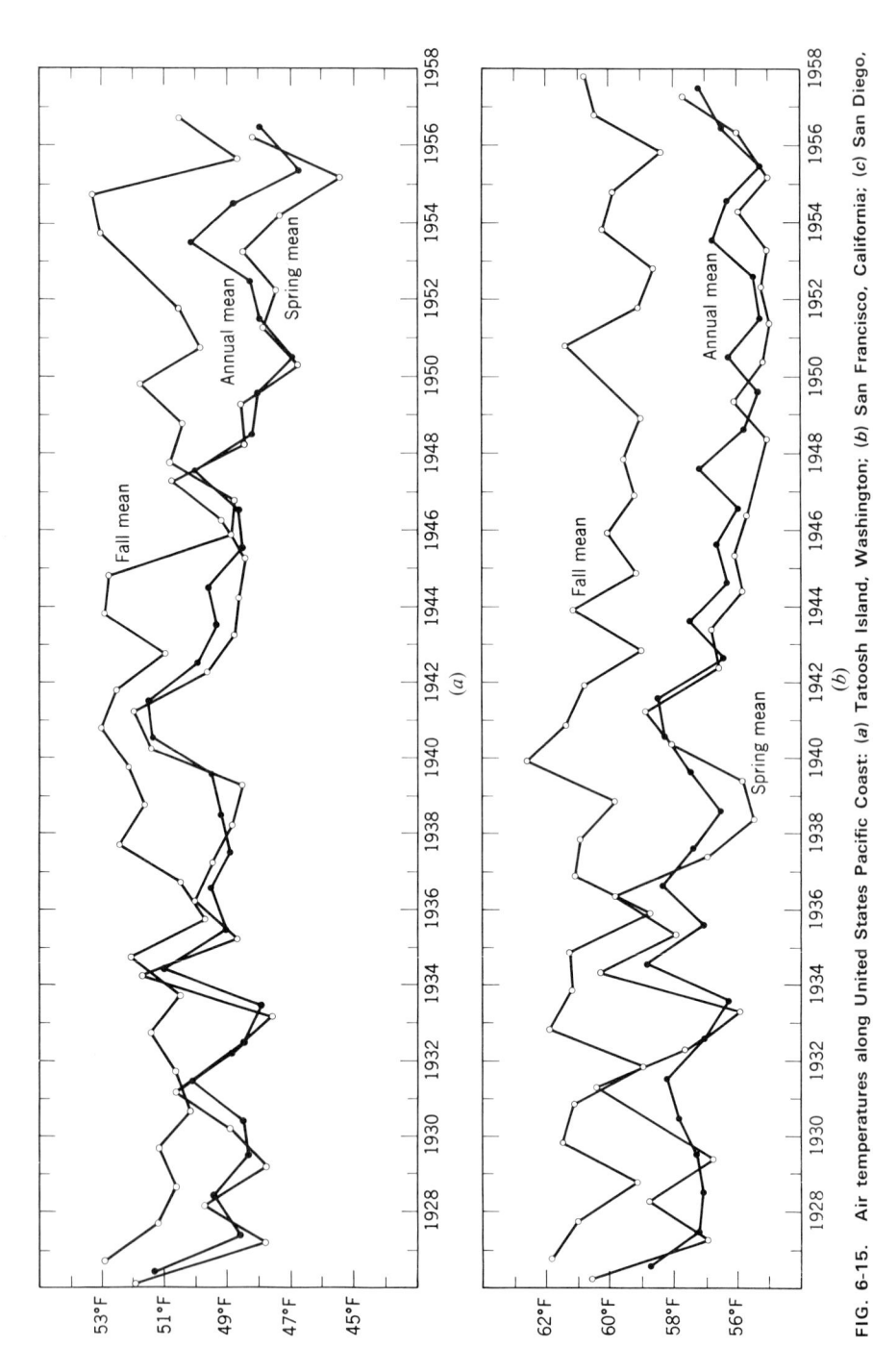

FIG. 6-15. Air temperatures along United States Pacific Coast: (a) Tatoosh Island, Washington; (b) San Francisco, California; (c) San Diego, California.

FIG. 6-15. (*Continued*)

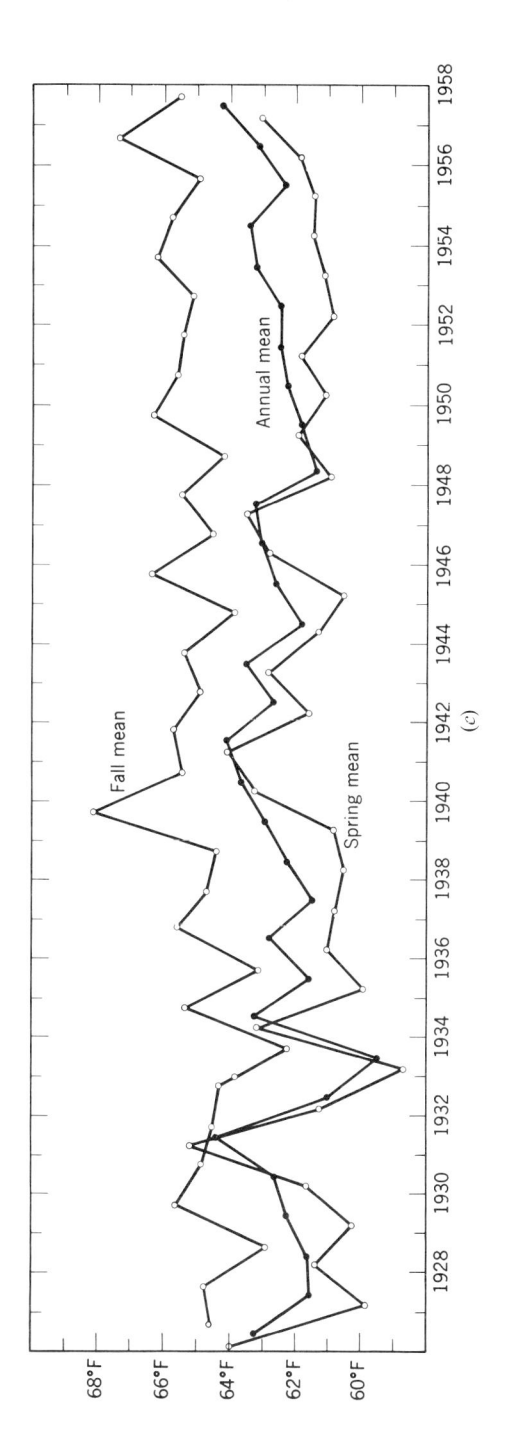

can be seen Fig. 16(e)–(h) to ascend again in swarms in the evening. We can visualize other important interactions of this nature.

Certainly there are places in the ocean where internal waves of diurnal frequency are common. Here the migrant zooplankters may also rectify the fluctuating current and be carried quite differently from any other component of the ocean—perhaps carried upstream in some cases, or maintained in certain areas, where they then act as a focus of distribution. Cases such as these and the one at San Isidro certainly are important factors in understanding the distribution of fishes and the development of fisheries. In the particular case I have shown in the figures the total food input into this area is in the order of 50 to 100 times the input of food material into most of the highly productive areas of the California Current.

I will give you one more example of these interdisciplinary freedoms and constraints of the ocean that are of vital importance to engineering intervention into it.

I referred previously to the fact that we now have the tools and the energy to intervene into oceanic processes by such processes as oceanic heating, but that we lack the understanding of the nature of climate changes and their causes. We undoubtedly could produce climate changes and perhaps we do, by the release of carbon dioxide into the atmosphere, but we have not the understanding to define the results of any given action. One way of going about attaining this understanding is to explore the nature of the changes in the ocean and in the associated climate changes. This has been the subject of considerable attention. A brief example follows: In 1958 there was a sudden warming of the ocean on the Pacific Coast. This terminated a period of unusual persistency of a moderately cold period starting sometime between 1946 and 1948. Figure 6-15 shows air temperatures along the coast. We see that prior to 1946, years were typified by very great fluctuation in mean temperature, with cold years, hot years, and "normal" years in random succession. There were also years with cold springs and hot falls, and vice versa. Beginning in about 1946, however, there was a series of cold years of uniform character. This same persistency is apparent in Fig. 6-16 which is a plot of the salinity anomalies at Scripps Pier (Sette and Isaacs, 1960), where we see that some process ceased in these years, and this or some other process that caused fluctuations did not start again until 1958. Some of this climatological behavior results directly from the interaction of the winds and ocean.

Figure 6-17 (Reid, Jr., 1960) shows a plot of the temperatures of the ocean off the California coast plotted against the strength of the north wind. These same remarkable changes and periods of persistencies are seen clearly here. The years 1931, 1932, and 1933 are examples spanning the entire range, whereas the years 1950 to 56 are clustered in a small group in the center of

FIG. 6-16. Salinity anomalies at Scripps pier, 1916–1959 (from Isaacs and Sette, 1960).

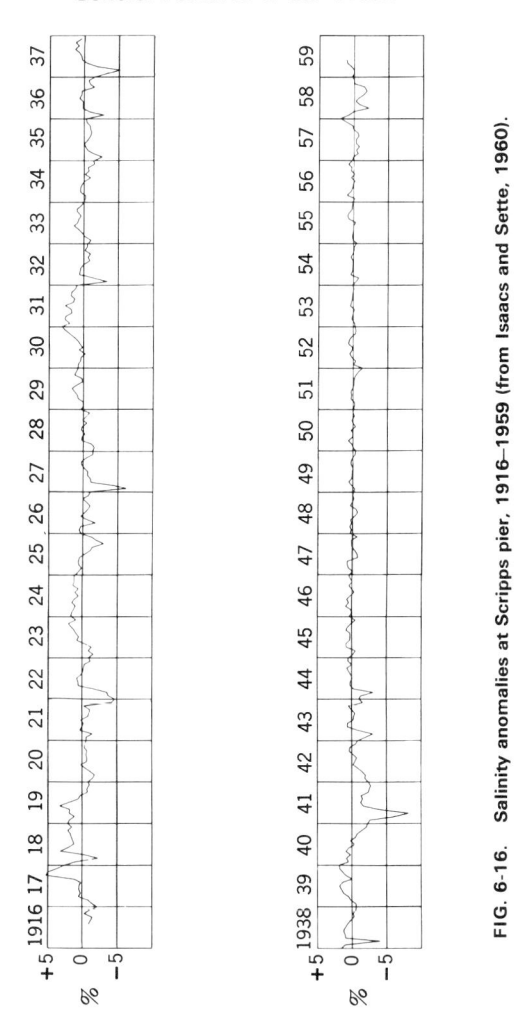

the chart. All this period of persistency that was characterized by a somewhat stronger than normal north wind and stronger than normal California current flow, upwelling, and cold, terminated abruptly on this coast in 1958. The earliest symptoms of the change can be traced back to the unusual winter of 1956–1957 in the North Pacific. The weather map of that winter (Fig. 6-18, Namias, 1960) shows the usual North Pacific low actually replaced by a high, and the low displaced far to the Asiatic side, pushing warm water up into the Gulf of Alaska. Warm water in winter is strictly unstable and produces instability in the atmosphere also. Thus there must have been great exchanges of heat and water vapor with the atmosphere in that period. Figure 6-19 (McGary, 1960) shows the unusual temperatures (6, 7, or 8

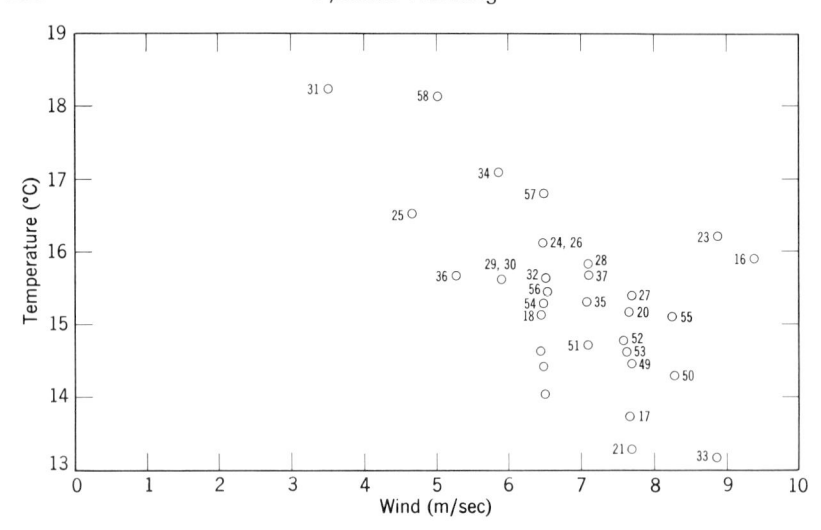

FIG. 6-17. Sea surface T versus north wind (from J. L. Reid, Jr., 1960). May seasurface temperature plotted against May wind for the years 1916–1938 and 1949–1958. Temperature is measured in the five-degree square 30–35°N, 115–120°W. Wind is computed from the pressure difference between 110 and 130° W along 30° N, and represents the component of wind from the north.

degrees above normal) in the Gulf of Alaska, caused by the transport of warm water from the Central Pacific into that region. I will not go into more detail, but this mass of warm water, plus perhaps an associated increase of warm water in the equatorial region, caused the development of a series of cyclonic storms moving across the North Pacific, and this whole process culminated in the winter of 1957–1958 with quite an opposed development— a great climatic low much more strongly developed than at any time in the immediate past, off the Oregon–Washington coast and in the Gulf of Alaska. This was the cause of the warming along the Pacific Coast with a sudden breakdown of all northerly components of the winds and the development of a strong countercurrent along this coast, a diminution of the California Current flow, and the invasion of tropical, subtropical, and Central Pacific organisms onto this coast. For instance, Fig. 6-20 is a distribution graph of the organism *Nyctiphanes simplex*, which had been carried thousands of miles out if its usual range by the remarkably narrow countercurrent.

In this brief sketch there are two points that I want to make. The first is that there are large and profound climate changes relating the ocean and the atmosphere and that these are also traceable by the changes they have imposed on the organisms. The second point is that if we could establish some highly resolved record of the past together with the relationships

General Features of the Ocean 189

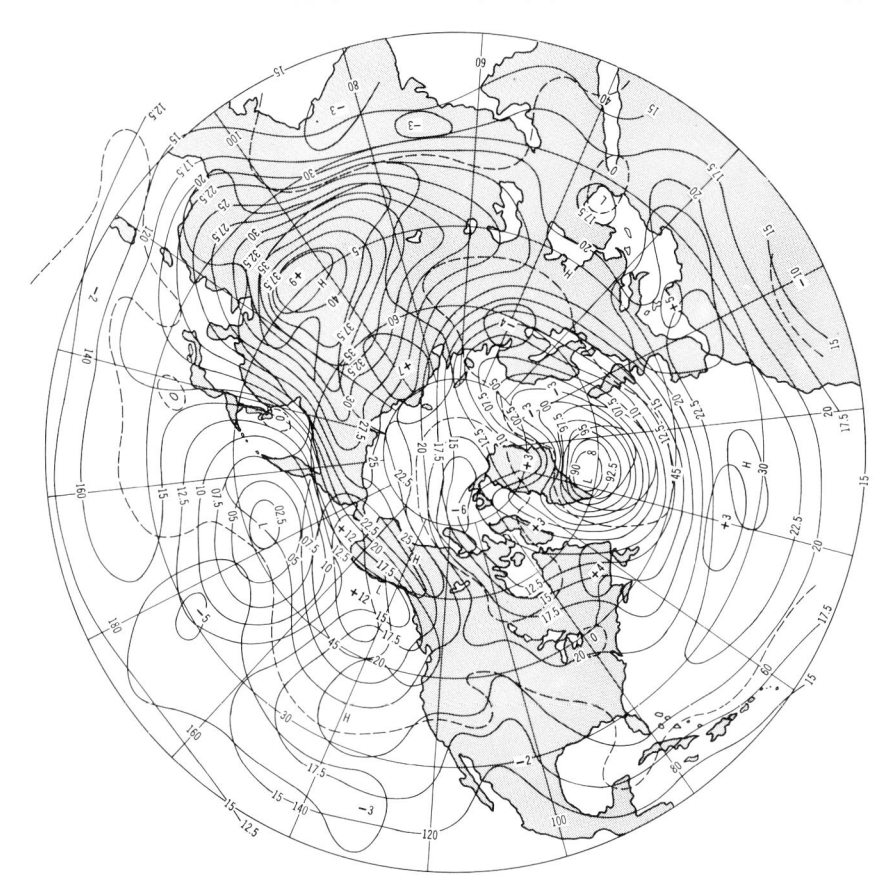

FIG. 6-18. Sea-level pressure field and departures from normal—winter 1956–1957—in millibars-1000 and millibars, respectively (from J. Namias, 1960).

between organisms and water masses, we would have a better idea of the nature of the oceanic circulation in past years, and a much greater insight into the nature of the processes that give rise to climate change. Now one place to look for this memory is in the sediments on the ocean bottom. There is nothing new in this—people have been looking at the ocean sediments for a long time for glimpses into the past, but there is now a possible entré that may permit a much better time resolution. In certain rather rare places along the coastal shelves there are basins and other bottoms that have conditions such that the sediments are not only rapidly deposited but virtually undisturbed by currents or organisms after they arrive there. These sediments therefore constitute a sort of annual memory of events. For instance, Fig 6-21 (McGowan, 1963) shows the distribution of the little marine pteropod that is characteristic of the Gulf of Alaska and of the

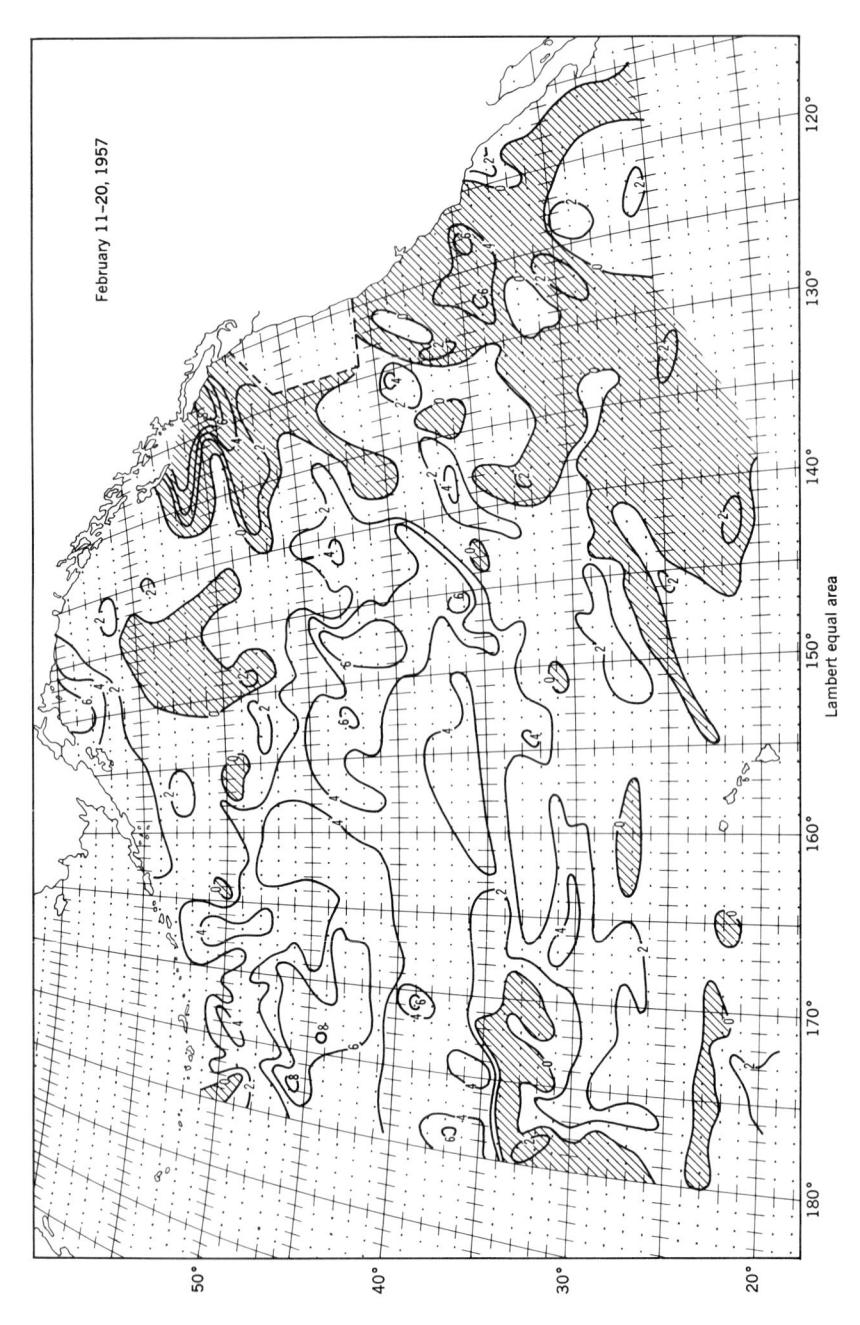

FIG. 6-19. Anomaly of sea surface temperature (°F) from 30-year mean (from J. W. McGary, 1960).

General Features of the Ocean   191

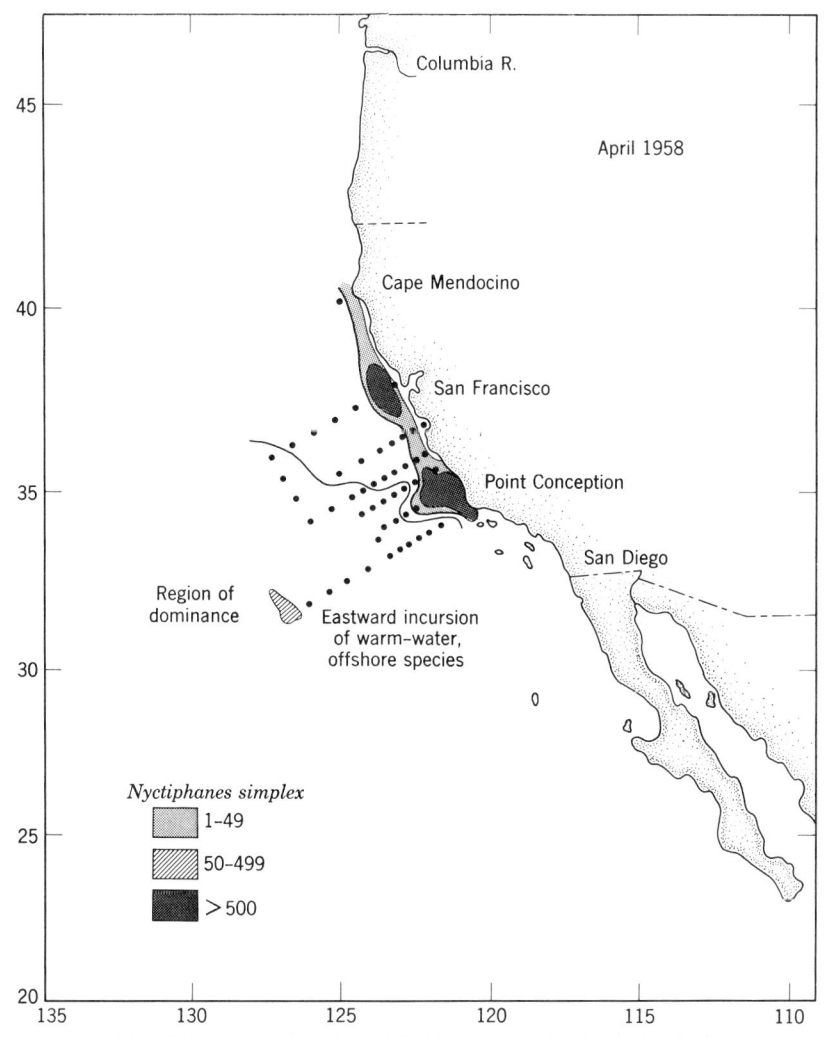

FIG. 6-20. Coastal invasion of *Nyctiphanes simplex* (by E. Brinton).

upwelling regions of the California coast. Its presence in Southern California waters indicates the strength of the California Current and a participation of northern water in it. If we now look into the sediments in the Santa Barbara Basin (which are rapidly reported and undisturbed) we find a great variation in the distribution of the shells of this tiny marine mollusc with depth (Fig 6-22). Each one of these blocks is a period of perhaps five years. Put together with other information, this gives us a possibility of reestablishing the short-term changes in the climate of the Eastern Pacific and perhaps the fluctuations of climate in the entire Pacific Basin. The entire column here represents a period of about 1500 years or so. There is evidence that climate changes in one part of the Pacific are associated with climate changes in other portions, and thus such a region as the Santa Barbara Basin is by no means as provincial as might at first be assumed. Assumptions can be made of related climate changes elsewhere around the Pacific Basin and possibly elsewhere around the world. We have recently discovered other sites of these rare sediments as far away as northern South America.

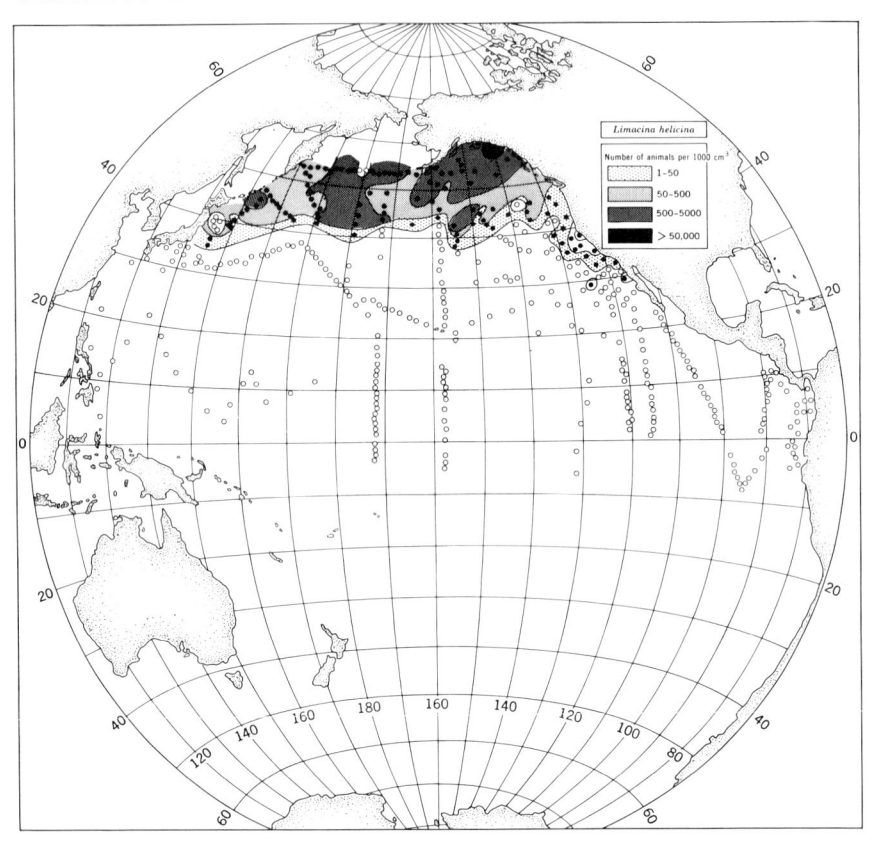

FIG. 6-21.   Distribution of *Limacina helicina* (from McGowan, 1963).

In these same sediments fish scales are very well preserved due to anaerobic conditions, and we also can examine the history of fish—in fact the outcome of this is one of the most important findings as far as the immediate problems of the fisheries in California are concerned. Two species of fish that have not been explored very thoroughly, the hake and the anchovy, are responsible for the dominant scales throughout these sediments.

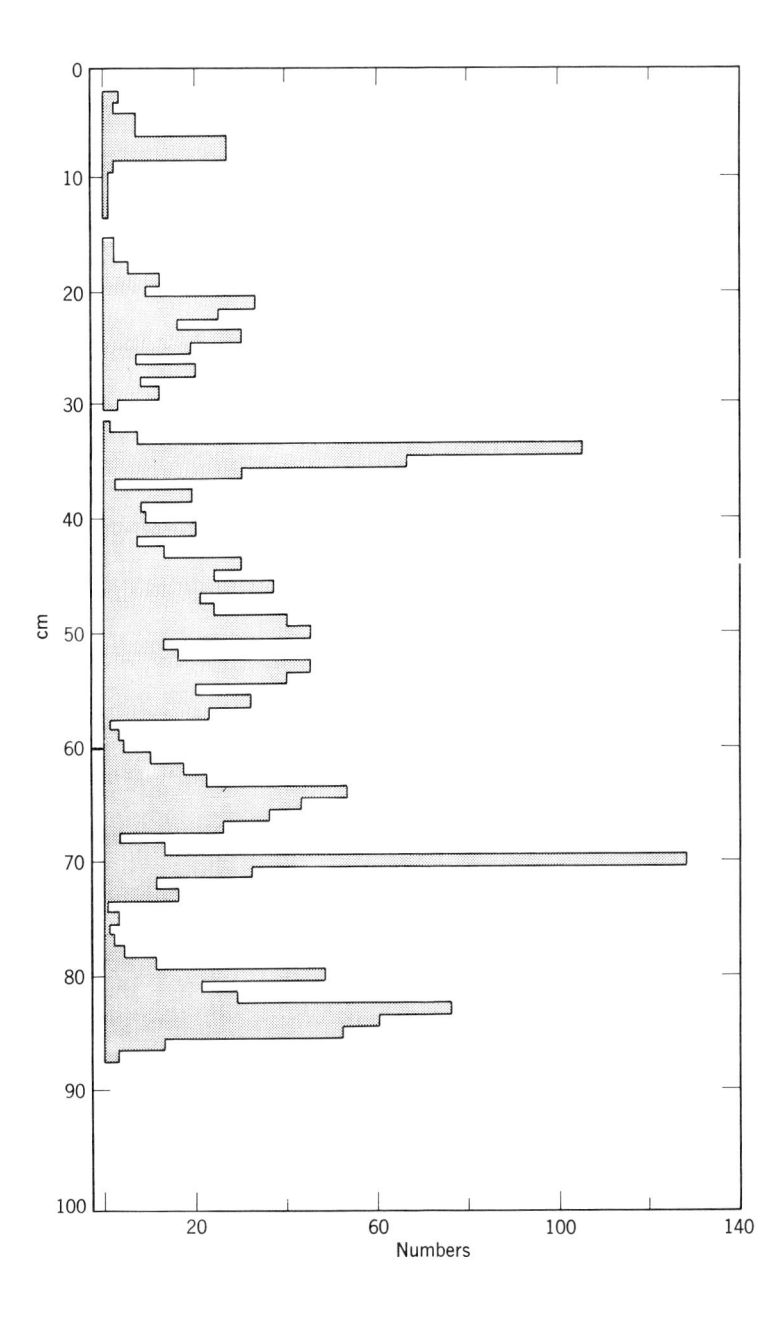

FIG. 6-22. Pteropod count in sediment core (by A. Soutar).

Figure 6-23, for instance, gives the distribution of anchovy scales with depth in the same cores. We see that this fish has been here almost constantly over the last 1500 years with the exception of the last 70 or 80 years. Figure 6-24 shows the scale count of what has been heretofore considered to be the normal inhabitant of the California waters, the sardine, and the sediments indicate that there have roughly been only two periods of abundance—the most recent period and a period some 700 years ago or so. Incidentally, in these same figures there are visible two periods in which the sediments have been disturbed. One of these apparently dates somewhere around the beginning of the nineteenth century when the only great undersea earthquake occurred along this coast—the San Nicolas earthquake of 1812. It is not impossible that the other disturbed period some 800 years ago was a similar event, in which case we would have tagged the occurrence of undersea earthquakes in the Southern California region. As we further investigate these sediments we may be able to determine the existence of terrigenous sediments carried into the sea by floods, and in this way develop a history of rainfall and floods of the last 2000 years.

These few cases exemplify the strong interdisciplinary aspect of the processes and problems in the sea. In the last section I would like to take up engineering and ocean engineering.

6.3 REQUIREMENTS FOR MAN'S INTERVENTION

I look at ocean engineering as a subject in which we are concerned with the purposeful and useful intervention into a new realm. This is a relatively new and involved field of human activity. Certainly at this time we should look carefully at the nature of the environment and the nature of the problems of the intervention. We should define ocean engineering as broadly and meaningfully as we can. If we thus define it as "promoting the purposeful and useful intervention into the sea," I believe that we will find that it is quite insufficient to continue to consider the field as some sort of transplanted marine homologue of terrestrial, mechanical, electrical, structural, agricultural, and systems aspects of engineering alone. A great number of factors, of which I will give you examples presently, argue that mere baptism with saltwater is wholly inadeqate. Even the usual structural, electrical, and mechanical problems are not only unconventional, but are largely unknown. Earthquake stresses, for example, can be handled on land rather empirically as horizontal or vertical accelerations affecting structures. These same types of earthquake shock take quite a different form under the sea. Since there is this great overburden of water, accelerations are not so important per se but rather they produce great brief overpressures as the ground is pushed against the water overburden. The

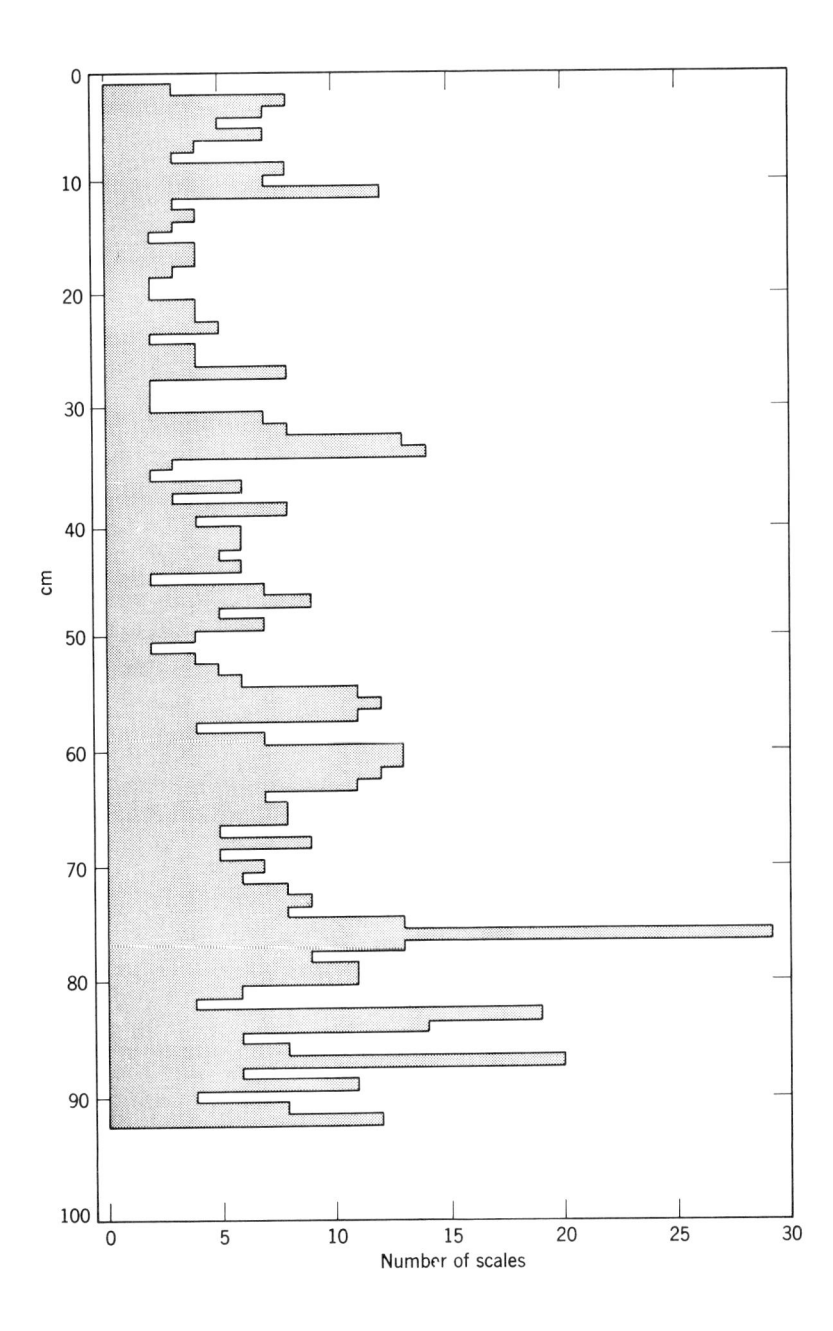

FIG. 6-23. Anchovy scale count in sediment cores (by A. Soutar).

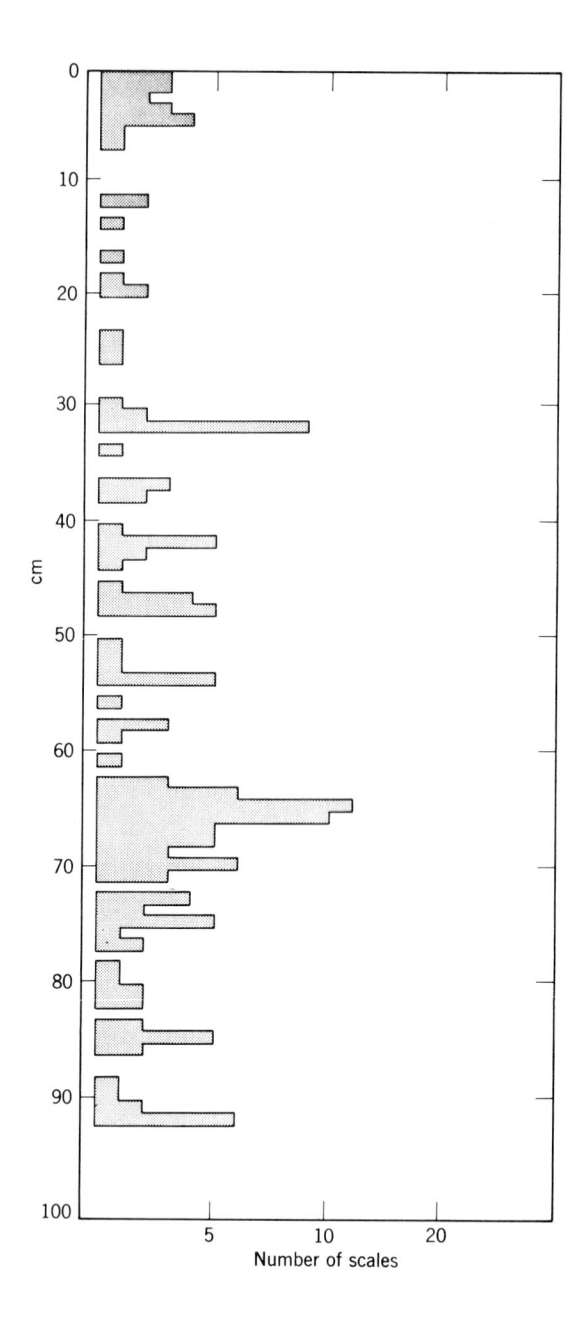

FIG. 6-24. Sardine scale count in sediment cores (by A. Soutar).

nature of sediment stability is practically unknown and underwater landslides take quite a surprising and vigorous form. These density flows are more like vast snow avalanches of the high-mountain regions or like the white-hot ash avalanches that occasionally have wiped out cities in the explosion of a volcano, such as Mt. Pelée. Such an undersea avalanche apparently occurred in 1929 in the Grand Banks region of the North Atlantic and swept widely for thousands of miles at high velocities out over the ocean floor ripping out cables and even tearing out sections of them. Corrosion is another example of an unknown circumstance. It would be easy to consider that all we had to do was to put the test sample in some laboratory tank and circulate seawater about it or allow seawater to become stagnant about it. The facts are that a piece of metal in the ocean is not subject to the open conditions of seawater but rather to the conditions underneath some bacterial plaque or fixed organism—a barnacle for instance. These are conditions that not only vary widely around the ocean but that are very difficult to duplicate in the laboratory.

Under some oceanic conditions, for example, we find that stainless steel is by no means stainless. This is merely one example of the manner in which marine biological factors are often controlling. Surfaces are fouled, and corrosion and fatigue result from the efforts of these organisms to find attachment. Animal noises, ever present throughout the oceans, are one of the obstacles to the use of underwater sound for detection, for communication, and for classification. Underwater cables are bitten and drilled, mooring lines are severed. Organisms, as we have already seen, cluster around underwater installations or underwater objects, make noise, excavate, set sediments in suspension, and alter conditions. Organisms are also the principal considerations in the disposal of atomic waste. In addition to the aspects already mentioned, the introduction of an object on the bottom attracts a community of organisms around it and shortens the pathway of radioactive materials escaping from containers into the organic realm of the ocean. In a more direct fashion, much of ocean engineering is concerned with fisheries or should be concerned with it, and it then must understand more about the behavior of fishes, their escape mechanisms, their behavior and the possible methods of their capture and acquisition. What I have said so far is that I think that ocean engineering must be far more broadly interdisciplinary then terrestrial engineering. The terrestrial engineer is usually horrified to find that he has to take into consideration the activities of uncontrolled creatures. A squirrel bites a cable and starlings enter a jet engine, a plant's roots invade pipes, termites attack buildings, but the terrestrial interference by organisms is relatively understood and rare. In the sea this is by no means the case. If a mechanism is placed on the bottom and is expected to operate for a month and then to emerge or to perform some other task, it will most likely we found that some hard organisms have become attached to it and some others have taken up residence in it and the thing works not at all.

Thus the ocean engineer finds himself in circumstances where biochemical or geological, or odd and unquantified physical phenomena (such as deep-sea combers) compromise his efforts.

Not only the scientific fields apply in a different way in marine and terrestrial efforts. On land, law, economics, and government policy are reasonably well formulated. In ocean-engineering efforts, however, these matters are by no means so well established. Because Dr. Chapman has treated in Chapter 5 the subject of legal constraints on man's activities, I will cite only some points which are essential to my discussion. The problems of rights of entrepreneurs, unlimited entry, and effectiveness of technology, have not been adequately handled by the United States. For example, our domestic fisheries utilize some of the most primitive acquisition techniques in the world. Any inventive person could make order-of-magnitude improvements on them. The result, however, is likely to be useless. Either the novel method already is unlawful, or, if successful, it soon would be made so. The price we pay for not solving the control of fishing and the entry problem is to have fishermen using methods scarcely evolved from past centuries. We are the only advanced nation in the world to so constrain our fishermen that in some cases they must place handles on the keels of their boats so that they can have something to hold on to when they capsize the miserable boats to which they are legally restricted. This is quite different from our treatment of farmers. Yet the fisherman is effective.

Basically, engineering maintains a puritanical (practical) viewpoint where production, transportation, or defense are the important human activities. However, in the United States, direction of development of aesthetic and recreational activities are becoming increasingly important. Witness, for example, the profound changes that snow machines have made on the economics of the Northeastern recreational areas. In like fashion on this coast, with the exception of defending stocks from foreign exploitation, sport fishing is more important than commercial fishing. Small-boat harbors deserve attention even more demanding than commercial harbors, and good surfing can change a beach community. I have yet to see, however, an engineering survey of the particular bottom configuration of Waikiki that gives rise to the wave refraction and wave transformation that makes that beach a superb surfing beach. Southern California is subject to the same swell from the Southern Hemisphere, and a properly configured offshore topography undoubtedly could create equally fine surfing conditions.

The superstable FLIP may make a fine platform for the study of underwater sound. Our minds immediately leap to the (antisubmarine-warfare) implications, but it also could solve the primary restraint to marine sport fishing—sea sickness.

Aesthetics will, I believe, play an increasing role also. The Bodega Head crises about the installation of a nuclear power plant, whose warm cooling-water discharge would have changed the local sea life, was an example.

I point out these latter examples as cases where ocean engineering must or should look more broadly at its problems than the traditional social economic problems with which engineering confronts itself on land.

Previously I have pointed out the strong scientific interdisciplinary demands of the sea. With such a spectrum of largely unexplored problems covering a range of scientific and human disciplines, all developing rapidly and all in a new medium, the most *basically educated* participants are required. Only when there is a large background of well-codified knowledge is the instruction of practitioners pertinent. Teaching the *practice* of ocean engineering at the undergraduate level can now only produce technicians in the future—not leaders, in my opinion.

It is now possible to identify some ocean-engineering problems whose solution would broadly enhance our abilities. These are in cases where the problem is sufficiently well understood and of sufficient general importance that its solution would eliminate serious constraints and thus greatly increase out freedom of action in the marine realm. For example, we have not yet achieved mastery of the ocean air. We are tolerated in crossing the sea, but the true marine aircraft as free of land ties as an albatross is not only feasible, but would constitute a powerful tool in antisubmarine-warfare, advanced seaborn nuclear deterrent, and in ocean science.

There is much opportunity even in relatively minor aspects of mastery. We have yet to improve on the Spanish of the sixteenth century on the fearsome business of putting a small boat over the side in a heavy sea. A pilot may ditch planes alongside aircraft carriers, but the real danger comes to the men being lowered to give him assistance. This is also a consideration in the servicing of a storm-tossed instrument platform. Soon arrays of these platforms, sensing and recording oceanographic and meteorological parameters, will be installed in parts of the oceans. In this business of the deep mooring of surface instrument platforms it seems that it should be easier to moor on a seamount or a shelf at 400 or 500 fathoms than in the deep sea at 2000 to 3000 fathoms. We are somewhat surprised to discover that this is not so! In deep water currents are weak, the deep sediments are often good holding ground, and the great lengths of mooring cable required provide great damping and elasticity to absorb and dissipate sudden loads. It now worries me to try to moor in only half a mile of water!

The full implications of superstable platforms have not yet been realized, I believe. Consider a circular pelagic harbor of superstable configuration. Deployed halfway between California and Hawaii, it could not only act as a refuge from an inclement sea but also as a site for one of the great nuclear power plants. Cooling water and nuisance-level atomic waste disposal would be no problem. What could be done with the immense power capacity of such plants in a remote ocean area? Of course, mineral extraction from sea water and fish flour concentrates are possibilities. But these would take only a tiny fraction. Is it possible to use unconventional modes of power transmission? After all, in the United States, the cost of generation is only a few percent of the charged cost of power. The rest goes to transmission, administration, billing, etc. Could power, wirelessly transmitted and receivable any place, eventually become the inherent right of all men from the fishing ship in the South China Sea to the farmer in Central Africa or the weekend camper in Antarctica?

What also are the opportunities of learning from the engineering method in studying natural phenomena? Special wave refraction makes some small-boat harbors absolutely unique in their safety. Depoe Bay, Oregon, in midwinter is sometimes the only enterable harbor between San Francisco and the Straits of Juan de Fuca. Curious nonlinear wave refraction gives it this peculiar and important quality. It is said that only an idiot can get himself into trouble on that entrance (I think this is not so, for I dismasted there myself, once).

Much work, these days, is done on desalination. All marine plants desalinate seawater. Some desalinate water that is three times as concentrated as seawater. Can these higher saltwater plants teach us how it is done—or by breeding can they "teach" crop plants to tolerate salt water—to alleviate the terrestrial fresh-water shortage, which stems primarily from its use in desert agriculture and from river pollution?

It is clear that organisms have solved many of the problems that face man in attempting to employ the sea for his needs. Echo location, temperature and bends control, navigation, homing, underwater communication, and propulsion are some of the problems that organisms have solved in various ways and from which we can learn. It often has been said that our submersibles would be far ahead had we sooner taken a curious and critical look at a tuna. My point here is that the ocean engineer must be intellectually free to learn from all sources of information.

There are many more matters that I would like to include in this discussion. I have not managed to get around to the use of engineering knowledge and methodology in the understanding of the sea, such as why the curious fact that a man perspires when ascending a hill rather than descending, has important significance to fish propulsion!

If any brief summary to my discussion is possible, it is this.

The ocean re-expresses great natural laws in a medium that is largely foreign to man. In this medium its components act and interact quite without regard for the specialties into which man has compartmented his thinking.

In this regard oceanography and ocean engineering constitute the archetype of a new science and technology where the analytical results of disciplinary science and fields of technology must be meaningfully put together in a system where the *interactions* generate the principal features. These may be only very weakly described by their components!

The successful ocean engineer will thus be fully aware and fully competent to understand the constraints and *opportunities* to which these interactions give rise. *His* accomplishments will *not* be mere waterproofed terrestrial transplants that emerge from abstract design criteria and which are immediately compromised by "real-world effects"!

REFERENCES

Bascom, W., 1961, *A Hole in the Bottom of the Sea*, Doubleday, Garden City, New York.

Carey, S. W., 1958, "A Tectonic Approach to Continental Drift." pp. 177–355. *Continental Drift – A Symposium, March 1956*, U. of Tasmania 1958.

Isaacs, John D., and Hugh Bradner, 1964, "Neutrino and Geothermal Fluxes," *J. Geophys. Res.* **69** (18), 3883–3887 (1964).

Isaacs, John D., and Richard A. Schwartzlose, 1965, "Migrant Sound Scatterers: Interaction With the Sea Floor," *Science* **150**(3705):1810–1813 (1965).

McGary, J. W., 1960, "Surface Temperature Anomalies in the Central North Pacific, Jan. 1957–May 1958," California Cooperative Oceanic Fisheries Investigations (CalCOFI) Reports VII. pp. 47–51.

McGowan, J. A., 1963, "Geographical Variation in *Limacina helicina* in the North Pacific," Systematics Association Publication Number 5, pp. 109–128.

Namias, J., 1960, "The Meteorological Picture 1957–1958," CalCOFI Report VII, pp. 31–41.

Putnam, W. C., *et al.*, 1960 *Natural Coastal Environments of the World*, U. of California Press, Los Angeles, California.

Reid, J. L., Jr., 1960, "Oceanography of the Northeastern Pacific Ocean During the Last Ten Years," CalCOFI Reports VII, pp. 77–90.

Sette, O. E., and John D. Isaacs., 1960, "Editors' Summary of the Symposium," CalCOFI Reports VII, pp. 211–217.

Vine, Fred, 1966, "Spreading of the Ocean Floor: New Evidence," *Science* 154(3755): 1405–1415 (1966).

CHAPTER 7

Physical and Hydrodynamic Factors

MYRL C. HENDERSHOTT

The ocean is driven by both thermal- and mechanical-energy fluxes. By far the largest energy input is solar radiant energy and 90% of this is absorbed in the upper 40 m of water. This supply, together with back-radiated heat and heat exchanged with the atmosphere, make up virtually the entire heat flux through the ocean (excepting only a small contribution through the sea floor, see Fig. 7-1). Viewed as a simple convective system, the ocean is not a very efficient converter of thermal energy into mechanical energy. This is due in part to the fact that the direct heating of the ocean is confined to a very thin layer near the sea surface. The situation is different in the atmosphere, which is driven from below by warming and injection of water vapor (because most of the thermal energy incident upon the earth is absorbed at ground or sea level). When the water vapor rises to the altitude at which it condenses, its large latent heat becomes available and is the primary source of kinetic energy in the atmosphere. This method of energy injection does not have an oceanic counterpart. Additionally, the earth's rotation inhibits direct flow from regions of high pressure to regions of low pressure so that the thermal-energy flux largely goes to maintain oceanwide gradients of temperature and salinity (via the mechanisms of evaporation and precipitation) in which large amounts of potential energy are stored.

The obvious motions of the sea, waves and currents (but not tides or tidal currents), derive their kinetic energy mostly from winds in the atmosphere. Because the density of water is so much greater than that of air, the transfer of kinetic energy across the air–sea interface is nearly a one-way street and hardly any energy returns to the atmosphere in this way (see Fig. 7-2). The wind energy that goes into the ocean drives the systems of wind waves and surface currents familiar to mariners but it does not do so independently of

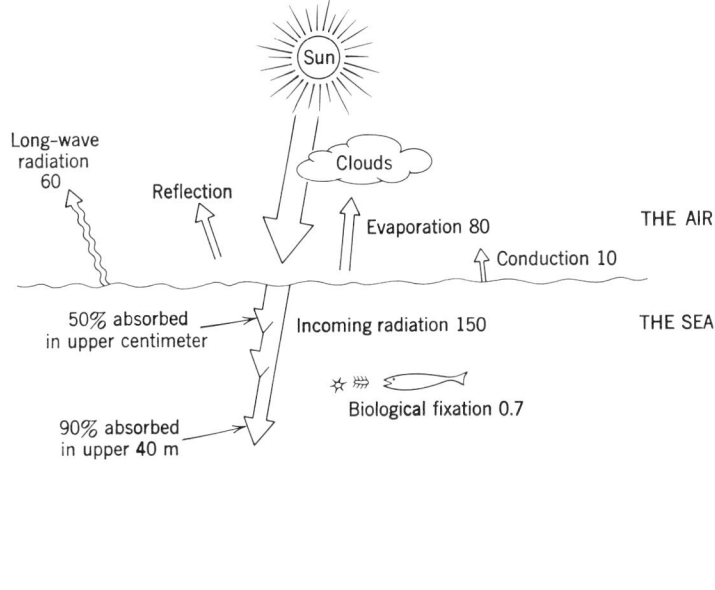

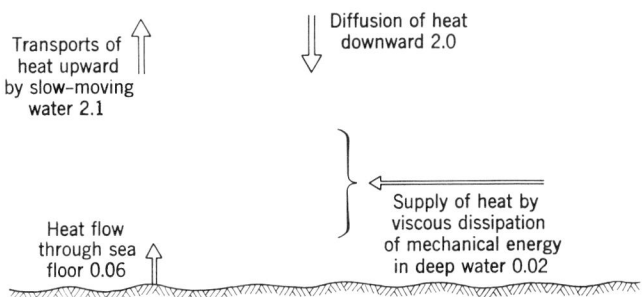

FIG. 7-1. Estimated average fluxes of thermal energy. All figures represent energy flow expressed in W/m² of sea surface.

the flux of thermal energy. The oceanwide gradients of temperature and salinity maintained by the thermal flux have a profound influence on the wind-driven motions and significant interaction between the thermal and mechanical fluxes probably occurs over a wide range of length scales.

The tides derive their energy from the rotational energy of the earth–moon–sun system and have traditionally been thought of as independent of other processes in the ocean in the sense that tides in fresh-water oceans on an isothermal earth with no atmosphere might not be too different from the actual tides. Much tidal energy is ultimately dissipated through turbulence in shallow seas but it is also possible that appreciable amounts are diverted into internal wave motion in the deep sea. This may represent an

204   Systems and Planning

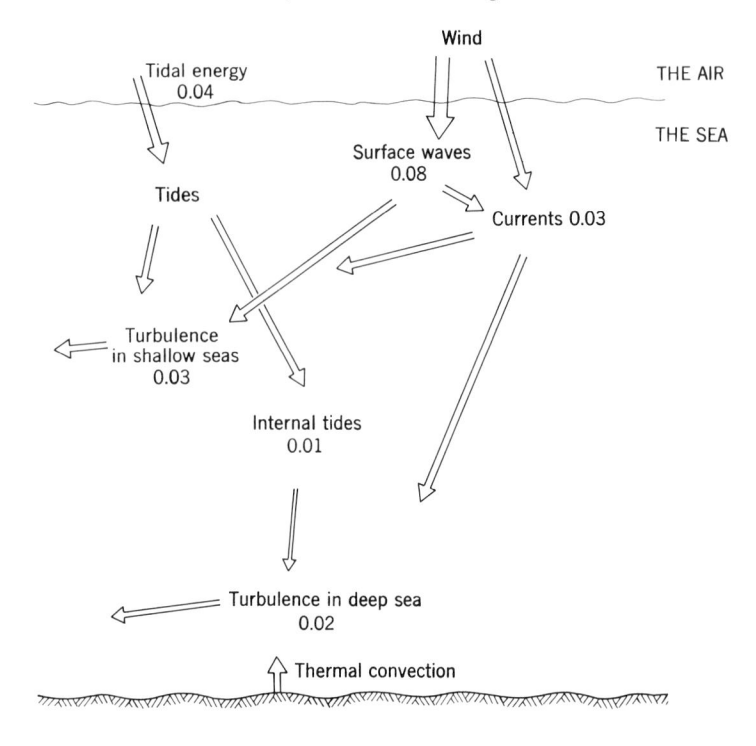

```
                              Wind
         Tidal energy                              THE AIR
            0.04
        ~~~~~~~~~~~~~~~~~~~~~~~~~~~~~~~~~~~~~~~~  THE SEA
                          Surface waves
                             0.08
           Tides
                                    Currents 0.03


           Turbulence
          in shallow seas
              0.03

                       Internal tides
                           0.01



                   Turbulence in deep sea
                           0.02

                        Thermal convection
```

FIG. 7-2. Estimated average fluxes of mechanical energy in W/m^2 of sea surface.

important source of energy for stirring and mixing in the deep sea, and, if so, the absence of tides might well alter the character of many oceanic phenomena.

A selected few of these phenomena are described in the following pages. Observations and theory have been combined to provide an easily visualized picture of each, but the descriptions are necessarily very schematic and are really intended only as prefaces to the serious reading of original sources.

7.1 SALINITY AND TEMPERATURE DISTRIBUTIONS; STRATIFICATION; THERMOHALINE CIRCULATION

The characteristic that distinguishes seawater is its salinity. The origin of the various dissolved salts in seawater evidently involves submarine vulcanism and the leaching of rocks—on the sea floor by the seawater itself and on land by precipitated water which then carries its dissolved mineral load to the sea. Salinity may be defined as the total amount of solid material in grams in 1 kg of seawater when all the carbonate has been converted to oxide, the bromine and iodine replaced by chlorine, and all organic material completely oxidized (Forch et al., 1902).

The salt content thus defined, actually a little lower than the amount of dissolved material per kilogram of sea water, is on the average approximately 35‰ (the symbol ‰ is to be read as parts per thousand). Over most of the ocean it generally falls between 33 and 37‰ although it may nearly vanish at the mouths of large rivers and is over 40‰ at the surface of, for example, the Red Sea (where evaporation is rapid). Values much higher than this have recently been observed at isolated locations on the floor of the Red Sea, possibly at submarine mineral springs. The greatest variations in salinity occur near the surface; values below 4,000 m generally are between 34.6 and 34.8‰

The upper portion of Fig. 7-3 shows mean values of evaporation and precipitation over the ocean from Antarctica to the Arctic according to Wüst (1954). The values are averaged over season and longitude. In the lower section of the figure the similarly averaged surface salinity as it varies with latitude is compared with the corresponding excess of evaporation over precipitation. The obvious correlation indicates that the salinity of surface water is strongly dependent upon the rates of evaporation and precipitation

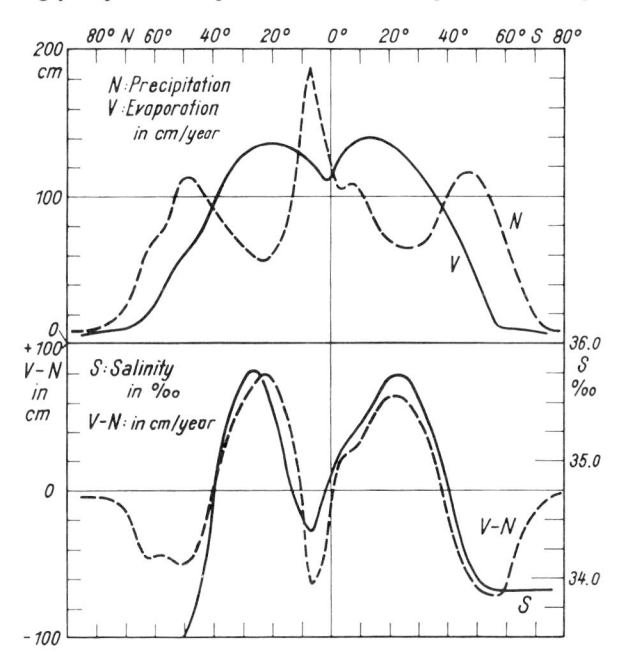

FIG. 7-3. Zonal distribution, in the annual mean, of precipitation *N*, evaporation *V*, *V–N*, and salinity at the sea surface, including the marginal seas (according to G. Wüst, 1954; from G. Dieterich and K. Kalle, 1963, *General Oceanography*, Interscience, New York).

at the sea surface. The two salinity maxima near 30°N and 30°S are near the latitudes of the atmospheric subtropical high-pressure cells over each ocean basin. The relatively clear air of these regions makes for high rates of solar heating and consequent evaporation, so that maxima of surface temperature and salinity tend to develop there.

The traditional method of measuring salinity has been to determine the concentration of halogen ions by direct precipitation with silver nitrate and then to use Dittmar's (1884) empirical rule—in seawater of any not too extreme salinity, the relative proportions of all constituents are nearly constant—to determine the total content of dissolved salts. The actual determination is made by comparing the amount of silver nitrate needed to precipitate the halogen contained in the sample itself with that necessary for a sample of normal seawater of the same volume (normal seawater is seawater having a precisely known halogen content). The ratio of volumes of silver nitrate thus obtained gives the ratio of halogen in the sample to that in the normal water. The halogen content in grams per kilogram is commonly called the chlorinity (although it reflects not only the chloride but also the bromide and iodide content as well) and is used to calculate the actual salinity by Knudsen's empirical relation (Forch et al., 1902):

$$\text{salinity} = 0.03 + 1.805 \times \text{chlorinity}.$$

The resulting salinity is generally considered to be within 0.02 to 0.04‰ of the correct value.

The concentration of dissolved material in seawater may be more sensitively determined by measuring the electrical conductivity of the water (Cox, Culkin, Greenhalgh, and Riley, 1962). Various conductivity bridges, conveniently operable on ships at sea, have been developed. All make use of an empirical salinity–conductivity relationship and may give salinities as precise as 0.005‰.

The temperature of seawater in the ocean ranges from about 30°C at the surface in tropical regions to about $-2°C$ in antarctic surface and bottom water. A representative, although by no means entirely typical, profile of temperature versus depth is shown in Fig. 7-4. Salient features of this profile common to most locations are (a) a warm, homogeneous surface layer which is produced by wind stirring and absorption of solar radiation and whose thickness ranges from tens of meters near the equator to perhaps 100 m in midlatitudes, (b) a *thermocline* region in which the temperature falls rapidly from the value in the mixed layer above to the low value characteristic of the deep water below, and (c) a deep region in which the temperature decreases only slowly with depth.

The thermocline is generally centered around the 8° or 10° isotherms, which lie at 300 to 400 m below the surface in equatorial regions and from

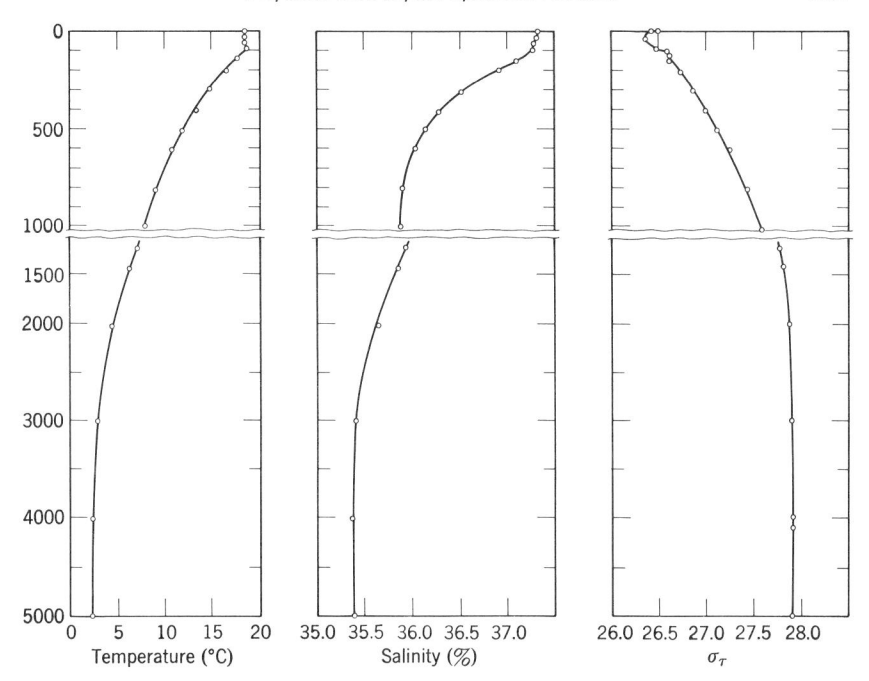

FIG. 7-4. Temperature, salinity, and the resulting values of σ_T at Michael Sars Station 144 (28°37' N., 19°08' W) (after H. U. Sverdrup, M. W. Johnson, and R. H. Fleming, 1942, *The Oceans, Their Physics, Chemistry, and General Biology*, (C), by permission of Prentice Hall, Inc., Englewood Cliffs, New Jersey.)

500 to 1000 m in subtropical regions. The relatively warm surface water and the transition water of the thermocline thus form but a small fraction of all the water in the sea, whose mean depth is close to 4000 m. The mean temperature of the oceans is about 3.8°C and even at the equator, a water column has an average temperature of only 4.9°C. The term "thermocline" is implicitly defined in a variety of ways in the oceanographic literature. In addition to the sense in which it is employed in this article, it is commonly used to denote the transition between the homogeneous mixed layer and the water just below. The temperature change at this transition depth is often particularly abrupt.

The temperature of water at the sea surface is influenced by incoming solar radiation plus evaporation, precipitation, and freezing, but the cause and effect relationship is not as clear cut as in the case of surface salinity. Average annual surface temperatures for the Atlantic and the Pacific are shown in Fig. 7-5 as a function of latitude.

The global distribution of such hydrographic quantities as temperature and salinity is generalized from some 40,000 hydrographic stations. At each

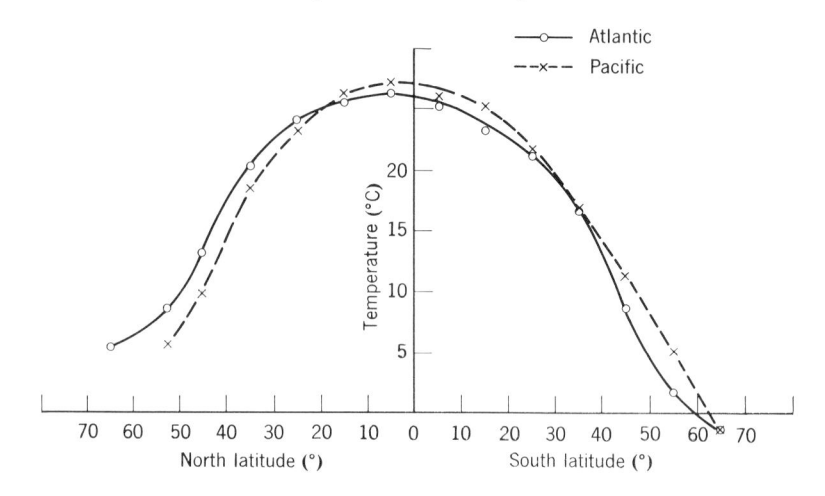

FIG. 7-5. Zonal distribution of sea surface temperature (10° averages) (after H. U. Sverdrup, M. W. Johnson, and R. H. Fleming, 1942, *The Oceans, Their Physics, Chemistry, and General Biology*, (C); by permission of Prentice Hall, Inc., Englewood Cliffs, New Jersey.)

station, values of temperature, salinity, and often such other quantities as dissolved oxygen, phosphate, and nitrate, have been determined at from 1 to as many as 18 or 20 different depths between surface and bottom. The traditional smooth interpolation of Fig. 7-4 between these measured points to obtain values at any depth appears to yield meaningful average vertical gradients but fails to reveal the recently discovered steplike microstructure which characterizes these fields on a small scale. The actual variation of temperature and salinity with depth is not smooth but rather proceeds in jumps of about 0.01°C or 0.1‰, respectively, every 10 m or so. The origin of this structure is the object of current research, but its observed coherent persistence at one location over appreciable lengths of time (several hours) is probably associated with the large-scale stable density stratification of the oceans established by the observed fields of temperature and salinity.

The density of seawater is determined by the salinity, the temperature, and the pressure. It decreases as the water is heated and expands, but increases as the content of dissolved salts increases. The pressure, which is primarily due to the weight of the water above, increases by 1 atm with every 10-m increase of depth so that it is quite large even at moderate depths. Consequently, the water is sufficiently compressed that the resulting increase in density must be taken into account if the density is to be determined to the accuracy needed for present-day oceanographic work (about one part in 10^5).

Because the density of seawater is always slightly greater than unity, it is customary to express the density ρ_{STP} of a sample at salinity S, temperature

T, and pressure P in terms of a quantity σ_{STP} defined by

$$\sigma_{STP} = 10^3 (\rho_{STP} - 1).$$

Thus a density ρ_{STP} of 1.02538, typical of seawater, is abbreviated as $\sigma_{STP} = 25.38$.

The calculation of density begins with Knudsen's (1901) empirical formula for the density σ_0 at 0°C and atmospheric pressure in terms of salinity:

$$\sigma_0 = -0.093 + 0.8149\, S - 0.000482\, S^2 + 0.0000068\, S^3.$$

This density σ_0 is then extrapolated to the required temperature, but still at atmospheric pressure, to obtain a density $\sigma_T = \sigma_0 - D$, which would be the density of the sample if it had been raised from its depth of origin to the sea surface without changing either the salinity or the temperature. The correction D has been tabulated by Knudsen (1901) on the basis of careful laboratory measurements. Finally, the σ_T value is extrapolated to the proper pressure to obtain the *in situ* density σ_{STP} of the sample at its depth of origin. The coefficients necessary for this correction have been tabulated by Ekman (1908).

The variation of σ_T with salinity and temperature is illustrated in Fig. 7-6. A noteworthy point is that, with increasing salinity, the temperature of maximum density (4°C for fresh water) is gradually lowered until, in seawater,

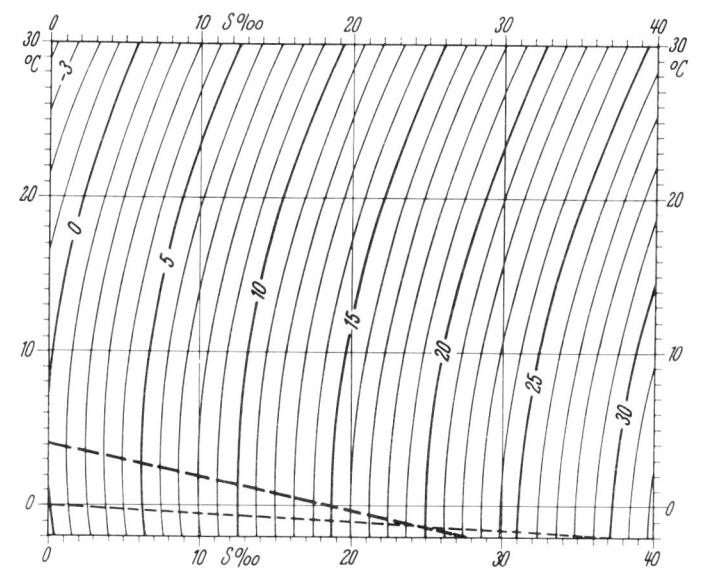

FIG. 7-6. σ_T for seawater as a function of temperature and salinity.---density maximum; ----freezing point (from G. Dieterich and K. Kalle, *General Oceanography*, 1963, Interscience, New York).

it is below the freezing point (which is also depressed from $0°C$ for fresh water to about $-1.9°C$ for saline seawater). Thus, contrary to the situation in most fresh-water lakes, freezing in polar regions of the sea usually occurs before the water has reached maximum density.

The mean density of surface water in the sea is 1.025 gm/cc, a σ_T value of 25.00. Values of σ_T at the surface generally range between 24 and 26, increasing to 28 or 29 in water from great depths. The σ_T profile of Fig. 7-4, computed from the accompanying fields of temperature and salinity is typical in this and other respects. Both σ_T and the actual density increase most rapidly in the thermocline, growing only slowly with increasing depth thereafter. Computed *in situ* densities are accurate to several parts in 10^5, although at very great depths, the goodness of the pressure correction to the coefficient of thermal expansion is so doubtful that the density gradient there may not be calculated with certainty (Eckart, 1960).

The density differences thus calculated are of great importance in spite of the relatively small departures of the density from its average value. The vertical distribution of density is nearly always such that if a parcel of water is raised (or lowered), it finds itself in lighter (heavier) water and tends to sink (rise back to its original position. The water is said to be stably stratified and any vertical motion or mixing is opposed by relatively strong buoyant forces. If an unstable gradient were ever to develop in the absence of motion, the heavier water on top would sink to the bottom until stability had been restored.

In as much as a parcel expands (is compressed) upon being lifted (depressed) to a region of lower (higher) pressure, the buoyant force is proportional not to the gradient of *in situ* density but rather to the gradient of what is termed the potential density. But this is very nearly equal to the density σ_T of the parcel if it is brought to the surface (where the pressure is atmospheric pressure) without changing either the temperature or the salinity. Mixing and motion in the sea, therefore, tend to occur along surfaces of constant σ_T and, because the oceans are very thin in comparison with their horizontal extent, these are nearly horizontal. On this account, water whose temperature and salinity are fixed at the sea surface by evaporation, etc., may creep below lighter water formed at another surface location, thus leading to a vertical distribution of temperature and salinity in which the values at depth reflect the characteristics of a surface region of formation that may be thousands of miles away.

Thus, in the antarctic winter, the formation of ice increases the salinity of the unfrozen water and it is cooled to such an extent that it sinks to the bottom. This cold and saline water then spreads northward throughout ocean basins along the sea floor and antarctic values of T and S may be obtained from deep water samples taken in subtropical waters. Somewhat to the north of the region where this freezing–sinking formation of bottom water occurs, the pattern of atmospheric winds is such as to force water to flow from both north and south towards a line, the Antarctic Convergence, which circles the Antarctic Continent on an irregular course between 50°S and 60°S. The water thus accumulated has nowhere to go but down; it thus sinks to form an intermediate water mass which spreads northward at a depth of 700 to 900 m. Because of the excess of precipitation over evaporation at these latitudes, the salinity of this water is quite low (Fig. 7-3) and the intermediate water appears in the salinity section of Fig. 7-8 as a tongue of low salinity prominent even north of the equator. A somewhat similar formation of bottom water occurs far to the north in the Atlantic, where salty water which has been brought north from tropical latitudes by the Gulf Stream mixes with cold surface water of arctic origin. This mixture is cool enough in winter to sink to the bottom of the North Atlantic Basin and even appears just above the Antarctic bottom water in the South Atlantic Ocean. There is no corresponding formation of bottom water in the North Pacific. The

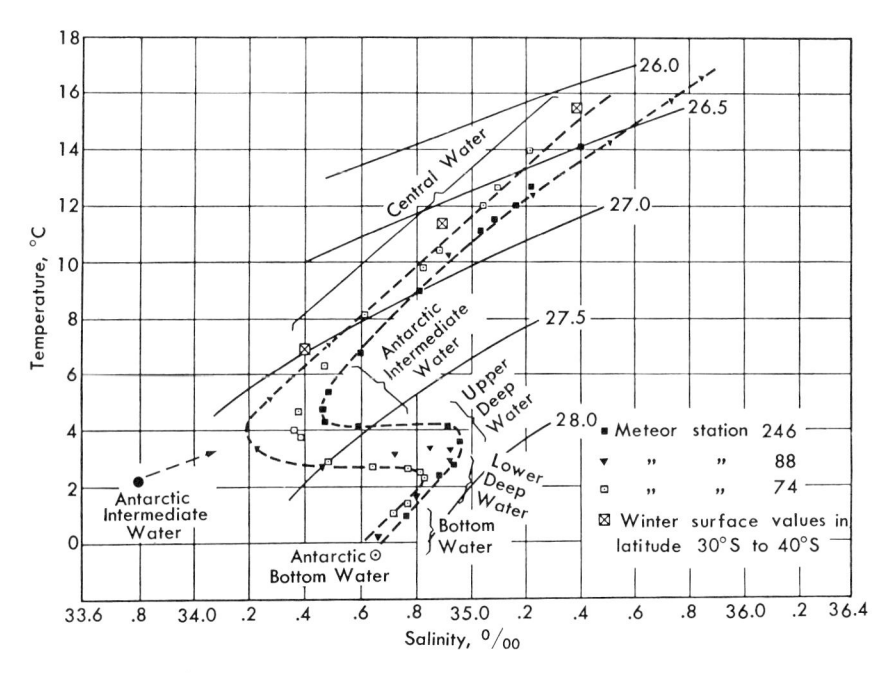

FIG. 7–7. Temperature (°C) and salinity (‰) of the water masses of the Atlantic Ocean as derived from Meteor Stations 74, 88, and 246. (from W. S. von Arx, 1962, An Introduction to Physical Oceanography. Addison-Wesley, Reading, Massachusetts. After H. U. Sverdrup, M. W. Johnson, and R. H. Fleming, 1942. *The Oceans, Their Physics, Chemistry, and General Biology*, (C); by permission of Prentice-Hall, Englewood Cliffs, New Jersey.)

deep and bottom waters of the Pacific and Indian oceans appear to flow northward from the region around Antarctica but the manner in which the circulation is closed is not known with certainty.

The hypothetical nature of these mechanisms must be emphasized strongly. Observations to confirm or refute their importance are very rare, partly because of the extreme difficulty of working in the antarctic winter. The sinking mechanism presumably operates seasonally but also might well be an intermittent and sporadic phenomenon or even one that occurs only during certain epochs of extreme climate.

An important visual aid in the type of water mass analysis leading to these hypotheses has been Helland-Hansen's *TS* plot, on which the values of temperature and salinity appropriate to a sample are plotted against one another. If the *TS* pairs obtained in a water column from surface to bottom are plotted on such a diagram together with *TS* pairs characteristic of different geographic locations and different depths, it is often possible to identify the water of the samples with water of known geographic origin. Such a diagram for data from one equatorial and two subtropical stations in the Atlantic is shown in Fig. 7-7. *TS* pairs typical of winter surface water in southern midlatitudes, Antarctic bottom water, etc., are indicated as well and their differentiation from one another is to be noted. The presence of Antarctic bottom water at the bottom of the three equatorial and tropical water columns is clearly evident, as is the mixing of the Antarctic intermediate water with adjacent waters.

North–south sections of T and S themselves over the entire length of the Atlantic appear in Fig.7-8. The processes outlined above suggest themselves with altogether typical subtlety, and the nature of the large-scale thermohaline (i.e., driven both by temperature and salinity gradients) flow implied by this model of water mass formation is, as noted above, only partly understood. For reasons related to those that cause the Gulf Stream to form on the western edge of the Atlantic rather than on the eastern edge, it has been postulated (Stommel, and Orans, 1960; Stammel, 1958) that bottom water formed at polar latitudes moves equatorward away from the formation region in narrow currents along the western edge of the ocean basins. This theory was in part prompted by Wust's (1955) careful analysis of hydrographic data in the Atlantic, which indicated that the flow of bottom water does not occur as a broad, uniform, gradual advance from the polar regions. Although not all details have been tested observationally, scattered observations (Swallow and Worthington, 1961) of a southward flow at the sea floor, not far from the northward-flowing Gulf Stream, suggest the equatorward flow of bottom water predicted in the theory. This characteristic has been realized in a similar laboratory model flow (Stommel, Arons, and Faller, 1958).

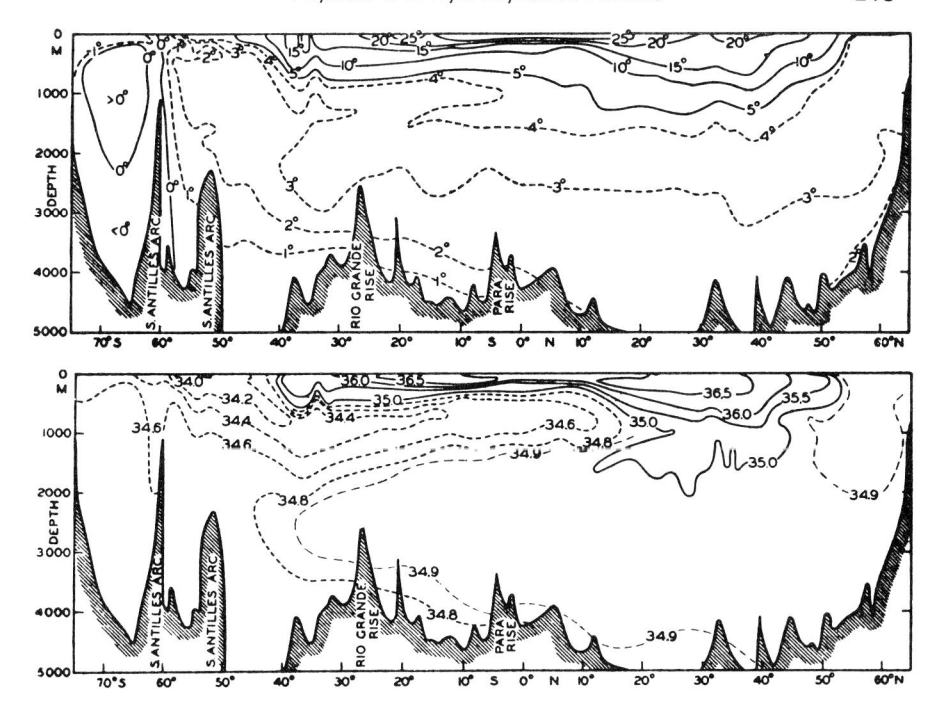

FIG. 7-8. Vertical sections showing the north-south distribution of temperature (°C) and salinity (‰) in the Atlantic (after Wust). (From H. U. Sverdrup, M. W. Johnson, and R. W. Fleming, 1942, *The Oceans, Their Physics, Chemistry and General Biology*, (C); by permission of Prentice-Hall, Englewood Cliffs, New Jersey.)

Since water sinks at high latitudes, it must rise elsewhere. The order of magnitude of this upward drift in mid and equatorial latitudes is about 1 cm/day and the resulting slow upward transport of water properties evidently balances the tendency of surface-produced maxima to diffuse downward. This makes stationary the thermocline that separates the warm surface water from the cold water below.

7.2 THE WIND–DRIVEN CIRCULATION

The major global systems of surface current are evidently driven by wind stresses transmitted across the air–sea interface, for the permanent ocean-wide gyres comprising these systems virtually coincide with the permanent zonal wind systems in the atmosphere.

The yearly averaged wind distribution over the oceans consists of a broad system of symmetric gyres, each one centered about one or two semipermanent high-pressure centers and each one somewhat flattened in the north–south direction. Over the northern and mid Pacific, for example, the westerly "roaring forties" appear as the upper boundary of a strong anticyclonic (clockwise in the northern hemisphere when viewed from above) gyre centered at roughly 30°N and stretching from California to Japan. The southern boundary of this gyre is the easterly system of trades. Similarly, most of the North Atlantic is covered by a single high-pressure cell about which the winds blow in a symmetrical, clockwise gyre with the easterly trades again forming the lower boundary.

The pattern of currents which this wind system would produce in an idealized rectangular ocean has been calculated by Munk (1950) and his result, outlined in Fig. 7-9, exhibits many elements on the observed circulation in the major ocean basins of the earth. The strong current at the western boundary corresponds to the Gulf Stream in the North Atlantic, the Kuroshio in the North Pacific, the Brazil current in the South Atlantic, the Mozambique and Agulhas currents in the Indian Ocean, and the East Australia current in the South Pacific. In the Southern Hemisphere, the current driven by the west winds of the "roaring forties" flows unobstructed

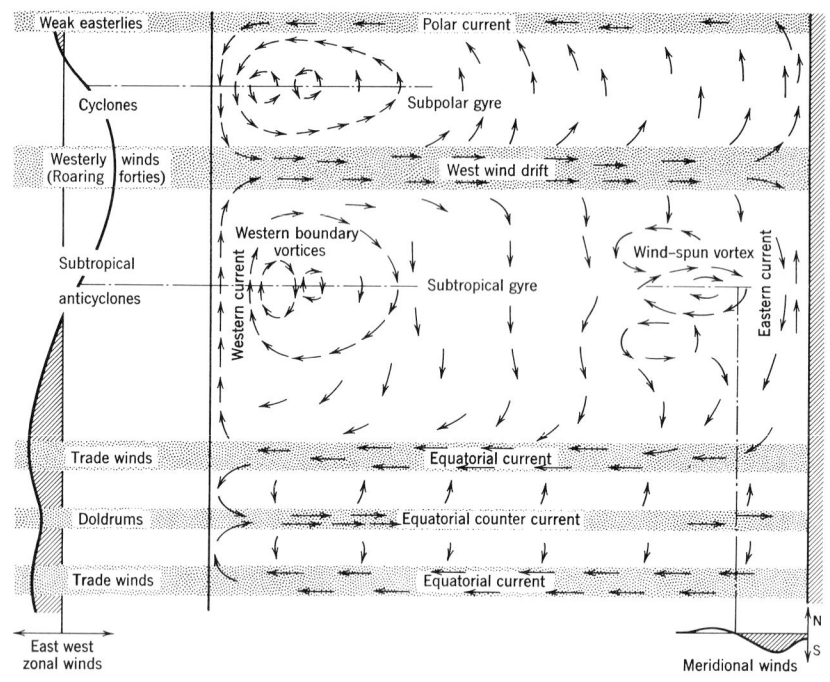

FIG. 7-9. Schematic presentation of circulation in a rectangular ocean resulting from zonal winds, meridional winds, or both. The nomenclature applies to either hemisphere, but in the Southern Hemisphere the subpolar gyre is replaced largely by the Antarctic Circumpolar Current (west-wind drift) flowing around the world [reprinted from W. H. Munk, 1950, *J. Meteor* 7, 89 (1950)].

by continental barriers and corresponds to the Antarctic circumpolar current.

This idealized circulation, driven by unvarying winds, corresponds to the persistent system of broad surface and near-surface currents in the ocean. Superimposed on this oceanic "climate," as Munk (1955) characterizes it, is a variable "weather" that changes drastically from week to week. Our understanding of this weather is about as well developed as is our ability to forcast rain next week, but the main features of the average circulation may be partly explained in terms of the driving wind and the basic physical processes possible in a thin, stratified ocean on a rotating spherical planet. We have already pointed out some of the consequences of the stratification and will now turn to a description of the remaining processes which comprise the dynamics of the mean flow.

If each high-pressure cell in the atmosphere is thought of as a hill with greatest pressure at the top, then the wind motion due to the pressure gradient is around the hill rather than down hill. Although the pressure forces on the air are outward, away from the center of high pressure, the resuting fluid motion is not outward, but rather occurs along lines of constant pressure. This "around the hill" type of flow is a result of the rotation of the earth and the consequent Coriolis force exerted on every object that moves over the surface of the earth. An explanation of the physics of this mechanism is beyond the descriptive scope of this article, but *geostrophic* motion of exactly this sort is observed to be an important element of the circulation in the atmosphere and in the oceans. A descriptive treatment of the Coriolis effect has been given by McDonald (1952). More detailed discussions may be found in such texts as that of Neuman and Pierson (1966).

It should be emphasized that this geostrophic relationship between pressure and flow is not one of cause and effect but is rather a dynamical constraint imposed by the rotation of the earth and the length and velocity scales of mean oceanwide flows. It does not constitute a solution to the problem of understanding the entire flow but rather allows us to calculate the velocity field once the pressure field has been determined.

In the ocean the pressure is nearly hydrostatic, that is, at any depth, the total pressure is nearly atmospheric pressure plus the weight of the overlying water column (per unit area). Horizontal pressure gradients, which give rise to horizontal movements of water parcels, consequently are produced only if there are horizontal variations in the weight of the water, that is, in the density field. The flows resulting from these pressure gradients will not be from regions of high pressure towards regions of low pressure, but rather will be along lines of constant pressure—geostrophic flows.

Now it is observed to be true that the major systems of current which are considered wind-driven ones may be calculated with good accuracy by measuring the density field (measuring temperature and salinity and calculating σ_{STP}), then using the hydrostatic relationship to calculate the horizontal pressure gradient, and finally calculating the resultant geostrophic flow. We must therefore conclude that the ocean responds to the average wind stress both by moving and by simultaneously adjusting its internal distribution of heat and dissolved salts in a way consistent with the geostrophic constraint. That this actually occurs in nature was first conclusively demonstrated by Sverdrup (1947) who showed that in the equatorial current system of the Pacific, pressure gradients calculated on the basis of the wind field do in fact nearly equal those calculated independently on the basis of the observed fields of temperature and salinity.

The situation must therefore be rather more complex than an intuitive picture in which surface winds are imagined to simply drag the water along behind them. Nonetheless, the final result is that each wind gyre does drive a geometrically similar gyre in the ocean directly below. These wind-driven flows do not extend from sea surface to sea floor undiminished in strength but rather are largely confined to the water in and above the thermocline. The strongly stable density stratification of the thermocline tends to inhibit the vertical motions that produce the wind-induced density adjustment accompanying wind-produced geostrophic currents. These currents consequently decay rapidly (although not completely) through the thermocline.

In laboratory experiments and in small lakes or inland seas, the effect of a steady wind is just to pile up water against the downwind side of the basin. This effect in the Baltic was in fact used to form the first estimates of wind stress on the ocean. However, details of the stress transfer process are not sufficiently well understood to allow extrapolation of the results to the open sea, where wind-generated wave systems absorb energy over thousands of miles before being deflected or destroyed at a coastline. An alternative procedure is to measure the wind profile over water and calculate the stress from the wind shear, but again no measurements truly representative of midocean regions have been obtained. In another method use is made of meteorological data only to calculate the stress by measuring the deviation of the wind from the geostrophic prediction. In spite of much effort in these and other directions the details of the relation between wind stress and wind speed have not yet been deduced and the basic physics of the air–sea interaction are still a subject of intensive inquiry.

An often-used approximation is: Wind Stress = C_d (air density) (wind speed)². Because the roughness of the sea surface changes as the wind changes, the drag coefficient C_d is not a constant but varies with the wind speed. When the wind speed is taken to be that measured at annemometer height (6 to 8 m) above the sea surface, C_d is the order of 0.003 and the values of stress which result from typical zonal winds are the order of 0.5 dyn/cm². But measured values of C_d vary from 0.01 to less than 0.001 and their variation with wind speed is not really known with any precision. A summary of various measurements may be found in Neumann and Pierson (1966).

If the earth did not rotate, this stress acting at the surface would simply drag the surface layer of water along in the wind direction, that surface layer would similarly drag along water immediately below, etc., so that the current at all depths would be in the direction of the wind although the speed of each layer would be smaller than that of all layers above it.

But on the rotating earth the surface layer is subject not only to the force of the wind but also to the Coriolis force, which tends to deflect the moving water to the right in the Northern hemisphere (to the left in the Southern). In theory, in the Northern hemisphere the result is that the surface layer moves 45° to the right of the wind. Below, under the combined influence of shear stress and the Coriolis force, each layer of water moves to the right of, but at a slightly slower speed than, the one above it. In an ocean of great depth, there is theoretically a level at which the motion is opposed to the wind, although the speed at this depth is very small. The direct frictional effect of the wind is thus confined to a thin near-surface layer of water, the frictional or Ekman wind-drift layer. The thickness of this layer has never been measured in the open sea but it is probably the order of 100 m or smaller. Below it, in theory, the wind stress is virtually zero. The net transport of water in the layer is 90° to the right (left) of the wind in the Northern (Southern) hemisphere—a relation between driving force and resulting motion reminiscent of the geostrophic relation.

The possible occurrence of such a flow in the ocean was first suggested to Ekman (1905), who developed the theory in detail, by Nansen's observation that sea ice in arctic latitudes consistently drifts to the right of the surface wind. Similar observations are still the best evidence for this flow in the ocean although recent observations (Katz, Gerard, and Costin, 1965) of the spreading of dye patches by near surface currents do suggest the theoretical flow pattern. The rotary structure at depth has been observed only where the sea surface is covered with a layer of drifting ice (Hunkin, 1966). But the Ekman layer has often been observed in laboratory experiments and is known to be an important mechanism in the flow of a rotating fluid.

Under a clockwise wind gyre in the northern hemisphere, water transported within the Ekman layer thus tends to accumulate at the center of the gyre. This results in an elevation of sea surface and a pressure maximum which is felt at depths far below those at which the wind stress has become negligible, that is, far below the Ekman lyer. Therefore water below the Ekman layer is not at rest but must move in response to the pressure maximum produced by the wind-driven surface accumulation of water. It does not move downhill, (outward from the region of high pressure) but, as in the atmosphere, moves around the hill (around the region of high pressure) in obedience to the geostrophic constraint. The motion of this deeper water is therefore a gyre around the high-pressure center, geometrically similar to the wind gyre above the surface. It extends from the bottom of the Ekman layer to the thermocline.

Associated with the convergence of surface water towards the center of the gyre is a vertical sinking of warm, relatively light surface water. This depresses the thermocline* (its rise towards both north and south from a depth of around 1000 m at the latitude of the convergence is clearly visible on the temperature section of Fig. 7-8) so that, below the thermocline, the pressure maximum produced by the elevation of sea level at the center of the gyre is partly canceled by the tendency toward a pressure minimum due to the shoaling off to the north and the south of the overlying layer of light water. The geostrophic flow consequently is much slower below the thermocline than above, although not necessarily vanishingly small.

In very schematic summary, the Northern-hemisphere circulation below a clockwise wind gyre consists of a surface convergence of water in the Ekman layer with a consequent increase in sea level at the center of the gyre. This generates a pressure maximum about which the water between the Ekman layer and the thermocline rotates in a geostrophic gyre parallel to the wind gyre. But the accumulation of surface water also results in a depression of the thermocline which greatly reduces the pressure maximum and corresponding geostrophic flow in the water beneath the thermocline. The net transport of water is in a gyre in the direction of the wind gyre but the flow is largely limited to water above the thermocline. This is the manner in which each of the gyres of Munk's ocean are driven. We see that to simply imagine the wind as dragging the water along is not a complete picture of the flow, even though it correctly predicts oceanic gyres coincident with wind gyres.

In nature, however, the centers of the oceanic gyres are always displaced to the west of the centers of the atmospheric ones so that currents along the western boundary are swift and narrow while those along the eastern boundary are slow and diffuse. Thus the Gulf Stream flows along the western boundary of the Atlantic, the Japanese Kuroshio along the western boundary of the Northern Pacific, and these swift currents have no similar counterparts on the eastern boundaries.

Stommel (1948, 1958) showed that the explanation of this asymmetric response is to be found in the tendency of fluid columns to conserve their

---

*A word of caution must be added concerning this interpretation of the isotherm deepening. Robinson and Stommel (1959) were able to produce this feature at mid-latitudes in a theoretical flow driven only by thermal gradients. The wind stress was identically zero.

angular momentum as they change latitude. A column of fluid initially at rest at the North Pole and brought from there to the equator would retain the rotation with the earth that it possessed at the Pole and so would appear to an equatorial observer to be spinning counterclockwise. A similar column initially at rest at the equator and brought from there to the North Pole would appear to a polar observer to be spinning clockwise.* We therefore expect water which flows from north to south in the eastern half of a gyre to accumulate counterclockwise spin whereas water flowing north in the western half gyre will accumulate clockwise spin. In the eastern half gyre the wind tends to spin the fluid columns clockwise so that the two tendencies nearly cancel whereas in the western half gyre, the fluid is spun clockwise both by the wind and by the accumulated spin gathered as it travels northward. These tendencies to speed up or slow down the spin of the fluid column are countered by frictional drag against adjacent fluid columns. Fluid friction may be visualized as the resistance of one layer of fluid to sliding along an adjacent layer. If the amount of sliding is small, as in a nearly uniform flow, the force of friction is correspondingly small, but if the sliding is extreme, as in a flow with regions of swift and slow motion close to one another, the force of friction is large. In the eastern half gyre, where the two tendencies to spin nearly cancel, the countering effect of friction need be but slight, and the flow may be broad and gentle. But in the western half gyre friction must counter the two tendencies as they are reinforcing one another, and, if this is to occur, the flow must be swift and narrow. So it is to be expected that the flow in the western half gyre will be more intense than that in the eastern half and this is what is observed. Each gyral flow in Munk's ideal ocean (Fig. 7-9) is therefore not symmetrical in the east–west direction but rather is most intense at the western edge.

Stommel himself was among the first to observe that, because the fluid itself carries spin (vorticity) from one place to another, the local balance of friction against the two spinning tendencies of wind and earth's rotation is incomplete so that the foregoing explanation is only a part, albeit an important one, of the entire story. Further development of this idea is summarized in his book (Stommel, 1965) and references may be found there. Significant work appearing subsequently includes Holland's (1966) numerical investigation of the effect of bottom topography in shaping the flow, and the model by Robinson (1965), which emphasized the important role played by the vertical structure.

The major western boundary currents such as the Gulf Stream and the

---

*In these statements we implicitly suppose that fluid particles at every level in the water column are constrained to move parallel to the earth's surface no matter where the column may be carried. In the ocean this important constraint is imposed by the stable density stratification.

Kuroshio are thus embedded in an oceanwide system of more or less closed gyres. They have, however, historically been studied as entities in themselves and it is appropriate to summarize some of the characteristics of the most intensively studied example, the Gulf Stream.

A major portion of the Gulf Stream is already present as a strong current (the Florida Current) in the Florida Straits. This flow, amounting to some 26×10^6 m³/sec, is fed by north equatorial surface water which has been driven into the Caribbean under the influence of the Trades. Along the east coast of Florida, this flow is joined by water which has travelled along the eastern side of the Antilles arc, and the augmented stream proceeds northward towards Cape Hatteras, growing to a total transport of order 70 to 90×10^6 m³/sec.* Along this portion of its path, the stream remains close to the 800-m isobath of the Blake Plateau but near Cape Hatteras, the lines of constant depth converge from the south and the stream is forced into rapidly deepening water. After this traumatic ejection the stream proceeds in an irregular and meandering path whose mean direction tends to be slightly east of the roughly diagonal lines of constant depth northeast of Cape Hatteras so that it gradually enters deeper and deeper water. The stream becomes more diffuse as it flows towards the mid-Atlantic ridge and the currents observed there are relatively broad and gentle.

A line between Chesapeake Bay and Bermuda intercepts the Gulf Stream shortly after it has passed Cape Hatteras. On such a traverse, the flow at the surface has the appearance of a jet of width 50 km or somewhat greater (between 10 cm/sec lines of constant speed on either side) with maximum speeds of up to 220 cm/sec. As one enters the stream from the shore side, the speed rises rapidly over 20 to 30 km to its maximum value and then decreases more gradually in the seaward direction. The profile is a skewed one with the maximum speed appreciably to the left of center. The change in speed is often so abrupt that it is visible from shipboard as a change in sea state.

---

*These and similar estimates of the transports of all the major currents rest on velocity profiles computed using the geostrophic assumption. But the velocities thus computed are not unambiguous. Rather, the geostrophic computation gives the rate of change of velocity with depth so that a measured or guessed velocity at one level is additionally necessary to reconstruct the actual velocity field. The transports derived in this way are very sensitive to the value of this constant of integration, which is usually given by specifying a depth at which the current is taken to be zero (level of no motion). Thus Fuglister's (1963) direct measurements of the deep flow beyond Cape Hatteras lead him to calculate a transport of nearly 150×10^6 m³/sec there, a value close to double the one formerly believed and close to double the one quoted above for the transport south of the Cape. The situation is further complicated by the possible presence of deep countercurrents in the vicinity of the surface stream and an entirely consistent description of the transport plan is not yet at hand. Concerning his large estimate, Fuglister observes that such a great transport would imply heretofore unanticipated movements of considerable magnitude elsewhere.

The velocity is not uniform with depth but decreases rather regularly from its maximum value at or very near the surface to values between 1 and 10 cm/sec at depths between 1500 and 2000 m. Direct measurements of the deep flow are rare and the profile is no doubt quite variable from time to time and location to location. The current does, however, seem to extend to the bottom when it is flowing over the Blake Plateau, and recent observations of Knauss (1965) support previous conjectures and indirect indications (Warren, 1963) that the bottom flow beneath the Gulf Stream is parallel to the surface flow in the region east of Cape Hatteras. This is in accord with the deep flows in this region reported by Fuglister (1963).

This simple picture is complicated by the system of deep flows which apparently combine to form a southerly countercurrent off the Blake Plateau. The existence of such a deep countercurrent appears certain near Cape Hatteras (Swallow and Worthington, 1961) but elsewhere its direction and location relative to the Gulf Stream have not been well determined.

Associated with the rapid downstream flow of the Gulf Stream are the large cross-stream pressure gradients demanded by geostrophy. The density gradients that produce these are largest in the high-speed core of the current and are most evident in the rapid rise of the thermocline from the Sargasso sea landward through the stream. This rise is clearly visible on both the temperature and salinity sections of Fig. 7-10. The Gulf Stream effectively divides the warm rather saline surface water of the Sargasso sea from the colder and fresher "slope" water on the shoreward side of the stream. Where the upward-sloping isotherms intercept the sea surface, rapid variations of surface temperature and salinity are observed in the cross-stream direction. Their magnitudes may be estimated from the figure.

Slightly seaward of the current maximum, the surface temperature tends to reach a weak maximum and then decay to the surface value appropriate to the Sargasso seawater. Associated with this tendency is the surface counter flow (opposite to the main flow) indicated on the velocity section of Fig. 7.10.

On many detailed sections across the stream, although not on the one of Fig. 7-10, thin (5 miles) filaments of cold and relatively fresh water, possibly due to runoff from the land, have been observed just landward of the current maximum. They have been traced to a depth of 400 ft and they were found on more than half of the June 1950 crossings of the stream over appreciable fractions of a 1200-mile length of stream east of Cape Hatteras (Ford, Longard, and Banks, 1952).

The Gulf Stream is nowhere an entirely steady current and variations in both its total transport and geographic position are known to occur. The relationship between these changes and possible changes in the Atlantic-wide circulation has not been established. The gyre of which the Gulf Stream

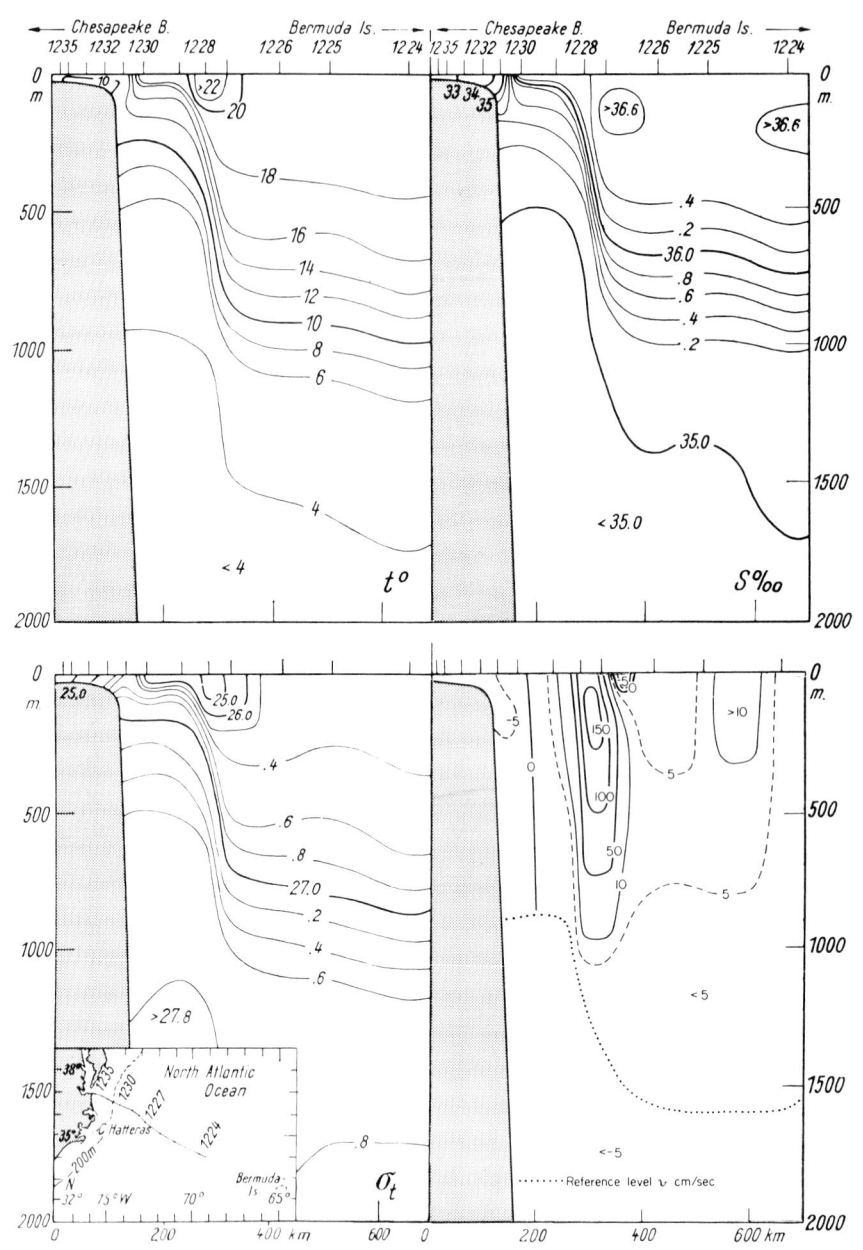

FIG. 7-10. Distribution of temperature (°C), salinity (‰) σ_T and of the geostrophic current at right angles to the section (no sign, NE current direction of stream; minus sign, SW current, counterflow). Location of stations noted on inset (from G. Dieterich and K. Kalle, 1963, *General Oceanography*, Interscience, New York).

is a part cannot be expected to follow daily or weekly changes of the driving wind because a change in the flow pattern would involve a simultaneous adjustment of the density distribution and the wind does not supply sufficient energy to do this rapidly. A comparison between the rate at which the wind supplies kinetic energy and the total potential energy stored in the density distribution shows that the stored potential energy would be sufficient to maintain the current system for 1700 days in the absence of wind (Stommel, 1965). But the theoretical possibility of long-period (weeks, months, years) wavelike responses of the ocean considerably complicates this simple estimate of response time to a variable wind stress, and a study of Veronis and Stommel (1956) suggests that some readjustment of the main thermocline depth may occur over intervals as short as a week.

Wertheim (1954) has made use of the submarine cable between Key West and Havana to measure the potential difference developed between these points as the Florida Current flows through the magnetic field of the earth. From these measurements the transport itself could be estimated and it showed irregularly occurring variations having magnitude comparable with the entire transport and taking place within a time as short as a month. Stommel (1965) analyzed meterological data related to the strength of the trade winds during the period of measurement and found that there were three sustained breakdowns of the Bermuda–Azores high during this period, each breakdown preceding a period of high flow of the Florida Current by about a month (but he cautions that exactly the opposite reaction has also been observed).

Fuglister's (1951) summary of surface currents in various areas of the North Atlantic shows pronounced seasonal variations of the surface velocity of the stream with the maximum occuring in early or midsummer, and Iselin (1940) using hydrographic data taken northeast of Cape Hatteras, found possible seasonal variations of the transport of the stream.

In addition to changes in the transport of the stream, which might be imagined to occur even if the path of the stream remained fixed, the stream itself moves bodily from one path to another. Off the east coast of the United States, south of Cape Hatteras, such variations do not greatly exceed the width of the stream itself, but after the stream has left the coast at Cape Hatteras, it develops wavelike meanders of path that are typically of length 100 to 400 km and in which the maximum displacement of the stream may be as great as 500 km. These meanders form, disappear, and re-form over periods of several weeks and a particularly sharp one may even cause an eddy to detach itself from the main body of the stream.

There is some evidence that the instantaneous curvature of the stream is determined primarily by the slope of the sea floor below so that the stream east of Cape Hatteras chooses a path consistent with its strength, the direction in which it left the coast, and the topography over which it is subsequently directed. In this view (Warren, 1963) small variations in the stream south of Cape Hatteras could lead to the marked changes in the path of the stream which constitute the meander pattern downstream of the Cape.

Although we recognize that the actual flow is not steady, the intuitive approach has been to believe that the flow averaged over, for example, a year would resemble the ideal flow described above. In so doing, it is implicitly supposed that the fluctuations of the flow are largely passive in their effect, that is, that their size, frequency, etc., are determined by the main flow and that their effect is only to enhance the processes of molecular diffusion by stirring. The mathematical expression of this assumption is to write fluid dynamic equations for the average flow itself but with coefficients of viscosity and diffusion that in principle depend upon the amplitude of the fluctuations. The solutions of these equations are the flows described above with their predictions of very slow wind-driven flow and even slower thermally driven flow in midocean water beneath the thermocline.

In order to examine these supposed flows of deep water, Swallow (1955) has devised a submergable float, which may be ballasted to sink to any desired depth where it will remain and drift with the current there. The float emits pulses of sound and may be tracked relative to a surface ship equipped with directional hydrophones. Measurements at several thousand meters in the Atlantic with these floats (Crease, 1962) do not confirm the existence of only slow drift currents but rather reveal the presence of rapid and rapidly fluctuating currents far below the wind-driven surface gyres. Amplitudes of 10 or 20 cm/sec changing by a similar amount during a week are evidently not uncommon. Needless to say, the mean flow described above is not readily detectable in the midst of such large fluctuations. Because these fluctuations are of such great amplitude and of such long duration, it is difficult to imagine that their role in the circulation is limited to the enhancement of diffusive processes. Consequently the relevance of the ideal flows envisaged above to the mean flow in the deep sea is not altogether clear. The work of Rossby (1947) and Longuet-Higgins (1964, 1965) indicates that the ocean is able to store large amounts of energy in planetary or Rossby wave fields whose characteristic periods of oscillation are always longer than several days and Longuet-Higgins suggests that the observed deep-water fluctuations are motions of this type. It is even possible that the fluctuations act as a source, rather than a sink, of energy for the mean flow (Webster, 1961, 1965).

In the foregoing discussion we have considered the manner in which the wind causes the water to flow in the open sea. We noted that there the transport of water within the Ekman layer is 90° to the right (left) of the wind direction in the Northern (Southern) Hemisphere. But if there is a coast nearby that intercepts this flow, water will tend to accumulate at the coast. This results in a pressure greater at the coast than seaward and the effect is twofold. First of all, by the geostrophic relationship, the seaward pressure gradient will drive a flow parallel to the coast in the direction of the wind and secondly, a sinking of surface water will take place near the coast. Depending upon the hemisphere, north or south, and upon the relative direction of wind and coastline, coastal water may tend to sink or rise. The motion does not usually extend to great depth on account of the stable stratification, but is restricted to the upper several hundred meters. The two cases of sinking and upwelling, both of which occur in the ocean, are illustrated schematically for the Northern Hemisphere in Fig. 7-11.

Upwelling of this kind is characteristic of the west coasts of continents where winds blow in a direction falling between seaward and equatorward along the coast. Direct measurements of the flow in regions of upwelling have not clearly revealed the circulation illustrated in the figure and upwelling is indicated primarily by surface-water properties (temperature, salinity, oxygen, etc.) characteristic of deeper waters offshore and by weakening of the thermocline near the coast. It is often accompanied by an unusual abundance of marine life due to the high nutrient content of the rising deeper water. When the wind pattern is interrupted, as off the coast of Peru during certain years, commercial fishing may be seriously disrupted.

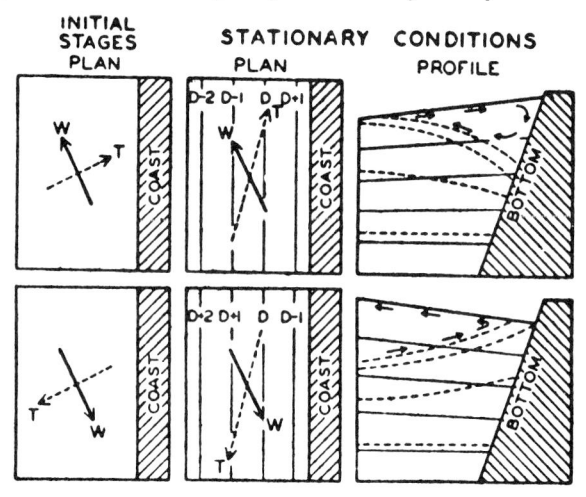

FIG. 7-11. Schematic representation of effect of wind toward producing currents parallel to a coast in the Northern Hemisphere and vertical circulation. W shows wind direction and T, direction of transport. Contours of sea surface shown by lines marked D, D + 1.... Top figures show sinking near the coast; bottom figures show upwelling (from H. U. Sverdrup, M. W. Johnson, and R. H. Fleming, 1942, *The Oceans, Their Physics, Chemistry, and General Biology*, (C) by permission of Prentice-Hall, Englewood Cliffs, New Jersey.)

We conclude this discussion of the mean circulation with a mention of a striking new discovery which must form an important part of any scheme of the circulation—the equatorial Cromwell Current. This is a narrow (500 km), thin (several hundred meters), equatorially centered ribbon of water which flows eastward, opposite to the westward surface drift, at about 100 m below the surface in both the Pacific and the Atlantic. Knauss (1960) estimates its transport in the Pacific to be about 40×10^6 m^3/sec; it is thus comparable with the Gulf Stream in magnitude. Its unanticipated discovery, coming in the midst of intense theoretical work on the general circulation, emphasizes the vast oversimplification of model flows such as the ones we have been considering.

7.3 WAVES; WIND WAVES; TSUNAMI; STORM SURGES; TIDES; INTERNAL WAVES; INERTIAL MOTIONS

We now turn to a description of the various motions in the sea commonly labeled waves. In steady flows such as the ones we have been considering, the flow of energy from one place in the ocean to another is accompanied by a flow of matter. Wave motion, on the other hand, is characterized by the transport of energy without a corresponding transport of matter. The steady flow emerging from a fire hose clearly carries energy, for it is capable of moving objects when it strikes them. However, an object floating at one end of a swimming pool will be moved by the disturbance created as a diver jumps into the water on the opposite end of the pool even though there is no actual flow of water from the diver to the object. The energy necessary to move the object reaches it without an accompanying steady flow of water. In a similar fashion, energy put into wave motion in the Gulf of Alaska may, several days later, do the work of demolishing a pier in Southern California even though no water has been transported from Alaska to California.

Wave motion is possible in a medium whenever the medium responds to any displacement with a restoring force that tends to cancel the displacement. The most obvious forms of wave motion in the sea are possible because the free surface of the water tends to oppose being hollowed out or raised up in a heap. The restoring forces are either gravitational or are due to the effect of surface tension. The nature of the motion depends upon the force that tends to disturb the free surface and upon which of the two restoring forces acts in opposition.

In idealization a surface wave has the form of an infinite set of alternating equally spaced crests and troughs progressing at a calculable speed (the phase speed). The distance from crest to crest (or trough to trough) is the wavelength λ (see Fig. 7-12). To an observer sitting still and watching the wave go by (perhaps from a ship), the local sea surface appears to rise and fall as successive crests and troughs pass by. The interval between successive

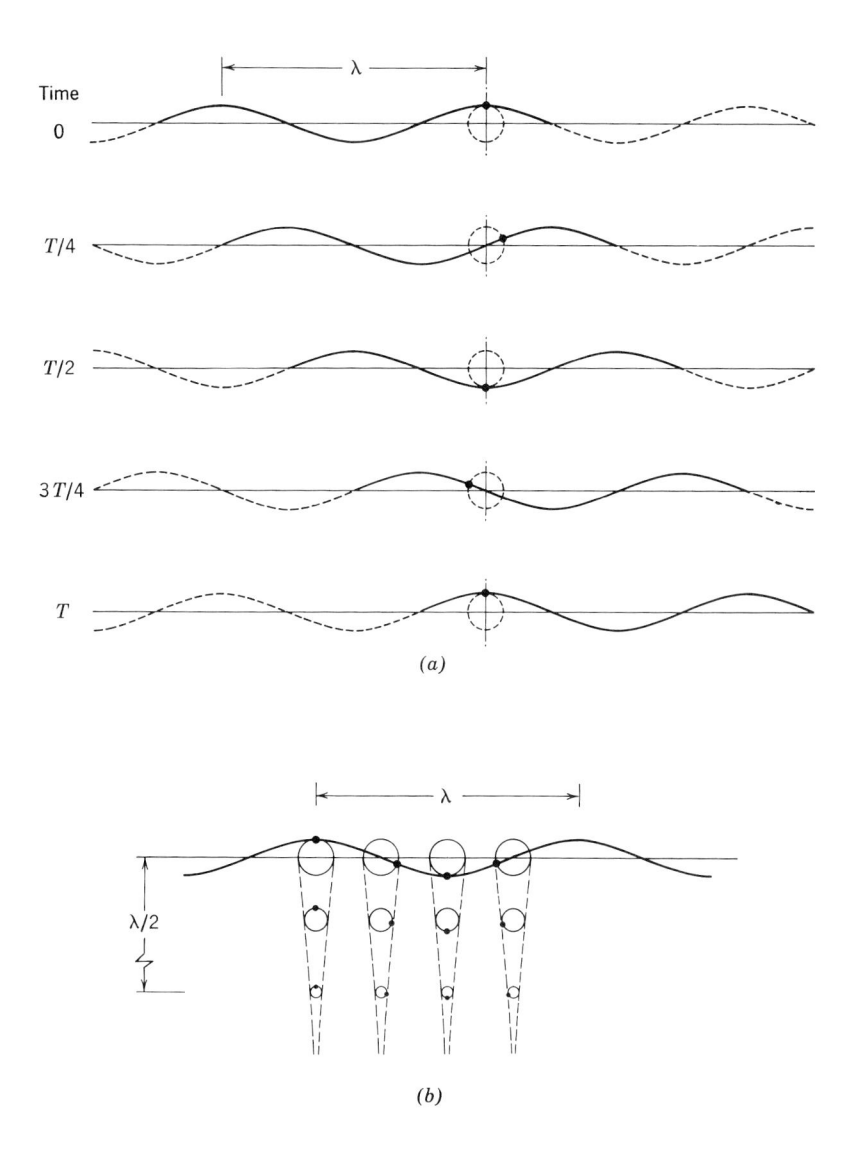

FIG. 7-12. (a) Instantaneous sea-surface profiles as a wave passes by. From the first sketch to the last, one wave period T has elapsed. Notice how the water particle marked by the heavy dot is carried forward, downward, backward, and upward so that its path during one wave period is the small circle in each sketch. (b) Cross section of a wave showing circular orbits or water particles at depth. Their diameter at a depth of half a wavelength ($\lambda/2$) is 1/23 of their surface diameter. Each water particle (heavy dots) traces out its circular path once every wave period.

227

times of high water is called the period T of the wave. Since a crest arrives every T seconds and since crests are spaced a distance λ from one another the crests travel with a phase speed c given by $c = \lambda/T$. The frequency $f = 1/T$ (in cycles per second) is also frequently employed.

Associated with this corrugated pattern which moves along the surface is a flow of the water beneath (see Fig. 7-12). As the crest approaches, water particles are carried up and forward. During the approach of the following trough, they are moved downward and backward. As successive crests and troughs pass by, the water particles are carried through small, nearly circular paths. But they do not continue to move steadily along with an individual crest or trough, so that the total amount of water moved forward in the direction in which the wave is going is almost zero.

Beneath the surface of water of great depth, the motion decreases rapidly. At a depth equal to half of the wavelength, the motion is only 1/23 of its value at the surface and at a depth of one wavelength, it is 1/530 of its surface magnitude. We may therefore expect significant differences in the properties of the wave depending on whether the water is much deeper than a wavelength or much shallower.

The chop, swell, and ripples which are the most obvious waves on the sea surface as viewed from a ship correspond to the first case. Their wavelengths are all much shorter than the ocean is deep and their motion therefore decays rapidly below the sea surface. Thus a submarine several hundred feet below the surface does not feel any but the longest of these waves and rides smoothly even in rough weather.

The force that generates these waves is, by ancient observation, the wind stress. However, a more detailed understanding of the relation between the wave field and the wind than this simple one of cause and effect is only now being developed.

The shortest of these waves are the ripples or capillary waves which are controlled mainly by the surface tension of the water (Cox, 1962). Their wavelengths are usually the order of millimeters and they have slow phase speeds, typically 30 cm/sec. They therefore appear as a fine corrugation on the slopes of longer waves, moving rather slowly. When the wind begins to blow over an entirely smooth sea, these capillary waves are the first to appear and it has been suggested (Van Dorn, 1953) that even when larger waves are present, the capillary wave roughness of the sea surface is the essential element in transferring energy from wind to water. The presence of a slick of oil or organic matter on the surface inhibits the formation of ripples very effectively (Cox, 1962) and causes those generated elsewhere to damp rapidly as they propogate into the slick. The visibility of slicks is mainly due to the absence of ripples.

The chop and swell, which cause the rocking of a ship at sea, exist by virtue of the gravitational stability of the undisturbed sea surface. The wavelengths of these waves range from capillary wavelengths to several hundred meters. Representative periods, wavelengths and phase speeds for such waves appear in Table 7-1.

TABLE 7-1

Period T (sec)	Wavelength (m)	Phase Speed c (m/sec)	Remarks
15	350	23.4	very long swell
10	156	15.6	
7	76	10.9	} short swell, "sea"
3	14	4.7	} and chop
0.074	0.017	0.23	slowest capillary wave

The phase speed of gravity waves increases with wavelength according to $c = \sqrt{g\lambda/2\pi} = gT/2\pi$ (where $g = 9.8$ m/sec² is the acceleration of gravity).

Heights and periods of chop and swell to be expected at sea depend in a dramatic manner upon the local wind, how long it has been blowing and the distance over open water over which the wind system extends. On the second or third of several windless days, the surface of the sea may be literally mirror smooth with only very long swell visible. Any slight breeze will immediately produce ripples that come and go with the wind. As the wind rises to about 3 knots (one knot = one nautical mile per hour = 0.51 m/sec), small wavelets about a foot high may appear. At 10 knots the wavelets are appreciably higher—several feet—and they occasionally break to form minute whitecaps. If the wind increases further, larger and longer waves appear, and the number of whitecaps grows. At wind speeds of 15 knots waves 10 ft high with plentious whitecaps are not uncommon and people at sea begin to complain of rough weather. Much higher wind speeds are not uncommon in stormy areas and then waves of 30 to 50-ft height may be expected. Occasionally, even higher waves are reported, the record being perhaps the 112-ft crest reported by the *U. S. S. Ramapo* (Lt. Comdr. R. P. Whitemarsh, reported in Bascom, 1959).

The shape of idealized surface waves of small height is sinusoidal. Crests and troughs are equidistant from one another and the wave viewed in profile is symmetrical about the mean sea level. With increasing wave height, this symmetry is lost. The troughs broaden and the crests become steeper, tending towards a very sharp tip. In a moderate sea, this lack of symmetry is visible and when it becomes sufficiently great, the crests of the waves erupt into a head of foam which is then blown down the slope of the crest by the wind. This production of whitecaps rapidly drains energy from the wave into turbulence of very small scale and it may be one of the principal ways in which wind energy is transmitted into stirring motions and steady flows of the water.

It is evident that the infinite sequence of alternating crests and troughs of the ideal surface wave is barely hinted at in a picture of the sea surface. Nonetheless, it is useful to consider the actual irregular shape of the surface at any instant to be a sum of many such ideal waves of all different lengths and heights and traveling in all different directions. This spectral characterization of the sea surface is now in common use, for it emphasizes just those statistical properties that distinguish one irregular and chaotic sea state from another. In a similar, but less complicated fashion, the irregular rise and fall of the surface at any one point may be resolved into a sum of periodic variations of various amplitudes and periods. When records taken at sea are thus analyzed, the results are as shown in Fig. 7-13.

This is a plot of the energy associated with waves of a particular frequency versus wave frequency. Each curve is an average resulting from analysis of many records at the indicated wind speed.

Clearly, the total wave energy (the area under each curve) increases rapidly with increasing wind speed. At each wind speed, the wave field is richest in waves having a frequency equal to the frequency at which the spectrum is sharply peaked. There are many waves of high frequency, but very few with frequency less than the frequency of the peak. From the various curves we may roughly estimate that

$$\text{wind speed} = \frac{1.35 \text{ m/sec}^2}{(\text{peak frequency})},$$

whereas for an individual wave we have, from the theory of surface waves,

$$\text{wind speed} = \frac{1.56 \text{ m/sec}^2}{(\text{wave frequency})}.$$

It is thus evident that waves at the peak frequency travel only slightly faster than the speed of the wind, that is, wave energy appears almost entirely as waves traveling at the speed of the wind or slower.

When a concentrated disturbance of the sea surface is produced, as in a strong local storm, it contains waves of all different periods (and hence of all different wavelengths) traveling outwards in all directions. As the disturbance progresses into previously undisturbed water, it sets this water in motion so that energy is propagated from the region initially disturbed into other regions.

When such a localized disturbance advances, it is observed not to move with the speed of the crests of the various waves of which it is composed. Rather, crests arise at the rear of the disturbance and move through it

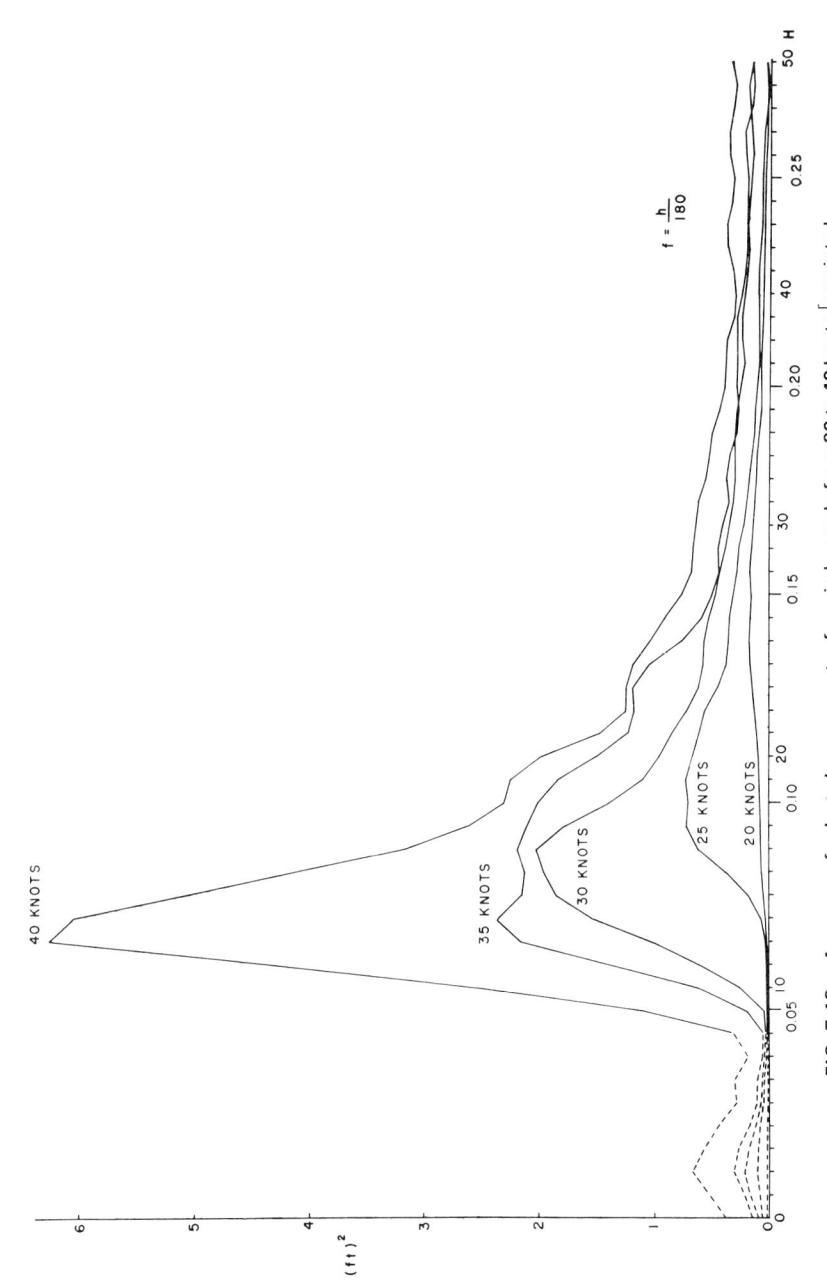

FIG. 7-13. Averages of selected wave spectra for wind speeds from 20 to 40 knots [reprinted from L. Moskowitz, 1964, *J. Geophys. Res.*, **69**, 5162 (1964)].

towards the front, where they decay and vanish. Thus the energy carried by the wave field does not move with the speed of the crests c but rather at a different so-called group velocity c_g. For the short gravity waves (of wavelength much less than the depth of the ocean) that comprise swell and chop, the group velocity is just half of the phase velocity; $c_g = \frac{1}{2}c$.

As we previously remarked, these waves are such that the phase speed of long waves is greater than that of short ones; the waves are said to be dispersive. This adjective characterizes the behavior of a localized disturbance composed of many such waves as it progresses from its point of origin. Initially, all of the component wavetrains, both short and long, are contained within the disturbance. But as time passes, the energy associated with long waves tends to outrun the energy associated with short ones so that the concentrated disturbance gradually spreads out and is finally dispersed.

Just this effect has been observed in records of waves from distant storms at sea. Because a group composed of waves of period near T travels with group velocity $c_g = gT/4\pi$, the time t required for the group to travel a distance D between storm and observer is $t = 4\pi D/gT$. Thus, at the observing station, the period T of arriving waves should vary inversely with time. This tendency is clearly obvious in the series of wave spectra of Fig. 7-14 taken off Cornwall, England in 1949. Moreover, since $f = 1/T = gt/4\pi D$, a plot of frequency of arriving waves versus time of arrival will yield a curve whose slope is $g/4\pi D$ so that the distance D to the storm may be estimated. Munk and his collaborators have made detailed studies of many aspects of this problem (Munk, Miller, Snodgrass, and Barber, 1963).

Short wind waves at sea are not stable because of their relatively great steepness and they tend to loose energy rapidly by formation of whitecaps, but a long swell such as that evident on the spectra of Fig. 7-14 has been observed to travel from a storm in New Zealand across the entire Pacific to a wave recorder in Yakatut, Alaska, with very little attenuation. Thus, in the absence of whitecap formation, most of the energy in the long swell is dissipated in turbulence formed as the swell breaks on a shoaling beach.

Wave motions whose wavelength is greater than the depth of the ocean manifest themselves in the sea primarily as tsunami waves, storm surges, and tides. Once more, the idealized wave is sinusoidal in profile but, because the wavelength is often comparable with the radius of the earth and the wave period comparable with or longer than a day, effects arising both from the sphericity of the earth and its rotation may be appreciable. This is particularly true of the tides. The phase speed c of these long gravity waves depends only on the depth D of the water and not on the wavelength or period, $c = \sqrt{gD}$. All long waves travel at the same rapid (200 m/sec in water 4000 m deep) phase speed. Consequently any localized disturbance

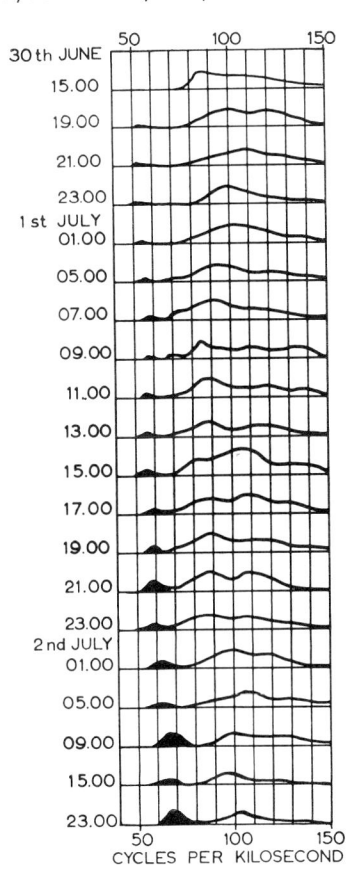

FIG. 7-14. Wave spectra taken at Cornwall, England, in 1949. The curves are smoothed representations of those in the literature. They plot mean amplitude (square root of power per unit frequency interval) against wave frequency. The isolated low-frequency activity attributed to the hurricane is distinguished in black. Its frequency shows some slight oscillation attributable to the effect of tidal streams near the Cornish coast, but the general trend toward higher frequencies is in accordance with the time and distance of a hurricane reported off Florida (from N. F. Barber and J. F. Tucker, 1962, *Wind Waves. The Sea*, Vol. 1, M. N. Hill, Ed., Interscience, New York).

remains localized, traveling with its component waves at the one phase speed, which is therefore identical with the group velocity, $c_g = c$.

A submarine earthquake or volcano may impulsively excite a "tsunami" composed of long gravity waves whose periods are of order 5 to 30 min and all traveling at nearly the speed $\sqrt{gD}$. The motion of the sea floor that generates the tsunami lasts only a few minutes so that, in a very shallow and uniform ocean of infinite extent, the disturbance arriving at a distant point would also last only a few minutes. However, the shortest-period waves in the tsunami have wavelengths small enough (though still several ocean depths long) that they travel about 10% slower than the longest period ones, which go at $\sqrt{gD}$. Thus a disturbance initially lasting a few minutes would be stretched out into one lasting about an hour when it had been traveling for 10 or 12 hr. More importantly, coastal and bottom irregularities scatter much of the energy of the tsunami out of the paths that would be predicted from the horizontal variations of $\sqrt{gD}$ into other directions and into other energy-storing modes of the ocean such as edge waves. The result is that the disturbance created by a tsunami "reverberates" throughout the ocean for days afterwards, decaying by about 37% during each subsequent day, and some unusual activity appears on wave records for as long as a week after the initial arrival (Munk, 1962; Miller, Munk, and Snodgrass, 1962).

The tsunami waves are barely noticeable at sea from shipboard since there, they are of such great length and small amplitude that the gradual rise and fall of the ship is imperceptible. But when they enter the shoaling water of coastal regions, they may crest to heights of 50 ft or more and occasionally become very much higher. The first indication of the occurrence of a tsunami may be either a rise or a fall of sea level, depending on the first motion of the earth at the source point. Successive high and low waters follow at intervals of 10 to 30 min and the rapid rush of water may do great damage to buildings and structures near the shore.

The record of sea level produced on the tide gauge at Arica, Chile, by the tsunami of 22 May 1960, generated during the great Chilean earthquake is shown in Fig. 7-15.

Appreciable variations of atmospheric pressure and of wind stress occur over large ocean areas with periods that range from longer than seasonal to less than a day. Prominent variations are associated with the development and motion of pressure centers and storms, and these may have periods ranging from weeks to a fraction of a day. When they occur over the ocean, they excite variations in sea level called storm surges.

The response to slow (weekly) changes in pressure is often the inverted barometer reaction giving a high sea level whenever the atmospheric pressure is low and vice versa. In this case, a slowly moving center of low pressure carries beneath it a mound of water, and if it crosses a coast, the sea level there will rise. This effect is very well illustrated in Fig. 7-16 where plots of sea level and inverse atmospheric pressure over a 3-month period are presented. But the situation is often further complicated by local winds which, in shallow regions, are especially effective in piling water onshore so that the day-to-day variation may be quite different from the inverted barometer response. Additionally, the possibility in theory that the ocean may respond to excitation at long periods in a nonequilibrium way ought not to be overlooked. Work by Longuet-Higgins (1964, 1965), both on the nature of the flow at long periods and on the nonequilibrium response, is particularly important. When the atmospheric fluctuations are more rapid, the response of the sea depends more upon the properties of long waves and is more difficult to describe succinctly. The shape of a bay or the width of a continental shelf may be such that a wave of one period is repeatedly reflected from shore to shore or from shore to shelf edge and back again over the same path. In this case the sea will be selectively responsive to forces that fluctuate at or near the resonant period of such a wave and, after the disturbing force has vanished, the resonant response will persist for a time depending upon the smallness of dissipative effects and of energy loss out of the resonant wave path. But the duration of the disturbance, the resonant periods typical of the local geography, and the damping time are generally all so close to one another that no simple relationships dominate the response, and prediction must generally be carried out by numerical techniques.

The complexity of storm-surge response is partly illustrated by the example of Fig. 7–17, in which sea-level records at various ports are shown during the passage of a hurricane along the eastern coast of the United States. The initial surge, which occurs approximately when the center of the storm is nearest but which may be either a high or a low water, is preceded by a gradual change in sea level and may be followed by one or more resurgences. Along the east coast, the slope of the nearshore bottom and the rotation of the earth each make possible the existence of resonant edge wave modes in which energy is trapped near the shore with only the possibility of flowing north or south. The resurgences visible on some of the records of the figure have approximately the period of such free edge waves traveling at the speed of the storm center and it has been suggested that these waves constitute a wake which travels along behind the storm (Munk, Snodgrass, and Carrier, 1956).

The phenomenon of the tides is the regular rise and fall of the sea surface, typically between 1 and 6 ft, observable every day at nearly every coastal location. Smaller values are found in enclosed seas and considerably larger ones may result from special configurations of the coastline and bottom nearby. The rise and fall cycle occurs twice per day in most locations and the times of high and low water are closely correlated with the passage of the moon over the local meridian. Since this occurs roughly once every 24 hrs and 50 min (a lunar day), the two daily high waters do not occur at fixed hours of the solar day but rather each occurs about 50 min later than the corresponding high water of the previous day. The lag between passage of the moon through the local meridian and the arrival of high water is constant at any individual port but varies widely from one port to another in a way

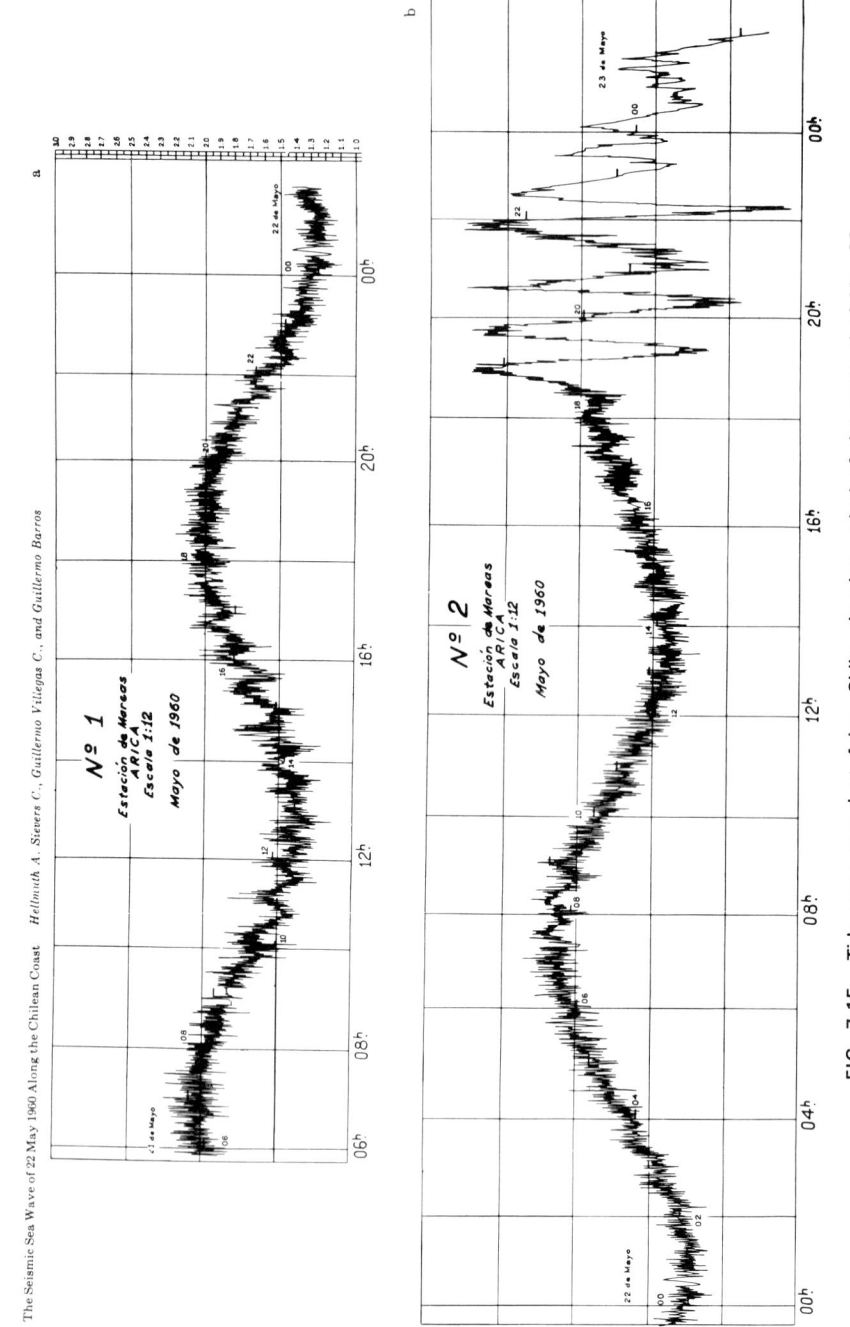

FIG. 7-15. Tide gage record at Arica, Chile, showing arrival of the tsunami of May 22, 1960. (reprinted from H. Sievers, G. Villegas, and G. Barros, 1963, *Bull. Seism. Soc. Amer.* 53, No. 6, p 1125, with permission of the Society.)

FIG. 7-15. (Continued).

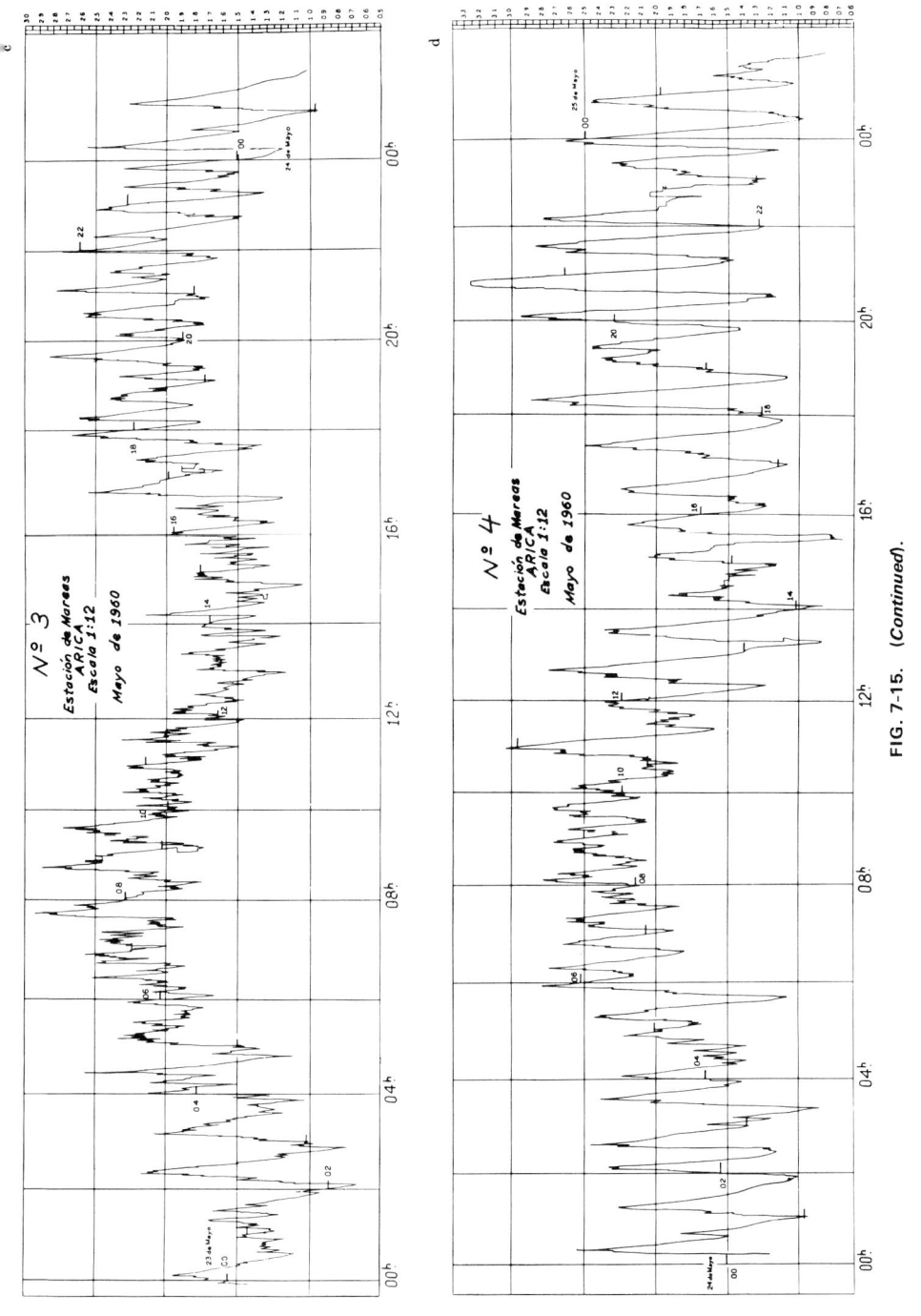

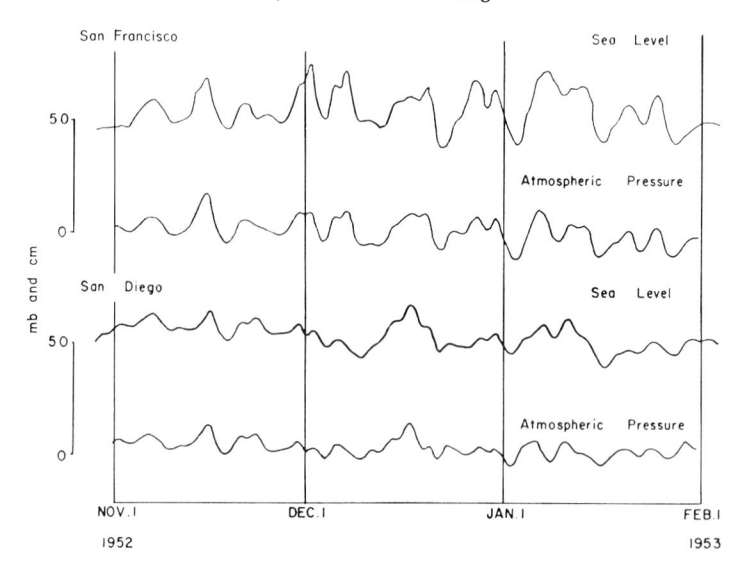

FIG. 7-16. Two comparisons of sea level and inverted barometer response (from P. Groen and G. Groves, 1962, *Surges. The Sea*, Vol. 1, M. N. Hill, Ed., Interscience, New York.)

depending upon the propagation of the tidal wave in the sea. When the moon is nearest the earth, the range of the tide is some 20% greater than when the moon is farthest away.

The tide-producing force is clearly the gravitational attraction of the moon and it may be wondered why there are two high waters per lunar day when the moon passes overhead only once per lunar day. The explanation is that both the earth and the moon rotate in nearly circular orbits about a common center of revolution which lies on the straight line joining their centers (it is in fact within the earth). Consequently, observers on the earth see both the force of lunar gravity and the centrifugal reaction to the circular motion of the earth. On the side of the earth nearest the moon, the gravitational attraction towards the moon is somewhat stronger than the opposing centrifugal reaction of the earth and the sea surface beneath the moon tends to bulge out towards the moon. On the opposite side of the earth, the centrifugal reaction away from the moon is greater than the gravitational attraction towards the moon and the sea surface again bulges out. There are therefore two high-water points on the earth, one directly beneath the moon and one on the other side of the earth. The mathematical development is given in many standard works, for example, Dronkers (1964). (The reader who pursues the matter will note that this verbal description glosses over a subtlety in the formulation of the centrifugal reaction.)

In this simple picture the two tidal bulges travel completely around the

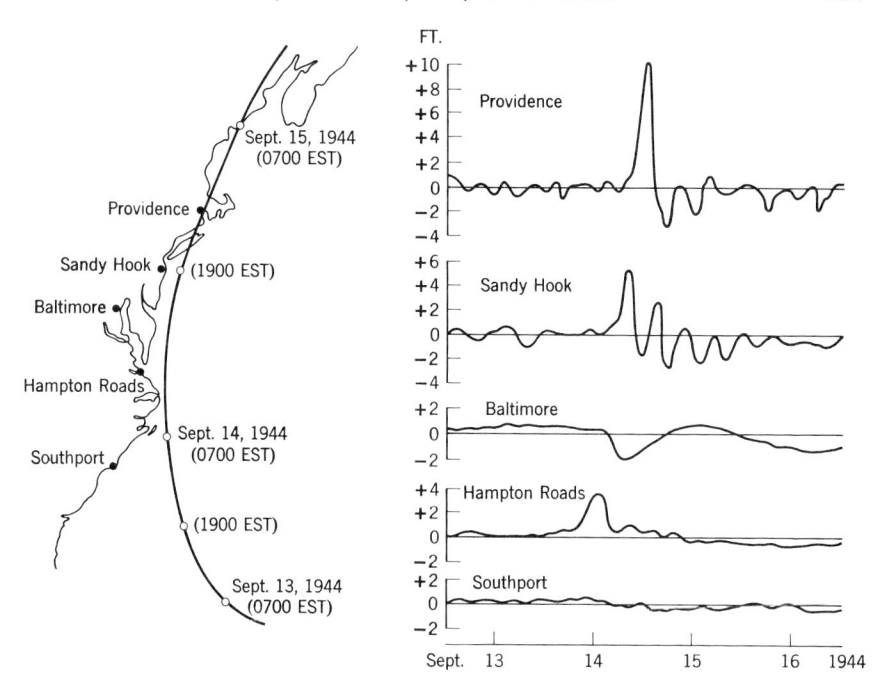

FIG. 7-17. Storm surge charts for the hurricane of September 1944 (from G. Neumann and W. Pierson, 1966, *Principles of Physical Oceanography*, (C); by permission of Prentice-Hall, Englewood Cliffs, New Jersey).

earth once every lunar day. They must consequently move much more rapidly than the swiftest free long waves are capable of doing. On a water-covered globe, this disparity would cause the bulges to arrive not when the moon is directly overhead, but rather at some other time. (But notice that they would still travel once around the earth every lunar day—the effect is not a slowing of the bulges). On the real earth this situation is further complicated by the irregular shape and depth of the great ocean basins, and the actual tides resemble those of the double-bulge model mainly in the arrival of two high waters daily.

The two daily high waters are generally not of equal height. They would be equal only if the moon were in the equatorial plane of the earth, as it is twice per lunar month (the period of rotation of the moon about the earth—more correctly, the period of rotation of the earth–moon pair about their common center). When the moon is north of the equatorial plane, the high water directly beneath it is in the Northern Hemisphere and the moon will consequently raise greater tides there than the opposite bulge, which is in the Southern Hemisphere and arrives 12 hr and 25 min later. This daily inequality thus goes through its own cycle, rising from zero (equality of the two high waters) to a maximum and returning to zero in half a lunar month.

Although the sun is much farther away than the moon, it is so much larger that its tidal attraction on the earth is nearly half as large as that of the moon. The picture of lunar tides presented above is therefore further complicated by the simultaneous presence of solar tides. But the existence of the two tide-generating forces makes possible a convincing explanation of the observed large spring tides (occurring once every half lunar month when solar and lunar tide-producing forces pull together) and small neap tides (occurring a quarter lunar month later when solar and lunar forces oppose one another).

The complexity of tide-producing forces makes for a tide much more complicated than a simple harmonic rise and fall of sea level. In a fashion similar to that outlined in our discussion of the spectral representation of swell, the rise and fall of water at a port or at sea may be regarded as a sum of many periodic fluctuations (the main difference being that a number of discrete fluctuations having only periods determined from astronomical calculations will usually suffice for the tide, whereas a continuum of fluctuations at all periods is necessary for the swell). Some of these components, together with their periods, are indicated in table 7-2 below:

TABLE 7-2

Tide and Symbol	Period
Principal lunar, M_2	12.42 hours
Principal Solar, S_2	12.00 hours
Larger lunar elliptic, N_2	12.66 hours
Luni-solar semi-diurnal, K_2	11.97 hours
Luni-solar diurnal, K_1	23.93 hours
Principal lunar diurnal, M_1	25.82 hours
Principal solar diurnal, S_1	24.07 hours
Lunar fortnightly, M_f	13.66 days
Lunar monthly, M_m	27.55 days
Solar semi-annual, S_{sa}	182.62 days
Solar annual, S_a	365.26 days
18.6 year	18.6 years

In identifying characteristics of the tide with corresponding features of the tide-producing forces, we have tacitly assumed that the water of the ocean comes to equilibrium with these forces instantaneously—without inertia. The ideal tide thus envisaged resembles the actual one mainly in the variety of periodic constituents to be expected. Investigations of this "equilibrium" tide began with Newton and culminated in Doodsen's (1921) analysis of tide-generating forces. The utility of these studies lies in the fact that the various periodic motions of the fictitious equilibrium tide have historically been the building blocks of practical predictions of real tides at ports and in harbors. Dronkers (1962) discusses variations of the basic scheme of harmonic analysis of observed tides at a location to produce predictions of the tide there, a plan common to every predictive scheme except that of Munk and Cartwright (1966), and gives detailed references.

However the ocean is always far from being at complete equilibrium with the tide-producing forces, and the inertia of the water is manifested in the properties of the long gravity waves which make up the tide. These waves differ considerably from the idealized wave on account of the sphericity of the earth and its rotation. Beneath the sequence of progressing crests and troughs that make up the corrugated sea surface in the ideal wave, the water moves forward under the crests and backwards under the troughs. Because the wavelength is so much greater than the depth of the water, the horizontal motion is nearly uniform from top to bottom and individual water particles move back and forth at the wave period along straight horizontal lines in the direction of wave travel. The ratio of horizontal displacement of fluid particles to vertical displacement of the sea surface is approximately the ratio of the wavelength of the tide (several thousand kilometers in the open sea) to the depth of the water (about 4 km in the deep sea). But in the ocean, the earth's rotation and the consequent Coriolis deflection of all moving particles at right angles to their direction of motion cause these straight-line paths to become horizontal ellipses. Thus as the crest–trough–crest sequence of a diurnal or semidiurnal wave passes by, the water particles go through horizontal elliptical orbits. Seen from above, the motion is clockwise rotary in the Northern Hemisphere and counterclockwise rotary in the Southern Hemisphere. Rotation in the opposite sense may be produced near the shore by rapid changes in depth, as in a shelf region (Larsen, 1966). The elevation of sea surface is usually less than 1m as the tidal wave passes so that the velocity of water particles estimated as above should be the order of 1cm/sec. A circular path traced out once in 12 hr would have a circumference of (1 cm/sec $\times$ 12 hr $\times$ 60 min/hr $\times$ 60 sec/min), that is 43,200 cm or a diameter of about 0.14 km. The diameter of tidal ellipses observed in the open sea is usually appreciably larger than this figure even though the estimate of surface elevation upon which it is based in a good one. We consider this discrepancy in the following discussion of internal waves. For now, it suffices to note that the paths of water particles executed with tidal periods are generally observed to be of the elliptical form described.

The rotation of the earth has a further effect on the propagation of the tidal wave, again closely bound up with the Coriolis acceleration. To understand this, consider a wave progressing parallel to a coast in the Northern Hemisphere with the coastline on the right-hand side of an observer facing in the direction of wave propagation. In the ideal wave water particles would move back and forth at the wave period in straight lines parallel to the coast. We have noted above that in the open sea, the Coriolis acceleration forces the particles to the right of their initial direction of travel, thus transforming the straight-line paths into elliptical ones executed in the clockwise direction. But now, the nearby coast prevents this cross flow from developing and constrains the water particles to move in a straight line parallel to the coast. In order for this to occur, the Coriolis acceleration must be balanced by a new pressure gradient force normal to the shore and this is manifested in an upward slope of each crest (plus a downward slope of each trough) towards the coast. An observer at the coast sees a series of crests and troughs move by parallel to the coast just as in the ideal case but if he goes out to sea, he sees that the farther away from shore he goes, the smaller the crests become and the shallower the troughs. The wave height is greatest at the coast and becomes very small far out at sea. This type of wave is called the Kelvin edge wave, after its discoverer (Lamb, 1932). Along a west coast (California), the Kelvin wave propagates only northward. It would go westward along a north coast (Alaskan Gulf), southward along an eastern coast (Japan) and eastward along a southern coast (Antarctica, if Antarctica were in the Northern Hemisphere).

Now consider such a Kelvin wave in a circular ocean. To an observer on the coast, the pattern of crests and troughs appears to progress along the coast in the direction explained above (counterclockwise in the Northern Hemisphere when viewed from above), but the motion becomes smaller towards the center of the circle as the crests and troughs become more and more gentle. At the very center there is no rise and fall of the tide and an observer there would see distant crests (near the shore) rotating about him. This rotation of crests and troughs about a midocean *amphidromic* point of zero tidal range is in striking contrast to the uniform progression of crests and troughs imagined in the ideal wave.

Extrapolation of coastal tides seaward by various means to obtain estimates of the deep-sea tide always indicates the existence of such amphidromic points in bodies of water the size of the Aral sea and larger and they are certainly a salient feature of the global tidal field. It must be added, however, that no actual deep-sea tidal measurements have ever been made near such a point nor have many been made anywhere else, and the necessary instrumentation for doing this is only now being developed.

Physical and Hydrodynamic Factors 243

Such an extrapolation of lunar semidiurnal coastal tides into the interior of the Atlantic is reproduced in Fig. 7-18. This figure is to be imagined as a quintuple exposure snapshot of the entire Atlantic basin with pictures snapped at 0, 3, 6, 9, and 12 hr after the passage of the moon over Greenwich. The solid lines labeled with hourly values are the positions of the tidal crests at each time. The wide variability of the lag between the passage

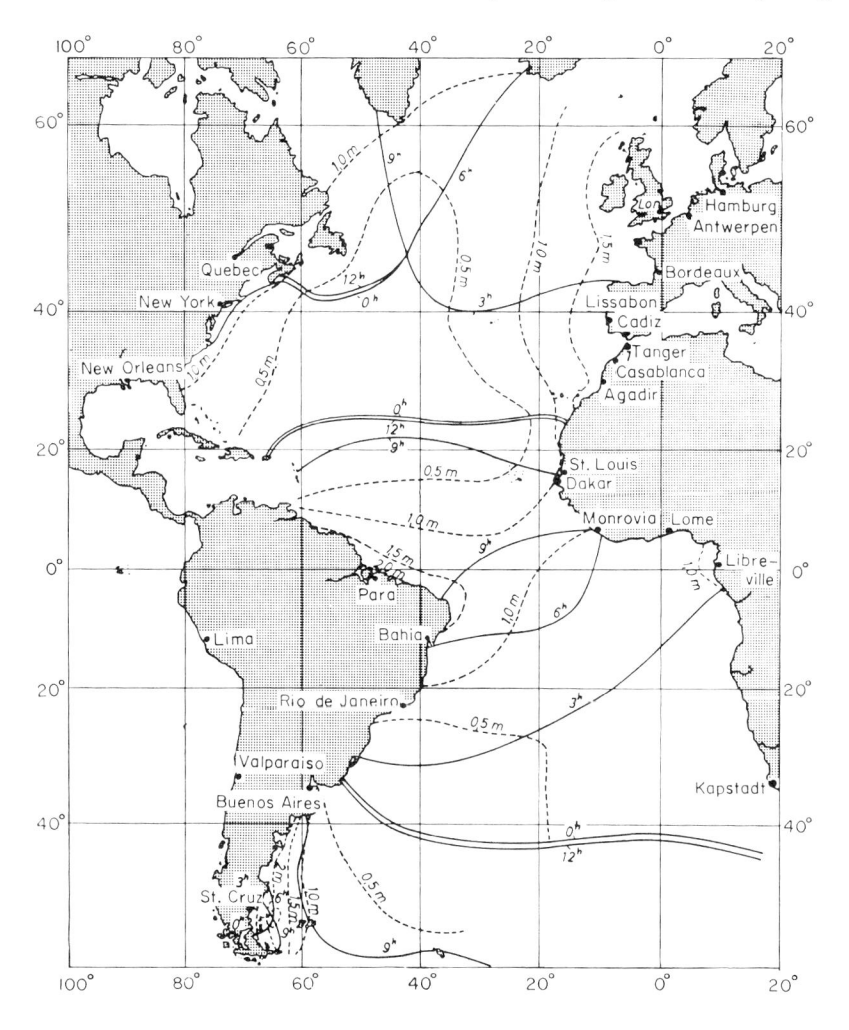

FIG; 7-18. Co-tidal and co-range lines for constituent M_2. Numbers on the full lines give time of high water after the moon's transit of Greenwich. The numbers on the broken lines give the range in meters. (From W. Hansen, 1962, *Tides. The Sea*, Vol. 1, M. N. Hill, Ed., Interscience, New York.)

of the moon overhead and the arrival of high water at various points on the same meridian is immediately obvious.

The crest progresses northward along West Africa at 0 hours, is moving along the coast of Western Europe at 3 hours, has reached Iceland by 6 hours (and is now traveling nearly westward), and is proceeding rapidly southward along the eastern seaboard of the United States by 12 hours. It then appears to swing very rapidly eastward along the Antilles arc, and join a new crest which has just crossed the equator (which, in theory, has the same ability to "trap" a Kelvin wave as would an equatorial coastline). The amphidromic point about which it rotates is evidently some 15 deg south of the tip of Greenland and, as may be seen from the dotted contours of equal amplitude, the tidal range there is evidently quite small. In the North Atlantic the tidal range is delineated by the contours of constant amplitude is largest near the shore and the equator, and the entire picture is qualitatively similar to that of the Kelvin wave in the circular sea. Larsen (1966) has found that the tide along the California coast may be accurately modeled as a single Kelvin wave traveling to the north and it is probable that the Kelvin-wave free modes of ocean basins are among the most important elements of the response of the ocean to tidal forces. They are not, however, the entire response and both motion analogous to simple seiching and response in the planetary waves of Longuet-Higgins must be considered.

Hansen (1962) has written a modern review of the tidal problem and further references may be found in his article.

We remarked previously that a midocean tidal range of 1m leads to an estimate of water orbital velocities of order 1 cm/sec but that observed tidal orbital currents are appreciably larger than this value. Typical observed diurnal and semidiurnal currents in the mid-Atlantic are shown in Table 7-3.

Note the large orbital velocities varying greatly with depth. This is at variance with the picture of tidal waves developed above, in which the tides were characterized simply as long gravity waves giving rise to uniform horizontal motions from surface to bottom.

This discrepancy indicates the presence of yet another type of wave together with the long gravity wave of the tide. Such waves are called internal waves, for they manifest themselves as vertical displacements which occur only within the body of the fluid and which nearly vanish at the surface. The presence of such a wave thus has a negligible effect upon the tidal rise and fall of sea level but, since its horizontal velocities are large and variable throughout the body of the fluid, the orbital currents of the tide may be entirely masked by those of the internal wave—as seems to be the case in the tabulated observations.

TABLE 7-3

Elements of the Semidiurnal and Diurnal Tidal Current Ellipses Based on Harmonic Analysis[a] of the Current Measurements at Station 385 of the *Meteor* Extending over a 60-hr Period from February 12 to 14, 1938[b,c]

Depth (m)	α_{12}[d]	α_{24}[d]	κ_{12}[e]	κ_{24}[e]	μ_{12}[f]	μ_{24}[f]	β_{12}[g]	β_{24}[g]	γ_{12}[h]	γ_{24}[h]
5	344	52	7.0	9.7	6.4	7.5	0.64	0.44	−	−
15	335	57	6.2	10.0	9.4	6.4	0.42	0.72	−	−
30	9	333	6.8	2.7	8.9	4.6	0.26	0.59	−	−
50	343	17	6.6	11.5	11.6	4.6	0.15	0.70	−	−
100	307	70	10.2	9.5	4.4	2.5	0.11	0.48	−	−
300	310	54	10.6	5.0	12.3	4.2	0.07	0.55	−	+
500	353	20	10.8	9.0	6.3	3.4	0.11	0.41	+	−
800	0	44	0.5	19.8	6.5	4.9	0.48	0.41	−	−

[a] After O. von Schubert (1944).
[b] 16°48′N, 46°17′W; depth = 2950 m.
[c] Indices 12 and 24 indicate semidiurnal and diurnal tidal currents, respectively.
[d] Current direction at time of maximum current velocity, in degrees.
[e] Time of occurence of maximum velocity, in Greenwich lunar time.
[f] Maximum velocity, in centimeters per second.
[g] Ratio of minor axis to major axis of the tidal current ellipse.
[h] Sense of rotation of tidal current ellipse; + = counterclockwise (*contra solem*), − = clockwise (*cum sole*).

Such waves may exist and travel in the ocean only by virtue of the stable density stratification that exists there. The restoring force is that of gravity, but it is now not a free surface which is being disturbed but rather the layered density structure within the body of the fluid itself. As we remarked in our earlier discussion of stability, a water parcel moved up or down is forced back towards its point of origin by the buoyant forces arising from the increase of density with depth. Just as in the case of the free surface, this means that the configuration is not only stable, but overstable, and it rebounds from any displacement in the oscillatory manner of a spring suddenly released from tension. The frequency at which this occurs may be calculated from the density structure and is called the Vaisala frequency, or Brunt-Vaisala frequency, or stability frequency. A plot giving typical values of this frequency in the ocean is reproduced in Fig. 7-19. The large maximum near the surface corresponds to a very stable seasonal thermocline. The smaller maximum near 800 m is associated with the permanent thermocline discussed earlier. In these thermocline regions, the density increases rapidly with depth and the stability is correspondingly great.

The sort of wave motion made possible by the stratification is most easily

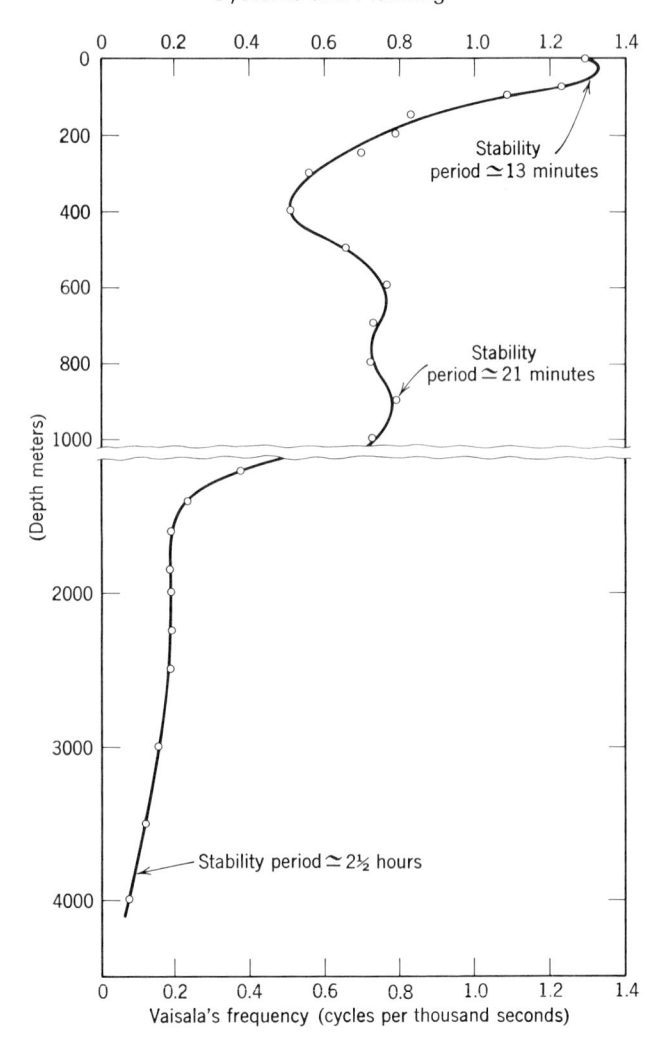

FIG. 7-19. A smoothed plot of Vaisala's frequency (in cycles per thousand seconds) in the ocean. (Atlantic Station 1227, April, 1932).

visualized in a model ocean composed of a lower layer of water with a thinner overlying layer of oil. This very crudely simulates the heavy deep water overlaid with a thin layer of light water above the thermocline—which becomes the oil–water interface. Just as ordinary waves exist at the real air–sea interface, so waves may travel along the oil–water interface of the model ocean. They exist because the oil is lighter than the water so that any depression of oil downwards is floated back upwards and vice versa.

Because oil is only slightly less dense than water, this restoring force is small and the waves travel much more slowly than those at the air–sea interface.

When the crests and troughs of such a wave travel along the oil–water interface, the oil–air interface is barely disturbed. When the wavelength is long compared with the depth of both layers, motion of the fluid above and below the interface occurs along straight lines parallel to the direction of wave propagation, as in the ordinary long gravity wave, but with this additional characteristic: flow in the oil is in the opposite direction to that in the water and the net transport of mass in one layer just cancels that in the other at every instant.

In summary, then, the vertical displacements are largest at the interface and nearly zero at the surface (exactly zero at the flat bottom), but the horizontal motion persists throughout the entire two-layer fluid, reversing its direction as one crosses the interface between the two fluids. The surface rise and fall of the astronomical tide in the oil–water ocean would closely resemble that in an ocean of homogeneous water even if, in addition, a strong internal wave were present in the oil–water ocean. But the horizontal motion of water particles would then be quite different in the two oceans.

The approximation of a sudden change of density between two otherwise homogeneous layers is not a sufficiently good one for the ocean since there, the density varies continuously even across the thermocline. More complete treatments—which correspond in some sense to approximating the actual continuous density structure as many small steps at each of which internal waves may occur—indicate that an infinite number of internal wave modes may exist in the ocean (Fjeldstad, 1933; Eckart, 1960). The frequency of each internal wave must necessarily be greater than the Vaisala frequency at some depth if it is to transport energy as a wave. The first mode resembles the one described above in the two-layer ocean with horizontal velocity in one direction at the surface and in the opposite one at the bottom. In the second mode, the velocity at the top is in the same direction as that at the bottom, but at middepths it is reversed. Higher modes have more reversals, mode number n having n of them, but in all cases the disturbance of the sea surface is very nearly zero. Some vertical profiles of horizontal and vertical velocity for the first few semidiurnal internal modes at a rather shallow station are shown in Fig. 7-20. The actual motion as a function of depth may always be represented as a combination of all such modes with varying amplitudes and phases.

Internal waves travel very slowly and therefore have much shorter wavelengths than surface waves of the same period. A first-mode internal wave of 24 hr period travels at about 5 m/sec and hence has a wavelength of about 430 km. This is very large for an internal wave and values less

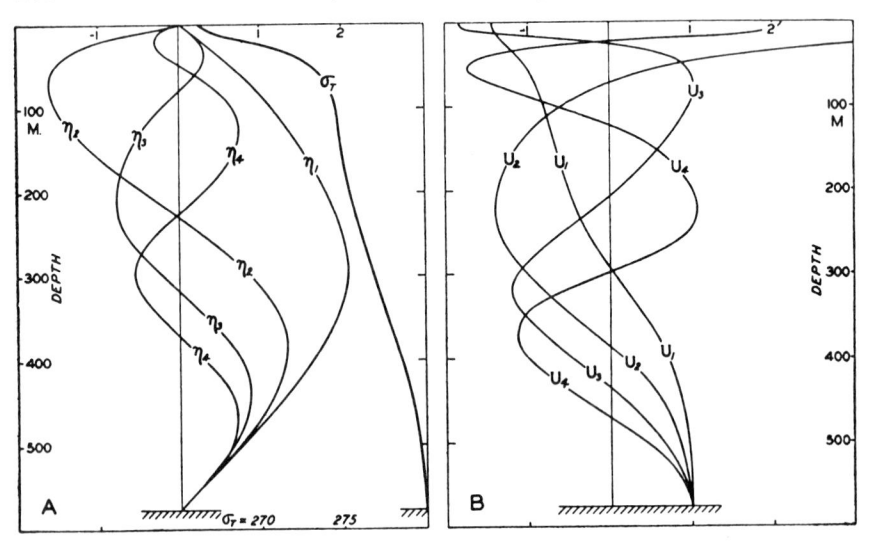

FIG. 7-20. (a) Variation with depth of the vertical displacements corresponding to internal waves of first, second, third, and fourth order at Michael Sars Station 115 (according to Fjeldstad). The density distribution is shown by the curve marked σ_T. (b) Variation with depth of the amplitudes of the horizontal velocities corresponding to an internal wave of first, second, third, and fourth orders (according to Fjeldstad). Vertical displacements and amplitudes are plotted on an arbitrary scale (from H. U. Sverdrup, M. W. Johnson, and R. H. Fleming, 1942, *The Oceans, Their Physics, Chemistry, and General Biology*, (C); by permission of Prentice-Hall, Englewood Cliffs, New Jersey).

than 100 km are much more typical for higher-mode 24-hr-internal waves. Waves of shorter period have shorter wavelengths and the most rapid ones, of period about 10 min, have wavelengths of only 100 m or less.

It is evident from Fig. 7-19 that Vaisala's frequency is so small at most depths that a wave of 15-min period would have a frequency greater than Vaisala's frequency everywhere except at depths between the surface and about 100 m below it. In this case, the motion of the 15 min-period wave is almost entirely confined to the layer between 0 and 100 m and no trace of it is to be found at any great depth. This trapping of short, rapid internal waves in thermocline regions is typical of midocean conditions. Indeed, a spectral analysis of fluctuations of temperature at a given depth generally shows relatively little motion at frequencies above the local Vaisala frequency.

Internal waves of tidal period seem to be very common in the ocean and they are often visible as semidiurnal vertical displacements of deep (to several hundred meters and occasionally deeper when apparatus permits observation) isotherms. A typical record is reproduced in Fig. 7-21. The

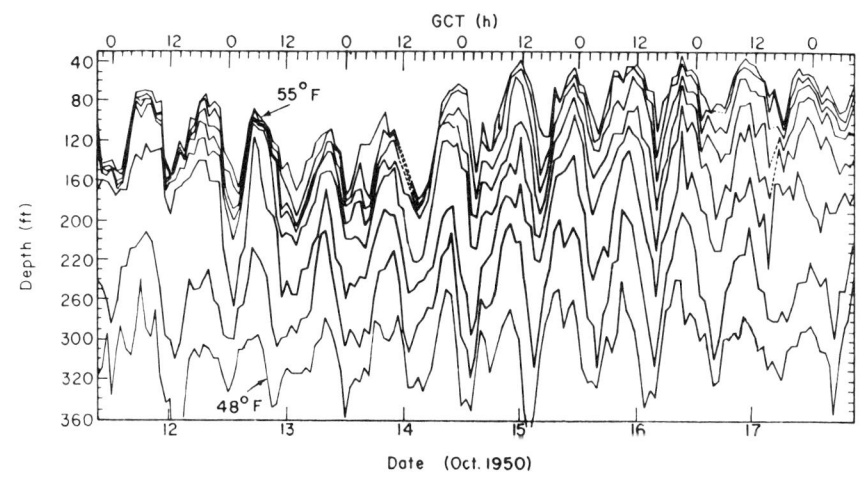

FIG. 7-21. Lunar semidiurnal fluctuations of large amplitude off the coast of California (from E. C. LaFond, 1942, *Internal Waves. The Sea*, Vol. 1, M. N. Hill, Ed. Interscience, New York).

semidiurnal period is clearly visible and vertical displacements of over 50 ft are indicated.

The origin of these internal waves has been a subject of considerable conjecture but no certain explanation has been advanced. The short-period fluctuations may in part be of atmospheric origin. Cox and Sandstrom (1962) suggest that an appreciable amount of the energy in the tides is diverted into the tidal internal wave field when the tide passes over bottom roughness of small (10 to 50 km) scale, and early experiments of Zeilon (1912) in the laboratory lead Rattray (1960) and others to suggest that the tide incident on the continental shelf radiates internal waves of tidal period. Recently, Sandstrom (1966) proposed a new theory of internal wave motion, which differs significantly from the traditional one outlined above and which makes necessary a re-examination of the mechanisms of Cox and Sandstrom and of Rattray. There is little doubt that tidal internal waves may be generated by these mechanisms, but their effectiveness remains to be established.

Internal waves in the sea are discussed by LaFond (1962) and Cox (1962) and detailed references may be found in their articles. The works of Rattray (1960), Cox and Sandstrom (1962), and Sandstrom (1966) represent important subsequent developments.

In the various types of gravity waves discussed above pressure gradients due to the slope of the free surface or of an internal surface of constant density are balanced by the Coriolis acceleration, at right angles to the motion of fluid particles, and by the inertial reaction of the particles, parallel to their motion. In the absence of Coriolis acceleration the displacement of a fluid column from its equilibrium position tilts the free surface or the surfaces of constant density. Gravity tends to restore them to their original level state and consequently the fluid column experiences a restoring force. The presence of a small Coriolis force alters only the details of the motion, but if the Coriolis acceleration exceeds the inertial reaction of fluid particles, the pressure gradient force does not tend to restore the fluid particles to their initial positions and gravity wave motion is no longer possible.

The inertial reaction is directly proportional to the wave frequency; the more rapidly the fluid particles must trace out their orbits, the greater the force necessary to make them do so. The Coriolis acceleration is proportional to twice the frequency of rotation of the earth; on a more rapidly rotating earth it would be greater. In tidal waves and internal waves, the fluid particles are constrained to move nearly parallel to the surface of the earth by the relative thinness of the ocean and its stable density stratification. This means that not the frequency of the earth's rotation, but rather some fraction thereof, is the relevant number to use in calculating the Coriolis force acting on fluid particles as such a wave passes by. At the pole, the full value must be used, whereas at the equator the Coriolis acceleration on a particle moving parallel to the earth's surface vanishes so that zero is the appropriate value. At intermediate latitudes, an intermediate value is appropriate. The value to be used at a given latitude is called the inertial frequency at that latitude. It varies from zero at the equator to once every 12 hr at the poles.

Clearly, there may exist waves for which the inertial acceleration of fluid particles exceeds the Coriolis acceleration near the equator but is less than the Coriolis acceleration near the poles. Their frequency must lie between twice the rotational frequency of the earth (in which case the inertial reaction dominates except just at the pole) and zero (when the Coriolis acceleration dominates except just at the equator); their periods must lie between 12 hr and infinity.

Such a wave may transport energy at lower latitudes but above the latitude where the inertial reaction and the Coriolis acceleration are equal, it loses this ability. At the crossover latitude, the frequency of the wave just equals the inertial frequency and the motion is most pronounced. A wave of period 12 hr or longer is thus trapped between the two latitudes where the inertial frequency just equals the wave frequency. A wave of 24-hr period is trapped between the latitudes 30° N and 30° S. Poleward of these latitudes, the motion decays. The trapping is most pronounced for internal waves on account of their short wavelengths and is scarcely noticeable for long surface tidal waves. The rotation of the earth thus makes possible a sorting out of long-period wave energy by latitude, with energy at each period manifesting itself as motion at the appropriate inertial latitude.

In theory, motion at the crossover latitude is particularly pronounced and observations of such rotary motion at the local inertial frequency are common at sea (Gustafson and Kullenberg, 1936; Knauss, 1962; Webster, 1963). There appears to be some enhancement of such motion at latitudes when the inertial frequency coincides with a tidal constituent (Reid, 1962; Hendershott, 1965).

7.4 WAVES ON A BEACH

When wind waves enter coastal water of depth less than about half of their wavelength, they begin to undergo a transformation which leads to phenomena unique to nearshore regions. Originally "deep-water" waves, they become "shallow-water" waves, which travel at phase speed $\sqrt{gD}$ (D is the local depth). Consequently, waves traveling into shallow water are retarded compared with those that enter neighboring water of somewhat greater depth, and it is meaningful to think of the refraction of waves by various features of bottom relief (Johnson, O'Brien and Isaacs, 1948; Arthur, Munk, and Isaacs, 1949; Munk and Arthur, 1952). It is easy to see that waves will generally be crowded into shallower regions so that wave activity will be greatest around exposed capes and off beaches before which the water deepens gradually. Conversely, waves are refracted out of deeper regions so that activity in bays and off beaches before which the water deepens rapidly will be minimal. These effects are well illustrated (Fig. 7.22) in a plot of successive positions of wave crests approaching a beach. The lines normal to the crests are the ray orthogonals. Where they converge, wave activity will be greatest. It should be emphasized that the focusing and shielding effects of offshore bottom topography are quite sensitive to the direction from which the waves are incident and a small change in wave direction may produce a dramatic change in wave intensity at certain points on the shoreline.

The only property of the waves that remains unchanged as they enter shoaling water is their period. Since their speed decreases as $\sqrt{gD}$, their wavelengths also decrease. Furthermore, as long as the waves are losing no energy by breaking, the motion of fluid particles must become more rapid as the waves enter shallower water. The waves thus become shorter, higher, and steeper. In so doing they lose their symmetry with respect to the undisturbed sea surface; troughs become broader and crests steepen. Finally, the wave becomes unstable and breaks. If the depth decreases slowly and evenly, a head of foam will spill out of the crest down into the trough ahead and the wave will lose energy as it travels until it is completely attenuated. If the depth decreases rapidly, the crest will steepen suddenly and fall forward in a plunging or curling breaker.

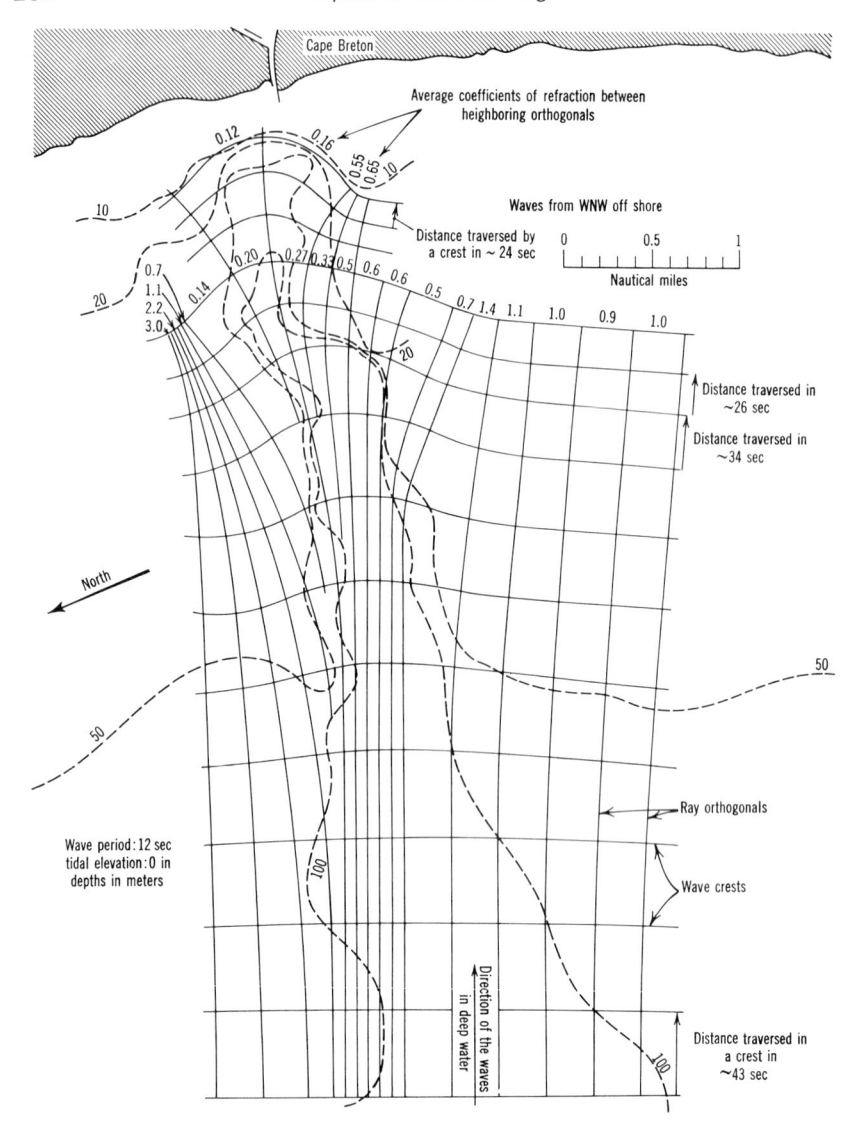

FIG. 7-22. Wave plan for the approaches to the trench of Cape Breton. (from H. LaCombe, 1965, *Cours d'Oceanographique Physique*, Gauthier-Villars, Paris).

Waves which steepen and possibly break as they approach the shore are a source of energy for various steady secondary nearshore flows. When waves strike the beach obliquely, the result is often a secondary flow parallel to the beach in the "downwave" direction (Putnam, Munk, and Trayler, 1949). Large amounts of beach material may be carried by this flow, and its interruption (by building a jetty seaward for example) is often accompanied by severe erosion of the beach area thus deprived of its source of replenishment for sand carried away by wave action. Another possibility is the formation of an intense and narrow jet flowing seaward, a rip current (Shepard and Inman, 1951). Rips usually do not extend far beyond the zone within which the waves are breaking, but their water velocities may be as great as 1 m/sec. After a heavy rainstorm, water at the beach is often colored with mud from shore runoff and rip currents are then readily visible as narrow plumes which extend out beyond the breaker zone and terminate abruptly in a region of rapid spreading.

7.5 ACOUSTIC PHENOMENA

Acoustic waves, like all others, exist by virtue of a restoring response of the medium to disturbance. The finite compressibility of seawater provides the restoring force in this case, and we close our discussion of wave motion in the sea by pointing out a few of the acoustic phenomena peculiar to the ocean environment.

In seawater of normal salinity, the speed of sound is primarily a function of temperature and pressure. It decreases as the temperature is decreased (contrary to its behavior in most fluids) but increases with increasing pressure. Thus temperature profiles of the sort usually encountered in the ocean combined with the hydrostatic pressure gradient lead to a vertical profile of sound velocity with a pronounced minimum at some intermediate depth, usually near 1000 m. Some typical profiles appear in Fig. 7-23. The

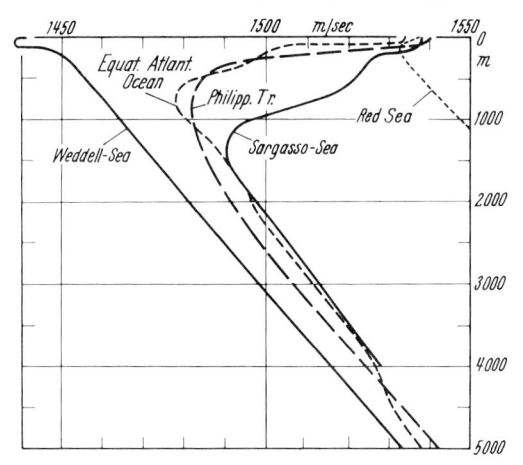

FIG. 7-23. Examples of horizontal velocity of sound in the world ocean (from G. Dieterich and K. Kalle, 1963, *General Oceanography*, Interscience, New York).

velocity decreases right up to the surface layer in the Weddell and Red Seas because the temperature structure there is relatively homogeneous.

This state of affairs leads to the existence of the so-called SOFAR (sound fixing and ranging) channel whose axis lies at the depth of the velocity minimum (Ewing and Worzel, 1948). An explosion at or near this depth radiates sound waves in all directions, but waves moving obliquely into depths of higher velocity (either up or down) are refracted back towards the depth of minimum velocity with the result that very little acoustic energy escapes from the channel. The detonation of several pounds of explosive in the SOFAR channel generates a signal detectable several thousand miles away by a hydrophone at the channel depth. The "direct" wave, which travels straight along the axis of the channel, arrives after any of the refracted waves, which traverse longer paths but at a greater speed, so that the sound actually heard begins as a rumble that gathers intensity and then cuts off suddenly.

Acoustic energy radiated at a depth where the speed of sound is smaller than in surface water will be refracted towards deeper water and a listener on the surface some horizontal distance away from the source may hear nothing at all. This fact finds its application when a submarine is trying to escape detection by the underwater listening apparatus of a surface ship.

Bubbles, marine organisms, irregularities of sea floor and sea surface, and density variations resulting from turbulence or internal waves all scatter acoustic energy. When the waves are repeatedly reflected from the ocean surface or bottom, there is considerable scattering of energy out of the path of multiple reflection. Bubbles are particularly effective scatterers at frequencies near their own acoustic resonant frequency. Energy absorption, as distinguished from scattering, differs from that in fresh water mainly in the loss of energy due to the dissociation of salts.

Deep-sea echo soundings consistently record the presence of one or more layers of scatterers at various depths above the bottom. In the deep ocean during the day, there are typically three such deep scatterings layers lying between about 900 and 2000 ft. Individual layers are of variable thickness, often between 40 and 60 ft. Because these layers migrate vertically twice per day—rising to the surface at night and descending at sunrise—they are believed to consist of small (but large enough to scatter effectively at the frequencies used) marine organisms which rise at night to graze in the plankton-rich surface waters. An interesting description of this phenomenon has been given by Dietz (1962); detailed references are to be found in Hersey and Backus (1962).

REFERENCES

*Indicates a textbook (including general works to which specific reference has not been made in the text.)
†Indicates an article of particular interest to the general reader.

Arthur, R. S., W. H. Munk, and J. D. Isaacs, 1949, "The Direct Construction of Wave Rays," *Trans. Amer. Geophys. Uni.* **33**. 855–865 (1949).
Bascom, W., 1959, †"Ocean Waves," *Scientific American* (August 1959).
Cox, C. S., 1962, "Ripples," *The Sea*, Vol. 1. M. N. Hill, Ed. Interscience, New York.
Cox, C. S., 1962, "Internal Waves, Part II," *The Sea*, Vol. 1, M. N. Hill, Ed. Interscience, New York.
Cox, C. S., and H. Sandstrom, 1962, "Coupling of Internal and Surface Waves in Water of Variable Depth," *J. Ocean. Soc. Japan* 20th Anniv. Vol.
Cox, R. A., F. Culkin, R. Greenhalgh, and J. P. Riley, 1962, Chlorinity, conductivity, and the Density of Sea Water. *Proc. Roy. Soc. (London)* **A 252**, 51–62 (1962).
Crease, J., 1962, "Velocity Measurements in the Deep Water of the Western North Atlantic," *Turbulence in Geophysics*, American Geophysical Union, Washington, D.C., pp. 3173–3176.
Defant, F., 1961, *Physical Oceanography*, Vol. I and II, Pergamon Press, London.
Dieterich, G., 1952, in Landolt-Börnstein, *Zahlenwerte und Funktionen*, Vol. 3, Part: Ozeanographie, Berlin.
Dietrich, G., and K. Kalle, 1963, *General Oceanography*, Interscience, New York.
Dietz, R., 1962, †"The Sea's Deep Scattering Layers," *Scientific American*, (August 1962).
Dittmar, C., 1884, "Challenger Report," *Phy. and Chem.* **1**, No. 11, 189–204 (1884).
Doodsen, A. T., 1921, "The Harmonic Development of the Tide-Generating Potential, *Proc. Roy. Soc. (London)* **A 100**, 305–329 (1921).
Dronkers, J. J., 1964, *Tidal Computations*, North-Holland, Amsterdam.
Eckart, C., 1960, *Hydrodynamics of Oceans and Atmospheres*, Pergamon, New York.
Eckart, C., 1961, "Internal Waves in the Ocean," *Phys. Fluids* **4**, 791, (1961).
Ekman, V. W., 1905, "On the Influence of the Earth's Rotation on Ocean Currents," *Arku. Mat., Astron. och Fysik*, **2**, No. 11, 1–53 (1905).
Ekman, V. W., 1908, "Die Zusammendruckbarkeit des Meerwassers," *Counseil Perm. Intern. Expl. Mer*, Pub. de Circonstance, No. 43.
Ewing, M., and J. L. Worzel, 1948, "Long Range Sound Transmission," in *Sound Transmission in the Ocean*, Geol. Soc. America, Mem. No. 27.
Fieldstad, J. E. 1933, "Interne Wellen" *Geofys. Publ.*, No. 10, No. 6 (1933).
Forch, C., M. Knudsen, and S. P. L. Sorenson, 1902, "Berichte uber die konstanten Bestimmung zur Aufstellung der hydrographischen Tabellen, *D. Kgl. Danske. Selsk. Skritter*, Raeke 6 *Naturv. og. Math.*, Afd. 12, No. 1.
Ford, W. L., J. R. Longard, and R. B. Banks, 1952, On the Nature Occurrence, and Origin of Cold, Low Salinity Water Along the Edge of the Gulf Stream. *J. Mar. Res.* **11**, 281–293 (1952).
Fuglister, C. W., 1951, "Annual Variations in Current Speeds in the Gulf Stream System," *J. Mar. Res.*, **10**, 119–127 (1951).
Fuglister, C. W., 1963, "Gulf Stream '60," in *Progr. Oceanography*, **1**, 265–373 (1963).
Groves, G. W., 1957, "Day-to-Day Variation of Sea Level," *Am. Met. Soc., Met. Monograph*, **2**, No. 10, pp. 32–45 (1957).
Gustafson, T., and B. Kullenberg 1933, Tragheitsströmungen in der Ostsee. *Medd. Göteborgs Oceanogr. Inst.*, No. 5 (1933).
Hansen, W., 1956, "Theorie zur Errechnung des Wasserstandes und der Stromungen in Randmeeren nebst Anwendungen," *Tellus*, **8**, 287 (1956).

Hansen, W., 1962, "Tides," *The Sea*, Vol. 1., M. N. Hill, Ed., Interscience, New York.
Harris, D. L., 1956, "Characteristics of the Hurricane Storm Surge," Tech. Paper No. 48, N. S. Dept. of Commerce, Weather Bureau.
Hendershott, M., 1965, "Inertial Oscillations of Tidal Period," Ph. D. thesis. Harvard University.
Hersey, J. B., and R. H. Backus, 1962, "Sound Scattering by Marine Organisms," *The Seas*, Vol. 1. M. N. Hill, Ed., Interscience, New York.
Holland, W. R., 1966, "Wind-Driven Ocean Circulation in an Ocean with Bottom Topography," Ph.D. thesis, University of California in San Diego.
Hunkin, K., 1966, "Ekman Drift Currents in the Arctic Ocean, *Deep-Sea Res.*, **13**, (4), 607–620 (1966).
Iselin, C. O'D., 1940, "Preliminary Report on Long Period Variations in the Transport of the Gulf Stream System, *Pap. Phys. Oceanography and Meteor.*, **8**, No. 1 (1940).
Johnson, J. W., M. P. O'Brien, and J. D. Isaacs, 1948, "Graphical Construction of Wave Refraction Diagrams," U.S. Navy Hydrographic Office Pub. No. 605.
Katz, B., R. Gerard, and M. Costin, 1965, "Response of Dye Tracers to Sea Surface Currents," *J. Geophys. Res.* **70**, 5505 (1965).
Kinsman, Blair, 1965, *Wind Waves*, Prentice-Hall, Englewood Cliffs, New Jersey.
Knauss, J. A., 1960, "Measurements of the Cromwell Current," *Deep-Sea Res.* **6**, 265–286 (1960).
Knauss, J. A., 1962, "Observations of Internal Waves of Tidal Period with Neutrally Bouyant Floats. *J. Mar. Res.* **20**, 111–118 (1962).
Knauss, J. A., 1965, A New Technique for Measuring Deep Ocean Currents Close to the Bottom with an Unattached Current meter, and Some Preliminary Results. *J. Mar. Res.* **23**, 237–245 (1965).
Knudsen, M., 1901, *Hydrographical Tables*, Blanco Luno; Second Edition, 1931.
Lacombe, H., 1965, *Cours d'Oceanographie Physique*, Gauthier-Villars, Paris.
Lafond, E., 1962, "Internal Waves. Part I." *The Sea*, Vol. 1. M. N. Hill, Ed., Interscience, New York.
Lamb, H., 1932, *Hydrodynamics*, Sixth Edition, Cambridge Univ. Press, New York.
Larsen, J. C., 1966, "Electric and Magnetic Fields Induced by Oceanic Tidal Motion," Ph.D. thesis, University of California in San Diego.
Longuet-Higgins, M. S., 1964, "Planetary Waves on a Rotating Sphere," *Proc. Roy. Soc.* (*London*) **A** Vol. **279**, 446–473 (1964).
Longuet-Higgins, M. S., 1965, "Planetary Waves on a Rotating Sphere. Part II," *Proc. Roy. Soc.* (*London*) **A 284**, 40–68 (1965).
Longuet-Higgins, M. S., 1965, "The Response of a Stratified Ocean to Stationary or Moving Wind System," *Deep-Sea Res.*. **12**, 923–973 (1965).
McDonald, J. E., 1952, †"The Coriolis Effect," *Scientific American* (September 1952).
Miller, G. R., W. H. Munk, and F. E. Snodgrass, 1962, "Long Period Waves over California's Continental Borderland II, Tsunamis," *J. Mar. Res.*, **20**, 31–41 (1962).
Moskowitz, L., 1964, "Estimates of the Power Spectrums for Fully Developed Seas for Wind Speeds of 20 to 40 Knots. *J. Geophys. Res.*, **69**, 5161–5179 (1964).
Munk, W. M., 1950, "On the Wind-Driven Ocean Circulation," *J. Meteor.* **7**, 79–83 (1950).
Munk, W. M., 1955, †"The Circulation of the Oceans," *Scientific American* (September 1955).
Munk, W. M., 1962, "Long Ocean Waves," *The Seas*, Vol. 1. M. N. Hill, Ed., Interscience, New York.
Munk, W. M., and R. S. Arthur, 1952, "Wave Intensity Along a Refracted Ray," *Gravity Waves*, NBS Circular 521, pp. 95–108.
Munk, W. M., F. E. Snodgrass, and G. F. Carrier, 1956, "Edge Waves on the Continental Shelf," *Science*, **123**, 127–132 (1956).

Munk, W. M., G. R. Miller, F. E. Snodgrass, and N. F. Barber, 1963, "Directional Recording of Swell from Distant Storms," *Phil. Trans. Roy. Soc.* **255**, No. 1062, 505–584 (1963).

Munk, W. M., and D. Cartwright, 1966, "Tidal Spectroscopy and Prediction," *Phil. Trans. Roy. Soc.* A **259**, 533–581 (1966).

Neuman, G., and W. J. Pierson, 1966, *Principles of Physical Oceanography*, Prentice-Hall, Englewood Cliffs, New Jersey.

Putnam, J. H., W. H. Munk, and M. A. Traylor, 1949, "The Prediction of Longshore Currents," *Trans. Am. Geophys. Union* **30**, 337–345 (1949).

Rattray, M., 1960, "On the Coastal Generation of Internal Tides," *Tellus*, **12**, 54 (1960).

Reid, J. L., 1962, "Observations of Inertial Rotation and Internal Waves," *Deep-Sea Res.* **9**, 283–289 (1962).

Robinson, A. R., 1965, "A Three-Dimensional Model of Inertial Currents in a Variable-Density Ocean," *J. Fluid Mech.* **21**, Part 2, 211–223 (1965).

Robinson, A. R., and H. M. Stommel, 1959, "The Oceanic Thermocline and Associated Thermohaline Circulation," *Tellus* **11**, 295–308 (1959).

Rossby, G. G., and Staff of Meteorology Department, University of Chicago, 1947, "On the General Circulation of the Atmosphere in Mid Latitudes, *Bull. Am. Meteor. Soc.* **28**, 255–280 (1947).

Sandstrom, H., 1966, *"The Importance of Topography in Generation and Propagation of Internal Waves*, Ph.D. thesis, University of California in San Diego.

Schubert, O. v., 1944, "Ergebnisse der Strommessungen und der ozeanographischen Serienmessungen auf den beiden Ankerstationen der Zweiten Teilfahrt," *Wiss. Ergeb. Deutsch. Atl. Exp.* (1937 and 1938). *Ann. Hydrographie und Marst. Meteorl.* 72 (1944).

Shepard, F. P., and D. L Inman, 1951, *Proc. First Conf. Coast. Eng.* Long Beach, California.

Sievers, H. A. (Part I–III), G. Villegas. (Part IV), and G. Barros (Preface), 1963, "The Seismic Sea Wave of 22 May 1960 along the Chilean Coast, *Bull. Seismol. Soc. Am.*, **53**, 1125–1190 (1963).

Stommel, H. M., 1948, The Westward Intensification of Wind-Driven Ocean Currents, *Trans. Am. Geophys. Union* **29**, No. 2 (1948).

Stommel, H. M., 1958, The Abyssal Circulation, *Deep-Sea Res.* **5**, 80–82 (1958).

Stommel, H. M., 1965, *The Gulf Stream*, Second Edition, University of California Press, Berkeley, California.

Stommel, H. M., and A. B. Arons, 1960, "On the Abyssal Circulation of the World Ocean, I." *Deep-Sea Res.* **6**, 140–154 (1960).

Stommel, H. M., and A. B. Arons, 1960, "On the Abyssal Circulation of the World Ocean, II." *Deep-Sea Res.* **6**, 217–233 (1960).

Stommel, H. M., A. B. Arons, and A. J. Faller, 1958, "Some Examples of Stationary Planetary Flow Patterns in Bounded Basins, *Tellus* **10**, 179–187 (1958).

Sverdrup. H. U., 1947, Wind-Driven Currents in a Baroclinic Ocean, with Application to the Equatorial Currents of the Eastern Pacific," *Proc. Natl. Acad. Sci.* **33**, No. 11 (1947).

Sverdrup, H. U., M. W. Johnson, and R. H. Fleming, 1942, *The Oceans*. Prentice-Hall, Englewood Cliffs, New Jersey.

Swallow, J. C., 1955, "A Neutral-Bouyancy Float for Measuring Deep Current," *Deep-Sea Res.* **3**, 74–81 (1955).

Swallow, J. C., and L. V. Worthington, 1961, "An Observation of a Deep Counter Current in the Western North Atlantic," *Deep-Sea Res.* **8**, 1–19 (1961).

Van Dorn, W. G., 1953, Wind-Stress on an Artificial Pond, *J. Mar. Res.* **12**, 247–276 (1953).

Veronis, G. and H. Stommel, 1956, "The action of variable wind-Stress on a Stratified Ocean," *J. Mar. Res.* **15**, 43–75 (1956).

Von Arx, W. S., 1962, *An Introduction to Physical Oceanography*, Addison-Wesley, Reading, Massachusetts.
Warren, B. A., 1963, "Topographic Influences on the Path of the Gulf Stream," *Tellus*, **15**, No. 1, 167–183 (1963).
Webster, F., 1961, "The Effect of Meanders on the Kinetic Energy Balance of the Gulf Stream," *Tellus*, **13**, 392–401 (1961).
Webster, F., 1963, A Preliminary Analysis of Some Richardson Current Meter Records," *Deep-Sea Res.* **10**, 389–396 (1963).
Webster, F., 1965, Measurements of Eddy Fluxes of Momentum in the Surface Layers of the Gulf Stream, *Tellus*, **17**, 239–245 (1965).
Wertheim, G. K., 1954, Studies of the Electric Potential Between Key West, Florida, and Havanna, Cuba. *Trans. Am. Geophys. Union* **35**, 872–882 (1954).
Wust, G., 1954, "Gesetzmässige Wechselbeziehungen zwischen Ozean und Atmosphäre in der zonalen Verteilung von Oberflächensalzgehalt, Verdunstung und Niederschlag," *Arch. Meteorl. Geophys. Bioklim.* **A 7**, 305–328 (Defant Festschrift) (1954).
Wust, G., 1955, "Stromgeschwindigkeiten im Tiefen-und Bodenwasser des Atlantischen Ozeans," *Papers in Marine Biology and Oceanography, Bigelow Volume* in *Deep-Sea Res.* 373–397 (1955).
Zeilon, N., 1912, "On Tidal Boundary Waves," *Kungl. Svenska Vetens. Akad. Handl.* **47**, 4 (1912).

CHAPTER 8

Chemical and Biological Factors in the Marine Environment

FRANCIS A. RICHARDS

This chapter is an attempt to describe those chemical and biological factors in the marine environment that must be considered for successful ocean engineering. Chemical and biological constituents of the ocean interact with each other and they will also interact with or alter manmade objects in the sea.

Some of the properties of the ocean are relatively uniform in both horizontal and vertical space, whereas others are more heterogeneously distributed. In general, the major dissolved constituents of seawater tend to be somewhat uniformly distributed, but living organisms and other particles are not.

8.1 THE CHEMICAL ENVIRONMENT OF THE SEA: THE NATURE OF SEAWATER

The water itself

Water itself is an anomalous substance with unusually high density, boiling point, heat capacity, freezing point, and other unusual properties that suggest that the formula H_2O is an inadequate representation of the water molecule. Probably H_2O units are polymerized by a variety of chemical bonding processes so that their effective average molecular weight is considerably greater than 18. The melting and boiling points of some other low-molecular-weight substances (Table 8-1) are much lower than those of water, presumably because their molecules are not so polymerized. Meteorologically and climatologically, the most important anomalous properties of water are its thermal properties: disproportionately large amounts of

TABLE 8-1
Boiling and Melting Points of Water and Some Compounds
that have Similar Formula Weights

Substance	Formula	Formula weight	B.P. (°C)	M.P. (°C)
Water	H_2O	18	100	0
Ammonia	NH_3	17.03	33.35	− 77.7
Methane	CH_4	16.04	− 161.49	− 182.48
Hydrogen flouride	HF	20.01	19.54	− 83.1

heat are involved in changing the temperature of water, freezing, or vaporizing it.

The widespread occurrence of water on the earth makes life as we know it possible; the high heat capacity of water moderates the effects of the uneven heating of the earth by the sun so that the tropics are not so hot, and the poles not so cold as they would be were the amount of water on the earth much less than it is or were its heat capacity much smaller—in effect, were its effective molecular weight much smaller.

The major solutes and salinity

Water is probably the substance most characteristic of earth among the planets, and yet the ocean forms but a thin film on the surface of the earth—a film about 1/2500th of the thickness (diameter) of the earth itself. Many of the characteristics and properties of ocean water are very close to those of fresh water, but the salt content of seawater makes it different in many ways that are of engineering importance. The chemical and physico-chemical differences are almost wholly attributable to some 11 *major constituents* that make up over 99.9% of the material dissolved in seawater. These are the anions Cl^-, $SO_4^=$, HCO_3^-, Br^-, and F^-; the cations Na^+, Mg^{++}, Ca^{++}, K^+, and Sr^{++}; and undissociated H_3BO_3 (Table 8-2). The ratio of these materials to each other varies only slightly in the open ocean, so that the salt that might be obtained by dehydrating seawater would have almost exactly the same composition no matter where in the ocean the water came from, providing it were reasonably remote from the mouths of major rivers and from sites of extensive freezing and thawing of sea ice. It is evident that the chemical properties of seawater are primarily attributable to the total concentration of these materials. This total contentration is closely related to the *salinity*, which is formally defined as "the weight in grams of the dissolved inorganic matter in seawater, after all bromide and iodide have been replaced by the equivalent amount of chloride and all carbonate converted to oxide." This definition was formulated by Knudsen (1901) and is a formalization of the notion of salinity rather than an

TABLE 8-2
Composition of Seawater. Concentration of Constituents in Seawater Having a Chlorinity of 19‰

	(g/kg)	(g/unit of chlorinity)
Chloride, Cl^-	18.890	0.99894
Sodium, Na^+	10.560	0.5556
Magnesium, Mg^{++}	1.273	0.06695
Sulfate, SO_4^{--}	2.649	0.1394
Calcium, Ca^{++}	0.4104	0.02106
Potassium, K^+	0.380	0.02000
	mg/kg	
Carbon (as HCO_3^- or CO_3^{--})	28	0.00735
Bromide, Br^-	65.9	0.00340
Strontium, Sr^{++}	8.1	0.00070
Boron, as H_3BO_3	4.6	0.00137
Silicon, as silicate	0.01–4.5	
Fluoride, F^-	1.4	0.00007
Nitrogen, as NO_3^-	0.01–0.80	
Aluminum, Al^{+3}	0.5	
Rubidium, Rb^+	0.2	
Lithium, Li^+	0.1	
Phosphorus, as PO_4^{-3}	0.001–0.1	

Trace Elements present in concentrations of 1 to 50 $\mu g/kg$: barium, iodine, arsenic, iron, manganese, copper, zinc, lead, selenium, caesium, uranium.

Trace elements present in concentrations of less than 1 μ/kg: molybdenum, thorium, cerium, silver, vanadium, lanthanum, yttrium, nickel, scandium, cobalt, cadmium, mercury, gold, tin, chromium, radium.

operational definition. Drying seawater to a constant weight is not a reproducible analytical procedure, so the salinity itself is never determined. Rather, some other constituent or property of the water is determined and, through the fixed relationships between the major constituents and other physicochemical properties, the salinity can be computed if desired. Many oceanographers state that the computation of salinity is unnecessary, and it is probable that the oceanographic use of salinity is more traditional than useful. However, the term does give the notion of total salt content, which has some conceptual use.

In the open ocean the salinity varies relatively slightly between approximately 32 and 37‰. Where rivers, which have essentially zero salinity, enter the ocean, all possible mixtures of seawater and freshwater will be formed, and the salinity will approach zero as the river is approached. In the Red Sea and the Persian Gulf evaporation exceeds precipitation and runoff, and salinities greater than 40‰ occur. In general, however, seawater contains about 3.5% salt.

In estimating salinity the classical procedure has been to estimate the chloride (plus iodide and bromide) content of the water by titration with silver nitrate. For oceanographic purposes, highly precise values of these variables are required, because they are used to calculate density which, in turn, is used in the calculation of currents by dynamic methods. For these reasons, highly standardized titration procedures using special glassware were developed by oceanographers and were formerly in universal use. However, the electrical conductivity of seawater at a fixed temperature can be related to the salinity, the chlorinity, or the density, and because electrical measurements can be made more precisely than the chemical determination, titration procedures have been replaced largely by conductivity measurements in most oceanographic ships and laboratories.

In inshore areas and estuaries salinity may be influenced by rivers or land runoff, and for many purposes less accurate salinity determinations may be adequate. Hydrometers or simple conductivity cells may be used, and continuous records from an instrument that records the salinity with less precision may be more useful than highly precise point observations. The salinity–temperature–depth recorder (STD) (Jacobson, 1948) is an early model of such an instrument.

In general the salinity of the oceans is reasonably uniform because the oceans are well mixed (Maury, 1855). Except at phase boundaries, all the processes in the ocean tend to mix it. Only at the interfaces between the water and the air, land, and solid matter in the water, including plants and animals, can processes that "unmix" or lead to a less uniform distribution, occur. Every part of the ocean is connected with and mixes with every other part of it. The consequence of this mixing is that the salinity of over 97% of the seawater in the world lies between 33 and 37‰ (Cox, 1965).

Properties of seawater that depend on the salinity

It is evident from the above that many of the properties of seawater can be defined as functions of the salinity, temperature, and pressure. These include the density, the temperature of maximum density, electrical conductivity, refractive index, speed of sound transmission, viscosity, the colligative properties (vapor-pressure lowering, boiling-point elevation, freezing-point depression, osmotic pressure), the specific heat (at constant pressure and at constant volume), the latent heat of evaporation, adiabatic temperature changes on compression and expansion, and the viscosity.

Density. The density of seawater can be determined as a function of the salinity and the chlorinity from the tables of Knudsen (1901) or expanded tables published by La Fond (1951). These tables give values for conversion between chlorinity and salinity, which may also be done using Knudsen's (1901) relationship, $S = 0.030 + 1.8050\ Cl$, in which S is the salinity as defined above and Cl is the chlorinity as determined by titration and defined as the total amount of chlorine, bromine, and iodine in grams contained in 1 kg of seawater, assuming that the bromine and iodine have been replaced by chlorine. Knudsen's tables also contain the corresponding values of σ_o and $\rho_{17.5}$. σ_o is $(s_o - 1)1000$, where s_o is the specific gravity of seawater at 0°C referred to distilled water at 4°C. It is related to the chlorinity by the equation

$$\sigma_o = -0.069 + 1.4708\ Cl - 0.001570\ Cl^2 + 0.0000398\ Cl^3.$$

The quantity $\rho_{17.5}$ is given by

$$\rho_{17.5} = \left(\frac{S_{17.5}}{S_{17.5}} - 1\right) 1000, \qquad [1]$$

where $S_{17.5}$ means the specific gravity of seawater at 17.5°C referred to distilled water at 4°C and $S_{17.5}$ is the specific gravity of distilled water at 17.5°C in proportion to distilled water at 4°C. It is related to σ_o by the equation

$$\rho_{17.5} = (0.1245 + \sigma_o - 0.0595\ \sigma_o + 0.000155\ \sigma_o{}^2) \times 1.00129.$$

Knudsen (1901) also published tables for determining σ_t, which is a measure of the density of seawater at the temperature t and at 1 atm. pressure. If the density at the temperature is s_t, then

$$s_t = 1 + \frac{\sigma_t}{1000}.$$

Knudsen's tables have been standard in the oceanographic world since their publication and were officially adopted by the International Council for the Exploration of the Sea, but their validity in reflecting precisely the world's seawaters is now in question (Cox, 1965). The relationships were based on precise analyses of a small number of seawater samples that were collected in a limited area. Cox (1965) has reviewed the objections to the relationships in Knudsen's tables, but the accuracy and precision of the relationships will probably be acceptable for all but the most precise calculations of physical oceanography. The relationships between $S(Cl)$, T, and σ_t are also available on nomograms published by the U. S. Naval Oceanographic Office as *Miscellaneous Publication 15530*.

Temperature of maximum density. Unlike pure water, which has its maximum-density at a temperature very near 4°C, the temperature at which the density of seawater is maximum decreases as the salinity increases, until at a salinity of 24.70‰ the maximum density is at the freezing point ($-1.332°$ C). At greater salinities, the density continues to increase as the temperature decreases until the freezing point is reached.

Electrical conductivity. Because electrical measurements can be made with a high degree of precision, it is now the general practice to determine the electrical conductivity of seawater as a measure of its salinity. Changes in temperature also alter the conductivity, so it is customary to make the measurements at a carefully controlled or carefully measured temperature. The generally used relationships among temperature, salinity (chlorinity), and conductivity are those of Thomas, Thompson, and Utterback (1934), although the absolute accuracy of these data have been questioned in recent years (Pollack, 1954; Cox, 1965). Weyl (1964) has reduced their data to an equation, according to which

$$\log K_S = 0.57627 + 0.892 \log Cl\ (\text{‰}) - 10^{-4}\tau\ [88.3 + 0.55\tau + 0.0107\tau^2$$
$$- Cl\ (\text{‰})\ (0.145 - 0.002\tau + 0.0002\tau^2)],$$

where

$\tau = 25 - t$ C, t is the Celsius temperature, and K_S is the specific conductance in millimhos per centimeter. The equation holds over a chlorinity range of 17 to 20%, from 0 to 25°C and at a pressure of 1 atm.

The refractive index. The refractive index of seawater is an optical property that varies with the salinity (chlorinity). Refractometers of high precision can be made, so it has been proposed that the refractive index be measured as a means of arriving at the salinity (chlorinity). The conductivity method now seems preferable, but Bein, Hirsckorn, and Möller (1935) made some use of refractive-index measurements to arrive at the density of seawater.

According to Utterback, Thompson, and Thomas (1934),

$$n_{D,25°} = 1.33250 + 0.000328\ Cl,$$

in which $n_{D,25°}$ is the refractive index of the D line of sodium vapor measured at 25°C. Miyake (1939) reported slightly different coefficients:

$$n_{D,25°} = 1.332497 + 0.000334\ Cl.$$

Cl is the chlorinity, in parts per thousand, in both equations.

Speed of sound transmission. This property is one of the most important variables of seawater, because sound is the only known form of radiant energy that is transmitted through great distances in water. In seawater, the speed of sound transmission varies with the salinity, the temperature, and the pressure (depth). It has been shown that the major part of the salinity effect is attributable to the magnesium ion, but because of the fixed relationship of the magnesium ion concentration to the chlorinity or salinity, widespread determinations of the magnesium ion are not indicated.

The speed of sound transmission can be computed from the density of seawater, its specific heats at constant volume and at constant pressure, and its compressibility. Calculated tables giving the speed as a function of the temperature, salinity, and pressure have been published by the Hydrographic Department of the British Admiralty (Matthews, 1939). Recently developed instruments permit the direct measurement of the speed of sound in seawater, and the U. S. Navy Oceanographic Office (1962) has published tables of the speed of sound based on direct measurements by Wilson (1960). These are probably more accurate than the computed tables.

The colligative properties. The colligative properties of a solution are those that depend on the number of particles of solute and not on the nature of the particles, which can be ions or molecules. (Other physicochemical properties depend not only on the number but also on the kind of particles. The colligative properties are the vapor pressure, the freezing point, the boiling point, and the osmotic pressure. In the case of aqueous solutions, these properties are generally compared with the corresponding value for distilled water.

Vapor pressure. The vapor pressure of pure water is a function of its temperature; values can be found in handbooks of chemistry and physics. Between 0 and 30°C, it increases from 4.6 to 31.8 mm of mercury. The vapor pressure and the chlorinity of seawater are related by the equation

$$e = e_o (1 - 0.000969\ Cl),$$

where e is the vapor pressure of the seawater, e_o is the vapor pressure of distilled water at the same temperature, and Cl is the chlorinity in parts per thousand.

Freezing point. The lowering of the freezing point (in centigrade degrees) of seawater arising from its salt content is related to the chlorinity (in parts per thousand) by the equation

$$\text{freezing-point lowering} = -0.0966\ Cl - 0.0000052\ (Cl)^3.$$

Boiling point. The *normal* boiling point of a liquid is that temperature at which its vapor pressure is 1 atm (760 mm of mercury). The following approximate boiling points of seawater have been determined:

Chlorinity (‰)	Boiling point (°C)
18.0	100.3 ± 0.1
20.4	100.7 ± 0.1
29.4	100.9 ± 0.1
48.3	101.3 ± 0.1
61.2	101.9 ± 0.1

Osmotic pressure. If pure water and a solution, or two solutions of different concentration, are separated by a semipermeable membrane (a membrane that permits the passage of water but not of the solutes), water will pass into the more concentrated solution and a pressure differential will be established across the membrane. This process is called osmosis; the pressure differential, the osmotic pressure. This phenomenon is vital in biological processes, and cells must be able to withstand osmotic pressure differentials across their walls or they will collapse or rupture. This is fundamentally why the salinity controls the distribution of many marine organisms; why fresh-water organisms are generally killed by exposure to seawater and vice versa; and the main reason we cannot drink seawater.

The osmotic pressure of seawater at 0°C can be calculated from its freezing-point lowering (in °C, see above) from the equation

$$OP_o = -12.08 \times \text{(freezing-point lowering)}.$$

At other temperatures, the equation

$$OP_t = OP_o \times \frac{273 + T}{273}$$

would apply. OP is the osmotic pressure in atmospheres and T the Celsius temperature. On increasing the chlorinity of seawater at 25°C from 10 to 22‰, the osmotic pressure increases from 12.87 to 29.33 atm.

Specific heat. The specific heat is the amount of heat (in calories) required to increase the temperature of 1 g of a substance by 1°C. The specific heat of seawater varies with the chlorinity and has different values at different temperatures. Two entities are to be considered, the specific heat at constant volume C_v and the specific heat at constant pressure C_p. The latter value is larger, because if a substance is heated at constant pressure, it will expand and do work against its surroundings, so a larger amount of heat energy is required to raise the temperature.

At 0°C and 1 atm pressure,

$$C_p \text{ (cal/g)} = 1.005 - 0.004136\, S + 0.0001098\, S^2 - 0.000001324\, S^3,$$

where S is the salinity in parts per thousand. C_v can be estimated from C_p by applying a term involving the temperature T, the coefficient of thermal expansion e, the density ρ, the true compressibility K, and J, the mechanical equivalent of heat:

$$C_v = C - \frac{Te^2}{\rho KJ}.$$

Latent heat of evaporation. It is assumed that the latent heat of evaporation of seawater (the amount of heat required to produce 1 g of water vapor at the same temperature of the water) is nearly the same as that of distilled water.

Adiabatic temperature changes. Seawater is slightly compressible, and when it is compressed work is done on it. If this is done without exchange of heat with the surroundings, the temperature of the water will increase. In the ocean the pressure is proportional to the depth, and at depths greater than 1000 m, the adiabatic temperature increase is appreciable. In most of the ocean the salinity effect on the temperature change is nearly constant, because the salinity at depths greater than 1000 m is nearly uniform. The effect increases with both the temperature and the salinity. In changing the salinity from 30 to 38‰ at 10°C (such as in the Mediterranean Sea) the adiabatic cooling associated with raising water from 1000 m to the surface would change from 0.118 to 0.130°C (Sverdrup, Johnson, and Fleming (1942)).

Viscosity. The viscosity of seawater increases as the salinity increases and as the temperature decreases. These variations have been investigated by Miyake and Koizumi (1948).

8.2 THE DISSOLVED GASES IN SEAWATER

Their nature and range

As a first approximation, oceanographers assume that at one time or another all the water of the oceans has been at the sea surface and at that time became saturated with air. The amounts of the air gases so dissolved (the equilibrium solubilities) are functions of the salinity and temperature. The solubilities of oxygen, nitrogen, and argon in seawater equilibrated with air have been reported by a variety of workers and recently reviewed by Richards (1965). The solubilities of the air gases differ, so their ratios in air-saturated seawater will differ from their ratios in air. Air is highly uniformly composed of 78.08% nitrogen, 20.95% oxygen, 0.03% carbon dioxide, and 0.93% argon (Glueckauf, 1951). However, if these gases were extracted from air-saturated seawater, the compositions of the mixtures would be as shown in Table 8-3, and would vary somewhat with the temperature and salinity at which the seawater became saturated. In addition to the marked differences in the ratios of these gases to each other, their absolute concentrations (except for molecular carbon dioxide) will generally be much smaller in air-saturated seawater than in air. Thus the concentration of oxygen in air is about 210 ml/liter, so the oxygen requirements of air-breathing animals are rather easily met, but in the marine

TABLE 8-3
Some Equilibrium Solubilities of Gases in Seawater Equilibrated with Air

	2°C			15°C			25°C		
Chlorinity (‰)	$O'_2{}^a$	$N'_2{}^b$	Ar'	$O'_2{}^a$	$N'_2{}^b$	Ar'	$O'_2{}^a$	$N'_2{}^b$	Ar'
4	9.22	16.50		6.75	12.61		5.54	11.81	
10	8.57	15.39		6.30	11.79		5.21	10.32	
15			0.405^c			0.298^c			0.248^c
16	7.96	14.28	0.416^d	5.90	10.97	0.312^d	4.90	9.43	0.261^d
		14.40^e	0.400^c		11.01^e	0.294^c		9.34^e	0.244^c
18	7.77	13.91	0.406^d	5.78	10.69	0.306^d	4.81	9.13	0.256^d
		14.05	0.389^c		10.79	0.286^c		9.17^e	0.237^c
19.12						0.282^f			
20	7.58	13.54	0.396^d	5.66	10.42	0.299^d	4.72	8.83	0.252^d
		13.71^e	0.379^c		10.56^e	0.279^c		9.00^e	0.230^c

Composition of Air Gases in Seawater Equilibrated with Air^g

	2°C				15°C				20°C			
Chlorinity (‰)	Total Gas	O_2(%)	N_2(%)	Ar(%)	Total Gas	O_2(%)	N_2(%)	Ar(%)	Total Gas	O_2(%)	N_2(%)	Ar(%)
16	22.78	34.94	63.22	1.83	17.22	34.26	63.93	1.81	14.50	33.79	64.41	1.80
18	22.23	34.96	63.21	1.83	16.88	34.25	63.94	1.81	14.24	33.79	64.41	1.80
20	21.69	34.95	63.22	1.83	16.52	34.26	63.93	1.81	13.97	33.78	64.41	1.80

^{a} O'_2 solubilities from Carpenter (1966).
^{b} N'_2 solubilities from Rakestraw and Emmel (1938).
^{c} From Rakestraw and Emmel (1938).
^{d} Calculated from N'_2 calculated according to Footnote e below and the Benson and Parker (1961) relationship, $N'_2/Ar' = 37.48 + 0.0522\,t$.
^{e} N'_2 solubilities calculated from the O_2 solubilities of Carpenter (1966) and the Benson and Parker (1961) relationship. $N'_2/O'_2 = 1.807 + 0.00396\,t$ for the chlorinity range of about 17.7 to 20‰.
^{f} From Köing (1963). For a chlorinity of 19.12‰, König also gives the following temperatures and corresponding Ar solubilities: 1°C, 0.359 ml/liter; 5°C, 0.329 ml/liter; 10°C, 0.306 ml/liter; 20°C, 0.245 ml/liter.
^{g} Italicized figures from the first part of the table were used to compute these percentages.

environment the concentration of oxygen is between approximately 5 and 10 ml/liter, so that marine organisms must have highly efficient organs for extracting oxygen, or lower metabolic requirements, or both, to survive the rigors of the marine environment. On the other hand, the availability of carbon dioxide is more favorable in the marine environment, where, in all ionic and molecular forms, its concentration is about 50 ml/liter, in contrast to about 0.3 ml/liter in air. We might assume that this difference is related to some of the differences in marine and land plants and their adaptations for carbon dioxide uptake and photosynthesis.

Oceanographic observations of the concentrations of the air gases in surface and near-surface waters of oceans suggest that the surface waters are approximately saturated with air, although there are regions where there are systematic departures from 100% of the saturation values of up to about $\pm$ 10%. In some regions the surface waters appear to be oversaturated with oxygen, probably as a result of oxygen photosynthetically produced in the upper layers diffusing into the surface layer. In other regions oxygen-deficient water may be carried to the surface, having lost its oxygen by the respiration of plants and animals at depth, a process that is discussed later.

The concentrations of oxygen, carbon dioxide, and nitrogen can be altered by biological processes. These processes alter the oxygen and carbon dioxide concentrations more extensively and more universally than they do the nitrogen. Oxygen and carbon dioxide are altered reciprocally by photosynthesis and respiration, which can be represented by the equation

$$CO_2 + H_2O \xrightleftharpoons[\text{respiration}]{\text{photosynthesis}} CH_2O + O_2.$$

In this equation the organic matter formed by photosynthesizing plants is represented simply as CH_2O, but this is not intended to be an accurate representation—the organic matter is obviously much more complex. Light energy and photosynthetic plants must be present for the reaction to proceed to the right; photosynthesis and the production of oxygen in the water are therefore limited to the upper layers of the sea that are penetrated by sunlight. In the clearest parts of the oceans, the central oceanic gyres, adequate sunlight for photosynthesis to exceed respiration penetrates to a maximum of about 100 m; in other parts of the oceans it penetrates much less, and the compensation depth (that depth at which photosynthetic production and respiratory consumption of oxygen just balance) may be within a few centimeters of the surface.

It should be pointed out here that quantitatively by far the most important marine plants are microscopic, generally one-celled organisms, the phytoplankton. These drift with the water and reproduce more rapidly when rapidly photosynthesizing. However, as their numbers increase, the absorption of light and consequently the depth of light penetration, increase and decrease, reciprocally. Thus, regions where there are large populations of photosynthetic plants are also regions where the depth to which photosynthesis can take place is limited.

It would be anticipated from the foregoing that the concentration of dissolved oxygen in the ocean would vary widely from the concentrations that might be expected from the solution of air, and such is the case (Fig. 8-1). In near-surface layers (generally the upper 25-50 m) oxygen

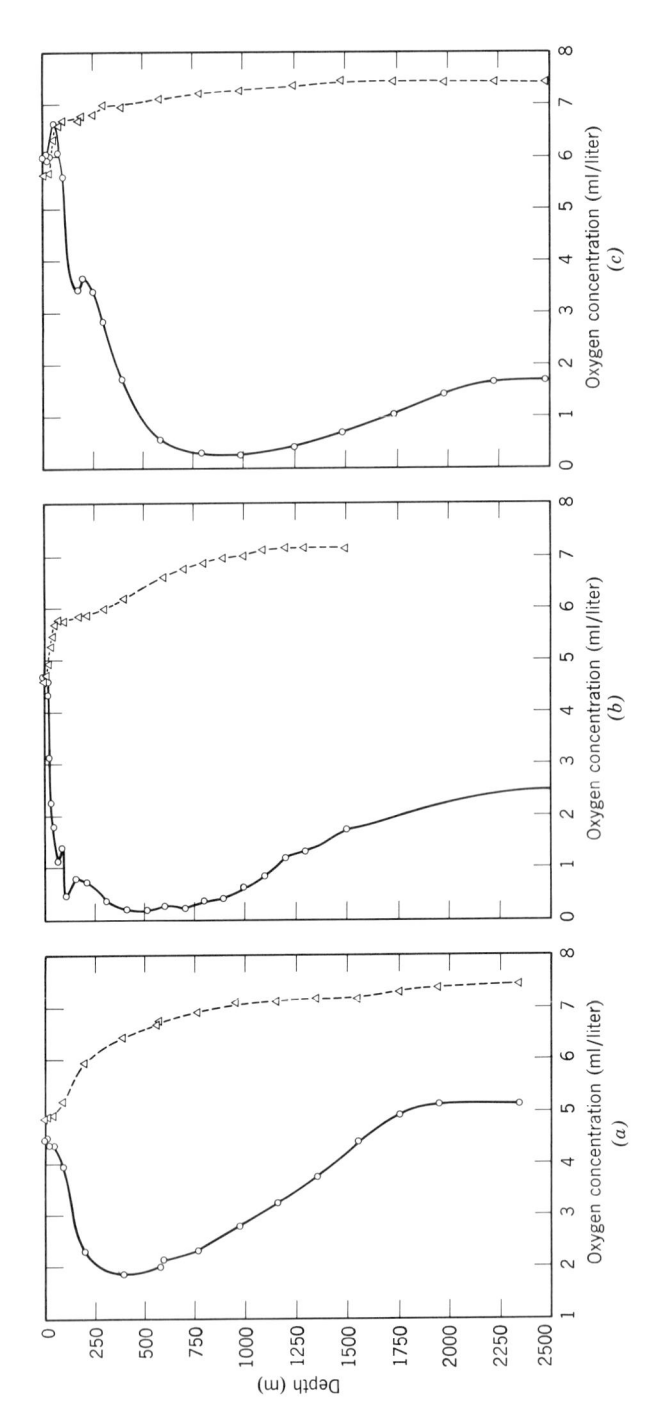

FIG. 8-1 Vertical distributions of dissolved oxygen (O) and of the amount of oxygen to be expected from the equilibrium solution of air (△). (a) A station in the tropical North Atlantic Ocean, R. V. *Albatross III* Station 44, 15° 25'N, 44° 43'W, Feb. 6, 1952. (b) An eastern tropical North Pacific Ocean Station, R. V. *Thomas G. Thompson* Station 1-57, 9° 9.5'N, 88° 59.7'W, Nov. 30, 1965. (c) An eastern North Pacific Ocean Station, R. V. *Brown Bear* Station 312-15, 48° 31'N, 126° 58'W, Sept. 16, 1962. The area between the two curves is an approximate measure of the amount of oxygen consumed from the water.

concentrations of up to twice those expected from the solution of air may be encountered; these apparently arise from photosynthesis. In mid-depths the oxygen generally decreases to some minimum value that may be only a small fraction of the equilibrium solubility. In the eastern tropical Pacific Ocean this intermediate zone of low oxygen is several hundred to over 1000-m thick, and the oxygen concentrations may be almost immeasurably low.

The deep waters of the ocean basins are generally renewed by waters that were formed by the sinking of cold waters in high latitudes. These waters have initially high oxygen and relatively low organic contents, so as they sink and flow to the north and to the south, they effectively ventilate the great depths of the ocean, which consequently never become oxygen free.

Although both oxygen and carbon dioxide in the ocean are drastically altered by biological processes, this is not the case with nitrogen and argon. Two biological processes are known that can alter the concentration of dissolved nitrogen—nitrogen fixation and denitrification. The former process is well known in land and fresh-water organisms and is the function of the nitrogen-fixing bacteria associated with the nodules on leguminous plants. It has recently been shown that marine blue-green algae of the genus *Trichodesmium* can fix nitrogen. This means that these plants may consume some of the free nitrogen dissolved in the water and produce various nitrogen compounds. There are no estimates of how this process alters the nitrogen economy of the ocean, but extensive blooms of *Trichodesmium* occur in the tropical oceans, so the process may be important to biological cycles in the sea.

Denitrification is a bacterial process during which oxidized compounds of nitrogen, nitrate, and nitrite ions are reduced to yield energy to the bacteria. It appears that the chief nitrogenous endproduct of this process is free nitrogen itself. However, the process probably only takes place in restricted marine areas, generally basins and fjords, where the circulation is so limited that stagnation occurs. As organic matter, in the form of the bodies or remains of plants and animals, rains into these waters, bacterial respiration proceeds at the expense of dissolved oxygen. Ultimately, all the oxygen can be so consumed, unlike in the open ocean, because the depths cannot be ventilated by the introduction of oxygen-bearing water. The evidence indicates that only after all, or nearly all, of the oxygen has been so consumed can denitrification begin. Thus this process that alters the nitrogen concentration is neither common nor extensive. The major example of an environment where denitrification can take place is the Black Sea, which is around 2000 m deep and becomes devoid of oxygen at depths of between 150 and 250 m. Other such environments have been discussed by Richards (1965). It appears that the quantity of nitrogen produced by denitrification in marine environments is only a few percent of that arising from the solution of air.

The above remarks concerning oxygen-free environments are concerned with the water column itself. Conditions under which sediments become anoxic are much more common. In sediments oxygen can be supplied only by slow diffusion, and if the material falling to the bottom is reasonably rich in organic matter, the sediments will be anoxic as a rule. Denitrification would be assumed to take place, but it would be quantitatively limited to the nitrate and nitrite ions trapped in the interstitial waters or produced there by the oxidation of ammonia released from the decomposing organic matter. However, the oxygen-free reducing environment that is common under the surface of sediments is of special engineering interest. The question of the anaerobic corrosion of iron in soil has been studied by Starkey and Wight (1945). The author, while working with the Turkish Naval Hydrographic Office, noted excessive corrosion of a galvanized plough steel wire cable that had been used for frequent samplings of the sulfide-bearing waters of the Black Sea.

Argon is the third most abundant gas in air, and König et al. (1964) have reported that water samples from various depths in the Pacific and Indian oceans contained within 3% of the amount of argon that would be expected from the solution of air. The same was found to be true for neon and helium, which are present in much smaller amounts in the atmosphere (18.2 and 5.2 ppm, respectively). Many other gases are present in trace amounts in the air and can therefore be assumed to be present in seawater in trace amounts, unless they are destroyed or altered by reaction upon solution. For example, ammonia, NH_3, would react with the water to form ammonium hydroxide, NH_4OH, which would in turn partially ionize to form ammonium and hydroxyl ions, NH_4 and OH_3.

The role of carbon dioxide in the ocean has been thoroughly reviewed by Skirrow (1965). It reacts with the water according to the equations

$$CO_2 + H_2O = H_2CO_3$$

$$H_2CO_3 = H^+ + HCO_3^-$$

$$HCO_3^- = H^+ + CO_3^=$$

At the pH of seawater, the most abundant carbonate species is the bicarbonate ion, HCO_3^-. Because of the above reactions, the amount of bicarbonate–carbon (the sum of the CO_2, H_2CO_3, HCO_3^-, and $CO_3^=$) is much larger than the concentration of CO_2, which, at the sea surface, is fundamentally governed by the partial pressure of CO_2 in the air, 0.0003 atm.

Gas solubilities and exchanges across the sea surface

In general the equilibrium solubility of a gas in seawater varies inversely with both the temperature and the salinity. Gases can and will exchange across the sea surface in response to a thermodynamic drive toward equilibrium, which is achieved when

$$p = P,$$

where p is the partial pressure of the gas in the gas phase and P its partial pressure in solution. The former is the product of the total pressure of the atmosphere and the mole fraction of the gas in the mixture.

The solubility of a gas in a liquid can be expressed by an absorption coefficient, α, which is the amount of gas that will be dissolved in the liquid when equilibrated with the gas when the pressure of the gas itself, without the vapor of the liquid phase, is 760 mm. In seawater the absorption coefficients are functions of the temperature and salinity, and decrease as both temperature and salinity increase. When they are multiplied by the mole fraction (approximately the volume percent divided by 100) of the gas in the air, the solubility of the gas in seawater equilibrated with a standard atmosphere is obtained (Henry's law). In studies of the solubility of gases in seawater, it has been the practice to equilibrate seawater with air and then to analyze the resulting solution.

Several workers have investigated the solubilities of the various air gases in seawater, notably Fox (1909), Whipple and Whipple (1911), Rakestraw and Emmel (1938), König (1963), and Truesdale, Downing, and Lowden (1955). For many years the values for the solubility of oxygen that were published by Fox were considered definitive, until new experiments by Truesdale et al. (1955) gave different results. Other investigators in turn questioned both sets of values. Carpenter (1966) published a set of values that he asserts are of higher accuracy and precision than those previously published, but cautioned that his methods for determining the dissolved oxygen content were also different from and more accurate than those of earlier workers. He concluded that determinations of dissolved oxygen carried out with most of the modifications of the Winkler method now in use are probably more comparable to some of the older oxygen solubility values than they would be to his newer, more accurate figures.

The above discussion is intended to introduce the concept that exchanges of oxygen across the sea surface—either solution of oxygen or its release to the air—take place in response to physical conditions, which are reflected in the partial pressure of the gas in the liquid and solid phases. The partial pressure of oxygen in air at the sea surface is relatively constant, varying principally with the total barometric pressure (assuming that at the surface of the sea the air is saturated with water vapor). Within the water, the partial pressure is proportional to the ratio of the oxygen content to the solubility of oxygen (O_2/O'_2). Both these factors may vary widely; O_2 in response to photosynthetic production and respiratory consumption, O_2 in response to temperature changes in the water.

The properties the dissolved gases impart to seawater

The ability of the sea to sustain forms of life higher than bacteria depends on its dissolved oxygen content. Although it has been shown that fish and other life occur in seawater that contains very little oxygen, oxygen is a biological requirement. On the other hand, there is little evidence of very direct correlations between the presence of fish and the abundance of oxygen. Correlations between the occurrence of fish and temperature and organisms on which fish feed are apt to be much stronger and more direct. Low concentrations of dissolved oxygen may alter, govern, or limit the presence of marine organisms, at least in special cases, but it is doubtful if variations in the concentrations of dissolved nitrogen and argon have any appreciable effects on the life of the sea.

From the point of disposing of organic wastes at sea, oxygen is essential to the efficient biochemical decomposition of such wastes. As pointed out elsewhere, the decomposition of organic matter at the expense of sulfate ions can proceed in the absence of dissolved oxygen, but the products, sulfides, are noxious and toxic.

8.3 THE *p*H, ALKALINITY, AND BUFFER CAPACITY OF SEAWATER

Their nature and range

Seawater is slightly basic, having a *p*H of 8.0 to 8.2 at the surface. The *p*H generally decreases to values of around 7.7–7.8 as the depth increases. Seawater contains an excess of cations of strong bases over the anions of strong acids, making it somewhat alkaline. The alkalinity arises primarily from the carbonate, bicarbonate, and borate ions—all anions of weak acids. Anions of other weak acids are present, such as phosphates, silicates, and in the special anoxic environments described earlier, sulfides, but their concentrations and the hydroxyl-ion concentration are generally too small to make measurable contributions to the alkalinity. The anions of weak organic acids, such as acetate, also contribute to the alkalinity, but their effect in the ocean has not been evaluated.

The *p*H, the alkalinity, and the solubility of calcium carbonate are intimately related in considerations of seawater. As carbon dioxide is added to seawater, as by respiration, it reacts with the water

$H_2O + CO_2$ (carbon dioxide) $\rightleftharpoons H_2CO_3$ (carbonic acid) $\rightleftharpoons$
$H^+ + HCO_3^-$ (bicarbonate ion) $\rightleftharpoons H^+ + CO_3$ (carbonate)

and the pH decreases because of the increased hydrogen ion concentration.

If solid calcium carbonate ($CaCO_3$) is present the reaction

$$CaCO_3 \text{ (solid)} + H_2CO_3 \rightleftharpoons Ca^+ + 2HCO_3^-$$

will tend to take place and the calcium carbonate will be dissolved. During the photosynthetic consumption of CO_2, the reverse processes will take place. In the course of a 24-hr day, photosynthesis and respiration can produce large changes in the CO_2 content of the water, so there will tend to be a daily cycle of these events. This cycle would tend to have a counterpart in the seasonal cycle in temperature and high latitudes, where photosynthetic consumption of CO_2 would predominate in spring and summer, respiration in winter.

Both temperature and pressure increases decrease the pH. There is also an independent pressure effect on the solubility of calcium carbonate (Revelle and Fairbridge, 1957) and, combined with the pressure effect on the pH, there would be a tendency for calcium carbonate (limestone) to dissolve from bottom sediments and be precipitated at the surface, provided water circulated from the surface to the bottom and vice versa. This mechanism may be approximated in the real ocean as exemplified by the precipitation of calcium carbonate on the Grand Bahama Banks. Currents bring somewhat deeper waters to the surface, where photosynthesis can proceed intensely. At the same time, the water is warmed by solar radiation, which has the effect of decreasing the solubility of CO_2 in the water and making it tend to escape to the atmosphere. The solar heating also decreases the solubility of calcium carbonate, which, unlike most salts, becomes less soluble as the temperature increases. The precipitation of calcium carbonate, principally as aragonitic oolites, follows.

The properties the *p*H, alkalinity, and buffer capacity impart to seawater

Primarily because of the carbonate system, seawater tends to resist changes in pH, that is, it is buffered. This buffering prevents wide fluctuations in the pH and also tends to fix the ratio of carbonate to bicarbonate ions. This presumably is biologically significant.

8.4 MICRONUTRIENTS IN SEAWATER

Their nature and range

Although the 11 major constituents and the atmospheric gases account for almost all the material dissolved in seawater, if other substances were not present the sea would be sterile and lifeless. Everything else present in seawater may be considered to be present in trace amounts, but some of these materials are essential to all marine life. *Micronutrients* may be defined as materials essential to the growth of plants or animals that are present in such small concentrations that they may be so completely stripped from the water by the growth as to limit or alter additional growth. Of these substances, nitrogen and phosphorus have been the most studied, but other inorganic trace elements, such as silicon, iron, manganese, molybdenum, zinc, cobalt, copper, vanadium, and organic compounds such as vitamins, may fit the definition. These materials are *nonconservative*, in contrast to the major constituents, in that biological activity may involve *major fractions* of the amounts of them present. On the other hand, most of the major constituents are also essential to life processes, but in general they are present in superabundant supply and insignificant fractions of their total amounts are incorporated into the bodies of plants or animals. The micronutrients generally occur in the parts-per-million or parts-per-billion concentration ranges.

Nutrient cycles in the sea

In the upper layers of the ocean that light penetrates, photosynthetic growth forms new plant material. This material is literally compounded from substances in solution in the water—the carbonate components and nitrogen and phosphorus compounds. During this process, substantial fractions or sometimes practically all of the micronutrients present are organically bound in the bodies of the plants. Once there, they can enter the food web in which the plants are eaten by herbivores, which are in turn eaten by carnivores and so on. During these *particulate* stages—when the nutrients are bound as living or dead solid matter—the nutrients can be moved through the water independently of the motion of the water itself by swimming and by gravity. These processes distribute the nutrients differently from the distribution of the major constituents, which are all distributed essentially identically. The net direction of transport of particles is downward, and below the depths to which light penetrates, the net biological process is respiration, by which the particulate matter is redissolved or returned to the inorganic form (remineralization). These processes constitute an irreversible downward transport of the nutrients, and only the motion of the water itself can return the nutrients to the upper, lighted layers where photosynthesis can proceed and the cycle can begin again. The same processes are common in lakes in the temperate latitudes. The summer warming of the surface layers results in a density stratification that presents an impediment to vertical mixing. The growth and sinking of organisms from the surface layers strip these layers of nutrients and effectively transport them to the deeper waters. The end of summer may be a season when little photosynthesis is taking place in such lakes, and photosynthesis cannot begin again until the autumnal cooling of the surface waters and the increased winter winds cause the lake to mix to the bottom and "overturn." The nutrients will then be almost uniformly distributed vertically, and the photosynthetic cycle can begin again in the spring.

In the temperate latitudes seasonal density stratification may develop in the ocean as described for lakes above. In such regions the nutrient cycle may take place seasonally, with annual refertilization of the upper layers by deep mixing of the water column in winter. Such is the case in the Gulf of Maine, which has been intensively studied by Redfield and his colleagues (Redfield, Smith, and Ketchum, 1937). Other parts of the ocean, such as the Sargasso Sea and the other central oceanic gyres, are permanently stratified and renewal of the upper layers by deep mixing cannot take place. These are sites of low rates of photosynthetic productivity, and the surface layers are nearly devoid of phosphates, nitrates, and silicates. One can then think of the nutrient cycle taking place in space, rather than time, in the open ocean. In regions of upwelling, generally along the west coasts of continents, the trade winds remove surface waters, usually to the west. These are replaced by somewhat deeper, nutrient-rich waters, and as long as the winds persist, these are sites of continuously high rates of nutrient supply. If the other factors for plant growth, such as the temperature and light intensity, are suitable, photosynthesis will proceed, corresponding to the springtime condition in the lake or seasonally stratified ocean.

Phosphorus cycle. The Atlantic Ocean may contain up to about 68 ppb of soluble compounds of phosphorus, whereas the Pacific Ocean may contain about 50% more. Because of the cycles described above, most of the inorganic form of the material is present at depths greater than the photosynthetic zone. In the depths to which light penetrates phosphorus is taken up, photosynthetically, by the growth of plants and incorporated into their bodies. At certain seasons of the year, most of the phosphorus may be thus bound into plant cells (Fig. 8-2). A small amount of the phosphorus may be excreted directly to the water by the plants, but in the main the phytoplankton cells will be ingested by herbivorous planktonic animals. These animals will assimilate a fraction of the phosphorus and eventually excrete some of it to the water as inorganic phosphate. Some of the ingested cells may pass through the animal's gut and be incorporated into fecal pellets that may sink to the bottom or be ingested by other animals. In either case much of the material eventually will be attacked by bacteria and further broken down or incorporated into bacterial cells. These in turn eventually

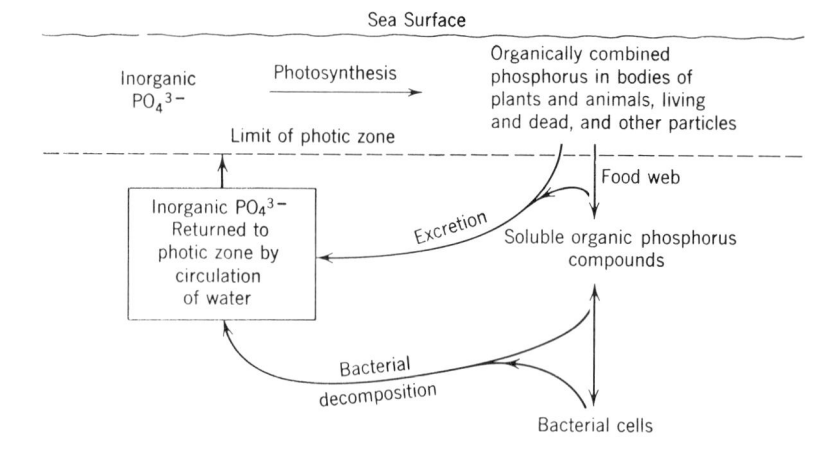

FIG. 8-2. Some aspects of the phosphorus cycle. The inorganic PO_4^{3-} can be converted into organic, particulate matter (living) only in the upper layers penetrated by light. The net tendency of the particles is to sink out of the photic zone, where the phosphorus is eventually returned to solution as inorganic PO_4^{3-} ions by respiration of plants, animals, and bacteria. It must then wait for the circulation of the water to return it to the photic zone, where it can re-enter the cycle. In lakes and seasonally stratified marine environments this return takes place in winter, when vertical mixing takes place. In the open ocean it takes place in areas of upwelling, which are highly productive as long as the upwelling continues.

must break down and release their phosphorus to the water, although a small fraction may be incorporated in the sediments.

Analytically, oceanographers distinguish between soluble and insoluble organic and inorganic phosphorus fractions. The definitions are operational—soluble is distinguished from insoluble on the basis of whether or not the material will pass through a specified filter. Inorganic phosphate is defined as that which will react with specified reagents to give a blue color; organic is that which will so react only after preliminary treatment to decompose the organic matter and release inorganic phosphate.

Nitrogen cycle. In describing the phosphorus cycle it was assumed that only one oxidation state of phosphorus (PO_4^-) need be considered. Such is not the case with nitrogen. The same distinctions between soluble and insoluble, organic and inorganic nitrogen compounds can be made as in the case of phosphorus, but nitrate (NO_3^-), nitrite (NO_2^-), and ammonium (NH_4^+) ions, all of which are soluble inorganic forms of nitrogen, must be considered. (We are here concerned with nitrogen compounds, not free, dissolved N_2.) Plants can take up nitrogen from the water any of these inorganic forms. We can then assume that when incorporated into plant cells, the nitrogen is in the amino form, $R–NH_2$, where R is an organic residue. We might further assume that ammonia (NH_3) is the first inorganic product in the breakdown of organic RNH_2 compounds (which may be complex proteins and soluble or simple amino acids formed by the breakdown of proteins). The NH_3 would react with the water to form ammonium ions, NH_4^+. These can then be attacked by nitrifying bacteria and oxidized first to NO_2^- and ultimately to NO_3^-. By far the most abundant form of combined nitrogen in the ocean is NO_2^-, and it is apparent that it is the most stable product in the nitrogen cycle. Thus, nitrite and ammonia are normally transient in the ocean, and large accumulations of them do not build up in the open ocean.

Although phosphorus compounds can be introduced to the sea almost exclusively from the leaching of rocks, nitrogen compounds can be formed by lightning discharges in the air and then directly introduced to the ocean. It has been mentioned that nitrogen fixation can be carried out by certain blue-green algae, adding to the stock of nitrogen compounds in the oceans, or nitrogen compounds can be destroyed during denitrification and lost as free N_2, depleting the stock.

Carbon cycle. The carbon cycle in the ocean follows a path similar to that of the nitrogen and phosphorus cycles. However, there are more than adequate supplies of inorganic carbon compounds to fulfill photosynthetic requirements, because the lack of nitrogen or phosphorus or possibly even silicates is certain to halt further photosynthesis before carbonate supplies are exhausted. However, during the cycle, inorganic carbon compounds in solution, CO_2, H_2CO_3, HCO_3^-, or CO^{2-}_3, will be taken up by photosynthesis and formed into organic compounds, and approximately 10% of the carbonate carbon may be so involved. Some of these organic compounds are respired directly by the plants, others enter the food chain to be ingested and either metabolized and eventually respired as CO_2, or excreted as organic compounds that may be passed along the food chain.

It is probable that few organisms are very efficient in recovering dissolved organic compounds from solution for their growth or respiratory needs. It has recently been demonstrated, however, that dissolved organic compounds may aggregate into particles that can serve as food for marine organisms. This aggregation may take place at the surface of bubbles or of other particles, and may form an important short circuit in the carbon economy of the sea. It is evident that the phosphorus and nitrogen in soluble organic compounds can be aggregated similarly.

8.5 OTHER MINOR CONSTITUENTS OF SEAWATER

Inorganic

Presumably all the naturally occurring elements and some of the manmade elements are present in seawater, but the concentration of many of them is minute, and little is known of their variations or distributions. In spite of their small concentrations, many of these elements may be biologically or geochemically significant, or they may have subtle catalytic properties. Goldberg (1963, 1965) has compiled lists of the abundance of the elements in seawater, but the lists contain little information on variations in time and space. Generally, these elements are difficult to determine analytically, and concentrations reported by different workers and from different laboratories are apt not to be in agreement or even comparable.

Several processes can alter the concentrations of trace elements or transport them through the oceans. If an element is introduced into the sea in solution, one process might be for it to form an insoluble compound with some component of seawater and precipitate out of solution. There is little evidence that this process has much control over the concentrations of trace elements. Krauskopf (1956) concluded that the sea was undersaturated with all of a group of 13 trace elements he studied. The most reasonable explanation seems to be that many of the elements are adsorbed or absorbed on particles and thus removed from solution.

Another important process involving trace elements is their biological uptake. It is known that many marine organisms can concentrate trace elements in their bodies or in specific organs. This uptake can be expressed as a concentration factor,

$$\text{C.F.} = \frac{\text{concentration in organism}}{\text{concentration in seawater}}$$

Many of the concentration factors that have been determined are large—many in the hundreds of thousands, particularly for elements that are present in minute concentrations. (Conversely, the concentration factors for the major constituents tend to be small—and may be less than unity.) It is evident that organisms will tend to scavenge trace elements from the water, effectively converting them to particulate forms in which they may eventually be deposited in the sediments on the sea floor.

An important consequence of the ability of organisms to concentrate trace elements in or on their bodies is concerned with radioactive trace elements. Nuclear explosions and the disposal of radioactive wastes at sea may introduce small concentrations of radioactive nuclides to the sea in solution. However, biological concentration processes can result in much larger concentrations of radioactivity in the bodies of marine organisms than are present in the waters in which the organisms occur. Thus radioactive chromium and zinc are scarcely detectable in the ocean water off the Columbia River, but there are readily detectable concentrations of them in some of the marine organisms and sediments of the region.

Organic

Many organic compounds or classes of organic compounds have been detected in seawater. Up to 1 or 2 ppm of organically combined carbon may occur in solution, with the highest concentrations occurring in the near-surface layers. The concentrations tend to vary seasonally in cycles that are intimately related to the biological cycles in the sea—most of the soluble organic matter probably arises as metabolic or breakdown products of plankton.

A large number of specific organic compounds or types of organic compounds have been identified. In a recent review of the subject Duursma (1965) lists carbohydrates, proteins and their derivatives, aliphatic carboxylic and hydroxy-carboxylic acids, biologically active compounds (vitamins, hormones, enzymes), humic acids, phenolic compounds, and hydrocarbons. Some of these compounds, such as thiamine and vitamin B_{12}, are known to be essential to life processes. It has been demonstrated that the concentration of carbohydratelike compounds can have an effect on the pumping rates of oysters. It has already been suggested that the aggregation of soluble organic compounds into particles may provide a source of food for zooplankton organisms.

8.6 PROCESSES ALTERING THE DISTRIBUTION OF CHEMICAL PROPERTIES IN THE SEA IN TIME AND SPACE

Exchanges across the boundaries of the ocean

Concentrations of chemicals (and heat) in the ocean can be altered by exchanges across the boundary of the ocean or by *in situ* (body) processes. The latter are diffusion, advection, and local time changes primarily attributable to biological processes.

Exchanges of water. Water is the most important substance that exchanges across the sea surface. Its exchange can take place by evaporation, precipitation, runoff, and the freezing and thawing of sea ice. When simple evaporation takes place we would expect only changes in the total concentration of the major solutes to result. The other processes will probably also carry along certain solutes and correspondingly alter the composition of the seawater. Precipitation may introduce nitric acid formed during lightning discharges and other materials, such as particles windborne from the land, combustion products, and so forth, to the sea. Runoff from the land will carry the products of the leaching of the land, the products of human activities, and other materials to the sea and thus markedly alter the composition of the ocean. In general seawater is dominated by the ions of sodium and chloride, whereas rivers are more commonly dominated by calcium, sulfate, and carbonate ions, so river runoff will tend to alter the ratios of the major constituents of seawater. However, the absolute concentration of ions in river water is much less than in seawater, so the effect is small. The freezing of sea ice will also alter the ratio of the major constituents in the water from which the ice is frozen. Although under ideal conditions, one might expect the freezing of ice to remove only water from the sea, this is never the case and pockets of brine are always trapped in the ice. The ratios of ions in this brine are not the same as in seawater, and they vary somewhat with the temperature and other conditions under which the ice is frozen. It can be seen, however, that the freezing and thawing of sea ice alters not only the salinity but also the ratios of the major ions.

Exchanges of solutes. Solutes can be exchanged across the sea surface primarily by the leaching of the land and by precipitation. Sea spray is blown ashore, there to deposit sea salt on the land. Nearly all of this salt, the so-called cyclic salt, is eventually carried back to the sea. This salt must be taken into account when one attempts to assess the leaching of the land from analysis of river waters. Large fractions of the sodium and chloride content of rivers are attributable to cyclic salts.

Solutes can be removed from the water and deposited in the sediments as the remains of organisms. The deposition of calcium carbonate and silica (as oozes) are important examples of this process. Other solutes, such as the trace metals, may be removed from solution by adsorption on inorganic or organic particles, or they may be incorporated into the bodies of plants or animals. Occasionally, the solubility of some inorganic phase may be exceeded and solutes removed by inorganic precipitation. These occurrences are probably rare, the most noteworthy being the precipitation of calcium carbonate on the Grand Bahama Banks referred to above. Sulfur may be removed from solution by the action of photosynthetic sulfur bacteria. This can take place when anoxic conditions have arisen and hydrogen sulfide has been formed. The photosynthetic sulfur bacteria can then use the hydrogen sulphate as an energy source in the photosynthetic formation of new protoplasm:

$$CO_2 + 2H_2S \rightarrow CH_2O + H_2O + 2S^o.$$

Exchanges of solids. Solids can be introduced into the ocean as the sediment burden of rivers, as dust carried to the sea, by subaerial and submarine volcanoes, by meteorites, and so forth. This solid material may be deposited into the sediments and subsequently undergo little or no physical or chemical change. In most cases, the particles probably are altered considerably in the marine environment by processes such as solution; aggregation; adsorption and absorption of other materials; and ingestion, maceration, and digestion by marine organisms. The particles finally deposited in the sediments may have little physical or chemical resemblance to the particles originally introduced from extramarine sources.

Exchanges of gases. The principal mechanisms by which gases are exchanged across the sea surface have been discussed. However, if the circulation of the sea brings into the surface layers waters from subsurface strata, gas exchanges will tend to take place because these upwelled waters will probably not be in equilibrium with the atmosphere, particularly with respect to oxygen and carbon dioxide but to a lesser extent with respect to the other gases. Heat will probably be transferred to or from the water, because it is not likely to be at the same temperature as the air. The warming or cooling will in turn alter the solubility of the gas and result in an exchange. For example, a parcel of upwelled water may contain just the equilibrium concentration of dissolved nitrogen for water of that temperature. If the water is then warmed by solar radiation, the nitrogen will become less soluble and the water will become oversaturated. A tendency for the nitrogen to be driven into the atmosphere will result. Such upwelled waters will generally be low in dissolved oxygen and will tend to take up oxygen from the air.

Body processes: conservative and nonconservative

Theory of the distribution of variables in the sea. Conservative properties of seawater have been defined as those properties that are not appreciably altered by *in situ* biological or chemical processes. Thus, although the major constituents such as potassium, sulfur, chlorine, and so forth are clearly required for life processes, they are present in such abundance that the fractions of them that are so involved are negligibly small. Another way of defining a conservative property is one that, except for boundary processes, whose local time change of concentration takes place solely by advection and diffusion. These same processes also alter the nonconservative, biologically (or chemically) altered properties, so that the local time change of their concentrations is the result of advection, diffusion, and *in situ* biological (or chemical) changes. Sverdrup *et al.* (1942) have formulated this as the equation for the distribution of variables:

$$\frac{\partial s}{\partial t} = \frac{A_x \partial^2 s}{\rho \partial x^2} + \frac{A_y \partial^2 s}{\rho \partial y^2} + \frac{A_z \partial^2 s}{\rho \partial z^2} \text{ (diffusion terms)}$$

$$- V_x \frac{\partial s}{\partial x} - V_y \frac{\partial s}{\partial y} - V_z \frac{\partial s}{\partial z} \text{ (advection terms)}$$

$$+ R \text{ (}in\ situ\text{ changes).}$$

In this equation S is any scaler property, or, in our case, a concentration. X, y, and Z are rectangular coordinates, X and y in the horizontal directions, Z originating at the sea surface and increasing downward. A_x, A_y, and A_z are diffusion coefficients, ρ is the density, and V_x, V_y, and V_z are velocities in the x, y, and z directions. The diffusive and advective terms alone will describe the local time change of concentration of a conservative property, but R must be added for nonconservative properties.

Gradients, velocities, and diffusion coefficients. In practice the above equation generally cannot be solved, and simplifying assumptions must be made to use it. The concentration changes with distance, $\partial s/\partial x$, $\partial s/\partial y$, and $\partial s/\partial z$ are concentration gradients. The horizontal gradients $\partial s/\partial t$ and $\partial s/\partial y$ are generally much smaller than the vertical gradients, $\partial s/\partial z$. In many cases the horizontal gradients can be assumed to be negligibly small and these terms are dropped, leaving a one-dimensional model concerned with vertical distributions. The usefulness of this kind of simplification becomes apparent if we consider the relative magnitudes of some typical horizontal and vertical concentration gradients. The dissolved oxygen distribution at two stations in the Gulf Stream might be considered (Table 8–4). Stations A and B are in the Gulf Stream system, where the horizontal gradients are apt to be unusually large, and thus could not be ignored, whereas Station C is beyond the major currents of the stream, currents are less marked, and the motion of water from Station B to Station C will alter the properties at C only slightly.

It is clear that the magnitude of the velocities in the advective terms will vary widely, and that the horizontal velocities will generally be larger than the vertical ones, at least in the open ocean. Currents of 5 to 6 knots have been measured in the Gulf Stream. Estimates of vertical velocities are much more difficult to make, but in general, they will be much smaller than horizontal velocities. However, because of the larger vertical concentration gradients, these terms may be important in governing the local time change in a concentration. This is particularly pertinent in regions of upwelling, where upward velocities may be small relative to horizontal currents but not negligible, and because of the large vertical gradients in the nutrients, the upward water motion provides a continuing source of nutrients to the photic zone so that these are regions of high plant productivity.

The diffusive terms A_x, A_y, **and** A_z. The magnitude of these coefficients of eddy diffusion is related to the distribution of density. If the density increases rapidly with depth, vertical mixing is hindered and A_z will be small, but the oposite is the case if the water column is only neutrally stable. Permanent and temporary pycnoclines occur in the ocean and represent layers where the resistance to vertical mixing is at a minimum. The pycnocline (Fig. 8-3) is that part of the water column where the density increases

TABLE 8-4

Horizontal and Vertical Oxygen Gradients In and Between Two Gulf Stream Stations (*Atlantis* Stations 4856, 4860) and In and Between Two Sargasso Sea Stations (*Atlantis* Stations 4863, 4865). Stations 4856 and 4860 are 39 km apart, 4863 and 4865, 37.12 km.

Depth (m)	Station 4856 O_2 (ml/liter)	$\frac{\delta O_2}{\delta z}$ (ml/m)	$\frac{\delta O_2}{\delta x}$ (ml/m)	Station 4860 O_2 (ml/liter)	$\frac{\delta O_2}{\delta z}$ (ml/m)	Station 4863 O_2 (ml/liter)	$\frac{\delta O_2}{\delta z}$ (ml/m)	$\frac{\delta O_2}{\delta x}$ (ml/m)	Station 4865 O_2 (ml/liter)	$\frac{\delta O_2}{\delta z}$ (ml/m)
1	4.32		1.5×10^{-6}	4.26		4.34		4.3×10^{-6}	4.50	
		1.5×10^{-3}			5.9×10^{-3}		9×10^{-4}			2.2×10^{-3}
100	4.17		1.7×10^{-5}	4.85		4.25		1.3×10^{-5}	4.72	
		7.2×10^{-3}			1.62×10^{-2}		1.8×10^{-3}			2.9×10^{-3}
200	3.45		5.6×10^{-6}	3.23		4.43		0	4.43	
		1×10^{-3}			3.5×10^{-3}		1.5×10^{-3}			1.4×10^{-3}
400	3.25		1.8×10^{-5}	3.94		4.13		5.4×10^{-7}	4.15	
		1.35×10^{-3}			5.1×10^{-3}		2.15×10^{-3}			1.85×10^{-3}
600	4.98		5.3×10^{-5}	2.92		3.71		1.9×10^{-6}	3.78	
		1.5×10^{-3}			6.4×10^{-3}		1.85×10^{-3}			2.45×10^{-3}
800	5.28		2.8×10^{-5}	4.20		3.34		1.4×10^{-6}	3.29	
		6.5×10^{-4}			4.9×10^{-3}		3.05×10^{-3}			1.55×10^{-3}
1000	5.41		5.9×10^{-6}	5.18		3.95		9.4×10^{-6}	3.60	
		6.4×10^{-4}			9.4×10^{-4}		2.9×10^{-3}			3.42×10^{-3}
1500	5.73		2.0×10^{-6}	5.65		5.40		2.4×10^{-6}	5.31	
		1×10^{-4}			4×10^{-4}		6.4×10^{-4}			8.2×10^{-4}
2000	5.78		1.8×10^{-6}	5.85		5.72		0	5.72	
		6×10^{-5}			2.2×10^{-4}		4×10^{-5}			1×10^{-4}
2500	5.75		2.6×10^{-7}	5.74		5.70		1.9×10^{-6}	5.77	

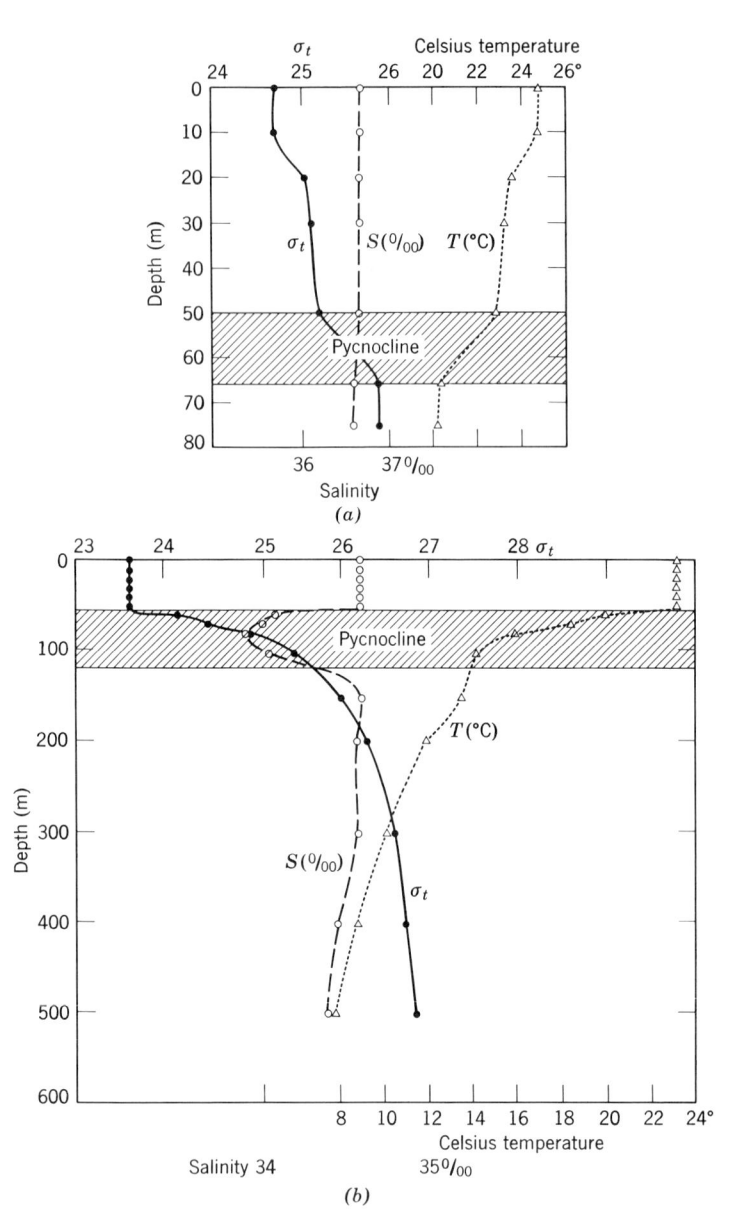

FIG. 8-3. Three examples of pycnoclines. (a) The Gulf of Cariaco on the Caribbean Coast of Venezuela, R. V. *Atlantis* Station 5673, 10° 28′N, 62° 28′W. Now 1,1958. The pycnocline, between about 50 and 65 m, arises solely from the thermal gradient. (b) R. V. *Thomas G. Thompson* Station 1–71 in the tropical eastern North Pacific Ocean, 24° 44.5′N, 113° 12.7′W, Dec. 6, 1965. (c) A station in Lake Nitinat, a fjord on Vancouver Island, British Columbia, R. V. *Hoh* Station 142-2, June 11, 1962. The low-salinity layer on the surface makes a major contribution to the pycnocline.

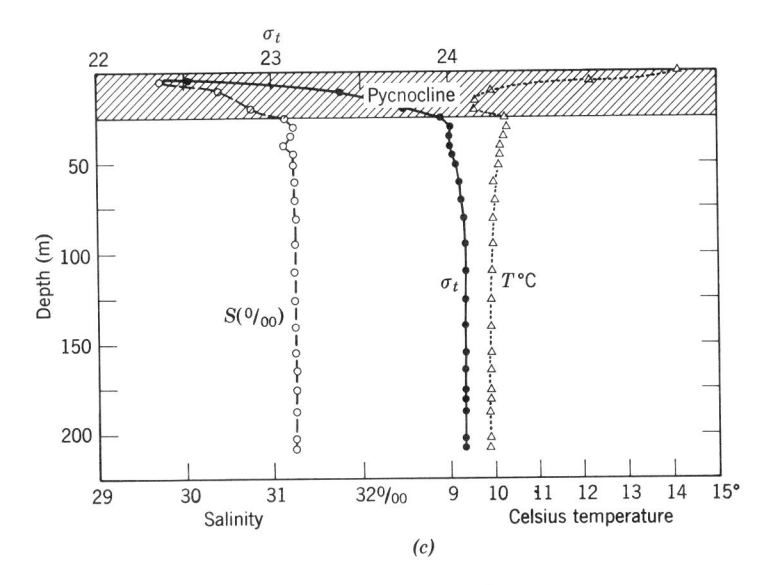

FIG. 8-3. (*Continued*).

rapidly with depth. It may be the result of a large vertical gradient in salinity, in temperature, or a combination of both. Below the pycnocline, the density increases more gradually with depth, so that vertical mixing should become easier.

The above remarks apply to eddy diffusion and infer the motion of water particles with their contained solutes or heat. Another kind of diffusion is molecular diffusion, in which the solute particles spread along concentration gradients without bulk motion or exchange of the water itself. Generally, molecular diffusion is too small to account for fluxes of solutes in the ocean, although ultimately this process is involved in eddy diffusion.

The task of evaluating eddy diffusion coefficients and velocities, particulary vertical velocities, in the real ocean is difficult. Riley (1951) has made such attempts by studying the distribution of salinity and temperature in the Atlantic Ocean and evaluating the equation for the distribution of variables from them. His analysis was continued by attempting to evaluate the biological rates of change (R) of dissolved oxygen, phosphates, and nitrates.

The advent of nuclear devices has made possible the tentative experimental evaluation of vertical eddy coefficients in the Pacific Ocean. Briefly, bomb testing introduced adequate amounts of radioactive material for their spread to be followed over a major portion of the ocean. Tracing the spread of strontium-90 and cesium-137 permitted Miyake et al. (1962) to estimate the value of A_z to be about 200 cm²/sec, although the spread of beta activity gave a value of 40 cm²/sec. The discrepancy in the numbers is perhaps not as important as the fact that the bomb tests were the first time man was able to introduce substances into the ocean whose spread could then be traced over vast areas of the ocean.

In situ processes. The concentrations of nonconservative constituents of seawater are altered *in situ* by processes that are primarily biological and biochemical. Purely inorganic chemical reactions may alter the concentrations of some of the trace metals, but such processes are poorly documented and probably are not quantitatively very important. Laboratory studies by Carritt and Goodgal (1954) were aimed at elucidating the processes of sorption (both adsorption and absorption) of phosphate ions on various kinds of particulate matter that occurs in the ocean, and Krauskopf (1956) studied the role of the solubility and adsorption in controlling the concentrations of trace metals in the ocean, but the part that these processes play in altering (or controlling) concentrations of various solutes in seawater is essentially unknown. The way that biological processes alter the concentrations of the carbonate components, dissolved oxygen, phosphorus compounds, nitrogen compounds, and silicates is somewhat better understood, and some reasonable estimates of the effects of biological processes on these concentrations have been made.

Biological cycles of C, N, P, Si, and O_2. In general, inorganic phosphate, ammonium, nitrate, nitrite, and silicate ions and carbon dioxide or carbonate or bicarbonate ions are converted, during photosynthesis, from inorganic materials in solution to organic compounds bound in the bodies of plants. When so bound, these materials are subject to swimming motions and the gravitational sinking of particles—influences to which they were not subject when in solution. It should be repeated that these constituents and the other nonconservative constituents of seawater are present in the oceans in such small concentrations that major fractions of their total amount (in the photic zone) may be so bound. This is the principal reason for their being distributed fundamentally differently than the conservative variables—R in the equation for the distribution of variables is of major importance with respect to nonconservative variables (essentially by definition), but is negligible with respect to major solutes. As a result, the distribution of the ions of sodium, magnesium, potassium, calcium, chloride, sulfate, and bromine are essentially identical to each other, but the distributions of the micronutrients are governed by different processes. Before discussing typical distributions of these constituents, some of the *in situ* processes that govern the distributions and the relationships among them are considered.

The individual cycles of nitrogen, phosphorus, and carbon have been discussed, but these cycles do not proceed independently in the ocean, and their effects on the distribution of nonconservative variables are correlated. When nitrogen, phosphorus, and carbon are converted, via photosynthesis, to plant tissue, they are taken out of solution in the ratios in which they appear in the plants, and changes in the concentrations of inorganic forms of these elements will take place in the same ratios. In the ocean, these ratios will be, on the average, 106:16:1, by atoms, or 40:7:1, by weight, and changes in the concentrations of these substances appear to take place in approximately these ratios. Because oxygen is evolved during photosynthesis and consumed during respiration in proportion to the consumption or production of carbon dioxide, changes in CO_2 will also be related to changes in the concentration of inorganic forms of nitrogen and phosphorus. These changes are reflected in Fig. 8-4, which shows the relationship between inorganic phosphate and nitrate ions at some stations off the Washington–Oregon coast. The diagram is typical of large areas of the world's oceans. The lower concentrations occur in the upper layers, the maximum concentrations at intermediate depths, and slightly lower concentrations at the great depths. It is evident that phosphate and nitrate occur in proportion to each other, and these proportions are close to those in which these elements occur in plankton.

The distribution of dissolved oxygen is also related to the distribution of nitrate and phosphate ions, but the relationship is somewhat more complicated than the linear nitrate–phosphate relationship—largely because the solution of air is an important source of oxygen in the ocean and oxygen can readily pass across the sea surface. Because of this, the relationship between dissolved oxygen and nutrient concentrations is better expressed in terms of the amount of oxygen consumed from the water than in actual oxygen concentrations. The apparent oxygen utilization (A.O.U.) is calculated from the formula

$$\text{A.O.U.} = O'_2 - O_2,$$

in which O'_2 is the amount of oxygen seawater of the same salinity and temperature would contain if just saturated with air at the sea surface, and O_2 is the amount of dissolved oxygen the water actually contains.

Plots of A.O.U. versus nitrate or phosphate are linear (or nearly so) over only part of the concentration or depth ranges. In some of the deeper parts of the ocean, the water contains more nitrate and phosphate than one would predict from a linear relationship between oxygen consumption and the respiratory remineralization of phosphate and nitrate. Redfield (1942) has called these excess quantities of nutrients "preformed." They can be accounted for if we consider that the waters containing them were "formed" at high latitudes. By "formed" is meant the acquisition of the water's

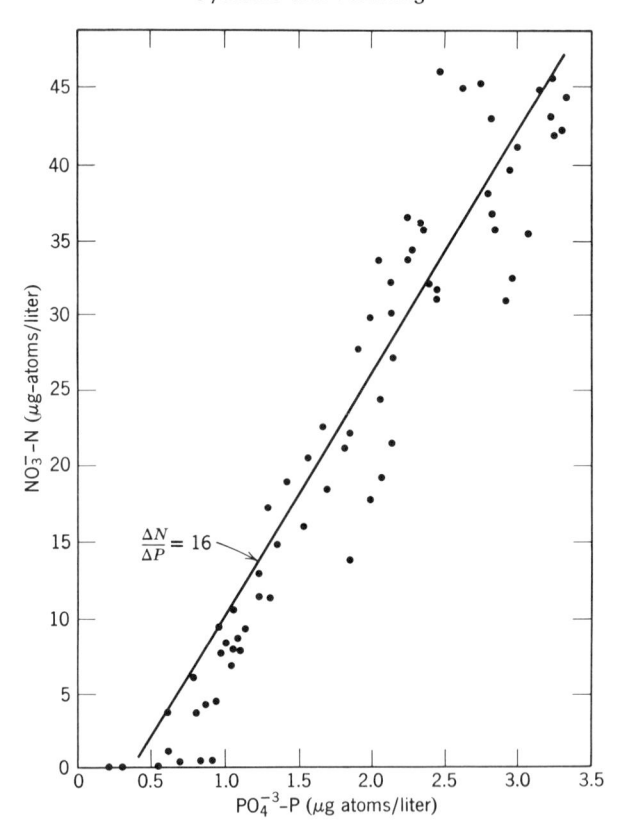

FIG. 8-4. The correlation between phosphate and nitrate concentrations off the coast of Washington and Oregon. The slope of the curve shown, $\Delta N:\Delta P = 16:1$, is that expected from the composition of plankton and not necessarily the best fit of the point shown. In general the low values are from the surface layers, the highest from intermediate depths.

characteristic salinity and temperature when last at the sea surface. These characteristics determine the density, which is high at high latitudes because of the low temperatures. This high-density water sinks below the sea surface and spreads out at greater depths toward lower latitudes, seeking its own density level. At the time of the water's departure from the photic zone, it can be assumed it was approximately saturated with dissolved oxygen, had an appreciable burden of organic matter in the form of plankton organisms, but still contained appreciable amounts of inorganic phosphate and nitrate. In the course of the water's travels from its site of formation, the organic matter decomposes and in so doing consumes oxygen and releases nitrate and phosphate ions in proportion to the amount of oxygen consumed. However, these amounts are in addition to those amounts present when the water mass was formed, so the total amounts of phosphate and nitrate present are not in proportion to the oxygen consumption.

The above discussion suggests that the distributions of nitrate and phosphate in the oceans should be closely related, and they are, but the distributions are not identical. This is presumably because the uptake and release of the two elements do not proceed exactly simultaneously. The dissolved oxygen will tend to be distributed in inverse proportion to nitrate and phosphate concentrations as pointed out above, the oxygen consumption will be distributed more nearly like the nitrate and phosphate, but there will be marked departures because of the preformed nutrients and also because of somewhat different rates of remineralization of nitrate and of phosphate and of oxygen consumption.

It should be pointed out that, although carbonates were included in the major constituents of seawater, they are not conservative. About one-tenth of the inorganic carbon may be involved in the biological cycle in one turnover. We might think of the nine-tenths as a reservoir of preformed carbonate, and its distribution would be governed primarily by diffusion and advection and be like that of the salinity. Superimposed on this distribution would be the biologically altered distribution that might involve about 20% of the total amount of the carbon in the system.

Silicates occupy a special case among the minor constituents. Because they are required for and consumed by the growth of diatoms and occur in such small quantities that diatom growth may essentially strip them from the water, they can be classed as micronutrient. However, once silicate has fulfilled the requirements of the diatoms, it is not required by most of the animals in the food web. As a result, the distribution of silicates seems to be closely related to biological events in the upper layers of the oceans, but the depths of the oceans contain much more silicate than we might expect from the concentrations of phosphate or nitrate there.

Distributions of biologically altered constituents. Typical vertical distributions of phosphate, nitrate, silicate, and dissolved oxygen are shown in Figs. 8-1 and 8-5. The surface layers of much of the ocean contain essentially the amount of dissolved oxygen that would be expected from the solution of air. In much of the ocean there is a wind-mixed, homogeneous surface layer that varies from a few tens of meters to 200 or 300 m thick. Characteristically, the nutrients and dissolved oxygen are uniformly distributed through this layer. Where the oceans are stratified near the surface, large vertical gradients in the nutrients and dissolved oxygen develop. Very low concentrations of the nutrients are characteristic of the surface layers in

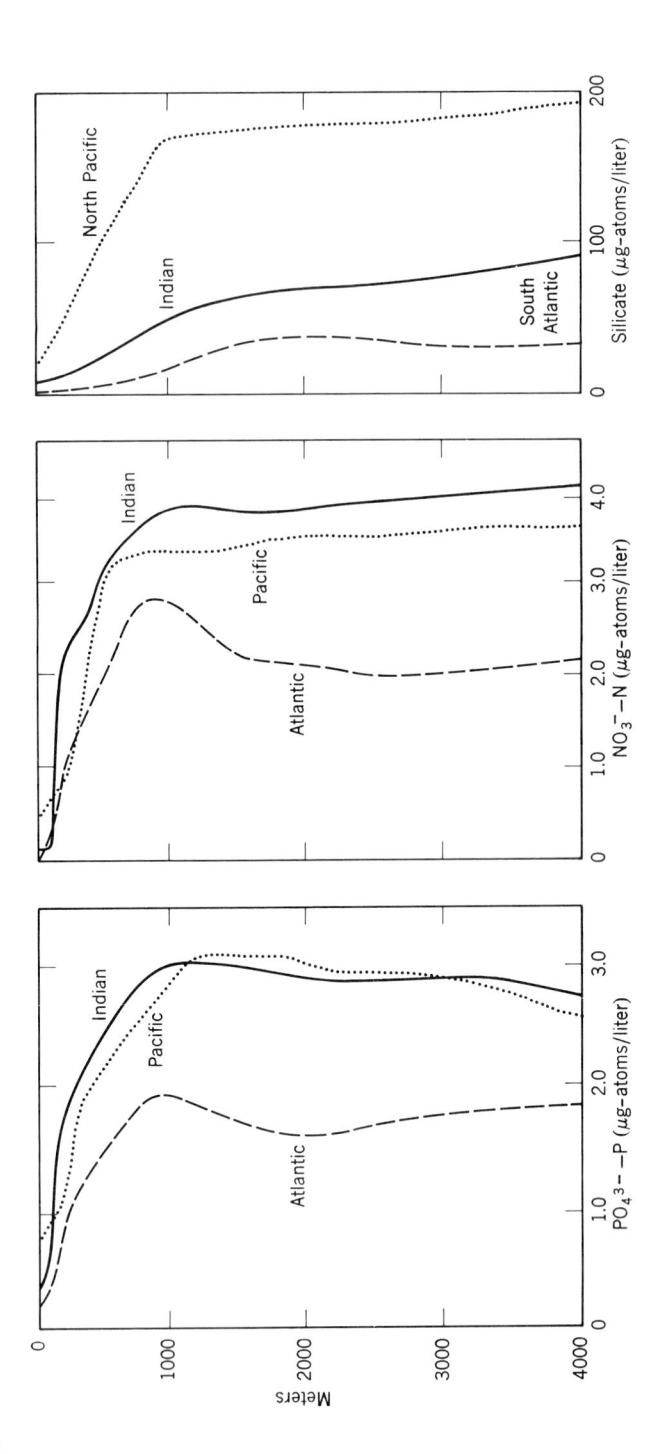

FIG. 8-5. Typical vertical distributions of nutrients (adapted from Sverdrup et. al., 1942, by L. K. Coachman).

these places, which may also be characterized by a subsurface oxygen maximum, evidently of photosynthetic origin. This maximum generally lies in the upper 50 m and may contain up to nearly twice as much oxygen as would be expected from the solution of air. Below this maximum or below the mixed surface layer, the oxygen concentrations generally decrease and the nutrients generally increase with increasing depth to their respective minima and maxima at depths of 600 to 1000 m. At greater depths the oxygen concentrations again increase and the nutrients decrease, owing to the circulatory replenishment of the deep waters by waters that were formed at the surface at high latitudes where they became rich in oxygen but somewhat poor in nutrients.

In general the Pacific and Indian oceans are somewhat richer in phosphates, nitrates, and silicates than the Atlantic Ocean (Fig. 8-5). At depths of 2500 m, as we proceed from the North Atlantic Ocean southward, there is a gradual increase in the phosphate (and presumably nitrate) content of the water through the Antarctic circumpolar ocean and then northward to the North Pacific Ocean, where about the maximum concentrations of phosphate and nitrate in the ocean occur.

Oxygen-deficient lobes of water extend westward from the west coast of Africa, from the Peruvian coast, and from the Pacific coast of Central America. The minimum oxygen concentrations in these lobes lie at an average of some 700 m, but the water columns in the Pacific oxygen-low regions are more devoid of oxygen than the corresponding region of the Atlantic Ocean.

As would be expected from the discussion of the inverse relationships between the nutrients and dissolved oxygen, the distributions of phosphate and nitrate tend to be similar and inversely to reflect the distribution of dissolved oxygen.

The distributions of the nutrients and dissolved oxygen vary seasonally, with maximum quantities of the inorganic forms of nitrogen and phosphorus in the upper layers in the winter. In winter the wind-mixed homogeneous surface layer will tend to be thick, and in relatively shallow regions this mixed layer may extend to the bottom. With the advent of vernal warming, stratification of the upper layers will be induced, and this is generally followed by increased activity of the phytoplankton—the so-called spring phytoplankton bloom. During the bloom, nutrients and carbon dioxide are converted, by photosynthesis, into the bodies of plants and animals and the nutrient cycles previously discussed are entered. Thus, in the upper layers, there will tend to be a seasonal cycle of the inorganic nutrients, with high concentrations in the winter and low ones in the summer. Such waters will tend to be stratified in summer, not in winter; the sinking and decomposition of the bodies of plants and animals and of fecal pellets into the lower strata will act as a one-way downward pump for the nutrients during the warmer part of the year, and these nutrients will have to be carried back into the upper layers by vertical mixing before they can re-enter the photosynthetic cycle.

The process by which the circulation of deep waters ventilates the deep ocean has been described, but there are systems, generally basins and fjords, in which the topography so limits the horizontal circulation that stagnation results. Under these conditions, the respiratory consumption of oxygen exceeds its circulatory replacement, and all the oxygen is consumed. As more organic matter showers into these systems, respiratory decomposition (oxidation) of organic matter will continue, but it will be a bacterial process and the source of oxygen will be ions in solutions, first nitrate (NO_3^-) and nitrite (NO_2^-) ions, and eventually sulfate (SO_3^-) ions: When the sulfate ions are reduced, hydrogen sulfide (H_2S) is produced. Such environments are typified by some of the Norwegian and British Columbian (Canada) fjords and by the Black Sea. The highly toxic hydrogen sulfide in these systems eliminates all forms of life higher than bacteria. The oxidative decomposition of organic matter probably proceeds more slowly than in oxygenated waters, and more organic matter tends to accumulate in the sediments than under oxygenated waters. The sulfides in the waters are probably highly corrosive to many metals, and the pH in these systems is somewhat lower than it is in comparable oxygen-bearing waters. Many of these conditions prevail in sediments underlying waters that contain oxygen, particularly in nearshore, relatively shallow systems where an abundance of marine life exists and furnishes large amounts of organic matter to the bottom. These reducing muds are generally black and odoriferous; dredging operations in such systems may introduce enough hydrogen sulfide into the water to kill organisms in it or even introduce enough hydrogen sulfide into the air to blacken lead pigmented paints and to create a general, if temporary, nuisance.

Some of the fjords that have been studied appear to contain anoxic water always, others contain them intermittently, whereas such conditions never arise in many basins and fjords. The kind of circulation in the basin and the rate of production of organic matter in the upper layers will determine which will be the case, because the balance between the rates of consumption of oxygen and its circulatory replacement is the fundamental control. Some of the fjords appear to flush annually, and the concentrations of sulfides that build up in them tend to be smaller than those in fjords that flush less frequently or perhaps never completely. Some fjords are known to flush only occasionally and then with damaging results, including extensive kills of fish and other animals.

8.7 SOME BIOLOGICAL AND CHEMICAL PROBLEMS OF SPECIAL INTEREST TO ENGINEERS

Corrosion

No attempt will be made to cover the corrosion of metals in the marine environment in this chapter, but some general statements are appropriate.

Corrosion is a term that is applied to chemical attack and destruction, generally of metals. Corrosion invariably starts at a surface, and may be a simple chemical reaction or may be more accurately termed an electrochemical process. Rusting of iron or steel would be an example of the former, whereas the solution of the electrochemically more active of two dissimilar metals in common contact with an electrolyte, such as seawater, constitutes the latter. Both processes are chemically complicated, and the reactions that take place will depend on a large number of chemical and environmental factors. The presence of small quantities of impurities in a metal may greatly alter its susceptibility to corrosion, and rates of corrosion can be expected to depend on the environmental temperature.

In general the susceptibility of metals to corrosion, particularly by such a complex medium as seawater, cannot be precisely predicted from theory, so exact knowledge of how metals will react with seawater is generally gained empirically by exposing test panels of the metal to seawater under various conditions.

The electrolytic solution of the more active of two dissimilar metals in common contact with an electrolyte can be more surely predicted from an electromotive series of the metals, in which the metals are arranged in decreasing order of electrochemical activity. Such series are contained in handbooks of chemistry and a series of structural metals has been published by La Que, May, and Uhlig (1955). In general, if two of the metals are connected and immersed in seawater, the metal higher in the series (more active) will tend to go into solution. However, if the metals are close together in the series, the rate of solution may be slow, or perhaps negligible. For example, the activities of yellow brass, red brass, and copper are so similar that their corrosion when in contact immersed in seawater might be negligible.

Where the use of dissimilar metals in contact with seawater is unavoidable, their electrolytic corrosion can be prevented by supplying sacrificial blocks of a highly active metal, such as zinc, which will be dissolved in preference to the less active metals; thus, zinc blocks are attached to the hulls of ships.

The prevention of corrosion is generally attempted by means such as the above or by providing a protective noncorrosive coating to the metal.

Marine fouling and boring organisms

Even the most casual observation of docks, pilings, or other structures in seawater will show the results of the activities of fouling organisms. Literally thousands of species of marine organisms will attach to structures and grow there. This is a matter of highly practical importance, particularly to marine architecture, because the layer or layers of fouling organisms on a ship's bottom can greatly reduce the speed of the ship. The weight of fouling organisms may alter the flotation characteristics of buoys and other aids to navigation, and the mobility of parts that may be engineered for undersea use at infrequent intervals could be greatly hindered by fouling organisms.

Extensive studies of marine fouling and its prevention were carried out by a group of scientists at the Woods Hole Oceanographic Institution during World War II. Their findings have been published in a book that should be of interest to anyone concerned with ocean engineering (Woods Hole Oceanographic Institution 1952).

In general fouling is a surface phenomenon and attempts to prevent it generally consist of coating the surface with a substance toxic to the organisms. Various organic substances, such as tar or creosote, may be used, but paints that contain certain metals that will slowly give up metal ions in solution are the most effective. Ions of silver, mercury, and copper are all effective in preventing fouling, but economy generally dictates the use of copper products. It should be stressed that the metal ions must be in solution. In one case a series of copper test panels was exposed to the sea, and after exposure, all the panels but one were free of fouling organisms. However, one of the panels had been attached to the supporting wood with galvanized nails. The zinc on the nails acted as a sacrificial metal and prevented the copper from going into solution, allowing the organisms to attach and grow on the copper plate.

The attachment of larger organisms to surface exposed to the sea appears to be preceded by the formation or growth of a bacterial slime coat that may be a necessary precursor for attachment.

Many of the organisms that attach to surfaces in the sea do so very firmly. For example, many barnacles actually penetrate the surface somewhat and can literally lift a protective coating from a steel surface.

In contrast to fouling organisms that grow largely on the surface, boring organisms can penetrate materials, generally wood, and can weaken structures to the point of uselessness simply by removing the material. There are many species of marine boring organisms, but two genera of boring mollusks, *Toredo* and *Bankia*, are particularly damaging. The latter includes the giant northwest shipworm of the Pacific coast of North

America. *Limnoria*, a genus of boring isopod, the gribble, is similarly damaging. Not even concrete structures are immune to attack by boring marine organisms, and several rock-boring mollusks are known.

In 1959 a symposium on Marine Boring and Fouling Organisms was held at the Friday Harbor Laboratories of the University of Washington. The papers were on the organisms and enzymes responsible, and on economic implications and evaluations of marine boring and fouling. The proceedings volume (Ray, 1959) should be a useful reference in this topic.

Pollution and disposal of wastes at sea

Increasing industrialization, increasing populations, increasing use of fertilizers and detergents, and increasing use of nuclear energy all lead to increasing amounts of the waste products of man's activity that require disposal. The sea has seemed to be a natural dumping ground for waste materials, but we now know that this is not acceptable practice in many cases. Some of the kinds of wastes man has disposed of in the sea or might consider so disposing of include domestic sewage, industrial wastes, radioactive wastes, heat, agricultural drainage, brines from saline water conversion or chemical recovery plants, and insoluble junk. Only a few salient features of some of these problems can be touched on here. However, a first principle is that if such wastes are mixed with sufficiently large quantities of seawater, their undesirable qualities will be diluted to unobjectionably low levels. The problem then reduces to one of dilution and that of possible reconcentration to objectionable or dangerous levels, generally by biological agencies.

Domestic sewage. The objectionable features of domestic sewage are its population of pathogenic bacteria; its high content of organic matter, which gives it a high oxygen demand; its high nutrient content, which gives it the potential of supporting large populations of algae and other plants which in turn may be almost as objectionable as the sewage itself; and the obvious aesthetic ones. The first and last of these can be overcome by proper treatment in biological sewage treatment plants, but the effluent from these plants is generally rich in nutrients (phosphates and nitrates), so that if the effluent is discharged into nearby inshore waters (which may also circulate poorly, as in many harbors), it may result in large and objectionable weed crops of algae and other plants. This has been the case in the Oslo Fjord, and serious attempts are being made there to treat the sewage by an electrolytic method that will produce an effluent that is not only bacterially unobjectionable but from which nitrate and phosphate have been removed to low levels.

An additional difficulty with disposing of domestic sewage and even the effluents from sewage treatment plants in the sea is that density of sewage is invariably less than that of seawater. Thus the sewage tends to float on or to the surface and, unless introduced into regions of strong currents, will mix with the seawater and be diluted only slowly. Even placing the sewage outfall considerably below the surface may not accomplish satisfactory dilution. It is therefore imperative that the current, wind, tide, and circulation patterns in general be thoroughly understood in planning outfalls for sewage or of effluents from treatment plants.

Industrial wastes. These can be of highly varied composition and present a variety of special problems. The wastes may be toxic to plants or animals. They may be highly acidic or highly basic. If they contain large quantities of organic matter, their biochemical oxygen demand may be objectionably high. Surface-active ingredients such as detergents may cause objectionable foaming or disrupt normal bacterial populations.

Some studies have been made of the dispersal of industrial wastes discharged at sea from a barge. It was found that this could be done in such a way that satisfactory dilution was accomplished. Very shortly after the barge passed, no chemical changes in the water could be noted. However, the effluent contained iron salts, and it is now evident that the sediments in the regions of these operations have become covered with iron compounds.

The disposal of petroleum industry wastes and wastes from the petrochemical industry may present special problems. Several studies have been made of the effects of these wastes and of the hydrographic regime on marine life, particularly along the Gulf of Mexico coast.

Radioactive wastes. These can be somewhat arbitrarily divided into high- and low-level wastes, depending on their activity. Radioactive wastes all have the characteristic of losing their radioactivity with time—some nuclides losing it quickly, others very slowly. A second consideration is that radioactive elements will enter the biological cycle and therefore the food web. If the radionuclides have long radioactive half-lives and are retained for long times in the body, they may present health hazards. The role of strontium-90 has caused particular concern, because it has a long half-life and may be incorporated in teeth and bones, substituting for calcium.

High-level radioactive wastes present complex problems in their disposal, and the author knows of no serious proposal to dispose of them at sea. In general, they will be impounded either permanently or until they decay to low levels. However, low-level radioactivity has been and is being disposed of at sea. Two methods are generally employed: direct discharge of very low-level material and the disposal at sea of low-level radioactive wastes that have been cast in concrete and, generally, encased in steel drums. The latter method is frequently used for the disposal of small quantities of radioactive wastes from medical studies and scientific research.

Low-level radioactive materials are introduced directly into the Celtic Sea from the Windscale atomic energy activity in England. The hydrologic regime in the Celtic Sea and the biological effects of the radioactive material have been carefully studied, and it has been concluded that no intolerable hazard is involved in the operation.

The Hanford plants in Washington State result in low levels of radioactivity in the Columbia River. River water is used to cool the reactors, and secondary radioactivity is induced in the zinc naturally present in the water and in chromium that is added to the water as chromium compounds to prevent fouling of the reactors. These materials are carried to the Pacific Ocean, where they are at nearly undetectably low levels in the water, but they are sufficiently concentrated in particulate matter to be detectable in the bodies of benthic organisms and in the sediments.

Brines from saline conversion plants (plants where fresh water is recovered from seawater) probably pose no great disposal problem. In general, the treatment will only about double the salt content and should not add foreign matter. The brines so formed will be somewhat denser than normal seawater and, if dispersed on the surface, will tend to sink and presumably be diluted readily by the ambient seawater. The effluent brines may be somewhat warmer than the seawater, but one would suspect inefficiencies in the operation if this were the case.

The sea as a source of commercial products

Marine products of commerce have been extensively discussed by Tresler and Lemon (1951). Most of the products they consider are either the major salts or biological products. Marine fishes and their byproducts are by far the most important, and will probably remain so for years to come. We might predict that the next most important product to come from the sea will be fresh water. Tresler and Lemon have listed soluble chemicals, red algae, brown algae, pearls, mother of pearl, coral, marine fishes, fish-meal oils, fish-liver oils, fish glues, isinglass, leathers, shellfish, turtles, whales, seals, and sponges as some marine products of commerce.

Mero (1965) has considered the mineral resources of the sea in some detail and has attempted to evaluate their economic potential and some of the engineering problems involved in their recovery.

The recovery of soluble chemicals from the sea is a thriving industry, and the recovery of ordinary salt from the sea is an ancient art. In modern times the production of both sodium and potassium salts from seawater has become important. Practically all of this country's magnesium metal and magnesia are now produced from seawater, as is a large fraction of the bromine.

Because of the vast differences in concentrations of the various soluble constituents, it is doubtful if seawater can be used as a commercial source of any but the 10 or 12 most concentrated solutes—even if the recovery of the minor constituents were accomplished incidentally. Some of the things that might be so recovered, such as silica and lime, do not demand premium prices, so that unless the seawater product has some special physical or chemical properties, its production will not be profitable.

Ocean water contains particulate solids in addition to the solutes. These may be organic (living or dead) or inorganic, and although some of the inorganic matter in suspension (particulate or colloidal) such as oxides of manganese and colloidal suspension of gold, might seem to be of commercial value, they are present in such small concentrations that their profitable recovery will remain impossible in the foreseeable future. The same can be said of the dead organic material, either dissolved or suspended in the water—it is too dilute for its recovery to become practical.

It is evident that, except for water itself and the major solutes, seawater holds little promise of supplying materials of economic importance, even in the future. However, it should be pointed out that the sea is the major repository of both deuterium and tritium hydrogen isotopes required for the release of thermonuclear energy. Since ultimately this will become the main source of energy for the earth as other energy sources are expended, it is to be expected these isotopes will eventually be recovered from seawater for these purposes, but these conditions are in the distant future.

Other than the major soluble constituents of seawater, the water, the deuterium and tritium of seawater, and the plant and animal life in the sea, the wealth of the ocean lies on and under the bottom. Mero has divided this realm into the marine beaches, the continental shelves, and the deep-sea floor. Several commercial products are now recovered from the first two of these environments, but the technology for exploiting the third is still in the research and development stage.

Marine beaches. Among the materials recovered or recoverable from marine beaches are diamonds, gold, tin, heavy minerals, silica, titanium minerals, and magnetite sands. Mero points out that, as ore sources, marine beaches have the advantage of being readily workable and, since the sands are already small particles, at least some of the initial milling stages have been accomplished by nature.

Continental shelves. Petroleum, gas, sulfur, and limestone are now being commercially produced on the continental shelves. Extensive deposits of phosphate nodules (phosphorite) are known to exist on the continental shelves, and in several instances it appears that the economics are or will soon become favorable for their commercial recovery for use in manufacturing fertilizers. For example, the state of California imports large quantities of phosphate fertilizers and a large fraction of the cost to the consumer is transportation costs. However, there are extensive deposits of phosphorite on the continental shelf off California. The engineering problems involved in mining this material do not seem too difficult, but the suitability of the material for the production of fertilizer has not been fully established.

The exploration of the structure of the earth's crust below the sea floor is being carried out by geophysicists. With this knowledge, it seems likely that ore deposits may be discovered under the sea floor on the continental shelves. With technological advances, the availability of large quantities of cheap nuclear energy, and greatly increased experience in man's living and working in the sea, these deposits can be expected eventually to be worked economically.

Deep-sea floor. The technological difficulties of mining the deep-sea floor are much greater than those of mining the continental shelves, but some engineers are optimistic that this can be accomplished in the relatively near future. One of the reasons for the optimism is the known existence of extensive beds of manganese nodules on the sea floor. Because these nodules are relatively rich in a variety of strategic metals and because they appear to be rather loosely lying on the sea-floor surface, there is some optimism that they can be recovered profitably. Analyses of these nodules show that they contain $1-20\%$ silicon, up to 1% titanium, $12-50\%$ manganese, $2-26\%$ iron, up to 2.3% cobalt, up to 2% nickel, up to 1.6% copper, and over 1 to 2.4% magnesium.

The oozes and sediments on the sea floor might eventually be used as sources of several strategic metals, but this development lies in the distant future when dry-land ore bodies have been exhausted and we would otherwise be forced to recover traces of the metals from igneous rocks. However, deep-sea sediments are generally richer in such metals as manganese, nickel, copper, cobalt, and molybdenum than are igneous rocks, so that when driven to extremes, the oceanic sediments might be more likely sources of these metals than would igneous rocks.

REFERENCES

Bein, W., H. G. Hirsckorn, and L. Möller, 1935, "Konstantenbestimmungen des Meerwassers und Ergebnisse über Wasserkörper," *Veroff. Inst. Meerskunde.*, *N. F. A. Geogr. Naturwiss. Riehe*, **28**: 1–240 (1935).

Benson, B. B., and P. D. M. Parker, 1961, "Relations Among the Solubilities of Nitrogen, Argon and Oxygen in Distilled Water and Sea water," *J. Phys. Chem.* **65**, 1489–1496 (1961).

Carpenter, J. H., 1966, "New Measurements of Oxygen Solubility in Pure and Natural Water," *Limnol. Oceanog.* **11**, 264–277 (1966).

Carritt, D. E., and S. Goodgal, 1954, "Sorption Reactions and Some Ecological Implications," *Deep-Sea Res.* **1**, 224–243 (1954).

Cox, R. A., 1965, "The Physical Properties of Sea Water," Vol, in J. P. Riley and G. Skirrow, Eds., *Chemical Oceanography*, Academic, New York, pp. 73–121.

Duursma, E. K., 1965, "The Dissolved Organic Constituents of Sea Water, Vol, *in* J. P. Riley and G. Skirrow, Eds., *Chemical Oceanography*, Academic New York, p. 433–475.

Fox, C. J. J., 1909, "On the Coefficients of Absorption of Nitrogen and Oxygen in Distilled Water and Sea Water, and of Atmospheric Carbonic Acid in Sea Water, *Trans. Faraday Soc.* **5**, 68–87 (1909).

Glueckauf, E., 1951, "The Composition of Atmospheric Air," in T. F. Malone, Ed., *Compendium of Meteorology*, American Meteorological Society, Boston, p. 3–10.

Goldberg, E. D., 1963, The Oceans as a Chemical System, Vol 2, in M. N. Hill, Ed. *The Sea*, Interscience, New York, pp. 3–25.

Goldberg, E. D., 1965, "Minor Elements in Sea Water," Vol. 1, in J. P. Riley and G. Skirrow, Eds. *Chemical Oceanography*, Academic, New York, pp. 163–196.

Jacobson, A. W., 1948, "An Instrument for Recording Continuously the Salinity, Temperature, and Depth of Sea Water," *Trans. Am. Inst. Elec. Eng.* **67**, 1–9 (1948).

Knudsen, M, 1901, *Hydrographical Tables*. G. E. C. Gad, Copenhagen.

König, H., 1963, "Uber die Löslichkeit der Edelgase in Meerwasser," Zeits. Naturforsch. **18a**, 363–367 (1963).

König, H., H. Wänke, G. S. Bien, N. W. Rakestraw, and H. E. Suess, 1964, "Helium, Neon and Argon in the Oceans," *Deep-Sea Res.* **11**, 243–247 (1964).

Krauskopf, K. B., 1956, "Factors Controlling the Concentration of Thirteen Rare Metals in Sea Water," *Geochim. Cosmochim. Acta* **9**: 1–32 B (1956).

LaFond, E. C., 1951, "Processing Oceanographic Data," (H. O. Pub. No. 614). U. S. Navy Hydrographic Office, Washington, D. C.

LaQue, F. L., T. P. May, and H. H. Uhlig, 1955, *Corrosion in Action*, International Nickel Co., Inc., New York.

Matthews, D. J., 1939, *Tables for the Velocity of Sound in Pure Water and Sea Water for Use in Echo-Sounding and Echo-Ranging* (H. D. 282, Second Edition), Hydrographic Dept., The Admiralty, London.

Maury, M. F., 1855, *The Physical Geography of the Sea*, Fifth Edition, Harper, New York.

Mero, J. L., 1965, *The Mineral Resources of the Sea*, Elsevier, New York.

Miyake, Y., 1939, "The Refractive Index of Sea Water," *Bull. Chem. Soc. Japan* **14**, 29–34 (1939).

Miyake, Y., and M. Koizumi, 1948, Measurement of the Viscosity Coefficient of Sea Water, *J. Marine Res.* **1**, 67–73 (1948).

Miyake, Y., K. Saruhashi, Y. Katsuragi, and T. Kanazawa, 1962, "Penetration of ^{90}Sr and ^{137}Cs in Deep Layers of the Pacific and Vertical Diffusion Rate of Deep Water," *J. Rad. Res.* **3**, 141–147 (1962).

Pollack, M., 1954, "The Use of Electrical Conductivity Measurements for Chlorinity Determination, *J. Mar. Res.* **13**, 228–231 (1954).

Rakestraw, N. W., and V. M. Emmel, 1938, "The Solubility of Nitrogen and Argon in Sea Water, *J. Phys. Chem.* **42**, 1211–1215 (1938).

Ray, D. L., Ed., 1959, *Marine Boring and Fouling Organisms*, Univ. Washington Press, Seattle.

Redfield, A. C., 1942, "The Processes Determining the Concentrations of Oxygen, Phosphate and Other Organic Derivatives Within the Depths of the Atlantic Ocean," *Papers Phys. Oceanog. Meteorol.* **9** (2), 1–22 (1942).

Redfield, A. C., H. P. Smith, and B. H. Ketchum 1937, The Cycle of Phosphorus in the Gulf of Maine, *Biol. Bull. Woods Hole* **73**, 421–433 (1937).

Revelle, R., and R. Fairbridge, 1957, "Carbonates and Carbon Dioxide," *in Treatise on Marine Ecology and Paleoecology*, Vol. 1, Ecology, J. Hedgpeth, *Ed.* Memoir 67, Geol. Soc. Am. p. 239–296.

Richards, F. A., 1965, "Anoxic Basins and Fjords," Vol. 1, in J. R. Riley and G. Skirrow, Eds., *Chemical Oceanography*, Academic, New York, pp. 611–645.

Richards, F. A., 1965, "Dissolved Gases Other Than Carbon Dioxide," Vol. 1. in J. P. Riley and G. Skirrow, Eds. *Chemical Oceanography*, Academic, New York, pp. 197–225.

Riley, G. A., 1951, "Oxygen, Phosphate, and Nitrate in the Atlantic Ocean," *Bull. Bingham Oceanog. Collection* **13**(1), 1–126 (1951).

Skirrow, G., 1965, The Dissolved Gases—Carbon Dioxide," in J. P. Riley and G. Skirrow, Eds., *Chemical Oceanography*, Academic, New York, pp. 227–322.

Starkey, R. L., and K. M. Wight, 1945, *Anaerobic Corrosion of Iron in Soil*, American Gas Association, New York.

Sverdrup, H. U., M. W. Johnson, and R. H. Fleming, 1942, *The Oceans, Their Physics, Chemistry, and General Biology*, Prentice-Hall, New York.

Thomas, B. D., T. G. Thompson, and C. L. Utterback, 1934, "The Electrical Conductivity of Sea Water," *J. Conseil*, Conseil. Perm. Intern. Exploration Mer **9**, 28–35 (1934).

Tresler, D. K., and J. McW. Lemon, 1951, *Marine Products of Commerce*, Reinhold, New York.

Truesdale, G. A., A. L. Downing, and G. F. Lowden, 1955, "The Solubility of Oxygen in Pure Water and Sea-Water," *J. Appl. Chem.* **5**, 53–62 (1955).

U. S. Navy Oceanographic Office, 1962, *Tables of Sound Speed in Sea Water*, (Supplement to H. O. Pub. 614). U. S. Navy, Washington, D. C.

Utterback, C. L., T. G. Thompson, and B. D. Thomas, 1934, "Refractivity–Chlorinity–Temperature Relationships of Ocean Waters," *J. Conseil*, Conseil Perm. Intern. Exploration Mer **9**, 35–38 (1934).

Weyl, P. K., 1964, "On the Change in Electrical Conductance of Seawater With Temperature, *Limnol. Oceanog.* **9**, 75–78 (1964).

Whipple, G. C., and M. C. Whipple, 1911. "Solubility of Oxygen in Sea Water," *J. Am. Chem. Soc.* **33**, 362–365 (1911).

Wilson, W. D., 1960, Speed of Sound in Sea Water as a Function of Temperature, Pressure, and Salinity, *J. Acoust. Soc. Am.* **32**, 641–644 (1960).

Woods Hole Oceanographic Institution, 1952, *Marine Fouling and Its Prevention*, U. S. Naval Institute, Annapolis, Maryland.

PART 2

SYSTEMS DESIGN— THE TECHNOLOGY

CHAPTER 9

Introduction

JOHN F. BRAHTZ

In introducing the subject matter of Part 2, the writer's aim is to extend the implications contained in Chapter 1 regarding benefits of systems methodology for ocean engineering, so as to include both planning and design.

By viewing the various categories of marine technology with the perspective required for design of large-scale systems, the reader can appreciate the unifying effect that engineering systems philosophy lends to both planning and design operations. Moreover, if generally recognized, this approach should provide a special impetus to ocean engineering as a new area for professional expression with its characterizing disciplines. In turn, such unification of professional thinking would have a significant effect on the achievement of national objectives for marine resources.

9.1 DESIGN OF OCEAN SYSTEMS

The morphology of large-scale systems design can be traced through a sequence of synthetical and analytical operations. The basic stages of design are first, feasibility studies; second, preliminary design; and third, detailed design. The systems approach to engineering design is primarily reflected in the first two stages wherein feasibility and preliminary analyses are conducted. This is supported by the fact that the philosophy of systems engineering is linked in essence to the study and definition of objectives, establishment of valid decision criteria, and finally system optimization. These aspects of systems engineering are also fundamental to the feasibility study and preliminary design stages.

The feasibility study, according to general practice, includes the following sequence of steps:

1. *Needs Analysis*: This step may include market analysis to determine economic justification, or sociopolitical studies, and military operations analysis to validate what appears primitively to be needed.

2. *Problem Definition*: This step provides for resolution of the engineering problem situation which stems from the needs analysis. Design objectives are defined and environmental factors are studied in consideration of effective criteria for decision making and full understanding of constraints on the problem solution.

3. *Concept Evaluation*: This step provides for the creative synthesis of plausible solutions to the problem which is defined in step 2. Admissible solutions are evaluated, judged, and selected for their relative merit in satisfying the basic need and related design objectives. Evaluation and judgment is usually made on the basis of technical feasibility, economic worth, and financial feasibility.

An operations analysis of the preliminary design stage would introduce and define several important steps in the design process which are not necessarily significant to the immediate purpose of this book. However, one aspect of the preliminary design stage particularly characterizes the theme of systems engineering. That single aspect is the optimization of a selected system according to carefully considered criteria. The system solution selected for optimization would be a product of the preceding feasibility study.

9.2 A READER'S VIEWPOINT

By maintaining a critical viewpoint based on an understanding of the requirements for large-scale systems design, the reader should expect his study of the whole of Part 2 to be constructive for the delineation of design goals within the category of his special interest in ocean engineering.

Part 2 identifies by chapters generally accepted major categories of marine technology. The categories are broadly defined; this is in fact tributary to the purpose of this book. Any suggestion of parochialism or artificial barriers to communication that might retard the development and interdisciplinary use of technology for engineering purposes has been conscientiously avoided wherever possible.

As indicated in the Chapter 1, the present technologies are being extended, and in certain cases extrapolated, to meet the information requirements for exploratory engineering developments. Consequently, it can be expected that voids in the body of engineering information that is needed for treatment of ocean-centered design problems will give rise in certain instances to new and definable areas of marine technology. It is partly for this reason that trends as well as present capabilities in the state of the art have been indicated for those major categories of technology which are discussed in Part 2.

By considering the morphology of engineering systems design as outlined above, the reader will readily discern the larger context in which the subject matter of both Parts 1 and 2 has been presented. Further, he will find that this approach to the study of Part 2 tends to reveal engineering objectives which promise to be most rewarding in terms of national, social, economic, political, and military goals. It is this writer's conviction that the reader will find that such an expanding viewpoint as suggested herein will be conducive to greater innovation and creativity in engineering design for the marine environment.

CHAPTER 10

Deep Ocean Installations and Fixed Structures

JOSEPH J. HROMADIK

Underwater installations and structures have been or will be directed for use as offensive and defensive systems, navigation arrays, maintenance facilities, supply depots, research stations, and the like. Offshore towers in shallow waters are commonplace, whereas sea-going floating weather stations as NOMAD and MAMOS are more recent. Prototype underwater fuel caches and stabilized platforms have appeared for strategic reasons. Bottom-installed well heads associated with drilling operations are standard practice. In addition various test arrays have been emplanted to various depths down to about 6000 ft and recovered, and still more recent are the habitable compartments at shallow depths of SEALAB and Conshelf experiments.

Construction in the ocean has been largely limited to relatively shallow depths, perhaps a few hundred feet. Further, the work has been done from the surface by lowering prefabricated components. Some remote manipulations have been accomplished, especially in completing oil wells on the ocean bottom, but still at diver depths. Much of this is history.

But ocean exploration and exploitation will continue in a variety of ways with new concepts adding to the old to an ever-broadening technology. This is man's quest for knowledge and recognition. The form, shape, and performance characteristics of future concepts are difficult to predict in terms of present knowledge and requirements, but it is certain that man will build in the sea environment. Although the ways for exploration and exploitation are varied, they can be reduced to three basic methods:

1. Man-in-the-sea techniques that expose the man diver directly to the environment;

Deep Ocean Installations and Fixed Structures   311

2. Man in the environment in vehicle, mobile or stationary, encapsulated in normal pressure breathing chambers—simple diving vehicles to complex habitats;

3. Man's capabilities in the environment through remote operations while he remains on the surface—simple oceanographic sampling to complex installations.

Although these three basic methods complement one another, each imposes its own distinct requirements on technology.

This chapter takes a condensed look at present knowledge and engineering capabilities applicable to deep-ocean installations and structures. The term installation is used to refer to complexes in the ocean that may be floating, submerged, or resting on the bottom, usually secured by an anchor system; fixed structures implies complexes fixed to the sea floor through some foundation system. Trends in the immediate future are also explored through concepts and attendant problems. The subject matter is restricted to installations and structures fixed to, or resting on the sea floor. Excluded are the ambient-pressure, manned habitats and floating structures that are discussed in other chapters.

To derive maximum benefit from this chapter, the reader is challenged to expand his thinking beyond this fragmentary presentation. The expansion of the subject matter has no foreseeable limits. It may begin with the Suggested Reading section at the end of this chapter and the various references cited throughout. Project into the future; but bear in mind that "Man tends to overestimate what he can accomplish in five years and vastly underestimates what he can achieve in thirty."[1]

10.1 PLANNING THE STRUCTURE/INSTALLATION

Interdisciplinary functions

Problems associated with the design and emplantment of structural installations draw on many disciplines. Consider some of the situations facing the designer.

As on land, the ocean bottom may vary from extreme regularity to steep inaccessable slopes. The bottom does have natural regular areas that may be used for sites, and these can be generally located by present-day bottom survey methods. Local irregularities however, often exist, and cannot always be detected. Moreover, the precision of the survey diminishes as depth increases; hence, a desirable site cannot be always assured.

Foundation investigations on bearing capacities, anchor-holding powers

---

[1] From an address by Roger Lewis, President General Dynamics delivered to the 63rd Anniversary Dinner Navy League of the United States New York Council October 27, 1965.

and settlement must be considered. Phenomena of mud slides, rock slides, turbidity currents, and unstable sediments challenge man's understanding.

Knowledge of the deep-ocean environment and its effects on materials of construction is important. Since little information is available on chemical, physical, and biological deterioration, the life expectancy of an underwater system cannot be accurately predicated. Existing materials will be used for initial construction. With the advent of new materials and alloys, vital structural components in high-degradation areas can be built with greater confidence.

Although both photography and television have proved to operate in deep-sea areas, sight distance is limited and a slight disturbance of the bottom often clouds the water, reducing visibility to zero. Thus, in some areas in the near future, underwater construction operations may have to rely on sight-unseen operations.

Since any constructing far below the surface with on-site control of operations does not appear feasible in the immediate future, reliance will have to be placed upon self-contained, prefabricated assemblies, with the control operations at or near the surface. In time submersibles will be performing many of the functions.

Areas remote from land will be difficult to locate, or establish a fix, with desired accuracies. Until navigational problems are resolved, successful search or relocation may depend on bottom features or buoys, none of which can be presumed permanent, except for major prominences.

Knowledge of the hydrodynamics and responses of deep-ocean installations to the forces imposed upon them are of prime importance to those who design and use the installation. These may be floating structures, submerged forms, or bottom-resting structures.

Drilling rigs and construction-type floating platforms undergo six degrees of freedom when disturbed by forces of winds, waves, and currents. The three translational motions are surging, swaying, and heaving, and the rotational motions are rolling, pitching, and yawing. A combination of these motions could capsize or heavily damage a vessel or platform.

Parameters that are of interest to the construction personnel aboard a moored barge in the open sea are those associated with lowering loads from such a barge to the ocean floor and returning them to the ship. These parameters are the forces in the moving cables, the displacements and tensions in the lowering lines, the degree of precision in placing the loads, the accelerations acting on the loads, and the magnitudes of impact on the ocean bottom.

The study of motions of ships at sea is a general problem of naval concern and has been devoted primarily to the problems of an advancing ship in head seas. The prime variables of concern in these studies are heave and pitch motions. There is a large background of information related to the hydrodynamic forces for zero speed of advance, but no complete treatment of this situation for a realistic ship has been completed. The problem is further compounded when the influences of mooring are included.

These are but some of the considerations to which ocean engineering must yield. Some established land-construction methods may be extended with modifications and reservation for use in the deep ocean. More likely however, due to the restrictive environment, a considerable portion of the land-construction practice will have to be abandoned, and operations and systems will have to be unique and capable of functioning under a much greater range of boundary conditions.

Design considerations[2]

The design considerations fall into four basic engineering requirements: (a) functional, (b) structural, (c) maintenance, and (d) habitability. The hostile ocean environment dictates careful planning and coordination of the engineering requirements for the design, fabrication, construction, and placement, whether they are offshore towers or completely submerged installations.

Functional. Accurately defining the purpose for the complex is essential. Installations and structures are accommodatious, space for facilities and equipment—and sometimes men—combined into a system to perform a useful function. The complex will be necessarily expensive because of the logistics involved and the forces of nature imposed on the system. Space must be optimized, balanced for mission efficiency. Unattended stations must be automated to alert of "go – no go" situations, and standby equipment must be available to take over in the latter. Manned stations must consider habitability, the transfer of personnel, and provisions for safety against mission hazards. Some functions may be more suitable in separate and isolated compartments.

Structural. Design criteria are preselected sets of conditions that must be met. It is important, and often difficult, to define the probability of the occurrence of any given set of conditions and the involved risk. Ocean structures are subjected to severe loading not normally encountered in terrestial structures—winds, waves, high hydrostatic loads, submarine land-mass movements, berthing impacts of servicing vessels, to name a few. The effect that these forces may have on the foundations, substructure, and super-

[2]McEntee, J. R., *Structures in the Deep Ocean*—Engineering Manual for Underwater Construction—Design, U.S. NCEL Technical Report R 284-4, in preparation under contract with Shell Development Company, Houston, Texas.

structure during all phases of operations must balance the risk of loss against the cost for a high factor of safety. Although it is neither feasible nor practical to eliminate risk entirely, it must be reduced to an acceptable level commensurate with the probability of occurrence of conceivable disasters. Moreover, each situation may subscribe to a new set of conditions, for which criteria must be adjusted. The alert designer capitalizes on latest advances in technology.

Maintenance. Maintenance of a deep-ocean structure can be optimized only if it is treated as a basic consideration during design, integrated with functional, structural, and habitability requirements. Structural geometry and choice of materials may depend more on maintenance requirements than the others. Unfortunately, little is known or understood of the phenomenon in maintenance. Perhaps, studies dealing with maintenance do not have the glamor and excitement to attract researchers. Also, time is a factor in maintenance studies, and a slow return on investments is not exactly a great inducement to research sponsors. An effective research program in this area would be most rewarding.

Habitability. Habitability is characterized by the conditions imposed upon the manning personnel performing the functional mission. Consideration must cater not only to mission and personnel needs, but also to personnel comforts and recreation—and a psychological feeling of security.

Factors affecting planning

Ocean construction will take place in an environment which alters the usual present-day concepts of construction operations. Factors having lesser significance in land-based constructions may take on major proportions at sea. Below is a listing of some of the factors that will be on most engineers' lists when they plan the structure.

1. *Natural*: geography; topography; geology; climate; weather; water condition.

2. *Construction*: equipment and limitation of availability of labor; source of materials and supplies, prefabrication and land-based operations; transportation to the site; on-site construction; scheduling.

3. *Legal*: Law, all types through maritime and international; jurisdiction of the environment; insurance; workman's compensation; safety of construction crew and manning crew.

4. *Efficiency of workmen at sea*: effects of sea state on operations; psychological factors on working crew; psychological factors on manning crew.

5. *Cost estimating all of the above.*

The offshore fixed tower

The prominent offshore fixed tower becomes daily a more commonplace structure. It provides an excellent example to illustrate the manner in which engineers attended design consideration.

Take for example, the U. S. Coast Guard's offshore light stations. By late 1965 the Coast Guard had installed five such structures,[3] three manned and two unmanned, to replace lightships; a sixth tower was under construction and two more were in the planning stage. The towers were installed at sites 10 to 30 miles off the Atlantic coast in water depths of 40 to 80 ft. Their function: navigational aid to ships. But why a fixed structure? Principally, economics. According to Ruffin (1965),[4] "the choice of a fixed structure over lightships was arrived at through economic studies of the initial cost of replacing worn-out lightships, of the maintenance costs for these ships and the manpower required for them as compared with the same costs for fixed structures. Besides its economic advantage, the fixed structure maintains its position and functional reliability in all kinds of weather."

Structural. The Coast Guard structures are four legged and of the fixed, template type, the predominate type used in the oil industry. The prefabricated structural template is positioned on the ocean floor and piling are driven through the template to provide support and stability. Subsequently, the prefabricated deck section is placed on top of the pile substructure.

Generally, the most significant forces acting on such a structure are those caused by waves. Also, they are probably the most difficult to ascertain. In consideration, the designer must arrive at the probability of occurrence of a certain design wave through forecast methods (hindcasting from meteorological and oceanographic records for an n-year storm) and translate it to a force diagram. A typical, theoretical internal force distribution diagram of an ocean wave is illustrated in Fig. 10-1; a schematic wind–wave force diagram is given in Fig. 10-2.

According to Ruffin (1965), after the Coast Guard decided on fixed structures, a fathometer survey was made of the selected area over a 1-mile radius. Borings were made at suitable sites, generally to 160 ft below the sea floor. The soil samples were analyzed and the results plotted for determining ultimate pile capacities. Considerations in designing the structure included 40- to 50-ft waves, and smaller waves that set up periodic motions; hurricane winds of 125 mph, equivalent to a pressure of 50 lb/ft^2 on a flat projection; vertical forces on the heliport of 20 kips plus 50% impact; loads

---

[3]These are discussed in detail in two articles in the November 1965 issue of Civil Engineering, published by the American Society of Civil Engineers, New York, New York. See Fowler (1965) and Ruffin (1965) in Suggested Reading.

[4]References cited by author and year are alphabetically arranged in the Suggested Reading Section.

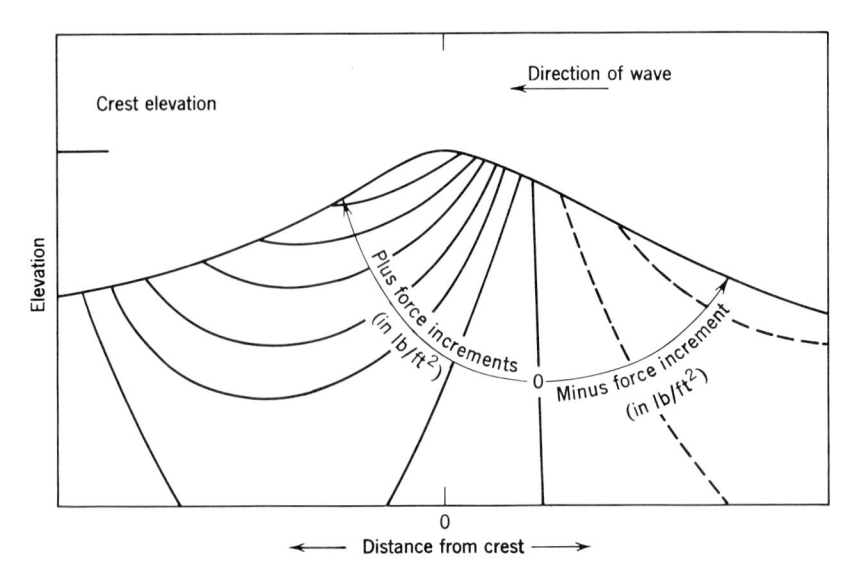

FIG. 10-1. Illustration of theoretical internal force distribution of an ocean wave.

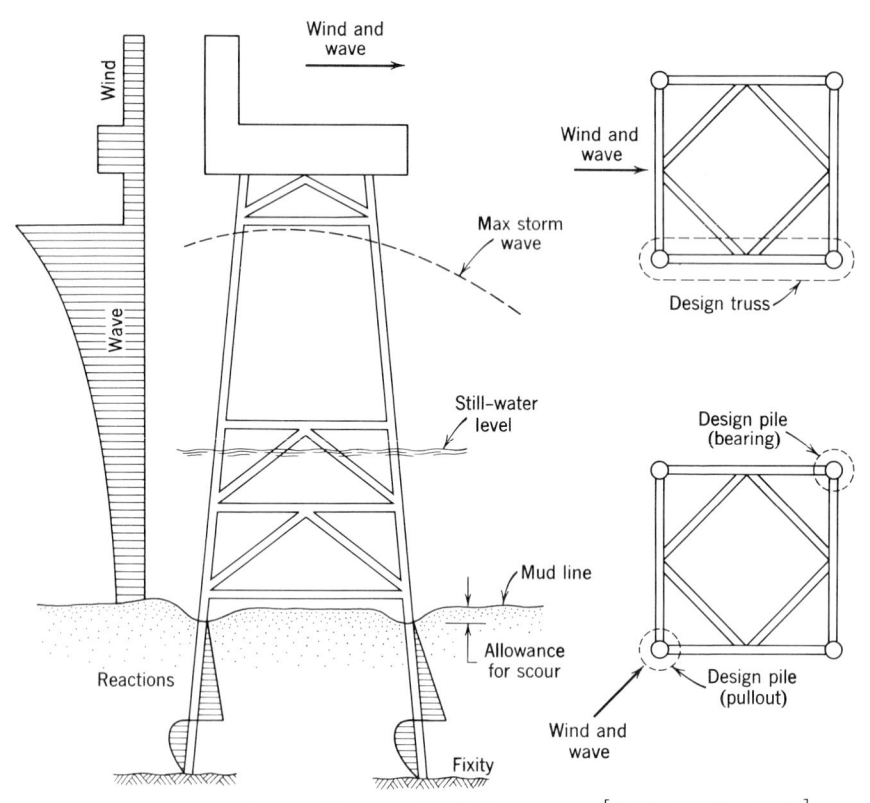

FIG. 10-2. Assumed external loads on USCG fixed towers. [Ruffin (1965)—ASCE]

on floors supporting 5000-lb generators, 13,500-gallon water tanks, and 8000-gallon fuel tanks; normal occupancy live loads, structure dead loads, storage and all types of equipment and furnishings. The maximum handling load of a typical station for the floating crane was the template, in excess of 200 tons. Member sizes for the Chesapeake Light Station, as given by Fowler (1965) are illustrated in Fig. 10-3.

Maintenance. Corrosion is the principal contributor to problems of maintenance. The offshore tower presents the greatest problem because it is exposed to four distinct and different corrosion zones, atmospheric, splash, submerged, and buried. Consequently, different protective measures must be employed in the different zones. Corrosion zones for a fixed tower are shown in Fig. 10-4; accepted protective practices are indicated in the figure.

According to Ruffin (1965), three grades of steel were used for the Coast Guard stations. Members exposed to the waves and atmosphere were of ASTM A-242 grade having a resistance to atmospheric corrosion four times that of ordinary carbon steel; framing protected from the atmosphere was made with ASTM A-36 grade. Stainless steel was used for exposed items, such as handrails, walkways, door and window framing, and so on. All exposed structural members were coated with a coaltar epoxy paint. The jacket (template) was protected cathodically by sacrificial aluminum anodes. The exterior walls of the deckhouse were of porcelain-enameled aluminum sandwich panels that should require but little maintenance.

Habitability. The U. S. Coast Guard Stations have a helicopter platform and living quarters for a six-man crew. Power is provided by diesel generators for the light beacons, for signals, radar equipment, and services. The unmanned stations are remotely controlled from shore establishments. Temporary quarters are available only for repair crews. The unmanned stations lack a heliport, but otherwise are equipped essentially the same as the manned station. Typical floor plans are given in Fig. 10-5.

10.2 SEA-FLOOR SOIL MECHANICS

It is reasonable to assume that at least partial support of structures and equipments will be afforded by the sea floor. To predict "how much," with any degree of reliability, we must turn to engineering-geology and soil-mechanics techniques.

Investigations

Soil investigations in shallower depths, as for offshore towers, can rely to a certain extent on current technology, as developed by the offshore oil

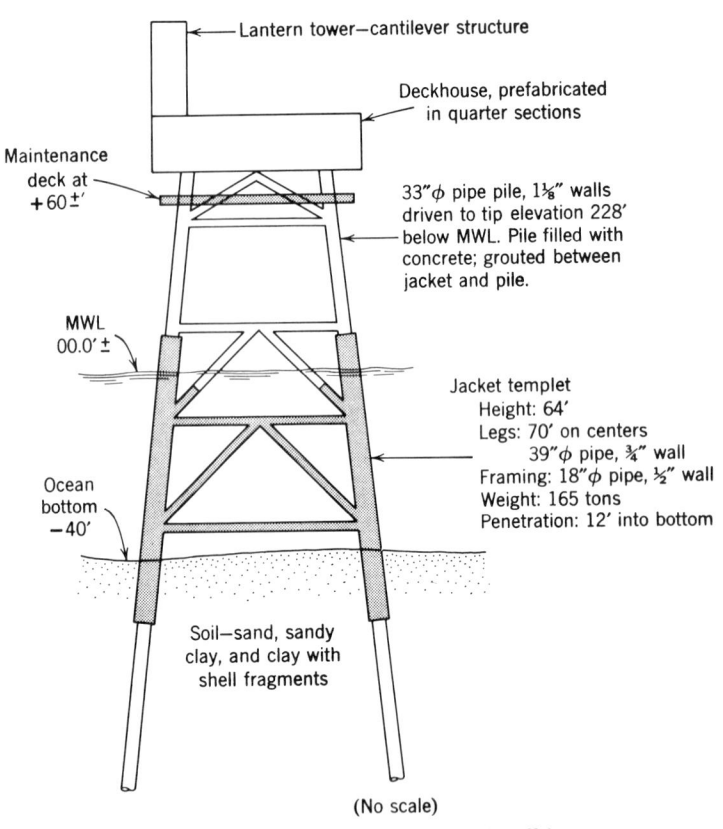

FIG. 10-3. Schematic diagram of Chesapeake offshore tower.

Labels on figure:
- Lantern tower—cantilever structure
- Deckhouse, prefabricated in quarter sections
- Maintenance deck at +60 ±'
- 33"φ pipe pile, 1⅛" walls driven to tip elevation 228' below MWL. Pile filled with concrete; grouted between jacket and pile.
- MWL 00.0' ±
- Jacket templet
  Height: 64'
  Legs: 70' on centers
  39"φ pipe, ¾" wall
  Framing: 18"φ pipe, ½" wall
  Weight: 165 tons
  Penetration: 12' into bottom
- Ocean bottom −40'
- Soil—sand, sandy clay, and clay with shell fragments
- (No scale)

Corrosion Zones	Acceptable Protective Measures	Chesapeake Tower Protective Measures
Atmospheric	Paints Encasements Coatings Special steels	Porcelain-enameled Aluminum sandwich panels
Splash		Special steels Exposed structural members coated with coaltar epoxy paint
Submerged	Encasements Coatings Cathodic	Jacket cathodically protected with sacrificial aluminum anodes
Buried		

FIG. 10-4. Corrosion zones for offshore towers.

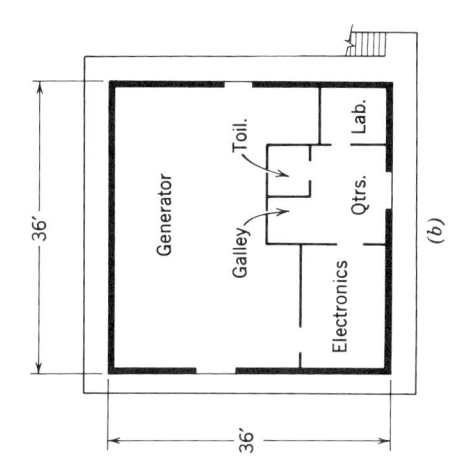

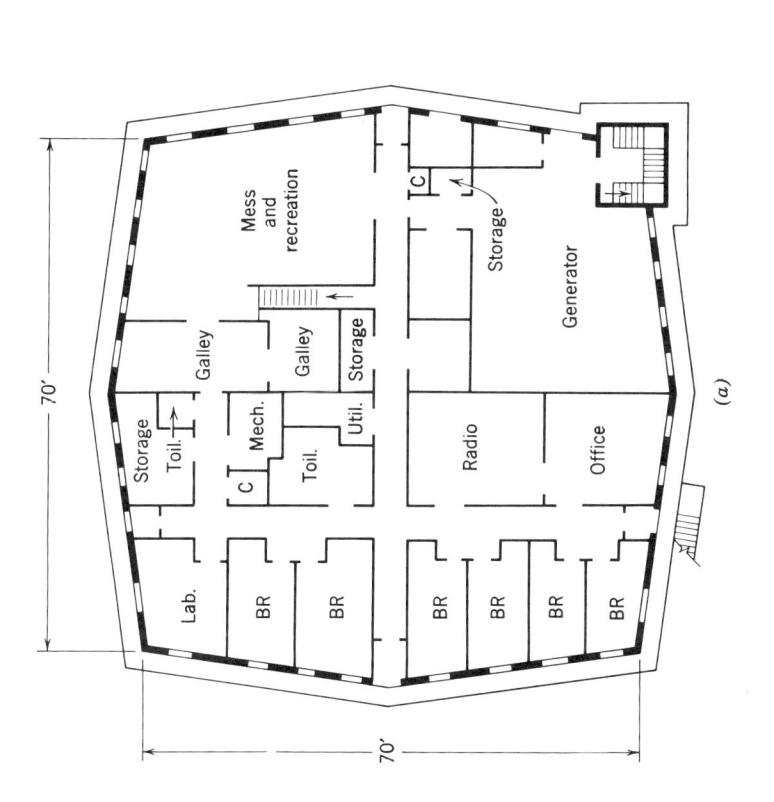

FIG. 10-5. Typical habitability layout of USCG offshore towers: (a) Manned; (b) unmanned [Ruffin (1965)—ASCE].

industry. But man's exploration into the deeper oceanic regions is presented with an appreciably different, and most difficult, set of circumstances. Usable exploration techniques are still in the formative stage and reliance on past experiences is an event of the future. Soil investigative techniques for these deeper regions have advanced little beyond the preliminary stage. There is a difference of opinion among notables in the field as to a suitable means of obtaining an undisturbed sample, most realistic testing method, and the translation of the findings into usable design criteria.

Until recently, examination of oceanic sediments as a foundation support material has provoked little interest. Heretofore, most sea-floor sampling was for oceanographic and geological purposes—of academic interest totally unrelated to foundation investigations. Man's desire to explore, and exploit, innerspace has presented him with a need to know, and the interest in sea-floor foundations is on the rise. To date at the U.S. Naval Civil Engineering Laboratory (NCEL), hundreds of sea-floor soil samples have been subjected to analyses for engineering properties. Tests essentially follow those standardized for terrestrial soils.

In general the analyzed sediments possessed varying properties. Near shore sediments were in most cases coarse grained in the sand-silt range and were high in carbonate and organic carbon. Computed bearing capacities in these areas would be high with probable settlement relatively low. In basin areas, sediments were almost always fine grained in the silt–clay range These areas would be characterized by low bearing capacities and high settlement values. Carbonate and organic carbon were present in lesser amounts than in the nearshore sediments. In all cases shear strength and bulk density increased with depth, whereas void ratio and water content decreased. Typical computer-processed data resulting from a core sample is given in Table 10-1.

Various types of coring tools are in common use today (Fig. 10-6). It is believed that the characteristics of the coring tool exert an influence on laboratory findings. Also in question is the degree of impressed disturbance resulting from the coring operation, and unknown changes that must occur in the sample being brought from great depths at high pressure to the surface, and the influence that transportation and storage may exert on the sample before testing. Moreover, the shallow penetration depths from which the core samples are taken are usually inadequate for foundation analyses. Only in rare instances do the penetrations exceed 10 ft. Understanding the behavior of sea-floor sediments may lie in perfecting techniques for *in situ* strength measurements.

Designers must learn how to use the ocean sediments for resting foundations—the same as they do with terrestrial soils. Engineering geologists are becoming concerned with predicting bearing values, stability and

Deep Ocean Installations and Fixed Structures 321

TABLE 10-1

****** CORE NO MH - 6 ******

LATITUDE 34 06.2 N	LONGITUDE	120 43.3 W		WATER DEPTH	395. FM	
INTERVAL (IN)	0-3	6-9	12-15	18-21	24-27	30-33
COLOR (GSA NO.)	5Y4/4	5Y4/4	5Y4/4	5Y4/4	5Y4/4	5Y4/4
ODOR	H2S	H2S	H2S	H2S	H2S	H2S
BULK WET DENSITY (GM/CC)	1.582	1.613	1.703	1.761	1.766	1.697
VANE SHEAR STRENGTH (PSI)	0.447	0.382	0.555	0.601	0.708	1.011
REMOLDED STRENGTH (PSI)	0.115	0.123	0.231	0.216	0.297	0.635
SENSITIVITY	3.9	3.1	2.4	2.8	2.4	1.6
WATER CONTENT (P)	62.1	60.5	48.7	46.8	44.8	36.6
SPECIFIC GRAVITY OF SOLIDS	2.608	2.694	2.720	2.699	2.795	2.759
DRY DENSITY (GM/CC)	0.976	1.005	1.145	1.200	1.220	1.242
VOID RATIO	1.674	1.681	1.375	1.247	1.294	1.222
POROSITY (P)	62.6	62.7	57.9	55.5	56.4	55.0
SATURATED VOID RATIO	1.620	1.630	1.325	1.263	1.252	1.010
LIQUID LIMIT (P)	39.6	44.0	40.9	35.4	36.0	31.6
PLASTIC LIMIT (P)	-	-	-	-	-	-
PLASTICITY INDEX	-	-	-	-	-	-
LIQUIDITY INDEX	-	-	-	-	-	-
COMPRESSION INDEX	0.27	0.31	0.28	0.22	0.23	0.20
CARBONATE CONTENT (P)	14.26	23.36	11.04	20.29	9.85	13.56
ORGANIC CONTENT (P)	2.14	1.60	1.45	2.06	1.03	1.05
ACTIVITY	-	-	-	-	-	-
PERCENT SAND	69.7	65.5	66.0	68.0	72.4	69.4
PERCENT SILT	12.6	16.3	16.5	16.4	13.5	17.0
PERCENT CLAY	17.7	18.2	17.5	15.6	14.1	13.6
MEDIAN DIAMETER (MM)	0.074	0.072	0.072	0.078	0.077	0.076
SEDIMENT TYPE	CLAYEY SAND	CLAYEY SAND	CLAYEY SAND	SILTY SAND	CLAYEY SAND	SILTY SAND
MICROSCOPIC ANALYSIS - PLUS 325 FRACTION, PERCENT						
QUARTZ	60	50	60	70	60	60
GLAUCONITE	10	5	5	5	10	10
SHELL DEBRIS	15	10	10	5	5	5
BENTHONIC FORAMINIFERA	5	5		5	5	5
PLANKTONIC FORAMINIFERA	5	20	20	10	15	20
SPONGE SPICULES	5	5				
RADIOLARIA	TR					
FISH TEETH	TR					
AGGREGATES		5	5	5	5	
CHARCOAL					TR	TR
ASPHALT				TR	TR	TR

REMARKS ****
GLAUCONITIC QUARTZOSE SAND APPARENTLY HOMOGENEOUS NEAR THE TOP, BECOMING BEDDED DOWNWARD. LOOKS FINER GRAINED THAN OTHER CORES IN THIS SERIES. BENTHONIC FORAMINIFERA VISIBILE. SOME DARK MOTTLES NEAR BASE. LIGHTER AND DARKER LAYERS REVEAL BEDDING NEAR BASE. IT IS ESSENTIALLY HORIZONTAL WITH SOME DISTURBANCE DUE TO SLUMPING, ORGANIC REWORKING OR DISTURBANCE BY PASSAGE OF CORER.

response to various loadings, in addition to age and rate of deposition of sediments. These latter characteristics may influence the engineering properties of marine soils.

The work at NCEL is directed not only toward improving coring methods to obtain undisturbed and longer samples, but also toward determining the influence of pressures on soil strength and developing techniques for *in situ* testing.

One task recently undertaken has as its objective the performance of shear-strength and consolidation tests on typical sea-floor samples in a high-pressure environment. Tests will be conducted in a pressure vessel with the pressures ranging from atmospheric to 10,000 psi. In this way it

FIG. 10-6. Ewing-type soil sampler.

may be possible to determine the effects of the deep-ocean environment, primarily that of elevated pressure, on the engineering properties of these sediments. The sheer strengths of samples will be measured using both vane shear and direct shear devices. The former will be measured using rotational speed of 6° per minute (clock motions speed) for tests on cohesive soils; the latter will accommodate both cohesive and noncohesive sediment. The consolidation tests will be performed with a consolidometer of the conventional fixed- or floating-ring configuration. A sufficient number of soil types will be used to obtain some representation of ocean sediments. The total number of tests to be performed on each soil will be adequate to allow analyzing the influence of the high-pressure environment, and to determine this influence for variations in pressure.

In another program[5] the NCEL has developed an *in situ* plate bearing device that will further man's knowledge of sea-floor sediments by conducting plate bearing tests directly on the sea floor. In a way, this device is an ocean installation. The unit is lowered to the sea floor on a tether line from a surface ship (Fig. 10-7). On the sea floor the device is supported on three pads. Each pad is articulated by means of a ball-and-socket joint;

[5]Kretschmer, T. R., "In-Situ Plate Bearing Device," Technical Report, U.S. NCEL, Port Hueneme, California (in preparation).

Deep Ocean Installations and Fixed Structures 323

FIG. 10-7. *In situ* plate bearing device.

rotation of the pad is prevented by a keeper. When tension is removed from the tether line, the load is transferred to a central piston that forces a given size bearing plate into the sea floor to induce a shear failure. The bearing plates are interchangable; various sizes of circular and square plates are available to evaluate various footing sizes.

Basically, the working mechanism (Fig. 10-8) consists of a large piston with a bearing plate attached at the lower end of the piston. The piston is supported and guided, and its rate of displacement controlled by three small hydraulic cylinders. The weight on the piston (and bearing plate) can be varied from the device's tare weight of about 300 lb to 5,000 lb by the addition of lead weights, housed in the central portion. An instrumentation system provides four channels of readout via an acoustic signal: monitoring plate displacement into the sediment, total load on the plate, and the inclination of the device from the vertical. The fourth channel is a calibration circuit. The instrumentation package has been tested to 6,200 psi hydrostatic pressure at temperatures as low as 0°C.

Sea trials with the device were conducted in depths from about 100 to 1,200 ft. More than 60 tests were performed. Typical data in soft sediments are shown in Fig. 10-9.

FIG. 10-8. Working mechanism of *in situ* plate bearing device.

Another ocean installation-type of device under development, and planned to be operational in late 1966, is the Deep Ocean Test Instrumentation Placement and Observation System (DOTIPOS). It too is a bottom rest structure that will be lowered from a surface ship, but in this case with coaxial cable. Design working depth is 6,000 ft. DOTIPOS will carry television and still cameras, lights, power source (15 kVA), electronic equipment and the experiment to be carried out on the sea floor. The structure will monitor a 12 ft by 12 ft central area for various experiments, such as *in situ* vane shear, sea-floor penetrations and dynamic loading of components under hydrostatic load. DOTIPOS will provide the power for the tests, manipulate the specimens and/or test devices, record test data, and provide a pictorial record of the tests.

However, investigations are only beginning and have not progressed to establishing firm criteria for the design of sea-floor foundations. Nevertheless, some generalities are possible in the interim: (a) for determining ocean-floor bearing capacity, laboratory test data alone is insufficient; (b) the effect of sampling procedures on *in situ* strength parameters is not fully understood because sediment properties are quite heterogeneous even

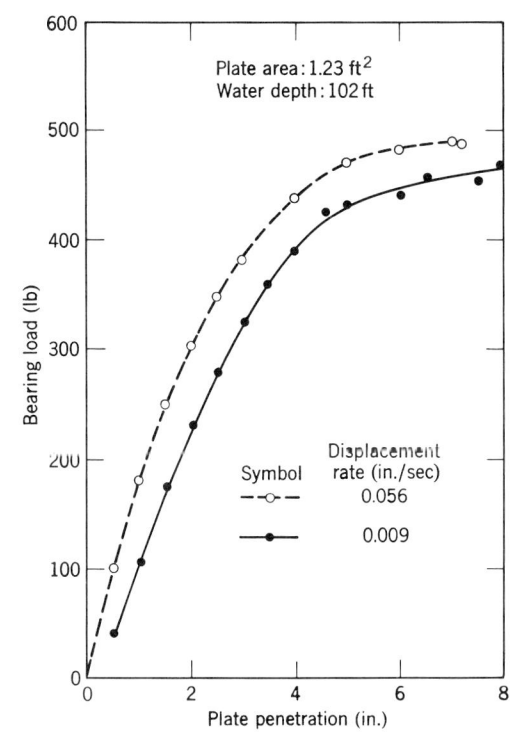

FIG. 10-9. Typical *in situ* plate bearing test results in soft sediments.

in a very limited area; (c) engineering judgment must be exercised and an adequate safety factor must be provided when sea-floor foundations are designed; (d) methods used for analyzing terrestrial soils may possibly be applied to marine sediments, but this possibility, should it prove well founded, neglects any effect a high-pressure environment may have on strength properties (high-pressure effect is currently being investigated); (e) in most cases, primary concern in foundation design and construction will be with safe bearing capacity based on tolerable settlement, rather than on the ultimate bearing capacity, which is the bearing stress necessary to rupture the soil mass beneath a footing.

Mass movements

The designer faces many problems in mass movements—submarine landsliding, slumping, and turbidity flow. Perhaps, the engineers' greatest challenge today is defining the problem. We need only to pick up a newspaper to realize that terrestrial land-mass movements, even though they can be observed and studied, are not well understood. Potential areas of seafloor movement and the prevalence of occurrence can possibly be defined.

A better understanding of the magnitude of forces involved would be a significant contribution to the designers of ocean structures. Until this is achieved, constructions will of necessity avoid potential danger areas.

Terzaghi (1956) in his paper on "Varieties of Submarine Slope Failures" offered these conclusions:

"Submarine slope failures can be divided into three large categories, failures involving the downward movement of patches of cohesive sediments perched on the steep slope of cohesionless material, spontaneous mass movements of short duration involving large quantities of sediments, and more or less continuous slumping on gentle slopes.

The rapid downward movement of a patch of cohesive sediment perched on the steep slope of a sand and gravel delta has been observed on Howe Sound. Slides of this category are minor details in the otherwise continuous growth of the deposit.

Spontaneous mass movements of short duration involving large quantities of material are quite common on both steep and gentle slopes above fine-grained, cohesionless or slightly cohesive sediments. They are caused by spontaneous liquifaction. Liquifaction can occur only on the condition that the structure of a large portion of the sedimentary deposit is metastable. Presence or absence of metastable structure can be ascertained either by field penetration tests or else by laboratory tests on perfectly undisturbed samples. The impulse for the liquifaction can be given by earthquakes, blasting operations, the seepage pressure exerted by the flow of the ground water towards the ocean during very low tides, and other agents tending to change the position of the solid particles with reference to each other. The liquifaction may start at one point and spread from there over larger and larger areas with a velocity of at least 10 miles per hour. Whenever the slope is steep, the mass movement assumes the character of a landslide. The slide material moves rapidly down and outward and covers the adjacent, gently sloping or horizontal ocean floor like an alluvial fan. If the slope is gentle, the ground movement assumes the character of a more or less violent agitation of the material, associated with a flow towards the lowest area and a corresponding change of the topography of the surface of the sediment. During the brief period of agitation, heavy objects sink into the liquified sediment and submarine cables resting on the agitated material are likely to snap. However, the distance through which the particles move may be relatively short and the velocity of their movement may be small compared to that of the propagation of the boundary between intact and liquified sediment.

Continuous slumping on gentle slopes occurs only on clay deposits. It is caused by the lag between sedimentation and consolidation. The excess hydrostatic pressures responsible for the slides can be measured by means of Bourdon gages attached to the upper end of piezometric tubes. Wherever the conditions for slope failures are satisfied, the clay movements assume the character of a more or less continuous process, whereby the seat of the movement shifts from place to place. The slide material flows down the slope and may even pass beyond the foot of the slope.

The effect of earthquakes or of explosions on slopes underlain by coarse-grained sediments or by clay is likely to be unimportant, whereas slopes on fine-grained, cohesionless, or almost cohesionless sediments may start to flow even under the influence of a mild earthquake shock or the vibrations produced by pile-driving operations or small blasts."

10.3 ANCHORAGE SYSTEMS

Most deep-ocean installations to date have been anchorage systems of one type or another. Anchoring of buoys seems to predominate, with systems such as the single-buoy taut line, two-buoy taut line, slack line, and the multileg. Ships and major structures have been anchored to various depths down to 20,000 ft. All such systems have used conventional or deadweight-types of anchors. Smith (1965) documents the current state of the art for buoys and anchorage systems. Included are the achieved installations of some of the major operations.

To improve holding power of anchors, the NCEL is studying various concepts of anchors for deep-ocean application. One of the tasks deals with evaluating propellent embedment anchors and the techniques and equipment necessary to their effective placement and employment. The propellent embedment anchor, also commonly termed explosive anchor, is a self-contained bottom-penetrating implement similar to a large caliber gun consisting of a barrel, a recoil mechanism, and the projectile, which is the anchor. These anchors have found limited application in shallow waters. Preliminary tests at the NCEL include testing vital components in a pressure vessel. The components of one anchor passed the test to 10,000 psi. In addition to shallow tests, five firings have been made in deep water, two each at nominally 1100 ft and 2500 ft, and one at 6000 ft.

In another study the application of these anchors to an anchorage complex for bottom-mounted structures is being investigated. The anchor complex, called a Padlock because it locks the bearing pads to the sea floor, is a tripodial frame terminating in articulated, steel bearing pads, each pierced by a vertical component carrying a propellent embedment anchor (Fig. 10-10). Each anchor is attached by cable to a battery-powered motor in the central portion. Once the complex is positioned on the sea floor, the three embedment anchors are fired. The cable arrangement for payout is shown in Fig. 10-11. The anchors are then tensioned by cables winding on drums

328   Systems Design—The Technology

FIG. 10-10. Sea-floor Padlock anchor with 6-ft bearing pads.

in the central compartment. This is accomplished by a rewind mechanism driven by the electric motor. The rewind mechanism has a 1356-to-1 gear ratio that is capable of developing a 10,000-lb line tension. The anchors are individually tensioned. Although the capacity of the complex depends upon the soil conditions, it is estimated that in a reasonably competent soil the complex will support 80,000 lb in bearing, a pullout force of 40,000 lb and an overturning moment of about 120,000 ft-lb. Shallow-water shakedown tests demonstrated that the complex is workable in the areas of major concern, such as cable rewind, cable payout, blast effect, and bearing-pad acceptance of activated anchors and cable.

10.4 SUBMERSIBLE TEST UNITS (STU's): THEIR INSTALLATION AND RETRIEVAL

An extensive research task that is underway at the NCEL is concerned with the long-term behavior of various materials of construction for underwater structures. The program includes the exposure of a variety of metallic

FIG. 10-11. Cable-payout arrangement on Padlock anchor. Cable secured with clips and plastic bolts that shear to allow cable to pay out.

and nonmetallic materials in ocean depths of nominally 6000 and 2500 ft for extended periods of time to determine the corrosive and biological degradation of construction materials exposed to an ocean environment. These depths were selected because (a) the 6000-ft depth represents a sea environment on the edge of a major basin, and although it is an unprecented depth for present construction-type operations, it appears to be attainable in terms of midrange objectives; (b) the 2500-ft depth represents the level of minimum oxygen concentration, offshore at NCEL. The work includes monitoring the environment at the sites. Table 10-2 summarizes the six units emplaced and retrieved while Table 10-3 defines the ocean environment for one period in time at one of the sites. The sites are geographically identified in Fig. 10-12.

An integral part of the over-all program was the emplantment and retrieval of relatively large installations. The STU itself is a bottom-rest structure designed to carry thousands of sample materials to the sea floor and expose them to the bottom environment. This section of the test describes the structure and rigging complex and the installation and retrieval

TABLE 10-2
Submersible-Test-Unit Data

STU	Depth (ft)	Weight in Air (lb)	Total No. Specimens	Materials Represented	Emplaced	Retrieved
I-1	5,300	7,000	1,282	396	Mar. 1962	Feb. 1965
I-2	5,640	5,600	1,521	429	Oct. 1963	Oct. 1965
I-3	5,640	5,200	1,367	398	Oct. 1963	Feb. 1964
I-4	6,780	6,710	1,852	567	June 1964	July 1965
II-1	2,340	7,350	2,385	603	June 1964	Dec. 1964
II-2	2,370	8,750	2,588	811	Apr. 1965	May 1966

techniques. Jones (1965) discusses these aspects in greater detail. Discussion of the corrosion aspects of materials is the topic of another chapter. Also see Reinhart (1965).

The structures

The first STU emplanted was in 1962. It was a towerlike structure on a 12-ft by 12-ft base (Fig. 10-13). The entire unit was fabricated of mild steel and coated with a vinyl paint system. Subsequent to the emplacement of STU I–1, an increase in program emphasis dictated an increase in payload not possible with the heavy towerlike structure, and a redesign was in order. Thereafter all structures were fabricated from aluminum-alloy angles and plates joined with stainless-steel hardware (Fig. 10-14). Polyvinyl chloride washers separated the aluminum members from the steel connectors. Finally, all voids in the joints were filled with a waterproof silicone

TABLE 10-3
Ocean Environment[a] at STU-I Series Site.

Sample Depth (ft)	Temp. (°C)	Salinity (‰)	Oxygen (ml/L)	pH	Eh (mv)
0	13.890	33.410	6.121	8.183	+164
1,716	5.458	34.269	0.362	7.619	166
2,346	4.833	34.336	0.299	7.604	166
2,981	4.048	34.446	0.480	7.590	180
4,193	3.208	34.543	1.050	7.615	146
5,384	2.357	34.594	1.467	7.734	179

[a] From oceanographic Data Report, NCEL Cruise A-601-1, Jan. 18, 1966, Station 7, Lat: N 33° 41'30", Long: W 120° 46'18", Sounding depth: 5500 ft.

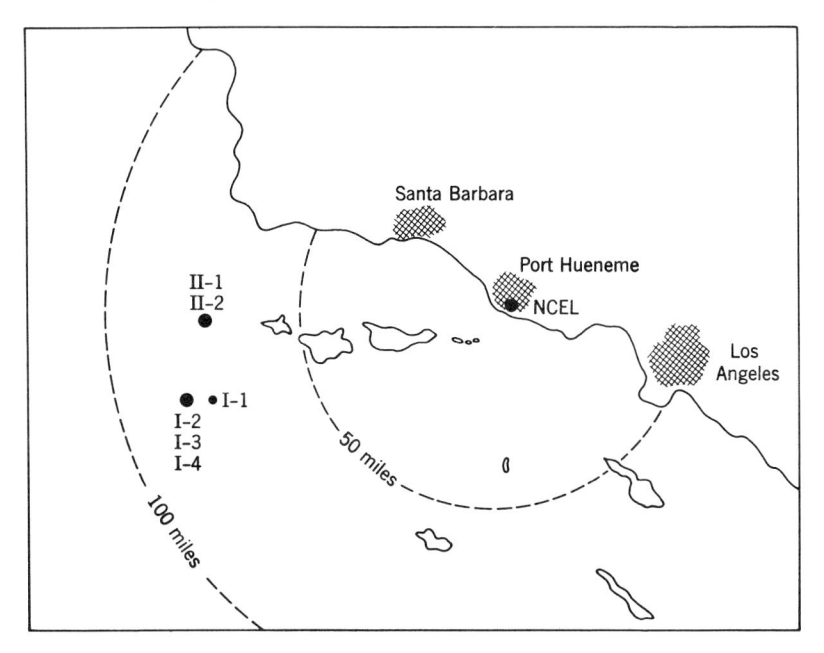

FIG. 10-12. Pacific Ocean sites for submersible test units.

sealing compound. The finished cagelike structure is shown in Fig. 10-15. It is approximately 12 ft high on a 12-ft by 12-ft base.

Provisions were made to suspend standard specimen racks in the long sides of the unit for exposure to the ocean water, and in the base, for exposure to the bottom sediments. The latter protruded below the bearing plates and were thus embedded into the sediments under the weight of the structure. The interior of the structure, as well as the ends, were used for other types of specimens that could not be mounted in standard racks, and for various instruments. To assist in retrieval, the structure's bearing plates were designed to eliminate the "mired-down" holding of the sea floor. The aluminum bearing plates that extended the full length of the structure (as may be noted in an earlier figure) were attached with magnesium bolts and steel nuts. The life of the magnesium–steel connections in sea water was but a few days. Thereafter the structure could be lifted from its bearing plates. Keepers were used to prevent lateral shifting of the structure on the bearing plates.

Rigging

The rigging was designed to ensure successful emplantment and retrieval. Basically, the over-all design for each unit was similar, consisting of a riser

FIG. 10-13. STU I-1 installation.

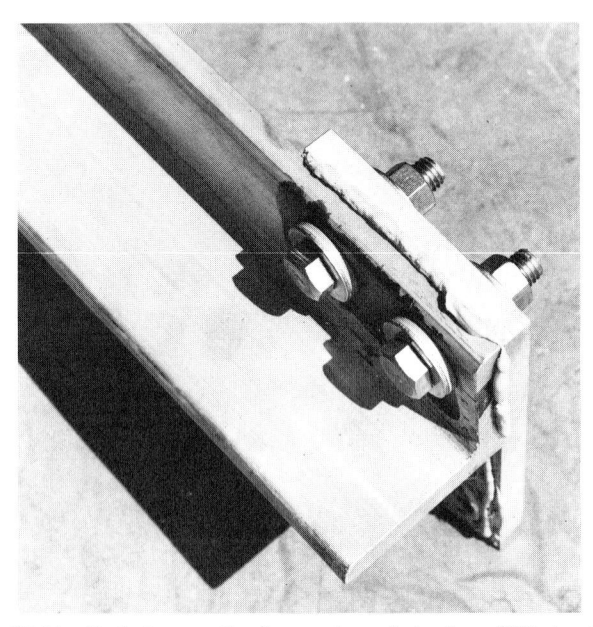

FIG. 10-14. Typical connection for members of aluminum STU structures.

Deep Ocean Installations and Fixed Structures 333

FIG. 10-15. Aluminum STU structure loaded with specimens. Buoys at top were used to increase payload.

line from a bottom anchor clump to a submerged buoy about 200 ft below the surface and a connecting line from the riser to the structure. In one installation two units (STU I–2 and I–3) were connected to a common riser. But individually each installation differed in details as necessary to conform to existing conditions.

Except for the STU I–1 complex that used black twisted nylon and polypropylene ropes, all other installations were made with braided black polypropylene rope. Under load the braided rope is nonrotating, as opposed to the twisted construction which tend to un'ay under tension. Two of the emplanted installations are shown schematically in Fig. 10-16 and 10-17; Table 10-4 lists the components used in each installation. According to Jones (1965) methods for retrieval for all STU's except the first were as follows:

1. A release mechanism actuated by coded acoustic command from a surface ship would disconnect the concrete sinker and allow the aluminum buoys to surface. Failure of either buoy, by collapse or flooding, would not

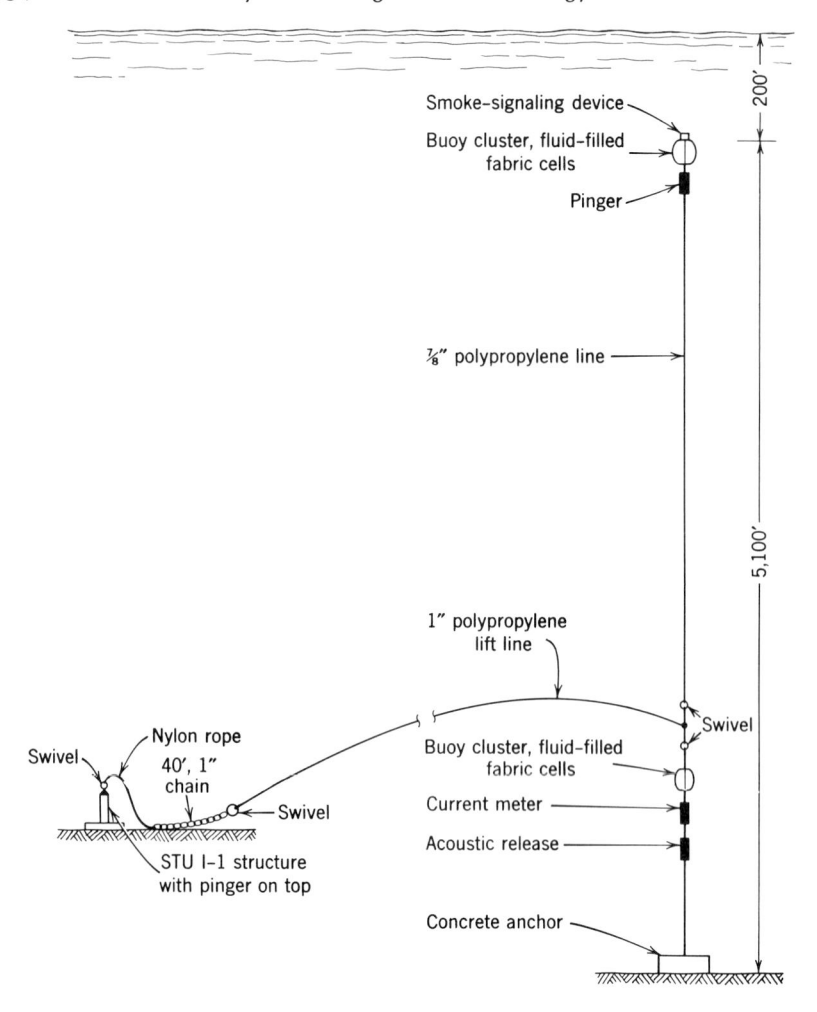

FIG. 10-16. Submersible test unit I-1 installation (not to scale).

prevent surfacing of the remaining buoy because each has 800 lb net buoyant capacity, sufficient to overcome the load of all components in the line.

2. A release mechanism (identical to the first except for actuating code) would, when actuated, allow 500 ft of 1-in.-diam braided polypropylene rope to be pulled from its bale as the upper buoy approached the surface.

3. A surface marker to allow the riser line assembly to be located and retrieved.

4. An inverted catenary formed by the lift line between the STU and riser line. If the STU and riser positions are known within a few hundred feet, a grapnel suspended from a ship can be towed normal to the STU lift line

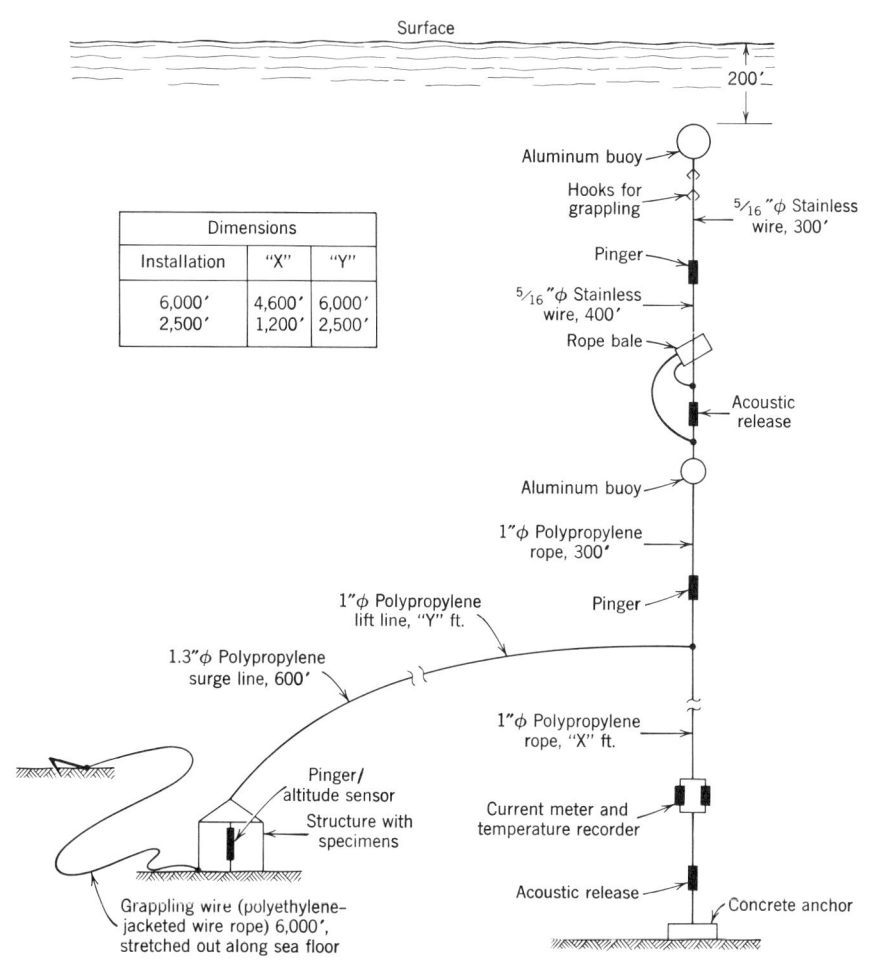

Deep Ocean Installations and Fixed Structures

Dimensions		
Installation	"X"	"Y"
6,000'	4,600'	6,000'
2,500'	1,200'	2,500'

FIG. 10-17. Typical STU installations as designed, excluding STU I-1 (not to scale).

catenary until contact is made, and then used to raise the line to the surface. The locating problem is simplified by the aluminum buoys, which are excellent sonar targets and appear clearly on fathometer records when the ship is within 200 ft radially from the buoys. The pingers also assist in positioning.

5. A hook assembly beneath the upper aluminum buoy would snag a weighted line suspended between two boats as the line is towed through the water.

6. A grappling wire (polyethlene-jacketed wire rope) laid on the sea floor would be snagged by a grappling assembly. This assembly consists of a Rennie chain grapnel and a sliding prong grapnel at the end of a length of

TABLE 10-4
STU System Components

	STU I-1	STU I-2[a]	STU I-3[a]	STU I-4	STU II-1
Structure material	Steel	Al 5086 H112	Al 5086 H112	Al 5086 H112	Al 5086 H112
Structure connections	Welded and bolt	Bolt, AISI 316	Bolt, AISI 316	Bolt, AISI 316	Bolt, AISI 316
Protective system	Vinyl paint	None	None	None	None
Surge line	Twisted white nylon, 1¼ in. diam	Black, braided 2-in-1 polypropylene, 1.3 in. diam	Black, braided 2-in-1 polypropylene, 1.3 in. diam	Black, braided 2-in-1 polypropylene, 1.3 in. diam	Black, braided 2-in-1 polypropylene, 1.3 in. diam
Lift line	Twisted black polypropylene, 1 in. diam	Black, braided 2-in-1 polypropylene, 1 in. diam	Black, braided 2-in-1 polypropylene, 1 in. diam	Black, braided 2-in-1 polypropylene, 1 in. diam	Black, braided 2-in-1 polypropylene, 1 in. diam
Shackles	Forged steel	Forged steel, Ni plated	Forged steel, Ni plated	Stainless steel, AISI 303	Stainless steel, AISI 303
Thimbles	Stainless steel, AISI 304	NiAl bronze, BuShips type	NiAl bronze, BuShips type	NiAl bronze and high-impact PVC	NiAl bronze and high-impact PVC
Swivels	4, Miller type	None	None	2, Miller type, stainless steel in wire of riser	2, Miller type, stainless steel in wire of riser
Lift line attachment to riser	Bottom, above concrete sinker	Beneath lower buoy	Beneath lower buoy	Beneath lower buoy	Beneath lower buoy
Riser line	Twisted, black, polypropylene, ⅞ in. diam	Black, 2-in-1 polypropylene, 1 in. diam; common to 2 STU's (1-2, 3)	Black, 2-in-1 polypropylene, 1 in. diam; common to 2 STU's (1-2, 3)	Black, 2-in-1 polypropylene, 1 in. diam, and AISI 302 wire, 7/16 in. diam	Black, 2-in-1 polypropylene, 1 in. diam, and AISI 302 wire, 7/16 in. diam
Sinker	Concrete	Concrete	Concrete	Concrete	Concrete
Buoys	Neoprene and fabric, fluid filled, with/ stainless-steel fittings	A17178, air filled	A17178, air filled	A17178, air filled	A17178, air filled
Surface float	None	Foam filled, 20 in. diam	Foam filled, 20 in. diam	None	None
Navigation aid	None	10 × 39 WE Coast Guard buoy	10 × 39 WE Coast Guard buoy	None	None
Pinger	Enclosed transducer	Enclosed transducer	Enclosed transducer	External transducer	External transducer
Pinger batteries	Mercury	Leclanche	Leclanche	Leclanche	Leclanche
Temperature recorder location	None	On structure	On STU I-2	Bottom of riser	Bottom of riser
Acoustic release	Failed (proprietary)	Failed (NOL type)	Failed (NOL type)	None available	None available
Grappling wire	None	½ in. diam, 7 × 19 ips, polyethylene jacketed	½ in. diam, 7 × 19 ips, polyethylene jacketed	½ in. diam, 7 × 19 ips, polyethylene jacketed	½ in. diam, 7 × 19 ips, polyethylene jacketed
Current meter location	Bottom of riser	Bottom of riser	Bottom of riser	Bottom of riser	Bottom of riser

[a] Originally placed with common riser line.

wire rope at least 1½ times the water depth; it would be towed behind the ship in a direction normal to the direction in which the grappling wire was placed. This method is an adaptation of the techniques used by cable repair ships when retrieving submarine telephone cables.

Emplacement and retrieval operations

Of the six units, it is intended to briefly describe the emplacement and retrieval of only STU II–1. In a way, they were all similar-yet dissimilar. Readers interested in greater details and in the operations involved with the other units are referred to Jones (1965) and NRR (1965).

Emplacement of STU II-1. The structure and all rigging was transported to the test site on board the deck of an auxiliary fleet tug. During its transport the structure loaded with its specimens was covered with a polyethylene sheet to protect it from the marine environment. The auxiliary fleet tug accompanied by an auxiliary power boat reached a fogged-in site in the early morning hours. The sea state was running at about a Beaufort 3. Because of the 2500-ft water depth, the vessel was not anchored, but allowed to drift.

With an assist from the auxiliary power boat to pull out lines, the 7300-lb structure was lifted, placed over the port side, and slowly lowered.

On lowering, the surge line was paid out around a capstan that was continuously wetted and the lift line was paid out through grooved rollers of a winch. Line tensions varied from 2000 to 11,500 lb during the first few hundred feet of lowering; the tension excursions reduced thereafter as the structure lowering continued. After about 3¼ hr the structure came to rest on the sea floor. Lowering during the last few hundred feet was slowed to insure a soft, upright landing. By the time the emplacement was completed, the sea became more severe. Consequently, the lift line was buoyed off and emplacement of the riser postponed. About one week later the operation was resumed; the riser with its instrumentation was attached to the STU line and lowered into place. This took about 2 hr. Finally, the riser's geographic position was tied into a navigational system.

Retrieval of STU II-1. The retrieval operation set out in a fleet salvage boat some 6 months after emplacement. Navigating to the general area was possible with the ship's gear, after which the ship "homed in" on the acoustic pingers—one attached to the structure and the other in the riser line. (See Fig. 10–17). Since each pinger had different characteristics, it was possible to discriminate between the signals and position the ship to pass between the structure and its riser. Once the ship was positioned, a weighted grapnel was suspended to a depth of 2000 ft. As the ship slowly passed between the structure and the riser, the grapnel engaged the lift line (line connecting the

FIG. 10-18. Grapnel engaged in jacketed wire rope of STU I-4.

structure to the riser). Contract with the lift line was indicated on a dynameter by an increase in tension in the grappling line. Thereafter the installation was systematically retrieved in about 4 hr. Once they were on board, the specimens were carefully washed (sprayed with fresh water) and sprayed with butyl alcohol to aid drying. The entire structure was then covered with a polyethylene sheet.

Of the six structures and installations retrieved by the NCEL, two were recovered by shallow grappling, engaging the inverted catenary formed by the lift line and the riser; three were recovered by bottom grappling, engaging the jacketed wire stretched out from the structure along the sea floor (Fig. 10-18); and one by bottom grappling, when the grapnel fouled on the nylon line attached to the structure (STU I-1).

10.5 FUTURE APPLICATIONS

Submerged or bottom-rest platforms, towers, and habitable complexes are seen for practically all depths for such functions as offense, defense, surveillance, power generation, storage, and research. This section considers some possible solutions.

Frame structure[6]

Picture a purely hypothetical case, in which, in some not-too-distant future, we may anticipate a requirement for a load-carrying structure to be supported by the sea floor. Factors of economy, handling, and design may

---

[6]After a feasibility study by P. A. Dantz and D. True at U.S. NCEL (unpublished).

Deep Ocean Installations and Fixed Structures 339

render buoyancy elements, to support the structure completely off the sea floor, unattractive or limited. To further the argument let us assume that our payload is 10 ft by 30 ft in plan by 50 ft high, and weighs 500 tons submerged. The problem is further defined by these assumptions or "ground rules":

1. The payload must be held to within 5° of vertical plumb and must be oriented to a predetermined azimuth after emplantment.
2. The structure must not have a water-depth restriction, but shall be in depths in excess of 300 ft; that is, below surface disturbances.
3. The structure must be capable of resting on the sea floor with slopes up to 1 in 10 in currents up to 2 knots.
4. A bearing capacity of 1000 lb/ft^2 is allowed.
5. The structure shall be as light as possible, consistent with structural adequacy, and shall have a factor of safety of 2 on maximum fiber stresses and buckling.
6. The material shall be structural steel having a minimum yield of 36,000 psi.

The basic structure selected for this problem was a truss-type truncated,

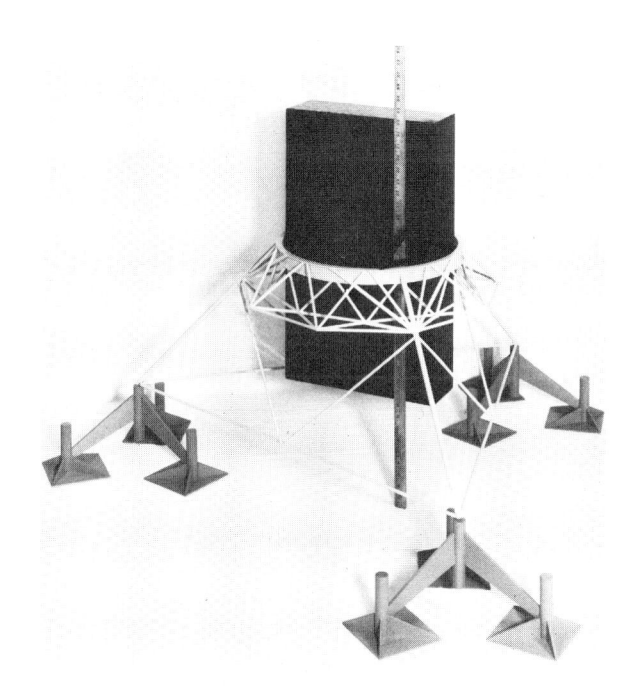

FIG. 10-19. Model of structural frame using Padlock anchors to secure to the sea floor.

equilateral pyramid resting on Padlocks, as modeled in Fig. 10-19. The selected geometry of the structure was as follows:

1. Support legs extend from the package support frame at 45°.
2. Height of the package support frame and center of gravity of the package is 30 ft above the base of the supporting legs.
3. The dimension in plan from the center of gravity of the package to the base of a support leg is 45 ft.

The design is controlled by the bearing pads and the anchor holding power. The critical loading occurs with the 30-ft-by-50-ft surface of the package normal to the current direction, one of the structure's main support legs is parallel to the current flow, and the current is down the slope with the structure resting on a maximum slope. Figure 10-20 illustrates the critical loading and selected geometry. The selected member sizes are given in Table 10-5.

The construction sequence for this structure may be somewhat as follows:

1. The prefabricated structure is towed to a preselected and surveyed site completely assembled.
2. All systems are checked and activated on surface at the site.

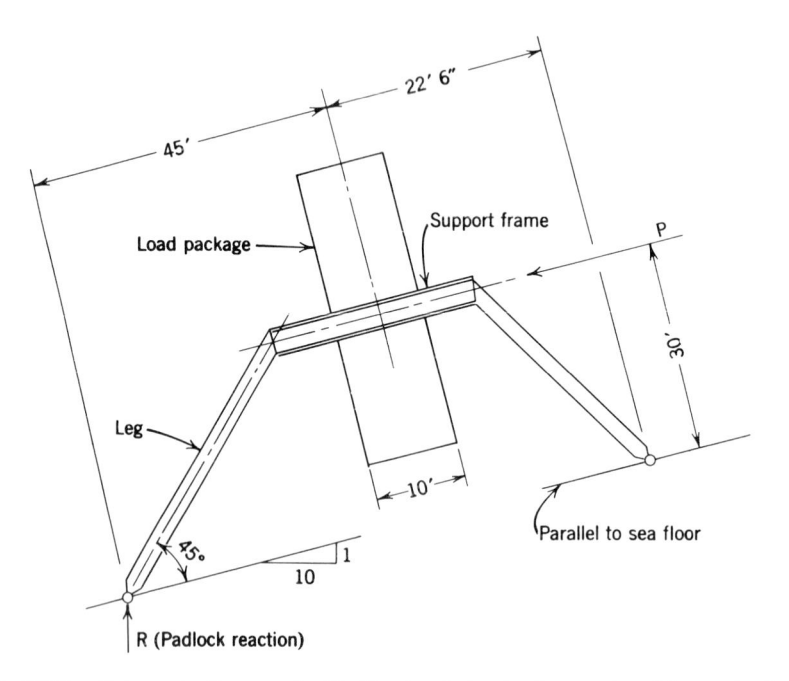

FIG. 10-20. Schematic diagram of critical load and structural geometry of frame structure.

TABLE 10-5
Member Sizes for Structural Frame

Component	Member Size
Support frame for package, main member	Built-up ring, 34-ft diam. in plan, with web and flanges, equivalent to 16×36 WF
Main leg	15-in. tublar, $\frac{3}{4}$-in. wall
Lower tension chord (at base of legs)	10-in. X-strong pipe
Struts (main truss)	8-in. XX-strong pipe
Horizontal members (main truss)	8-in. X-strong pipe
Secondary bracing	Not sized
Bearing pads	12-ft square or $13\frac{1}{2}$-ft diam., circular
Anchors	16-kip vertical holding[a]

[a] Based on assumption that sea floor is frictionless plane and that resistance to sliding is provided by embedment anchors.

3. The structure is lowered to the sea floor and maneuvered into position.

4. Upon command, a preprogrammed device activates the anchors and subsequently the rewind mechanism to lock the structure to the sea floor.

5. The package is lowered and guided into its support frame.

6. Finally, the package is plumbed and oriented in azimuth through systems built into the package support points.

Construction submersibles are seen as a definite asset, if not a necessity, to the many phases of the operation.

Deep-ocean reactor-placement study

An ocean reactor-placement study is reported by Quirk (1965). His work documents the results of a study pertaining to handling of heavy weights in the ocean, specifically, the problem of placing a 155-ton nuclear reactor on the ocean floor at depths to 20,000 ft is considered. The study includes two approaches: (a) an all-surface system where all the work is accomplished from the surface and (b) a vehicle-assist system, wherein submersibles are used to assist in the handling and placement of heavy components. It is concluded from the study that basically the all-surface system is better, with the role of the submersible reduced to suited tasks, such as connecting power cables, surveying the site, and inspecting and directing the work.

Manned habitats

A logical solution for early manned habitats, where life support will be a normal sea-level atmosphere, appears to be in winch-down concepts. Essentially, this type of structure consists of a sea-floor anchor with a flexible

342 Systems Design—The Technology

line connected to a hull structure with positive buoyancy. The structure is lowered to any desired elevation in the water column by winching against the anchor (Fig. 10-21).

This type of construction offers many advantages. It will not require a major technical break through to achieve by the early 1970's. The concept offers the potential of operating at any depth, down to design depth, to perform experiments for extended periods of time. It can be designed to be

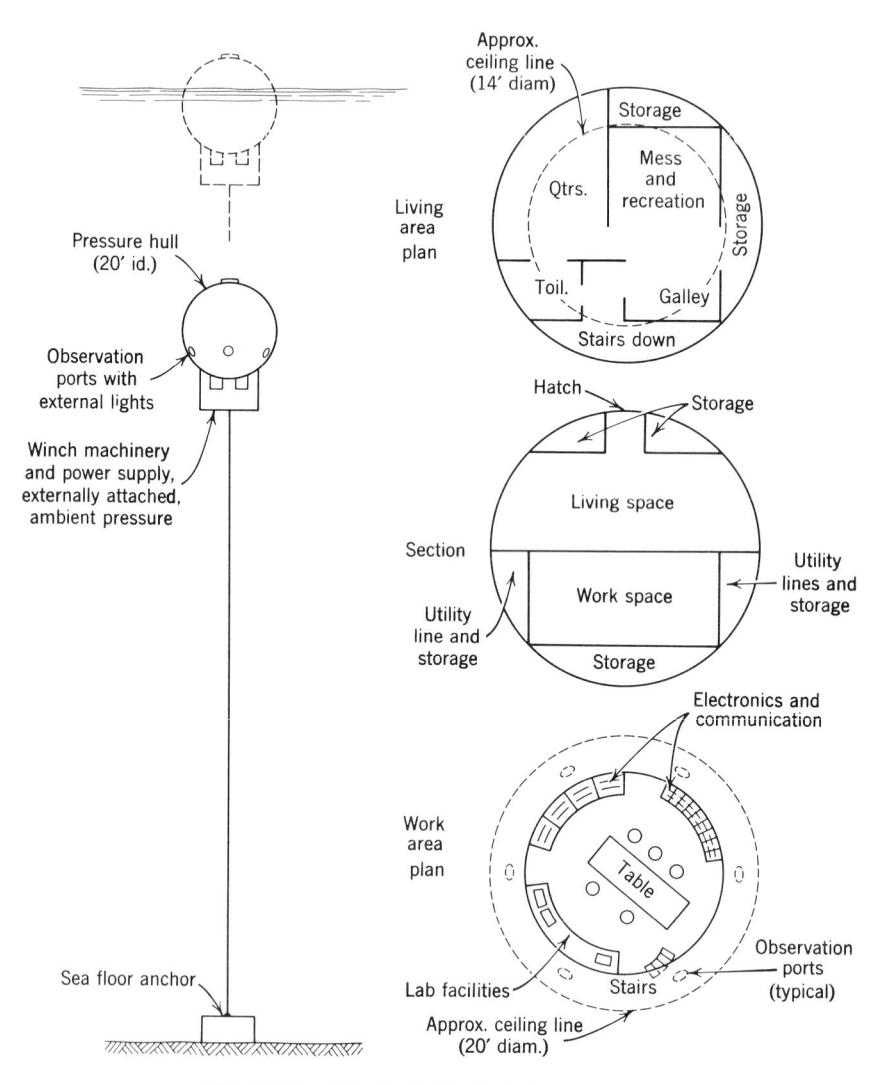

FIG. 10-21. Spherical hull, winch-down concept.

entirely self-contained, thereby eliminating the requirement for surface tenders. Positive buoyancy is possible by taking on or discharging ballast. Because of the buoyancy control, anchoring forces may be minimized, and major foundation problems virtually eliminated. Personnel transfer for crew change over could be attained by returning to the surface or to shallow depths where transfer vehicles could be used. Moreover, the installation is not permanent. Once it has accomplished its mission, it can easily be relocated.

A single research station with a five- to six-man crew for a mission duration of several weeks could be accomodated in a 20-ft spherical pressure hull. As shown in Fig. 10–21, an upper and lower level would provide ample living and work space with sufficient space remaining for storage. Winch machinery and the power could be attached external to the hull, exposed to ambient conditions. If more space is required to accommodate a large crew or more equipment, a cylindrical spar-type hull with hemispherical ends or a toroidal-type hull might provide the answer—or the coupling of two or more spherical hulls. One may even visualize a large complex with several interlocking hulls, completely self-contained, being preassembled on shore, floated to the site and winched down to the desired depth. For transfer of personnel and supplies, a spherical hull with mating hatches could serve as an elevator. The possibilities are many.

Naval Edreobenthic Manned Observatory.[7] The two-man Naval Edreobenthic Manned Observatory (NEMO) planned for the near future is designed on the winch-down principle. It is planned for use as an observatory–laboratory. Its spherical pressure hull of acrylic plastic will be entirely transparent. Facilities for the crew (a central structure with back-to-back seats), mission equipment, and the life-support system will be contained within the hull; the winch, power supply, lights, and ballast are attached to the underside of the hull (Fig. 10-22). Plans call for a 10-ft hull with a 1000-ft working depth. To minimize cost and fabrication complexity, the hull is made up of 12 spherical pentagons.

According to Moldenhauer et al. (1966), the three unusual aspects of the NEMO hull are (a) use of acrylic for an external pressure hull, (b) the modular construction utilizing spherical pentagons, and (c) the use of metallic inserts for hatches attachments and feed throughs in an otherwise completely plastic hull.

The metallic inserts, located top and bottom are connected by spring-loaded tie rods. These metal zones not only provide attachment points for the topside hatch and underneath equipment, but also attachment points for securing the interior equipment and for the lifting of the entire structure

---

[7]The NEMO concept was originated by James G. Moldenhauer and Richard G. McCarty, Physicists at the U.S. Naval Missile Center, Point Mugu, California.

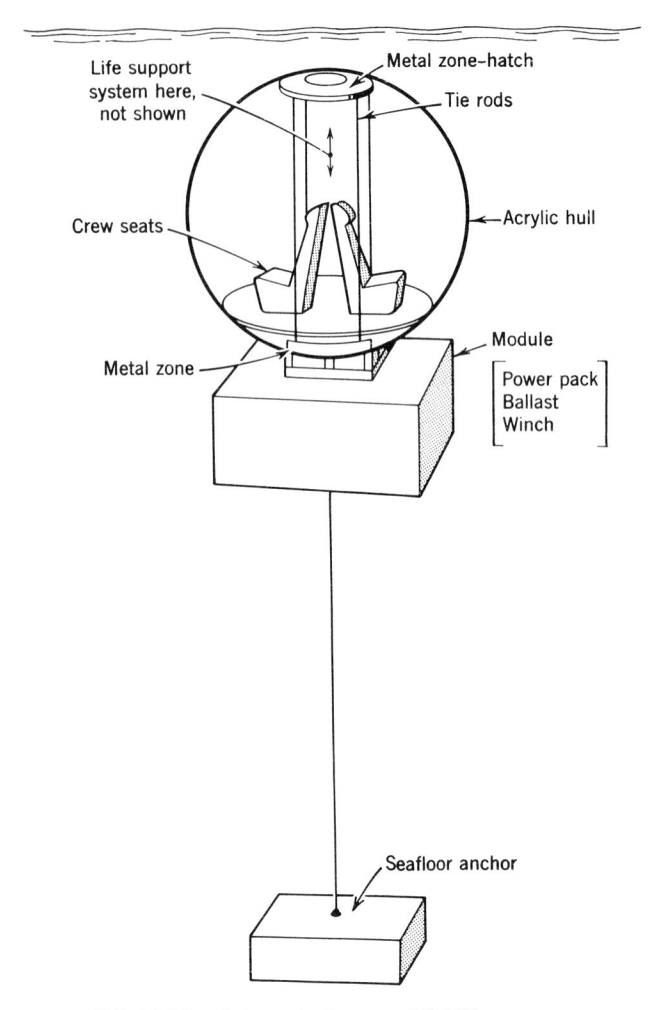

FIG. 10-22. Schematic diagram of NEMO concept.

from the water. The concept, without precedent, is being studied through the testing of 15-in. models (Fig. 10-23) in a simulated environment in pressure vessels.[8] Results from short-term tests have indicated a model-collapse depth of about 3600 ft. NEMO is seen as a new class of undersea vehicles.

Concrete structures. To this day, little consideration has been given to the use of concrete for pressure-hull structures; yet in the shallow to medium depths, say 3500 feet or less, buoyant hulls are possible. Exploratory experiments in the use of concrete are in progress at the NCEL. Sixteen-in. model

---

[8]See Chapter 16.

Deep Ocean Installations and Fixed Structures 345

FIG. 10-23. Fifteen-in. acrylic model of NEMO.

spherical hulls, with 1-in. wall thickness, were tested to failure in pressure vessels. The models imploded at simulated depths of 7000 to 7200 feet.

The advocates for concrete cite many advantages, such as lower material costs, resistance to corrosion, ease of formation, massive walls eliminate buckling problems associated with thinner-wall hulls and because of the massive wall construction, tolerances in sphericity are not as critical. Although the hulls may be massive, they would still be buoyant to medium depths.

Presently the principal disadvantage in using concrete for pressure hulls is a "lack of knowledge." Questions confront the engineerings, a few of which are—what stress patterns will be generated by surface discontinuities created by windows and hatches, and the intersection of surfaces?; will laboratory-scale models predict the behavior of full-scale ocean structures?; what are some of the fabrication, transportation and emplacement

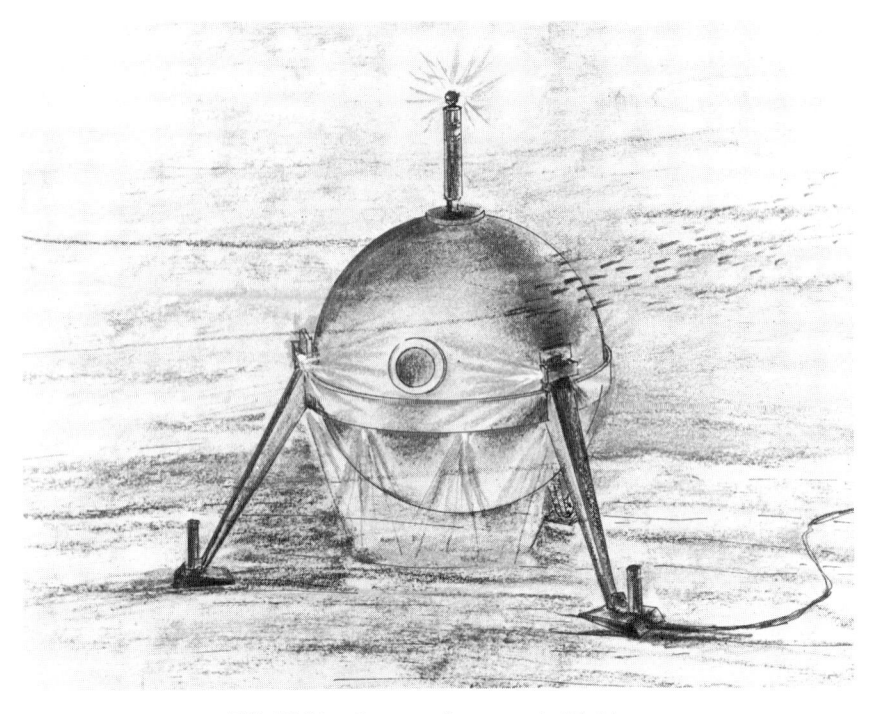

FIG. 10-24. Concept of concrete-hull habitat.

problems? I leave these thoughts and the concept of Fig. 10-24 with the reader.

10.6 LONG-RANGE FORECAST

Just what is in store for ocean technology in the next 20 years? A poll of engineers and scientists had this to say about ocean constructions.

In the next two decades towers and platforms will continue to play a dominant role in exploitation of natural resources. These will be of a more stable design to cope with the adverse elements of the environment. Surface and bottom structures at practically all depths to 20,000 ft will serve the gamut of military and nonmilitary needs. These will be surveillance systems for detection, tracking, interrogation and warning, submarine refueling and repair stations, underwater missile offense and defense systems, power-generating stations, storage and supply depots, navigational towers, research stations, and political occupancy. Some of the complexes will be inhabited, perhaps to depths of 12,000 ft.

The environmental data gathering and prediction system will be available for military and nonmilitary applications. A widespread network of collecting stations will feed data to central processing points, where the data will be computer-compiled and analyzed. Much as weather data is processed today, current synoptic and forecast charts will be prepared and disseminated.

Seafaring construction will use oceanographic services to plan and expedite on-site construction activities. The successful outcome of many construction ventures may well depend on this service.

The preliminary site-selection process will be greatly simplified by more accurate, detailed bathymetric maps and improved underwater photographic techniques. The availability of factors to enable conversion of sediment compositions to a parameter of strength, and increased information on the factors that govern sediment dispersion, will enable engineers to plot worldwide charts of sea-floor strengths, such as is now done with select physical and chemical measurements. The next 20 years will undoubtedly bring improved or perfected penetrometers, probes and signal-generating systems for sea-floor and subbottom investigations. The accuracy of physical analysis and evaluation of ocean-bottom sediments will be greatly improved by development of new apparatus and equipment and refined methods and techniques.

Studies of the occurring frequency of mass movements will permit these phenomena to be better understood. Predictions of occurrence, even with little accuracy, will provide an opportunity to observe them. Methods of defining to define the magnitude of the forces of landslides and turbidity currents and the effects of scour will be of increasing importance.

Much progress is expected in the capability to understand, predict, and design for the hydrodynamic behavior of deep-ocean installations and equipment. Advanced statistical methods which more accurately and realistically describe the effects of waves and other natural forces will be developed. Highly developed economical dynamic mooring systems should be available within the next two decades.

Tensile strengths of 400,000 psi for metals will be available with other properties more suitable for deep-ocean applications. Improved casting techniques and advanced welding processes will insure stronger and better alloys that are immune to stress corrosion cracking. Working stress on the order of 10,000 psi is forecast for concrete.

Nonmetals with properties of transparancy, flexibility, noncorrosivity, water and biological resistance, and insulation will be available within two decades. Structures made of these materials will be capable of withstanding the hydrostatic pressures of 20,000-ft depths, although problems of economics of raw materials and production costs must be overcome.

Virtually all major constructions designed for bottom use in the deep ocean seem destined for nearly complete prefabrication and bottom assembly techniques relevant to successful placement; semiautomatic assemblies seem to be the key developments needed. New techniques for which there is a demonstrated need in construction methods used in deep water will be developed.

If the current interest in oceanic exploration is pursued, a whole family of bottom swimming vehicles, and perhaps manned or unmanned tracked vehicles, will undoubtedly be developed. For those required construction techniques involving loose-material handling, successful applications probably will have been made within this period. Concreting, drilling, and trenching in particular, appear to be both useful and rather easily accomplished, if a need is demonstrated.

High-capacity hydraulic tools using local ocean pressure for major assembly of prefabricated structure, furnishing power for short-time use, are foreseen. Specialized safety techniques and warning devices of all sorts probably will have been devised or adapted to the environment.

Finally, for the adventurous fun lovers, underwater recreational facilities will be provided.

REFERENCES

Anderson, M. P., 1949, "Sea-Going Construction Plant Drives Huge Pipe-Piles to Close Tolerances," *Civil Engineering* **19**, (May 1949), 30–33.

Barnes, S., 1964, "Man's Race to the Oceans Bottom," *Machine Design* **36** (March 12, 1964), 136, and **36** (March 26, 1964), 154.

Berres, D. S., 1955, "Habitability of Naval Ships," Society of Naval Architects and Marine Engineers, Paper No. 10 for November 9–12 Meeting.

Booda, L. L., 1966, "Industry Bees Swarm at NEL," *Undersea Techn.*, **7**, 723–25 (July 1966).

Bruce, R. N., 1956, "Prestressed Precast Platform Built in Gulf," *Civil Engineering* **26** (July 1956), 441–443.

Cousteau, J. Y., 1964, "At Home in the Sea," *Nat. Geograph.* **125**, 4.

Craven, J. P., 1962, "Advanced Underwater Systems," *Mech. Eng.*, **84** (October 1962), 42.

Eichber, R. L., 1963, "Construction at Sea," *Undersea Techn.* **4**, 12.

Fowler, J. W., 1965, "Construction of the Chesapeake Light Station," *Civil Engineering*, **35**, (November 1965), 76.

Heezen, B. C., and M. Ewing, 1952, "Turbidity Currents and Submarine Slumps, and the 1929 Grand Banks Earthquake," *Am. J. Sci.*, **250**, 849.

Jones, R. E., 1965, "Deep Ocean Installations," *Trans. Joint Conf. Exhibit, Ocean Science and Ocean Engineering*, **1**, Washington D.C., and Technical Report R-369, Design, Placement, and Retrieval of Submersible Test Units at Deep-Ocean Test Sites, U.S. NCEL, Port Hueneme, Calif. May 1965.

Kuenen, P. H., 1948, "Turbidity Currents of High Density," *International Geology Congress*, 18th Session.

Link, E. A., 1964, "Tomorrow on the Deep Frontier," *Natl. Geograph.*, **125**, 6.

Moldenhauer, J. G., K. Tsuji, D. T. Stowell, and J. D. Stachiw, 1966, "NEMO—A New Undersea Observatory," *Undersea Techn.* **7**, 6, p. 39.

Naval Research Review (editorial staff), 1965, "The Recovery of STU I-1," *Naval Res. Rev.*, **18** (June 1965), 16.

Newmark, N. M., 1956, "The Effect of Dynamic Loads on Offshore Structures," *Proc. Eighth Texas Conference on Soil Mechanics and Foundation Engineering*, September 14 and 15, 1956, Austin Texas, University of Texas.

Quirk, J. T., 1965, "Deep-Ocean Reactor Placement Study," ASME Publication 65-WA/UNT-11.

Reese, L. C., and L. P. Johnston, 1963, "Criteria for the Design of Offshore Structures," The First University of Texas Conference on Drilling and Rock Mechanics, Austin, Texas.

Reinhart, F. M., 1965, "First Results—Deep Ocean Corrosion," *Geo. Marine Techn.* **1**, 9, 295.

Ruffin, J. V., 1965, "Steel Offshore Towers Replace Lightships," *Civil Eng.* **35**, (November 1965), 72.

Rutledge, P. C., 1956, "Design of Texas Tower Offshore Radar Stations," *Proc. Eight Texas Conference on Soil Mechanics and Foundation Engineering*, September 14 and 15, 1956, Austin Texas, University of Texas.

Smith, J. E., 1965, Technical Report 284-7, "Structures in Deep Ocean—Engineering Manual for Underwater Construction, Buoys and Anchorage Systems," U.S. NCEL, Port Hueneme, California, October 1965.

Spilhaus, A., 1964, "Engineering the Oceans," *Undersea Techn.* **5**, 5, 130.

Stetson, H. C., and J. F. Smith, 1938, "Behavior of Suspension Currents and Mud Slides on the Continental Slope," *Am. J. Sci.*, 5th series, **35**, 1–13.

Terzaghi, K., 1956, "Varieties of Submarine Slope Failures," *Proc. Eighth Texas Conference on Soil Mechanics and Foundation Engineering*, September 14 and 15, 1956. Austin Texas, University of Texas.

Wallace, G., P. Morrissey, CDR and D.L. Hayen, 1961, "Argus Island," *The Navy Civil Engineer*, **2**, 6, 34.

Wiegel, R. L., 1964, *Oceanographical Engineering*, Englewood Cliffs, N.J., Prentice-Hall, 1964.

CHAPTER 11

Vehicles and Mobile Structures

OWEN H. OAKLEY

The ocean covers about three-quarters of the earth's surface to a mean depth of roughly 13,000 ft. Of this area a little over 8% lies at depths of less than 1000 ft. This includes the continental shelves, which are of particular interest and importance because of the potential mineral wealth they contain and the food sources contained in the waters which cover them. Ninety-eight percent of the ocean floor lies at depths less than 20,000 ft, and the deepest spot in the ocean, in the Marianas Trench off Guam, has a depth of 35,900 ft. These depths are small compared with the earth's diameter; thus the ocean is in a sense only a thin slick of water on the earth's surface. Table 11-1, gives the distribution of ocean floor area as a function of depth.

Seawater weighs about 64 lb/ft^3, so that for every foot of depth a pressure of 0.444 psi is produced. At 35,900 ft the pressure is about 16,000 psi. Special structural-design methods and the use of high-strength materials are required to produce efficient vehicles capable of operating under such loads.

The corrosive nature of seawater is well known to the marine designer. It has a special significance in the construction of deep-diving submarines because most of the very high-strength metals, which would otherwise be highly desirable, are susceptible to stress corrosion in seawater.

Another characteristic of seawater that is important in the design of deep-diving submarines is compressibility. As seawater compresses with depth and becomes more dense, so does the pressure hull of the submarine. Depending upon material and design depth, a submarine pressure hull may be more or less compressible than water. Flotation materials and other elements of the submarine also have ranges of com-

TABLE 11-1
Distribution of Ocean-Floor Area
As a Function of Depth from
"The Deep Submersible" by
R. D. Terry (1966)

Depth (ft)	Ocean-Floor Area at Less Than Indicated Depth (%)
1,000	8.5
2,000	10.5
3,000	11.9
6,000	15.9
9,000	20.5
12,000	34.5
15,000	60.0
18,000	90.0
21,000	99.0
24,000	99.5
36,000	100.0

pressibility depending upon the materials employed. The difference between the compressibility of seawater and the net compressibility of the submarine gives rise to buoyancy changes with depth which constitute a special control problem for deep-diving craft. Thermal-expansion characteristics, both of seawater and the submarine, and variations in salinity of the water produce similar buoyancy variations which also contribute to the control problem.

The propensity of water to develop surface waves under the influence of wind or the passage of a ship affects the design of surface ships possibly more than any other single consideration. Wind-induced waves produce motions that affect the handling of gear at sea and the stomach of man, and the waves incident to propelling a ship through the water absorb energy and influence the amount of propulsive power that must be provided.

Seawater transmits light and other forms of electromagnetic radiation poorly but is an excellent medium for the transmission of sound. Thus underwater communication and navigation are heavily dependent upon sonic devices.

Economic and national-interest factors also provide a kind of environment which shapes the development and evolution of marine vehicles and craft. Man's increasing need for food and minerals has resulted in the development of special craft of many types for studying the ocean and the ocean floor and for tapping the resources they contain.

The evolution of marine vehicles is being strongly influenced by the competition of aircraft in ocean transportation. As larger, more efficient aircraft are produced, they become more economically competitive with ships. For high-value cargoes, transocean air freight promises to encroach deeply into the market now controlled by ships. For passenger transport, the airplane is already the principal type of carrier. In 1964, for example, about four times as many passengers crossed the Atlantic by plane as by ship. The possibility exists that air cushion vehicles (ACV) may insert themselves between aircraft and ships in competition for high-value cargoes, but this cannot be evaluated until the technological promise of the ACV is clearer. For transoceanic passenger service, it is unlikely that the ACV will lure many people away from the rapid and relatively comfortable passages by air which are now available and which should improve in the future. Ships, on the other hand, appear likely to remain the principal means for mass transport of bulk cargoes across the ocean. The production of larger and larger "super tankers" indicates the direction of evolution in bulk carriers dictated by economics.

Finally, the influence of naval requirements upon the development of ships and submarines should be noted. The capability to fight all-out and brushfire wars, to provide means of deterrence that are credible and of low vulnerability, and to extend national power into the deep ocean, all require continuous agressive and imaginative development of naval ships and craft on a wide front.

Ocean engineering is taken to refer to the exploration and exploitation of the ocean and the ocean floor rather than to the activities of commercial ocean transportation or naval uses of the ocean. Also, in a discussion of this type it is not feasible to cover the technology in depth. For these reasons the present discussion will be confined to a description of typical ocean-engineering working craft and a discussion of the more significant problems peculiar to their design.

11.1 SURFACE SHIPS

Oceanographic research ships—requirements

The study of the ocean embraces a wide variety of scientific endeavors. The three principal areas, physical oceanography, marine biology, and meteorology, are not neatly separable, and oceanographic ships usually must be designed to accommodate the needs of all three. All require laboratories of one sort or another, shops, space to carry scientists and technicians, office space for the inevitable paperwork, and, of course, the ubiquitous computer. In addition, facilities for working over the side, taking bottom samples or cores, water samples, bathythermic measurements, etc., require suitable deck areas and gear for handling the scientific instruments. Arrangements for visual observations either from a viewing chamber attached to the hull, via closed circuit television from a camera lowered at the end of a cable, or from a small submarine transportated on the deck of the ship, all call for special consideration in the design of oceanographic research ships.

Of all the characteristics of oceangoing research ships, sea-worthiness is probably the most important. The amount of useful work that can be accomplished on a voyage depends upon the number of good working days on station. The ability to maintain good speed in bad weather in transiting to station is obviously desirable. Similarly, the ability to continue work in the face of worsening weather and when forced to discontinue, the ability to ride out a storm without being forced too far off station so that work is resumed promptly as soon as the weather has moderated sufficiently, help to maximize useful working time.

Good freeboard and flare forward assist in keeping decks dry when headed into a seaway. Freeboard in way of the working area, however, is unfavorable for handling gear over the side. Research ships are, therefore, usually built with forecastle decks and good sheer to provide large freeboard forward and low freeboard aft in the working area.

Relatively deep draft forward is likewise desirable in order to minimize lifting out of the forefoot and slamming. Fine, vee-shaped sections forward also aid in minimizing slamming.

Flare, both forward and aft, introduces damping in pitch and heave and is desirable from this point of view. Large, wide, nearly horizontal areas, such as may exist in a ship with a broad transom stern, however, can produce annoying slamming in a quartering sea.

Closely related to the ability to remain at sea is the matter of ship motion.

Sea-induced motions limit the performance of surface ships and floating structures as regards crew effectiveness, passenger comfort, equipment operation, weight handling, in fact, virtually every factor in ship operation. There is no practical way of avoiding motions completely in a floating body, but they can be reduced, much as mechanical vibrations are reduced, by avoiding resonances, introducing damping, and by actively opposing the motion inducing forces or moments.

First, let us look at the exciting forces. Waves on the ocean surface have been reasonably well correlated with the winds that cause them in terms of wind velocity, duration of blow, direction, and fetch. It is still a most complex phenomenon to describe simply so that engineering analysis can be made. Power-spectra techniques have proved highly useful in dealing with the concept of sea state. Probably the most widely known are the

Neumann spectra. These describe a fully developed sea as a function of wave height squared versus wave frequency. Figure 11-1 illustrates two Neumann spectra, for sea states 4 and 6; it also gives Neumann's empirical equation for his spectra.

Ship response to the seaway is described by terms which refer to each of the six degrees of freedom in which the ship can move—heave, surge, sway, roll, yaw, and pitch. The motion in any particular degree of freedom can also be described in spectral form. Linear theory provides reasonably accurate predictions even though the motions are relatively large. The key is the development of the "transfer function." This can be done by calculation using theoretical ship-motion equations and estimated or measured empirical values for damping and force coefficients. It can be done more practically and with better accuracy using the results of model tests in waves. By towing a model at a given speed in regular waves of various discreet frequencies and dividing the measured response at each frequency

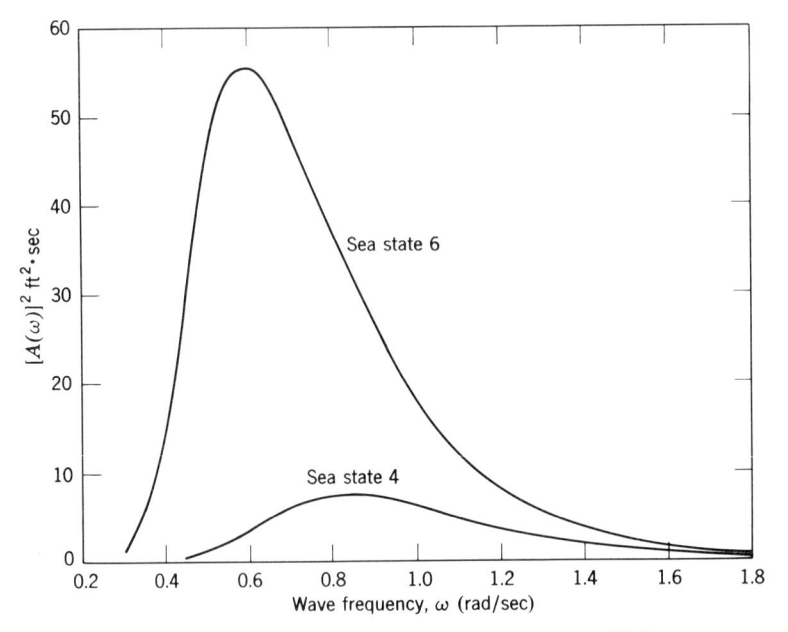

FIG. 11-1. Neumann spectra, sea states 4 and 6. $[A(\omega)]^2 = \dfrac{51.6}{\omega^2} e^{-2g^2/v^2\omega^2}$

where $[A(\omega)]^2$ = energy spectrum,
ω = wave frequency,
g = 32.2 fps^2,
v = wind velocity (fps).

For sea state 6, v = 44 fps = 26 knots;
for sea state 4, v = 32 fps = 18.6 knots.

by wave height, a curve of transfer function factor versus wave frequency is developed. Each ordinate on the transfer function curve is then multiplied by the one at the corresponding frequency in a Neumann (or other) spectrum for a particular sea state to give an ordinate of the ship response spectrum for a given ship speed. The process is illustrated in Fig 11-2 for the pitching characteristics of a destroyer-type hull. Useful estimates of heave and pitch can be made by this method. It may also be used for estimating derivatives of the motions such as hull-bending moments amidships. From the spectral analysis it is possible to predict the average value of, for example, the $\frac{1}{3}$ and $\frac{1}{10}$, or the $\frac{1}{100}$ highest pitch angles, or stress peaks.

Much thought has been expended in devising means for reducing motions. For normal ship forms not much can be done about stabilization in heave since other factors preclude radical changes in natural heave period. Flare at the waterline can provide some damping but even here practical considerations intervene to prevent maximizing this effect.

Pitch damping has been attempted using fixed fins at the bow, and, although pitch amplitudes can be reduced, slamming and vibration are apt to be encountered. No fully successful installations have been accomplished.

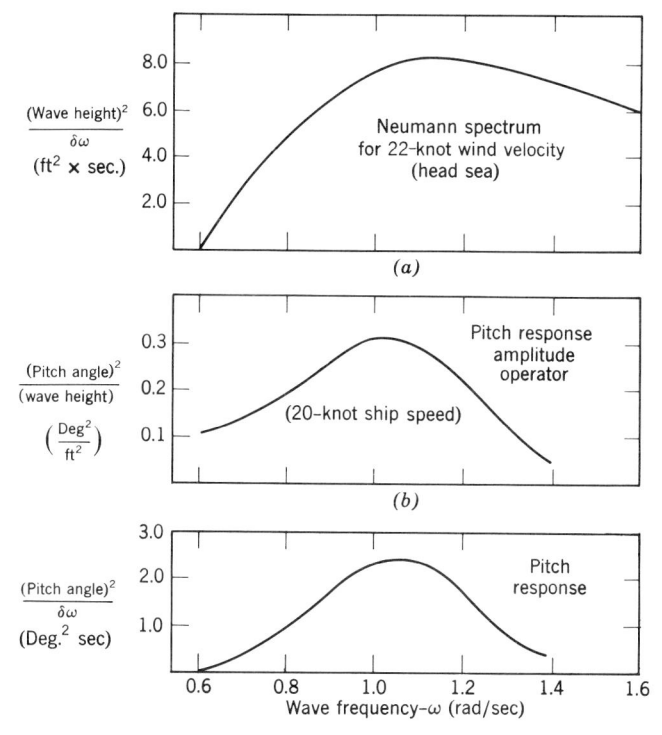

FIG. 11-2. Estimated pitch spectrum for destroyer type hull in head seas.

The natural pitch period of a ship is given by the formula:

$$T_p = 2\pi \sqrt{\frac{J_s + J_w}{g \times W_s \times GM_L}}$$

where J_s = longl. mass moment of inertia of the ship about its c.g. and
$\quad\quad\quad\quad = r_s^2 W_s$
$\quad r_s$ = longl. gyradius of ship
$\quad W_s$ = weight of ship
$\quad J_w$ = longl. mass moment of inertia of entrained water about c.g. of ship = $r_w^2 W_w$
$\quad r_w$ = gyradius of entrained water
$\quad W_w$ = weight of entrained water
$\quad GM_L$ = longl. metacentric height.

Modification of the natural pitching period of the ship can be affected to a very limited degree in design by adjusting the shape of the water plane which affects GM_L, and locating weights within the ship so as to control the magnitude of the longitudinal radius of gyration which affect J_s. Considerations of powering and transverse stability prevent extreme variations in water plane area and shape, and the longitudinal radius of gyration is most difficult to change appreciably within practical limits of arrangement. Large bulbs at bow and stern can effect measurable increases in added mass moment of inertia J_w, with attendant increase in natural period of pitch.

The aim of changing the natural period in heave or pitch is, of course, to remove it from a high energy part of the wave spectrum. Radical changes in natural period require drastic changes of hull form, weight distribution or dimensions. These are usually dictated by other considerations such as stability, powering or economics so that in reality not much can be done to modify natural heave or pitch periods appreciably.

Modification of the hull form to affect the phase relationship between heave and pitch can change the total amplitude of motion at the ends of the ship. The longitudinal location of the center of buoyancy, which can often be adjusted appreciably in design without serious penalty in other characteristics, can influence the heave – pitch phase relationship significantly.

In a head sea a ship captain is most apt to reduce speed when the violence of pitching causes bow emersion and slamming. His option is to slow or to change course. In the latter case rolling may become extreme. Stabilization in roll is far more practical than in heave or pitch. Roll stabilization is accomplished by a variety of devices. The best known is the ubiquitous bilge keel, which reduces the response to wave forces by increasing the damping. Since wave frequencies and ship roll frequencies lie in the same band, rolling is a near resonant phenomenon, and damping, therefore, is highly effective in reducing amplitude. Bilge keels have the advantage of requiring no actuating power or sensing devices; their principal disadvantage is that they increase the propulsive power required slightly. All other modes of reducing roll amplitudes involve opposing the rolling motion by applying out of phase moments to the hull.

The simplest means of doing this involves the use of tuned passive antirolling tanks. One form of such tanks is the flume type which is usually arranged to extend across the ship about as shown in Fig. 11-3. The principle of operation of these tanks is similar to that of the dynamic damper in a mass spring system; this is also illustrated in Fig. 11-3. In such a system the tank is "tuned' by adjusting the depth of water in the tank to have the same natural frequency as the natural roll frequency of the ship. As the response characteristics show, the system is not effective off resonance and may even increase roll amplitude slightly, but is quite effective in opposing the near-resonant components of the sea state. The mass of the active fluid, "m" in the mechanical analogy, need be only 1 or 2% of the mass of the ship. A fair degree of success has been obtained with these tanks, and their use is on the increase. Principal disadvantages is that they use valuable space in the ship. Active tanks depend upon pumping the active fluid from side to side

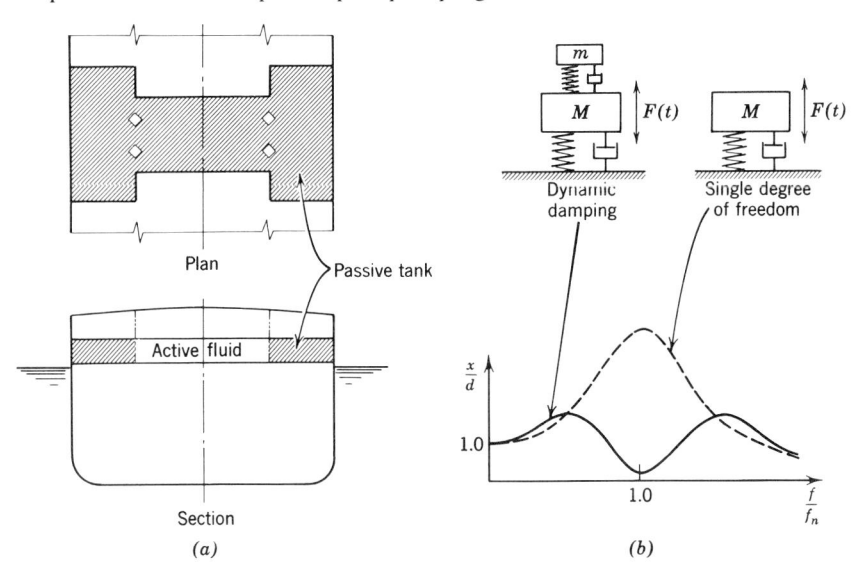

FIG. 11-3. Action of passive antiroll tank. (a) Passive-tank installation, schematic diagram. (b) Response characteristic of analagous dynamically damped single-degree-of-freedom system.

under the control of a sensing system which is designed to anticipate and oppose rolling motion. These installations are naturally more expensive than passive tanks and have, therefore, not been widely used.

Active fins, as the name implies, develop antirolling moments by means of pairs of radially disposed hydrofoils usually located at the bilges near amidships. The rolling moments induced by a pair of these hydrofoils are controlled by a motion-sensing device that actuates the foil "steering" machinery. Although complex, these devices do not require excessive amounts of space and have proven quite effective. Since the lift force developed by a hydrofoil is a function of the speed, it is obvious that these devices are of little use at low speeds or when dead in the water. Righting moments induced by passive and active tanks, on the other hand, are independent of speed.

Gyro stabilizers oppose rolling movements by precession forces and require sensing and control mechanisms for actuation. Their effectiveness is limited, however, and although once fairly common are little used today.

Oceanographic research ships

To satisfy the nation's growing need for oceanographic research ships the U.S. government has sponsored the construction of a fleet of AGOR's of standard design for use by government and private laboratories. It has also funded the construction of other oceanographic ships for private institutions under grants. The ships produced under this program are the first modern oceanographic research ship designed, so to speak, "from the keel up."

The AGOR-3 class was designed by the U.S. Navy to serve the needs of a mixture of government and private laboratories.

As illustrated in Figs 11-4 and 11-5, these ships have an upper or forecastle which extends aft about three-quarters the length of the ship. On the starboard side, the shell is carried up to the upper deck only as far back as amidships, whereas on the port side the shell extends to the upper deck all the way back to the three-quarter point. An L-shaped working deck is thus provided with a long working area at the starboard rail. This is useful for laying out coring gear or other long overside equipment. The deck house is displaced to port and houses laboratory areas which are thus readily accessible from the working deck. The presence of the deck house also provides a degree of protection for the starboard rail area when the port side is to weather.

The AGOR-3's have diesel electric propulsion. Two 600-hp diesel generators supply power to a single 1000 hp dc electric propulsion motor, which drives the propeller directly. A retractable, trainable bow thruster driven by

FIG. 11-4. AGOR-5 class.

a 175-hp motor is provided to aid in position keeping. Quiet operation was emphasized in the design, the main-propulsion diesel generators and other major items of machinery are sound isolated. A 300-kW gas-turbine-driven generator is located in the stack for propulsion in the quiet mode.

A compensating fuel system is employed.

The ship is provided with passive antiroll tanks of the type illustrated in Fig. 11-3. Principal characteristics of the AGOR-3 are given below:

Length, over-all	(ft)	208
Beam	(ft)	37[a]
Draft	(ft)	14.25
Displacement	(tons)	1,370
SHP[b]		1,000
Speed	(knots)	13
Range at 12 knots (miles)		12,000
Accommodations:		
Crew		22
Scientists		15
First of class in service		1,962

[a]Applies to AGOR-3 – 7 class; increased to 39ft on AGOR-9, 10, 12, 13.
[b]SHP–Shaft Horse Power.

The design and construction of *Atlantis II* overlapped that of AGOR-3, and were the responsibility of the Woods Hole Oceanographic Institution operating under a grant from the National Science Foundation.

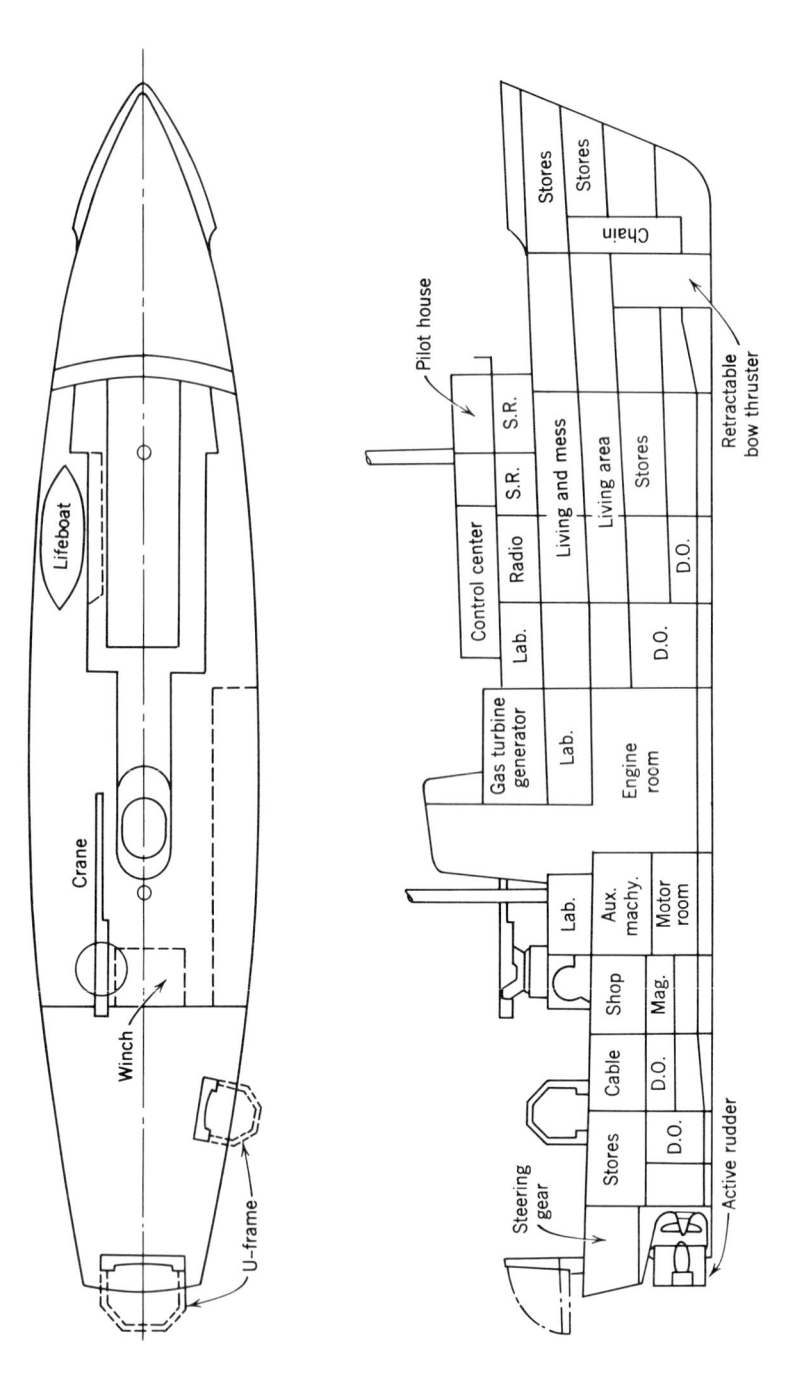

FIG. 11-5 AGOR-3 class.

Vehicles and Mobile Structures 361

Atlantis II, like AGOR-3, is a forecastle deck ship with a similar unsymmetrical main deck working area featuring a long rail working space to starboard (Fig. 11-6).

A special effort was made in this ship to provide maximum flexibility to accommodate the varying requirements for laboratories, office space, and berthing from voyage to voyage. To this end laboratory areas are arranged to permit ready conversion to office or berthing spaces. Work tables are made portable and pads are provided in the deck, bulkheads and overhead to permit the attachment of furniture, joiner bulkheads or scientific equipment.

As in AGOR-3, quietness was given careful consideration in the design. Propulsion is by steam reciprocating, unaflow, engines. This type of propulsion was considered superior to a sound-isolated diesel electric installation from the point of view of low noise and vibration in the continuous low-speed operating range. Some penalty was paid relative to the diesel in terms of oil needed for endurance and a slightly larger ship resulted from the choice.

Twin screws were selected instead of a single screw for reasons of reliability. The dividend in better maneuverability was an incidental benefit.

FIG. 11-6. *Atlantis II*.

A 250-hp bow thruster mounted in a transverse duct is also provided.

A passive antiroll tank is located below the second deck slightly forward of amidships. So located, the tank cannot be as effective as though it were placed higher in the ship. However, a more efficient use of space results by not giving over "prime real estate" to the stabilizing function.

A bow observation station is provided in the form of a bulbous bow.

The principal characteristics of *Atlantis II* are given in the following tabulation:

Length, over-all	(ft)	210
Beam	(ft)	44
Draft	(ft)	16
Displacement	(tons)	2200
SHP		1400
Speed	(knots)	13
Range at 12 knots	(miles)	8000
Accommodations:		
Crew		28
Scientists		25
Year in service		1963

The AGOR-3 class was criticized by the oceanographers for a number of shortcomings. The principal ones stemmed from the fact that the design was not flexible enough to satisfy the diverse requirements of the several laboratories and the design was simply too small for all the requirements laid upon it.

To overcome these objections a new design was undertaken by the Navy to replace the older design commencing with AGOR-14. The design was closely coordinated with the oceanographic community and a number of interesting features were developed.

In order to satisfy the spectrum of requirements laid on by the prospective users the concept of offering optional equipment suits was invoked. With this scheme a basic hull, machinery, and living arrangements were designed to which could be added a selected combination of some twelve basic options and six special options. The number of options allowable in any single combination is limited by the ability of the design to accept the weight, space needed, or cost.

The ship generally resembles the AGOR-3 and *Atlantis II* in that it is a forecastle deck type with the usual L-shaped working deck aft (Figs. 11-7 and 11-8). As in *Atlantis II*, provision is made for easy conversion of laboratory space to office or other use by means of temporary bulkheads and portable fixtures. Provision also is made for the stowage of portable vans containing special equipment, or fitted out as additional berthing accom-

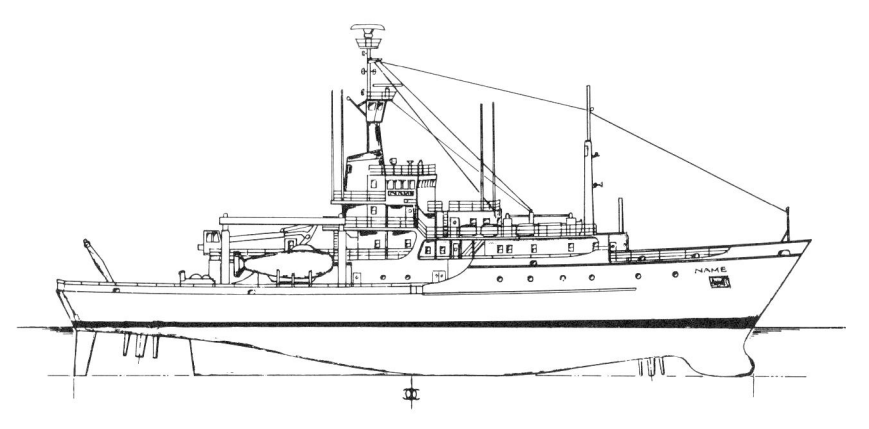

FIG. 11-7. AGOR-14 class.

modations. These would be secured on the main or upper decks aft where provision is made to supply them with electrical, water, and other necessary services.

A special feature of this design is the use of cycloidal propulsion. As indicated in the illustrations, two cycloidal propellers are employed, a large one aft as the primary propulsion unit, and a smaller one forward to act as a bow thruster unit. Both can be employed for propulsion. By directing the thrust of both propellers athwartships a thrust of about 30,000 lb can be developed to hold the ship broadside against wind and sea. This capability is expected to be useful to permit holding position and heading when working on the bottom with cables overside.

The propulsion units are driven by a single 2000-hp diesel engine with shafts taking power off each end of the engine.

As in *Atlantis II*, a bow observation station is provided in the form of a bulb.

A passive antiroll tank is provided amidships between the main deck and the first platform level.

The principal characteristics of the AGOR-14 and 15 are as follows:

Length, over-all	(ft)	245
Beam	(ft)	46
Draft	(ft)	15
Displacement, loaded	(tons)	1,915
SHP		2,000
Propeller diameter	(ft)	forward 10.5; aft 7.9
Speed	(knots)	12
Range at 12 knots	(miles)	10,000

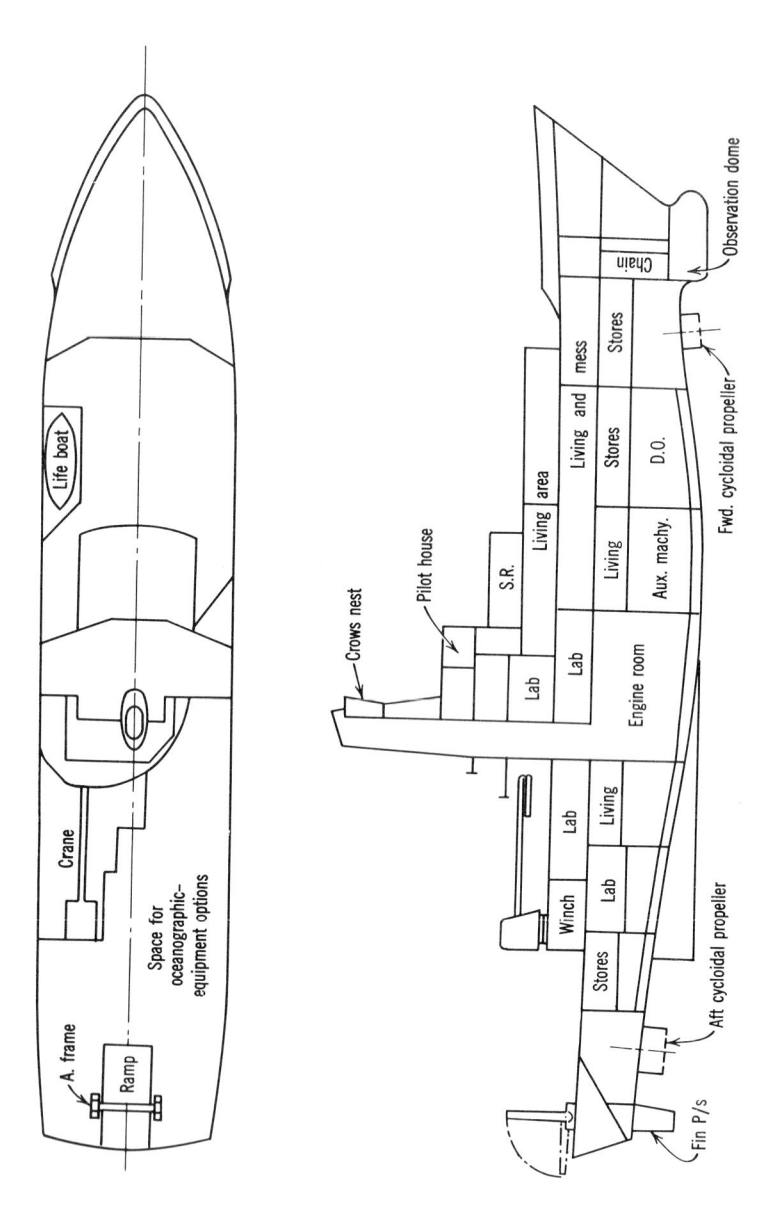

FIG. 11-8. AGOR-14 class.

Accommodations:
  Crew   25
  Scientists   25
Date to enter service   1968.

Basic Options
1. Laboratory equipment
2. Demolition-charge magazine
3. Trawl & hydro winches
4. A-frame
5. Portable vans (5)
6. Internal well
7. Articulated cranes (2)
8. Cafeteria mess
9. Biological seawater system
10. Underwater observation ports
11. Pressure transducer spaces
12. Stern ramp.

Special Options
1. Research submarine handling
2. Towing machine
3. Workboat
4. Antiroll fin (¢ daggerboard)
5. Drill tower and 4000 ft of pipe
6. Storage batteries

To satisfy the need for a larger oceanographic research ship, the Navy undertook the design of a catamaran, intended for construction in the 1967 shipbuilding program. At the present writing the design is still being developed.

The virtues of the catamaran are that it provides the large deck area so badly needed in oceanographic ships, and it allows work in the water to be carried on in the protected space between the hulls. This is particularly desirable for handling small research submarines which are expected soon to become a usual part of the equipment of oceanographic ships.

The rolling motion of a catamaran, although less in amplitude than in a ship of normal form, is apt to be quicker, so that the maximum acceleration at the deck edge may be about the same. The location of the working area close to the centerline will minimize motion and acceleration and a substantial net improvement is expected.

The principal problem in the design of catamarans is the matter of the strength of the structure connecting the two hulls. Since there is little historical data on which to base quasistatic comparative calculations, it is necessary to make realistic estimates of the stresses in a seaway. By running model tests in waves, the rigid-body response and hence the stresses in the structure can be related to the seaway that causes them. The elastic characteristics of the hull may be important if any of its natural frequencies lie in the range of wave frequencies in the seaway spectrum where the energy is high.

Figure 11-9 shows AGOR-16 as envisioned in the preliminary design. The principal characteristics of the AGOR-16 (catamaran) are given below:

Length, over-all	(ft)	246
Beam, over-all	(ft)	80
Space between hulls	(ft)	32
Beam of one hull	(ft)	24
Draft	(ft)	21
Displacement, loaded	(tons)	3,100
SHP, total		4,560
Speed	(knots)	15
Range at 13.5 knots	(miles)	6,000
Accommodations:		
Crew		44
Scientists		25

Special-purpose ships

The catamaran AGOR was a modification of the design for a submarine rescue ship (ASR) completed by the Navy for construction in the FY 1966 shipbuilding program. This craft is somewhat larger and more powerful than the AGOR, which it otherwise resembles rather closely.

The principle features of note in the ASR are the provisions for handling the Deep Submergence Systems Project (DSSP) rescue submarine and the saturation diving arrangements.

The small rescue submarine replaces the McCann rescue bell in the new ASR. This all but eliminates the necessity for the elaborate and precise mooring arrangements required in the older ASR's. It also permits rescue operations in water too deep for mooring.

The submarine is handled through a well in the structure between the two hulls. Raising and lowering are accomplished by means of an elevator arrangement. On deck the submarine is positioned athwartships by means of a 60-ton bridge crane betwen the forward and after deck houses. The operation of the rescue submarine is described later in the chapter.

The principal characteristics of the ASR are tabulated below:

Length, over-all	(ft)	251
Beam	(ft)	86
Space between hulls	(ft)	34
Beam of one hull	(ft)	26
Draft	(ft)	19
Displacement, loaded	(tons)	3,410
SHP, total		6,000

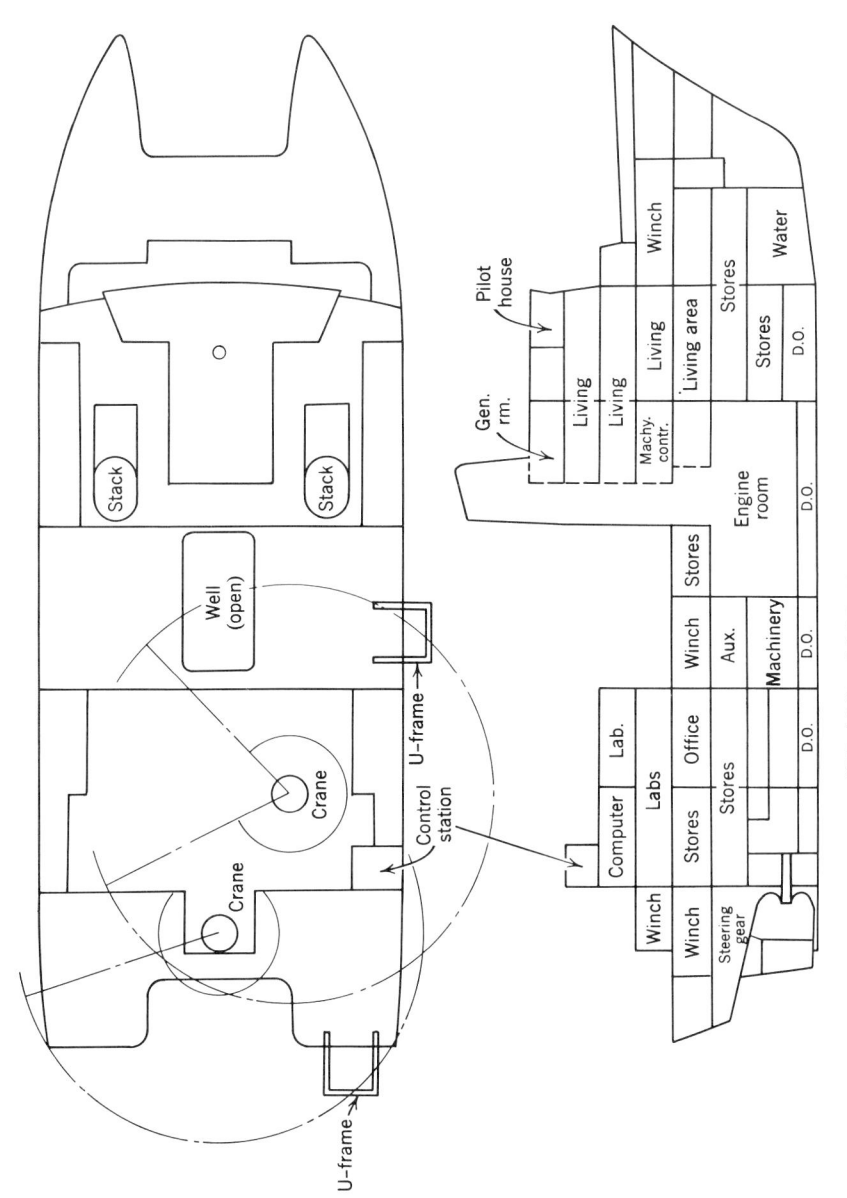

FIG. 11-9. AGOR-16, catamaran.

Speed	(knots)	15
Range at 15 knots	(miles)	10,000
Accommodations		163
Date in service		1969

Another use of the catamaran hull has been exploited by the oil industry in the form of a mobile drilling ship the *E. W. Thornton*, used for boring exploratory wells in deep water. This ship is pictured in Fig. 11-10 and its principal characteristics are as follows:

Length, over-all	(ft)	278
Beam, over-all	(ft)	105
Beam, each hull	(ft)	37
Space between hulls	(ft)	31
Depth of hull	(ft)	34
Draft, maximum	(ft)	20.75
Displacement	(tons)	8,500
SHP, total		3,000
Machinery		diesel electric, twin screw
Speed	(knots)	12
Complement		86

11.2 FLOATING PLATFORMS

Certain ocean-engineering activities require much more stable platforms than can be provided in a normal ship configuration. Oil-drilling rigs are a good example of this requirement. Other activities such as the investigation of sound propagation in the ocean require platforms of little motion and low background noise.

As noted earlier, one way to achieve low motions in the presence of a spectrum of exciting forces is to "detune" the system so that its natural frequency lies outside the spectrum of the forcing function, or at least in a region of low energy. In a spring–mass system having a single degree of freedom such as is illustrated in Figure 11-11, it can be seen that minimum motion results when the system has a low natural frequency relative to the frequency of the exciting force. Detuning in this direction calls for a "soft" spring, that is a low spring constant k, relative to the mass M, and results in a response to the right of the resonant peak in the figure.

For a floating body in the heaving mode, the spring constant is the rate at which buoyancy changes with immersion, the "tons per inch" of the naval architect. Thus, to increase the natural period, the water-plane area must be reduced. The extreme case of such reduction results in the spar buoy.

FIG. 11-10. *E. W. Thornton.*

In it the mass is large relative to the water plane, which, for a cylindrical shape is a measure of the spring constant k, thus making for a low value of natural frequency, f_n. Both the water-plane area and the draft of the buoy affect the forcing function in that it is the variation in pressure "seen" by the area of the bottom of the buoy which induces vertical motions. Inasmuch as the orbital motions of the water particles decay exponentially with depth, the effective wave amplitude at the bottom of the buoy, and hence the pressure variation, is less than near the surface. By making a very large buoy, appreciable reduction in the forcing function for all but the largest ocean waves is possible. It should also be apparent that by varying diameter to make the top smaller than the bottom further modification of the forcing function can be accomplished. Damping is primarily viscous and important only in the vicinity of resonance. In a spar-buoy design that under special conditions may experience near-resonant conditions, it would be quite feasible to increase the damping by the use of horizontal plates. This would also have the effect of increasing the effective mass and so increasing the natural period of the system by the entrained water effect.

Systems Design—The Technology

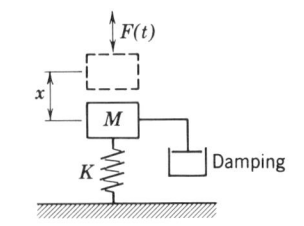

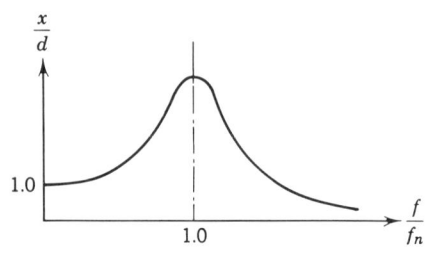

FIG. 11-11. Response characteristics of a single-degree-of-freedom system, $f_n = \frac{1}{2\pi}\left(\frac{K}{M}\right)^{\frac{1}{2}}$ (approximately),

where f_n = natural frequency,
M = mass,
K = spring constant,
x = amplitude of vibration,
F = amplitude of exciting force,
f = frequency of exciting force,
d = static deflection of spring under steady force, F.

FLIP (Floating Instrument Platform), an oceanographic research platform developed and operated by Scripps Institution of Oceanography, is designed along the lines of the foregoing discussion. FLIP is 355 ft long with a draft of about 300 ft and has a light displacement of about 600 tons. As the name implies, FLIP is designed to be towed from place to place in the horizontal position and then, by ballasting, "flipped" to the vertical working position, as illustrated in Fig. 11-12 and 11-13.

The same principles for achieving minimum motion that were used in the design of FLIP have been used in the design of large stable floating platforms. In order to support the large horizontal deck areas that characterize these platforms, multiple cylindrical "legs" are provided.

Figure 11-14 shows SEDCO-135. This is a huge floating oil-drilling rig for offshore work. The platform is triangular in shape and measures about 300 ft on each side. At each corner there is a large vertical "leg" at the bottom of which is a huge footlike tank. This arrangement provides the small water plane relative to the large underwater volume that results in a low natural frequency in heave and pitch.

FIG. 11-12. FLIP.

The rig weighs about 9000 tons, and has a draft of about 135 ft. It is positioned over the drilling site by nine anchors with $2\frac{1}{2}$-in. wire-rope mooring cables. In the floating mode drilling can be done at depths of 600 ft.

The rig can also be used as a fixed platform in water depths up to about 135 ft. In this mode the rig is ballasted to press the "feet" firmly against the bottom.

FIG. 11-13. FLIP.

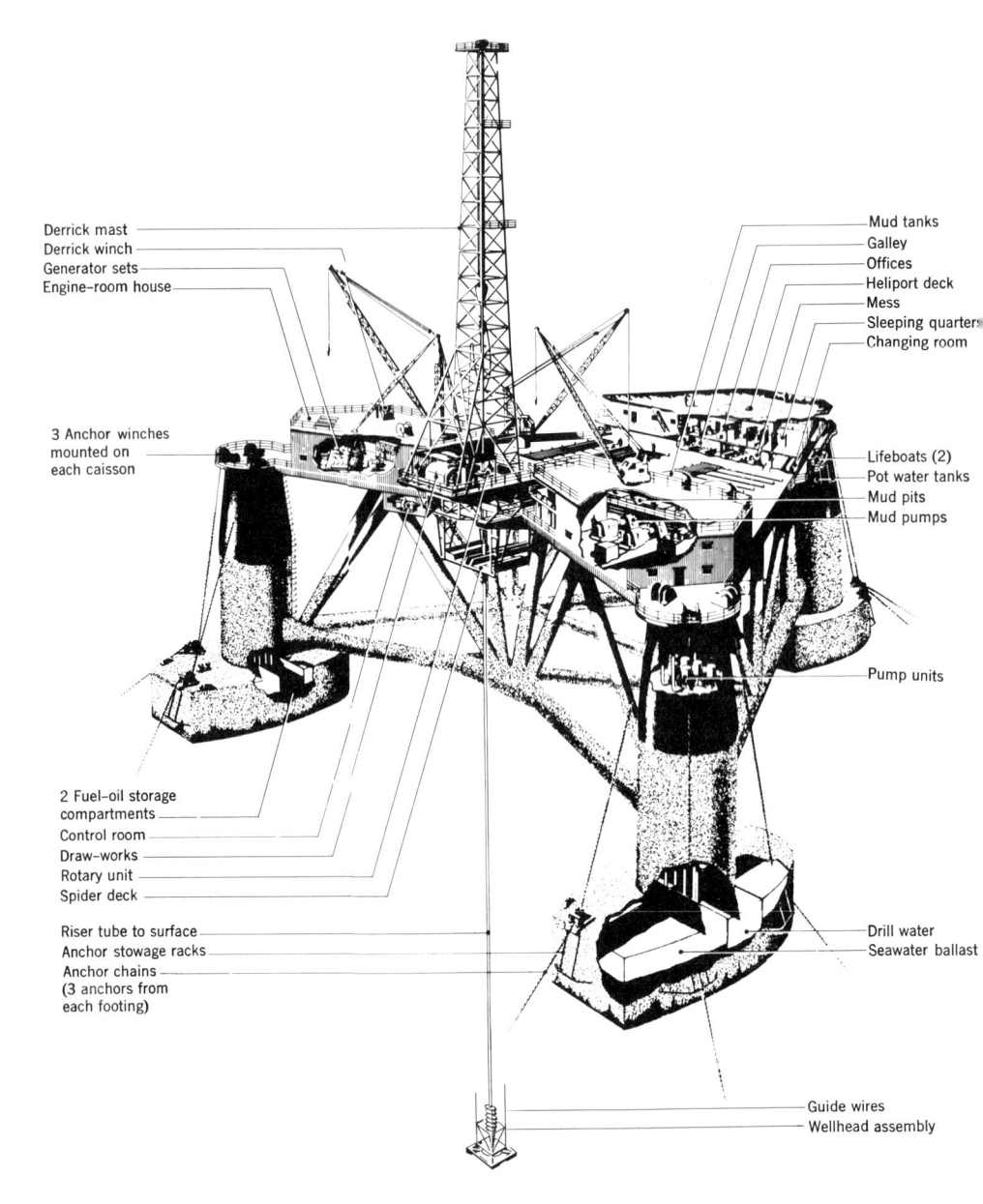

FIG. 11-14. SEDCO-135.

Vehicles and Mobile Structures 373

One of the most ambitious drilling schemes ever undertaken was the Project MOHOLE, which derives its name from the Mohorovičić Discontinuity separating the earth's crust from the mantle. Although the project was terminated before construction of the MOHOLE platform was completed, the design and engineering work that went into the effort substantially advanced the state of the art for large floating platforms. The MOHOLE concept envisioned the drilling of an exploratory hole through the earth's crust to the upper mantle. The decision to do this in the ocean rather than in way of a land mass was made because the crust is only about half as thick under the deep portion of the ocean as it is under the continents. Drilling was to have been done in a depth of water of about 14,000 ft and to have penetrated 16,000 to 19,000 ft below the ocean floor.

Figure 11-15 shows the configuration of the MOHOLE platform. The six "legs" provide the small water plane and the large horizontal cylinders

FIG. 11-15. MOHOLE platform.

provide most of the buoyancy. The rig is self-propelled with an installed power of 15,000 SHP. When underway it floats at a draft of 29 ft; in this condition about 6 ft of the horizontal cylinders are exposed. The displacement in this condition is about 17,000 tons. In the drilling mode the rig is ballasted to a draft of 60 ft and displaces about 23,000 tons. Position over the drilling site is maintained dynamically by means of six trainable thruster units which are lowered from the bottoms of the horizontal cylinders. Each thruster rated at 750 hp.

11.3 SUBSURFACE CRAFT

Examination of the ocean from the surface has traditionally been done by means of water and bottom samples taken at the ends of long cables lowered from surface ships. Much information about the world that lies below the ocean surface has been gleaned by this method, but much also has had to be inferred by scientific intuition. Underwater photography and, more recently, closed-circuit television have improved the information flow. None of these, however, permits the controlled and selective study of the depths afforded by manned underwater vehicles.

Submarines for exploration and research are undergoing a population explosion. In addition to their obvious utility for scientific purposes, these craft have proven useful in the location and salvage of objects in deep water. The recovery of the hydrogen bomb lost in 2800 ft of water off Spain in April 1966 was materially aided by the availability of the research submarines *Alvin* and *Aluminaut*.

One of the effects of the *Thresher* disaster was to direct attention to the problem of rescuing the crew of a submarine sunk at less than its collapse depth. Fortunately the occasions when such capability is needed have been infrequent and are likely to remain so. Nevertheless, the extremely limited capability for rescue beyond depths of a few hundred feet is clearly inadequate in the face of the trend toward deeper-diving submarines, both naval and civilian.

For many years the McCann rescue chamber has been the only means available for removing crews trapped in sunken submarines. The chamber is handled over the side of a rescue ship and hauled down to the submarine by a cable streamed from the escape hatch of the submarine. This operation is dependent upon positioning and mooring the submarine rescue vessel (ASR) over the submarine, proper streaming of the haul-down wire or the use of divers, and the cooperation of the elements, for it cannot be carried out in anything but a low sea state.

Small submarines as rescue vehicles largely eliminate the dependence of the operation on the foregoing conditions because of flexibility of operation inherent in an untethered, maneuverable vehicle. The Deep Submergence Systems Project (DSSP) has undertaken the development of a rescue system based on the use of small rescue submarines. Prototype development is underway at this writing and will be ready for operational evaluation in 1969.

Small submarines are becoming increasingly in demand for servicing undersea installations such as cables, oil wells, and manned sea habitats. As the tempo of underwater activity quickens, more and more of these craft will be needed as the "workboats" of ocean engineering.

Shallow-diving submarines

The continental shelves lie at depths less than 1000 ft. The waters over these shelves contain most of the ocean's food resources, and beneath the shelves lie the most accessible of the undersea oil and mineral deposits. It is therefore reasonable to expect that much of the activity of ocean engineering will take place in way of the continental shelves.

The technology for the design and construction of shallow-diving submarines is relatively well established, as is exemplified by Ashera II (Fig. 11-16). The problems of submergence to depths of several hundred feet are

FIG. 11-16. Ashera (Star II).

much less critical than those associated with diving depths of thousands of feet.

The very proximity to shore makes the logistic support problem easier. Also, the time required to go from the surface to operating depth is a small part of the total time of a mission.

Cousteau's undersea habitation "Conshelf Two," at a depth of 35 ft included a "garage" for the *Souscoupe*, a small submarine with a diving depth of about 1000 ft. This submarine also serviced "Conshelf Three," which was located at a depth of 328 ft.

For most ocean-engineering uses, high submerged speed and long range are not required. Low-speed control and manipulative capability are the characteristics principally needed. There are, however, a few tasks related to the ocean sciences that require speed and endurance. One of these is the concept of a "fish chasing" submarine which could follow schools of fish or sea mammals to observe their habits. Such a submarine has been studied for the U.S. Bureau of Fisheries but no decision to construct one has been reached.

The high speed and maneuverability needed in a submarine of this type requires a hull form of low resistance such as was developed for the experimental submarine *Albacore* (Fig 11-17). *Albacore's* hull form is a body of revolution having a length to diameter ratio (l/d) of just under 7. This ratio results in a hull form of minimum resistance. Figure 11-18 illustrates

FIG. 11-17. Albacore.

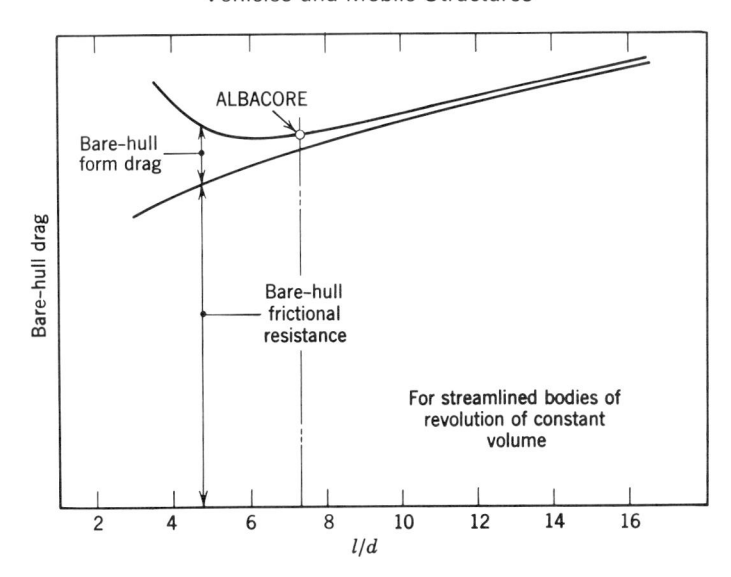

FIG. 11-18. Drag versus l/d.

the relationship of l/d to total resistance for bodies of revolution having constant displacement. At the very low l/d's, eddy-making resistance is large. For an $l/d = 1$, that is, a sphere, eddy-making resistance is by far the greatest part of the total drag, whereas wetted surface and hence frictional resistance is at a minimum. For large values of l/d, long slender streamlined shapes, eddy making is only a few percent of the total and frictional resistance predominates. As shown in the figure there is a low point in the total-resistance curve which for submarines of normal roughness and appendages occurs at an l/d between 6 and 7.

The low submerged drag of *Albacore* is obtained at the price of surface performance. Wave drag on the surface is high; there is virtually no working deck area and cleats and fairleads are minimal in number and retractable to reduce drag submerged.

The size of a submarine designed for long endurance is in part dictated by the requirements of life support of the crew. Mission times of 8 or 10 hr do not require much in the way of accommodations, sewage disposal or feeding arrangements, other life-support features may also be minimal. As soon as a requirement for a longer mission is imposed, a jump increase in crew size and the life-support features occurs. The crew must be enlarged to provide relief for the operators and very likely the larger ship required to hold the additional people and increased equipment and stores will require still more people to handle it.

If an endurance of more than a few hours at, say, 20 knots is required, the conventional diesel electric submarine with lead-acid or silver–zinc batteries is clearly unsuitable.

Fuel cells hold the promise of much greater ranges, but at present these are in the early developmental stages. Only very small fuel-cell plants have been tried in submarines. The development of a fuel-cell propulsion system in the thousand-horsepower range is estimated to require about five years of well-supported effort. It is expected, however, that in the low power levels, fuel cells will be in use in small submarines in a few years.

Nuclear power is the only developed submarine propulsion system capable of providing at the same time high submerged speed and long endurance.

TABLE 11-2
Characteristics of Shallow-Diving Submarines

	Ashera (Star II)	Auguste Picard	Submarine PC-3B	Souscoupe
Diving depth (ft)	600	1000	600	1000
Length (ft)	17	93	22	10
Beam (ft)	8.7	18.6	3.5	10
Weight in air (tons)	3.8	164	2.5	3.5
Speed (max)	4	6	5	1.5
Endurance (hr)	10	8	8	4
Crew/Passengers	1/1	3/40	1/1	2

Table 11-2 lists the principal characteristics of several typical research submarines that might be classed as "shallow divers," if we may define "shallow" as 1000 ft or less.

Deep diving craft

Submarines and other craft designed for deep diving share all the problems of their shallow-diving relatives plus a few more. These additional problems arise principally from the fact that water is both heavy and compressible.

The great pressure to be resisted requires that the pressure hull be made exceedingly strong and the weight of these structures becomes a dominant factor in design. It is obviously desirable to use high-strength and lightweight structural materials to the extent these are suitable and available.

The compressibility of water and the net compressibility of the submarine together produce a special operating problem in the maintaining of a balance between weight and buoyancy as depth is changed.

In shallow-diving submarine it is possible to drive a propeller by a motor located within the pressure hull. The pressure at great depths precludes the use of seals around rotating shafts. The friction losses in such seals would be excessive, but the truly disqualifying factor is leakage. In a small deep submarine even a small amount of leakage is unacceptable. The propulsive thrust on an internal thrust bearing would be augmented by the load caused by the hydrostatic pressure pressing the shaft inwards. At great depth this is likely to exceed the propeller thrust many times over.

To avoid these problems, deep-diving submarines are equipped with external motors for propulsion and thruster units. Direct-current electric motors are the most likely candidates because they can be run directly off the batteries (or fuel cells). Because of the commutation problem, they must be encased in oil-filled containers compensated to sea pressure. At present such motors are available in sizes up to about 15 hp. It is expected that dc waterproof motors as large as 100 hp will eventually be developed.

Alternating-current motors, on the other hand, need not be oil encased and can be built in large sizes. Installations in the multihundred horsepower range are in service today. In order to produce alternating current from a battery (or fuel cell) power source, a motor generator or static inverter would be required. These are heavy items and would have to be located within the pressure hull. The losses associated with the conversion of dc to ac would be felt in terms of larger ship size.

Buoyancy-weight relationship. Deep-diving submarines are normally "weight critical," that is, the size of the craft depends upon the weights to be supported, and hence the buoyancy to be provided, rather than the volume required to house items which must be carried inside the pressure hull. The usual means of providing buoyancy, that is, the one used in conventional submarine design, is to make the pressure hull large enough to furnish the requisite displacement. This, of course, will only work if the density of the pressure hull is sufficiently less than that of water. Where the pressure hull density is greater than that of water, a flotation material such as gasoline or syntactic foam must be used. In this case it is best to the pressure hull to just hold all that must be protected from sea pressure, and provide the needed additional buoyancy by flotation material. *Trieste II* is a good example of this situation. When the pressure hull density lies between that of water and that of a suitable flotation material, it is also advantageous, from the standpoint of craft size, to minimize the size of the pressure hull and use the flotation material to provide the buoyancy.

A submarine is said to be "volume critical" when the volume of pressure hull required to house the items which must fit in it is greater than that needed for buoyancy. In this case lead ballast must be added in order to achieve the neutral-buoyancy condition in which submarines operate when in "diving trim."

The weight of a pressure hull to resist the hydrostatic pressure of deep submergence is a substantial part of the total weight of a submarine. For this reason high-strength materials and favorable structural shapes are employed to minimize the weight penalty. The most favorable shape for a pressure hull is a sphere. Stiffened cylinders and cones and even prolate spheroids are also used where these shapes are advantageous as regards arrangement and external shape. For the same enclosed volume and design depth, these shapes are heavier than the sphere.

Penetrations require reinforcements that are costly in weight, as do intersections of pressure-hull elements, for example intersections of spheres and/or cylinders.

The ratio of the weight of a pressure-hull structure to that of the water it displaces is known as the "hull fraction." In Fig. 11-19 the hull fractions for spheres of a high-strength steel and titanium are given as a function of design operating depth. These spheres were designed using methods developed by the David Taylor Model Basin (DTMB) and presented in DTMB Report 1985.

Based on the hull fractions presented in Fig. 11-19, Fig. 11-20 was developed to illustrate the relationship of size to design operating depth for two families of small submarines. In one family (dashed lines) the pressure hull was sized to just contain the necessary working space and equipment. The necessary additional buoyancy was provided by a flotation material having a specific gravity, or hull fraction equivalent of 0.68 relative to seawater. All submarines were designed for constant speed payload and endurance.

An examination of Fig. 11-20 shows that for depths less than 10,000 ft (for the HY-150 hulls), a slightly smaller craft results by sizing the pressure hull

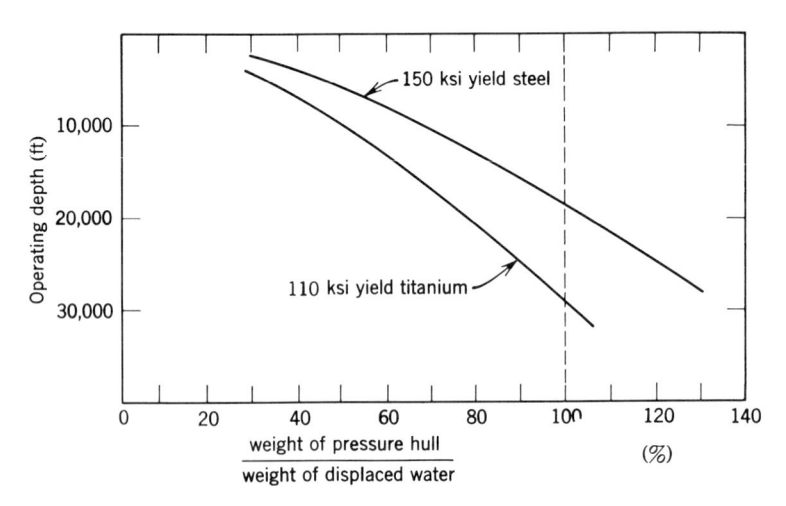

FIG. 11-19. Hull fraction versus operating depth, spherical pressure hulls.

Vehicles and Mobile Structures 381

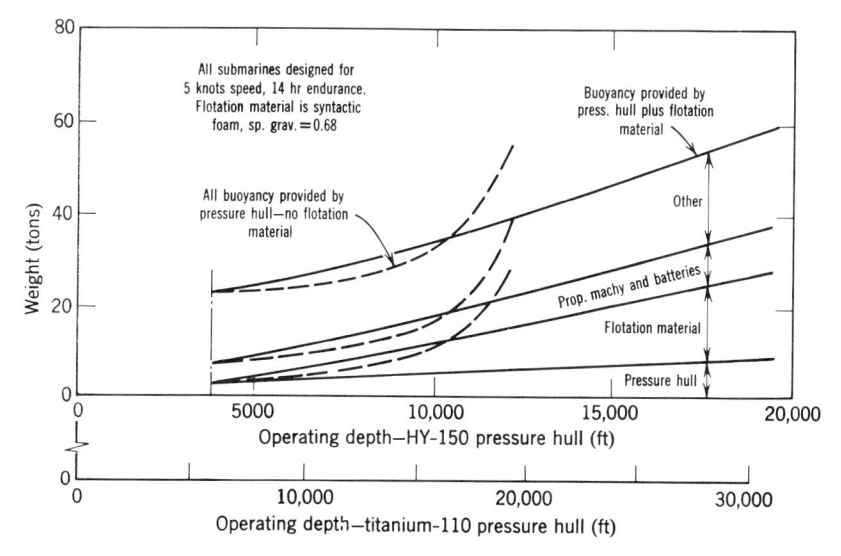

FIG. 11-20. Weight of deep SS versus operating depth, for small submarines.

to provide all buoyancy. For greater depths, it becomes progressively better to size the pressure hull to provide the minimum needed internal volume and use flotation material for the remaining buoyancy.

Referring again to Fig. 11–19, it will be noted that at 10,000 ft the hull fraction of the HY-150 sphere is 0.68, exactly that of the flotation material, which explains why the solid and dashed curves of Fig. 11-5 cross at this depth. The same thing is true for the titanium designs at a depth of about 17,000 ft.

Beyond the crossover point, the craft without flotation material increase in size with increasing rapidity with depth.

Materials. From the foregoing it is apparent that a strong need exists for pressure-hull materials having high strength and low densities. Additionally, corrosion resistance, toughness, and good fatigue strength are greatly to be desired. Repairability is another virtue which, today, is practically synonymous with weldability.

The following are among the more likely candidate materials for pressure hulls and buoyancy materials:

HY-80: A quenched and tempered nickel – chrome alloy steel having a yield strength of 80,000 psi. It is weldable, tough, and currently in wide use, particularly by the U.S. Navy, which was responsible for its development.

HY-100: This is a modification of HY-80 coming closer to "STS," the

Navy's old ballistic steel, in physical properties. It has a yield strength of 100,000 and resembles HY-80 in respect to weldability and toughness. The ALVIN sphere is made of HY-100.

HY-130–150: This is a developmental steel sponsored by the U.S. Navy which shows excellent promise. It should be in service within the next few years.

Maraging Steels: The maraging steels appear capable of development in the 180,000–200,000 psi yield range while still retaining toughness. Above this the toughness and susceptibility to stress corrosion appear to be limiting for marine applications.

Titanium: Titanium alloys having strengths in 110,000–120,000 psi yield range offer highly favorable strength-to-weight ratios. Some difficulty with stress corrosion has been encountered but modified alloys appear to have overcome this problem without serious compromise for strength.

Glass-Reinforced Plastics: High strength-to-weight ratios are possible with these materials. However, uncertainties relating to strength and fatigue life in seawater at high pressure require further development.

Glass: Structural glass has the most favorable strength-to-weight ratio of all the candidate materials. In compression the yield strength of glass can be well over 300,000 psi, but it has very little strength in tension. Under extreme hydrostatic pressure a spherical glass hull can resist considerable bending caused by a concentrated load because of the compressive stress bias induced by the hydrostatic loading. Means of prestressing the surface layers in compression are available that can improve resistance to concentrated loads when not submerged. To date only small-scale tests of spherical grass hulls have been accomplished.

One of the best-known flotation materials is gasoline. It has specific gravity of 0.68 relative to seawater and, being liquid, can fit into unaccessible spaces that would otherwise have to be left free flooding. It also has the advantage of being removable by pumping when it is desired to lighten the craft for lifting out of the water. The disadvantage of a liquid flotation material is that it can be lost if the containing hull is ruptured. Gasoline has nearly double the compressibility of water which, as will be seen, can present a considerable problem in depth control. The usual precaution in handling flammable liquids, of course, apply and this must also be counted a disadvantage. All of the bathyscaphes have used gasoline for flotation.

Of the solid flotation materials, syntactic foam appears to hold the most promise. This is a plastic lightened with tiny hollow glass spheres ("microballoons"). A specific gravity of 0.68 associated with a crushing strength of about 10,000 psi has been attained. Syntactic foams having great strength and specific gravities of 0.5 or less are expected to be available in the near future.

Mention has been made of the problems of depth control caused by the compressibility of the pressure hull and other elements of the submarine relative to water. For an ideal "thin" spherical shell, the compressibility (defined as the change in unit volume per unit change in pressure) is

$$\frac{1}{V}\frac{dV}{dp} = \frac{3\sigma_y(1-\mu)}{Ep_y}$$

where V = volume
p = pressure
p_y = pressure at which stress in the shell equals yield strength (approximate collapse depth)
σ_y = yield strength of shell material
E = Young's modulus
μ = Poisson's ratio.

For this relationship it can be seen that compressibility varies directly as σ_y and inversely as E and design depth. Thus high-strength materials will result in pressure hulls of greater compressibility than lower-strength materials. In the case of glass, which has a σ_y of the order of 300,000 psi or more and an E of only about 9×10^6 psi, the compressibility of a spherical shell will be about $6\frac{1}{2}$ times that of a shell of HY-150 steel designed for the same depth and having $\sigma_y = 150,000$ psi and $E = 29 \times 10^6$ psi.

The compressibility of other elements of the submarine such as flotation material, oil seal for external batteries, etc., may increase or decrease the net compressibility of the craft as a whole. In a shallow-diving submarine, the usual method of compensating for buoyancy changes caused by compressibility, thermal expansion and contraction, changes in salinity, and changes in consumable load, is by pumping out or taking aboard seawater in the variable ballast system. At great depths the weight and power required for pumping against the external pressure practically eliminate this method of buoyancy control from consideration.

Some typical deep vehicles. Deep-diving vehicles are principally used for oceanographic research and exploration. The development of craft for undersea exploration has a long history but relatively little progress was made until comparatively recent years.

In 1934 Dr. William Beebe and Otis Barton descended to 3028 ft in their "bathysphere." This vehicle, if it can be so classed, consisted of a steel sphere 4 ft 9 in. in diameter fitted with viewing ports and suspended from a tending ship by a $\frac{7}{8}$ in. cable. The depth to which such craft can be lowered is ultimately limited by the weight of the sphere and cable because of stress in the cable. Ship motion affects this powerfully and imposes severe limits on operation. The amount of useful observation which can be accomplished is small indeed, because of the lack of mobility.

A major advance in mobility for the deep-ocean investigation was made by the Picards who designed and constructed the first "bathyscaphe." Borrowing from their free ballooning experience, they suspended the heavy pressure-proof spherical "gondola" from a large float filled with gasoline. This fluid, having only about two-thirds the density of water, provided the necessary buoyancy to support the gondola, batteries, motors and other weights. As in the case of the free balloon, altitude was controlled by dropping solid ballast (in this case iron shot) to rise, and venting off gasoline to descend. In January 1960 *Trieste I*, with Jacques Picard and Lt. Don Walsh (USN) aboard, made a dive to the deepest spot in the ocean, in the Marianas Trench off Guam, which is 35,900 ft deep.

Trieste II, a modification of *Trieste I*, is pictured in Fig. 11-21. The heavy sphere and forward short tubs are located near the bow and the batteries and after shot tubs are located in the stern. The gasoline is located amidships. This arrangement allows the longitudinal centers of weight and buoyancy to be located in the same vertical line as required for fore-and-aft balance. Variations in trim due to loading are compensated by a mercury trim system. The gasoline float is subdivided into several tanks, each small enough so that if the float is punctured on a bulkhead the resultant loss of two tanks of flotation fluid can be compensated by dropping all shot ballast. Shot release is controlled by electrostrictive valves located at the bottoms of the shot tubs. Both the shot tubs and batteries are jettisonable to lighten the ship for emergency ascent. On the surface, water ballast tanks are blown dry to provide freeboard and reserve buoyancy for handling and towing. Submerged, the craft is propelled and maneuvered by two electric motors that give it a speed of 2 to 3 knots.

The characteristics of the bathyscaphes are given in Table 11-3.

Figure 11-22 shows *Aluminaut*. This craft is considerably more maneuverable than the bathyscaphes. It resembles what may be considered a more conventional submarine in that the pressure hull is sufficiently less dense than water and provides all the buoyancy needed to support the weight of the craft. As in *Trieste II*, a mercury trim system is employed, but in this case it is inside the pressure hull rather than externally located. Iron shot in external saddle tanks with electrostrictive release valves is used for depth control. As mentioned earlier the pressure hull of *Aluminaut* is less compressible than water and gains buoyancy with depth. This permits an operating technique wherein the craft is ballasted on the surface just heavy enough for it to be in equilibrium at the desired operating

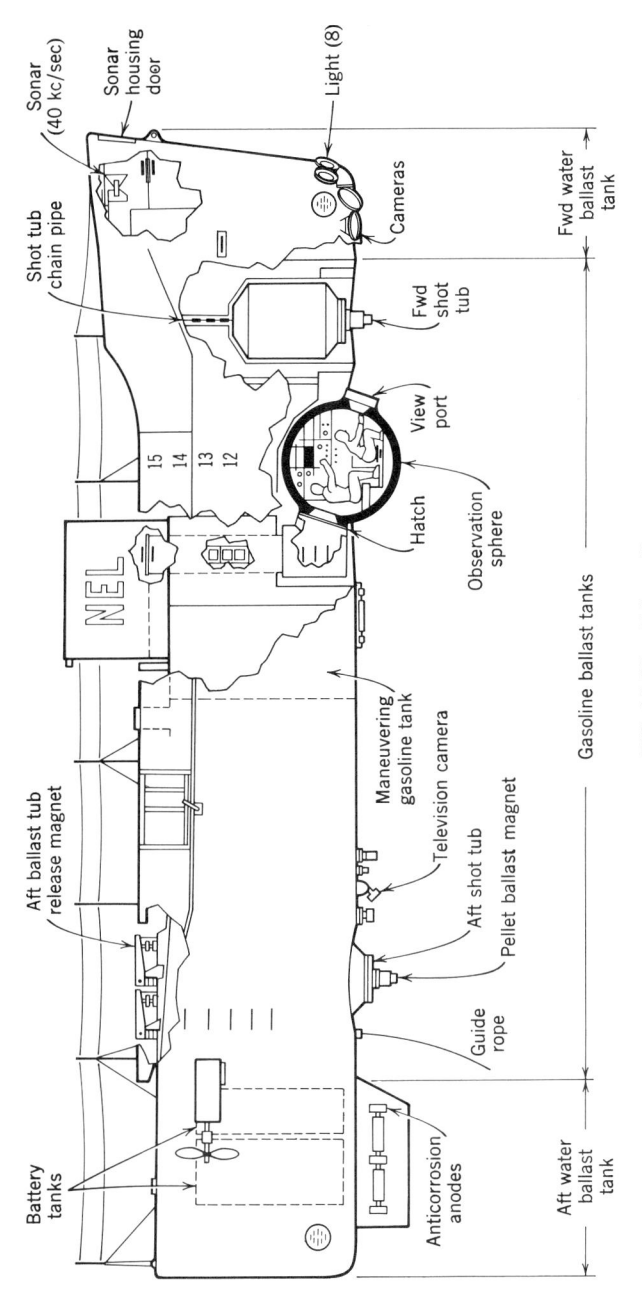

FIG. 11-21. *Trieste II*.

TABLE 11-3
Characteristics of Bathyscaphes

		Trieste II	Archimede	FRNS-3
Diving depth	(ft)	20,000/35,800	35,800	13,500
Length	(ft)	67	69	53
Beam	(ft)	15	13	16.6
Displacement—submerged	(tons)	220	200	100
Pressure-hull o.d.	(ft)	7.2	7.87	7.2
Pressure-hull thickness	(in.)	3.54/4.72	5.91	3.5
Pressure-hull material		Forged steel	Forged steel	Cast steel
Flotation material		gasoline	gasoline	gasoline
Speed	(knots)	2–4	3	1
Range	(miles)	14 at 2.4	7.5/15	3 to 5
Complement		2–3	3	2

depth. Shot is dropped to rise to an intermediate depth or to surface. Fine control in a vertical sense can be accomplished by the vertical axis propeller mounted amidships. *Aluminaut* is designed for an operating depth of 15,000 ft. The pressure hull is constructed of forged aluminum channel shaped rings machined to precise shape and bolted together. The ends are closed by forged, machined hemispherical end caps also of aluminum. *Aluminaut* participated in the location and recovery of the hydrogen bomb.

Alvin, shown in Fig. 11-23 approaches the conventional submarine more closely than *Aluminaut* in that it makes use of a variable water ballast system for control of weight–buoyancy equilibrium. This gives it a high degree of flexibility of operation. *Alvin* is a small craft weighing only 15 tons compared with 81 tons for *Aluminaut* and 220 tons for *Trieste II*. It has a diving depth of 6000 ft.

The pressure hull, which is spherical, although light enough to float its own weight, does not have sufficient buoyancy to support the batteries and other weights external to the sphere. The needed additional buoyancy is supplied by syntactic foam flotation material in which are embedded light titanium spheres. Vertical equilibrium is achieved by taking water into or pumping it out of spherical pressure-proof ballast tanks. At 6000 ft, given a submarine that does not require excessive compensation for compressibility, etc., the pumping requirements for a variable water ballast system are acceptable. At greater depths the system costs too much in weight and power to be considered. Where it can be used, however, it provides the most flexible type of variable ballast system. Propulsion and control are by three hydraulic motors driving steerable shrouded propellers, a large one at the stern and two smaller ones port and starboard just forward of amidships. Hydraulic pressure is supplied by an electrically

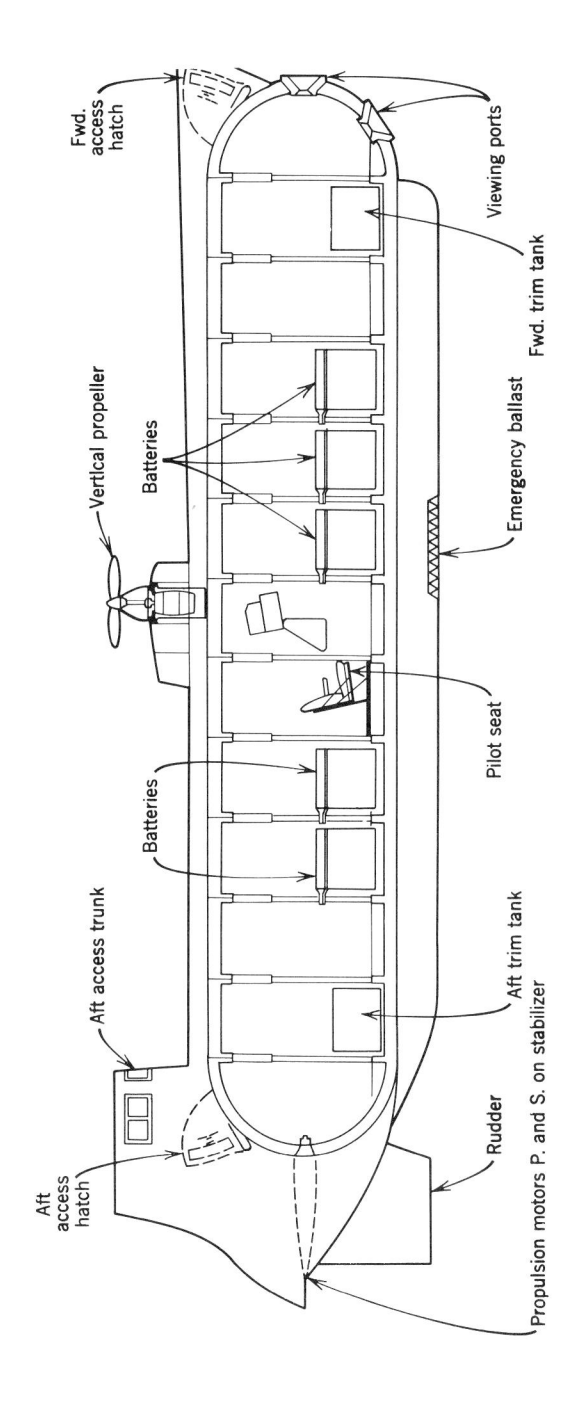

FIG. 11-22. *Aluminaut.*

FIG. 11-23. *Alvin*.

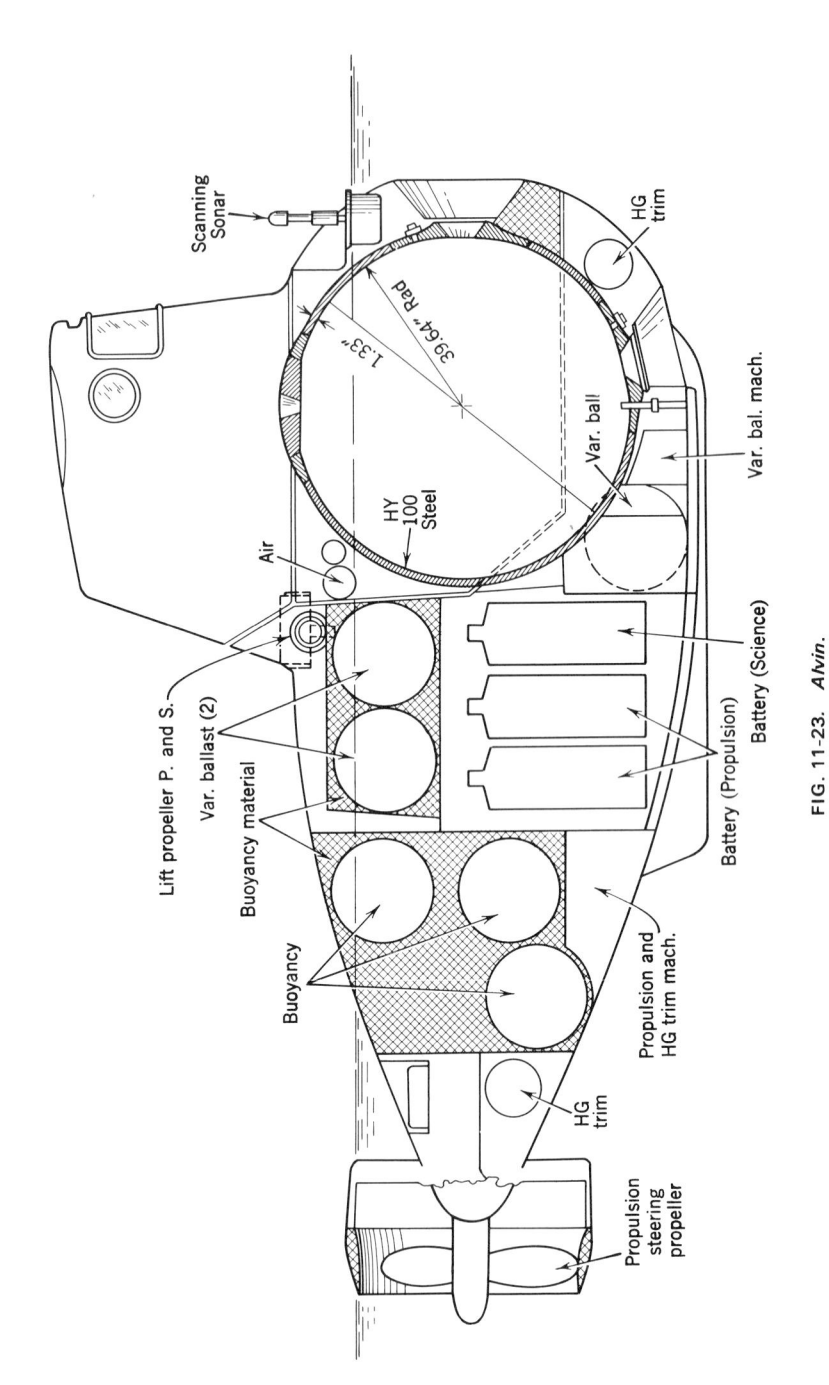

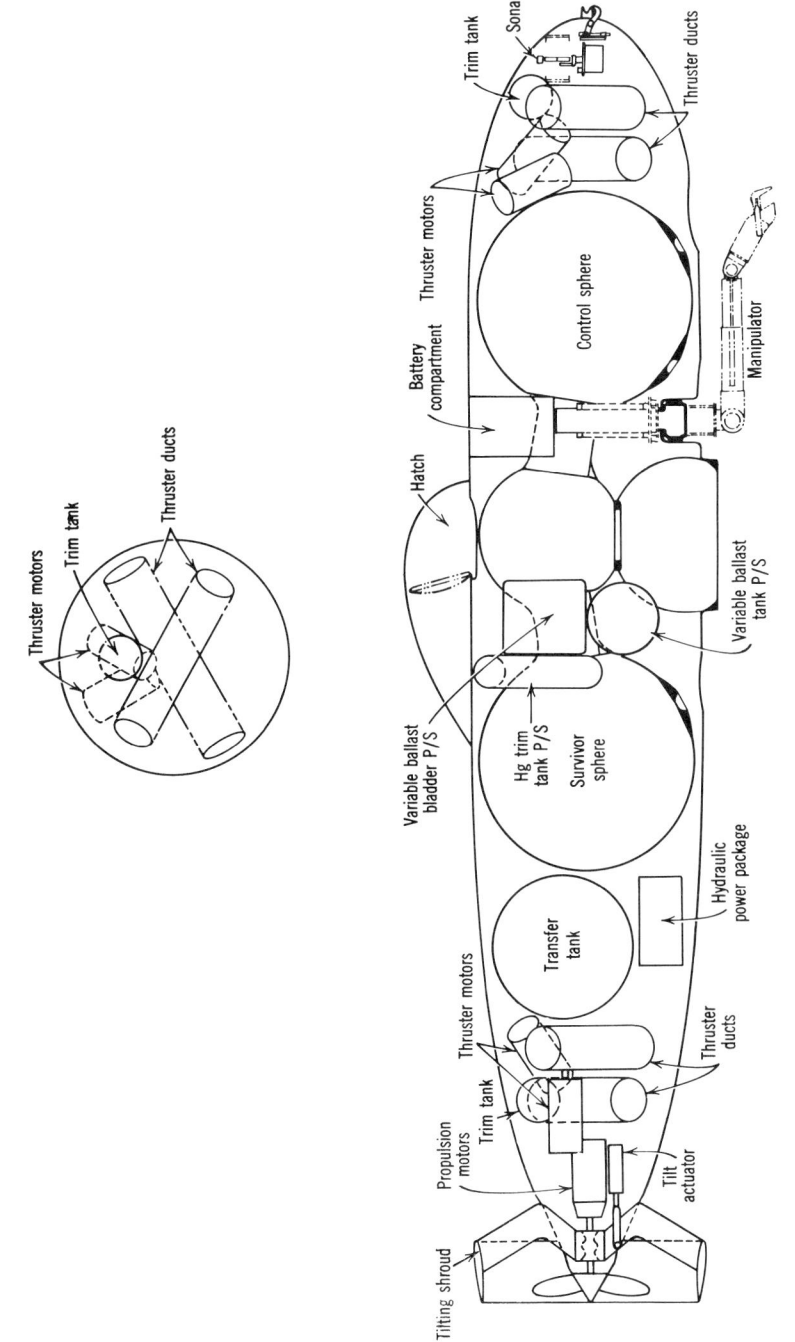

FIG. 11-24 DSRV.

driven pump. As in *Trieste II* the batteries are encased in an oil bath and compensated to sea pressure.

A unique safety feature is incorporated in this craft. The fairwater is jettisonable and the sphere is releasable from the rest of the craft by means of a connection at the bottom of the sphere. The sphere, being lighter than seawater, can float to the surface.

As a result of the loss of *Thresher*, the Navy re-examined its submarine-rescue capability and devised a new kind of rescue system based on the use of a small submarine in lieu of the McCann rescue chamber. The new system will be operable at considerably greater depths than the old one. The concept of operation envisions the deep submergence rescue vehicle (DSRV) being brought to the scene of a disaster either on an ASR or "piggyback" on a large combatant submarine. In the latter arrangement the DSRV is secured over a hatch of the host submarine. At the scene of the disabled submarine the DSRV is manned and released to locate it while the host submarine hovers nearby. The DSRV can seat on the same hatch ring as the rescue chamber and secure itself by engaging the bail of the hatch cover. Transfer of crew members from the rescue submarine to the host submarine is done submerged in order to avoid the wave action on the surface. This will permit rescue under adverse weather conditions in which surface operation would be impossible.

The DSRV is pictured in Fig 11-24 and its principal characteristics are listed in Table 11-4.

TABLE 11-4

Characteristics of Deep-Diving Submarines

		Aluminaut	Alvin	DSRV
Operating depth	(ft)	15,000	6,000	3,500
Length	(ft)	51.25	20	49
Beam	(ft)	10	7.5	7.92
Displacement-Submerged	(tons)	81	15	34
Pressure-hull diameter	(ft)	8.1	6.83	7.50
Pressure-hull thickness	(in.)	6.5	1.33	0.738
Pressure-hull material		Al 7096 T6	HY-100	HY-140
Flotation Material		—	Syntactic foam and metal spheres	—
Speed—submerged	(knots)	3–8	2–6	3–5
Range—submerged	(miles)	80 at 3	25 at 2.5	

REFERENCES

Abkowitz, M. A., L. A. Vassilopoulos, and F. H. Sellars, 1966, "Recent Developments in Seakeeping Research and Its Application to Design," *Trans. SNAME*, ABS, 1966.

Rules for Building and Classing Steel Vessels, American Bureau of Shipping, New York.

Arentzen, E. S., and P. Mandel, 1960, "Naval Architectural Aspects of Submarine Design," *Trans. SNAME* **68**, 622–692 (1960).

Arnott, David, ed., 1965, *Design and Construction of Steel Merchant Ships, SNAME*, New York.

Chadwick, J. H., Jr., 1955, "On the Stabilization of Roll," *Trans. SNAME* **63**, 237–280 (1955).

Covey, C. W., 1964, "*Aluminaut*," *Undersea Techn.* **5**, 9, 17–23 (1964).

Craven, J. P., and H. Bernstein, 1965, "Materials for Deep Submergence-Development Dilemma," *Mater. Res. Std.* **5**, 551–556 (1965).

Dermody, J., J. Leiby, and M. Silverman, 1964, "An Evaluation of Recent Research Vessel Construction in the United States," *Trans. SNAME* **72**, 404–444 (1964).

Fisher, F. H., and F. N. Speiss, 1963b, "FLIP – Floating Instrument Platform," *J. Acoust. Soc. Am.*, **35**, 1633–1644 (1963).

Gibbs and Cox, Inc., 1960, *Marine Survey Manual for Fiberglass Reinforced Plastics*, McGraw-Hill, New York.

Gibbs and Cox, Inc., 1962, *Marine Survey Manual for Fiberglass Reinforced Plastics*, New York.

Hobaica, E. C., 1964, "Buoyancy Systems for Deep Submergence Structures," *Naval Eng. J.* **76**, 733–741 (1964).

Johnson, H. F., 1946, "Development of Ice-breaking Vessels for the U.S. Coast Guard," *Trans. SNAME*, **54**, 112–151 (1946).

Kieran, T. S., 1964, "Predictions of the Collapse Strength of Three HY-100 Steel Spherical Hull Fabricated for the Oceanographic Research Vehicle *Alvin*," DTMB Report No. 1792 (March 1964).

Korvin-Kroukovsky, B. V., 1961, *Theory of Seakeeping*, Soc. Naval Architects and Marine Engineers, New York.

Krenzke, M. A., 1965, "Structural Aspects of Hydrospace Vehicles," *Naval Eng. J.* **77**, 579–606 (1964).

Krenzke, M. A., K. Ham, and J. Proffitt, 1965, "Potential Hull Structures for Rescue and Search Vehicles of the Deep Submergence Systems Project," DTMB Report No. 1985, February 1965.

Krenzke, M. A., and T. S. Kiernan, 1965 "The Effect of Initial Imperfections on the Collapse Strength of Deep Spherical Shells," DTMB Report No. 1757, February 1965.

Lank, S. W., and O. H. Oakley, 1959, "Application of Nuclear Power to Icebreakers," *Trans. SNAME* **67**, 108–139 (1959).

McKee, A. I., 1959, "Recent Submarine Design Practices and Problems," *Trans. SNAME* **67**, 623–652.

Mandel, P., 1960, "Hydrodynamic Aspects of Deep-Diving Oceanographic Submarine," 3rd U.S. Office Naval Res. Symp. Naval Hydrodynamics.

Marks, W., 1963, "The Application of Spectral Analysis and Statistics to Seakeeping," *Tech. Res. Bull.* 1–24, SNAME, New York (September 1963).

Oakley, O. H., 1965, "Design Considerations for Manned Deep-Sea Vehicles" *ASTM-Mater. Res. St.* **5**, No. 11 (1965).

Pellini, W. S., 1964, "Status and Projections of Developments in Hull Structural Materials for Deep Ocean Vehicles and Fixed Bottom Installations," Naval Res. Lab. Rpt. NRL-6167, November 4, 1964.

Pellini, W. S., R. J. Goode, P. P. Puzak, E. A. Lange, and R. S. Huber, 1965, "Review of Concepts and Status Procedures for Fracture-Safe Design of Complex Welded Structures

Involving Metals of Low and Ultra-High Strength Levels," U.S. Naval Res. Lab. Rep. NRL-6300, June 1965.

Puzak, P. P., A. J. Babecki, and W. S. Pellini, 1958, "Correlations of Brittle-Fracture Service Failures with Laboratory Notch-Ductibility Tests," *Welding J. Res. Suppl.* (September 1958).

Rossell, H. E. and L. B. Chapman, eds., 1939, *Principles of Naval Architecture*, SNAME, New York.

St. Denis, M., and W. J. Pierson, 1953, "On the Motions of Ships in Confused Seas," *Trans. SNAME* **61**, 280 (1953).

Shumaker, L. A., 1964, "*Trieste II*," *Naval Eng. L.* **76**, 513-519 (1964).

Terry, R. D., 1966, *The Deep Submersible*, Western Periodicals, North Hollywood, California.

Vasta, J., A. J. Giddings, A. Taplin, and J. J. Stilwell, 1961, "Roll Stabilization by Means of Passive Tanks," *Trans. SNAME* **69**, 411 (1961).

Wah, Thien, 1960. A Guide for the Analysis of Ship Structures, U.S. Department of Commerce, Office of Technical Services, U.S. Government Printing Office, Washington D.C.

Walsh, J. B., and W. O. Rainnie, Jr., 1963, "*Alvin*, Oceanographic Research Submarine," ASME Paper 63-WA-160 for meeting November 16–22, 1963.

CHAPTER **12**

Instrumentation and Communications

JAMES M. SNODGRASS

The material to be presented is in no way an attempt to present a complete picture of oceanographic instrumentation, since this would be a very voluminous document and well beyond our present concern. For the sake of convenience, we will further limit our consideration of instruments mainly to those developed at the Scripps Institution of Oceanography, and even at this level by no means all instruments are to be considered.

Before looking at present-day instrumentation, I suggest we turn the clock back approximately 200 years and examine an early oceanographic program. This was long before the Office of Naval Research or the National Science Foundation were available to support oceanographic research and there is no evidence that any "grant" was involved. It is almost certain that the researcher financed the project essentially from his own pocket. This early oceanographic investigator was none other than Benjamin Franklin. Figure 12-1 represents the Gulf Stream of the Atlantic Ocean presented in a publication of Benjamin Franklin's which was made shortly after he became Postmaster General of the American Colonies (Brown, 1938). Admittedly there is some controversy with respect to this work but, by and large, it is the result of measurements made by Benjamin Franklin. During the Revolutionary War Franklin made several trips back and forth across the Atlantic, and undoubtedly learned something on these trips that caused him to undertake his research. Franklin was a very ingenious and resourceful gentleman and it is significant for our purposes that he introduced some very simple but effective tools for the study of oceanography. The tools he used were a thermometer and a bucket, probably a wooden bucket, with which he took samples of water used for his temperature measurements. With these very simple tools he was able to

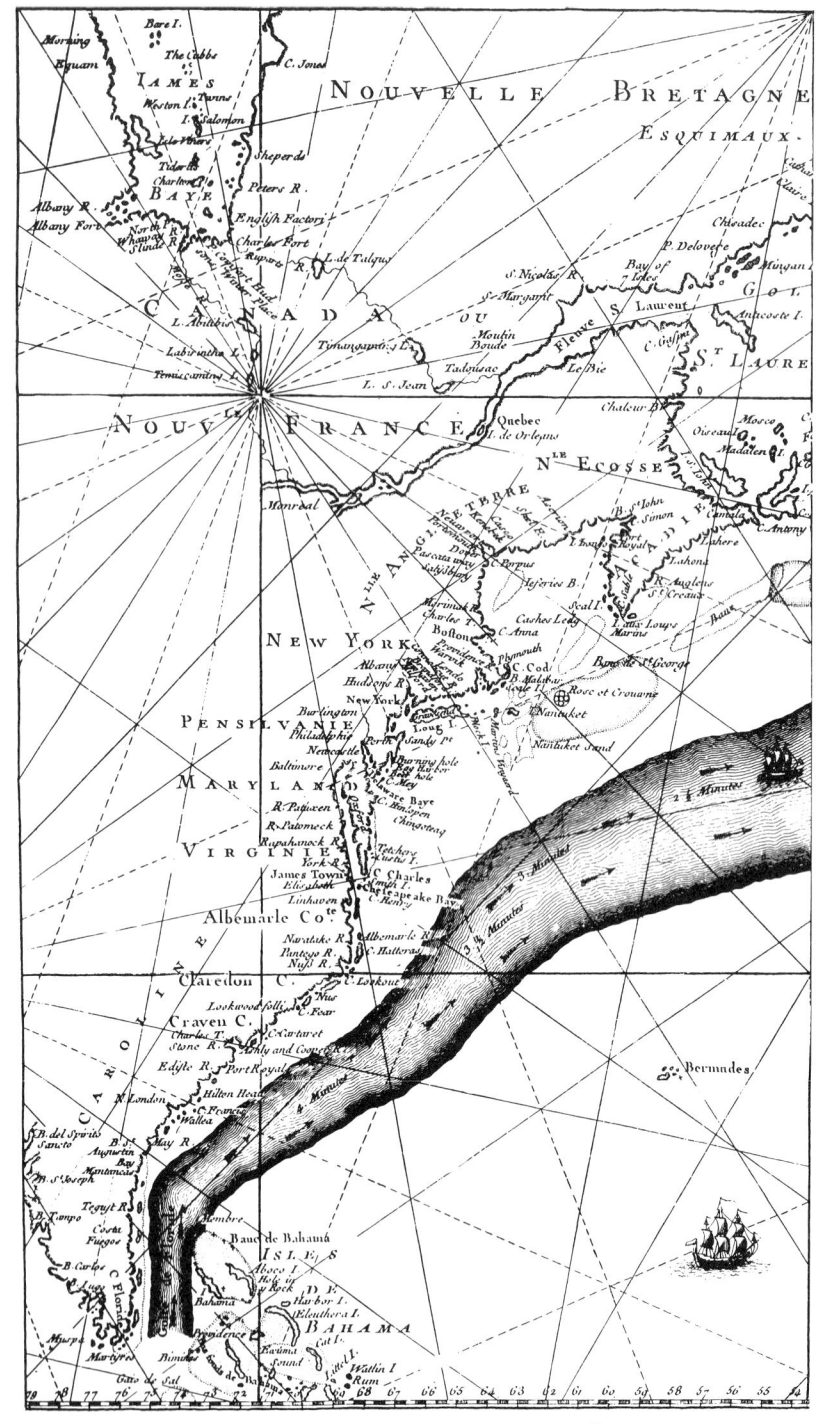

FIG. 12-1. Gulf Stream, as published by Benjamin Franklin. [R. H. Brown, "De Brahm Charts of the Atlantic Ocean, 1772–1776," *Geograph. Rev.*, 28, 124–132 (1938).]

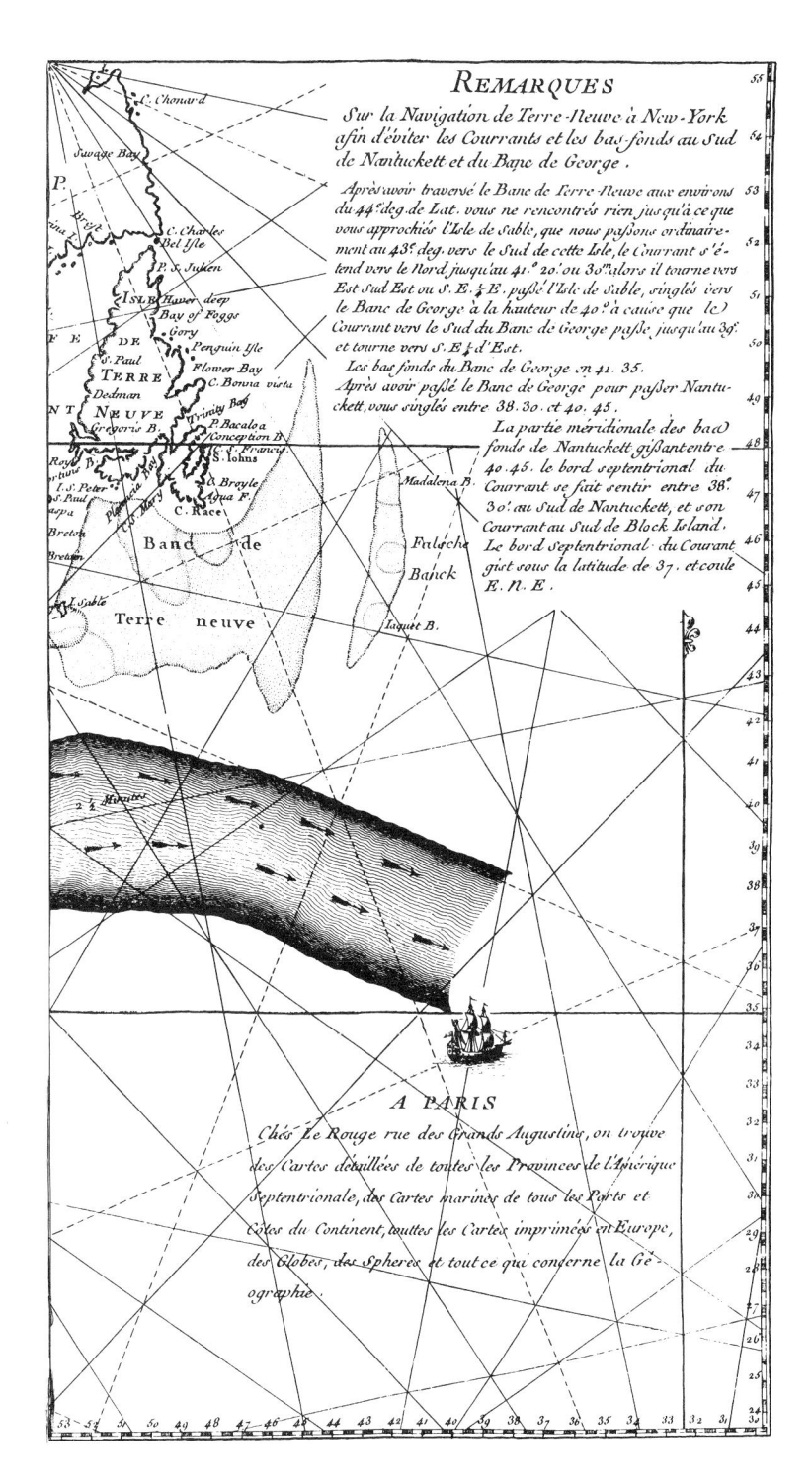

FIG. 12–1. (*Continued*)

define the boundaries on the Gulf Stream, essentially as illustrated. As a result of his temperature studies on the Gulf Stream, Franklin developed his principles of "thermometrical navigation" which if practiced by mariners should effectively speed their ships.

Now, turning the clock forward to very nearly the present time, Fig. 12-2 illustrates the application of modern tools to oceanography. As we closely examine the illustration it becomes evident that, rather than improving during the intervening years, the "modern" bucket temperature may be significantly inferior to that obtained by Benjamin Franklin. Certainly Franklin's wooden bucket would be considered a significant improvement, from the thermodynamic standpoint, to the galvanized iron bucket illustrated. Furthermore, since we are being critical at the moment, it is evident that the bucket is in the sunshine and exposed to the wind.

Until the advent of World War II oceanography proceeded at a rather leisurely pace. The antisubmarine-warfare program during World War II forced the rapid development of underwater acoustics. Since the behavior

FIG. 12-2. "Modern" bucket temperature (used by permission of U.S. Coast and Geodetic Survey).

of sound in water is intimately related to the physical properties of water such as density, temperature and salinity, oceanography received a tremendous boost. However, the actual instruments employed to make measurements in the ocean were rather crude, technologically, in contrast to much of the other instrumentation of the day involving electronics. After World War II many young men who had become acquainted with oceanographic techniques felt that there must be a better way of obtaining oceanographic data and felt that electronics would certainly be of assistance. It was at the Woods Hole Oceanographic Institution that the first major efforts to introduce electronics into oceanography occurred. Funds were obtained and electronic instruments duly built. Unfortunately, practically without exception, these instruments failed, not necessarily because of faulty electronics or conceptual aspects, but because the engineers did not properly understand the marine environment and the consequent packaging problem. Simple leaks proved to be disastrous. Since research funds were scarce, there was naturally considerable resistance to the introduction of these "new-fangled" instruments. In fact, after repeated failures, it was suggested that "The ideal oceanographic instrument should first consist of less than one vacuum tube." As this was in pretransistor days, it left little latitude for electronics. Such opinions and attitudes nearly served to sweep electronics out of oceanography.

When I first came to the Scripps Institution of Oceanography in 1948, it was quite evident that the word "electronics" was in many ways what might be considered a dirty word and, more often than not, preceded by an uncomplimentary adjective. Mechanical instruments had been used in oceanography literally for many decades and were performing creditably. Since mechanical instruments had achieved relatively high reliability, there was little or no inclination to turn to electronic instrumentation, which, aside form being relatively fragile, was at the time quite temperamental and admittedly unreliable. It was not until approximately 1950 that electronics was successfully introduced into oceanography at, both the Woods Hole Oceanographic Institution and the Scripps Institution of Oceanography. The decade from 1950 to 1960 was in many ways a period of very slow growth from the standpoint of electronics. The advent of solid-state systems and increased "know how" are responsible for much of the present success. At the Scripps Institution of Oceanography one particular instrument was in many ways responsible for the electronic "breakthrough." However, even in the case of this instrument, a bottom sediment temperature-gradient recorder, which is described in section 12.7, more than a year lapsed between its introduction and any significant influence.

12.1 INSTRUMENT CLASSIFICATION

There are, of course, many ways in which instruments may be classified, but for the sake of convenience, we may arbitrarily divide them into two large classes as follows: (a) pressure-protected instruments; (b) pressure-equalized instruments.

Pressure-protected instruments

The term "pressure protected" is descriptive of this class of instruments, since it means that the instrument components, electronic or electrical, are protected from the pressure of the ocean depths by means of a strong, rigid, often heavy-walled instrument case. The internal atmosphere of the instrument is an exceedingly important environmental factor and both its pressure and composition need to be selected with care. In general, pressure-protected instruments enjoy relatively narrow environmentals with temperature under operating conditions rarely below 1°C and no higher than 35°–38°C. However, since instruments are often carried on the deck of ships, the nonoperating environmental temperatures may be quite low, for that matter relatively high, particularly if the instrument is left on the deck, exposed to the tropical sun. The relative humidity of the instrument atmosphere is extremely important, for example if an instrument is opened at sea, under conditions of high relative humidity, with the instrument remaining open for any length of time or if the chassis is removed from the instrument case and then subsequently closed up and lowered into the ocean. Under these conditions as the instrument goes deeper the temperature decreases so that the moisture in the instrument atmosphere is condensed and collects at the lowest point in the case. It is not unusual to encounter condensation volumes of several cubic centimeters. Then, as the instrument is raised from deep water, the temperature increases; if the instrument is hauled up rapidly into warmer surface layers the case tends to act as a boiler and, since the interior electronic chassis in general may make poor thermal contact with the case, it remains cold and condenses the moisture "boiled off" from the warm instrument case. Silica gel may, of course, be introduced into the instrument case to absorb excess moisture but even here the silica gel must be carefully packaged. The difficulty is that if the silica gel package permits the passage of the very fine silica dust, the dust may be disastrous to precision mechanical systems. Whether silica gel is used or not, it is recommended that the instrument cases after closure be flushed with either anhydrous air or nitrogen, or even helium, if good convective transfer of heat is required. Another difficulty that is common with pressure protected instruments is the fact that both mechanical and electrical seals are required. The mechanical seals may be required simply to effect instrument closure. If electrical leads are required through the instrument case, special electrical lead-throughs will be required. There is now a relatively wide selection of excellent electrical lead-throughs available on the market. There are even some that may be connected and disconnected underwater very successfully.

Pressure-equalized instruments

As the descriptive term implies, the pressures inside and outside the instrument are very nearly identical regardless of depth or ambient pressure. In general this means that the instrument must be filled with a suitable fluid, usually a dielectric. The pressure between the filling fluid and the ocean is equalized by means of a suitable designed, highly compliant membrane or coupling mechanism. In some cases fluid filling is eliminated and a rigid dielectric material fills the unused space. However, care must be taken regarding the effects on electronic components of the forces that can be transmitted or developed by the rigid materials. Pressure-equalized instruments have two major advantages: there are, in general, no sealing problems either mechanically or electrically because the pressure differential is practically zero; and there is very good heat dissipation because the filling fluids conduct heat readily to the instrument case, which is in contact with the best possible thermal sink, the ocean itself.

Problems of filling fluids

As in the case of the atmosphere of the pressure-protected instrument, the filling fluids in the pressure-equalized instruments are extremely important and must be selected carefully and optimized for the particular application. The temperature coefficient of expansion of the fluid is very important and serious damage can result if it is not taken into consideration. Here again it is necessary to know the extreme ranges of environmental temperatures to which the instrument will be exposed. It is convenient to remember that most hydrocarbons have very nearly the same temperature coefficient of expansion, 1 cc/L/°C. The compressibility of fluids due to pressure change is also exceedingly important. Most hydrocarbons tend to change volume about 5% at 20,000 psi. However, silicone oils are much more compressible and change approximately 12% at 20,000 psi. It is important to realize that as an instrument is lowered into the ocean the temperature coefficient and compressibility tend to work in the same direction, that of decreasing the volume of fluid. With these considerations in mind it becomes obvious that the designer needs to minimize the unoccupied volume. In fact it sometimes becomes desirable to introduce various rigid solid shapes simply to take up what would otherwise be an inconvenient fluid volume. There is another disadvantage of fluid-filled instruments that should not be ignored, if we take compassion on the field technician or engineer. Before a fluid-filled instrument can be worked on at sea, the fluid must be drained from the instrument. It was Allyn Vine of the Woods Hole Oceanographic Institution who first pointed out that one of the unhappy aspects of repairing fluid-filled instruments at sea. In the case of the old R/V *Atlantis*, in heavy weather the only place to work on an instrument and hold it steady was on a bunk. Vine simply refused to work on a fluid-filled instrument on his bunk, since they are so messy.

Pressure effects on components

We must not forget that in the pressure-equalized instruments all of the components must be capable of operating at the designed depth of the instrument. Furthermore, some of the contaminates such as moisture in the filling fluids may cause serious problems. A glass electron tube with a $T3\frac{1}{2}$ envelope will successfully withstand pressures in excess of 15,000 psi. Under high pressure any trace of moisture in the dielectric fluid surrounding the tube envelope will rapidly migrate inside the tube, making it fail because of water vapor and consequent gasping. If the filling fluid is completely dehydrated or a nonmoisture-permeable envelope surrounds the tube, it will operate successfully at these high pressures. If it is necessary to do optical recording, flashlight bulbs in the $G3\frac{1}{2}$ size will withstand pressures over 15,000 psi. In the case of flashlight bulbs, approximately 20% will usually be found to leak, although those that do not leak are quite reliable. Here again, water vapor must be excluded. Oscillograph recording is practical under very high hydrostatic pressures with a fluid-filled system since we have available oil-filled galvanometers. Photographic film will operate very successfully in a water-white fluid, such as mineral oil. It is only necessary to remove the oil from the film with carbon tetrachloride or other suitable solvent before developing. It should be noted that the more modern prefocused flashlight bulb does not stand high pressure readily, as does the approximately round $G3\frac{1}{2}$. Resistors such as those employed in electronic circuits may or may not be affected by high pressure, depending on their construction and material. Figure 12-3 illustrates a large change in resistance of carbon composition types of resistor under high pressure (Buchanan, 1961). The carbon film or tin oxide shows no significant effect due to pressure. It might be anticipated that wire-wound resistors would be stable under high hydrostatic pressures. However, this is not necessarily true; there

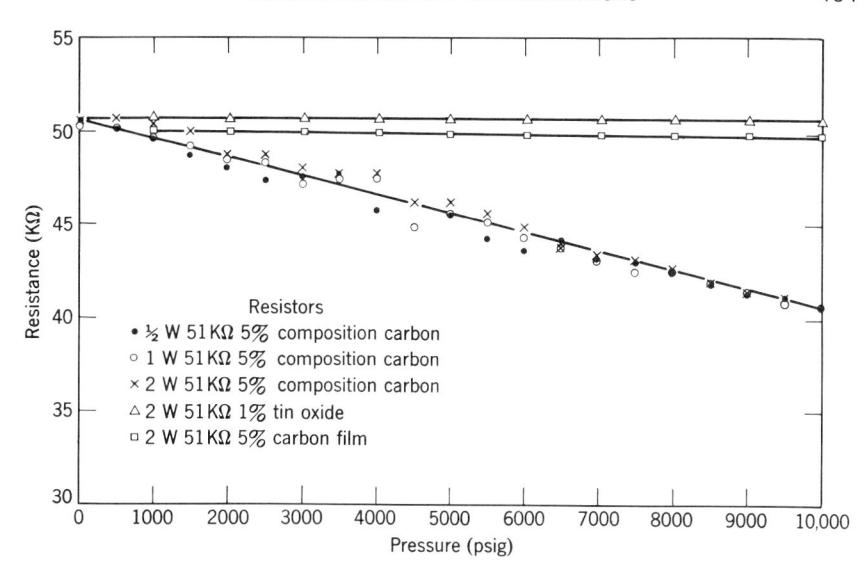

FIG. 12-3. Effect of hydrostatic pressure on three types of resistor. [C. L. Buchanan and Matthew Flato, "Putting Pressure on Electronic Circuit Components", *ISA J.* (November 1961).]

may be significant effects depending on the way in which the resistor is wound and impregnated. If there are gas inclusions within the body of the resistor or between some of the windings there may be significant pressure effects. Capacitors such as those employed in electronic circuits vary a great deal with regard to their ability to withstand satisfactorily high hydrostatic pressure. This is particularly true in the case of high-value capacitors of the electrolytic wet or dry types. Rather striking effects may sometimes be observed in the case of this type of capacitor. For instance, if we devise a circuit to measure leakage resistance as a function of hydrostatic pressure, leakage may be observed to increase up to, for instance 2000 psi. In fact the capacitor may actually short out at 7000 or 10,000 psi but if the pressure is continued upward a point may be reached at which the leakage resistance begins to fall and may occasionally fall lower than its initial atmospheric value. It is interesting that after a cycle of this sort is completed the capacitor may remain at an improved low leakage value, even though the pressure is reduced to atmospheric. Magnetic elements such as transformers or especially ferrites, though not necessarily directly affected by hydrostatic pressure, may be affected because of the medium in which they are mounted, if they are potted in a rigid plastic compound, which can introduce nonsymmetrical strains. Passivated silicon transistors are apparently insensitive to very high hydrostatic pressures. This comment does not, however, apply to the conventional case since these will readily collapse at low pressures. It is necessary to fill the transistor with a suitable fluid, if it is to be used at high pressure. Many types of relay, stepping switch, snap-action switch, and so on, will operate successfully in fluid-filled media. Perhaps it is important to point out that at present it is necessary to test every component being used under the pressure conditions it will be required to withstand.

Plastic cast and oil-filled photocells

The pressure-equalized construction may often introduce significant economies and simplification. Figure 12-4 illustrates a Weston Model 856 Selenium photovoltaic cell that has been fluid filled and potted in a rigid epoxy resin. The photocell as received from Weston is normally filled with helium. The helium is removed and replaced with either silicone oil or a water-white mineral oil and resealed. Electric leads are attached with a suitable strain relief such as the spiral steel helix, and a large blob of silicone grease spread over the rear surface of the photocell. The entire unit except the photocell window is then potted in the rigid epoxy. After the epoxy is set a small hole about 0.05 mm in diameter is drilled from the rear surface of the resin into the silicone grease. The silicone grease with its narrow duct communicating with sea water acts as the compliant pressure-equalizing element. A photoelectric cell mounted in this fashion is capable of operating successfully at any depth in the ocean without further protection. It should be noted that it is now

FIG. 12-4. Photoelectric cell cast in epoxy resin, requiring no machine operations.

possible to buy directly from Weston photocells that are filled with fluid in place of helium.

12.2 POWER SOURCES

Electric power in one way or another is required for various electronic systems. This is true whether the electronic system is operating in a buoy or is on the end of a cable connected to a ship or on the bottom of the ocean. Power sources are many and varied at the present time and are constantly being improved, modified, and added to. It is impossible to make generalizations in this area. It should, of course, be obvious that the exigencies of a particular situation will dictate the optimum power source.

Hydrocarbons

There are only two hydrocarbons at present that are considered practical power sources in the marine area: fuel oil and propane. Of the systems employing the direct combustion of propane, the most common employ thermal electric converters. Propane may be burned either as an open flame or in a so-called catalytic burner. A question that is somewhat unresolved in the marine field concerns the application of the propane-fueled thermoelectric systems when applied to oceanographic buoys or other environments very near or close to the sea surface. A problem here is what happens to all the sea salt, which, of necessity, must be removed from the air or must pass through the burner system as it breathes air. It is estimated that a simple 10-W propane-fueled thermoelectric unit will have to pass somewhere near 200 g of salt in the form of sodium chloride in the course of one year. Other salts are also involved but in substantially less quantity than sodium chloride. Presently, no long-term tests are known to have been carried out that appear to be adequate for life prediction purposes. Since most thermoelectric propane-fueled units employ a high-conductivity liner, usually made of an aluminum alloy, there is also reason to anticipate difficulties, for in some burners chlorine may be generated from the sea salt. The gasoline internal combustion engine is easily modified to burn propane, and there has been significant work done in this area which has given valuable data in terms of reliability. An incidental difficulty in the modification of gasoline engines to propane is that the lubricating oil that is conventionally employed in the gasoline engine and is formulated with various additives will gradually thicken. Thus, if propane is used in place of gasoline for fuel, it is necessary to use a different lubricating oil for this application. The propane-fueled prime mover may be either employed directly for electric power or used to charge electric storage batteries.

Solar power

Most of the satellites employed for scientific purposes have made use of multiple photoelectric cells mounted in panels. In fact the space program might have been delayed had we not had such solar panels available. It is frequently suggested that solar panels should be employed on ocean buoys. At this juncture perhaps it is advantageous to point out that the solar panels employed in the space program never had to contend with sea gulls. An ingenious attempt was made to solve the gull fouling problem by mounting the solar panels so that they were very nearly awash and could be continuously washed by wave action. However, the engineer who devised this solution was not cognizant of another factor in the marine environment, fouling by marine growth. In most areas of the ocean panels of this type would become essentially inoperable after only a very few weeks. However, assuming that we might solve the problem of the fouling of solar panels, we should not lose sight of the fact that there are many areas on the globe where they will be unsuited, such as at very high latitudes where little solar energy is available, as well as an extensive portion of the ocean in the equatorial regions that have a very heavy overcast. This essentially means that we would be limited to selected areas in the temperate zones.

Wave power

The possibility of utilizing waves to generate power has intrigued man for centuries. Unfortunately, little or no success has been achieved, even up to the present time. The only known successful wave-powered machine to operate for a reasonable period of time was devised by S. J. Savonius in 1931 to pump salt water up the cliffs to the aquarium at Monaco (Savonius, 1931). In a way, of course, wave energy has been used for whistling buoys for several decades. However, the problem of obtaining a significant amount of power still remains to be solved for the open-sea conditions.

Potential energy

There is another source of energy that should not be lost sight of which is almost uniquely available in the marine environment. This is the potential energy represented in the unoccupied space of a closed container as it is lowered into the ocean. The increasing pressure gradient may be utilized to generate power. Though this has been used for many years, it has recently been elaborated by Carl Holm, (1965–1966) an engineer for North American Aviation who proposed an ingenious system to use potential energy of this type for useful work.

Electric

Primary batteries. The classical dry cell (Leclanché), within its limitations, may be used to power instruments lowered into the ocean. For design purposes, it should be realized that even under low discharge rates the battery will lose at least 25% of its nominal capacity at temperatures between 0°C and 1 or 2°C. High-rate discharges become quite impracticable at the low temperatures. There is little significant direct pressure effect on the dry cells and it is only necessary to keep them out of contact with saltwater by imbedding them in a suitable material or fluid. In spite of the best efforts by manufacturers to seal the dry cells so that they do not lose moisture, they continue to do so over a period of time. Since voids in the case will cause physical distortion under high hydrostatic pressures, it is advisable to obtain batteries as fresh as possible from the manufacturer, if they are to be used at high pressures. The so-called mercury batteries have special problems. The mercury battery, as in the case of the dry cell, is not pressure sensitive, but has a high sensitivity to low temperature. An example of this problem is illustrated in Fig. 12-5, which shows a typical discharge curve over a period of several hours of a standard Mallory RM-42 mercury battery delivering power into a 3.0-Ω load at 8°C ambient. It is quite obvious that the moment the battery begins cooling the voltage begins dropping. In fact, at about $4\frac{1}{2}$ hr, we may observe the first rather erratic potential change which is repeated with emphasis at about $9\frac{1}{2}$ hr. The only relatively flat portion of the discharge curve occurs between approximately $12\frac{1}{2}$ and $14\frac{1}{2}$ hr, though it is suspected that this is basically a happenstance, and at around 15 hr the waves in the curve correspond to the on–off cycle of the refrigerator. On removal from the refrigerator, after about 20 hr, the potential rapidly returns to somewhere near normal. The erratic character of the discharge curve is due to mechanical problems related to the construction of the battery. Internally, there are four cylindrical stacked zink anodes held together by pressure exerted on them by a compliant slug of neoprene. At somewhere around 8°C, the neoprene loses its elasticity and compliance and therefore ceases to exert adequate pressure on the zinc anodes which then may develop an erratic contact. It should be noted that the manufacturer does not recommend this battery for operation at the 8°C

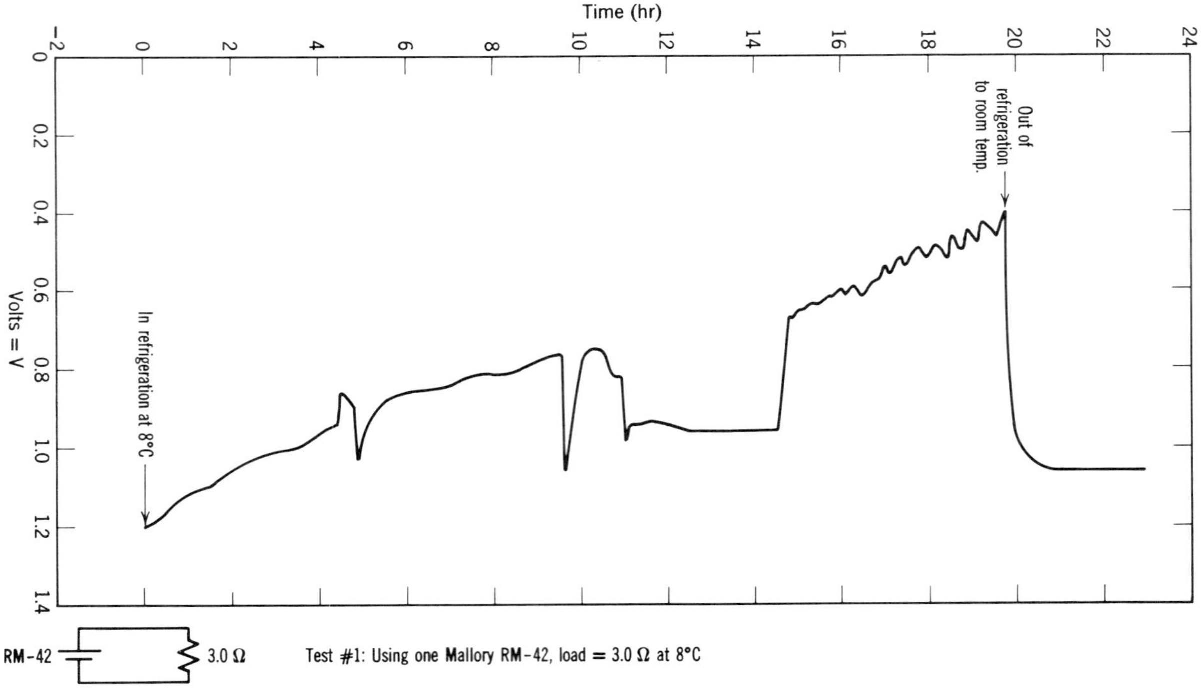

FIG. 12-5. Low-temperature voltage discharge curve for RM-42 mercury battery. 3-Ω load with battery at 8°C ambient temperature.

temperature. If it is desirable to use mercury batteries at low temperature there are some three models available for this type of service. Since mercury batteries represent rather high wattages per unit volume it is sometimes possible to use some of their electric power to heat a protective jacket, providing excellent thermal insulation. In certain specialized applications it may also be possible to employ chemical exothermic jackets.

Secondary batteries. The most familiar of the secondary batteries are the lead–acid batteries, which are the common storage batteries employed in the automotive vehicles. The lead–acid battery, properly pressure equalized and protected from saltwater is little affected by the temperatures encountered in the ocean or the pressures. In fact a lead–acid type of battery has been successfully tested by the U.S. Navy Electronics Laboratory in San Diego to 80,000 psi. Lead–acid batteries are rather easily pressure equalized by filling the cells completely full of sulfuric acid of the correct specific gravity and adding a suitable extra volume to allow for compression and cooling. The terminals may be easily painted with battery compound to keep saltwater from the connecting terminals. The lead–acid cell, since it is in such common use, permits some rather interesting applications of cost-effective studies. For instance, the University of California obtains a state price on ordinary conventional automobile storage batteries which is quite low. In some applications when it is necessary to obtain low-voltage electric power for an oceanographic instrument and at the same time have a fairly heavy weight or anchor attached, a lead–acid battery combines both of these functions rather economically. Although it may make strong Scotsmen wince, this is often true in those cases when it is necessary to expend the battery. The automotive type of storage battery, which we have been discussing, is not recommended when it must operate for long periods of time at low temperature and high pressure since there is a rather high degree of internal self-discharge. Lead–acid batteries specially designed for long-charge retention are available on the market known as "low self-discharge" types. The manufacturer states that these retain approximately 80% of their charge after a 1-year period at 70°C. A manufacturer of bottom-mounted acoustic beacons has reported obtaining at least 3 years of service from storage batteries of this type. In situations that require higher ampere-hour ratings per cubic inch the silver–zinc storage battery is perhaps one of the best. This battery may be utilized in the marine environment when repackaged as suggested for the lead–acid battery. Nickel–cadmium batteries may be employed quite readily in the marine environment.

Nuclear power. Nuclear power is, of course, rather attractive for use in the marine environment particularly for powering deep-sea electronic devices that must operate for a prolonged period of time. However, at present they remain relatively expensive and may have problems with respect to the weight of shielding required. Different isotopic fuels have different shielding requirements. The matter of shielding itself is one that is difficult to properly delimit since once an instrument is in place on the ocean bottom shielding appears to be a minor requirement. However, the matter of getting it in place and the initial handling certainly does require considerable care from this standpoint. There is also the possibility that an isotopic fuel system might be accidentally fished up from the bottom either by fishermen or by scientists. Here again, for those applications requiring extremely long operating time, it is quite possible that isotopic fuels will be the most attractive.

12.3 ELECTRICAL NOISE IN THE SEA

Perhaps it is well to explain what is meant by electrical noise in the sea since it is by no means obvious. Electric noise, for our present purposes, may be defined as electric potentials occurring both in frequency and amplitude that may be detrimental to the operation of electronic instruments and sensors in the sea. There are electric signals in the ocean which correspond to all of the known ocean-wave frequencies. A typical electrical potential pattern generated by ocean waves is illustrated in Fig. 12-6. The emf is picked up very simply by means of electrodes approximately 100 ft apart. It is obvious that there is a wide range of frequencies in evidence. How are these emf's generated? The orbital motion of the waves represents a moving electric conductor since the ocean is made up of saltwater. This electric conductor is moving in the earth's magnetic field and, as a consequence, generates a voltage which is easily picked up and recorded. In the surf zone higher frequencies are to be found, as well as greater voltage amplitudes. In the case of exposed metal structures in the surf zone these wave-generated potentials undoubtedly contribute to the general corrosion problem. The range of frequencies in the ocean is considerable. There are definite contributions from the tides and surf zone that can generate frequencies of several cycles per second. The magnitudes of potentials to be encountered in long telephone cables are perhaps somewhat surprising. In the case of telephone cables running between New York and Newfoundland, which employ thermionic repeater amplifiers, the filaments of the amplifiers are connected in series and powered by means of a constant-current power supply, that possesses a relatively high voltage compliance. The compliance which the power supply must handle is measured in tens of volts. There are other sources of emf generated outside of the oceans; some of them are generated terrestrially and some extraterrestrially. At the low-frequency end, there are micropulsations of the earth's electromagnetic field, which occur in the frequency range 0.1 to 10 cps. Other signals of terrestrial origin are known as

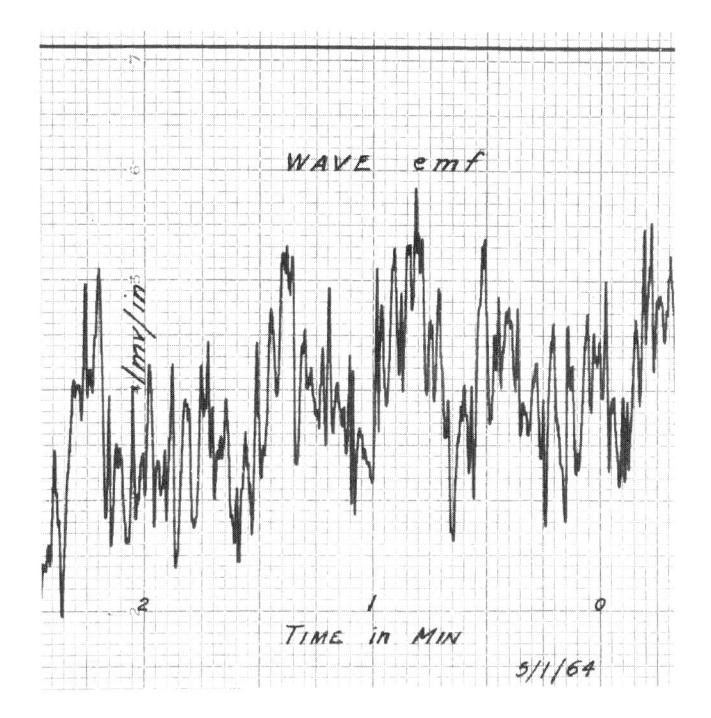

FIG. 12-6. Open-sea wave EMF; horizontal scale, two large divisions per minute; vertical scale, one large division per millivolt.

whistlers or atmospherics (sferics), which occur in the frequency range from 300 to 30,000 cps. There are other significant sources of very low frequencies (VLF) generated in the magnetosphere (Bleil, 1964). Atmospherics have their origin in lightening discharges, often thousands of miles away. A lightning discharge shock excites the space between the surface of the earth and the ionosphere. The natural frequency of this "cavity" is in the VLF and extra low frequency ranges. This electromagentic energy is propagated as though it were in a leaky waveguide. Part of this energy leaks out of the waveguide into the ocean, penetrating to different depths, depending upon the electromagnetic frequency. Since these frequencies are electromagnetic in nature, we cannot hear them unless we convert them by means of a simple amplifier to audio frequencies. A pair of electrodes suitably spaced in the ocean may be simply coupled to the input terminals of a high-gain audio amplifier with the amplifier driving headphones or loudspeakers and it is very easy to "hear" the atmospherics. Since over the surface of the earth there are several hundred lightning discharges occurring every minute the atmospheric noise may be very nearly continuous. The major sources of lightning discharge are, in general, the

410   Systems Design—The Technology

Indian Ocean and the Amazon Valley. Figure 12-7 shows the azimuthal variation of atmospherics as received by a 100-ft long electric dipole which was towed behind a submarine on the surface. The dipole was steered by turning the submarine through a large-diameter circle. Although it is quite true that certain fish do generate electromagnetic signals, it should be pointed out that some investigators have confused atmospherics with fish noises. Atmospherics tend to undergo extensive diurnal changes in magnitude. They begin approximately sunset local time and tend to reach a peak midway between sunset and sunrise. At sunrise they are usually rapidly fading out and are heard infrequently during daylight hours. It is interesting to note that one of the first clues to the failure of a shield or insulation in a shipboard sonar system transducer lead is when one begins hearing atmospherics after sundown.

12.4 EXPENDABLE INSTRUMENTS

"Expendable instruments" is perhaps not the proper term for what is meant, because expendable may imply a nonvoluntary consumption or actual sacrifice of equipment. The term "disposable" might be preferred, since it seems to imply a somewhat happier disposition of the instruments. However, the term "expendable" has now come into general use and will be the term used hereafter in this discussion.

Azimuthal variation
1Oct. 53 Baya, reel #4

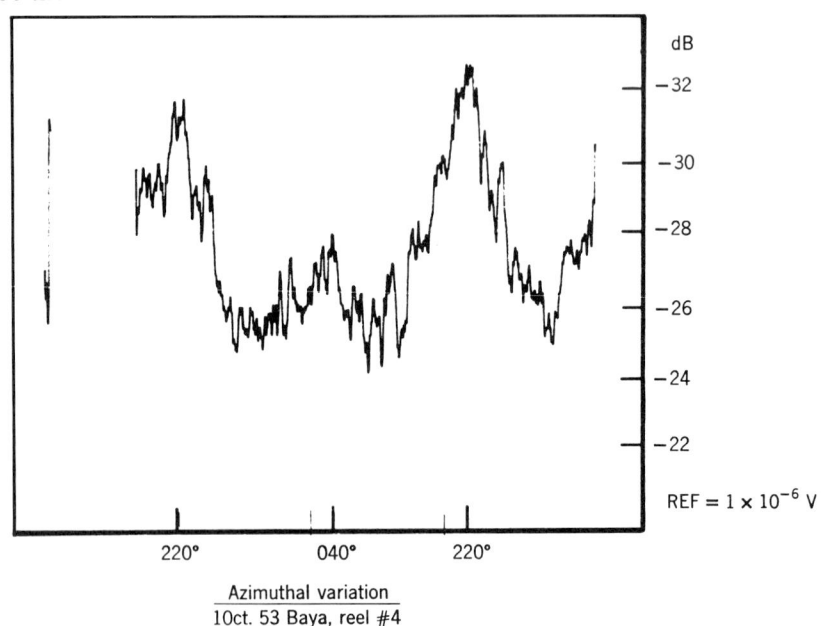

REF = 1×10^{-6} V

FIG. 12-7. Azimuthal variation of atmospherics.

Origin — historical

It is a rather entertaining and interesting exercise to speculate on a possibly significant origin of expendable instruments. It is doubtful, however, that this would be more significant than entertaining since even primitive man may well have found it convenient to utilize tools that might well have been defined as disposable instruments. For instance, if we were in a small boat and no convenient flotsam were close at hand, and if we wanted to know what the relative speed of the boat was with respect to the water, we could toss over a wooden chip or other suitable object as an index. In any case, during World War II and perhaps before, aircraft, particularly seaplanes, made use of air-droppable smoke floats in order to determine the speed and direction of surface winds before attempting an open-sea landing. A relatively sophisticated expendable instrument that has been used for a good many years is the meteorological balloon, in particular, the radiosonde. These are rather sophisticated miniature radio transmitters that transmit temperature, barometric pressure, etc., to ground radio receivers. Practically every country on earth makes use of these meteorological balloons daily. In fact they are used by the tens of thousands. Even though millions of dollars are involved, no one questions the cost basis of the data collected. It is considered sufficiently important to expend the required money. During World War II, expensive air-dropped sonobuoys were developed for the purpose of tracking submerged submarines acoustically. Initially, these cost somewhat over $100 and had a life expectancy of approximately 20 min. Another interesting air-droppable device is the automatic weather station, particularly as it is used in the Arctic and Antarctic for sensing local surface and meteorological data and transmitting it by radio. Oceanographers apparently have inherited a problem in taking certain kinds of data at sea, since it is often necessary to stop ships and lower various devices on cables in order to obtain information as a function of depth. This is enormously inconvenient and expensive as well as time-consuming.

Nighttime water-transparency measurements

For many years the method of making "measurements of the absorption of the visible portion of the optical spectrum in seawater was by lowering a white disk of standard size (30 cm) known as the Secchi disk, and observing the depth at which the disk disappeared from sight" Sverdrup et al. (1942). Aside from being slightly awkward to handle and, in appreciable currents, requiring the use of a fairly heavy weight, quite acceptable data were obtained by this means. However, no measurements of this sort could be made at night or late in the day. A practical and simple solution to the problem of nighttime transparency measurements is illustrated in Fig. 12-8, which presents a night "Secchi disk." Aside from the facts that a Secchi disk, cannot be used at night, and this device does not resemble a Secchi disk, we should not be bothered, since the instrument performs excellently. The construction of the expendable light source was very simple. A small flashlight bulb was soldered directly to the positive terminal of a D-size flashlight cell and an electric insulated wire soldered to the other terminal of the bulb. The battery was then pushed partially out of its cardboard case and a hole was punched with a paper punch near the bottom through which a bared portion of the stranded wire was inserted and wrapped, as shown in the illustration. To prevent accidental turning on, the devices were placed over wooden blocks, which prevented the battery from sliding back into place. These were then packed by the dozens in boxes for transportation and later use. To turn the device on it was simply picked up and the cell squeezed together manually; the zinc terminal contacted the bare wire, turning on the light. For making the trasparency measurement the device was dropped in the water alongside a ship proceeding slowly. The instant it hit the water a stopwatch was started and the moment the light source disappeared the watch was stopped. This time interval could be used to determine the depth of the light source when it disappeared since the sinking rate was reasonably uniform. During some of the survey programs at the Scripps Institution of Oceanography, these night "Secchi disks" were made by the hundreds and proved to be quite successful.

FIG. 12-8. Expendable night "Secchi disks".

Temperature-depth measurements

For our present consideration, insofar as we are concerned with instruments, we consider oceanography as it relates to the extracting of data from the oceans. A major factor tending to bring the data collection problem to a head is the increasing interest, both scientific and national in the potentiality of the oceans. There are very real pressures directed toward the collection of more data, more accurately, and more expeditiously. It has become increasingly evident that classical methods and techniques are in many ways grossly inadequate to meet present demands, let alone the future.

It appears advisable, especially in the case of the collection of routine data, to examine the problem of data collection from the basis of cost effectiveness, as far as the means to be employed are concerned. To put the situation somewhat differently, we might restate the problem as one of simply cost per data point. This does not imply that the scientist carrying out research programs should necessarily be constrained on this basis but it would certainly seem that, as mentioned (Snodgrass, 1961) previously, routine data collections are certainly involved.

In examining the over-all data problem it would appear that the most effective attack could be made from the standpoint of the data most commonly used or required. It is most important that the community of users be in general agreement as to the specific parameter to be measured. Not only should they be in agreement as to the parameter, but also in agreement as to required accuracy, range, etc.

In seeking to satisfy the above requirements, one parameter tends to stand out as being almost universally used and that is temperature as a function of depth. This measurement is especially useful to the oceanographer studying dynamical seasonal changes within the ocean as well as assisting in the measurement of current structures and boundaries. The measurement is especially applicable in the field of antisubmarine warfare (ASW) since it provides an index as to the behavior of sound within the ocean. In the equation for the vertical velocity gradient of sound, temperature occurs as a dominant term (Horton, 1959). The bathythermograph (BT) is the instrument that has been used for many years to measure temperature as a function of depth. The BT is a wire-connected mechanical device that makes an *in situ* recording. A special winch is employed to lower and raise the instrument at sea. Figure 12-9 illustrates a present-day BT and some of its associated accessories. An analog plot of temperature as a function of depth is recorded on the approximately 1 in $\times$ 1$\frac{1}{2}$ in. smoked slide which is shown sticking out from the side of the instrument just about the first inch on the ruler. The photograph shows the slide being inserted since, in actual use,

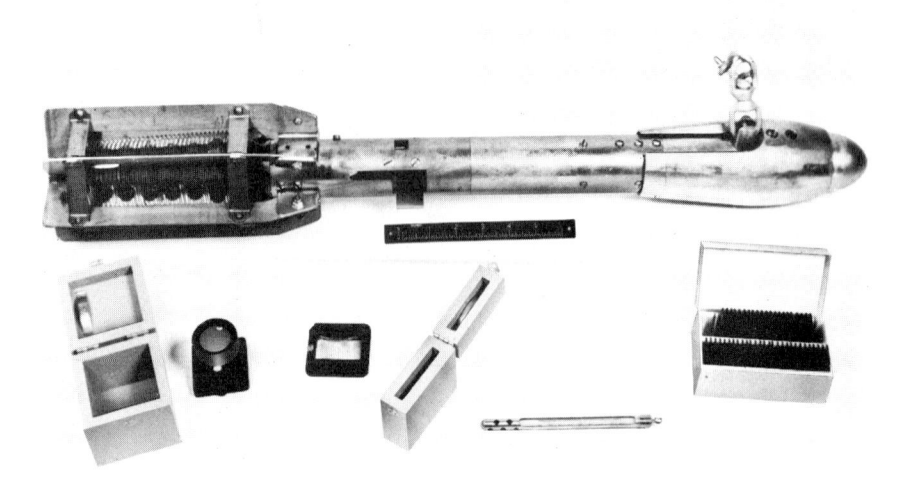

FIG. 12-9. Bathythermograph showing cable attachment swivel and additional nose piece which is added to obtain greater depth. In order, from lower left, are: box for viewing ocular, viewing ocular, calibration grid, calibration grid carrying case, thermometer, and box of smoked slides.

it is wholly contained within the body of the instrument. As an interesting bit of technical information, it should be mentioned that the smoke employed on the glass slide is obtained from burning skunk oil. Many types of deposit, such as the smoke from camphor or kerosene, were tried, but none were able to provide a smoke film that would not wash off in water, with the exception of the skunk oil.

The details of the operation of the BT are more clearly portrayed in Fig. 12-10. The pressure element that drives the depth axis of the record consists of a highly compliant metallic bellows loaded by a helical spring. The bellows is directly coupled to a piston head, which carries the smoke glass slide. In order to prevent rupture of the metallic bellows due to accidentally excessive pressure the bellows are partially filled with a suitable oil. The temperature-sensitive elements consist of several feet of thin-walled hollow tubing wound within the tail structure, and connected to a spiral Bourdon tube driving the stylus arm which traces the temperature displacement in accurate form on the smoke slide. The thin-walled copper tubing, as well as the Bourdon tube are entirely filled with xylene which possesses a rather high temperature coefficient of expansion. We thus have all of the elements for supplying the analog X–Y plot.

An actual record obtained from the BT is reproduced in Fig.12-11. Only the graphic trace itself is recorded on the glass slide and the coordinates must be applied by superposition of a calibration grid and viewed or projected

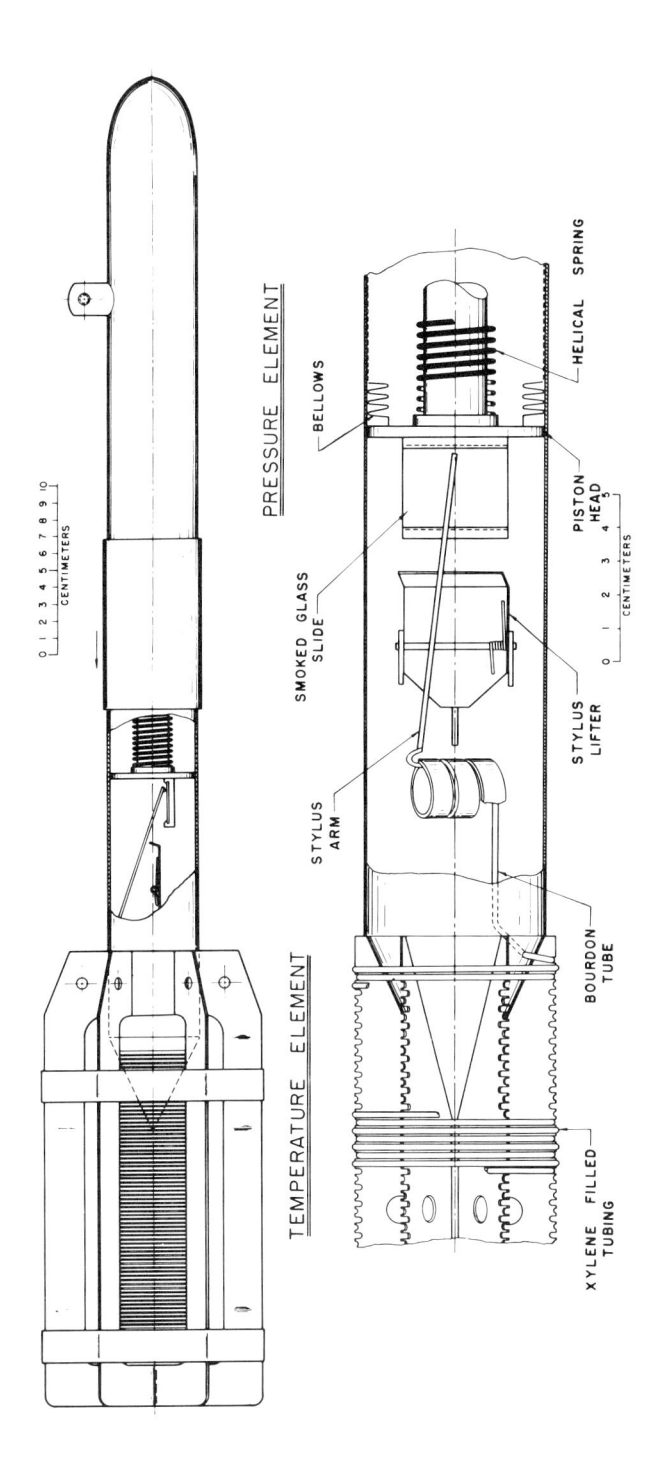

FIG. 12-10. Drawing of bathythermograph.

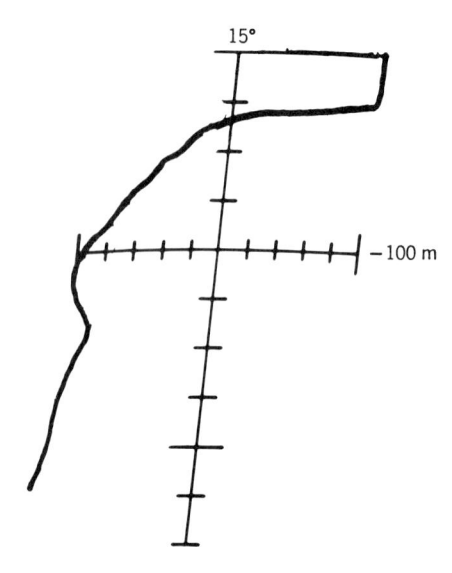

FIG. 12-11. Bathythermograph trace: Vertical scale, depth marked in 25-m increments. Temperature scale, marked in 1°C increments. Mixed layer surface to 25 m at which point thermocline begins. Note inversion layer at about 140 m deep. (Size of original may be determined by the fact that the temperature increments are approximately 1 mm apart.)

through a suitable optical system. For the purpose of this illustration, the artist has supplied depth and temperature coordinates. The horizontal scale is the temperature scale, each division corresponding to one degree centigrade; and the vertical scale is marked in 25-m increments. The size of the original may be estimated from the fact that the divisions on the temperature scale are approximately 1 mm apart. The advent of thin-metallic-film technology has made possible the major improvement in the optical and mechanical properties of the coating supplied to the glass slide. The thin metallic film has now displaced the skunk-oil smoke and lil' Abner's Skunk Works has "lost its contract" and may be considered a victim of technological obsolescence.

The bathythermograph over the years has perhaps been more extensively used than any other single oceanographic instrument. The BT was originally conceived and developed by Dr. Athelstan F. Spilhaus in 1937 (Spilhaus, 1938). The BT was improved early in World War II and has remained essentially unchanged for more than two decades. The BT has been a most prolific instrument from the standpoint of records, since literally millions of bathythermograms have been recorded.

Problems of the bathythermograph

Particularly in postwar years, there has been a significant degradation of the performance of the BT. Both temperature and depth errors have become more frequent. At one time the BT was severely troubled with hysteresis. It may be of interest to explain how one manufacturer "eliminated" hysteresis in the BT. An ingenious scheme was introduced to lift the stylus from the slide just as the BT began to be hauled back to the surface. Thus the hysteresis was demonstrably "eliminated."

The cost of the BT has steadily increased from approximately $100 in 1945 to over $530 a short time ago. Repairs also are very expensive. At the Scripps Institution of Oceanography about six years ago, it cost over $12,000 simply to repair and recalibrate the bathythermographs used in research programs.

There is a marked inability of the bathythermograph to hold its calibration. In any case the instrument at present is not sufficiently rugged to withstand the necessary rigor of the shipboard environment.

The present BT's are severely depth limited (Snodgrass, 1966). The depth limitations are of two types: The first is apparently due to a structural limitation in that if, for instance, a 900 ft BT is lowered to 900 ft, it suffers a significant error in depth calibration. For this reason, normally the 900 ft BT's are not used below 400–450 ft. The other type of depth limitation is due to the fact that for many applications it is desirable to go deeper than the 900-ft design depth.

In many ways the BT must be considered to be sea-state limited. This is because in heavy weather with decks awash it is often not safe to have personnel operate the winch and handle the BT on the open deck. This, of course, is a function not only of the sea state itself, but of the size of the vessel being employed.

The ship employing the BT is necessarily speed limited during the time of taking BT measurements. "Officially" the BT was designed to be used from ships at 15 knots. However, it is quite true that certain destroyer personnel during the war became reasonably proficient in operating from ships at even 20 knots. However, at any speed over 15 knots there is a distinct hazard, especially when high waves are present. Under high-speed operation, the BT has been known to wrap itself about the ship's funnel, behaving somewhat as a tethered lethal missile. Fortunately there is no known record of anyone being seriously injured under such conditions. However, the general difficulties outlined above, as well as in this section, have resulted in slowing ships at least to 6 knots and in the case of research vessels often stopping them in order to take BT's. The necessity to slow or stop vessels in order to take a BT tends to seriously complicate the problem on the basis of cost effectiveness since very real additional costs are involved for getting a ship back up to speed after slowing or stopping and also this adds to the total time required for the ship to perform its mission.

Due to the nature of the bathythermograph it is almost impossible to tell visually whether the instrument has suffered any damage that might tend to make the data inaccurate. In general difficulties of this sort can only be determined by disassembly or actual recalibration in the laboratory.

Development of the expendable bathythermograph (XBT)

As we have seen, the mechanical bathythermograph, though used over a quarter of a century, possesses many highly undesirable characteristics. In any case it appears obvious that there is great latitude for improvement.

Essentially, it appears evident that it would be very much worth expending considerable effort to solve the problem of expeditiously measuring temperature as a function of depth. Having reached this decision, it is well to establish some basic assumptions as guidelines:

1. The instrument should place no constraint on the ship's speed.
2. The parameters being measured, in this case temperature and depth should be available essentially instantaneously on board ship.
3. The instrument must be of small size for reasons of logistics, principally that of stowage, since large numbers of the instruments will undoubtedly be required and ships, either of the research category or military, are always chronically short of stowage space.
4. Requirement under 2 above inherently implies that a telemetry system of some type will be required.

Within the constraints developed in the above paragraph, let us proceed to examine various avenues of approach that might lead to likely solutions of the problem:

1. radio telemetry;
2. direct (dc) electrical field or ultralow frequencies (ulf);
3. optical telemetry;
4. acoustic telemetry;
5. "Hard-wire" telemetry.

Possibilities 1 and 2 face some serious problems with respect to the ocean itself. Had we set out to devise a medium which would be less useful for the transmission of electromagnetic energy, it is doubtful that we could have done better than nature herself in the creation of salt water (Snodgrass, 1958). Radio frequencies are so rapidly attenuated within the ocean that it becomes very difficult to endow the small device with adequate power to transmit useful energy of the required bandwith over the necessary distances (Baños and Wesley, 1953, 1954, 1966). Optical telemetry also is doomed to little success, due to the frequent excessive turbidity and absorption of optical energy to be encountered in the ocean. Here too, there would be problems of carrying adequate power sources within a very small device. Acoustic telemetry thus far remains a possible approach. However, even here there are problems that must be considered which relate to the propagation of sound within the ocean. In the upper layers of the ocean, sound rarely travels in a straight path since it is refracted by the changing vertical velocity structure. This problem becomes especially acute if there is a significant horizontal range involved. To some extent, the problem of the horizontal range could be minimized by towing hydrophones well astern. Unfortunately, even with the precaution of towed hydrophones, there are problems relating directly to the noise generated by high-speed ships, such as destroyers, that would be imposing severe problems of noise-limited operations. There are, however, possibilities of combining radio telemetry and acoustic telemetry so that a sinking temperature probe might transmit to a floating surface buoy, which would, in turn, receive the acoustic signal and retransmit it to ship or plane by means of radio. The major difficulties with this type of composite system are the high cost and difficult logistic problems. "Hard-wire" telemetry remains our sole hope. Development of hard-wire telemetry in the earlier phases of its development was plagued by the impossibility of obtaining fine wires which had insulation which was essentially pinhole free. Several years ago, "good" insulation, in the case of number 28 copper wire, permitted approximately six pinholes per thousand feet. The problem was not solved by simply running the wire through the insulation equipment a second time since the reason for the existence of the first pinhole was often due to dust or contamination on the surface of the wire and the consequent surface tension prevented the hole from being closed on the second pass through the insulating coating equipment. The sober fact was that statistically it was evident that the problem could not we solved with the techniques then being employed for wire coating There was also a great deal of reluctance on the part of naval personnel to support attempts at hard-wire telemetry since the general prevailing opinion was that the highly turbulent wake of the destroyer would tear apart the fine wires that appeared to be necessary. Fortunately, the insulation problems have been solved and the required fine wires are now available routinely. It is also indeed fortunate that the skepticism of naval personnel with regard to the hazards of the destroyer wakes proved to be unfounded. The availability of fine reliably insulated hard wire permits application of more or less conventional telemetry to the transmission of information from the sensing device to the ship. The designer now has a selection of single and multiple conductor leads with and without utilizing sea return circuits. Either FM or dc systems are applicable.

If a designer is interested in working out the details of a perhaps hypothetical expendable BT (XBT), what sort of a cost figure might be considered as being competitive with the existing mechanical BT? It was very difficult to obtain a figure that could be applied to the cost of taking a conventional BT slide and not even approximately valid data were available on this subject until about 1960–1961, at which time it was estimated that an unprocessed BT slide delivered on deck of the ship might cost approximatley $25. This does not take into consideration labor but only considers the original cost of the bathythermograph and its amortization, plus the maintenance of the required shore-based calibration and repair facilities and the prorated cost, etc., of the shipboard BT winch, cable, etc. (Miller-Columbian Reporting Service, 1961). In any case the figure of $25 served as a target below which it would be decidely advantageous for the production cost of an XBT. It was originally felt that, unless the cost was much lower than $25, the market would be somewhat dubious. However, it was expected that the volume consumption curve would tend to rise very steeply if the costs were very much below this figure. It also seemed possible that though the unit cost of taking an XBT record might be lower than the mechanical BT, if they were easy to use, a large consumer such as the U.S. Navy would not find the total annual cost of taking temperature data to be any less, since many more records would be taken.

Well over a dozen manufacturers were interested in the possibilities of developing an XBT. However, only three have succeeded in developing what might be considered a successful instrument. The companies are General Motors–Packard Electric,[1] Hytech Marine Products,[2] and Sippican Coporation.[3] The systems developed by all three companies have certain elements in common. First, all are designed to pay out wire both from the ship and the rapidly sinking plummet. Second, all obtain depth of sensor by measuring time since sinking rate was determined to be quite dependable and uniform. Third, thermistors were used as the temperature-sensing circuit element. Fourth, power to actuate the sensor, etc. is supplied from the shipboard portion of the electronic system. FM telemetry systems are utilized by General Motors and Hytech, whereas Sippican employs a dc-powered telemetry technique [Knopf, Cook (1965)]. Both General Motors

---

[1] *General Motors Corporation*, Defense Systems, Inc., Santa Barbara Laboratories, Box T, Santa Barbara, California

[2] *Sippican Corporation*, P.O. Box 87, Marion, Massachusetts 02738

[3] Bissett-Berman Corporation, Hytech Marine Products, "G" Street Pier, San Diego, California 92101

and Sippican have succeeded in obtaining high reliability with ship speeds from 0–30 knots.

Operational costs and logistics

In order to appreciate the significance and utility, as well as the economy of the XBT, it is desirable to consider the situation applying to two different-speed task forces employing destroyer screens. We will assume that the mechanical BT is being used and that the speed of the first task force is 15 knots and the second task force is 25 knots. (Note: The destroyer must slow to 6 knots and take a mechanical BT, then catch up to the advancing task force.) See Table 12-1.

TABLE 12-1

	15 Knots	25 Knots
Fuel oil	$18	$ 85
BT Cost	$25	$ 25
	$43	$110

The fuel-oil cost, as mentioned in the table, is simply for fuel oil that is required over and above that which would have been used for steady-state running of the destroyer. In other words this is the additional consumption of fuel oil utilized to accelerate the destroyer back to speed and to maintain an adequate speed to overhaul the task force. Furthermore, the costs are figured on the basis of the cheapest fuel oil, namely Bunker C No. 6. It is thus evident that the cost of the bathythermograph record itself may be the smaller part of the operational expense. However, there are some rather serious aspects to this type of operation. For example, in the case of the 25-knot task force, the destroyer will have to run for approximately 5 hr in order to catch up. Furthermore, during this period, from the moment the destroyer is detached to take the BT until it succeeds in catching up, it is not available to perform its primary task of screening the task force. There has been no attempt to estimate what this period of unavailability costs. As though this situation were not bad enough, it is further downgraded because of the fact that the actual BT data obtained is, in a sense, "ancient" since it will have been obtained many miles behind the position of the task force. It should be quite obvious that, if the destroyer had been using an XBT it would not have been necessary for it to leave station in the task force screen and the data which it would obtain would be current both as to place and time (Snodgrass, 1966).

It must be admitted that there has been an unexpected bonus obtained as a byproduct of the XBT development. The bonus is represented by the increased accuracy of both the temperature and depth measurements of the XBT in contrast to the conventional mechanical BT (MBT). Temperature and depth measurements have been improved by roughly a factor of 3 in accuracy and depth resolution by a factor of 5–10. The depth range, even at speeds of 30 knots for the XBT, has gone from 900 feet (MBT at 6-knot ship speed) to 1500 feet. Research models of the XBT have been successfully employed to depths of 6000 feet (Arthur D. Little, Inc., 1966).

It appears to be a significant fact that the electrical and mechanical problems of the XBT are, for all practical purposes, solved. However, there are problems remaining that were present even in the early stages of the XBT development and these are of somewhat more subtle nature. For instance, they essentially involve people and psychological reactions to the philosophy and concepts represented by the XBT. To properly appreciate the nature of these problems, it should be realized that since the reliability of the MBT is, at best, dubious, such successes as were obtained were dependent on careful handling and careful calibration, with calibration being affected as often as practicable. With this background it is easy to understand the reluctance of the oceanographer to trust an instrument which he does not calibrate nor for that matter is the instrument susceptible to calibration after it leaves the factory. The facts are that modern electronic production techniques involving careful application of quality control in reliability engineering have made it possible to assign a confidence figure to the performance of each XBT, which is something that we have never been able to do in the past. Even in the event that a single XBT is used and a temperature idiosyncrasy appears in the trace, we now have an extremely high probability that the record is valid. However, should this be questioned, it is simple to use an additional XBT for confirmation (Snodgrass, 1966).

A practical XBT

The XBT terminology in fact refers to a complete system, only a portion of which is, strictly speaking, of the expendable, or throw-away type. However, since the latter portion of the system is, essentially, the key to the success of the system as a whole and further since it, in effect, incorporates essential engineering features in which we are specifically interested, we will confine our consideration to this portion. Partly because more information is available on the model of XBT developed by the Sippican Corporation, as well as the fact that it represents a very fine bit of engineering, we will focus our attention on the Sippican XBT design. After what might well be considered an inordinate amount of much hard work, plus unusually thorough testing (at least as far as oceanographic instruments are concerned), dictated by strict quality control, a truly fine system, especially the sensor and related package, has been developed. Additional attention should be given to the paying out of wire because this is in essence, the fundamental principle of operation. Historically, there is the background development of the wire-guided torpedo and from the submarine. Although a relatively delicate wire is employed for the electric control link, it nevertheless possesses high reliability and rarely breaks. The basic principle involved is the paying out of wire under the slightest tension developed by the relative motion either between the torpedo and the water or between the submarine and the water. Therefore the wire itself is not pulled through the water and, as a consequence, the forces that the wire is required to withstand are minimal. It is essentially the wire payout principles utilized in the wire-guided torpedo that are applied so successfully to the XBT. In the case of the XBT two separate spools of wire are joined by a continuous length of wire (no splices), with one spool remaining on shipboard paying out wire as the ship advances and the second spool housed within the body of the sensor plummet package. The second spool pays out wire freely permitting very nearly freefall conditions for the plummet. However, the destroyer under way at high speeds represents a condition that applies little, if at all, to the case of the wire-guided torpedo. This is the existence of large numbers of vorticity cells within the destroyer wake. The forces on the wire, due to these vorticity cells exert considerable localized stress on the wire which it must successfully withstand. Figure 12-12 displays the elements of the primary sensing and telemetering portions of the XBT system just described.

The manner in which the components are assembled within the canister package is quite evident in the cutaway canister shown in Fig. 12-13. The safety pin is shown inserted through holes in the tail fin of the plummet.

An enlarged view of the plummet is illustrated in Fig. 12-14. The significant hydrodynamic properties of the plummet, which pertain to the sinking rate, are exceedingly sensitive to minuscule variations in the shape and symmetry of the tail fins. Therefore manufacturing tolerances are surprisingly tight.

The XBT may be launched by means of very simple equipment. Figure 12-15 shows an early design of a launching tube lashed temporarily to a ship rail stanchion. The "breech loading" inboard portion is clearly visible. Recent sea trials have demonstrated that the launching tube need not be mounted on the open deck but may be mounted in a suitable location below the main deck with the lower end only piercing the ships hull above the water line.

Insofar as the XBT has made a most successful beginning from the stand-

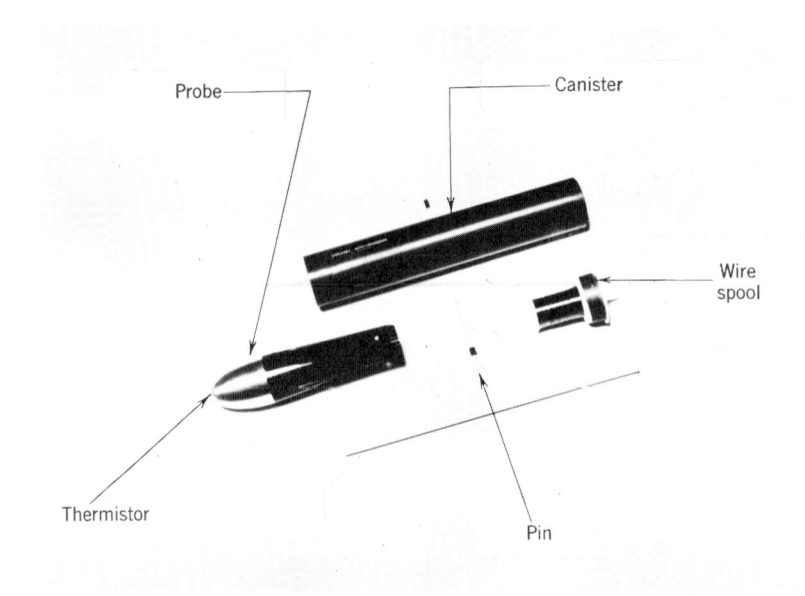

FIG. 12-12. Disposable portion of XBT system. Exploded view showing probe, plummet, and location of thermistor sensor. The wire-spool unit is attached to the canister and remains with it on deck. Connecting pins are on the right (top) of the wire spool, which make connection to a suitable jack fitting in the breech of the launching tube.

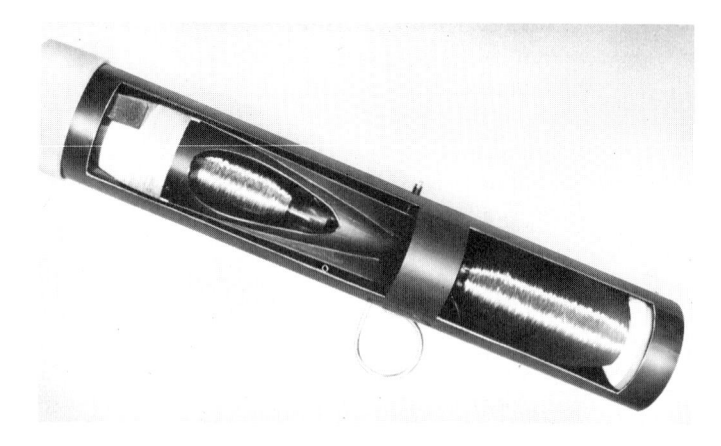

FIG. 12-13. Cutaway view of canister with assembled plummet and shipboard wire spool. The safety pin is shown inserted through holes in the tail fin of the plummet. For clarity, the wire spool in the plummet has been revealed by a cutaway section.

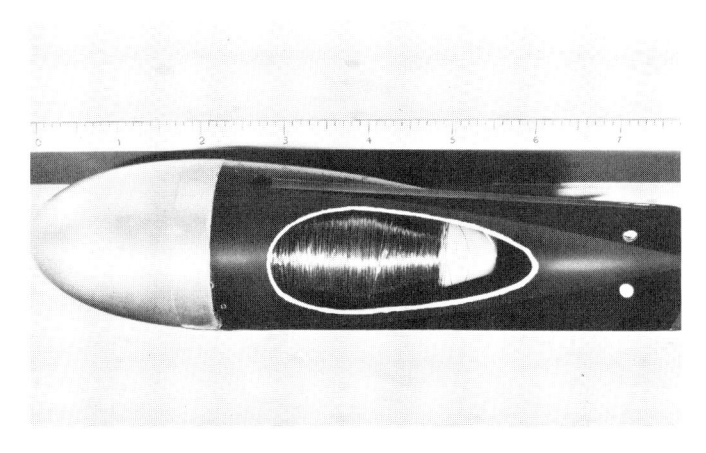

FIG. 12-14. Probe or plummet, the rapidly sinking component with temperature sensor of the XBT system. The sinking rate is approximately 7 m/sec. The length of the probe is 8in. and the diameter 2 in.

point of attacking the problem of extracting data from the oceans, it would seem advisable to examine future extensions of this concept. The major engineering problems of telemetering data from rapidly sinking sensor packages to ships under way, even at high speeds, may be considered essentially solved. Therefore we may now examine some of the next steps to take, if other parameters are to be measured. Unfortunately the guidelines

FIG. 12-15. Early developmental model of XBT probe launching tube. Launching tube is clamped temporarily to ship rail stanchion. "Breech loading" assembly on upper right of tube.

outlined under "nighttime water transparency measurements" now become slightly less definite, because no other parameter that is normally measured approaches the near universality of temperature. However, other desirable parameters which we might consider are oxygen, salinity (conductivity), light, and sound velocity.

Microelectronics appears to offer one of the best means of continuing the development of expendable instruments. In the past, oceanographic instruments employing electronic circuits have rarely been able to utilize the best in electronic circuitry either due to limitations of cost, bulk, or size of components. For the purpose of considering expendable types of instruments, it would be anticipated that the quantities would be relatively large and therefore a broad base would exist for amortizing the cost. It would also seem that this would offer a means of optimizing circuit performance since little difference would be reflected in unit costs, even if a dozen active elements were employed in place of three or four.

In the case of microelectronics it is perhaps interesting to examine the various tradeoffs employed between programs involving outer space and inner space. In the case of outer space, it has often been quite justifiable to pay high unit costs for very small numbers of microelectronic assemblies since reliability, small size, and precision were required and this appeared to be the best way to accomplish the purpose. However, in the case of inner space, the same requirements exist but it would appear that, since large quantities are involved, the basic cost would be spread over the large base and we should obtain fairly reasonable unit costs. These requirements, of course, are also basically compatible with logistics of stowage of large numbers of units on shipboard, etc., as well as making it possible to perform multisensing functions within a relatively small space.

12.5 THE OCEAN AS AN ELECTRICAL FILTER

In Section 12-4 we referred to the fact that a relatively simple electrode and amplifier system can pick up a wide range of electrical noises from the sea. The attenuation of electromagnetic energy in the ocean is highly frequency dependent. Baños (1966) has published an exceedingly complete treatment of the attenuation of electromagnetic energy in seawater. L. Liebermann (1962) disposes of the problem of electromagnetic attenuation in seawater very deftly. In spite of the rather exhaustive and thoroughgoing theoretical works on the subject of electromagnetic attenuation in seawater, as Liebermann points out, there persists a great reluctance to "believe" the theoretical studies and repeatedly we hear that someone has found a "window" in the ocean that is relatively transparent to electromagnetic radiation. Model experiments that have been carefully made confirm theoretical figures. However, as mentioned previously, even the model tests have not been accepted by some of the confirmed skeptics. It seemed advisable several years ago, in view of this chronic skepticism, to carry out a careful series of full-scale experiments that, in any case, would not be capable of being criticized on the basis of the model experiments. Such a series was carried out by Snodgrass and Rudnick (1954) which confirmed the theoretical figures, etc. as developed by Baños. In spite of the rather high attenuation of electromagnetic energy in the sea its frequency-dependent nature was permitted some significant applications. The British were especially active during the period of World War II and immediately thereafter in the development of what has been called "leader cable." Leader cable consists of an insulated electric conductor laid on the bottom in relatively shallow water to assist in guiding ships (Brock-Nannestad, 1964). To accomplish this, a relatively low frequency of between, for instance, 50 and 400 cps is impressed on the electromagnetic conductor and the ocean. The ship then, if it is equipped with the proper type of pickup coils, can maneuver on the basis of signals intercepted so that it can position itself while under way directly over the cable. During the latter phases of the German submarine campaign in World War II, in which German submarines were based in Brest, France, leader cable was installed and the submarines were able to get in and out of Brest simply by running submerged over the desired cable.

12.6 THE BALL BREAKER: A DEEP-WATER BOTTOM SIGNALLING DEVICE

Isaacs and Maxwell (1952) have developed an exceedingly interesting and useful device for telling the shipboard winch operator when an instrument reaches bottom in deep water. Before the advent of telemetering techniques involving acoustic systems or cables containing electric conductors it was difficult, if not impossible, to tell when an instrument lowered through several thousand meters of water actually reached bottom. We might assume that a suitable dynamometer connected into the cable would be quite capable of giving an index of arrival of the instrument at the bottom. However, this is by no means easy since, even with an instrument weighing several hundred pounds, the dynamometer may range from tensions of several tons to zero under certain conditions of sea and swell. Assuming that we had adequate time to look at a graphic recording of the dynamometer record, it is quite possible that relatively reliable determination of instrument arrival on the bottom could be worked out. However, it is imperative that fairly immediate information be obtained in order to avoid paying out excess cable, which would almost certainly tangle or kink at the very least, and, at

428 Systems Design—The Technology

the worst, break, causing the loss of the instrument. The ball breaker is connected in series, with the instrument lowered in such a manner that the slackening of tension when the instrument reaches the bottom causes a fairly heavy weight to fall, which, by means of a specially designed tip, causes the violent rupture of a small glass sphere. The glass sphere is approximately 3in. in diameter and is filled with air, simply at atmospheric pressure. Therefore as it is lowered through the water, since it is hermetically sealed, it represents an increase in potential energy due to the increasing hydrostatic head. At great depth, this represents a relatively large amount of potential energy and the sudden rupture and subsequent violent implosion initiates a very strong acoustic signal. This signal may be rather easily picked up on shipboard by means of either the ships existing fathometer transducer or by means of a special hydrophone lowered over the side of the ship. The clean acoustic signal produced by this implosion is shown in Fig. 12-16. The signal, once heard, is clearly distinctive and not

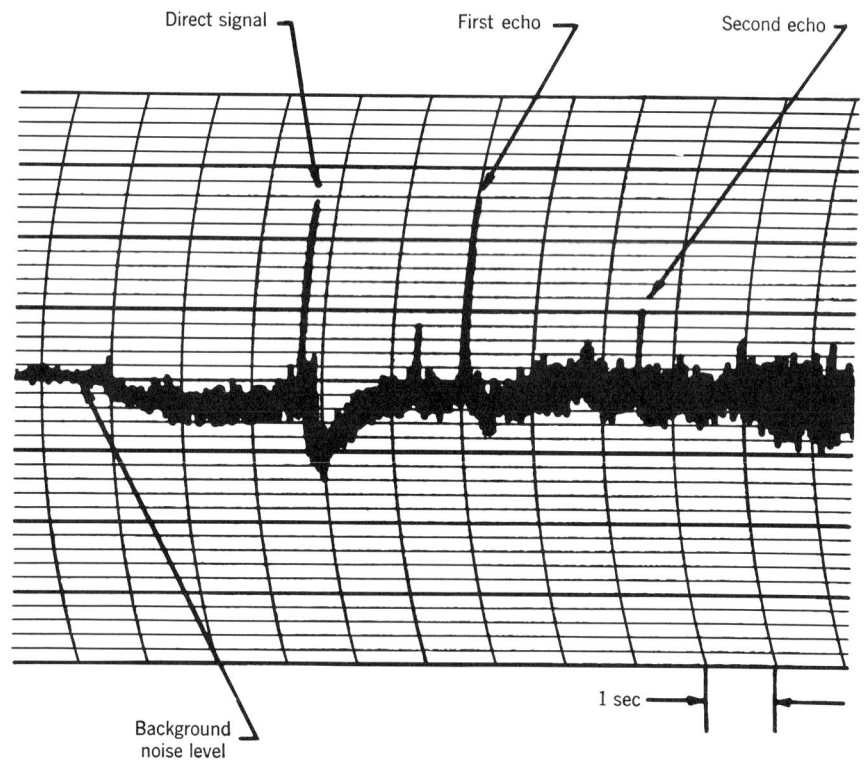

FIG. 12-16. Implosion of glass sphere on bottom at a depth of 1000 fathoms. [From J. D. Isaacs and A. E. Maxwell, "The Ball-Breaker, A Deep Water Bottom Signalling Device," *Sears Foundation J. Marine Res.* 66 (1952).]

likely to be confused with other miscellaneous acoustic noises. The ball breaker itself is illustrated in Fig. 12-17. Much more is not known of the physical properties of glass in compression when subjected to high hydrostatic pressures. It was originally thought, that it would be very easy to cause the glass sphere to rupture when it was subjected to high hydrostatic pressures. On the initial tests of the ball breaker, a hardened steel wedge was fastened to the lower portion of the 60-lb falling weight. Statistically it was found that only about 20% of the trials resulted in implosion of the glass sphere. In an attempt to improve the reliability of the instrument the steel wedge was serrated again hardened. The result, however, was little different than the original 20% success. The next test was made with a conical hardened steel point. During the test no sound was heard and when the ball breaker was hauled back to the surface it returned in the tripped position; the point had penetrated about half of its length into the glass sphere and the glass sphere was about $\frac{1}{5}$ full of water. It was evident that the glass had not been brittle under these conditions and had actually been able to seal itself about the conical tip. The solution leading to a reliable means of rupturing the glass is illustrated

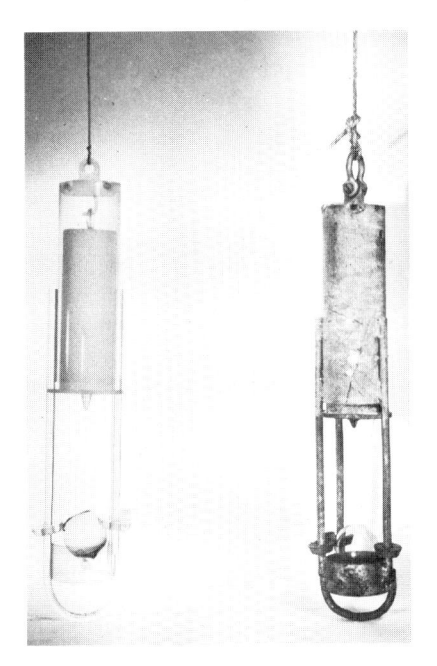

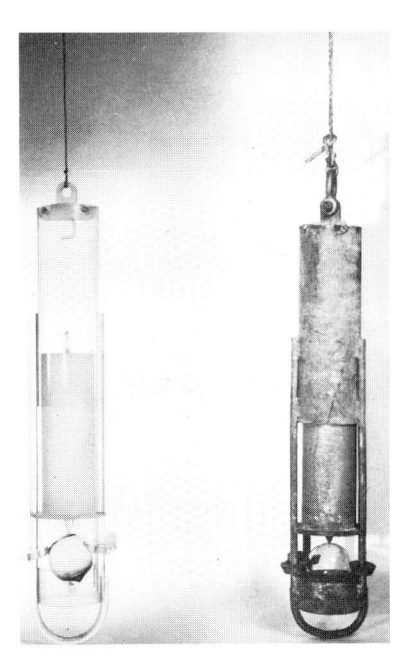

FIG. 12-17. Ball-breaker mechanism. The left pair illustrates the cocked position and the right pair demonstrate the tripped position. Of the two ball-breakers shown, the one is plastic so that the internal construction is clearly visible, whereas the right one of each pair is the seagoing model.

in Fig. 12-18. The basic secret of success was to provide a means of punching through the glass but not supplying a basis for support of the glass. The simplest configuration supplying this need was a fluted configuration of the pointer tip. The implosions are often accompanied by such a violent release of energy that the tips, though very rugged, are shattered. It was found that the standard machine countersink, which is available at the local hardware store at only a relatively small cost, performed every bit as well as the very expensive, specially ground and hardened tips. The ball breaker exploits a very interesting situation in that the deeper the water it is used in the more the potential energy available for conversion into sound. This additional energy apparently comes very close to completely compensating for the increased attenuation of the sound, due to the increased distance. Therefore the signals heard from the imploding glass sphere, at least to a first approximation, all seem to have the same received acoustic intensity independent of depth.

12.7 BOTTOM SEDIMENT TEMPERATURE-GRADIENT RECORDER [Revelle and Maxwell (1952) and Von Herzen et al. (1962)]

The temperature-gradient recorder was developed to measure the *in situ* temperature gradients in the sediment of the ocean bottom. The problem that was being studied was the heat flux through the bottom sediments. In order to do this it is necessary to measure both the *in situ* temperature gradient and the thermal conductivity of the sediments. The instrument that was developed occupies at least a small niche in the history of oceanographic instruments, specifically oceanographic electronic instruments. The instrument is a complete, self-contained, null type, self-balancing potentio-

FIG. 12-18. Detail of fluted tip for rupturing glass. [From J. D. Isaacs and A. E. Maxwell, "The Ball-Breaker, A Deep Water Bottom Signalling Device," by *Sears Foundation J. Marine Res.* 65 (1952).]

metric recorder. The instrument's success on the Scripps Institution of Oceanograph's MIDPAC Expedition in 1950 served to spawn an entire new generation of electronic precision oceanographic measuring instruments. The instrument was developed in an environment of skepticism and criticism of electronics, such as existed in the early part of the last decade, as previously discussed. The instrument to be described, as well as two others, were built from scratch for a definite ship sailing date deadline within a period of six weeks. The three instruments all operated on different bases and each instrument performed as designed. Since a severe time limitation existed both for design and construction, it was necessary to resort to off-the-shelf components and subassemblies where available. In the light of modern technology it is sometimes difficult to appreciate the state of the art in oceanography over 15 years ago. Nevertheless, there are still features of this instrument of interest at the present time. Since it would occupy an inordinate amount of space to go into a detailed discussion of the design, we will confine ourselves to obtaining a general understanding of the instrument and its function. In view of the skepticism that existed in 1950 when the temperature-gradient recorder was first used, it is perhaps not at all surprising that when temperature-gradient values approximately two times greater than those which were anticipated were recorded, there was general disbelief and an unwillingness to accept the values. It was not until over a year later, that another, almost identical instrument was being used in the same area in the vicinity of the Marshall Islands, that an identical high-temperature gradient was recorded and the instrument's earlier results were than credited as being valid.

Probe schematic

Figure 12-19 presents a simplified sketch of the thermal probe that punches into the sediments. The upper and lower pairs of thermistors designated by the letters UU and LL, respectively, are maintained in good thermal contact with the tube wall by means of a compression spring. The remaining portions of the hollow tube are sealed off from sea pressure by means of O rings and packed with absorbent cotton and filled with oil. The upper and lower sensing pairs are spaced roughly 2m. apart. The pressure-protected environment was selected for the thermistors because it was not known how they would be affected by the high hydrostatic pressure. A footnote in the Western Electric catalogue of the period stated, "May be pressure sensitive", on the basis of which it seemed prudent to avoid any deliberate error which might be introduced into the measurements were the thermistors subjected to the ocean pressure. However, it is of interest to know that somewhat over a year later, while visiting the Bell Telephone Laboratories in New York, it was learned that the footnote referred to was

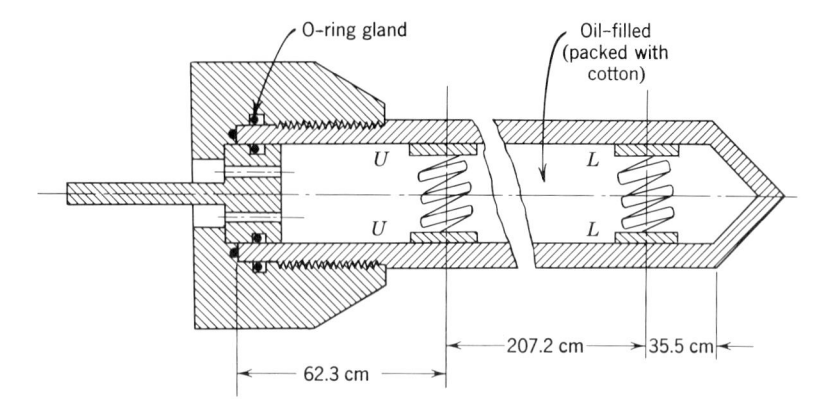

FIG. 12-19. Simplified sketch of probe. *UU* and *LL* represent the upper and lower pairs of temperature sensitive thermistors.

inserted simply because it had been observed that if the thermistors were clamped in a vise and the vise tightened "an effect could be observed." Insofar as the type of thermistor elements being employed were subsequently found to be only negligibly affected by high hydrostatic pressures, more recent probes have not been pressure protected.

Circuit schematic diagram

The circuitry employed in the temperature-gradient recorder is straightforward and essentially conventional. The simplified diagram is shown in Fig. 12-20, where the circuit functional elements are easily identified. The upper and lower pairs of temperature-sensitive thermistors identified by UU and LL are connected as four active elements in full Wheatstone bridge. The bridge operates on dc and the null, self-balancing potentiometer is driven by the dc servomotor to null the bridge output voltage. The unbalanced voltage between the bridge and the potentiometer is presented to a mechanical vibrator inverter and the resulting ac is then amplified by a hearing-aid amplifier to drive a phase-sensitive detector which powers the dc servomotor. The recording stylus is mounted rigidly to the same mechanical assembly that holds the potentiometer balancing slider.

Electronic chassis (early model)

The electronic chassis of the temperature-gradient recorder has a quite straightforward construction. Figure 12-21 illustrates the disposition of the larger electronic components. The construction details going from right to left are as follows: circuit test jacks for final check out after instrument

Instrumentation and Communications 433

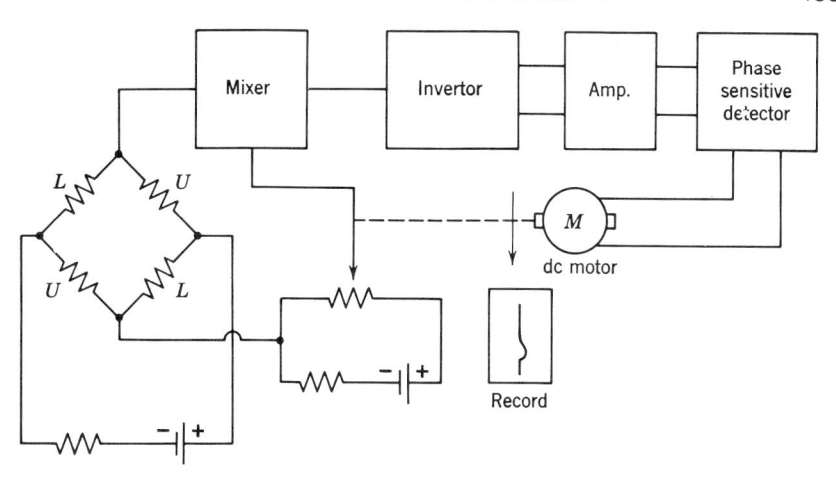

FIG. 12-20. Circuit schematic of heat-flow instrument.

is inserted in pressure-protective case; dc powered time-delay switch, adjustable from 15–120 min.; main on-off switch and range-calibrating adjustments; the second compartment comprises low-voltage batteries for dc time delay and hearing-aid type of amplifier shown on horizontal top desk; note microswitch driven by cam on chart paper drive which is mounted on bulkhead under amplifier; the third section comprises stripchart recorder with slide wire, slider, and recording stylus; the fourth compartment contains the dc servo-operated balancing motor with batteries; the fifth compartment and rear of the fourth bulkhead houses the

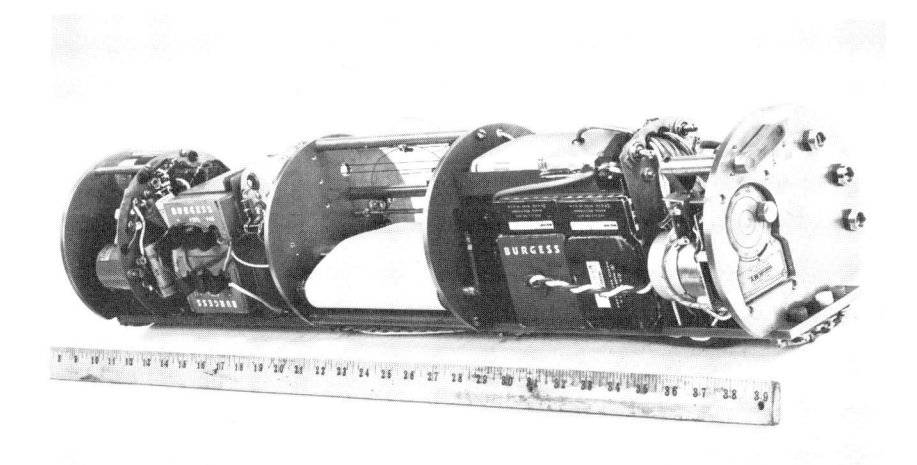

FIG. 12-21. Electronic chassis of bottom sediment temperature-gradient recorder.

phase sensitive detector. The time-delay system was necessary in order to conserve batteries during the lowering of the instrument to the bottom. The timer was set so that the instrument would come on while the probe was held stationary in isothermal water a few meters off the bottom for a calibration check point on recorder. The microswitch referred to, which was mounted on the rear of the second compartment bulkhead, applies a calibrating signal with every revolution of the strip-chart paper drive. This was later found to be unnecessary and was discontinued. The strip chart itself is an electrically sensitive paper known on the market under the proprietary name of Teledeltos Paper and is manufactured by the Western Union Telegraph Company. The potentiometer slider and strip-chart recording stylus are driven by means of a silk-thread transmission system from the dc balancing servo motor. There was no effort to incorporate limit switches in this model because it was found that the motor pulley would conveniently slip at torques which were well within the safe range of the motor gearing. It is apparent that even with the slippage, there was no differential action between the recorder balancing slider and the stylus, since they were mounted on the same movable portion of the system. The temperature range of the instrument as normally employed was approximately 0.5°C across the 6 in. width of the strip chart. The sensitivity was such that it responded to differences of less than 0.002°C. As previously mentioned, the amplification was supplied by means of a conventional vacuum-tube type of hearing-aid amplifier. The reason for selecting the amplifier was due to the fact that the performance of these amplifiers was rather well known and they possessed a high degree of reliability. The gain was more than ample for the task and in general they performed very creditably.

Electronic chassis (newer model)

The normal evolutionary processes in instrument development are quite evident if we compare Fig. 12-21, the first of the temperature-gradient recorders, with a somewhat later model illustrated in Fig. 12-22. Electrically and functionally the instruments are very nearly identical. In some ways the design had been improved but at least in one significant aspect it may be considered slightly inferior. The newer model is easily distinguished by the fact that now plug-in amplifiers and potted circuits are much in evidence. The balancing slide wire in the older model which was quite literally, both in the electrical and physical sense, a linear slide wire has been replaced in the newer version with a helical ten-turn slide wire and the recording stylus is driven by a fine thread lead screw. This "innovation" is substantially more expensive than the earlier model and mechanically con-

Instrumentation and Communications 435

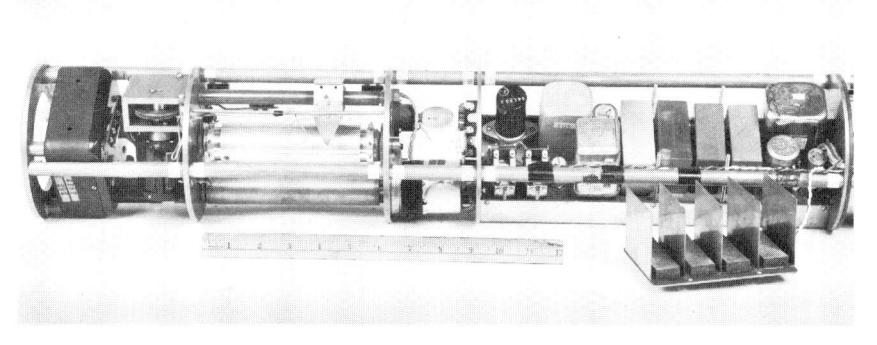

FIG. 12-22. Temperature-gradient recorder (new). Note especially the center section with the flashbulb angle recorder as explained in the related text.

siderably more involved. A most interesting feature was installed on both instruments but was not visible in the photograph shown of the older device and may be seen in the exact center section of the photograph (Fig. 12-22). The device to which I wish to call to your attention looks very much like a small lamp bulb but is, in reality, an M2 photoflash bulb. The purpose of this device is to serve as an angle recorder. The problem is that should the probe as it enters the bottom sediments depart significantly from vertical, the fact would need to be known since a cosine correction term would need to be applied. Though it is not shown in the photograph, in use the flash bulb is dipped into a mixture of beeswax and paraffin, to which has been added a substantial amount of lampblack. The angle recorder is activated the moment the flashbulb is fired. A substantial amount of the heat released from the flash bulb is absorbed by the lampblack in the beeswax, which assists the beeswax-paraffin mixture to melt and run to the lowest point of the bulb, at which position it "freezes" into a drop. By this technique, it is possible to measure the angle of orientation of the probe within $\pm 3°$. In practice, the angle indicator is fired by means of a microswitch which is activated by the stylus carriage as it moves across the strip chart from the position corresponding to the zero temperature differential. The angle recorder just described operates with an exceedingly high reliability and represents the survivor of many aborted alternative approaches.

Closing instrument case in tropics

As mentioned earlier under "Instrument classifications" precautions must be taken to avoid electronic instrument damage due to condensation of moisture. Figure 12-23 shows a scientist bolting up the pressure–protected instrument case, prior to lowering for a measurement. It is evident from the

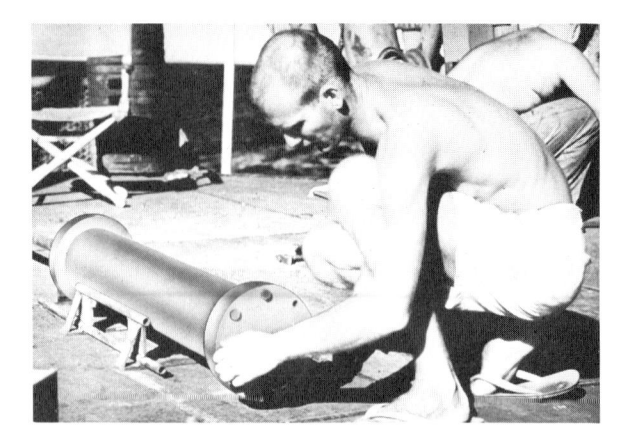

FIG. 12-23. Closing the pressure-protected case in a tropical environment. The instrument case has just been purged with anhydrous nitrogen to remove all significant traces of atmospheric moisture.

figure that the locale is definitely tropical. Unfortunately the photographer, lacking directions to the contrary, cropped the illustration, omitting what was intended to be its most significant point; namely, the presence of a hose for purging the instrument case with anhydrous nitrogen before tightening the final seal. In order to facilitate the instrument purging with anhydrous nitrogen a tube fabricated as part of the chassis leads the entire length, and it is this tube to which the purging hose is connected. With precautions of this type no visible moisture will condense either on the case wall or on the chassis. It is most important that the purging operation become a definite part of the instrument preparatory ritual.

Figure 12-24 illustrates one of the techniques employed in putting relatively heavy instruments into the water. The technician on the left of the photograph is holding a snubbing line in his left hand in order to control and restrain the motion of the instrument. Clearly visible, attached above the temperature gradient recorder case, is one of the ball–breaker mechanisms referred to previously.

Record obtained by temperature gradient recorder

A sample of a characteristic record made by the bottom sediment temperature gradient recorder is illustrated in Fig. 12-25. The scale shown in original dimensions as 30 cm also corresponds to 30 min, since the chart moved 1 cm/min. The lower trace on the record identified as "recorded reference line" was placed on the original record by a fixed–position stylus and serves as a point of reference, should the paper feed accidentally permit

Instrumentation and Communications 437

FIG. 12-24. Bottom sediment temperature-gradient recorder suspended over ship stern preparatory to lowering. Note ball breaker attached to instrument case.

the strip chart to move laterally. The next tracing designated at its beginning "zero temperature difference" is the temperature gradient as recorded by the instrument. As mentioned under "Expendable instruments" the zero reference mark was obtained by permitting the instrument to hang freely a few meters above the bottom in what was assumed to be reasonably isothermal water. The rapid rise, shown in the record, is due to the liberation

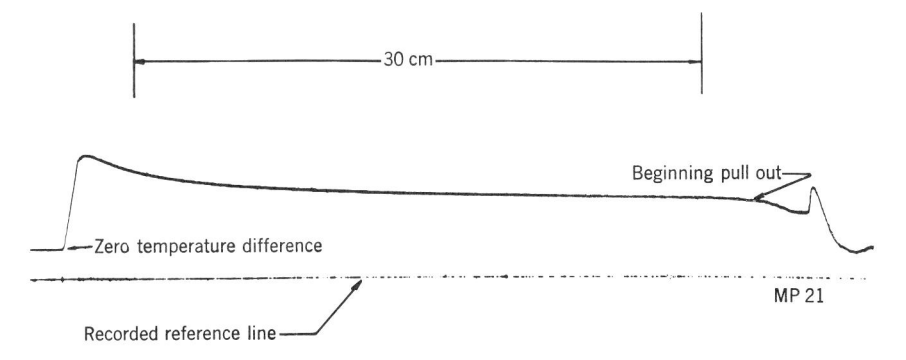

FIG. 12-25. Experimental record obtained with thermal gradient probe. Time scale 30 c equal 30 min. See text for details.

of heat by the probe as it penetrates rapidly into the bottom sediments. Since the lower pair of temperature sensors located nearest the tip of the probe moved the farthest, they also experienced the largest amount of heat developed by friction as the probe moves into the sediment. By examining the record, it is quite evident that the heat generated by the friction of the probe is significantly above the undisturbed thermal gradient and that it is necessary for this excess heat to "leak" of into the sediment for several minutes, during which time the record begins to approach the original undisturbed geothermal gradient asymptotically. The record is especially interesting at the moment when tension is again applied to the steel cable to pull the probe out of the sediments. At the point on the curve designated "beginning pullout," it is evident that the upper pair of temperature sensing thermistors in the probe begin to sense an increase in temperature, which is due to the motion of the upper portion of the probe without a corresponding displacement of the lower portion of the probe. This may be explained by the assumption that the initial tension imparted to the probe results in a pull toward the side which results first in bending the probe and then where the rapid rise in temperature also is shown, the probe as a whole begins to move and the lower pair of temperature sensors again experience rapid heating due to the friction of the probe in the sediment. An essential feature of the instrument design is unfortunately lost in the photographic reproduction. The original temperature tracing inscribed by the recording stylus is no means as broad as shown in the photographic copy. In actual fact, the apparent width of the line is due to the almost constant motion of the stylus back and forth in a small amplitude technically known as "dithering." As a first reaction to the statement that the servo system driving the potentiometer balancing contact dithered, we might assume that the servo system was poorly designed. However, it should be pointed out that, at least in this case, this was a design requirement. It was originally anticipated that a record roughly similar to the one being discussed would be obtained. Therefore, since the temperature would be falling approaching a final asymptotic value, this would mean that should the servo system fail to function and simply trace a straight line, we would have no way of knowing whether this was the final asymptotic value or whether the servo system simply failed to function. Therefore, in designing the servo system to dither, we were sure that if the trace showed the dither, the servo system was operating satisfactorily and such proved to be the case. The dithering of the servo system was insured by designing the system with a short time constant and an excess of sensitivity in the servo loop circuit. A substantial amount of theoretical work has been done to investigate the effects on the undisturbed geothermal gradient of the presence of the metallic probe (Ferris, 1953). The essential conclusion that may be drawn from the theoretical work is that the axial perturbations are negligible in spite of the thermal conductivity of the steel probe. It has also been shown that the radial disturbance, though significant, is brief, and that after only a few minutes there is no significant remaining error.

12.8 BUREAU OF SHIPS TELEMETERING CURRENT METER

The name that has been used to identify the ocean current meter in some respects characterizes its time of conception. Though the Bureau of Ships had long since been superseded by the U.S. Naval Ship Systems Command, for convenience, the older title is used in this text. Some of the specifications placed on the design requirements are as follows: the instrument should be designed for telemetering the observed current data and transmitting it for immediate readout on deck; both current direction and instrument depth are also to be telemetered; the output should be reasonably linear; high sensitivity is required, as well as an ability to respond to extremely weak currents; the instrument should be designed to ignore vertical components of current since it is desired to use the instrument for plotting vertical current profiles; if electric current is required for the operation of the underwater portion, it should be obtained from on deck. The Bureau of Ships' current meter was designed and constructed early in 1957. The selection of a current sensor was somewhat difficult but after preliminary investigations and tests, a mechanical sensor was selected. The original requirement for the Bureau of Ships was simply for visual indicating instrument readout on deck and not for recording. However, it seemed prudent to provide for making an analog record while the instrument was being raised and lowered so that current direction and magnitude could be plotted as a function of depth. Subsequent history validates the latter assumption.

Mechanical ocean current sensor

Having decided on a mechanical current sensor, there remained two essential choices. There are rotors that rotate about a horizontally disposed axis. The most common of this type are propellers. The propeller type was discarded since it normally requires a means of insuring its being properly aligned with the current to be measured and is quite sensitive to off-axis error. The other type of rotor is that which is designed to rotate about a vertical axis. Examining the vertical rotor, there appear to be two major types: the first known technically as Rauschelbach rotors (Stommel, 1954) or more commonly "S-rotor"; another rotor operating on a vertical axis is the Savonius rotor (Savonius, 1931). Examining the characteristics of these two rotors further: The Rauschelbach or S-rotor has a serious limitation in that it tends to stall quite easily and once stalled cannot be restarted by the force of the current alone. Also, while it does rotate, it rotates with a highly nonuniform angular rate. Stommel of Woods Hole Oceanographic Institution (Von Herzen et al, 1962) has, nevertheless, succeeded in utilizing the S-rotor to measure ocean currents when it is suspended beneath a small buoy. It is felt that one reason that the stall characteristics did not apparently trouble Stommel's experiment is due to the fact that the S-rotor would be almost constantly undergoing a fair amplitude of forced motion due to the motion of the buoy itself. The forces exerted by the buoy on the S-rotor certainly involve asymmetric forces which can easily set the rotor in motion past its stall point. An exceedingly attractive feature of the Savonius rotor, on the other hand, is that it does not stall over a wide range of velocities. However, the Savonius rotor also suffers from the limitation of a changing angular rate of revolution even in a uniform current. If the torque curve for the Savonius rotor is plotted, it is evident that it is a reasonably symmetrical ellipse. If we section the Savonius rotor into two equal portions and rotate them 90° with respect to each other and reassemble, we have, to a large extent, retained all of the original advantages of the Savonius rotor with few, if any, of the disadvantages insofar as it now possesses a rather precise uniform angular rate of revolution and furthermore still does not tend to stall. Fortunately, it is also a highly linear device. Figure 12-26 shows the two types of rotor we have been discussing. In the case of the Savonius rotor, the solid curved vanes represent the situation on one deck and the dotted curves represent vane positions on the second deck. It is perhaps worth mentioning at this point that there is obviously an error in terminology presently being employed with respect to the rotor resulting from the changes discussed, which were introduced into the original Savonius configuration. In order to avoid confusion it is suggested that the rotor resulting in the change be properly designated as a "modified" Savonius rotor.

Problem solution via materials engineering

Having made a decision as to the current-sensing rotor to be employed, we must now examine how we are going to make it responsive to low or weak currents. The problem confronting us, of course, is an old one, namely that of undesirable bearing friction due essentially to the weight of the rotor. As a first step toward reducing the effective weight of the rotor, the rotor was fabricated from polyethylene (density approximately 0.93) and the rotor balanced and weighted with metal slugs in order to make it neutral buoyant in saltwater. The rotor performed excellently but it could not be kept operational due essentially to the fact that polyethylene has a very

Instrumentation and Communications 441

(a)   (b)

FIG. 12-26. Current-sensing rotors that have been used for measurement of oceanographic currents (a) Rauschelback rotor; (b) Savonius rotor.

strong "plastic memory" and it tended to distort soon after completion of fabrication. Also, the polyethylene posed problems of fabrication since it could not be readily cemented. In attempts to eliminate the problem of plastic memory in regard to forming the rotor blades, sections were cut from proper radius polyethylene extruded tubing. Here again, problems were experienced since the tubing inevitably had strians internally and distorted almost impossibly the moment machine tools attempted to shape it to required tolerances. Fortunately, a commercial product known as Cycolac, as manufactured by the Marbon Chemical Company was available. It has a specific gravity of 1.01 which means that it will float comfortably in sea water with an average density of 1.025. Cycolac is a high-impact material and molds easily. It is easily assembled by cementing and possesses little or no plastic memory. With care in assembling, rotors fabricated from Cycolac maintain their dimensions indefinitely.

Advantages of "weightlessness"

Perhaps it is well to pause a moment and further examine what we have achieved in the way of a significant payoff by introducing a rotor-sensing system for the measurement of oceanographic currents that is, for all practical purposes, weightless in seawater. It should be noted that although it is weightless it is not without inertia. If the current meter is not being raised or lowered, it should be obvious that in the case of a neutral buoyant rotor that the axial load is zero. Furthermore, the radial bearing load approaches zero as the ocean current driving the rotor approaches zero. In many respects this is a rather unusual situation, and one we may use advantageously. As a convenient result of the neutral buoyant characteristic of the rotor, "stiction" or starting friction is essentially nonexistent. This fact is further borne out in practice insofar as the rotor current threshold is so low as to be practically indeterminate.

Rotor-bearing details

The rotor bearings are made of $\frac{1}{8}$ in. diam burnished stainless-steel axles with a thin disk of tungsten carbide press fit into each end. The axial thrust of the tungsten carbide insert is taken up by a $\frac{3}{16}$ in. sapphire ball. The radial thrust is absorbed by a ring bushing of sintered Teflon. The bearing is purposely fit loosely and approximately 2–4 thousandths "slop" is used in standard practice.

Limitations of commercial rotors

Practically all of the commercial modified Savonius rotors on the market have come from the same source and are fabricated from a high-impact modified polystyrene, which has considerably higher density than Cycolac. As a consequence of the increased weight of the rotor, substantial bearing loads are a permanent feature. Normally, these bearing loads do not begin to show any difficulty with currents above 0.1 knots. In practice the rotors are formed from two shapes: a vane and a disk. These are cemented together for assembly. It is recommended, however, that the grooves not occupied by the rotor blades be filled (Gaul et al., 1963).

Assembled current meter

The most conspicuous feature of the current meter is its large orienting fin. It is required to rotate the entire instrument in order to sense the current direction which is detected by means of a compass inside the instrument case. In the field the current meter is suspended from an electrical swivel so that

Instrumentation and Communications

it is free to rotate in any direction. A large yoke surrounding the instrument holds the instrument proper on a trunnion, which is placed so that the torques about the trunnion, due to horizontal current motion, are balanced. This insures the current meter remaining vertical at any current from zero to approximately 6 knots. Although it is true that, in Fig. 12-27, the instrument is not vertical, this is due simply to the fact that it is out of water and the heavy fin of Cycolac exerts the disturbing force. However, when the instrument is in water, it is vertical. The large fin is open top and bottom so that there is a minimum of drag due to raising and lowering of the instrument.

Chassis of underwater unit

Figure 12-28 illustrates the internal structure of the current meter when removed from its pressure-equalized case. Since the instrument is in a pressure-equalized case, the pressure-sensing transducer, known as a

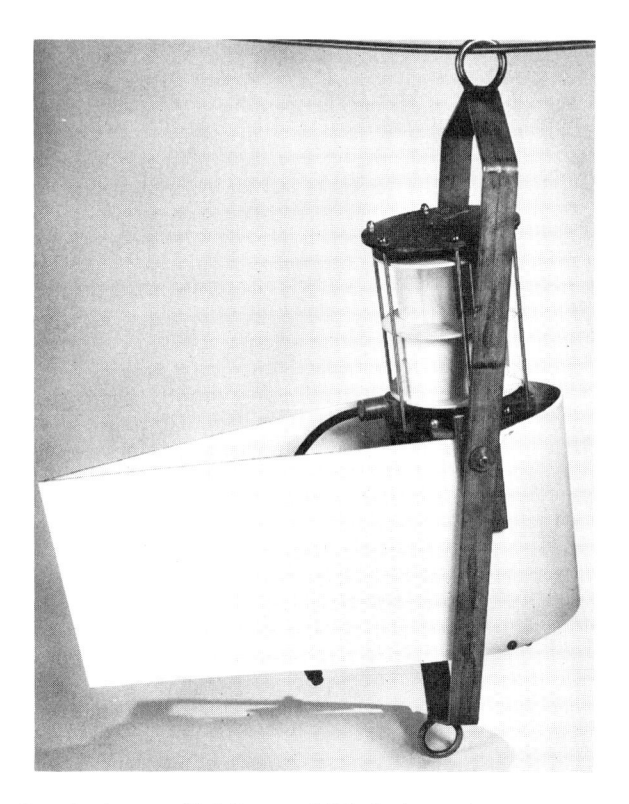

FIG. 12-27. Completely assembled Bureau of Ships' telemetering current meter. Instrument is supported on trunnions attached to a yoke, which in turn is attached to an electric swivel.

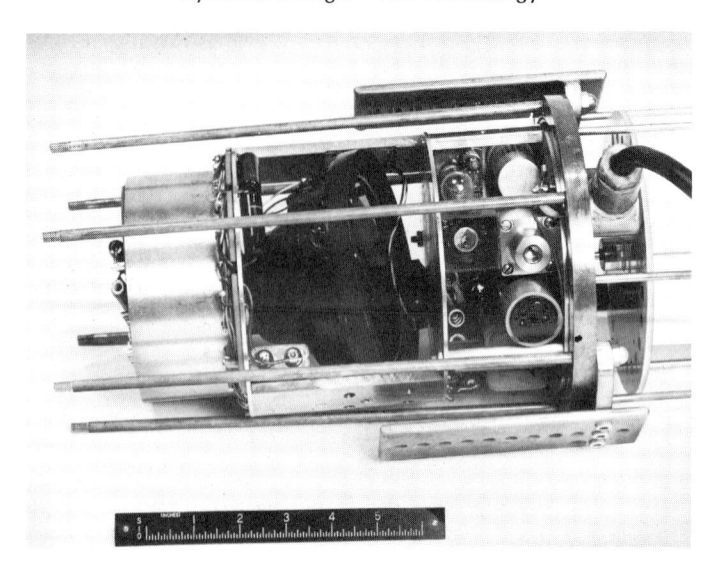

FIG. 12-28. Underwater chassis, Bureau of Ships' telemetering current meter. From left to right: toroid power transformer; mariner's compass modified for current direction; electronic deck. FM telemetry utilizes standard IRIG subcarrier frequencies.

vibrotron, is simply coupled directly to the fluid which fills the instrument case. The direction element consists of a modified standard mariner's small boat compass in which the compass card is replaced by a spiral-edged opaque disk. A light source is placed on one side of the disk and two balanced photoelectric cells are placed below the disk edge. The photoelectric cells serve as photoelectric resistors in the proper arms of a Wien-bridge oscillator. Thus an FM signal is transmitted whose frequency corresponds to the compass heading. The rotor pickoff mechanism is visible to the right of the figure where the 1-cm diam brass shim stock slugs are attached to the lower deck of the modified Savonius rotor. These slugs pass close to but not in contact with a thin electric coil which is part of a blocking oscillator. Thus, as each slug passes the pickup coil, the oscillator is gated. All three parameters go through separate isolating amplifiers and are then combined at a mixer terminal and finally amplified together for telemetering to the surface. The frequencies employed are those commonly used as subcarrier oscillators under the IRIG telemetry standards (Nichols Ranch, 1956). To the left of the photograph, at the left of the instrument chassis is located a toroidal power transformer. In operation the instrument employs two electric conductors to carry power down from the surface and transmits the IRIG frequencies back to the surface on a phantom circuit across the power leads and the sea return circuit.

Instrumentation and Communications 445

Deck control and readout unit

The on-deck portion of the telemetering current meter consists of two portions: a power unit and the control and indicating unit. These two constitute the top and bottom of a water tight carrying case. In practice, the two portions of the carrying case are separated and interconnected with the necessary power and signal cables. Figure 12-29 illustrates the control and readout unit. The readout in the deck unit is by means of specially calibrated D'Arsonval hermetically sealed and ruggedized 0–100 microameters. All of the indicating instruments are nonlinear except the current-direction instrument. The reasons for the nonlinearity will be discussed in the following section, "Current-speed indicating dial." The indicating instruments reading from left to right are as follows: current 0–0.03 knots; second instrument, range 0–0.3 and 0–3.0 knots; the third instrument depth in two ranges 0–100 ft and 0–1000 ft; the fourth instrument, direction 0–360°. The resolution of all indicating instruments is infinite. A recorder output jack is provided which supplies linear dc signals for 0–10 mV x_1, x_2, y recorder. With this provision, both current magnitude and direction can be recorded simultaneously as a function of depth. Insofar as the dc current output from the indicating control is linear, the dc signals could be integrated readily or,

FIG. 12-29. Deck control and indicating unit for Bureau of Ships' telemetering current meter. Current magnitude, depth and direction are conveniently read out on the indicating meters. Both current range and depth scales may be changed and noted on the instrument panel.

446 Systems Design—The Technology

since the basic information is contained in a pulse train, it would be feasible to simply count the pulses over the integrating period. We might well inquire why two current indicating instruments are necessary. The reason is that at the lower current velocities that the instrument is quite capable of responding to, the modified Savonius rotor requires about 6 min to turn one revolution. This in turn means that there are only four signal pulses per minute supplied to the control-readout unit. Additional circuitry has been added so that both the on-pulse and the off-pulse supply positive signals of identical energy to the indicating instrument. In order to smooth the meter deflections for ease of reading the instrument case is completely filled, without leaving any bubble, with silicone oil of approximately 500 centistokes viscosity. At current velocities above 0.03 knots it is not necessary to resort to fluid damping, hence the second meter for indicating currents. Newer designs of current meters supply on the order of 250 pulses per revolution of the modified Savonius rotor and there is therefore no problem in smoothing pulses in order to obtain steady indication.

Current speed indicating dial

Figure 12-30 has been selected to demonstrate the nonlinear indicating characteristic of the current meters. All of the nonlinearity is achieved by

FIG. 12-30. Nonlinear scale of indicating dials employed to facilitate reading over at least two current decades without shifting scale ranges.

Instrumentation and Communications 447

special shaping of the magnetic pole pieces. With this nonlinear feature incorporated, it is possible to read currents covering three decades in magnitude. A full-scale deflection of the instrument, which corresponds to a 3-knot current, requires 100μA. Half-scale deflection corresponding to 0.5 knots or $\frac{1}{6}$ of a full scale equals 16.66μA or $\frac{1}{6}$ of 100μA, which is full-scale deflection. One result of the nonlinear scale is that it may be read with approximately equal ease at any point with approximately the same percentage accuracy.

Shallow-water current profile

Under "Deck control and readout units" it was stated that the deck control unit supplied an analog output signal that could be used in an x–y recorder; Fig. 12-31 illustrates such a function plot. The locus of the

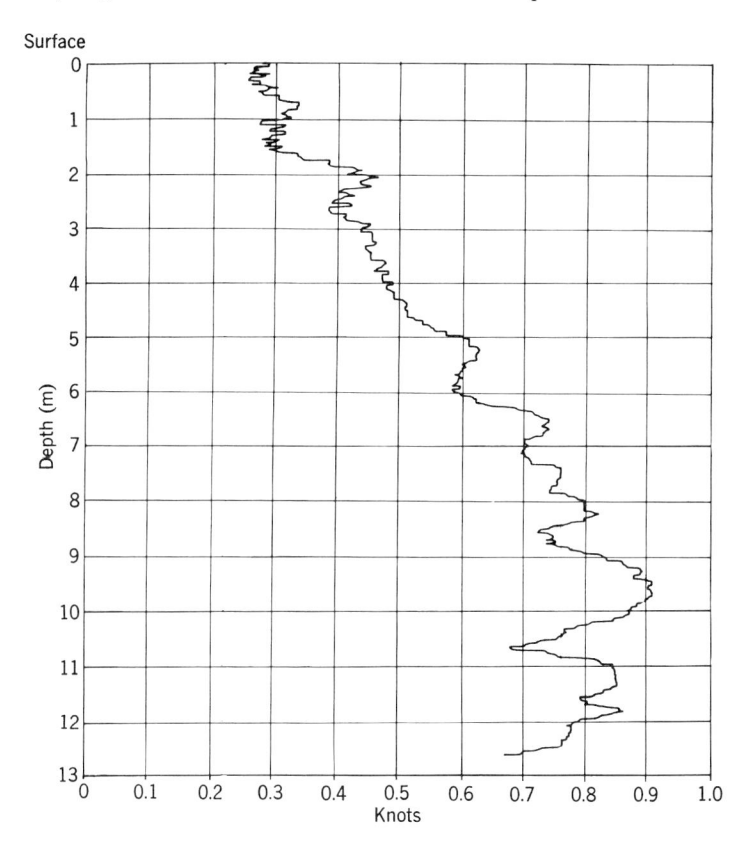

FIG. 12-31. Vertical current profile recorded in shallow water near Ballast Point in San Diego Bay.

measurement was in San Diego Bay, near Ballast Point. The detail visible in the vertical current profile is quite interesting, especially when it is realized that the two maxima, the one located at about $9\frac{1}{2}$-m depth and the other between 11- and 12-m depth are currents flowing in different directions. Unfortunately, a two-point recorder was not available at the time this record was made.

Tow-tank current-meter calibration

The accurate calibration of a current meter, especialy at the low velocity ranges, is rather tedious and in any case difficult. We were indeed fortunate that the Convair Division of General Dynamics Corporation in San Diego has an excellent tow tank, approximately 350 ft long, 13 ft deep and 12 ft wide. Since large-scale eddies take a very long time to die down, it is necessary to have the tank undisturbed for several hours before use. It should be pointed out that, as a matter of planning, it is advisable to begin calibrating at the lowest velocity that is anticipated will be of interest since, should we begin at the other end of the velocity scale, the tank is essentially useless for a good many hours before it can be used for the low velocities. One of the simplest ways of checking water for motion is to toss into the water a few crystals of potassium permanganate, which leave a well-defined dye track. Any deviations from a straight line will be clearly visible. This, of course, indicates persisting eddies. Figure 12-32 illustrates the method in which the current meter was mounted for tow purposes. The basic towing equipment

FIG. 12-32. Calibrating setup for Bureau of Ships' telemetering current meter in Convair's tow tank. Current meter is slung beneath shallow-draft supporting raft.

provided by Convair was not considered to be very reliable in ranges appreciably below 1 knot. Therefore a special towing arrangement capable of working down to 0.001 knot was provided. As shown in the illustration, the current meter is suspended from a very lightweight float, which is restrained toward the rear by means of a silk line wound on a small miniature winch driven by a constant-torque fractional horsepower motor. The raft is towed through the water by a bead chain, which is kept from touching the surface of the water by being threaded through the hole of hollow floats, as may be seen in fig. 12-32. The floats must be spaced so that the bead chain does not touch the water. Should the bead chain touch the water, the surface tension of the water forces the bead chain into a catenary curve, which adds such an excessive amount of compliance to the system that uniform motion cannot be achieved. During the tow operation the floats move down the pool with the bead chain and raft and stack up in a row near the bead chain puller. Fortunately the homemade puller was able to operate at sufficiently high speed that we had an overlap in the two tanks' towing power system so that the higher velocities could be readily checked. As mentioned earlier, the task of calibration is rather prolonged and actually 2-3 days may be required to calibrate one rotor. In the case of the commercial rotors which are somewhat heavy, it is rather interesting that the curves obtained are, for all practical purposes, identical with a neutral buoyant rotor at speeds above 0.1 knot. However the departures below 0.1 knot may be rather large. With the exception of the errors contributed by bearing friction, the rotors perform so nearly identically that individual calibrations are no longer considered mandatory.

Calibration curve

For the sake of convenience, the calibration curve shown in Fig. 12-33 has been plotted on log-log paper. The current meter is capable of operating over slightly more than four decades of current magnitude. The graph stops at approximately 2 knots but it has since been extended along the same slope to 5 knots.

12.9 ACOUSTIC BACKGROUND NOISE

The electrical background noise in the sea was referred to previously. There is also a very significant acoustic background noise in the sea. Both of these must be considered in instrument design, especially if the equipment has high sensitivity. Here, too, as in the case of electrical background noise, there are both animate and inanimate sources. The acoustic background noise may occasionally be of such intensity that

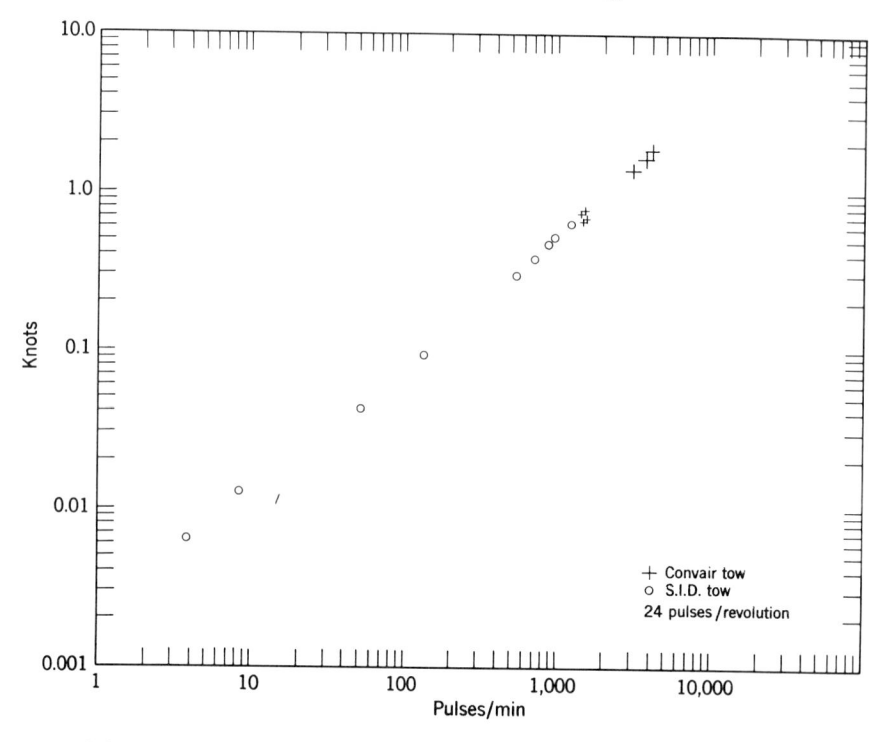

FIG. 12-33. Calibration curve for Bureau of Ships' telemetering current meter.

manmade systems working in the field of underwater acoustics may be severely noise limited. Continuing the parallel with the electric background noise, the acoustic noise spectrum covers a very wide range of frequencies. There are, for instance, frequencies below the tidal frequencies and at upper end well over 100,000 cps. It is quite obvious that a heavy surf pounding into a pebble beach is an excellent noise generator. Meteorological disturbances thousands of miles away generate noise which has been used to track the disturbance itself (Groves et al., 1966). Volcanos, both active and apparently inactive, frequently are excellent noise generators (Richards and Snodgrass, 1956). Major seismic disturbances rather obviously generate noise but at the lower end of the intensity scale, we find that seismic disturbances are apparently going on practically continuously. Perhaps one of the best examples of animate noise sources are fish (Wenz, 1964). The porpoise is well known as a noisemaker; less well known, however, is the whale (Shevill and Watkins, 1966). The ubiquitous snapping shrimp, *genera crangon*, and *synaltheus*, make such loud noises that they may be heard above water, in spite of the air-water acoustic mismatch. These shrimp radiate a nearly "white" noise between 12–26 kHz. The author has used patches of snapping shrimp as convenient calibrating sound sources to check out destroyer sonar equipment. The shrimp may be used for checking the sonar tuning for resonance and also the training in azimuth. The research scientist, as well as the submarine sonar operator is intensely interested in classifying and identifying underwater sounds. This has occupied many individuals for over 25 years. This is an exceedingly complex field when we realize that sometimes it is difficult to defferentiate between animate and inanimate sound resources such as, for instance, volcanos and toad fish (Wenz, 1964).

12.10 INTEGRATING RADIANT-ENERGY RECORDER

The oceans are in many ways vast reservoirs of food. At the very low end of the food chain are the organisms which utilize incident radiant energy photosynthesis. Thus the productivity of the sea is critically dependent upon the incident radiant energy. Fortunately, the organisms engaging in the photosynthetic process do not seem to care about the instantaneous value of the incident radiant energy, but seem to function as though they were integrating the available radiant energy. Thus it is evident that a suitable recorder of incident radiant energy does not have to have a short time constant providing the sensor is able to perform as an integrator of short-period variations in the magnitude of the incident radiant energy.

A pyrheliometer is easily able to perform the short time integrating as

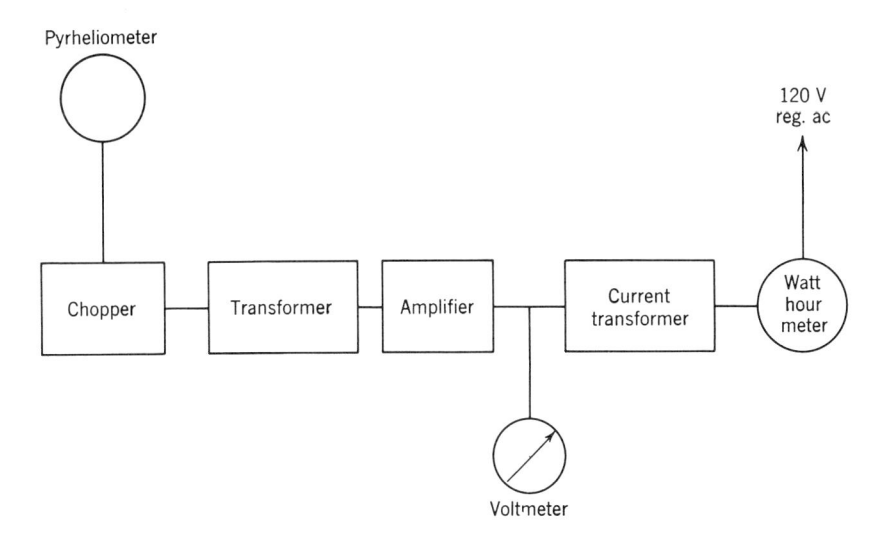

FIG. 12-34. Integrating radiant energy recorder, block diagram.

required. The Eppley Pyrheliometer[4] has been used for the precision measurement of radiant energy for many years. The pyrheliometer is a multijunction thermocouple with both the hot and cold junctions enclosed in the same glass envelope. Figure 12-34 is a block diagram of the functional components of the integrating radiant-energy recorder. Since the output of the pyrheliometer is dc, it must be inverted to ac for ease of amplification. An electrically driven chopper inverts the dc signal which is fed through an impedance-matching transformer to an amplifier. An ac voltmeter across the amplifier output indicates the near instantaneous values of the incident radiant energy. The amplifier output also drives a high impedance winding of an output current transformer the secondary of which drives the low-impedance winding of a conventional watthour meter. The potential coil of the watthour meter is energized from a small constant-voltage transformer. The current transformer itself is a small, 200-W General Radio Variac. The slider is moved from the Variac and the amplifier output is connected across the full Variac winding. The low impedance secondary consists of about 5 turns of No. 14 insulated wire wound over the toroidal Variac windings. Some precautions are in order if the pyrheliometer is to be mounted at the top of the ships mast as it must be gimballed and shock mounted. Good filters must be connected in series with each lead to prevent accidental burnout by currents induced from the ship's radio transmitting antenna.

12.11 PANEL INTEGRATING RADIANT-ENERGY RECORDER

Figure 12-35 shows the indicating instruments employed for the total accumulated incident radiant energy and the instantaneous value. The watthour meter has been adjusted to read directly in gram calories/cm^2 of ocean area. The voltmeter is calibrated to read in gram calories/cm^2 min. The watthour meter is a standard 5-A instrument with a low ratio geartrain. We often tend to forget that the watthour meter is a very fine computing machine that possesses very high reliability.

12.12 INTEGRATING IRRADIANCE METER

The problem, as presented, was to measure light levels in kelp beds down to depths of 100 ft. It was necessary to integrate the total luminous flux over a variable time period of days and weeks. Some of the limitations were that the instrument should have high precision, it should be rugged and capable of being easily handled by scuba divers. Another most important requirement was that it should be inexpensive. Incidentally, this latter requirement

Instrumentation and Communications 453

FIG. 12-35. Integrating radiant energy recorder, panel. Note pyrheliometer on top of cabinet.

is all too common and serves, nevertheless, to be quite a stimulus towards seeking a possible solution to the problems presented. The problem was ultimately solved by applying a rather old method of measuring electric current. In fact, the tool used is known as the silver voltameter and is identical in principle to that used by physicists to define the ampere. The ampere, or unit of current, is defined as the number of milligrams of silver plated out per second. The sensor selected was a photoelectric cell of the photovoltaic or self-generating type (Weston Photronic Cell, Model 856RR). The electrical impedance of the silver voltameter is very close to the optimum for the photronic cell to make its output independent of temperature over the anticipated environmental range. In practice the silver electrodes are weighed on an analytical balance before and after the experimental period. Thus the change in number of milligrams of silver is proportional to the total luminus flux received. The exact relationship is obtained by a very simple calibration. The photronic cells are manufactured to quite close tolerances and it is necessary to calibrate only one unit. Greater precision may, obviously, be obtained by inserting a simple resistance network between the photronic cell and the silver voltameter, to make all the units perform identically. MacRadyen (1949) has described a similar adaptation of the silver voltameter to record subsoil temperatures over a long period of time. The silver voltameter system may be used to integrate any dc signal that has a linear relationship to the quantity being measured. The irradiance meters are placed in black opaque plastic bags until ready for use. Scuba divers carry several attached to their belts. The meters are secured to the giant kelp by means of a lanyard threaded through a hole in the base of the meter. A large cork ring (Fig. 12-36, K) holds the assembly in the vertical position. Since these integrating irradiance meters have been used, there are now products on the commercial market that can do the same or better task and permit somewhat easier readout. One of these is known as the "E" cell and is manufactured by the Bissett-Berman Corporation in Santa Monica, California. It is substantially smaller than the silver voltameter described but is distinctly easier to use. The completely assembled integrating irradiance meter is illustrated in Fig. 12-37. All of the external nonoptical portions of the instrument are fabricated from polyvinyl chloride (PVC).

12.13 DUAL RADIANCE METERS: IN SITU RECORDING

Over the years a fairly large number of highly successful photometric devices have been built, mostly for use in biological research. The larger portion of these instruments have been devised to telemeter data directly to the surface as well as permitting control signals to be sent from the surface to the recording instrument to change optical filters, sensitivity, etc. These instruments have ranged from relatively crude, rugged devices employing in the simplest fashion photosensitive transducers (Boden et al., 1960) to fairly involved circuits with photomultipliers (Holmes and Snodgrass, 1961). Interesting observations have been gleaned along the way—for example, the fact that the ocean does not always get uniformly darker as the depth increases (Boden and Kampa, 1957). The reason for this is the existence of luminescent organisms. A rather specialized instrument has been devised to obtain additional data on the optical characteristics of some of these evasive luminescent organisms. One of the difficult problems in making measurements of the luminescent organisms is that they are highly susceptible to mechanical stimulation. For instance, if an organism should bump part of the equipment, such as the optical window, it tends to give off a flash of light. The problem was to determine what their natural and, insofar as possible, undisturbed state of activity really is. In order to do this, it was necessary to devise a means of restricting the observational field. This was done by mounting a pair of radiance meters, each possessing a fairly narrow restricted optical cone of sensitivity. These instruments were

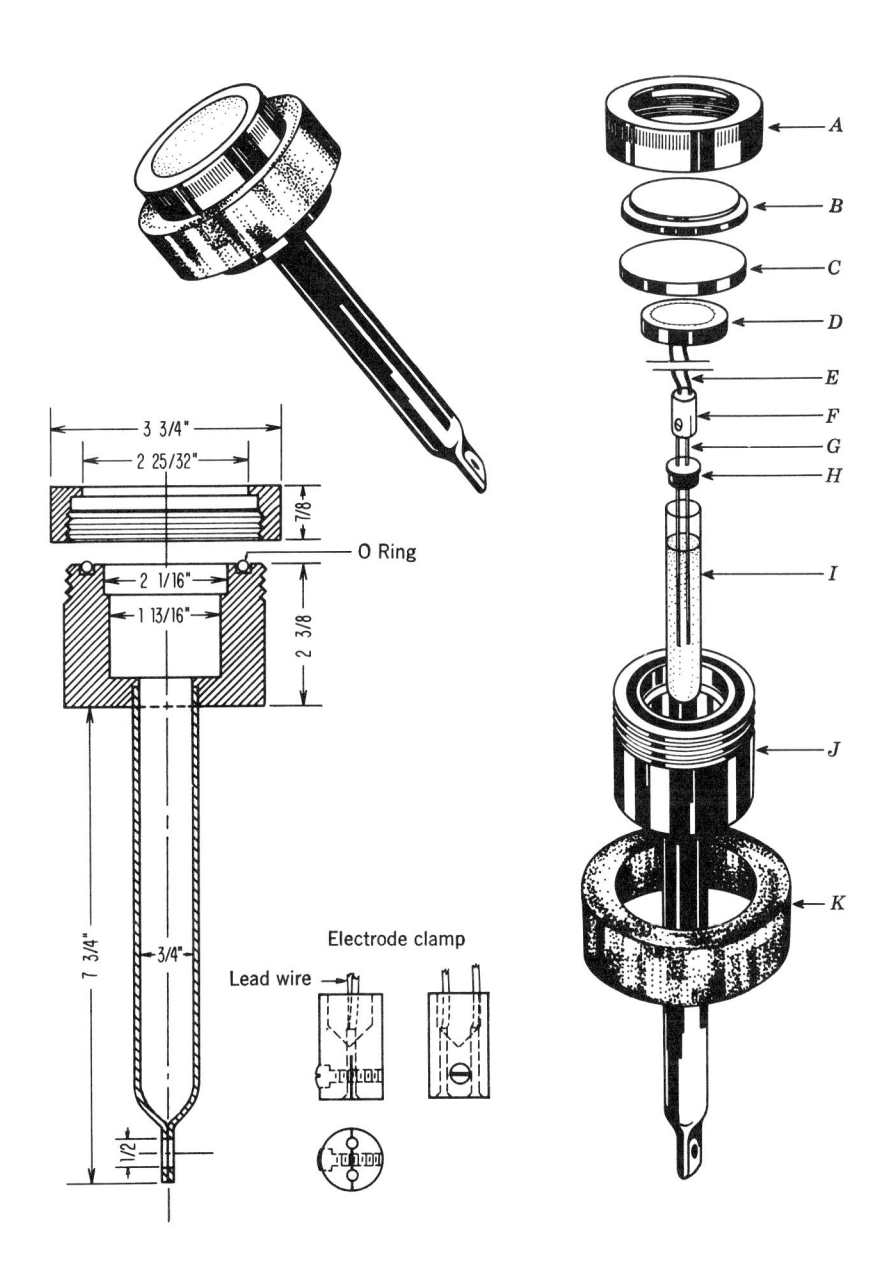

FIG. 12-36. Integrating irradiance meter. *A*, assembly ring; *B*, optical diffusion disk functioning as Lambert collector; *C*, optical filter disk; *D*, photronic cell; *E*, connecting leads; *F*, electrode clamp; *G*, silver electrodes; *H*, bacteriological type stopper; *I*, test tube holding electrolyte; *J*, body of instrument with O-ring seal; *K*, cork ring.

456   Systems Design—The Technology

FIG. 12-37. Completely assembled integrating irradiance meter.

mounted so that the optic axes are at 90° to each other, such as illustrated in Fig. 12-38. The duplicate instruments have optical beam widths of approximately 5° each. They have a common volume of approximately 1 m from the end of their optical tubes. The total common volume is approximately 1 liter. Statistically, it is evident that each instrument would independently see many more flashes of light than is represented by those flashes which would be seen by each identically. It is evident that the determination of an identical single source of light seen by each cannot be made on the

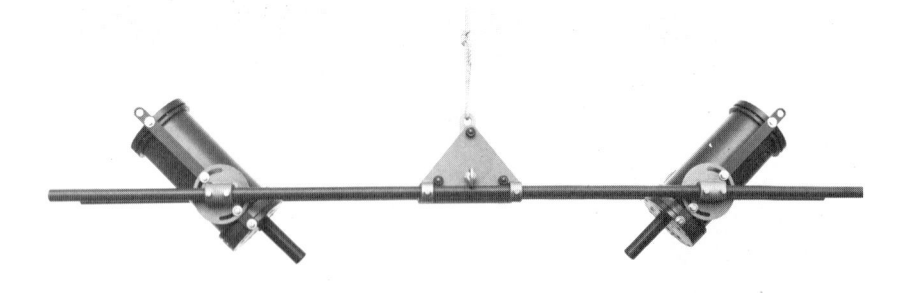

FIG. 12-38. Dual radiance meters, *in situ* recording. Radiance meters shown suspended in mounting bracket. The optic axes are 90° apart and focus on a common volume about 1 m from the optic tubes. The common volume is approximately 1 l.

basis of amplitude alone, since light output from the luminescent organisms varies as a function of angle. Arbitrarily, it has been assumed that coincident recordings would be assumed to be from the same light source. Statistically, of course, this assumption will have some errors. Electrically, the two radiance meters are interconnected such that the signal from its companion is recorded in the form of an AM signal, whereas its own observation is recorded in terms of an FM signal. Wire recorders capable of continuous 3-hr operation are employed. In the application which was planned for the radiance meters, it was not possible to telemeter the electrical signals to the surface for recording, though this can be done as there is nothing in the design which prevents this and the electrical signals are available externally (Boden et al., 1965).

Records from dual radiance meters

As was mentioned above, the luminous background noise in the ocean tends to vary with depth. Figure 12-39 illustrates differences between 70- and 131-m depths. Relatively speaking, the shallower water shows the lowest degree of activity. Record *B* at 131m shows distinct activity in both the AM and FM channels. The AM channel is the upper and the FM channel the lower of the two traces. The records presented in Fig. 12-39 were obtained from magnetic-tape playbacks from the original material. It becomes evident from these illustrations that the degree of time resolution is inadequate to establish coincidence or lack of coincidence of the light flashes. However, Fig. 12-40 does present clear resolution. The use of magnetic tape for work of this sort permits a great deal of latitude in data readout. For instance, by varying the relative speeds between the magnetic tape and the graphic recorder, it is possible to, in a sense, "expand time" to any reasonable degree. For instance, Fig. 12-40 represents a time expansion of approximately a factor of 8. The time marks at the bottom of the figure represent 125 m sec in terms of real time of the original recording. Here again, the upper trace is from the AM track and the lower trace is from the FM track. On the basis of our previous assumptions, traces marked *A* and *A* are coincident and represent the same luminescent individual; whereas flash *B*, which appears only on the AM track, would not be considered as being within the common 1-liter optical volume or it may have been in the volume but its azimuthal characteristics were such that it was completely blocked out from the radiance meter at a 90° angle. As is so often the case, an improved research tool often raises more questions than it succeeds in answering, and the present tool is no exception. If it is considered legitimate to set arbitrary relative luminescent thresholds, it would be quite feasible to apply state-of-the-art coincidence counting techniques to the AM and

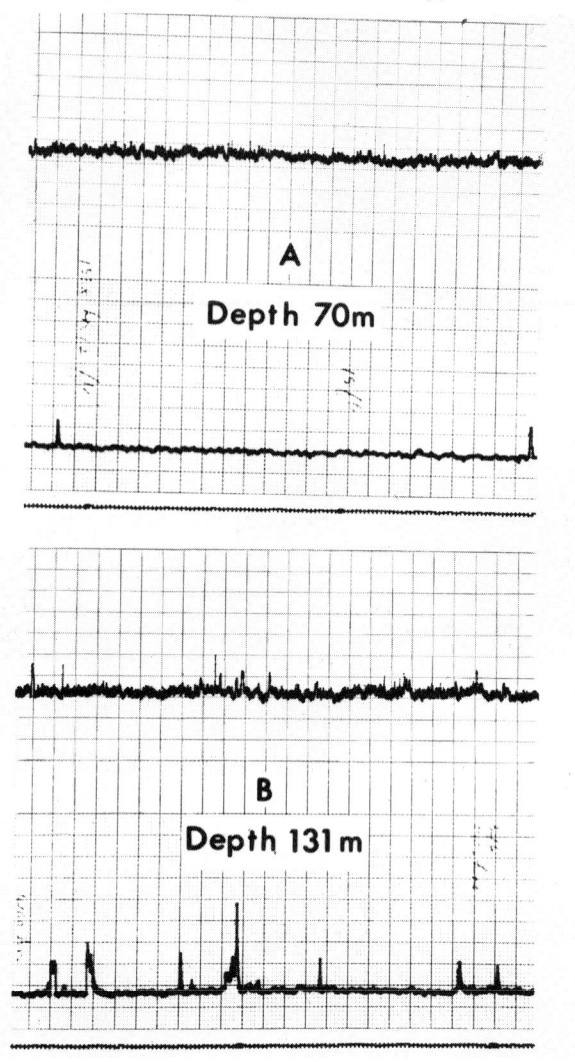

FIG. 12-39. Variations of luminous background noise between 70- and 131-m depths.

FM recorded tracks. However, a close examination of the record emphasizes the fact that any arbitrary assignment of relative threshold does, in fact, ignore a lot of probably pertinent data.

Electronic decks of dual radiance meters

The dual radiance meters were built as pressure-protected instruments, basically because of the difficulties of rebuilding or obtaining the wire

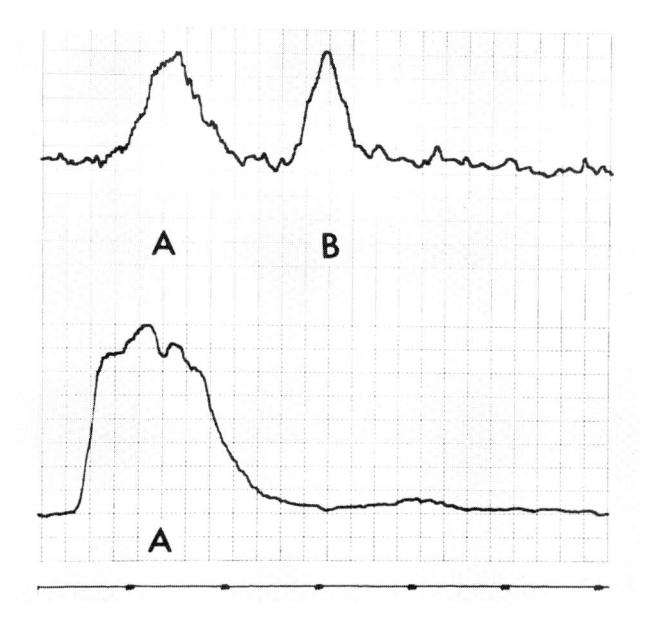

FIG. 12-40. Dual traces from dual radiance meters, illustrating "time expansion." Traces marked AA would be considered to be coincident and from the same individual, whereas a trace marked B would be either an individual outside the common optical volume or one whose orientation made it invisible to the one radiance meter.

recorders in a form that would make them practical to use in a fluid-filled medium. Several problems had to be considered, concerning the wire recording, the requirements for transcription to magnetic tape, and the ultimate derivation of the desired information. In the first place, means had to be insured to either control or correct the wire recorders for variations in speed during recording periods since, in the case of the FM track, frequency was related to the intensity of the observed light source. Since photomultipliers were to be employed, there was an additional need to obtain the high voltage necessary to operate the dynodes. There was also the problem of insuring a minimum disturbance in the FM signal from temperature changes. The high-voltage requirement and the stability of the FM signal were resolved by utilizing the wide-range voltage controlled oscillators (VCO's). To obtain the wide frequency span required, beat-frequency techniques were employed. To minimize the temperate difficulties, all of the VCO's were designed with approximately the same temperature coefficient and, to further insure their seeing the same temperature, they were mounted in thermal contact with a heavy aluminum plate. Figure 12-41 shows the relative disposition of the electronic components. The high voltage for the photomultiplier was obtained from a 24 kc/sec

FIG. 12-41. Electronic decks, dual radiance meter. Wide range VCO's mounted on circuit board with high-voltage transformer (secondary windings consist of 4 pi coils).

frequency from one of the VCO's which was stepped up by a specially made transformer employing four pi-type secondary windings. The pi type of construction was necessary due to the necessity of avoiding high distributed capacity in the secondary windings in view of the relatively high frequency of 24 kc/sec. Since 12 instruments were required, printed-circuit techniques were resorted to, in order to save time and to further contribute toward making each instrument functionally identical. Figure 12-42 illustrates a partial assembly with the circuit board from Figure 12-41 shown mounted on its metal plate the rectangular circuit board plugged in and shown vertically is the remainder of the electronic circuitry. The plugs protruding from near the base of the vertical circuit board are plug-in terminals for the wire recorder. The entire units are powered by mercury batteries specially designed for low-temperature operation (refer to Section 12-2). It was found necessary to test all completed instruments, as well as critical subassemblies in environmental chambers over the temperature range 25–1°C.

12.14 COMMUNICATIONS

Communications, as such, are playing an increasingly important role in the field of science. Oceanography is critically dependent on communications from the standpoint of obtaining data from remote ocean observing

FIG. 12-42. Electronic decks, dual radiance meters. Partially assembled instrument showing VCO's in contact with large aluminum thermal sink. Large rectangular control circuit board plugged in to base assembly.

platforms and ships. The oceanographer has been confronted with special difficulties in obtaining radio frequencies for telecommunications purposes. As early as 1957, it began to become apparent that very real problems existed in the obtaining of radio frequencies for oceanographic research purposes. The red tape necessary for obtaining radio frequencies proved, in general, to be insurmountable by the oceanographer. The problem was further complicated when the last major radio frequency allocations were accomplished by the International Telecommunications Union (ITU) resulting in the 1960 Radio Regulations. Requests for radio frequencies for telecommunications purposes, after 1960, were met with the very simple answer that there were no radio frequencies allocated for this type of service (oceanographic communications). The United States initiated the first effort to resolve the communication problem in oceanography. In 1960 and 1961 the National Academy of Sciences Committee on Oceanography became concerned with the problem and through their Panel on New

Devices initiated a study of the problem. Further, in 1962, the National Academy of Sciences Committee on Radio Frequency Requirements for Scientific Research became interested in the oceanographic telecommunications problem and further, the then new international organization, known as the Intergovernmental Oceanographic Commission appointed a Working Group on Communications to consider the problem. The major underlying facet of the difficulties faced in obtaining specific radio frequencies for use in oceanography and data collection from the marine areas is the fact that high-frequency (hf) radio is necessary due to the long distances and requirements for 24 hr communications. Further complicating the problem is the fact that the hf portion of the electromagnetic spectrum (3–30 MHz) is the most congested in the world. Furthermore, practically all users of high frequencies are struggling to obtain more frequencies. This, of course, puts a natural resource as a commodity in very short supply. Since the competition for radio frequencies is intense, it may be readily appreciated that the request for radio frequencies by oceanographers is somewhat less than welcome. It is true that satellites in various configurations with instrument buoys have been proposed for the collection of data from the ocean areas; however, this appears to be substantially in the future, as far as any serious consideration of realistic data collection systems are concerned. Much more development work and inventing is necessary before such systems can become operational; whereas, in the case of hf, nothing further need be developed. Since 1961, there have been numerous meetings, both national and international seeking ways to obtain radio frequencies for the collection of data from the ocean areas. The World Administrative Radio Conference (WARC) for Maritime Mobile radio frequencies, September 1967, had an item on its agenda which provided for consideration of the resolution of the oceanographic telecommunications problem.

System proposals

As a point of departure in considering the needs of the oceanographers for telecommunications high frequencies, it is necessary to consider the general characteristics of a final system and also how its needs might be resolved. Since high frequencies are being considered, the system plans must take into account practical means of compensating for changes in radio propagation involved in the day–night ionospheric changes. Essentially, the problem is providing a means of telecommunications for obtaining data from remote unmanned observation platforms such as buoys and the transmission of the data to suitable data-processing points and ultimately to users on a world-wide basis (Snodgrass, 1963a; 1963b; 1964; 1966; Haydon et al., 1962; 1963; Terry, 1965). Figure 12-43 presents in block form the

Instrumentation and Communications 463

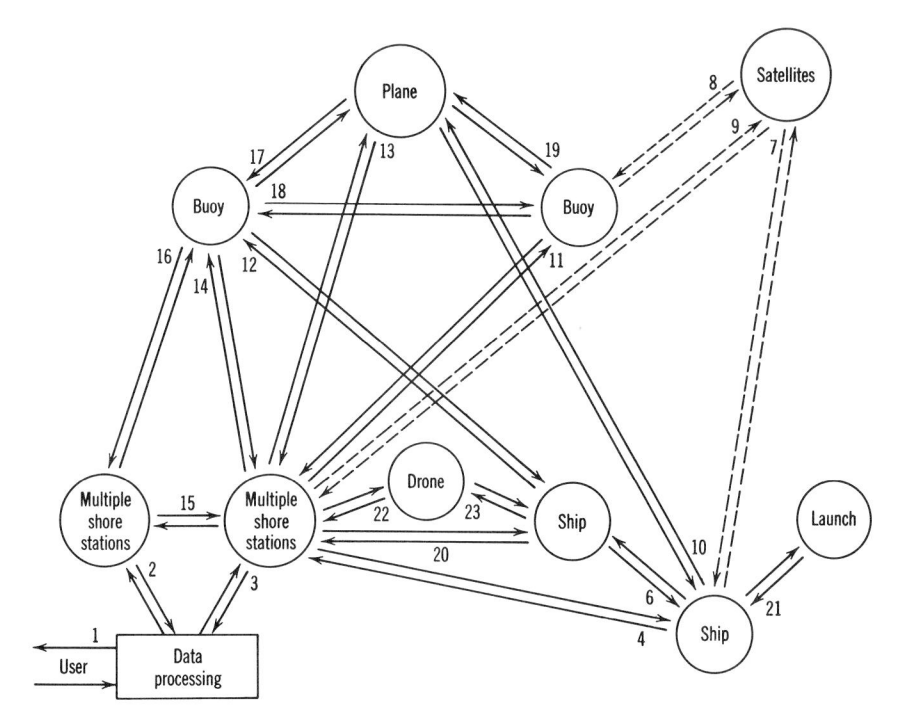

FIG. 12-43. Systems engineering block diagram, oceanographic communications.

essential elements for any practical telecommunications network. Some technical points are involved in that it is to be noted from the diagram that all of the communications paths are two-way. This is dictated by the necessity for employing interrogation systems in contrast to automatic-clock-programmed transmissions. Such requirements are technically sound because they permit a much more efficient utilization of the radio frequencies than could possibly be the case if the system were clock programmed. Also, if interrogation is used, the system can be "reprogrammed" as the operating requirements change.

The maritime mobile spectrum

As mentioned previously, it is necessary to provide 24-hr-a-day data transmission from each observing platform in the ocean. Due to the changes in the ionosphere within a given 24-hr period, it appears that more than one radio frequency per buoy may be required. Furthermore, it would seem practical, if we are considering world-wide data-collection network, to assign identical radio frequencies to buoys on a world-wide basis. This, with a properly programmed system, would facilitate the orderly and efficient use of the assigned radio frequencies. There are two present services listed in the 1960 Radio Regulations that employ identical frequency bands on a world-wide basis. These are the Maritime Mobile Services and the Aeronautical Mobile Services. This, of course, is done for obvious reasons, insofar as both ships and planes range over the globe crossing over boundaries which are used for purely regional radio assignments. Figure 12-44 which is a small portion of the entire hf radio spectrum shows the position of one of the Mobile Maritime bands between 6.200 and 6.525 MHz. The International Frequency Registration Board (IFRB) suggested in 1962 that perhaps the oceanographers should seek frequencies from the Maritime Mobile bands for their telecommunications purposes. It is largely on the basis of this recommendation that the IOC has based its thinking and argument.

Data channel proposals

On the basis of discussion with experts in radio propagation, recommendations have been advanced as to means of utilizing a 3.5-MHz slot within each of the six Maritime Mobile exclusive frequency bands for the transmission of automatic data (see Fig. 12-45). It has been proposed that 10 channels of 300 Hz be used for data channels[4]. There would be also be an upper and lower buffer channel 250-Hz width. Several technological factors lead to the recommendation of 250-Hz data channel. These are, among others–decision to limit transmission from the buoy to 100-W power; to limit the transmission to a rate of 100 bits/sec. The buffer channels are provided to reduce difficulties due to the possibility of relatively high-powered stations causing adjacent channel interference. However, the buffer channels can be occupied for data transmission of rather special sorts. It is desirable to incorporate, within the rf assignments for the collection

[4]The Maritime Mobile World Administrative Radio Conference (WARC) meeting in Geneva, October–November, 1967 officially designated radio frequencies for the exclusive transmission of oceanographic data. The precise frequencies are indicated below and are available for use with buoys under FCC authority.

Limits	Oceanographic data transmission	Limits
4162.5	4162.9–4165.6 10 frequencies spaced 0.3	4166
6244.5	6244.9–6247.6 10 frequencies spaced 0.3	6248

(*Continued on next page*)

Instrumentation and Communications

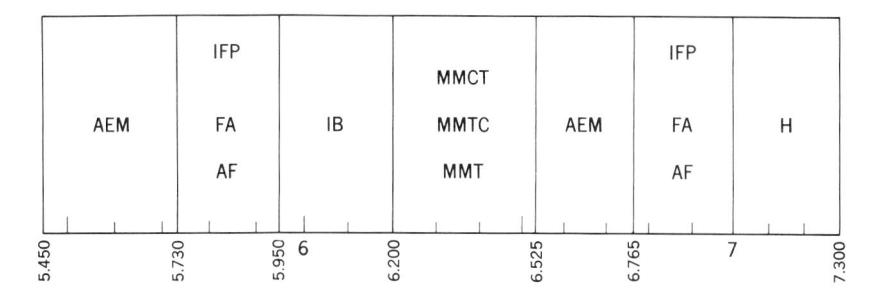

AEM	IFP FA AF	IB	MMCT MMTC MMT	AEM	IFP FA AF	H

5.450 5.730 5.950 6 6.200 6.525 6.765 7 7.300

FIG. 12-44. MMCT, MMTC, MMT: Position of one of six Maritime Mobile exclusive frequency bands in the high-frequency region.

of marine data, provisions for tsunami or tidal-wave warnings and hurricane warnings. Since the essential information from this type of warning message does not require a large number of bits, low data rates and narrow channels are practical; these could be combined with relatively sophisticated signal-to-noise detection and processing techniques at the receiving station to insure high reliability of the circuits. The receiving stations utilized for picking up the warning types of transmission would, of course, operate 24-hr a day. It is further assumed that radio transmissions of data from ships could also conform to the data-channel requirements. In planning efficient use of the data-channels, it is obvious that it would be necessary to resort to computers to assist in the assignment of frequencies on a world wide basis in order to avoid interference. It is suggested that "synchronized" interrogation be employed in order to avoid adjacent channel interference within the data channels. The reason for this suggestion of synchronous interrogation is that, if the high-powered interrogation stations all transmitted the interrogation signals at the same moment in

limits	Oceanographic data transmission	limits
8328	8328.4–8331.1 10 frequencies spaced 0.3	8331.5
12,479.5	12,479.9–12,482.6 10 frequencies spaced 0.3	12,483
16,636.5	16,636.9–16,639.6 10 frequencies spaced 0.3	16,640
22,160.5	22,160.9–22,163.6 10 frequencies spaced 0.3	22,164

time on a worldwide basis and then stood by for reception of data from the relatively weak 100-W buoy transmitters, this would in large part avoid serious problems that would otherwise arise if interrogation were carried out in a nonsynchronous manner. Insofar as time may now be determined with very high precision at any point on the globe, this seems to be a practical constraint to place on the system.

Use of space diversity to improve reliability

It was previously mentioned that consideration would have to be given to the problems introduced by the propagation of high frequencies, due to

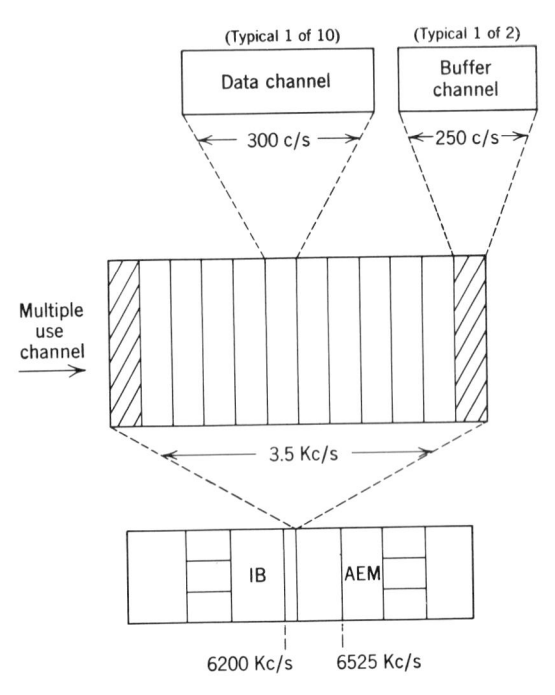

FIG. 12-45. Proposed relation of data channels to multiple use channel in one of the HF Maritime Mobile exclusive frequency bands.

variations of the ionosphere in the day–night changes. One way to assist in minimizing this difficulty, even though a single frequency might be used, would be to employ widely spaced interrogation-receiving stations. It would be advisable to obtain as great an east–west spacing as practicable. Such a hypothetical situation is shown in Fig. 12-46 in which a buoy is shown off the coast of Japan with possible interrogation stations at Tokyo, Vladivostok, Glinka, and Midway Island. This disposition of interrogation stations is selected on the assumption that statistically the over-all probability of being able to receive a transmission from the buoy is significantly enhanced if we provide a significant east–west spread of interrogation stations. Thus, if one station were unable to contact the buoy, another station might be able to do so with ease. The task of actually studying such a system is properly carried out with the assistance of a computer. Studies of this type, which are based on mathematical models of radio propagation have been carried out by the Central Radio Propagation Laboratory (ITSA) at Boulder, Colorado. A sample page of computer readout is presented in Fig. 12-47. The geometry of the system is illustrated in Fig. 12-46, which presents data that has been further

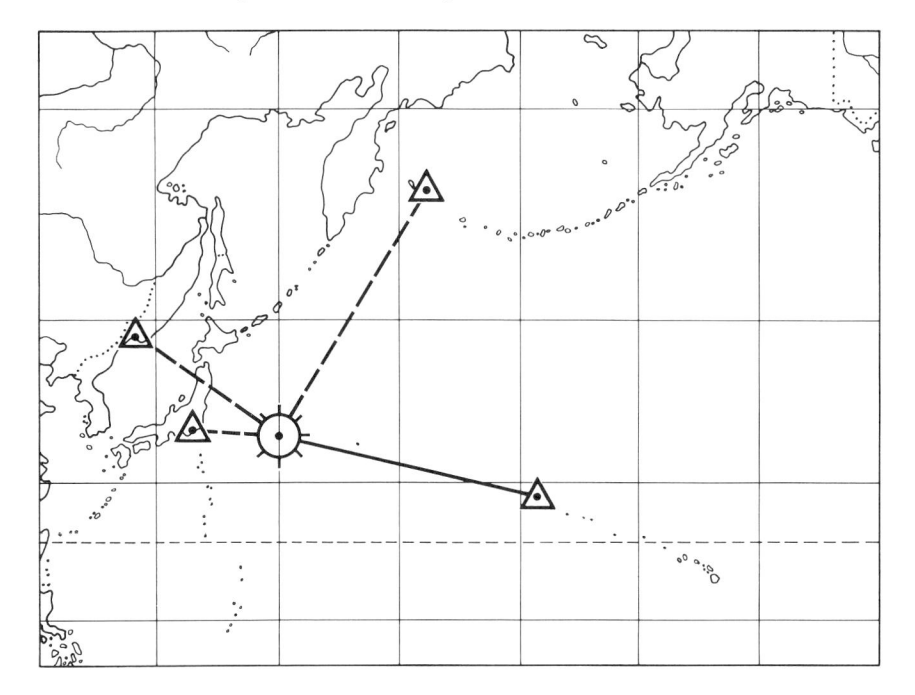

FIG. 12-46. Hypothetical scheme illustrating space diversity of multiple interrogation stations in relation to a given ocean buoy.

improved upon in a study published by the National Security Industrial Association in which the over-all composite reliability of the system diagramed was calculated (Terry, 1965). The essential important results of this show that, under the initial conditions imposed, no more than two frequencies are necessary on a year-around basis in order to obtain extremely high reliability of reception of data from the buoy position off Japan. It should be pointed out that one of the constraints placed on the

1. ② 2. MARCH 3. SSN= 10 4. TK 8.005
5. TRANSMITTER   RECEIVER   BEARINGS   10. N.MILES
6. 35.00N 7. 150.0   8. 35.40N - 139.40E 9. 275.7 89.6   520.1
11. VERTICAL   8H   ODEG   11 A NOISE= 3   ANT= 12DB
12. PWR= .10KW   17. OPERATING FREQUENCIES   13. REQ.S/N= 45DB

14.GMT	MUF	FOT	3	5	7	9	11	13	15	20	25	30	FOT	
1	10.9	10.9												16
			1E	2F	1F	1F	1F	00	00	00	00	00	1F	MODE 18
			11	46	26	26	26	0	0	0	0	0	26	ANGLE 19
2	11.9	11.1	0	0	78	97	99	0	0	0	0	0	99	RELIABILITY
3	12.7	11.1												
			1E	2F	1F	1F	00	00	00	00	00	00	1F	MODE 20
			11	46	26	26	0	0	0	0	0	0	26	ANGLE
4	12.8	10.8	0	1	82	98	0	0	0	0	0	0	99	RELIABILITY
5	12.1	10.2												
			1E	1F	1F	1F	00	00	00	00	00	00	1F	MODE
			11	25	25	25	0	0	0	0	0	0	25	ANGLE
6	11.2	9.5	0	56	96	99	0	0	0	0	0	0	99	RELIABILITY
7	10.2	8.7												
			2F	1F	1F	00	00	00	00	00	00	00	1F	MODE
			44	24	24	0	0	0	0	0	0	0	24	ANGLE
8	9.0	7.6	36	97	99	0	0	0	0	0	0	0	99	RELIABILITY
9	7.5	6.4												
			1F	1F	00	00	00	00	00	00	00	00	1F	MODE
			24	24	0	0	0	0	0	0	0	0	24	ANGLE
10	6.1	5.2	91	97	0	0	0	0	0	0	0	0	97	RELIABILITY
11	5.2	4.4												
			1F	00	00	00	00	00	00	00	00	00	1F	MODE
			25	0	0	0	0	0	0	0	0	0	25	ANGLE
12	4.9	4.2	91	0	0	0	0	0	0	0	0	0	94	RELIABILITY
13	4.9	4.2												
			1F	00	00	00	00	00	00	00	00	00	1F	MODE
			26	0	0	0	0	0	0	0	0	0	26	ANGLE
14	5.0	4.2	87	0	0	0	0	0	0	0	0	0	91	RELIABILITY
15	5.1	4.4												
			1F	00	00	00	00	00	00	00	00	00	1F	MODE
			27	0	0	0	0	0	0	0	0	0	27	ANGLE
16	5.1	4.4	87	0	0	0	0	0	0	0	0	0	92	RELIABILITY
17	5.0	4.2												
			1F	00	00	00	00	00	00	00	00	00	1F	MODE
			27	0	0	0	0	0	0	0	0	0	27	ANGLE
18	4.7	4.0	91	0	0	0	0	0	0	0	0	0	94	RELIABILITY
19	4.2	3.6												
			1F	00	00	00	00	00	00	00	00	00	1F	MODE
			26	0	0	0	0	0	0	0	0	0	26	ANGLE
20	4.6	3.9	93	0	0	0	0	0	0	0	0	0	97	RELIABILITY
21	6.4	6.3												
			2F	1F	1F	00	00	00	00	00	00	00	1E	MODE
			46	26	26	0	0	0	0	0	0	0	11	ANGLE
22	8.3	8.3	0	81	98	0	0	0	0	0	0	0	95	RELIABILITY
23	9.6	9.6												
			1E	2F	1F	1F	00	00	00	00	00	00	1E	MODE
			11	46	26	26	0	0	0	0	0	0	11	ANGLE
24	10.3	10.3	0	3	90	98	0	0	0	0	0	0	92	RELIABILITY

FIG. 12-47. Single page of computer output for Tokyo with telemetering link as shown in Fig. 12-46. Supplemental notes: 1, Computer sequence no.; 2, month; 3, sun-spot level (activity) 10 = low, 100 = high; 4, computer code for receiver location (TK 8,005 = Tokyo); 5, 6, 7, transmitter: latitude and longitude; 8, receiver—latitude and longitude; 9, bearing of receiver from transmitter and bearing of transmitter from receiver (computed); 10, nautical

Instrumentation and Communications

computer program was the assumption that we wished to receive a signal, having a signal-to-noise ratio of at least 45 dB. This means that if the predicted signal level should be less than 45 dB that the computer would print a zero. This does not mean that no signal was present but it simply means that 45 dB was arbitrarily selected as a cutoff point for the program. Also since the computer does not utilize three digits, it does not recognize percentages of reliability higher than 99. The beginning and the general character of the day–night frequency shift characteristics are easily visible in the shift of the maximum operating frequency from right to left as time increases down the page and return from left to right after 2000 GMT. The computer, properly programmed, is obviously an exceedingly powerful tool in assisting in the solution of radio propagation problems and the design of a suitable network.

Hemispheric plots of available signal to noise

The computer is an extremely versatile tool and the output may be programmed to different formats. Figure 12-48 utilizes the same kind of input data as presented in Fig. 12-47, but employs a totally different printout. In this case every pair of numbers corresponds to the value of the available signal to noise at a particular latitude and longitude over an entire hemisphere, the coordinates of which are visible on the margins. For con-

---

miles between transmitter and receiver (computed); 11, transmitter antenna constants (assumed vertical whip with top 8 m above water); 11A, noise figure for latitude of receiver; 12, transmitter antenna input power; 13, receiving antenna: assumed 12 dB gain. Required signal/noise of 45 dB, that is, ratio of total signal power-to-noise in a one-cycle band; 14, GMT—Greenwich Mean Time; 15, MUF—Maximum Usable Frequencies; 16, FOT—Optimum Traffic Frequencies; 17, operating frequencies 3–30 MC/sec for which data are computed; 18, MODE—Propagation path. Layer of ionosphere and number of reflections; 19, angle of received signal above horizontal; 20, RELIABILITY—Percent of the days of the month (2) that a signal will be received with a signal-to-noise ratio of 45 dB (13) or better, at a given hour GMT (14) with a specified sun-spot level (3) and at selected operating frequencies (17). *Note A:* The computer calculates the MUF (15) and the FOT (16) for each hour GMT (14). Every two hours GMT (14) the MODE (18), ANGLE (19) and RELIABILITY or S/N (20) are calculated for each of the Operating Frequencies (17) and the Optimum Traffic Frequency. (Right-hand column of data.) *Note B*: Zeros will automatically appear to the right of the last printed RELIABILITY or S/N value if the calculated received signal level is below a signal-to-noise ratio of 45 dB (13), that is, it would be zero if the S/N ratio were 44 dB, etc. *Note C*: The left-hand columnar data represents calculations of the S/N ratio of the received signals. The values given represent the monthly median of the hourly median signal-to-noise ratios to be expected under conditions set forth for RELIABILITY (20). *Note D*: Values are calculated for all conditions which yield reliability figures of 1 or above. In addition S/N ratios are calculated for any operating frequency (17) below a frequency for which a reliability is given. This explains some of the minus levels shown.

470 Systems Design—The Technology

```
                    AVAILABLE SIGNAL-TO-NOISE
        TRANSMITTER 33.00N - 72.00W        POWER  0.10KW
        FREQUENCY  5.0CMC/S      DEC   SUNSPOT 10   GMT  6
        VERTICAL   8H  -0L  -00EG                ANT= 12DB

   80   23  23  23  23  23  24  24  24  24  24  24  24  23  23  23  23  23  23  23  80
   70   23  23  24  24  24  25  26  26  27  27  27  26  26   +   +   +  23  23  23  70
   60   23  23  24  24  25  46  49  49  50  50  50  49  48  26  25  26  25  24  24  60
G  50   42  42  43  43  45  49  50  52  56  57  55  51  53  52  50  49  46  44  40  50  G
E                                                                                        E
O  40   37  43  43  43  45  49  49  54  61   +  61  57  54  52  50  48  47  45  45  40  O
G                                                                                        G
R  30   44  43  43  43  44  49  49  55   +   +  62  57  54  51  50  48  46  45  43  30  R
A                                                                                        A
P  20   43  45  45  47  43  46  49  52  57  58  55  55  53  50  49  46  44  42  40  20  P
H                                                                                        H
I  10   42  43  45  46  48  50  51  48  48  48  47  51  51  49  47  42   +  39  39  10  I
C                                                                                        C
    0   41  42  44  45  46  48  49  42  40  36  35  39  43  43  46  43  40  39  38   0
  -10   39  41  43  44  44  45  47  45  37  33  33  36  41  43  38  36  38  38  37 -10
L                                                                                        L
A -20   40  40  41  43  43  44  46  46  40  37  33  34  39  38  38  42  39  37  36 -20  A
T                                                                                        T
I -30   39  39  40  42  43  44  45  44  43  39  37  33  36  38  42  41  38  37  35 -30  I
T                                                                                        T
U -40   38  40  39  40  41  42  43  44  44  36  35  37  43  42  41  38  37  35  33 -40  U
D                                                                                        D
E -50   38  39  40  39  40  41  41  42  42  42  42  41  41  40  39  37  36  34  35 -50  E
  -60   35  38  39  39  40  39  39  40  40  40  40  40  39  38  36  35  34  32  35 -60
  -70   35  35  33  37  39  39  39  40  35  35  35  35  36  35  34  34  32  35  35 -70
  -80   31  31  30  30  36  36  36  36  36  36  36  36  36  30  30  31  31  31 -80
        160     140     120     100    80      60      40      20      0      -20
                              GEOGRAPHIC LONGITUDE
                                      -29-
```

FIG. 12-48. Computer printout of available signal to noise over hemisphere with buoy located in North Atlantic.

venience, the equator is the horizontal row of figures running across the center of the plot. The buoy position shown is different from Fig. 12-47 in that it is in the North Atlantic, latitude 33.00 north, longitude 72.00 west. As it is somewhat difficult to visually relate the data points to geographic coordinates relating to land masses, an artist has prepared a somewhat more revealing illustration, shown as Fig. 12-49, where the conditions are essentially identical to those presented in Fig. 12-47. In many ways it is quite dramatic that the signal from a relatively low-powered buoy is so strong and covers such a wide area. It is most interesting that quite a strong signal is obtainable over the Indian Ocean from a buoy well off the coast of Japan. It is also evident that there is little useful signal to be obtained under the initial conditions presented as one goes further east than Honolulu.

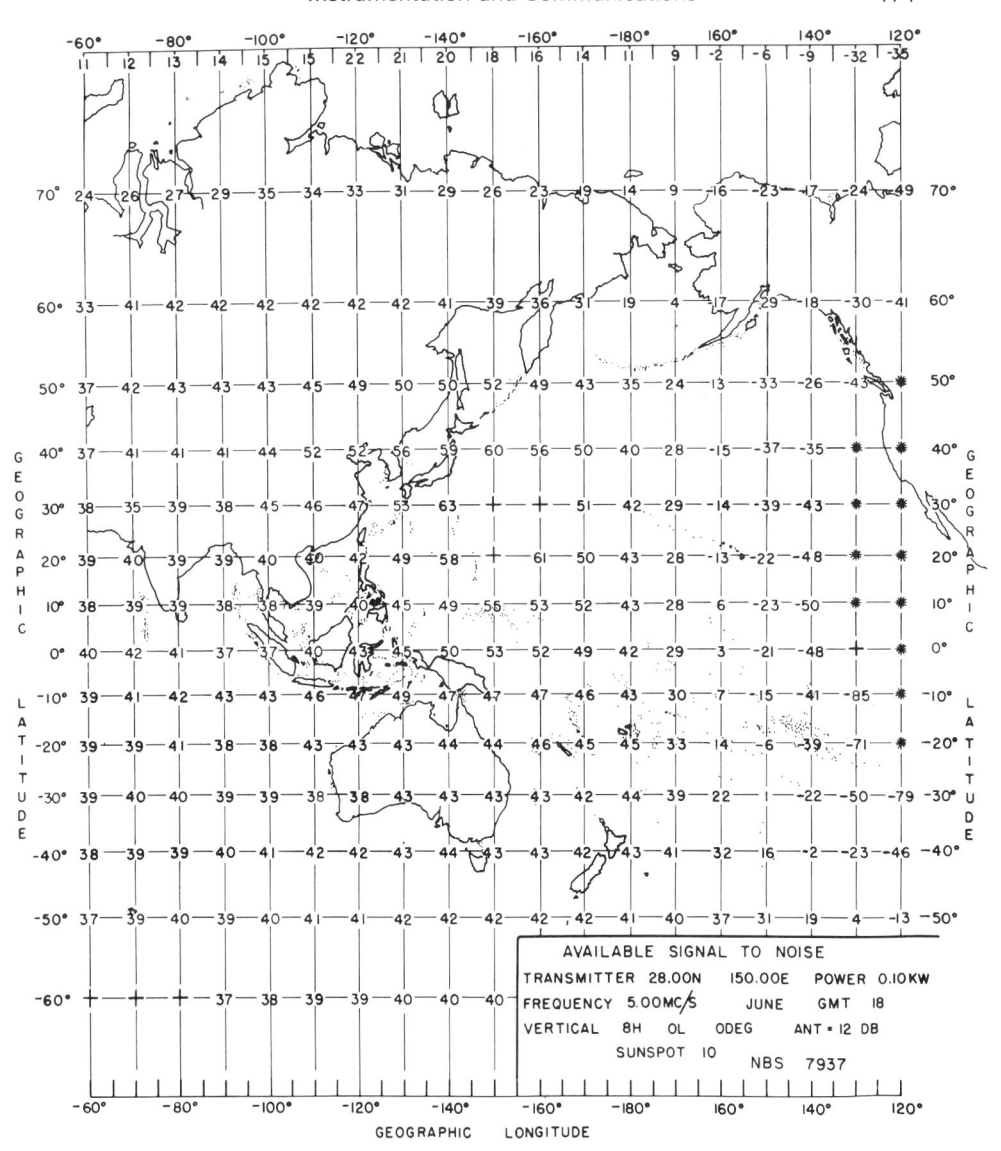

FIG. 12-49. Portion of hemispheric plot of available signal to noise superimposed upon local geographic details.

12.15 THE "GIANT" BUOY

As a result of a study and model testing by the Convair Division of General Dynamics Corporation under an Office of Naval Research contract, a full-size, ocean going buoy for the collection of meteorological and oceanographic data has been fabricated. The Convair buoy, which has been nicknamed the "Giant" buoy, is some 40 ft in diameter and is designed to be wholly self-supporting for a minimum of one year at sea without maintenance. It is capable of being anchored in water of any depth and it is designed to have a high probability of surviving extreme storm and hurricane conditions. Two of the "Giant" buoys have been built, one of which has already been through two hurricanes and has performed admirably during the entire time. Details of the construction and operation of the buoy have been published in *Geo-Marine Technology* (Devereux and Jennbury, 1966; Brown, 1966). Figure 12-50 shows the much-decorated buoy as it was launched by the shipyard. When we see the size of the men on the buoy the nickname does not seem inappropriate. It is perhaps well to comment on some of the fortuitous byproducts of the really large "Giant" buoy. For many years, one of the major difficulties of oceanographers, on both the east and west coasts of the United States has been the fact that their buoys have been subject to all types of vandalism and consequent damage. Among the types of vandalism experienced are damage as a result of shooting with high-powered rifles. Perhaps as an extreme case, however, is a buoy belonging to the U.S. Navy and located off Point Conception, in which vandals cut off some of the hardware with cutting torches. There is no question that the

FIG. 12-50. "Giant" buoy, as launched by shipyard at Jacksonville, Florida. Note size of men.

Instrumentation and Communications 473

adult vandals present serious problems and in view of the fact that no maritime law at present applies to the protection of buoys, the problem may well persist. It is, therefore, most interesting to report that in over a year's usage the "Giant" buoy has yet to show any evidence of successful vandalism, other than the shooting out of its beacon light. It is evident that it has not been boarded. There are perhaps several reasons for this, all of them having to do with its sheer bulk. If one attempts to board the buoy at sea from a small boat, such as a skiff, it is obviously very hazardous in view of the hull geometry of the buoy and its high freeboard. There are no handholds provided and there would be a distinct element of very real risk involved. On the other hand, attempting to board the buoy from larger craft present significant hazards to the boarding vessel. The massive buoy is substantially solid and in a seaway is capable of inflicting damage on any

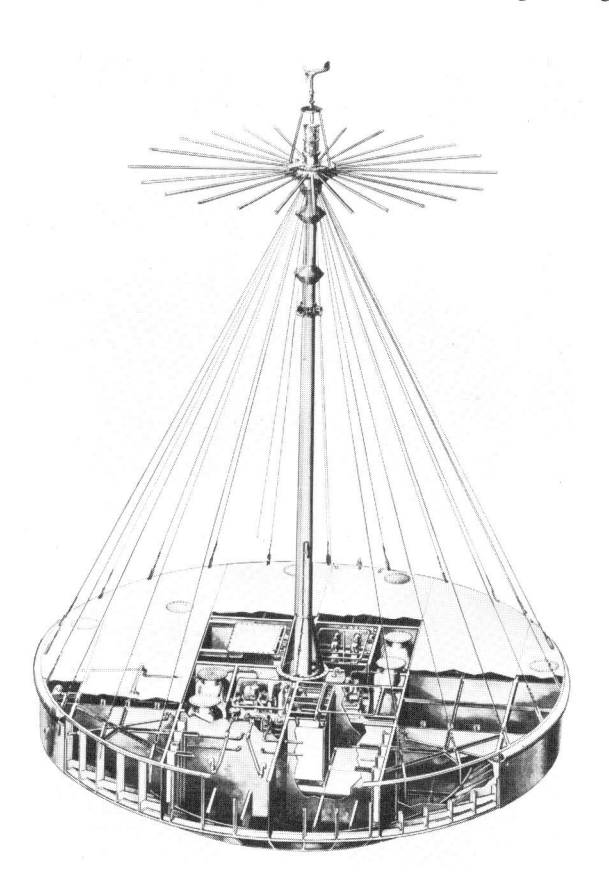

FIG. 12-51. Artist's sketch of Convair "giant buoy".

craft that should come close alongside. In any case, whatever the reasons, it appears that a "Giant" buoy does possess a high degree of immunity from vandalism, which is an exceedingly significant situation. The internal structure of the buoy is highly compartmented, as shown in the artist's sketch in Fig. 12-51. The buoy is powered by two independent propane-fueled internal combustion engines charging a bank of storage batteries. The entire below-deck area is sealed and filled with an anhydrous atmosphere. The mast is a multifunctional element that serves both to support the radio antenna structure which is a disk–cone radiator and as an inlet and outlet snorkel for the internal combustion engine and, of course, support for the warning lights and foghorn. Figure 12-52 shows the "Gaint" buoy being towed to sea in its work dress, which is a great contrast from its "party" dress, shown in Fig. 12-50. The second "Giant" buoy is expected to be installed off the Pacific Coast of the United States early in 1967, at which time it will begin transmissions utilizing high frequencies. It will be actively studied from the standpoint of radio propagation and other essenial operating characteristics.

FIG. 12-52. "Giant buoy" being towed to sea.

REFERENCES

Baños, A., Jr., and J. P. Wesley, 1953, 1954, Scripps Institution of Oceanography Contribution Reference 52-33 (September 1953) and Reference 54-31 (August, 1954). "The Horizontal Electric Dipole in a Conducting Half Space," (Note: new and revised edition, *Dipole Radiations in the Presence of a Conducting Half Space*, A. Baños, 1966, Pergamon Press, Long Island City, New York.

Bleil, B. F., ed., 1964, "Natural Electromagnetic Phenomena below 30 Kc/s," *Proc. NATO Advanc. Study Inst., Bad Hamburg, Germany, July 22-August 2, 1963*, Plenum Press, New York.

Boden, B. P., and Elizabeth Kampa, 1957, "Records of Bioluminescence in the Ocean," *Pacific Sci.*, *11* (April, 1957).

Boden, B. P., Elizabeth M. Kampa, and J. M. Snodgrass, 1960, "Underwater Daylight Measurements in the Bay of Biscay," *Marine Biol. Assoc. U.K.* **39**, 227-238 (1960).

Boden, B. P., Elizabeth M. Kampa, and J. M. Snodgrass, 1965, "Measurements of Spontaneous Bioluminescence in the Sea," *Nature* **208**, 1078–80 (December 11, 1965).

Brock-Nannestad, L., 1964, Unclassified NATO Technical Report No. 25, 2nd ed. "Bibliography on Electromagnetic Phenomena with Special Reference to ELF (Extra Low Frequency) 1-3 Kc/S," Saclant, ASE Research Center, La Spezia, Italy (October, 1964).

Brown, N. C., 1966, "Monster Buoy's Sensor Package," Reprinted from *Geo-Marine Tech.* **2**, V-198–V-210 (June, 1966).

Brown, R. H., 1938, "De Brahm Charts of the Atlantic Ocean, 1772-1776", *Geograph. Rev.* **28**, 124-132 (1938).

Buchanan, C. L., and M. Flato, 1961, "Putting Pressure on Electronic Circuit Components", *ISA J.*, **8** (November, 1961), 38.

Devereux, R. S., and F. D. Jennings, 1966, "A Special Report: ONR Ocean DATA System," reprinted from *Geo-Marine Tech.* **2**, **V-105-V-125** (April 1966), 8-29.

Ferris, H. G., 1953, "The retical Analysis of a Temperature Pr be Developed for Ocean Bottom Studies," Scripps Institution of Oceanography Contribution Reference 53-54 (May, 1953).

Gaul, R. D., J. M. Snodgrass, and D. J. Cretzler, 1963, "Some Dynamical Properties of the Savonius Rotor Current Meter," *Marine Sciences Instrumentation*, Vol. 2, Instrument Society of America, Plenum Press, New York, p. 115.

Groves, G., K. Hasselman, G. Miller, W. Munk, W. Powers, F. Snodgrass, 1966, "Propagation of Ocean Waves Across the Pacific," *Phil. Trans. Royal Soc. London*, **A259**, 431–497 (May).

Haydon, G. W., D. L. Lucas, and R. A. Hanson, 1962, 1963, "High Frequency Radio Propagation Predictions for Data Transmission from Ocean Buoys," NBS Report No. 7256 (June 13, 1962); "High Frequency Radio Propagation Predictions for North Pacific and North Atlantic Ocean Buoys," NBS Report No. 7284 (July 19, 1962); "High Frequency Radio Propagation Predictions for Oceanographic Data Stations at Selected Locations," NBS Report No. 7937 (August 15, 1963).

Holm, Carl, 1966. Marine Anchor, Patent No. 3,280,782. Oct. 25, 1966.

Holmes, R. W., and J. M. Snodgrass, 1961, "A Multiple-Detector Irradiance Meter and Electronic Depth-Sensing Unit for use in Biological Oceanography," *J. Marine Res.* **19**, 40–56 (March 15, 1961).

Horton, J. W., 1959, *Fundamentals of Sonar*, Chapter 3, United States Naval Institute, Annapolis, Md.

Isaacs, J. D., and A. E. Maxwell, "The Ball-Breaker, a Deep Water Bottom Signalling Device," *Sears Found. J. Marine Res.*, p. 66 (1952).

Knopf, W. C. and H. A. Cook, eds., 1965, Marine Sciences Instrumentation, Vol. 3, pp. 85–117, Plenum Press, N.Y.

Liebermann, L. N., 1962, "Other Electromagnetic Radiation," *The Sea*, Vol. 1, Interscience, New York, pp. 469-75.

Arthur D. Little, Inc., Cambridge, Massachusetts, for the Department of the Navy, 1966. "Expendable Bathythermograph (XBT) System Evaluation for Tactical Sonar Application," Naval Ship Systems Command, Nobsr-93055, Project Serial Number SF 101-03-21, (ADL Report No. 4150866), p. 7 (August, 1966).

MacRadyen, A., 1949, "A Simple Device for Recording Mean Temperature in Confined Spaces," *Nature 164*, 965-66 (December 3, 1949).

Nichols, M. H., and L. L. Rauch, 1956, *Radio Telemetry*, 2nd ed., Wiley, New York, Appendix 14, p. 436.

Proc. Government-Industry Oceanographic Instrumentation Symp., Washington, D.C., August 16-17, 1961, p. 232, Miller-Columbian Reporting Service, Washington, D.C. (1961).

Revelle, R., and A. E. Maxwell, 1952, "Heat Flow Through the Floor of the Eastern North Pacific," *Nature* 170, 199 (August 2, 1952).

Richards, A. F., and J. M. Snodgrass, 1956, "Observations of Underwater Volcanic Acoustics at Barcena Volcano, San Benedicto Island, Mexico and in Shelikof Strait, Alaska," *Am. Geophys. Union. Trans.* 37, 97-104 (February, 1956.)

Rudnick, P., and J. M. Snodgrass, 1954, "Field Measurements for an Electric Dipole Submerged in the Sea," Scripps Institution of Oceanography Contribution Reference 54-30 (May 14, 1954).

Savonius, S. J., 1931, "The 'S' Rotor and its Applications," *Mech. Eng.* 53, 333-338 (1931).

Schevill, W. E., and W. A. Watkins, 1962, *Whale and Porpoise Voices, a Phonograph record*, 24 pp., 35 text-figs., phonograph disk, Woods Hole Oceanographic Institution, Woods Hole, Mass.

Schevill, W. E., and W. A. Watkins, 1966, "Sound Structure and Directionality in Orcinus (Killer Whale)," *Zoologica*, Scientific Contributions of the New York Zoological Society, **51**, 71-76 (Summer, 1966).

Snodgrass, J. M., 1958, "Bathymetry Telemetry," *Proc. Nat. Telemetering Conf., Baltimore, Md.*, pp. 139-146, Institute of the Aeronautical Sciences, New York.

Snodgrass, J. M., 1961, "Introducing Oceanography," *ISA J.*, **8** (August, 1961), 75.

Snodgrass, J. M., 1963a, "Problems of the Oceanographer in the Space Age—1963," *Intern. Telemetering Conf. Proc., London, England, September, 1963*, **1**, 182-191 (1963).

Snodgrass, J. M., 1963b, "The Requirements for Radio Communication Facilities and Frequency Channels Formulated by the International Oceanographic Commission," *Radio Tech. Comm. for Marine Services, Symp. Papers*, **1**, 11-21 (May 14-16, 1963).

Snodgrass, J. M., 1964, "Oceanographic Communications and Telemetry Involving Buoy Systems," *Trans. 1964 Buoy Tech. Soc., March 24-25, 1964, Washington, D.C.*, pp. 285-300.

Snodgrass, J. M., 1966a, "Radio Telecommunications for Data Telemetering from Ocean Bouys," *Proc. Radio Tech. Comm. Marine Services Meeting, March 9, 1966, Williamsburg, Va.* (1966).

Snodgrass, J. M., 1966b, "Undersea Instrumentation Reliability: Where Away?", *IEEE Trans. Aerospace Electron. Syst. AES-2*, 6, 631 (November, 1966).

Spilhaus, A. F., 1938, "A Bathythermograph," *J. Marine Res*, **1**, 95-100 (1938).

Stommel, H., 1954, "Serial Observation of Drift Current in the Central North Atlantic Ocean", Scripps Institution of Oceanography Contribution Reference 54-26 (1954).

Sverdrup, H. U., M. W. Johnson, and R. H. Fleming, 1942, *The Oceans*, p. 82 Prentice-Hall, Englewood Cliffs, N. J.

Terry, R. D., ed., 1965, "Communications," North American Aviation, Inc., El Segundo, California; *Ocean Eng.*, 3, 58-67, National Security Industrial Association (1965).

Von Herzen, R. P., A. E. Maxwell, and J. M. Snodgrass, 1962, "Measurement of Heat Flow Through the Ocean Floor," *Temperature, Its Measurement and Control in Science and Industry*, Vol. 3, Reinholt, New York, Part 1, pp. 769-777.

Wenz, G. M., 1964, "Curious Noises and the Sonic Environment in the Ocean," *Marine Bio-Acoustics*, W. N. Travolga ed., Pergamon Press, New York.

CHAPTER 13

Undersea Ambient Environmental Habitation and Manned Operations

GEORGE F. BOND

Historically, in his efforts to gain dominion over the seas of this planet, man has sought the means of deeper and longer ocean penetrations, incorporating the capability of free-ranging activity. His lack of success in this endeavor may best be underscored by the fact that in the year when an astronaut was able to emerge from his capsule into the near-vacuum of space more than 100 miles above the earth, the free-diving aquanaut could descend only about 100 m beneath the surface of the ocean.

The reasons for this disparate state of affairs may seem obvious on preliminary examination. Considering the weight of seawater, a pressure differential will be imposed on the man diving to only 33 ft of depth that will exceed that to be encountered by the astronaut on the surface of the moon. Again, comparing decompression problems, it can be seen that through proper denitrogenation, the astronaut can be guaranteed protection against bends in all save the most severe accidental circumstances, for example, rupture of his space suit during Extra-Vehicular Activity (EVA). Such is not the case with the aquanaut exposed to high ambient pressures of inert gases over varying periods of time. For this man there is virtually no sure protection against bends; and the cure of this serious disorder does not consist of simply returning to atmospheric pressures of sea level, but rather requires application of added pressure, often many times that of his original exposure. Finally, it might be surmised that the environmental hazards of the aquanaut exceed those of the astronaut by virtue of the fact that the funding of the two programs is so disproportionate as to assume antipodal relations.

So much for the superficial comparison of the two programs of environmental extremes. A re-examination of the situation discovers some commonalities that cannot be denied, and may one day spark the drive for combined research into the areas of both inner and outer space. In this context, let us examine briefly the commonly shared problems of the astronaut and the aquanaut.

To begin with, both explorers face the major hazard of decompression sickness, or bends; and, although this hazard is minimized for the astronaut, the fatal end results of its occurrence stand as a stark reality. Next, it should be remembered that the astronaut and the undersea swimmer are almost equally weightless, and hence subject to Newton's third law of motion; and, although the latter is working in a medium about 850 times denser than that of the astronaut on EVA, the former must probably exert greater comparable effort to overcome the resitance of an inflated space suit. Again, with both extremes of ambient-pressure exposure, a critical feature is that of the inspired breathing mixture, its monitoring and control. The astronaut has a relatively simple single-gas oxygen supply, whereas the aquanaut requires a multiple-gas mixture, synthesized in proportion to pressure and often with extremely low percentages of the oxygen component. Although these differences exist, control of the breathing mixture remains a common denominator between the two space programs. In further extension of this comparison we readily see the identity of problems that exist with respect to crew selection and training, small-group interaction, captive-atmosphere contaminant control, sensory deprivation and/or alteration, stress phenomena, and a host of other problems not directly related to the engineering obstacles that harrass both programs.

Still the question persists: Why, until so recently, has man made virtually no progress as a free agent in the ocean depths—or, more specifically, on the continental shelves of the world? The answer is easily understood once the constraints of conventional diving techniques are clarified. These conventional diving practices call for the undersea worker to depart from sea level, descend to his working depth, and then to return to the surface, the total sequence constituting a single working dive. Such dives, although satisfactory for brief bottom stays with short periods of work, are quite inadequate for the purpose of exploration or exploitation of the continental shelves. The physical factors which so severely limit the usefulness of this type of diving largely stem from the established laws of Boyle, Dalton, and Henry. Thus, since the amount of inert gas driven into solution in the blood and tissue fluids of the diver has a direct relation to the partial pressure of the breathing gas and to the duration of exposure at depth, the deep-sea diver will rapidly accumulate a dissolved gas load in his body, proportional to the depth and time of his dive, that will require equivalent and insupportable increases in decompression times for safe return to sea level. To further compound the problem, the ratio of required decompression time to useful time on the ocean bottom increases with depth in a nearly exponential rather than linear manner. Thus, a single 200-m dive to the outer margin of a given continental shelf, with only *one minute* of bottom stay, would require more than 12 hr of staged decompression for the diver, who could accomplish virtually no useful work during his 60-sec visit. From this it might logically be concluded that working dives below 200 ft are rarely worthwhile in terms of useful human endeavor.

Faced with this situation, diving physiologists sought an oblique approach to the problem. In 1957 the novel concept of "saturation" diving was propounded. This new philosophy of undersea manned existence was quite compatible with the physical laws that so constrain the conventional diver. In essence, the saturation diving system postulated that the undersea worker should be provided with a sea-floor house, pressurized to ambient pressure with a suitable breathing-gas mixture. From this habitat, the divers could exit directly into the surrounding ocean environment, doing many hours of useful work at this depth, and returning to the safety and comfort of the underwater house as required. Since no substantive change of pressure would be involved, there would be no decompression penalty for these excursions. Rather, the decompression of the "saturated" diver would be accomplished in a single linear ascent to the surface, following days, weeks, or even months of useful work on the ocean bottom. This concept, if workable, would seem to offer the needed breakthrough in manned undersea venture.

Following the guidelines of the saturation-diving concept, animal and subsequently human experimental work was vigorously pursued. Within six years it became evident that proper gas mixtures could be provided and adequate decompression schedules promulgated, to permit prolonged human exposures at considerable depths on the ocean floor. Shortly thereafter, volunteer Navy subjects participated in Operations SEALAB I and SEALAB II, conclusively demonstrating that, within the state of present physiological and engineering art, man could indeed become a useful inhabitant of inner space.

In summary, more than 10 years ago there arose a requirement to place be possible for a saturated Aquanaut to live and do useful work on the a deep-sea diver at the outermost limits of the continental shelf, there to remain and do useful work for long periods of time. The best solution to this problem lay in the area of saturation diving, which postulated that the ocean-bottom dweller would become totally saturated by the gas which he breathed. This man, now known as an aquanaut, could do many hours or days of useful work at saturation depth, playing the decompression penalty only at the end of his ocean bottom stay. In this manner, it might ocean bottom for 30 days or more, with a proportionally short period of decompression. Such is the philosophy of saturation diving. The purpose of this introduction has been to acquaint the reader with a historical background of the story of the Man-in-the-Sea program. Since the design of this narrative is to provide a cursory examination of the engineering problems involved in saturation diving, the following rubrics will be directed to this important aspect of the undersea program. In a condensed presentation the following parameters will be surveyed: (a) the external and internal environment, including the habitat; (b) the aquanaut himself; (c) the aquanaut equipment; and (d) predictions for the future of man's invasion of the inner space of our world's oceans.

13.1 THE EXTERNAL ENVIRONMENT

The ocean environment, or more specifically, that of the waters covering the continental shelves of this globe, is clearly hostile. The threatening aspects of this undersea world relate to the water temperature and opacity and to the dangers of the marine biota customarily found in this environment. Each of these environmental hazards will have individual or synergistic effects on the Aquanaut and/or his equipment. In this context, some major problems will be examined.

Although the adverse effects of the submarine environment have been most publicized relative to the aquanaut himself, it is probable that hardware malfunction related to this exposure may be of equal importance.

Generally speaking, the Man-in-the-Sea program has a history of unhappy experiences with respect to immersed items of hardware. As might be anticipated from the unique nature of the program, much of the equipment required for safe undersea existence was not originally designed for other than sea-level, dry-environment reliability. When such hardware components are now exposed to an environment which is chemically corrosive, highly pressurized, 800 times denser than air, and populated by marine life with fouling and other destructive characteristics, severe engineering problems must be anticipated. Judging from the experience of the SEALAB series, Murphy's law of the sea can be postulated: virtually all equipment placed in the ocean will become (a) fouled, (b) lost, or (c) seasick.

13.2 THE INTERNAL ENVIRONMENT

As we proceed with a program to give man a living and working capability at the outer reaches of the continental shelf, at about three times the depth of SEALAB II operations, it is prudent to recognize the formidable problems that face the physiologist and engineer alike. The physiological unknowns, although still considerable, have at least been under scientific attack for many decades, and show promise of resolution in due time. In the case of the engineering fraternity, however, the situation is less promising. Until very recently there was no incentive to examine the engineering problems associated with hyperbaric synthetic-gas atmospheres. Equipment requirements for manned exposures to such a situation were limited to a handful of U.S. Navy recompression chambers, offering no real inducement to engineering development in this area.

This is no longer the case. A priority program now calls for development of a free-ranging, manned capability to a depth of 100 fathoms within less than five years. Thus the challenge to all engineering disciplines is not only of ominous proportions, but is time compressed as well. The purpose of this cursory inspection is to identify some of the critical engineering problems with which we are faced in the Man-in-the-Sea program.

In the course of the two SEALAB experiments it was possible to rely on modified shelf items of hardware to meet the engineering requirements of the program. As greater depths are sought, such a makeshift arrangement will no longer be possible. As we proceed to 15 or many more atmospheres of ambient living pressure, a considerable program of engineering development must be anticipated.

Two basic factors become apparent as we examine the new engineering requirements imposed by the exigencies of the Man-in-the-Sea program. The first problem relates to the increased ambient pressure required for saturated existence at great depths. Here, we must be assured that all hollow vessels are constructed to tolerate the pressure differentials involved. This is one of the simpler problems, involving only such items as light bulbs, oscilloscopes, and Bourdon tubes, as a start. Next we must take into account the inevitable collapse of all cellular insulating material, with resulting loss of efficiency. Finally, it must be noted that compression of the atmosphere results in an increase in molecular availability, which, in turn, will greatly change the factor of heat transfer.

By far the major problem posed by the synthetic atmosphere, however, relates to the use of helium as the inert-gas component of the atmosphere. Bitter experience has repeatedly shown us that virtually no shelf equipment, either mechanical or electronic, will function reliably in a helium atmosphere. Since practically all of the common items used in undersea living are designed to function at sea-level pressures, and in an air atmosphere, it is clear that extensive modification or—more often—developmental engineering will be required.

In considering the engineering problems encountered in undersea habitations, it is vital to bear in mind the enigmatic and frustrating properties of helium gas. At sea level, it exhibits a thermal conductivity equal to 6 times that of air; and, at increasing depths, with resultant mounting molecular availability, this problem tends to escalate. Thus electronic gear that is thermally dependent in any respect will be partially or totally disabled in the exotic He—O_2 atmosphere of the habitat. This in itself, is not too disturbing, and should be amenable to careful engineering design. Nonetheless, unexpected problems will continue to confront the engineer who is not totally oriented to the properties of helium.

To cite a few examples of thermal conductivity problems, I look to some dangerous incidents that occurred during the SEALAB II experiment. First, a high-voltage line chafed through, and began arcing toward the habitat. The circuit breakers, thermally conditioned for air conductivity, did not work; it was only through heroic measures that all electrical circuits to the habitat were broken. Next, we found that the Calrod burners of space heaters and stoves did not turn red when activated. Because of this there was a traumatic tendency for the aquanauts to touch the coils in order to determine whether they were functional. The resultant contact burns were generally treated with success.

Finally, the thermal conductive properties of helium are such that the habitat must be heated to temperatures in excess of 85°F to insure shirt-sleeve comfort to the aquanaut, who functions as a radiating blackbody.

If the thermal conductivity of the helium gas is a troublesome factor, surely the other physical characteristics of the problem child are of equal weight. It is generally known that helium is a remarkably pervasive gas, capable of passing through any material, including glass, given sufficient time and pressure differential. This property poses a very real hazard to undersea-habitat designers, since it must always be assumed that, given sufficient time, helium will saturate any gas-filled cell, and will there remain to exert all of its remarkable physical properties. Exposed TV cameras will be invaded, with resultant loss of focus control; insulative cladding of freeze boxes and of the habitat proper will be first crushed, then invaded by helium, suffering tremendous loss of efficiency. Even such a prosaic item as the pancake is not exempt: the internal bubbles of the flapjack are made up of helium, and will combine forces with the external atmosphere to prevent thorough cooking of the interior of this item.

Even this is not the sum of the helium problem. Because of its Reynold's number, so far out in right field with respect to our more conventional inert gases, all engineering systems dependent on turbulent-versus-laminar flow must be re-examined; and, in the case of gas-flow appliances, the equation becomes more complex, since gas density must also be considered. In another system, not amenable to engineering development, the human voice is so distorted by helium as to require intervention of an electronic helium voice unscrambler. In conclusion, it is well to remember that the kinematic viscosity of this gas is a factor in all calculated gas-flow equations.

The internal equipment of the future sea floor habitat continues to pose engineering problems, all related to the increased ambient pressure, and to the peculiar characteristics of either helium or hydrogen, whichever inert-gas diluent is to be used as the major atmospheric component. In this connection it would be useless to discuss each internal component since every piece of operating equipment without exception will be radically affected by the presence of lighter inert gases and high ambient pressures. This is a problem which has been largely ignored by the engineering fraternity. As a result of this understandable neglect, virtually no currently available equipment will work in the SEALAB of the future.

If we consider the needs of the man in the SEALAB habitat, it is necessary to provide the temperature–humidity profile best fitted to assure comfortable living conditions in the underwater house. To date, these parameters have not yet been determined. Control of the atmospheric components is likewise poorly developed. As greater depths are attained, increasingly elegant control of the atmosphere of the habitat is required. If we wish to maintain the O_2 partial pressure not to exceed 200 mm Hg, and a CO_2 partial pressure of not more than 10 mm Hg, it will be necessary to develop more reliable sensors, to meet the greater depth requirements of the program. When we add to this the problem of increased gas density with associated problems of heat transfer, the engineering situation is further complicated.

13.3 THE HABITAT

Generally speaking, the design of the habitat for operation at our ultimate chosen depth should pose no severe problems. It must be borne in mind that a primary requisite for the habitat of the future is achievement of autonomy for the ocean explorers. In that connection, the last remaining deficiency is that of a power source to be located on the ocean floor, and engineered to meet the requirements of a fully functional undersea habitat. Obvious candidates for this power source are modified nuclear-reactor–turbine combinations, radioisotope sources, and fuel cells. Each of these systems has points of attraction, and obvious areas of weakness. A well-planned engineering tradeoff study should help to identify the power source of greatest promise. It should be noted however, that a hydrogen–oxygen cryogenic system may best fit the needs of ultimate deep living. This concept of life support would involve two cryogenically stowed gases operating in a system to provide power via a fuel cell, yield a good breathing mixture, and power a catalytic burner to burn atmospheric contaminants, and provide the heat requirements for the habitat.

At this point the engineering requirements for the ultimate fixed undersea habitat may be simply stated. The habitat must be capable of withstanding an internal pressure of 25 atm; it must be engineered, within and without, to cope with a synthetic atmosphere of either helium–oxygen or hydrogen–oxygen; and, finally, an automonous ocean-floor power source must be provided.

Looking ahead, however, some thought must be directed to the development of a mobile undersea habitation, since full utilization of the saturated diver concept will require mobility, for access to all continental shelves of the world. Obviously, the ideal vehicle for this venture would be a nuclear-powered submarine, configured in such a fashion that a large compartment of the vessel's pressure hull would be capable of independent pressurization to ambient sea depth with a synthetic breathing mixture. This compartment would be occupied by aquanauts, who would have the capability of exiting from the bottomed submarine to do many days of work, whether it be salvage or other engineering/scientific tasks of military significance. In this case power supply would be no problem, nor would bottom stay be limited; in addition, the factor of weatherproof undersea operations could lend considerable importance to the development. At the present time it seems most desirable to examine the tradeoffs between new construction and modification of an existing nuclear-powered submarine, with careful consideration of the probable assigned missions of such a mobile SEALAB unit.

13.4 THE AQUANAUT

For the purpose of this chapter, little need be said about the physiological effects to date observed on our subjects. Physiologically, we have found few changes of consequence in our aquanauts during their stay in the habitat. Outside, in the hostile environment of the ocean, we have no means of monitoring this diver, nor do we have any reliable means of communication. These are problems of undersea engineering and biotelemetry which must be solved by engineers, not physiologists. The problems are formidable, and I am not hopeful of quick solutions.

Briefly, it can be stated that so far the only significant physiological findings have related to the levels of those blood enzymes which indicate physical or emotional stress. In the case of the aquanauts these stress indicators showed a sharp rise immediately upon commencement of the exposure in SEALAB. After about five days of bottom stay, however, the levels tended to return to normal. Thus the *nonspecific stress effects* represent an abnormal physiological finding that required close attention but probably will not result in damage to the aquanauts. Two more physiological hazards must be carefully monitored, although to date neither has presented a problem. These are the potential hazards of damage to bones resulting from inadequate decompression schedules, and possible damage to lungs, due to improper selection of breathing-gas mixtures. It seems certain that both of these latter hazards can be avoided by prudent management of the saturation dive and subsequent decompression.

13.5 AQUANAUT TEAM SELECTION AND TRAINING

The selection of personnel to participate in any undersea venture is always difficult; but this problem is considerably compounded when we seek to select a fairly large group of aquanauts who must live and work together on the ocean floor and under very hostile conditions for long periods of time.

During the early phases of the Man-in-the-Sea program this selection process was simple, albeit unscientific. Since the number of participating personnel was relatively small, selection was made by the principal investigator on the basis of maturity, physical condition, and personal observation of deep-diving ability.

With expansion of the program, this system could no longer be tolerated. A new technique, based on fairly reliable criteria of aquanaut selection, was established. Utilizing the results of psychophysiological test procedures used and refined in the early human laboratory tests and in subsequent SEALAB series, a preliminary protocol of test procedures was developed. This system, although as yet quite imperfect, offers a reasonably good array of selection criteria.

What physical and psychological qualities are to be sought in the aquanauts of the present and future? Although the answers to these questions will not be immediately forthcoming, some clues have been derived. First, it has been established that the successful aquanaut will have a tremendous interest in scientific matters, regardless of his background. This interest covers not only matters of oceanographic and marine ecological focus, but extends to embrace almost any area of physical, physiological, or philosophical science. To further complicate the selection scenario, test results indicate that the more successful aquanauts favor physical fitness, uninhibited night life, undersea living, and wood carving.

From the point of view of the principal investigator, these selective test criteria present a very real problem. First, it is required that all aquanauts be divers *par excellence*; second, they must be unusually motivated for the job; finally, they should ideally be in perfect physical condition. Whereas these criteria can be met without difficulty in the case of military personnel, some of these conditions must be waived when civilian scientists are involved in the experimental exposures, since these individuals are generally included in the program on the basis of specialized and desirable professional skills. Thus it is frequently necessary to qualify, train, and utilize aquanauts who cannot meet the rigid physical criteria established for the military members of the aquanaut team.

If some compromise must occasionally be accepted with respect to physical standards, the team-training protocol permits no such exceptions. The training schedule of all aquanauts, civilian and military alike, must be identical and concurrent, to guarantee the safety and efficiency of the underwater operation.

Proper aquanaut training will inevitably spell the success or failure of the program. Currently, the training schedule calls for a continuous cycle of instruction amounting to nearly 12 months. Within this time period the aquanaut will complete qualification as a first-class diver (hardhat); go through a special underwater swimmer's course in use of MK VI and associated swimmer's gear; participate in a number of saturation-diving experiments, as inside-chamber occupant and outside controls operator; and devote at least two months to classroom instruction in subjects ranging from general oceanology to mathematical determination of decompression schedules and development of gas-flow formulas. In spare time, of course, the aquanauts must continually maintain open-sea diving proficiency, while learning every detail of the habitat complex in which they will ultimately live beneath the sea.

The importance of this rigorous training program must be emphasized. To the aquanaut the ocean environment is much more than a liquid medium to be penetrated from time to time, with a reasonably available means of escape to the safety of sea level pressure and dry sod. After a few hours of bottom stay, the ocean dweller must accept the sea floor and surrounding waters as a natural environment, since rapid return to the surface would be fatal. In order, therefore to survive and perform meaningfully in this water environment, the aquanaut must of necessity learn as much as possible about the ocean and its population. Inevitably, he must come to identify himself completely with the total environment of the hydrospace in which he lives; and, in due time, he must come to appreciate the ocean as a living, coherent entity, possessing a complex system of internal physiology, geology, and whether totally unlike any dry-land systems. As this understanding becomes more complete the efficiency and safety of the aquanaut will increase geometrically; and, in time, the undersea worker will induce the hostile environment to work for him. This is the ultimate goal of the training program.

Although it is certain that the psychological stresses of undersea living are severe, it would appear that, with proper selection techniques, actual personnel breakdown and loss of group effectiveness is not likely to occur. It is important that valid criteria for personnel selection be established and tested in anticipation of an expanding program.

13.6 AQUANAUT EQUIPMENT

Discussion of aquanaut equipment will herein touch only on a few select areas that are critical to the undersea program and will require considerable developmental effort. These areas include communications, protective clothing, visibility aids, work tools, Scuba equipments, and propulsion units.

Communications continue to be the most unsatisfactory of all systems in use for aquanauts. Communications from the habitat to surface require interposition of an electronic helium speech "unscrambler" to correct in part the distortions and ablations caused by the helium-oxygen atmosphere. Communications from the habitat to the outside swimmer are possible, for a limited range, at the cost of considerable power. Voice communications from diver to habitat, diver to diver, or diver to surface, however, cannot presently be achieved with available equipment. The problems here involved are formidable the diver's mouthpiece impedes articulation; exhalation noises provide interference; helium speech distortion precludes intelligible conversation; and omnidirectional broadcasting poses a heavy power requirement. It is quite possible that conventional modes of underwater voice communication must be abandoned altogether.

In the cold environment of continental-shelf waters a major problem to the aquanaut is that of body-heat loss. Here the expected loss to the surrounding water is magnified by the thermal conductivity of the helium breathing mixture. At the present time thermal protection is provided by conventional foam-rubber wet suits, which are virtually useless under conditions of SEALAB operations since the gas cells of the suits become filled with compressed helium, and ambient water temperatures are 45° F or lower. Currently, development is underway with electrically-heated suits that can be powered by batteries, or via an electrical cable in the case of a tethered diver. Additional investigation now includes fluid heating with tube-lined underwear, possibly utilizing a shielded radioisotope heat source. Although all of these developments are promising, to date no completely satisfactory thermal protective garments have been produced.

The problems of decreased visibility in undersea situations deserve considerable attention. In view of the fact that in most continental-shelf areas at depths of our contemplated operations visibility may range from a few inches to 20 ft for unaided vision, it is certain that portable and fixed sources of illumination will be required. These sources should be compact, of long life, and capable of producing as much as 5000 W. At the present time the best state of the part has rarely been applied to diver lighting problems. Consideration must be given to pulsed high-intensity lights, polarized sources, underwater low-luminescent sources, and image itensification.

The undersea worker is faced with many unusual problems, very few of which have demanded solution until quite recently. The conventional hardhat diver has always used whatever standard suit of tools happened to be at hand. This has been the case for two obvious reasons. In the first place the tasks assigned to conventional divers have historically been unsophisticated and capable of accomplishment with use of crude tools. Secondly, conventional hardhat suits are designed so that the diver can make himself "heavy," thus remaining fixed in the vicinity of the job, and not easily dislodged. With the development of Scuba gear and the advent of the untethered underwater worker, the situation has changed radically. With freedom of motion and vastly increased manual dexterity, the task that could be performed by the diver expanded in order of complexity. For these sophisticated technical jobs, an entire new suit of tools must be developed.

Basically, two factors are to be considered in all future developments. First, the Scuba diver-technician, unless anchored in place with restraining devices, is subject to Newton's third law of motion. Therefore all tools to be developed must have torque-free characteristics if they are rotary in character, or of special design if they are to be of the impact type. Some progress has been made on this over-all problem, but a great deal of work remains to be done in developing tools and diver-fixation devices and in determining the exact profile of man's undersea work capability.

Very little will be said here concerning the diving gear of aquanauts, since this is a familiar subject to all, and most engineering details are readily available. For aquanauts working at depths of only 7 atm it has been found that two presently available types of diving gear can efficiently be employed. For the free diver, the MK VI semiclosed diving apparatus can be used for about 50 min without refill. This Scuba gear, charged with a predetermined $He-O_2$ gas mixture, is so engineered that roughly every third exhalation is dumped from the breathing bag to the surrounding water. Flow rate of the gas to be inhaled is a preset value, determined by a simple formula, in which absolute depth and gas composition are the two critical variables. For work that can be performed by a tethered diver, a Hookah system is most desirable, since this imposes no time limitations on the user. With such a device, the respirable atmosphere of the SEALAB habitat is pumped to the diver in one hose, and his exhaled breath is returned to the habitat, where the CO_2 scrubber system (LiOH or molecular seive) handles the carbon-dioxide accumulation.

So much for the 7-atm situation. At the 20-to-30-atm depth levels envisioned for the future, a safe, completely closed-circuit apparatus must be developed for the free diver, and new development in compressors will be required for the Hookah equipment. In the case of the closed-circuit gear a completely reliable device for oxygen monitoring and flow control is mandatory. To date, and after more than 15 years of experimental effort, such a device is not available. With respect to the Hookah gear, compressor development is not yet adequate to answer the problems posed by the increased density of the gas to be pumped to the diver, who may be required to range several hundred feet deeper than his habitat-placed gas-supply system.

The least efficient means of traversing the earth's surface is seen in the mode of an underwater swimmer. At best, a fully-garbed aquanaut swimmer can only hope to maintain a speed rate of about 0.8 knot. Obviously, some devices to provide swimmer propulsion are required. Three sources of locomotion are possible: the "wet-diver" propulsion unit; the dry-hull small submarine; and a tractor-type bottom-crawling device. Space does not permit an evaluation of the developmental problems related to each of these equipments. It is sufficient to say that probably all three types of propulsion units will be required for future Man-in-the-Sea efforts.

13.7 THE FUTURE

It is not likely that man will be content to develop a deep-diving capability of only 20 to 30 atm. Without question the commercial and scientific returns of the continental slopes and submarine canyons will demand that we extend man's free-ranging capabilities to 100 atm—perhaps ultimately to the 12,000-ft depths of the abyssal plains, although this greater reach may be of no more than experimental and historical value. For the purpose of this discussion, we treat the depths of 1000 to 3000 ft as practical goals, reserving the greater depths as speculations for narrative consideration.

It seems certain that within the next three years we will have developed the necessary technology to dive to a depth of 1000 ft. In order to accomplish this goal a tremendous amount of energy, time, and money must be devoted to engineering and physiological research and development. Yet it can be done, since we may predict with confidence that the breathing mixture at this depth will be 0.65% oxygen, and the balance helium. Given this controlling factor, the engineering problems and developments can be tailored to reach the desired end result of a system which will function at this level of atmospheric control.

Below 1000 ft, and jumping to the 3000 ft level, physiological factor will control. Helium can probably no longer be used as the inert-gas constituent, since it now becomes narcotic, and we must move to the left, and bitter end, of the periodic table of elements. Obviously, there remains now only one inert gas which can be used as an atmospheric diluent. This gas is hydrogen; and, although Zetterström made significant dives to 14 atm using a hydrogen–oxygen mix two decades ago, we are not yet prepared, from the point of view of physiology or engineering technology, to devise a diving system that would use such a mix.

Let us speculate on man's ultimate free-diving depth capability—an obviously unknown figure. For the purpose of final discussion, however, consider a depth of 12,000 ft, which is close to the average depth of the earth's oceans. It is *possible*, though not necessarily probable, that man may one day swim freely at this depth. It is feasible, with proper surgical and engineering techniques, to devise a penetration of the human lung airway through a hole in the windpipe. Through this hole, an oxygenated solution of Ringers solution, suitably buffered to accomodate metabolic CO_2 output, could be pumped through the lungs, supplying the metabolic needs of the deep diver. The system would be engineered to maintain a constant pO_2 of 200 mm Hg in the circulating solution, and the rate of pulmonary circulation appropriately compensated through use of a miniature pump. Next, we flood the remaining head sinus cavities of the aquanaut with any fluid, and put him on an elevator for the deepest of all dives. Theoretically, this diver can walk the abyssal plains of our seas for almost 2 hr under these conditions, and be returned to the surface without requirement of decompression, and with only transient, curable damage to his body. This is a way-out projection; but so was saturation diving 10 years ago.

13.8 CONCLUSION

This brief and inadequate treatment of the problems, engineering and physiological, that are foreseen for the Navy's Man-in-the-Sea program, is not designed to provide other than the most meager guidelines for ocean engineers who elect to address themselves to such problems. By way of apology, it must be pointed out that the author of this chapter is medically-oriented, with no academic background in engineering. Nevertheless, by virtue of nearly a decade of investigation in the saturation-diving concept and coupled with an equal time span of wrestling associated engineering problems, it is hoped that some provocative inputs have been made available to the engineering disciplines which will ultimately determine the success or failure of man's drive to dominion over the seas of the world.

SUPPLEMENTAL READING

Human Factors in Undersea Warfare, 1949, Committee on Undersea Warfare, National Academy of Sciences, National Research Council, Washington, D.C.

Man's Extension Into the Sea: Trans. of the Joint Symp. 11–12 Jan. 1966, Washington, D.C., Marine Technology Society, Executive Bldg. No. 828, Washington, D.C., 20005.

Nelson, L. B. *Implications of* "Man-In-Sea" Program, ASME Paper 66-MD-15 (Presented at Conference May 9–12, 1966).

Proc. First Underwater Physiology Symp., Jan. 10, 11, 1955, sponsored by Committee on Undersea Warfare, National Research Council, National Academy of Sciences Publ. 377.

Proc. Second Underwater Physiology Symp., Feb. 25, 26, 1963, sponsored by Committee on Undersea Warfare, National Research Council, National Academy of Sciences Publ. 1181.

Proc. Third Underwater Physiology Symposium, March 23, 24, 25, 1966, sponsored by Committee on Undersea Warfare, National Research Council, William & Wilkins, Baltimore, Md.

CHAPTER 14

Deep-Ocean Work Systems

WILLIAM H. HUNLEY

Deep-ocean work systems may be defined as those systems designed or used for the purposes of finding and extracting natural resources from the waters or the beds of the oceans; finding and recovering manmade objects lost, sunk, or stranded; or the construction, emplacement, servicing, or removal of structures and equipment in deep water. Although the list of such systems is long, some of the most interesting are those systems related to undersea mining, offshore oil operations, diving and salvage operations, and manipulator-equipped deep-submergence vehicles.

14.1 UNDERSEA MINING

Undersea mining has been going on at a modest scale for over 4000 years but has reached the point of being economically profitable only in the past few years. Oceanographic research and exploration and the search for offshore oil and other materials over the past decade, have brought to light sufficient mineral reserves to supply our needs for some metals and other raw materials for hundreds or thousands of years. A list of some of these is shown in Table 14-1.

Several of these minerals are now being mined in the shallower waters of the continental shelves.

Mining systems currently in use

Undersea mining systems now in operation (Fig. 14-1) are variations of four basic types; bucket–ladder dredges, surface-pump hydraulic dredges, wire-line dredges (drags and grabs), and airlift hydraulic dredges.

TABLE 14-1
Major Undersea Minerals

Type	Locations	Water Depth (ft)
Diamonds	Continental Shelf, Africa	0–300
Platinum	Continental Shelf, Alaska	0–300
Gold	Continental Shelf, Alaska, Russia	0–300
Tin	Continental Shelf, Indonesia; Thailand	0–130
Iron	Continental Shelf, Japan	0–100
Glauconite	Outer Continental Shelf	30–6000
Phosphorite	Continental Shelf, California	110–11,400
Manganese nodules	Deep-Sea Floor, Atlantic, Pacific	340–23,000
Calcarous ooze	Deep-Sea Floor, Pacific	340–12,000
Siliceous ooze	Deep-Sea Floor, Pacific	12,500–17,500

Bucket–ladder dredges can be operated efficiently to depths of about 150 ft, but can dredge only in protected waters or in calm weather. The stiff ladder required to support the bucket chain in position against the mining face cannot accommodate large motions of the mining platform.

Surface-pump hydraulic dredges are operational to about 200 ft of depth, depending upon the size and density of the material being dredged. Friction losses in the suction pipe make it difficult to maintain a velocity sufficient to keep large or very dense materials in suspension. Suspension ladders are used on most present hydraulic dredges to support the suction pipe and maintain the suction head in contact with the mineral body. This introduces the same rough-water operating problems as for the bucket-ladder dredge, and limits operations to protected water or calm weather.

Drag-and-grab dredges have proven feasible to depths of about 500 ft for recovery of commercial minerals, and have of course been used to greater depthts for scientific surveys and bottom sampling for scientific pruposes. The cost of operation of a drag-or-grab bucket is greatly influenced by cycle time, which varies directly with depth. Because of the flexibility of the system, these dredges can be operated in much rougher water than stiff ladder–bucket and hydraulic dredges.

Airlift dredges offer considerable flexibility for recovery of loose, unconsolidated material ranging in size from finely divided ooze to cobbles and small boulders. They have been used extensively for applications ranging from dredging diamondiferous gravels, to archaeological excavations, to tunneling under a sunken ship for placement of lift slings, in water depths from a few feet to about 1000 ft. Because of the flexibility of the dredge column, usually made up of sections of pipe joined by sections of

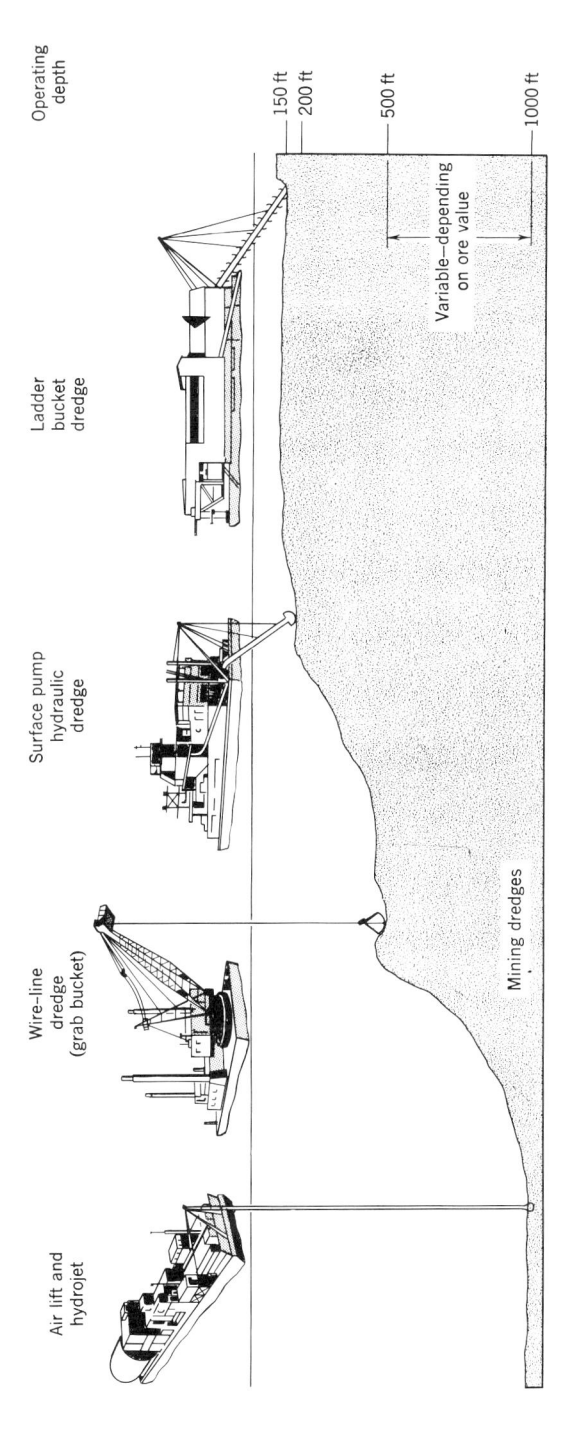

FIG. 14-1. Present undersea mining dredges are variations of four basic types. Two types, the air lift and wire-line dredges are useful at depths beyond 200 ft. (After Wilson).

flexible rubber hose, the airlift can be operated from a barge or ship in exposed locations without damage to the equipment.

Sea diamond mining (Wilson, 1965; Hess, 1965; Webb, 1965)

Perhaps the most spectacular of the offshore mining operations is the mining of diamonds in the open sea off the southern coast of Africa. The diamonds occur in sand and gravel deposits and are now being mined in the area of the West Coast just north of Cape Town. Although it has been known for the past half century or more that diamonds existed on the sea bottom in this general area, successful prospecting was first accomplished in 1961 when the Marine Diamond Corporation began operations with the tug *Emerson K*. Other previous attempts at prospecting in the area failed due to the lack of the two most critical elements required for the process: ultra-precise navigation equipment and suitable dredging equipment.

Navigation. The tremendous amount of material that must be processed to separate the diamonds from the waste dictates that the processing be done on-site, which means dumping the tailings from the processing plant back onto previously mined areas in close proximity to the bottom area being worked. An extremely precise navigation system is required both to fix the location of promising sites during prospecting surveys, and to map mined-out areas with sufficient precision to avoid reworking the tailings from previous operations. Navigation precision far greater than anything presently available for broad ocean navigation must be obtained.

The close inshore locations of the diamond-bearing areas allows the use of recently developed microwave navigation instruments such as the Hydrodist system, whose line-of-sight range of about 25 miles, and requirement for fixed measured baseline between base stations, precludes its use over most of the open ocean. Using the microwave system, the ship's distance from two shore stations can be measured by measuring the time of travel of the radio beam from the shore station to the ship and back. Ships' positions can be plotted continually with an accuracy of plus or minus 3 ft.

Mining platform. Operations in the South African sea-diamond mines have shown a rectangular barge to be superior to ship-shaped hulls for this application. The advantage of the barge-type hull stems from several important factors:

1. less costly construction;
2. compartments that are more watertight and have a greater resistance to damage in case of grounding;
3. large deck areas for equipment;
4. low freeboard, allowing opertions closer to the waterline.

To maintain accurate position and to provide precise maneuverability over the mine field, the barges are equipped with four anchors, streamed one from each corner with about 3000 ft of cable, allowing coverage of an area approximately 4200 ft^2 between sets.

Dredging equipment. Early dredging operations by the Marine Diamond Corporation were carried out using airlifts and hydrojet lifts to pump the diamond-bearing gravel aboard the barge for processing. Since the loose material on the bottom overlies the heavier diamonds, the suction end of the dredging device must be large enough to pass large boulders continuously without stoppage and small enough to enter gullies and crevices in the bottom.

Air lift. The air lift (Fig. 14–2) consists of an open tube with its lower end close to the bottom. In the diamond-dredging application, compressed air at 100 psig is injected into the airlift tube near the lower end. The air, expanding and rising in the tube, decreases the density of the contents of the tube and "floats" the column of water out of the top of the tube and into the processing plant.

The inrush of water at the lower end of the tube entrains the loose gravel, cobbles, and boulders from the bottom and, since the water column rises at a speed greater than the terminal velocity of the gravel falling in water, carries it entrained to the top. Periodic high-pressure air blasts are introduced into the column to keep the contents agitated. High-speed water jets are fitted at the suction end of the lift to aid in loosening the aggregate and in keeping it suspended and fluid at the intake.

Hydrojet lifts. The hydrojet lift is arranged in essentially the same fashion as the air lift, but uses the kinetic energy of high-velocity water jets to impart motion to the water column in the lift and provide the head to pump the water column and its entrained gravel onto the barge.

The air lift has an efficiency advantage over the jet lift, since it requires only 0.7 hp/ton/hr dredged, whereas the jet lift requires 1 hp/ton/hr dredged. The jet lift has the advantage, however, of being useable in series with surface hydraulic pumps.

Centrifugal surface pumps require only $\frac{1}{3}$ hp/ton/hr, but cannot work to depths below 60 ft because of friction losses. This limitation can be overcome by using hydrojet boosters to boost suction velocity and decrease suction head on the pump. It appears that dredging depth could be extended well beyond the 100 ft of water in which the diamond-mining barges now operate by use of several stages of hydrojet boosters in series.

Dredging-column suspension. With jet-lift, air-lift, and centrifugal-pump dredging, one of the major problems was suspension of the dredging column. The average height of the sea swell running in the area of the diamond deposits is 12 ft, and wave height frequently runs as high as 30 ft.

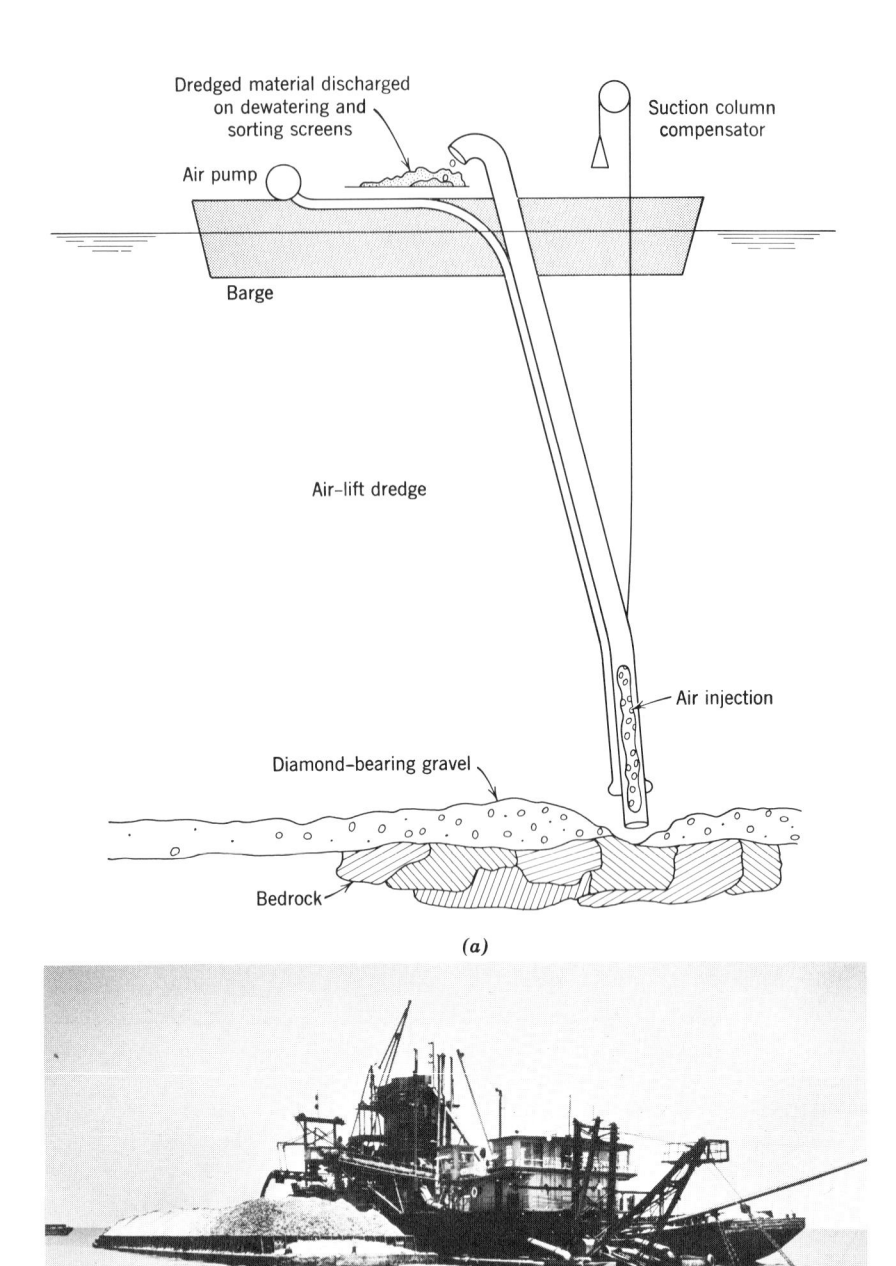

FIG. 14-2. (a) Flexible-column air-lift dredges are used extensively in sea-diamond mining where wave action would destroy stiff ladder dredges. (b) Haden Dredge No. 9, a surface centrifugal pump dredge, recovers shells for cement making from Galveston Bay. (*Maritime Reporter* Photo.)

Dredge columns made up of lengths of flexible hose to add "give" to the column were tried. This proved unsuccessful during prospecting operations, and counterbalanced compensators were developed to control the dredge column. Suspended from the counterbalanced hoist rig, the lift column is suspended in a vertical position and kept in contact with the gravel bed despite the rise and fall of the barge.

Costly equipment. The equipment required for ocean mining is understandably more costly than equipment to perform an equivalent operation on land. The offshore mining vessel must provide not only the mining and processing equipment, but also complete living facilities for a crew large enough to man the operation around the clock. Since ocean-mining sites are usually remote from populated areas, transportation facilities must be provided for transportation of relief crews, food, and the tools and supplies required to maintain the dredge and processing plant at the mining site.

The cost of a mining barge of the type being used by the Marine Diamond Corporation runs close to $1\frac{3}{4}$ million dollars (Table 14-2). The profits from using such a barge, capable of dredging about 85 tons of solids per hour, may be seen by comparison with the corporation's smaller Barge No. 77. Number 77, converted from a marine pipeline barge, and equipped to dredge about 55 tons of solids per hour and process up to 30 tons/hr through her concentrating plant, was put into operation in 1962. She was driven aground and wrecked less than a year later. She recovered over 51,000 carats of gem-quality diamonds worth about $1,770,000.

TABLE 14-2

Cost of Diamond-Mining Barge[a]

Barge hull	280,000
Mooring and positioning equipment	280,000
Dredging equipment	500,000
Processing plant	300,000
Services	200,000
Accommodations	150,000
Navigation and communication	10,000
	$1,720,000

[a]Compiled from Webb (1965).

Tin (Wilson, 1965)

One of the earliest undersea ventures, mining of offshore tin deposits, is receiving widespread attention as the free world's need for tin increases.

Indonesia has nine seagoing dredges in operation, recovering tin ores using bucket-ladder dredges, in 60 to 100 ft of water.

Off Thailand, tin-bearing cassiterite sands are found in sediments at 90 to 130 ft and are mined by bucket-ladder dredges and grabs.

The tin-mining barges have concentrating plants on board which process the ore on site to reduce transportation requirements.

Iron, gold, and platinum (Wilson, 1965)

Magnetite iron sands are common in many beach sands throughout the world. Most deposits do not contain enough iron to make mining for iron alone profitable at present. It appears, however, that some may profitably produce iron as a byproduct or coproduct of the recovery of other minerals. Mine tailings can be a valuable byproduct in themselves when a requirement exists for fill or beach sand in close proximity to the mine site. Tailings from the Ariake Steel dredge in Ariake Bay off the Island of Kyushu, Japan, are being used as landfill under a government contract, turning waste disposal to a profit. Rich iron-bearing sands in this area are recovered, originally with a grab-bucket dredge and today with a hydraulic dredge equipped with a cutter to loosen the sand and gravel. Exploitation of iron sand deposits, long known to exist in shallow water, again had to await recent developments in technology, in this case a successful and profitable method of smelting the iron sands.

Gold and platinum are produced in Alaska in very shallow waters, using hydraulic suction dredges, bucket line dredges and drag line buckets. Underwater gold is mined on a small scale, but prospecting permits have been issued for large offshore areas, in water down to 300 ft deep. Platinum mines in the Goodnews Bay area have provided 90 percent of the U.S. platinum production from deposits in very shallow water, and discoveries have been made in deeper water.

Phosphorite (Hess, 1965)

Phosphorite deposits occur in nodules and slabs on the continental slopes of many countries in water depths from 110 to 11,400 ft. The deposits off California contain about 29% phosphate, only 2% lower than the phosphate rock mined in Florida and shipped to California for agricultural use. The Collier Carbon and Chemical Company mined phosphorite from the 40-mile bank deposit off San Clemente Island in 1962 and 1963, using a clam-shell bucket in about 400 ft of water, but this operation was abandoned, reportedly because live ammunition, left from Navy practice firing, was coming up along with the phosphorite.

Glauconite (Hess, 1965)

Glauconite (green sand) occurs in marine sediments on the continental shelf, most abundantly on the crests and upper slopes of ridges and sea mounts. A hydrous potassium–iron–silicate mineral, it is used as a soil conditioner and water softener. Large deposits off the Pacific coast in 60 to 2400 ft of water could be mined for use as a soil conditioner and potash fertilizer ingredient, but is not yet being exploited.

Manganese nodules and crusts (Wilson, 1965; Hess, 1965)

Manganese nodules and crusts offer an unparalleled challenge for large-scale sea mining. This material is found distributed over the deep floor of the large oceans at depths from 300 to 23,000 ft. The deeper deposits—6000 to 18,000 ft deep—in the Pacific are relatively higher in copper, cobalt, and nickel content than the shallower deposits. Atlantic deposits and the shallower Pacific deposits tend to be higher in manganese, with relatively high iron content. Manganese nodules have not yet been mined extensively from the sea, but several companies and the Bureau of Mines are doing exploration and studies of the feasibility of mining them. Although in seemingly inexhaustible supply, manganese nodules are considered low-grade ores because of the variety of minerals contained, and the difficulty of separating them by present methods. Refining processes capable of separating the various component metal ores cheaply is the key to success of sea manganese-mining ventures, if mining equipment can be developed to recover the nodules economically.

Future undersea mining systems

Present undersea mining operations are all relatively low-volume production operations, most of which are mining precious metals, gemstones or high-value metal ores in 400 ft of water or less. In most instances, only increased size or number of units is required for expanded production in the shallower waters. The Marine Diamond Corporation is expanding its mining fleet and increasing productivity by introducing more efficient dredging equipment. Its converted LST, *Diamantkus*, is equipped with three 16-in. airlift dredges and three 14-in. jet lifts capable of dredging to 400-ft depths. Up to 300 tons of gravel per hour can be handled through the *Diamantkus*' concentrating plant, to produce an average of about 1000 carats of diamonds daily.

As technology improves, we expect future undersea mining operations to expand and to move into deeper water, probably to 20,000 ft. This will give access to practically all of the presently known underwater mineral deposits. Large-scale operations will be required to make deep mining economically feasible.

Continental-shelf deposits of hardrock ores

The continental shelves are submerged continuations of the land masses of the continents and have a similar geologic formation, so they can be expected to contain commercial concentrations of the same ores and minerals that occur on dry land.

Outcrops on the bottom could be mined by blasting and recovering the ore by dredging. Inshore operations of this type, however, would raise questions of pollution and destruction of sea life by the blasting.

Subbottom hard-rock ore deposits may be recovered by sinking a steel shaft and cementing it into the bottom, in the manner of a large diameter oil well casing, Fig. 14-3. A platform mounted at the surface would provide living accomodations for off-duty mine operators, and a stowage facility for the ore. It would also serve as a loading platform for ore carriers used to transport the ore to processing plants ashore. Deep-water oil production platforms already provide a design basis for such a facility.

Phosphorite and manganese nodules

Phosphorite and manganese nodules can be mined using drag dredges, hydraulic dredges, or airlift dredges. Drag dredges as presently developed can be used to recover the nodules, but become an expensive method in very deep water because of the cycle time. With a 300-ft/min line speed, round-tip cycle time for a drag dredge operation at 20,000 ft would be approximately $2\frac{1}{2}$ hr. Use of a drag dredge for mining phosphorite nodules in depths of 400 ft, with round-trip cycle time on the order of 5 min would be more feasible.

Since the air lift obtains its suction by a decrease in density of the pipe contents with respect to the water surrounding the pipe, the pumping force is, in effect, applied by pressure difference at the bottom of the dredge pipe, rather than by suction at the upper end. Therefore the depth at which the air lift can be operated is not limited by the friction head in the pipe to the same extent as is the suction dredge. This feature, plus the advantage of continuous production, makes the air lift dredge the most likely candidate for deep dredging of loose materials on the outer continental shelves and in the deep ocean basins.

To operate at depths as great as 20,000 ft, nearly 4 miles of pipe and hose, 16 in. or more in diameter must be connected up and lowered.

Deep-Ocean Work Systems 503

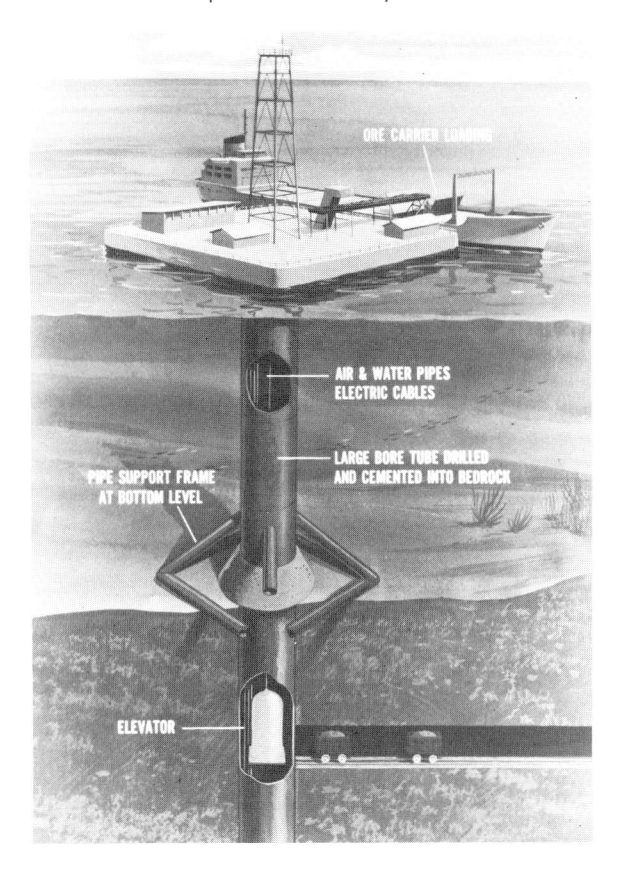

FIG. 14-3. Undersea hardrock ores may be mined through steel shafts, cemented into the sea floor. (U.S. Navy artist's concept.)

Equipment similar to that used for handling drill pipe in offshore oil-drilling operations and intermediate floats to support the weight of the dredge column as shown in Fig. 14-4 will be required. Once the dredge column is made up and streamed, a bottom-resting support or sled will keep the suction head in the proper position with respect to the bottom. However, to maintain the dredge column in the near vertical position required for maximum dredging efficiency, while towing the dredge column with its lower end in contact with the bottom, will require a very stable dredging platform.

Staging of the air injection at intervals along the column will be required to keep the contents agitated as they rise. Since it will not be necessary to aerate the entire length of the column to provide the buoyancy necessary to induce flow, it may prove more economical to inject air only in the upper

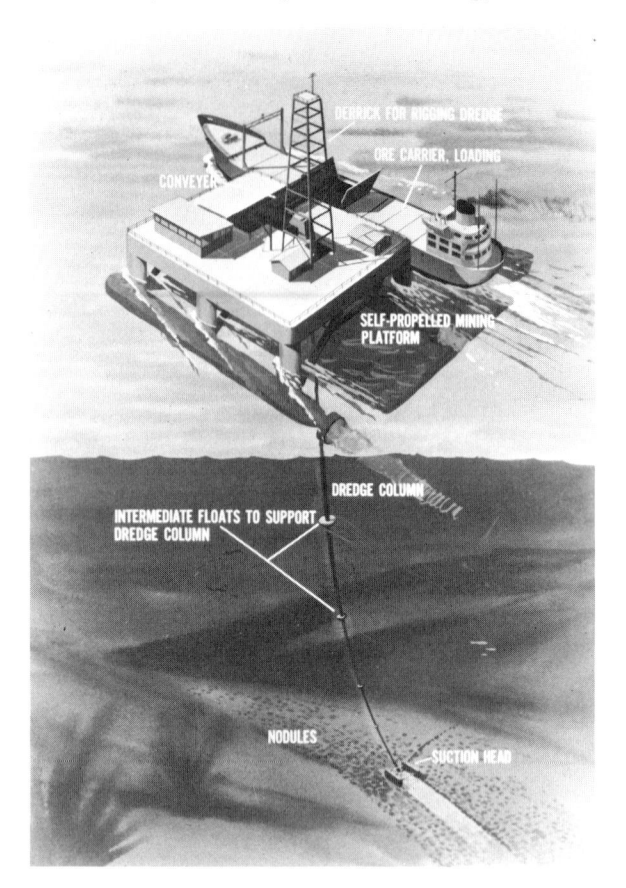

FIG. 14-4. Hydrojet boosted air-lift dredge concept for collecting nodules from the deep-sea floor. Very stable, semisubmerged dredge platform serves as ore stowage and living space; could dredge continuously in bad weather. (U.S. Navy artist's concept.)

portion of the column and use hydrojets to provide the necessary boost and agitation at the lower levels. Deep-submergence pumps, located on the dredge column adjacent to the booster jets, would eliminate the power losses of water flow through pipes from the surface, providing a more efficient lift.

Rigging and unrigging the dredge column will be too expensive an operation to be performed routinely to accommodate changing weather conditions. A low water-plane, semisubmerged platform, similar to an offshore oil-drilling platform, equipped with a propulsion system capable of propelling it and towing the dredge column at slow speed, could maintain dredging operations in all but very severe weather.

Developments of any type of deep-ocean mining will require development of surface and bottom navigation in addition to the mechanical mining equipment. Precise navigation will be required to plot the position of the surface vessel and the relative position of the mining dredge head to maintain position of the dredge head in the orebody. This will be more important as areas are extensively worked. Very low frequency (VLF) navigating systems now available for broad ocean use give position accuracy of 2 to 15 miles, contrasted with the 2- to 3-ft accuracy of the microwave system used by the Marine Diamond Corporation. Underwater television is a possibility if it can be kept moving ahead of the "dust cloud" stirred up by the dredge head, and if runs are long enough to allow the mud to settle before the next pass.

Surface storage of ore or processing on site will be necessary because it will not be feasible to use the mining vessel to transport the ore to shore. Provisions will have to be made to hold the ore at sea until a sufficient quantity is accumulated to load the ore carrier. Transportation costs could be minimized by processing or concentrating the ore on site.

Transfer of ore between the mining vessel and the transport by normal dockside equipment will not be possible in bad weather. Preprocessed ore could be pumped as a slurry, using transfer techniques similar to present Navy fueling at sea.

14.2 OFFSHORE OIL-WELL DRILLING AND COMPLETION

As demands for petroleum products increased the search for oil moved underwater. The first underwater wells were in shallow, protected waters, near producing oil fields. Platforms were erected over the water or drill rigs mounted on barges that could sit on the bottom and the wells were drilled and completed like dry-land wells. Exploration moved farther offshore, and mobile platforms were developed that could be moved to the drilling site and set on the bottom by extending movable legs. These legs are retracted high above the rig while it is being moved into position as a floating barge and are extended by hydraulic or mechanical jacks which force the legs into the sea bottom and raise the work platform above the reach of storm waves.

Mobile rigs have been built for successively greater depths, and four rigs capable of operating in 300 ft of water began operation in 1965. More are going into operation in 1966.

When bottom-supported rigs appeared to be approaching a practical depth limit of 200 to 300 ft, new methods and new equipment were needed to extend oil operations onto the deeper slopes of the continental shelf. Exploration and drilling in deeper water required drill rig support independent of the bottom.

Floating-rig support

Floating-drill-rig support developed in two types of craft: semi-submersible deep-moored platforms, and ship-hulled drilling vessels held in position by deep moorings or equipped for dynamic positioning.

Many configurations were investigated and tested in model form to determine the hull form best suited to the requirements of a floating drill platform. Investigation of various configurations indicated that for minimum response to sea motion, inherent stability, and ability to maintain location over drill hole for long-term, on-site operation, the semi submersible, triangular, or rectangular platform, connected by vertical members with a very small waterplane area to deeply submerged horizontal cylindrical buoyancy chambers, is best. The *Bluewater 1* rig continued drilling operations for Shell Oil Company in 22-ft seas and stayed on location in 28-ft seas and 65-mph hurricane winds. It was secured in position over the drill hole by an eight-legged moor, each leg of which consisted of one-half mile of 3-in. line, a 10-ton anchor, and a spring buoy to take up shock.

Exploratory drilling requires a more easily mobile platform. In tradeoffs between mobility and platform stability, the dynamically positioned ship-hulled drilling platform comes out very favorably. Though not as steady a drilling platform as the semisubmerged rig, the dynamically positioned ship can dispense with tug services and with the time-consuming operation of placing a deep-mooring array. Since it is unnecessary to anchor, positioning the ship is independent of water depth, at least within the depth limits of interest in oil-well drilling.

Shell's 136-ft long, 450-ton drilling ship Eureka has been moved and repositioned by using a dynamic positioning system for drilling in as many as nine locations in a single day. Drilling operations have been carried out in water depths from 30 to 4000 ft with waves 20 ft high and winds of 40 miles/hr.

The Automatic Positioning Equipment on *Eureka* uses a gimbaled tiltmeter to sense the perpendicularity of a taut wire attached to an anchor on the bottom. Tiltmeter signals are used to control the direction and thrust of propellers to maintain position and heading of the ship. Station-keeping accuracy within 2 or 3% of water depth can be maintained on automatic or manual control in weather within the limits of the ship's propulsion power system.

With the taut line attached to the wellhead, the ship may be positioned automatically above the wellhead for re-entry. Or the taut line may be attached to an underwater manipulator vehicle, allowing the surface support ship to follow directly above the swimming vehicle as it searches for, locates, and attaches to the wellhead.

Wellhead completion

Drilling practices on floating drill rigs have followed those used on bottom-standing rigs. Completion of wellheads, however, has required development of new techniques for use below the normal 250-ft working depth of the hardhat diver. Completion techniques have followed three general paths of development:

1. remote drilling and completion systems;
2. manipulator-operated systems;
3. off-bottom wellheads with diver completion.

The remote drilling and completion systems, such as the Shell RUDAC system, are based on use of the guide structure on the bottom to guide specially designed underwater wellhead equipment and blowout preventers into position to be locked down and hydraulically sealed by remote control from the surface. The guide structure for RUDAC is connected to the conductor pipe, which is cemented into the ocean floor and serves as a guide for the drilling tools and as a return for drill mud to the surface platform. Hydraulically operated production control valves are designed to be failsafe and are controlled electrically from the surface.

Manipulator-operated systems have been developed in which the wellhead completion tasks normally performed by a diver are performed by an underwater manipulator. The first of these, the MOBOT system, was developed for Shell Oil Company by Hughes Aircraft Company. The MOBOT, Fig. 14-5, and its replacement, ROBOT, Fig. 14-6, were designed to work in conjunction with a modified land-type wellhead.

Development of the Shell MOBOT system was approached by modifying a land-type wellhead to simplify assembly, disassembly, and operation by tasks to the point that they could all be performed by a powered socket wrench rotating on a horizontal axis. The usual vertical wellhead flange bolts were replaced by horizontal radial locking screws. The MOBOT then became a swimming, controllable, horizontal socket wrench. The powered socket wrench, lights, and television camera were mounted on an extensible mast.

The MOBOT is supported by its power cable and is propelled toward the wellhead by twin propellers. MOBOT can locate a wellhead at long range by sonar, which guides it into television range so that the operator can see the wellhead on the TV monitor. MOBOT engages a circular track mounted on the wellhead, and is positioned radially by its powered wheels. Any bolt or valve that requires operation or adjustment can be reached by extending the hydraulic mast vertically, or by extending the socket wrench horizontally All of the bolts, nuts, and valve stems that require actuation by the MOBOT

FIG. 14-5. Shell MOBOT was used to perform underwater oil wellhead assembly and operation tasks, using a hydraulically driven socket wrench. (Shell Oil Co. photo.)

FIG. 14-6. ROBOT, Shell's successor to MOBOT, was designed for wellhead completion and operation at depths of 1000 ft. (Shell Oil Co. photo.)

manipulator are provided with specially shaped heads to assure easy mating with the flexibly mounted socket-wrench head.

A claw-type grip is provided that can be fitted in place of the socket wrench for such tasks as replacing snap-fitted hydraulic hoses on the blow-out preventer controls, attaching a re-entry guide line to the wellhead, or for wielding brushes and other tools.

A newer version of the underwater manipulator, ROBOT, built for Shell by the Ventura Tool Company, is simpler and more rugged than MOBOT and is designed to operate at depths greater than 1000 ft.

The manipulator-operated system has been used by Shell in drilling operations off Point Reyes, California, in 363 ft of water and in the Molino gas fields in 240 ft. One of the primary advantages of the manipulator-operated system is that the actuation and control equipment for the wellhead is almost entirely contained in the manipulator itself. Since the actuating equipment is used only during drilling or re-entry of the well, expensive control and actuation equipment need not be left with the wellhead on the sea floor.

Off-bottom wellheads

Although a degree of success has been achieved in the use of remote drilling and completion systems and manipulator-operated systems, the equipment required is expensive and its use has not become universal. Many wells in water too deep for divers to work at the bottom are being completed by providing bottom-mounted structure to raise the wellhead far enough above the bottom to allow normal diver access for wellhead completion tasks.

Production facilities

Production facilities for deep-water wells have not kept pace with drilling and completion techniques. Bottom-supported production platforms are feasible and are used in water depths to about 300 ft, but for wells in deeper water, production facilities are still in the developmental stage. When deep-water wells are located close to shallower areas, oil and gas may be piped to collection facilities in shallower water. Floating collection tanks, bottom-mounted collection tanks, and stowage in washed out salt-domes in the underlying strata, are among the possibilities being investigated, but as yet no universally satisfactory solution has been found. Most deep-water wells are still being capped after completion to await later development of production facilities.

Present trends

Development of manipulator systems and vehicles aimed at underwater oil-field work continues. Autonetics Division of North American Rockwell is developing two concepts, for 300- and 1000-ft depths, each equipped with manipulators fitted with a variety of tools designed to be interchanged at the submerged work site. A number of other undersea vehicles are already in existence or are being developed. Some of them will certainly find use in undersea oil-field work.

Current developments in diving indicate that in the near future it will be feasible for divers to perform wellhead completion work at greater depths than is now the practice. During the French Conshelf III experiments in 1965, the saturation-diving oceanauts performed wellhead completion work on a commercial-type oil wellhead, including plugging and reopening the well at a depth of 370 ft. It has been reported that these tasks were performed in less than the expected length of time, and the divers, who presumably had practiced on land, considered the task somewhat lighter under water because of the buoyant effect on the tools.

Atlantic Refining Company, using the Global Marine Exploration Company center-well drill ship *Glomar V*, drilled and completed a well off the coast of Libya in 1965 in 525 ft of water. The wellhead completion work was reportedly done on the sea floor by hardhat divers, working with oxygen–helium breathing gas.

CJB-Divcon Ltd., a British company formed by Constructors John Brown Ltd. and Divcon International (U.K.) Ltd., to develop, build, and operate submersible work chambers, are currently building two such chambers (Fig. 14-7) for use at depths to 600 ft, and are designing others to be capable of use at 1000 ft.

The submersible work chambers have two compartments separated by a pressure-tight hatch. The upper compartment, kept at atmospheric pressure, is occupied by a control technician and a relief diver. The lower compartment is pressurized to ambient sea pressure and is occupied by the working diver, in modified hardhat diving dress, who enters and leaves by the lower hatch. The upper compartment can be pressurized in about 4 min to allow the relief diver to go to the aid of the working diver if required. Under normal conditions, neither the relief diver nor the technician would require decompression upon surfacing, since their compartment is maintained at atmospheric pressure.

The ADS IV system (Fig. 14-8) completed in March 1966 by Ocean Systems, Inc., has a personnel-transfer capsule that mates to the connecting trunk between the two two-man deck decompression chambers. Upon completion of testing, this system was shipped to Norway for use in oil-drilling operations off the Norwegian Coast in depths expected to reach 425 ft.

Deep-Ocean Work Systems 511

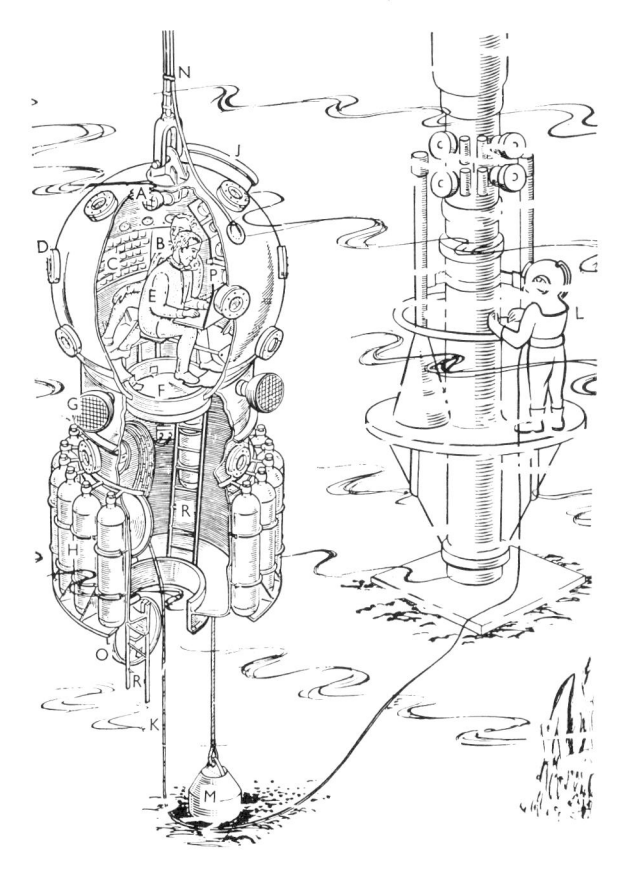

A CO₂ scrubber
B Standby diver
C Sodasorb cartridges
D Observation ports
E Technician
F Inter-communication hatch
G Floodlights
H Helium & oxygen cylinders
J Double access hatch
K Diver's gas supply hose
L Working diver at wellhead
M Ballast weight
N Umbilical cord
O Double access hatch
P Control panels
R Access ladders

FIG. 14-7. CJB-Divcon diving chamber. Diving chambers such as this British concept (CJB-Divcon), used in conjunction with deck decompression chambers and oxyhelium breathing mixtures, will speed up deep-diving operations. (*Shipbuilding and Shipping Record* Sketch.)

Although diving techniques have not yet evolved to the point where working dives 600 ft deep are routinely feasible, the use of saturated diving techniques and the submersible work chambers, or personnel-transfer capsules as the Navy calls them, will result in considerably more effective use of divers on oil-well projects. Among the advantages that will result are

1. Rapid ascent and descent. With the use of the personnel-transfer

FIG. 14-8. ADS IV System Personnel Transfer Capsule mates to access trunk between decompression chambers for transfer of divers under pressure. (Ocean Systems, Inc. photo.)

capsule to maintain the divers at sea-bottom pressure, the divers can be brought to the surface rapidly and undergo decompression on deck in the transfer capsule, or in a separate decompression chamber. This will eliminate the lengthy idle periods while the drill rig waits for the slow ascent of the diver after completion of diving work.

2. Diving in rough weather. The work chamber can be lowered from the fixed or stable drilling platform through surface conditions that would prohibit surface diving. This allows work to proceed in areas such as the North Sea, where wave heights of 6 to 9 ft are prevalent.

With operating costs of an offshore drilling rig running $900/hr and up, the savings in time and money possible through the use of the submersible work chamber would be considerable. For this reason, if near-future evaluation proves this technique successful, it can be expected to become almost universal for oil-field diving operations.

14.3 DIVING WORK SYSTEMS, SALVAGE AND RECOVERY

Until recently most diving work has been done by divers in hardhat dress, wearing 190 lb of equipment and trailing an air hose and a lifeline. Although the hard-hat diver suffers considerable lack of mobility, this handicap can be turned to an advantage. He can wield a sledge hammer, pull a wrench, or operate a power drill without excessive difficulty in remaining in place. When the operation requires, the hard-hat diver can adjust the amount of air in the suit to control his buoyancy and fix his position, and can then exert considerable force on the tool being used.

The rapid growth of free-swimming diving since World War II, and more recently the development of saturation diving, have added tremendously to the potential capability of divers to do work under the sea, and the trend now is more and more toward the light-configuration swimming diver, working at or near neutral buoyancy to maintain mobility.

The Navy's SEALAB III is planned for installation in 430 ft of water during 1968. Divers will work from the habitat for 45 days, with excursion dives planned to depths of 600 ft. At this point we will have made the continental shelf, out to the limit U. S. sovereignty, accessible for work and exploitation by man in the sea. Doing so, however, will have created more problems that must be solved if we are to work efficiently at that depth. The tools and equipment presently in use will be grossly inadequate to meet the needs of the swimming saturation diver working nearly a fifth of a mile deep in the sea.

Tools and work equipment: past practice

Tools for the hardhat diver have not differed significantly from those used by the topside mechanic. Most hand tools were ordinary mechanic's tools. A few have been modified to fit into tight spots or to operate special fittings. The power tools used have been conventional pneumatic shop tools: drills, impact wrenches, chipping hammers, grinders, and brushes, modified slightly for underwater use.

Welding and cutting. A notable exception is underwater cutting and welding tools developed expressly for underwater work. Although developed in 1908, underwater flame cutting did not see practical use at deep depth until 1926 during the salvage of the submarine S-51. After experiencing difficulty in removing fittings and clearing debris, the Navy developed an oxyhydrogen torch which was used successfully at a depth of 132 ft and which materially aided in the S-51 salvage operations. Oxyhydrogen cutting was the commonly used method for underwater work until 1942 when wartime necessity forced development of the electric arc-oxygen process to speed the work of salvaging war-damaged ships.

The arc-oxygen process has the advantage of being faster because no preheating is necessary, more positive because no flame adjustment is required, and simpler because only one gas is required. It has one distinct disadvantage, however, in that the diver must be completely protected from shock. Navy practice requires heavy rubber or rubberized canvas dress and gloves to insulate the diver from the water, the work, and his tools, and to insulate him from any metallic parts of his dress, particularly from his metal helmet.

Both the oxy-hydrogen and the arc-oxygen methods depend upon oxidation of the metal to effect the cut. Intense heat is directed to a small spot on the metal to be cut by an oxyhydrogen flame, or by an electric arc. When the metal has reached the proper temperature, a jet of pure oxygen impinged upon the hot metal causes it to burn very rapidly. These processes work only on those metals such as plain carbon and low-alloy steels which oxidize very rapidly. To cut nonferrous metals and the less easily oxidized steels, a third method, the metallic-arc method is used. This continuous melting process has the advantage of requiring no fuel gas for its operation.

Falcon nozzle. The Falcon nozzle is another unique device developed to provide a high-velocity water-jet nozzle that can be used by divers for jetting away unconsolidated bottom material without being jet-propelled across the bottom in the process. Consisting of a high-velocity "suicide nozzle" equipped with a ring of rearward thrusting jets to neutralize the thrust of the main jet, this device was used with considerable success to cut away mud and sand from beneath the battleship *Missouri* to aid in refloating her after she went aground on Thimble Shoal.

A similar device was used more recently by Swedish Navy divers in conjunction with an air lift to remove loosened material and to cut tunnels under the ancient Swedish warship, *Vasa*, at depths of 100 to 130 ft. The Swedish divers spent several months tunnelling through the clay beneath the ship which had capsized and sunk in Stockholm harbor in 1628. Steel cables were then threaded through the tunnels and used to lift *Vasa* to the surface, after more than 300 years on the bottom.

Tools and equipment for the future

The swimming diver, suspended weightless in the water, requires tools and equipment of a different sort if he is to perform with efficiency. If he swings a heavy sledge, Newton's third law takes effect and he must swim back into position for the next blow. Maintaining sufficient pressure on a drill to make the turning bit cut requires constant swimming effort or a handhold that can be used to hold onto while pushing the drill. A window-washer's belt was tried in SEALAB II (Fig. 14-9) but attachments are required to secure the ends.

Swimming diver's tool problems. As the depth increases, pneumatic tools and surface-powered and supplied cutting and welding tools become less practical. The air to power the pneumatic tools must be compressed to

FIG. 14-9. *Sealab II* aquanaut John Reeves tries special air-powered diver's drill designed by Battelle Memorial Insitute. Belt helps diver maintain position on test stand when underwater. (Official U.S. Navy photo.)

higher pressure to overcome sea pressure at the point of use. Longer hoses cause greater frictional pressure drop between the compressor and the tool and the air hose ties the system to the surface. To achieve maximum effectiveness from a diver work system, the working diver must be as free as possible from encumbering and entangling lines and hoses, and dependence on surface support for power and work supplies must be minimized. A number of new and modified tools and equipment systems are being developed to provide the freedom and power required by the deep diver.

One of the most frequent tasks required of a diver is the attachment of patches and lift gear to objects or structures to be lifted or refloated. In the past, when the lift was to be attempted, or when an airtight patch was required for containment of air to provide buoyancy, welding or drilling and bolting have been required; both are time-consuming underwater tasks.

Stud guns. To speed these tasks and reduce the physical exertion required of the diver, fastening tools and devices using velocity-power studs are being developed. Following the same principle as stud guns used extensively in shore construction work for driving fastenings into masonry and metal, these tools (stud guns) fire a hardened steel stud from a smooth-bored, breech-loading gun barrel, using conventional gunpowder and primer.

The NUD-38 underwater stud gun, Fig. 14-10, fires $\frac{1}{4}$ in. diam studs capable of piercing mild steel plate up to $\frac{1}{2}$ in. thick for fastening patches in place. Handheld by the diver, this stud gun can be used at depths to 200 ft, and development is continuing to make it usable to maximum diving depth.

A heavier tool, the Model D, Fig. 14-11, was redesigned for improved safety and tested in SEALAB II. The Model D fires a $\frac{3}{8}$-in. diam stud capable of penetrating a 1-in.-thick HY-80 steel plate, and was used to attach a pressure-tight patch to a curved HY-80 steel plate, Fig. 14-12, during the SEALAB II tests. The Model-D gun can also be used to fire a hollow stud through mild steel plating up to $\frac{1}{2}$ in. thick to provide an attachment for blowing compressed air into a compartment to provide buoyancy during salvage.

Power-stud-attached lift pad. The attachment of lifting lines is one of the most difficult tasks in salvaging a sunken ship, particularly on a smooth hull with few accessible strong attachment points, such as a submarine. A new attachment-and-lift system presently under development will make this task easier in the future.

The velocity power-lift pad, Fig. 14-13, a new concept tested for the first time in SEALAB II, promises a quickly and easily attached lift point applicable to any area where the plating is strong enough to develop the holding power of the studs. The test lift pad is in the form of a cross with two stud driver barrels attached to each arm of the cross and a freely rotating lift eye at the center. For attachment, the pad is placed against the hull or object to which it is to be attached and is held in contact by four magnets. The eight stud drivers are then fired, either separately or simultaneously,

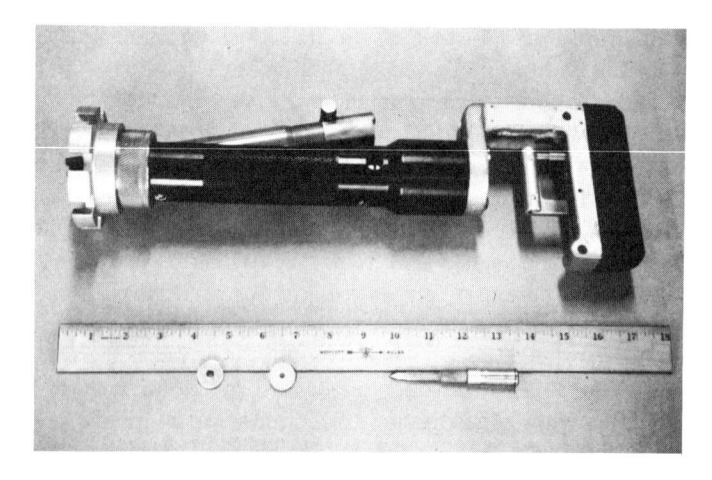

FIG. 14-10. Navy NUD-38 model studgun fires $\frac{1}{4}$-inch diameter studs into steel structure at depths to 200 ft. (Mine Safety Appliance Co. photo.)

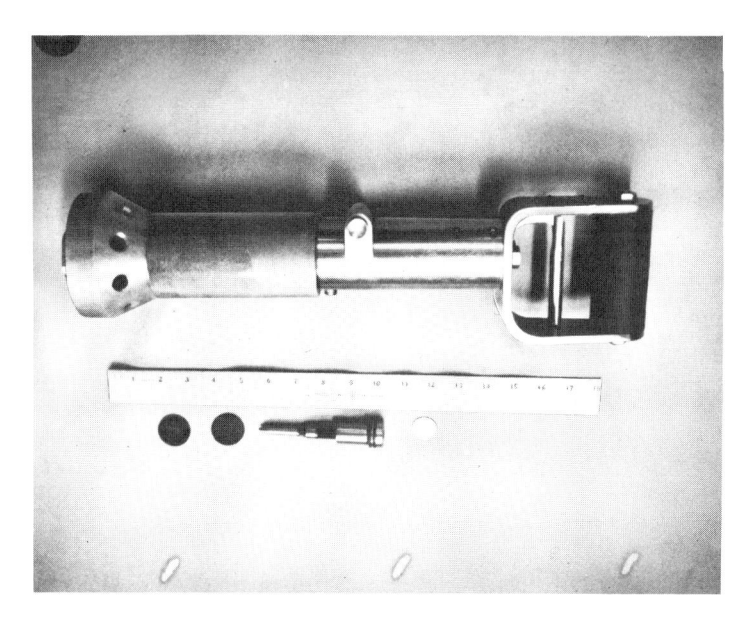

FIG. 14-11. Navy Model D underwater stud gun fires $\frac{3}{8}$-in. diameter studs through 1-in.-thick HY-80 steel. (Mine Safety Appliance Co. photo.)

FIG. 14-12. Aquanaut Billy Meeks fires a stud from Model-D studgun into 1-in.-thick steel plate in *Sealab II*. (Official U.S. Navy photo.)

driving the studs through pre-drilled holes in the pad and into the plate against which it rests.

The stud-driver barrels on the lift pad are adapted from the Model-D underwater stud gun, and are provided with firing mechanisms that can be operated mechanically by use of a lanyard, by a diver, or if required for

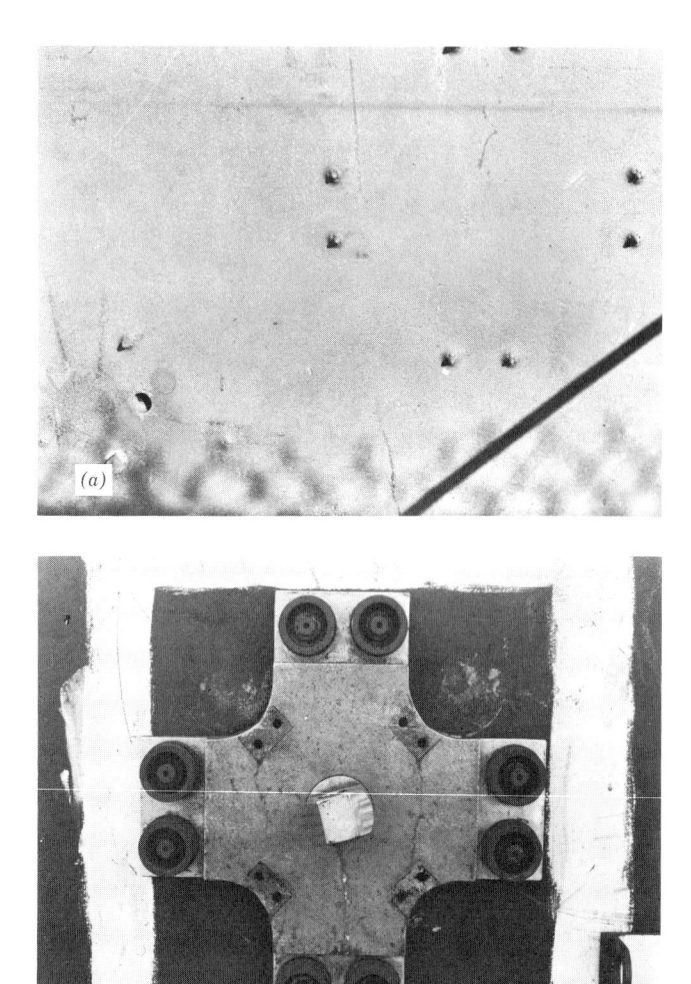

FIG. 14-13. (a), (b), and (c): Ten-ton capacity velocity-power stud attached lift pad tested in *Sealab II*. Magnets hold pad in place while stud-driver barrels (one still attached to pad) fire studs into structure to be lifted. Studs penetrate 1-in. HY-80 steel. (Official U.S. Navy photo.)

use with a manipulator-equipped vehicle at depths beyond diver capability, by remote electrical or mechanical control. Although recoil with the handheld stud gun is very light and the actuation of the propellent almost unnoticeable to a diver, additional precautions have been taken to eliminate recoil and gas escape with the multibarrel attachment pad. Copper crushing coils absorb the recoil and the barrels are sealed with a pressure-tight diaphragm at the muzzle and with a gasket at the breech to prevent entry of water into the barrel. A captive piston drives the stud down the barrel and is stopped at the muzzle, sealing the propellent gases in the barrel so there is no noticeable pressure wave generated when the studs are fired. Thus the use of the stud attachment pad may be extended to depths beyond diving depths, where the possibility of pressure waves generated by the collapse of escaping propellent gas bubbles might be a source of worry, if not of actual danger, to the vehicle operator.

The attachment pad studs, fired at a depth of 200 ft in SEALAB II tests, penetrated the 1-in.-thick HY-80 steel test plate, (Fig. 14-13). Though the pad was designed for a nominal 10-ton lift, it resisted a pull of 50,000 lb without separating from the plate. A 25-ton capacity velocity power lift pad is being developed, and testing is planned during SEALAB III.

Sealbin salvage pontoon. In SEALAB II tests, the velocity power-lift pad was used in connection with a newly developed Sealbin salvage pontoon, an adaptation of the U.S. Rubber Company's Sealbin container.

Though originally conceived for use with a remotely connected surface lift cable, the velocity power-stud lift pad provides an almost ideal com-

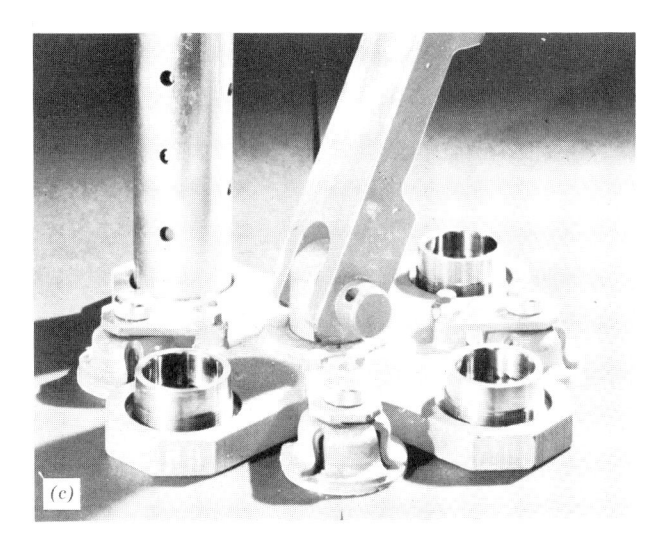

Fig. 14-13. (*Continued*)

panion piece to the Sealbin salvage pontoon. The pad can be prerigged to the collapsed pontoons and attached at will to any accessible area of the structure to be raised, providing distributed loading over the structure when the pontoons are inflated.

Hunley-Wischhoefer recovery system. The velocity-stud-attached lift pad, when used in conjunction with the remote lift-cable coupler, Fig. 14-14, (a) and (b), and the messenger buoy, Fig. 14-15, will provide a rapidly attached lift capability with a minimum of diver effort. For this operation, the attachment pad would be attached by a short lift pendant to a male connector, through the center of which is threaded a messenger line. The messenger line is stowed in the buoy, which is ballasted to make it neutrally

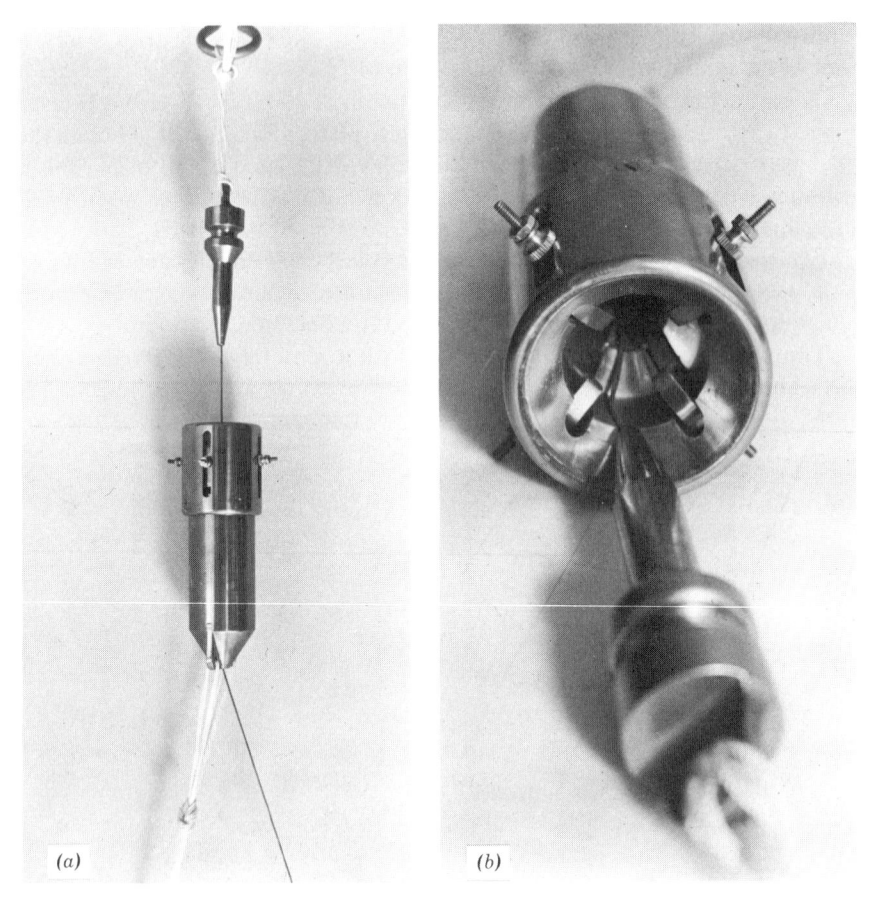

FIG. 14-14. (a) and (b) Female coupling device on lift cable is lowered down messenger line until pawls lock into groove in male coupling attached to object to be lifted from deep water. (Official U.S. Navy photo.)

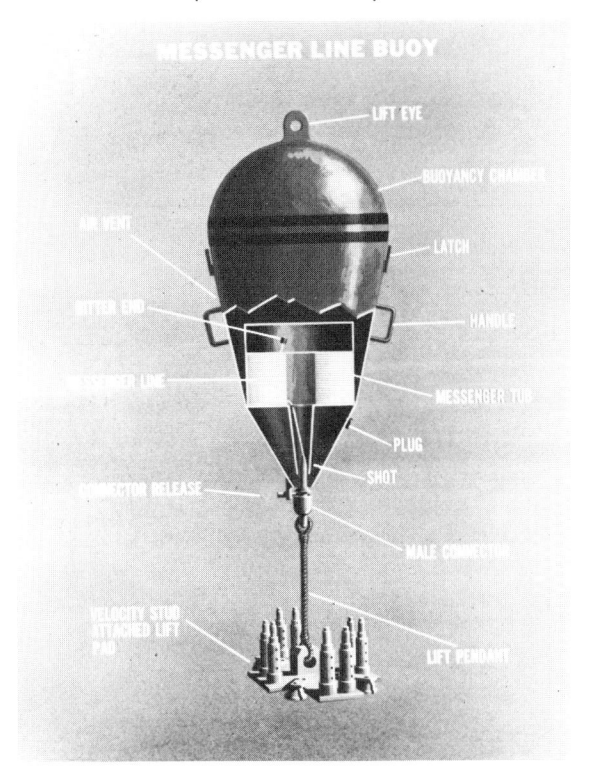

FIG. 14-15. Messenger buoy supports velocity-power stud attached lift pad while it is moved into position. After studs are fired into object to be salvaged, buoy is released to take messenger line to surface. (Official U.S. Navy photo.)

buoyant. After the lift pad has been attached to the structure to be lifted, the messenger-line buoy released to the surface, and the buoy retrieved, the messenger line will be threaded through the female connector to a winch and placed under tension. The female connector, attached to the heavy lift cable, can then be lowered until it latches onto the male connector, Fig 14-16, connecting the lift cable to the load to be raised. This system has been built in model form and is under development for possible testing in SEALAB III.

Foam-in-salvage. Another new tool in the diver's kit is foamed-in-place polyurethane foam, which the Navy calls foam-in-salvage. Developed and first used by the Murphy-Pacific Marine Salvage Company to remove the sunken barge *Lumberjack* from Humboldt Bay in late 1964, it was further developed for the Navy and tested at a depth of 200 ft in SEALAB II.

The foam gun (Fig. 14-17) consists of a mixing chamber, flow controllers, trigger-operated valves, and a nozzle to introduce the mixture into the compartment. Resin and catalyst are fed to the gun under pressure in

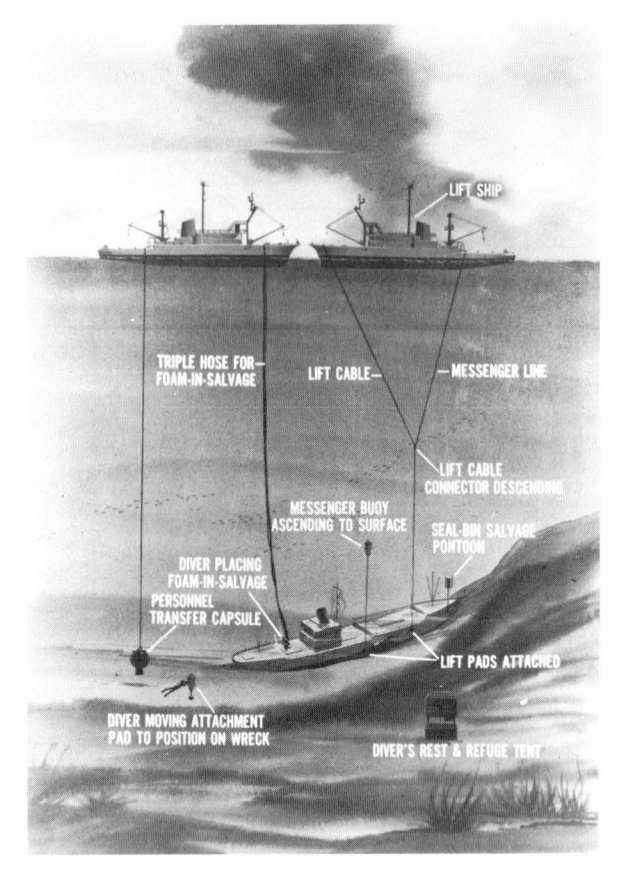

FIG. 14-16. Artist's conception of salvage operation using newly developed salvage tools and techniques.

FIG. 14-17. Foam-in-salvage gun for mixing and placing polyurethane foam underwater. (Official U.S. Navy photo.)

separate hoses, each hose being equipped with a flow regulator to assure equal flow of the foam components. A third hose conducts solvent for use in purging the foam gun after use.

The resin and catalyst are forced through baffles in the mixing chamber, Fig. 14-18, assuring thorough mixing of the components for complete reaction. Upon release from pressure (reduction of pressure) in the mixing chamber, a low-temperature Freon dissolved in the components becomes gaseous, starting the foaming process as the components mix. The heat of reaction continues the initial foaming and vaporizes additional high-temperature Freon to complete the foaming process. The tolyene diisocyanate mixture used in SEALAB II tests hardens in about 1 min and continues to cure for several hours.

Strength of the foam increases with curing time. It is therefore necessary to allow the foam to cure long enough to assure development of enough strength in the bubble walls to contain the gas as the object being salvaged rises and the presssure decreases. Premature raising of the foam to the

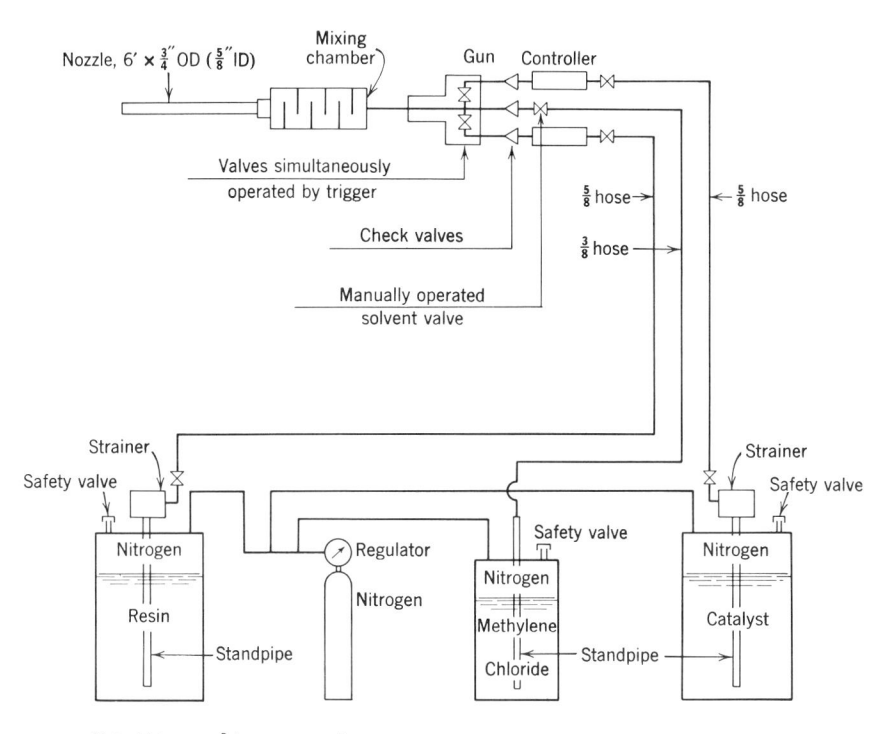

FIG. 14-18. Schematic diagram of foam-in-salvage polyurethane foam-generating equipment. (Murphy Pacific Salvage Co. sketch.)

surface results in rupture of the foam as shown in Fig. 14-19 and partial loss of gas and buoyancy.

Foam-in-salvage has been used successfully in salvaging several vessels in relatively shallow water, and it played an important role in refloating the U.S. Navy destroyer *Frank Knox* after she went aground in the Western Pacific. During SEALAB II, it was tested at a depth of 200 ft to float an aircraft hulk (Fig. 14-20). It promises to become a very useful tool for the salvage forces.

Heavy salvage lifts. Present heavy salvage lifts from deep water are performed by the use of salvage pontoons, or by the use of heavy lift ships, which use the rise of the tide to provide the required lifting force. Two ships of this type, *German Ausdauer* and *Energie*, Fig. 14-21, were used to raise the British submarine *Truculent*, which sank in 66 ft of water in the Thames Estuary.

The Navy's new salvage tug (ATS), in addition to its towing and diving

FIG. 14-19. Foam-in-salvage showing damage from rising to surface before foam was completely cured. (Official U.S. Navy photo.)

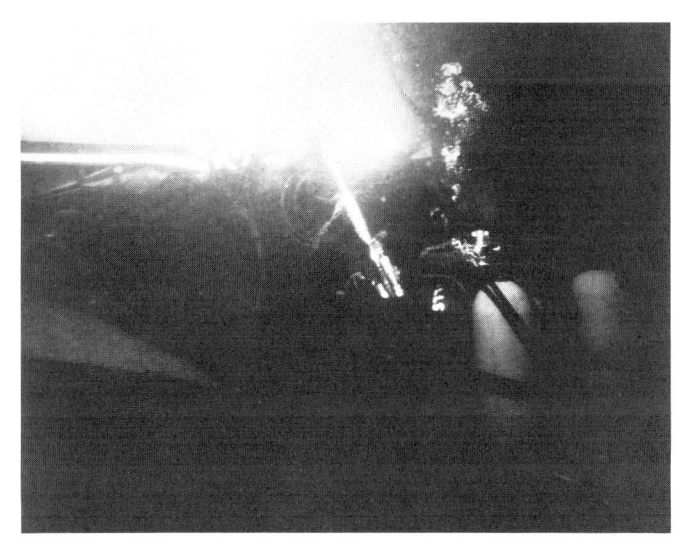

FIG. 14-20. Aquanaut Paul Wells places foam in aircraft fuselage during SEALAB II experiments. (Official U.S. Navy photo.)

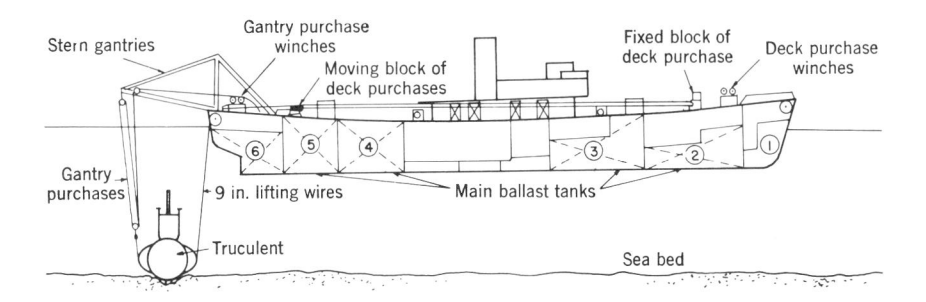

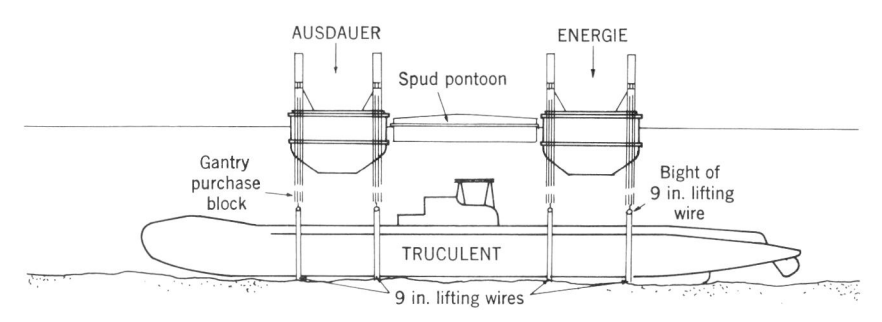

FIG. 14-21. Large lift barges. *Ausdauer* and *Energie* lifted the British submarine *Truculent*, sunk in the Thames Estuary.

525

support capabilities, will be equipped for lifting loads up to 300 tons as a tidal lift ship.

Rosenberg hydrodynamic winch. The lifting force for raising heavy loads to the surface from deep depths in future salvage operations may be provided by a new device being developed at the Navy Electronics Laboratory under patents issued to E. N. Rosenberg. Termed a "hydrodynamic winch," the device, Fig. 14-22, is in effect a combined salvage pontoon and heavy lift winch. It consists of a large cylindrical hull, divided internally by a number of longitudinal bulkheads into wedge-shaped compartments radiating from a central tube. At each end of the device, a counterbalanced power pump, and control station is mounted in bearings that permit the power pump, and control complex to remain in an upright position while the main structure rotates. In operation, a system of valves allows flow of water from one of the wedge-shaped compartments, Fig. 14-23, to the pump chamber, from which it is pumped via a transfer manifold to a higher chamber, the weight of which produces a rotating moment on the main structure. Continuation of the process causes the floating winch to turn about its longitudinal axis, winding the hoisting cables in grooves on its outer surface.

The hydrodynamic winch is projected in several sizes with lift capacities ranging from 10 to 10,000 tons and is presently undergoing small-scale model testing by the Navy Electronics Laboratory, (Fig. 14-24) (see Table 14-3). Operating alone or combined with a more conventional barge, it could provide the controlled heavy lift required to lift a large ship from the bottom.

Heavy lift attachments. Heavy-lift capabilities are being developed under projects previously discussed and under other projects concerned with development of salvage lift capabilities. At depths attainable by divers, these lift systems can be attached to objects to be lifted, using existing attachment points such as shafts and deck fittings, or by tunneling under the object to be raised and passing lifting slings underneath it. This technique has been used very successfully, employing the Falcon nozzle on a high-pressure water hose to wash away the mud, clay, or sand under the ship to form tunnels for the placement of lift slings.

At depths beyond the reach of divers, however, there is still no way to attach lifting lines to large, heavy objects so that they may be raised. The velocity power-stud attached lift pad, although attachable by a deep-diving vehicle, does not appear at this time to lend itself to lifts of more than a few hundred tons because of the multiplicity of attachments that would be required to develop a heavy lift. This problem will require more attention as a deep salvage capability is developed and the solution will require further developments in deep-submergence vehicles with manipulator capabilities.

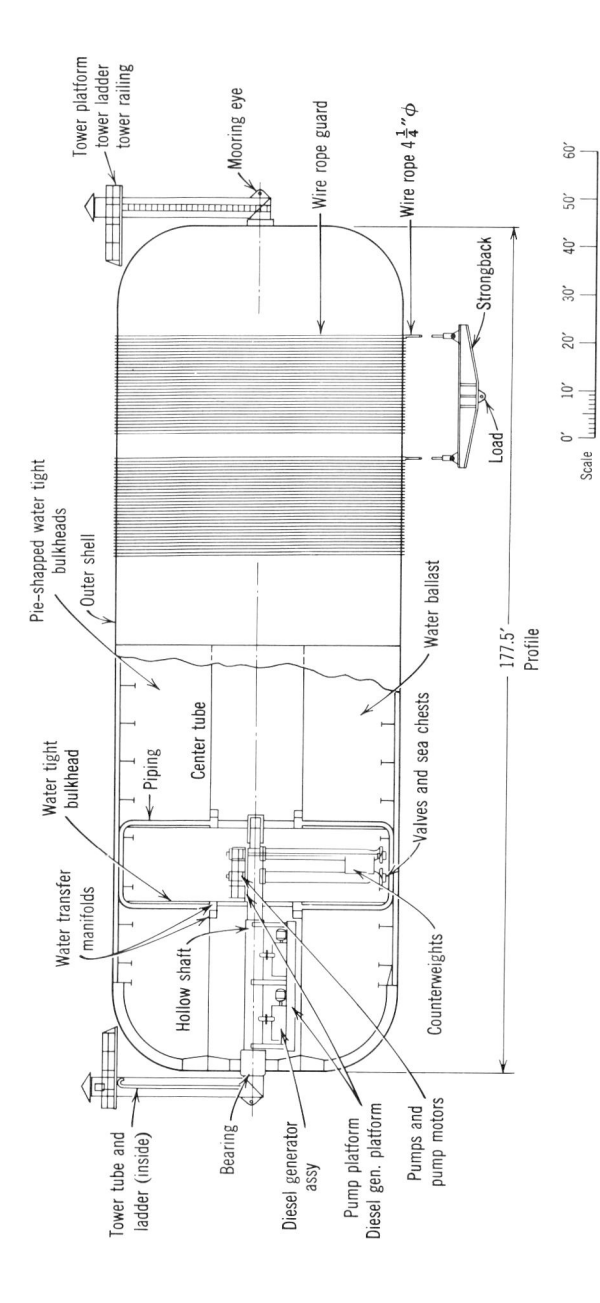

FIG. 14-22. The Rosenberg hydrodynamic winch, a floating cylinder, will lift heavy weights by winding lift cables on its outer surface.

528 Systems Design—The Technology

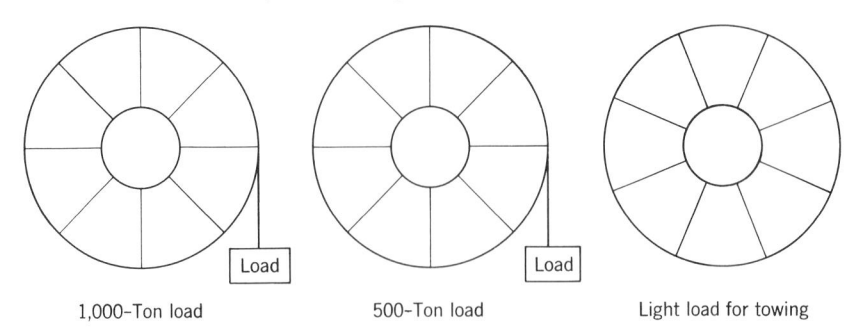

| 1,000-Ton load | 500-Ton load | Light load for towing |

FIG. 14-23. The Rosenberg hydrodynamic winch is rotated by pumping water between wedge-shaped longitudinal compartments to produce a turning moment.

14.4 VEHICLE-MANIPULATOR SYSTEMS

Deep-ocean work systems utilizing deep-submergence vehicles as an extension of man's capability to perform mechanical tasks underwater are still new. There is still a great deal to be done to produce a system or systems that will enable us to carry out a variety of tasks at will in the depths of the sea. Evolution of these vehicle-manipulator systems has taken place, and is continuing, along paths toward the accomplishment of three general purposes. These are to recover things we have lost in the sea; to discover new things about the sea through research; and to perform

FIG. 14-24. Sections of model of Rosenberg hydrodynamic winch, showing valves in pump plenum. (Official U.S. Navy photo.)

TABLE 14-3
Rosenberg Hydrodynamic Winch
Estimated Characteristics

		Capacity (tons)				
		10.	100.	1,000.	5,000.	10,000.
Length	(ft)	38	82	178	304	382
Diameter	(ft)	13	29	62	106	134
Draft	(ft)	8	18	38	66	83
Natural period of oscillation						
Roll	(sec)	6	9	14	18	20
Pitch	(sec)	5	7	10	14	15
Heave	(sec)	4	4	9	12	13
Hoisting Line diameter (four each of steel)	(in.)	0.4	1.4	4.2	9.5	13.4
Factor of safety in line			1.8	1.8	1.8	1.8
Total cost	(millions of dollars)	0.07	0.7	6.2	31.2	62.5

mechanical tasks such as emplacing, monitoring, servicing, repairing, or moving structures and pieces of equipment in deep water.

Most of the vehicles so far developed have capabilities in more than one of these areas. It may, therefore, be best to group them for discussion according to the type of vehicle used as the platform for the mechanical work system. There are at present four basic types of deep submergence vehicles.

Bottom-crawling remote-controlled vehicles

The remote-controlled underwater manipulator (RUM) vehicle, Fig. 14-25, was built for Scripps Institution of Oceanography on an ONTOS self-propelled rifle chassis, modified for use as a bottom-crawling vehicle capable of operating underwater at depths to about 20,000 ft. Intended for use in placing and tending equipment in a benthic laboratory, the RUM was designed to operate up to 5 miles away from a shore or surface control station, receiving electric power and control signals over a cable laid from a reel on the vehicle as it traversed the bottom. Lights and television cameras, shown in Fig. 14-25, were provided to enable the control operator

FIG. 14-25. RUM was built to crawl over the bottom at depths to 20,000 ft. Propulsion, lights, sonar, and manipulator were powered and controlled over coaxial cable from up to 5 miles away. (Official U.S. Navy photo.)

to guide the vehicle and to see and control mechanical tasks being performed by the manipulator.

Operations with the RUM vehicle focused attention on several problems that are inherent in bottom-crawling vehicles. Difficulties were frequently encountered because of unevenness or low bearing strength of the bottom. When operating in areas where finely divided sediment was present, the cloud stirred up by the progress of the vehicle over the bottom obscured vision and made television monitoring ineffective.

Surface-tethered manned vehicles

Oldest of all the types of deep-submersible vehicles, the surface-tethered manned vehicle traces its lineage from the diving bell through William Beebe's Bathysphere of the 1930's, which was the first truly *deep* submersible. Though it had a long head start on the other types of undersea vehicles, the surface-tethered manned vehicle suffers from several inherent shortcomings, and has been built in very small numbers.

In spite of sharply limited mobility, limited search capability, and limited operational depth, however, the surface-tethered vehicle can perform admirably under favorable conditions. Although most tethered vehicles have been designed for research work and observation, as was the Bathysphere, the Japanese *Kuroshio II* and *Yomiuri* are both equipped with manipulators for sample collection.

Recoverer I, Fig. 14-26, on the other hand, was built by the International Underwater Research Corporation for use as a work vehicle, and is equipped with a pair of the most powerful underwater manipulators yet fitted to a deep-submergence vehicle. These manipulators were designed for use in heavy salvage work, and provide a lift or horizontal force capability of 500 lb each at an outreach of 10 ft from the shoulder pivot. Intended for heavy-duty service on a vehicle having ample reserve buoyancy and tethered by cable from the surface, the manipulators were made rugged to reduce possibilities of damage in use. The structural linkages are made of rolled structural steel shapes, and all joint motions except grip rotation are powered with standard hydraulic cylinders. Continuous grip rotation is provided by a hydraulic motor.

Hydraulic power is provided by a hydraulic power plant inside the

FIG. 14-26. *Recoverer I*, 1000-ft depth surface tethered salvage vehicle has two 500-lb capacity manipulators.

pressure shell, Fig. 14-27, driven by electric power from the surface. The hydraulic oil flow through the pressure shell to the manipulator actuators at 3000 psi is controlled by hand-operated valves mounted on a manifold at each operator's position. Each manipulator is controlled by a separate operator.

Recoverer I was used continually for several months in diving operations in efforts to raise a 165-ft sunken fishing vessel in approximately 200 ft of water off Cape Lookout, North Carolina. In the process of clearing away nets and rigging, preparing the ship for raising, and attaching the flotation containers, the manipulators were used to perform a variety of mechanical tasks.

Experience with the direct-control hydraulic system has been very favorable. The operator develops a feel for direction and speed of arm

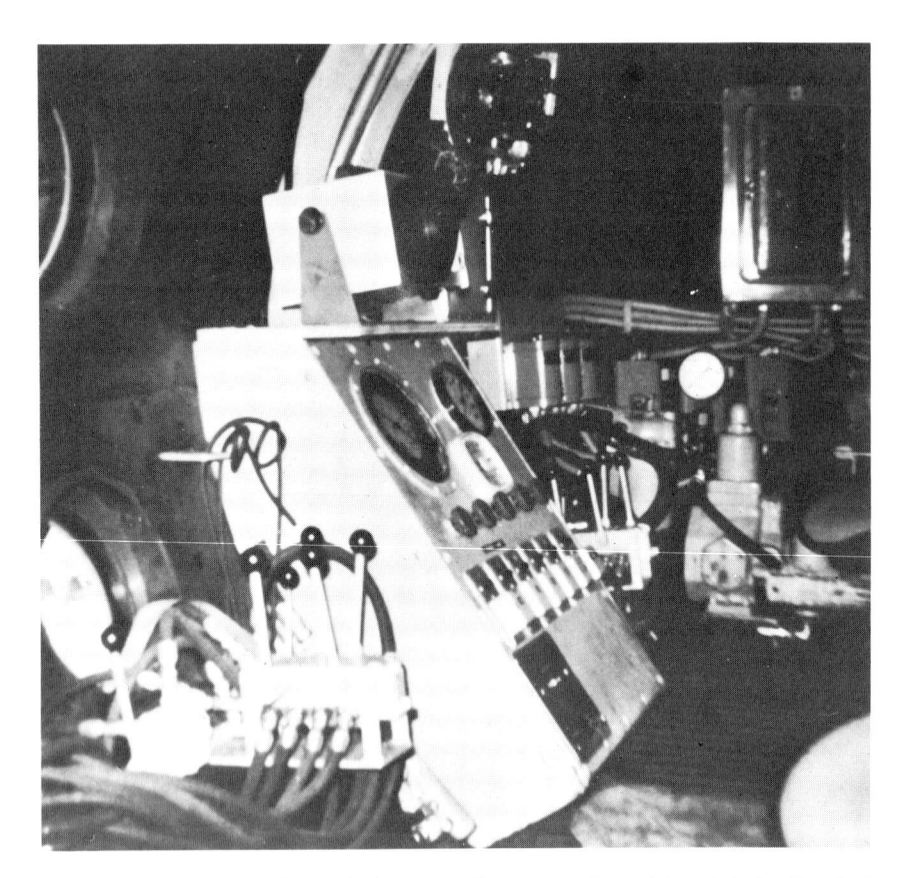

FIG. 14-27. *Recoverer I*, manipulators receive power from internal hydraulic plant. Manipulators have individual control through manually operated valves.

motion corresponding to the feel and sound of the hydraulic valve, which aids materially in controlling the manipulator.

It has also become quite evident that it is next to impossible to do work with the manipulators while the vehicle is suspended free or hovering neutrally buoyant at the work site. Some attachment to the object being worked on or to the sea bottom or other firm object is necessary to provide a sufficiently steady platform and the required reaction forces to prevent the manipulators from moving the vehicle.

These observations were borne out by subsequent experience when attempts were made to use *Recoverer I*, now operated by Pan American Undersea Research & Salvage, Inc., in the recovery of aircraft wreckage from Lake Michigan in about 250 ft of water. Attempts to use the tethered *Recoverer I* for search and recovery operations were unsuccessful primarily because of lack of control of position of the vehicle. Operating suspended from the surface vessel, it was not possible to maintain optimum height above the bottom for proper visibility in the relatively turbid water. Occasional contact with the bottom further aggravated the condition by stirring up mud. When wreckage was sighted, it was impossible to stop the surface support ship before the vehicle had been carried beyond visual contact with the item sighted. Even if the ship had been able to stop immediately, the vehicle would have continued to move until it had reached equilibrium under the boom-head suspension point.

It may be concluded that although a tethered vehicle can be usefully employed for general observation while moving, it must be suspended stationary or, preferably, firmly attached at the worksite for mechanical work with the manipulators.

Surface-tethered remote-controlled vehicles

Several surface-tethered remote-controlled vehicles have been developed, primarily for the purpose of finding and recovering objects from deep water, although some have been used for research purposes, for underwater cable inspection and for similar tasks. Typical of these vehicles is CURV, Fig. 14-28.

CURV was developed and is operated by the Naval Ordnance Test Station, Pasadena, California. Although it is a member of the "older generation" of undersea work vehicles, CURV is by no means obsolete, as was demonstrated recently when CURV handily recovered the lost nuclear device from a depth of about 2,900 ft in the Mediterranean off Palmares, Spain.

CURV, which weighs approximately a ton, is operated from a control console on the surface support ship which transports CURV to the recovery

534 Systems Design—The Technology

FIG. 14-28. CURV vehicle is tethered to surface ship for power and control. It is guided remotely by sonar and television. Hydraulically actuated claw can be released and buoy line used to recover objects too heavy to be recovered by the vehicle. (Official U.S. Navy photo.)

area. After the general location of the object to be recovered is determined, the support ship is anchored as nearly above the object as possible, and CURV is lowered over the side, trailing its control and power cable.

Propelled by two aft-thrusting and one vertical propeller, each powered by its own 10-hp, 440-V ac pressure-compensated motor, the vehicle swims toward the recovery objective. The course and speed of the vehicle can be very finely controlled by use of the variable autotransformer motor controls. Using the Sonar display at the console, the operator continues to guide the vehicle toward the object until it becomes visible on the television monitor. In addition to the FM high-resolution Sonar with active and passive modes, the SLAD-503 acoustic instrumentation system provides locator, altimeter and depthometer functions to assist in controlling the vehicle.

Once the recovery object has been identified by television, the vehicle is maneuvered into position to attach the hydraulically actuated claw. When the claw is attached the vehicle can be surfaced with the object, if the object is sufficiently small. If the object is too large or heavy to be brought up with the vehicle, a recovery buoy is released, which takes a nylon guideline, attached to the claw, to the surface. The claw is then released by means of a quick-disconnect mechanism and ejected from the vehicle. CURV is returned to the surface and the recovered object is hoisted to the surface by separate means.

In addition to Sonar and television, CURV is equipped with a 35-mm deep-sea camera for documentation of recovery operations or for use in preliminary search and identification.

Free-swimming manned vehicles

Free-swimming manned vehicles comprise the largest group of underwater work vehicles. Most of the deep-submergence manned vehicles in existence today have one or more manipulators, or some means of doing a mechanical task, even if it is only to grasp a rock or a shell to bring it to the surface.

The work system involving a free-swimming manned submersible vehicle includes much more than just the manipulator system, or even the complete vehicle. The surface support ship is as much, or more, a part of the work system as is the case with a tethered vehicle which depends upon a surface ship to hold it suspended in the deep. Considerations for safety of life alone require a large assortment of equipment and several attendant personnel. Servicing the tools and equipment to be used in performing a task underwater add to these. The fact that the vehicle is *not* attached to the ship by a supporting cable, that the vehicle must make contact with the surface support ship and be brought aboard for servicing and maintenance, generates one of the stickiest problems in the whole system. Although space will allow only brief treatment here of the problems involved in handling vehicles on and off the mother ships, this is an area which requires a great deal of consideration in the design of a vehicle system, and should be one of the prime criteria in the design of the vehicle itself.

Trieste. *Trieste* was built by Swiss scientists August and Jacques Picard to prove the feasibility of extreme deep submergence. After her acquisition by the U.S. Navy, she was piloted by Navy Lt. Don Walsh and Jacques Picard to the record depth of 35,800 ft in Challenger Deep. Shortly thereafter, in 1961, she was provided with the first underwater manipulator ever fitted to a manned deep submersible.

To provide a manipulator for scientific sampling and for performing other tasks at the bottom, the Navy turned to industry to draw on the knowledge accumulated over the previous two decades of nuclear hot-cell manipulator development.

To take advantage of this background, and to meet the requirement for quick delivery within a rather limited budget, the fully developed and proved General Mills Model 150, Fig. 14-29, was modified for underwater use. With a rated capacity of 50 lb at an outreach of about 39 in., the Model 150 offered an attractive addition to the capabilities of *Trieste*. A mounting was provided, pivoted to the forward ballast shot tub. It was equipped with a hoist, which could lower the manipulator into operating

FIG. 14–29. Small, modified Gen. Mills Mod. 150 nuclear hot-cell manipulator on *Trieste* was first on manned deep submarines. (Official U.S. Navy photo.)

position where it could be seen from the pilot's view port, and retract it out of the way when not in use. This arrangement also provided a safety release in the event that the arm became entangled in an obstruction on the bottom, since the shot tub was suspended from a magnetic release device and could be released from inside the pressure sphere.

This safety feature proved the undoing of the first manipulator. The release was inadvertently tripped shortly after installation and the manipulator was lost before extensive use data could be obtained.

A second Model 150 was procured and delivered in 1962. Experience with the second one brought to light a number of problem areas, some resulting from unexpected effects of adapting a manipulator designed for an air environment to operation in a saltwater environment, and some from changes in operating procedure. The Model 150 manipulator procured for *Trieste* retained the aluminum–magnesium structural materials used in the hot-cell version. Because of the size and nature of *Trieste*, it remained in the water for relatively long periods of time, making the manipulator inaccessible for maintenance. Long immersion led to corrosion of the fastenings and castings. Corrosion also developed at points where the surface got scraped, and developed around fastenings as a result of gradual soaking of seawater through the paint film.

Additional problems arose as a result of the necessity for towing *Trieste* to its operating site. Exposed tubing carrying electric wires on the exterior of the manipulator members loosened when water drag loads due to towing and wave action exceeded loads anticipated during design. Resultant leakage and damage to the wires caused corrosion and malfunctioning of the manipulator.

Difficulty was also experienced in getting proper commutation with the drive motors operating in the pressure-compensated oil bath. The system was found to be particularly sensitive to changes in the brand or composition of the oil.

The Model-150 manipulator was reworked before being refitted to *Trieste* in 1963 for the series of dives in the *Thresher* disaster area.

The reworked manipulator was used in the now-famous dive from which *Trieste* returned to the surface with a piece of pipe from *Thresher*.

Trieste II. After the Atlantic operations in 1963, the float structure of *Trieste*, designed primarily for vertical ascent and descent, was replaced by a more streamlined float better suited to towing and to propulsion for search over the bottom. In the spring of 1964 *Trieste II* was outfitted for further operations in the Atlantic.

When outfitting *Trieste II* for the 1964 series of dives, it was apparent that a larger and more powerful manipulator was required to recover larger objects from the bottom. A lift capability of 500 lb at an outreach of approximately 10 ft was required, with a capability of grasping objects at least 8 in. thick. The only existing manipulators designed and built for use on a manned submersible and capable of handling the loads required, were the manipulators built by the International Underwater Research Corp. for use on their tethered vehicle, *Recoverer I*. Although the through-the-hull hydraulic piping of the 1000 ft depth *Recoverer I* was unsuitable for use on deep-diving *Trieste II*, the manipulator linkages were easily adaptable. A remotely electrically controlled version of the *Recoverer I* manipulator was built by American Car & Foundary Co. and International Underwater Research Corp. and installed on *Trieste II* in May 1964 (Fig. 14-30).

Because of the danger of entanglement with bottom obstructions, the manipulator was fitted to a releasable base held in place in the float structure by a chain leading up through the float to an electromagnetic release device similar to those used to secure the ballast shot tubs. A switch arranged to preclude inadvertent operation was provided for the operator to break the circuit and release the manipulator.

FIG. 14-30. 500-lb capacity *ACF/IURC* manipulator was fitted to *Trieste II* for *Thresher* search. (Official U.S. Navy photo.)

The hydraulic power plant for the manipulator was housed in single pressure-compensated oil-filled tank, mounted on the top of the float structure for ease of access for maintenance. Hydraulic lines for each manipulator motion were led over the side of the float and to a fixed plate mounting at the edge of the manipulator recess in the bottom of the float, arranged in such a way that the hose fittings will shear off if the manipulator is released. The pump motor and solenoid valves are controlled electrically from the portable control box in the pressure sphere.

This manipulator remained on *Trieste II* throughout the Atlantic operations in 1964. Although it might be expected that corrosion problems would result from use of structural steel for a large part of the mechanism after such long immersion, no serious problems resulted from this cause.

A third manipulator was developed by the Navy Electronics Laboratory, Fig. 14-31, and fitted to *Trieste II* during its outfitting period in Boston in May 1964. Intended primarily as a device for recovering samples of various types from the bottom, this manipulator was unique in that it was actuated by a seawater hydraulic power system.

The manipulator consisted of a telescoping boom mounted at one end on the *Trieste II* float structure, and supported at the outer end by two cables led down through opposite sides of the float from two cylinders mounted on top. The two cylinders actuated a sheave arrangement multiplying the cylinder motion to allow the boom head to be lowered about 12 ft below the bottom of the float and swung through an arc of 30 deg.

Deep-Ocean Work Systems 539

With this cable arrangement and the 3-ft hydraulic extension of the telescoping boom, a bottom area about 5 ft wide and 3 ft fore and aft could be reached.

The boom was equipped at its outer end with a hydraulically actuated clam-shell grab. The clam-shell jaws were shaped to facilitate taking samples of mud or sand, and were fitted with interlocking teeth to pick up objects too large to fit within the jaws. Spring-loaded covers on the sides of the jaws covered cutouts which permitted recovery of long slender items such as pipes or rods.

Hydraulic power for actuation was supplied by a standard submersible deep-well multistage centrifugal pump capable of pressures 175 psi above ambient. Specially developed electrically operated saltwater valves were used to control the various motions of the manipulator.

The structural elements of the boom and its power system were built chiefly of stainless steel. All mating surfaces with sliding contract were made of glass-filled Teflon and stainless steel, both in the hydraulic cylinders and the solenoid control valves. It was found that the glass-filled Teflon on stainless steel had a very low coefficient of friction (lower even in seawater than in fresh water), and there was no apparent tendency to stick and slip, or to chatter.

While this manipulator was undergoing its first tests at the bottom off Boston, the weather on the surface became quite rough. While *Trieste II* was being towed back into Boston through a choppy sea, the manipulator

FIG. 14-31. Navy Electronics Laboratory manipulator, built for use on *Trieste II*, had unique seawater hydraulic power system. (Official U.S. Navy Photo.)

boom became unhooked from its stowage fitting and the violent motion caused the support cables to part and it was lost. No new manipulator has yet been built on this principle, although the seawater hydraulic system worked well. This type of power system should be developed further.

Alvin. *Alvin*, Fig 14-32, the 6000-ft depth vehicle built by the Applied Science Division of Litton Systems (formerly Electronics Division of General Mills, Inc.) has proven its usefulness as a work vehicle in the search and recovery operations to recover the nuclear device mentioned above in the discussion of the CURV vehicle.

To provide surface support facilities for *Alvin*, Woods Hole Oceanographic Institute built a catamaran, Fig. 14-33, on two 96-ft long oval minesweep pontoons. The tricky problem of launching and recovering *Alvin* in a disturbed sea was approached by providing a novel free-swinging elevator between the twin hulls. This elevator has been used to recover *Alvin*, surfaced, in sea state 4, and is designed to be lowered below the

FIG. 14-32. *Alvin*, 6,000-ft depth vehicle built by Litton Systems, Inc., for the Office of Naval Research and operated by Woods Hole Oceanographic Institution, found lost H-Bomb off Spain. (Official U.S. Navy photo.)

FIG. 14-33. *Alvin* is maneuvered between twin floats of catamaran barge, hoisted by elevator to top deck for service and transportation. Vans (upper right and left) house shops and instruments. (Woods Hole Oceanographic Institution photo.)

surface for recovery from a submerged position of lesser wave action. For surface recovery, *Alvin* is maneuvered between the hulls and into position over the elevator platform under its own power, but with the guidance of six lines tended manually from the pontoon decks. Once seated on the elevator, *Alvin* is lifted to the upper deck and rolled into the servicing deck area on a carry-all type hoist.

The heavy, low-lying catamaran has a response to waves so similar to the surfaced *Alvin* that the two move together during pickup operations so that elevator contact causes very little shock to the vehicle.

Alvin's support shops and equipment are housed in portable vans on the upper deck of the catamaran, making it an easy matter to transfer *Alvin* and its support complex by air to another ship in a distant area, as was done for the hydrogen-bomb recovery operations in the Mediterranean.

The manipulator for *Alvin* is an example of the direct application of nuclear hot-cell manipulator design background to the design of a manipulator system for use at deep depth in the oceans. The *Alvin* manipulator is operated by electromechanical drives controlled by switches on a portable control box in the pressure sphere. The manipulator drives and linkages are similar to those used in the nuclear hot-cell manipulators, but have been designed specifically for use under the sea. Arm pivot and rotation drives are powered by dc electric motors through gear trains and chain and sprocket drives equipped with slip clutches to protect the motors and the manipulator linkage against overload damage. The force exerted by the parallel jaw grip is adjustable from 0 to 100 lb through electrical control of the clutch in the grip drive.

All external parts of the manipulator are made of stainless steel to resist corrosion and the interior is completely filled with oil compensated to ambient sea pressure to prevent leakage of seawater into the mechanism.

The manipulator is mounted to a bracket on the forward port corner of the structural frame of *Alvin*, and is retained in position by a safety release mechanism which can be withdrawn at will by operation of a switch mounted on the operator's control panel. The switch is covered so that two deliberate operations are required to release the mechanism to drop the manipulator.

Alvin had just completed deep-dive trials to her 6000 ft design depth in the Carribean when the need arose for her use in Spain. During test dives, the manipulator was used to collect shells and other objects on the bottom, and was used, in conjunction with a coring tool designed to be held in the grip, Fig 14-34, to take mud and sand cores.

Although the *Alvin* manipulator has worked well, experience indicates that it would be desirable to have two manipulators to perform work at the bottom. One manipulator could then be used to hold the work while the other performed the task.

Others. A number of other vehicles have been equipped with manipulators. Costeau's *Diving Saucer SP-300* and the Westinghouse *Deepstar 4000* are each equipped with a stiff-armed manipulator designed primarily for sampling and specimen gathering. A pair of 200-lb-capacity manipulators has been built for the Reynolds *Aluminaut*. *Star III*, General Dynamics' latest deep submersible has a single 150-lb-capacity manipulator. The *Deep Quest* vehicle, under construction at Lockheed, will be equipped with a manipulator system for mechanical operations.

In development or construction

Beaver. *Beaver*, Fig. 14-35, a work vehicle concept under development by Autonetics Division of North American Rockwell, has several advanced features applicable to vehicle work systems. Its work equipment includes two hydraulic work manipulators with interchangeable power tools to perform mechanical tasks, and two hydraulic positioning arms that work together with an anchor to hold the vehicle in position while work is performed.

FIG. 14-34. *Alvin* can carry four coring tubes in an external rack. The manipulator removes a tube from the rack, obtains a sample by forcing the tube into the bottom and returns the loaded tube to the rack. (Woods Hole Oceanographic Institution photo.)

The work manipulator has been built in prototype and has been in use in Autonetics in-house manipulator studies and in a manipulator system development study for the Bureau of Ships. It has nine motions to allow positioning the tools within a hollow hemisphere of 2 ft inner radius and 6 ft outer radius. A hook-type grip is used to handle materials (50 lb capacity in any position) or to hold tools not fitted to be engaged in the ball detent wrist socket. In addition to the usual vertical and horizontal shoulder pivot, vertical elbow and wrist pivot, and grip open or close and rotate motions, the work manipulator has horizontal wrist pivot, and can extend the wrist socket 3in. in a straight line to facilitate tool engagement, drilling, etc. It is also capable of rolling the entire arm 90 deg about its longitudinal axis.

Actuation of the pivot joints in the manipulator and rotation of the

544 Systems Design—The Technology

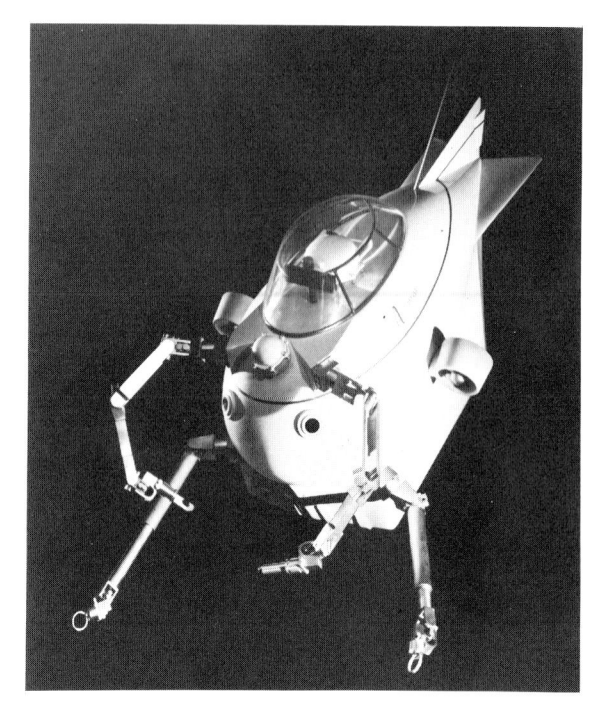

FIG. 14-35. Autonetics' *Beaver* concept model has two work manipulators with tools that can be changed at the work site. Two additional arms, with simpler articulation, will hold the vehicle in position while manipulators are used. (Autonetics photo.)

hand socket are all accomplished by piston-driven rack and pinion rotary actuators. These actuators, because of their inherent slow leakage rate, provide a hydraulic lock with a very low power-off creep rate.

The controller, Fig. 14-36, is arranged to provide spatial correspondence between the operator's hand motions and those of the tool in the manipulator.

Extension and retraction of the hand telescope is controlled by motion of the top button on the control handle, and the button on the front of the control handle controls the opening and closing of the grip. The power on/off switch is located on the control box, as is the switch controlling the roll motion of the manipulator about its horizontal axis, and the high–low rotation-speed control. The other four switches on the control box serve to invert the control hookup so that when the arm is rolled 90 deg, the spatial correspondence between the control handle and the manipulator motions will be maintained.

The 90-deg roll motion of the arm provides considerably increased facility for performing mechanical work tasks, because the vertical bend

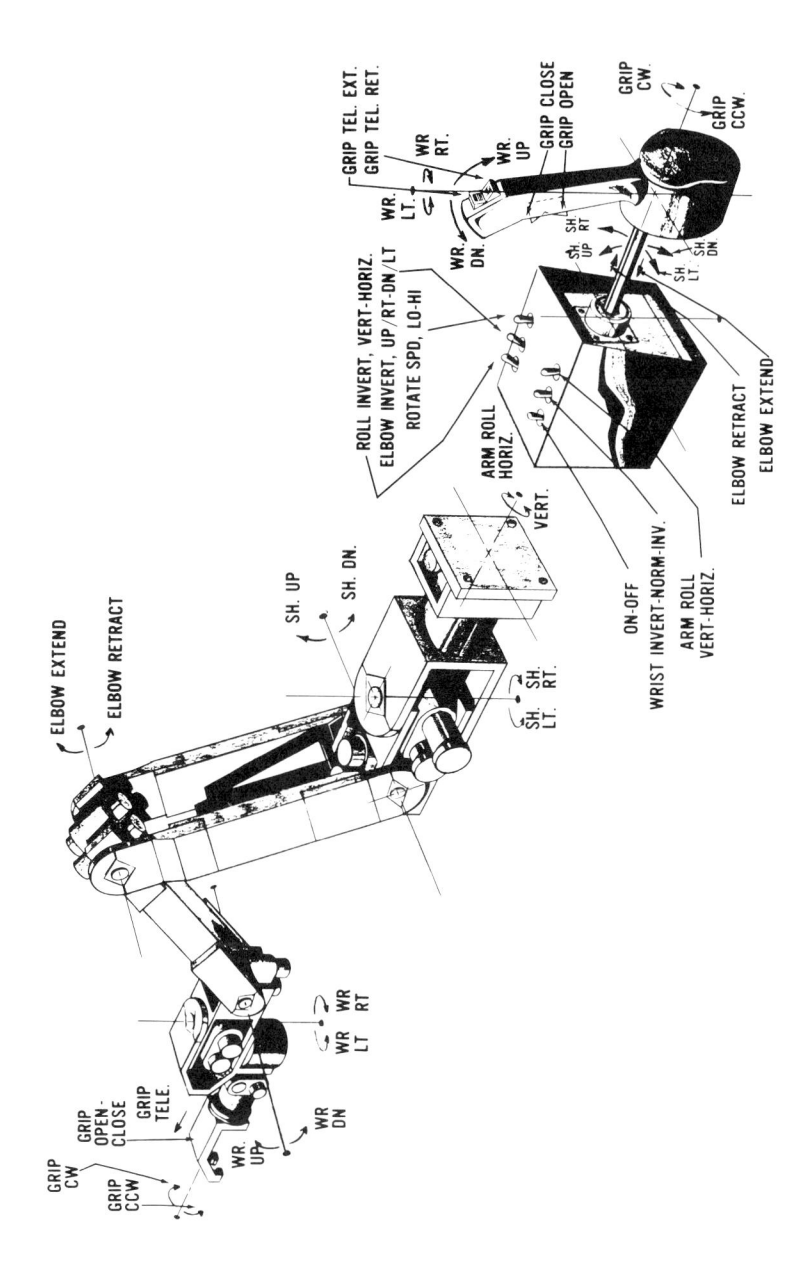

FIG. 14-36. The controller for the Autonetics manipulator provides spatial correspondence between the motions of the operator's hand and the motions of the manipulator. (Autonetics photo.)

FIG. 14-37. Remote change tools for the Autonetics *Beaver* manipulators will be carried in a rack for changing at the work site. Tools are, left to right: water pump-jet, impact wrench, high-speed hydraulic motor, powder-actuated stud driver and punch, hook grip and drill chuck. (Autonetics photo.)

at the elbow now becomes a horizontal bend and the manipulator can reach around, through, and behind obstructions, or can bring both grips or tools close to the observation port for better visibility of the work being performed in turbid water or poor light.

The ball detent wrist socket can be remotely disengaged to allow remote interchange of tools, so that any tool carried in the vehicle tool rack may be interchanged for any other while the vehicle remains at the underwater work site. The tool equipment being evaluated in the test system at present includes (Fig. 14-37):

1. hook grip;
2. powder-actuated stud gun and punch;
3. high-speed hydraulic motor which powers (a) a water-jet pump, (b) an impact wrench, and (c) drill chuck for drills, a wire brush, and grinding wheels.

AUTEC. AUTEC vehicle, Fig. 14-38, is now being developed by Electric Boat Division of General Dynamics for the Navy's Ship Systems Command, for use as a work vehicle on the AUTEC Range. AUTEC, 25ft long, weighing about 16 tons dry and capable of submergence to 6,500 ft, is slightly larger than *Alvin*, but small enough to be very maneuverable. This feature proved essential during *Alvin*'s search for the lost hydrogen bomb off Spain. As a submersible range work vehicle, AUTEC will be used to place, inspect, test, and retrieve electronic equipment and systems on the ocean bottom, assist in salvage work, perform site surveys, including core sampling, and perform or assist in oceanographic research operations. To perform the mechanical aspects of these operations, AUTEC will be equipped with two fully articulated manipulators and an initial set of tools including the ones just shown for *Beaver*, Fig. 14-37, plus parallel jaw hand, prosthetic hand, circular grapple, three-jaw orange-peel grab, and cable cutter.

Because of the small size and low inertia of AUTEC, it will be necessary to provide restraining forces to hold the vehicle in position during most manipulative tasks. For this purpose, an anchor is provided to assist in holding position. When the vehicle cannot sit on the bottom during the performance of manipulator tasks, it will use one manipulator as a holding device, and steady itself with the anchor if sufficiently near the bottom.

Tool racks will be provided to hold the tools for each manipulator so that they may be interchanged remotely at the work site.

The self-propelled AUTEC barge, Fig. 14-39, is being built and equipped

FIG. 14-38. Artist's conception of Navy's AUTEC vehicle. AUTEC will have too fully articulated manipulators and remotely interchangeable tools. (Concept by J. Claffey.)

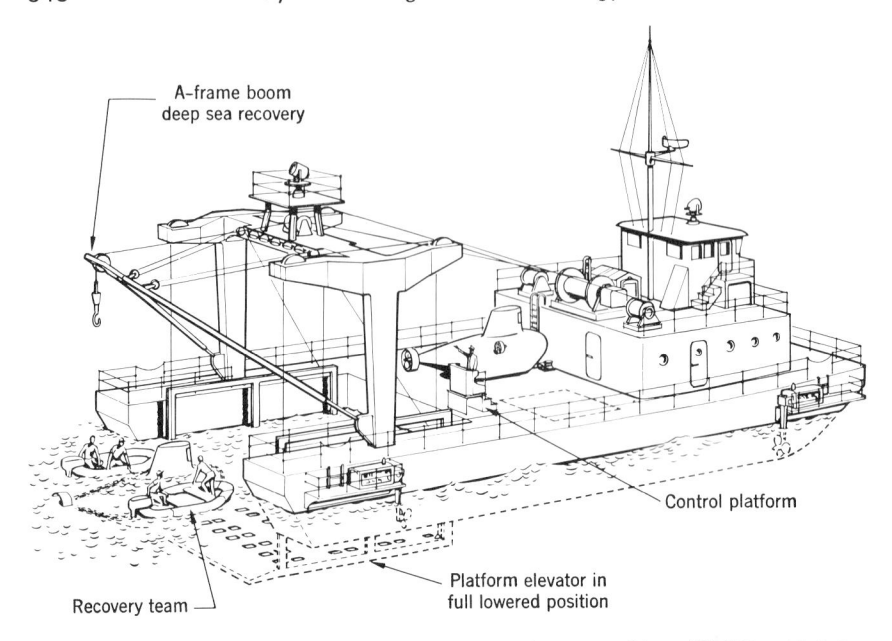

FIG. 14-39. AUTEC vehicle support barge will have elevator to pick up AUTEC and hoisting facilities for unmanned deep testing of deep submersible vehicles.

as a mother ship for the AUTEC vehicle. Powered by harbormaster diesel outboard drives, the barge will be able to move in any direction to maneuver into position to recover AUTEC with the elevator built into the bow of the barge. Service and support equipment will be housed in portable vans to allow quick and easy changes of instrumentation without lengthy tie-up of the barge and vehicle.

Deep-submergence rescue vehicle (DSRV). The prototype DSRV is being developed by Lockheed Aircraft Corp. under the management of the Navy Deep Submergence Systems Project office.

The primary mission of the DSRV is to rescue personnel from a disabled submarine. To accomplish this mission it must be able to locate the submarine escape hatch under very low visibility conditions at depths beyond the reach of present rescue equipment, maneuver into position, and mate to the submarine hatch with a pressure-tight seal. To assure the capability of the DSRV to perform the mating operation, a manipulator will be provided to connect the vehicle haul-down line to the submarine hatch bail, to cut and remove the messenger buoy cable which will be stretched across the seating plate around the hatch where the DSRV skirt seal must seat, and to remove dirt, debris, or damaged structure from the seating area. This work must be completed before the DSRV skirt can be seated and sealed to the submarine, and it must be done in the shortest possible time to assure maximum probability of a successful rescue mission.

The single, bottom-mounted manipulator is located so that it can be stowed within the vehicle fairing to reduce drag in transit but can reach into the vehicle skirt to assist in connecting the haul down line. Conceptual studies of the manipulator include a combination tool, Fig. 14-40, for use during the rescue mission which combines a prosthetic hand containing a cable cutter, capable of cutting the messenger buoy cable, and a rotary device which may be fitted with either a rotary brush or a centrifugal pump jet for clearing away dirt and debris on the hatch seating plate.

The DSRV will be air transportable to assure rapid availability of the rescue capability at any point in the world. To assure the capability of performing rescue operations in any kind of weather, the rescue vehicles

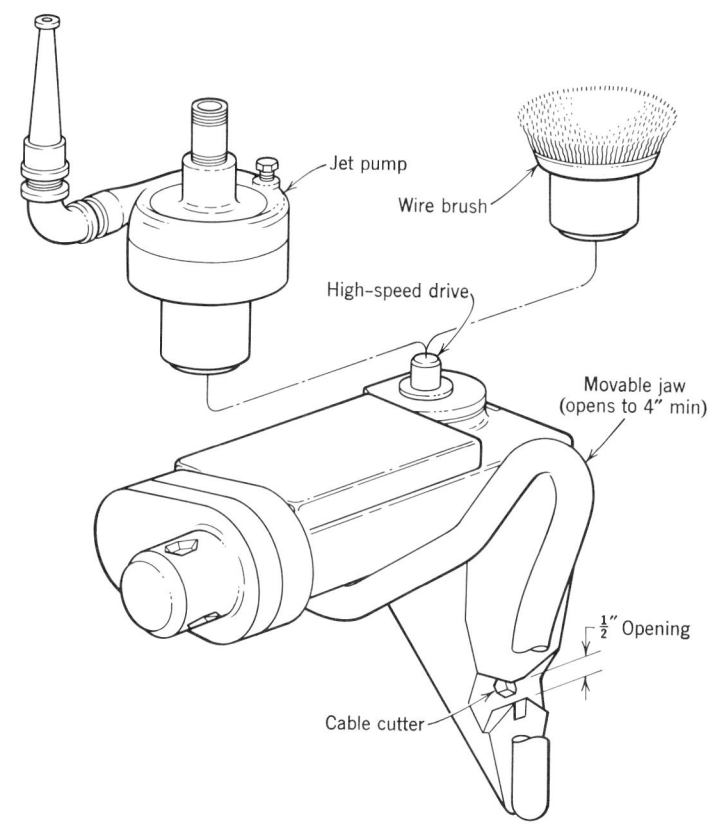

FIG. 14-40. The Deep Submergence Rescue Vehicle (DSRV) manipulator will be equipped with tools to cut cables, to clear away debris, and to clean the seating plate around the submarine escape hatch.

will be operable from a submarine, submerged below the effects of surface weather. Riding piggy-back over the after escape hatch, the DSRV will be carried to the rescue area, will leave the mother submarine, mate with the escape hatch on the disabled submarine, take on survivors, return to and mate with the mother submarine. The rescuees can then be transferred to the mother submarine to be cared for while the rescue vehicle returns to the aid of the stricken submarine.

Two DSRV's will be carried by the new submarine rescue ships, the twin-hulled ASR.

Trends for future development

Very little complex mechanical work has yet been done by undersea vehicles. All of the vehicles presently operating are equipped to do a fairly simple mechanical task such as attaching a claw type recovery attachment (CURV), to perform a task specifically designed for and in conjunction with the vehicle (MOBOT and the Shell modified wellhead), or to perform simple sampling or grasping operations (*Deep Star*). Work with these vehicles has brought out, and recent studies have reinforced, several principals which should be applied to future development of underwater vehicle work systems:

1. Definition of the tasks to be performed must be the first step in the design of any successful manipulator system. This is just as applicable to a general purpose manipulator system as it is to a special purpose one such as the Shell MO system. It is frequently desirable to design a work system to perform a series of different tasks, but a completely successful system is unlikely to result unless all of the different tasks are identified, and their peculiar requirements built into the system.

2. For performing mechanical tasks such as monitoring, operating, or assembly and disassembly of equipment, the vehicle-manipulator system and the equipment to be worked on should be developed as one system. This approach, followed in the development of the Shell MO system, results in the maximum simplification of the over-all system, assures that task requirements will be compatible with the manipulator and vehicle, and offers the opportunity to incorporate the more complex and, hence, expensive control and actuation functions in the manipulator vehicle. Providing control and power to open and close valves in the vehicles reduces the cost of the equipment which must remain in place on the ocean floor for long periods of idleness.

3. The design of the manipulator system must integrate the manipulator mechanism, lighting, view ports, TV, controls, and human engineering factors to make the vehicle and the operators integral parts of the manipulator

system. The most important problem in any underwater vehicle manipulator performed task is keeping the operator constantly and accurately informed of what is happening and of the relative positions and velocities of the work tool, manipulator, and vehicle. The fewer variables, the operator must contend with, the greater his ability to observe accurately and to make correct solutions. Therefore, the greater the degree of fixity in the relative positions of the vehicle and the object being worked on, the easier it is for the operator to observe the operations of the manipulator and relate these operations to the performance of the task. When this degree of fixity cannot be attained because of the position or conditions under which the vehicle must function, a complex dynamic control system may be necessary to allow use of vehicle manipulator positioning.

4. For deep-depth, multiple task mechanical operations, interchangeable or multipurpose tooling is required to avoid expensive repetitive surfacing for tool or equipment changes.

5. Some means of fixing the vehicles with respect to the work object is necessary when performing tasks that require movement of weights or cause reaction forces.

6. For safety, manipulators or other work devices must be designed to break loose or be jettisonable in case of entanglement.

AUTEC and other vehicles now under development will be equipped with manipulators having many of these features. AUTEC vehicles will also have remotely interchangeable tools. The Submarine Rescue Vehicle (DSRV) will have a combination tool capable of performing several tasks during submerged mating operations, and integrated control of the vehicle and manipulator is under development.

One of the most difficult problems in developing a manipulator work system is definition of the work tasks to be performed under water. As other areas of ocean engineering develop, new needs and new tasks will require further developments in manipulator systems. Tailoring the system to the task and integrating the equipment to be serviced, the vehicle, the manipulators, and tools into an over-all system to assure compatibility of all components and tasks, is the best approach to assure success.

REFERENCES

Dyment, R., 1966, "Cachalot Diving System Aids Repairs on Utility Dam," *Compressed Air Magazine* v. 71, No.6 (June 1966), pp. 4–8.

Gaber, N. H., and D. F. Reynolds, Jr., 1965, "Ocean engineering and Oceanography—From the Businessman's Viewpoint," *Trans. Joint Conf. Marine Tech. Soc. and Am. Soc. Limnol. and Oceanography, June 1965, Washington, D.C.*

Graham, J. R., 1965, "Mooring Techniques in the Open Sea," *Marine Tech.*, p. 132 (April 1965).

Hess, H. D., 1965, "Esploiting Hydrospace," Part IV, *Eng. Mining J.*, **166**, 8 (1965).

Holmer, E. C., 1966, "Offshore Oilwells Go For Deep Water," *Undersea Tech.*v. 7, No.1 (January 1966).

Hunley, W. H., and G. Houck, 1965, "Existing Underwater Manipulators," ASME Paper 65-UNT-18. Condensed in *Mech. Eng.* (March 1966).

Mavor, J. W., Jr., 1966, "Ten Months with *Alvin*," *Geo. Marine Tech.*, **2**, 2 (1966).

Rogers, L. C., 1965, "Libyan Offshore Test Drilled in 525 ft. of Water," *Oil Gas J.*, **63**, 10 (1965).

Talkington, H. R., 1965, "Underwater Engineering Experiments with NOTS YFU-53," *Trans. Joint Conf. Marine Tech. Soc. and Am. Soc. Limnol. and Oceanography, June 1965, Washington, D.C.*

Terry, R. D., 1964, "The Case for the Deep Submersible," Autonetics Division of North American Aviation, Inc.

Thornburg, R. B., 1965, "Petroleum's Pandora, Offshore Oil Drilling and Associated Technologies," *Trans. Joint Conf. Marine Tech. Soc. and Am. Soc. Limnol. and Oceanography, June 1965, Washington, D.C.*

Webb, W., 1965, "Technology of Sea Diamond Mining," *Trans. Joint Conf. Marine Tech. Soc. Limnol. and Oceanography, June 1965, Washington D.C.*

Wilson, T. A., 1965, "Exploiting Hydrospace," Parts I, II, and III, *Eng. Mining J.*, **166**, 5, 6, 7 (1965).

"Deep-Diving Cachalot is Operational," *Armed Forces Management*, **12**, 3, 14, (December 1965).

"Man's Extension Into the Sea," *Trans. Joint Symp. January 1966, Marine Tech. Soc. Washington D.C.*

"California and Use of the Ocean," 1965, University of California Institute of Marine Resources. IMR Ref. 65–21 (Oct. 1965).

"A Special Report on the Sea," Staff Report *Industrial Research* Vol. 8, p. 42 (March 1966).

"Diving Chamber to Increase Operating Depths," *Shipbuilding and Shipping Record* (January 20, 1966).

CHAPTER 15

Oceanographic and Experimental Platforms

F. N. SPIESS

Many seagoing activities cannot be carried out effectively because of interference arising from motion of the vehicle from which the work is being done. Marine technology thus must concern itself with minimizing this interference. Several approaches can be used, depending on the nature of the work to be accomplished and the constraints of time and place. Since the most violent sea forces occur close to the surface one appropriate line to follow is simply to go below this region, using a submarine or similar craft as the supporting vehicle. Although in many instances this is very appropriate, such a course of action has inherent disadvantages through loss of easy ability to communicate, to transfer personnel from one craft to another, to obtain oxygen (particularly for engine operation) and to make observations concerned with the atmosphere or outer space. Craft supported by their own positive buoyancy at the surface do not face these problems and are generally simpler to build and operate.

Minimization of motion for work done on or from a surface ship can be approached in various ways: by reducing the motion of the ship itself, by design of special subsystems which reduce the motion at some particular place or by measuring the motion and subtracting its effect from the observations being made. The latter two approaches are combined for example in instrumentation for measurement at sea of small (a part in a million) spatial variations in the acceleration of gravity. Here the basic sensing element is mounted in a gimbal system and in addition the disturbing acceleration components are measured and their effect subtracted off. An example of a motion controlling subsystem is the servo-controlled winch which can hold an object at a fixed depth or maintain fixed wire tension in spite of motion of the ship on which it is mounted. These methods, however, are more a part of the instrumentation than of the vehicle itself.

Stabilization of conventional ships has been a topic of concern for marine engineers for many years, resulting in utilization of gyroscopes, bilge keels, active fins and various tankage systems on many ships. A far less well-developed field is that of use of unconventional hull configurations as a means for reducing motion. Such an approach cannot easily be undertaken if it is essential that one move rapidly and efficiently from place to place, but there are many situations in which this is not the controlling element. In particular there are growing numbers of instances in which one wants to work at a fixed sea-floor location or move with a single patch of water, and it is to this family of problems that the present chapter—stable platforms—is addressed.

In this context "platform" means a special sort of ship that does not require a high degree of mobility. It can, therefore, have an unconventional form dictated primarily by the work to be done on station rather than giving strong concern to efficient travel between ports and working locations. "Stability" is also a word that requires some definition—it clearly refers to minimization of motion of the ship, but is often in naval architecture given an absolute meaning. In that sense it means the ability of the craft to keep from capsizing when given a displacement from equilibrium. In this present discussion, however, we will be considering the opposite extreme— that is an attempt to keep the ship from being displaced at all by the action of wind and water. This goal can naturally never be fully achieved but at least we can often reduce the motion to be within the tolerances appropriate to some particular uses to which the craft will be put.

That there is a need for stable platforms (as here defined) is clearly determined by looking into a few specific problems. In so doing we are led to ignore, perhaps, a broad common gain made in every instance with regard to habitability. A man is not at his best when being thrust about by the deck on which he stands, yet this does happen often to those engaged in research at sea. For example, Fig. 15–1 is a photograph of the Scripps Institution's research ship *Horizon* in moderately heavy weather in the mid-North Pacific. The picture was taken by a crew member on board our ocean-going stable platform, FLIP (Fig. 15–2), where vertical motions, were reduced to an inch or two. On FLIP personnel were barbecuing steaks on their charcoal grill while on *Horizon* those who felt like eating could barely keep the dishes on the table. Not only eating but thinking is affected, of course, thus careful conduct of sophisticated experiments is enhanced by use of such craft.

Habitability alone is not enough to justify such specialization for scientific

Oceanographic and Experimental Platforms   555

FIG. 15–1. R/V *Horizon* in moderately heavy weather.

or engineering applications; the specific needs of the work must be considered. These needs fall broadly into three categories:

1. need to hold instruments in accurately known positions or orientations;
2. provision of a stable reference system (position and/or orientation);
3. minimization of motion of or relative to remote parts of system being used in the water or on the sea floor.

As an extension of this advantage it appears possible that, in the not-too-distant future, consideration will be given to establishing semipermanent communities on the sea—as fishing or vacation resorts or in support of commercial activities. Stable platforms of the sort discussed below may be the most appropriate modules to use as basic structures for such enterprises.

The first of these provided the prime motivation for construction of FLIP and SPAR both of which are discussed below. In each instance the need was to be able to carry out underwater acoustics research in which the locations of the receiving hydrophones were very accurately known not only relative to one another (which any rigid suspended structure could provide) but relative to the surrounding temperature structure of the water and to a reference system at the sea surface. Temperature (or sound velocity) sensors and acoustic receivers are fixed at various depths on the ship structure and appropriately accurate interpretations of the measurements can be made.

The second category of application has led to studies by various government agencies and commercial organizations of the design of satellite

FIG. 15-2. FLIP in vertical attitude.

tracking and meteorological observation stations at sea. Here the requirement is to provide stations on the ocean for nonmarine purposes. Existence of such stations could provide data from regions (particularly in the southern hemisphere) where we do not now have an assured capability of maintaining shore stations. The ability to provide a nearly fixed reference system has applications in oceanography also. For example, this function plus ability to hold the sensing elements in known places on the structure led to the use of FLIP in the summer of 1963 as a midocean wave-measuring station with capability to observe long-period swell components having amplitudes as small as a few centimeters superposed on waves a few meters in height.

The third category may well turn out to have the most important applications. The availability of a stable point from which equipment can be lowered into the sea has already led to work on three research programs—investigations of acoustic background noise in the ocean using deep suspended arrays of hydrophones, studies of the directional nature of internal waves using servo-controlled isotherm followers, and research on the fine structure of the deep scattering layer with high-resolution Sonar equipment suspended in the layer itself. Of perhaps more importance is the capability which could be produced to maintain position stably relative to a particular point on the sea floor in order to be able to drill or do other work on the bottom. This is already a field being actively pursued in shallower parts of the sea and the drilling platform that was studied for the MOHOLE project provides an example of how these techniques could be extended into deep ocean application.

A moment's reflection will reveal that, even in the brief discussion above we have encountered a wide enough variety to indicate that minimization of particular categories of motions may have varying importance in different situations. The several categories which must be considered can be divided roughly into two classes—one in which the oscillating component dominates (as in surface-wave induced motions) and one in which the steady component is the principal feature (as in ocean currents). Within each of these we must concern ourselves with changes both in position and in orientation with rotation about or displacement along horizontal axes being differentiated from those related to the vertical axis. The six oscillatory motions involved have standard names in marine terminology:

- roll—rotation about horizontal axis through ship's longer horizontal dimension
- pitch—rotation about horizontal axis through ship's shorter horizontal dimension
- yaw—rotation about vertical axis through ship's longer horizontal dimension
- heave—vertical displacement
- surge—horizontal displacement along ship's longer horizontal dimension
- sway—horizontal displacement along ship's shorter horizontal dimension.

Whereas in one particular application or another one or more of the oscillatory motions may be bothersome, the corresponding steady motions are not in general of comparable concern—those corresponding to heave and roll (or pitch) being treated quite differently from those corresponding to yaw, surge and sway. In fact, ship designers have rather naturally concentrated on complete elimination of steady vertical displacement since they place great emphasis on having their products remain afloat. Similarly steady rotations are taboo and even steady torques about the horizontal axes (producing list or heel arthwartships or trim down by the bow or stern) are controlled carefully in conventional ship design. Thus the principal steady-state problems that are new are those of combatting horizontal forces and torques about the vertical axis.

Our concern here will be first with a brief discussion of the environmental factors against which we must fight to achieve stability followed by a treatment of certain general guiding principles. From this we move to a brief description of some stable platforms which have actually been built or for which extensive design studies have been made, and finally we move to a more detailed look at the design, construction, and operation of one particular craft—FLIP.

15.1 ENVIRONMENTAL CONSIDERATIONS

Three gross aspects of the marine environment are of concern in any discussion of the motion of surface ships. Forces applied by both the air and the sea are naturally involved; thus wind, surface waves, and currents must be considered. All three obviously occur in some sense randomly but have been investigated enough to make clear the existence of regularities of both geographical and seasonal nature. Average patterns for wind and near surface currents are summarized on the pilot charts published by the U.S. Navy Oceanographic Office.

Winds of high velocity may overtake a craft of low mobility in tropical waters (in hurricane or typhoon regions) as well as in high latitudes (40° or more). With such a wide range of occurrence it seems essential to take these into consideration in all design work. This does, however, bring to the fore the point that the need for good stability to carry out a job should be considered separately from the actually overriding need to produce a safe seaworthy ship. Thus in stating or reviewing requirements it is essential to distinguish these two aspects. The MOHOLE drilling platform criteria provide a good example of this duality. Since it was expected that drilling operations would take place in the trade-wind region, in areas off Hawaii where tropical storms do not occur, it was felt that position keeping for drilling purposes would be adequate if it could match a 33 knot (17 m/sec) wind. Since the craft might, however, have to traverse areas of heavy weather en route to the drilling site the safety requirements was set at 140 knots (70 m/sec) with gusts to 200 (100 m/sec). These latter numbers provide an ample safety factor against almost all storms. This is an area in which the designer may have a little more ability to relax in these times. Tracking of developing tropical storms using satellite information is finally putting even the operator of a relatively slow craft in a position to avoid meeting this source of trouble.

Waves on the sea surface are generated by the wind but propagate to great distances from the regions in which they are produced. Except for nonlinear effects associated with breaking of waves, the seaway can be thought of as a superposition of sinusoidal components of various frequencies, propagating in various directions. The range of periodicities in this composite is from a second or so to close to 20 sec. (Korvin-Kroukovsky, 1961). The distribution of energy among these various components varies smoothly across the spectrum with a principal peak in the 5 to 15 sec range depending on wind speed, duration of the blow and distance (fetch) over which the wind is effective. The fully developed sea for a 30-knot wind will peak at about a period of 13 sec while a 20-knot wind will produce about an 8.5 sec peak. In a given area there may be subsidiary peaks in the spectrum toward the long period (low frequency) end representing the swell due to large storms far away. For example in the North Pacific there is, a large fraction of the summer, a set of small spectral peaks in the 15- to 20-sec range due to severe storms occurring between 40° and 60° south latitude (Snodgrass et al., 1966). An actual observed wave-power spectrum is shown in Fig. 15-3. Although the sea may have some apparent regularity in direction of propagation, it usually shows considerable angular spread in its energy, leading to peaking or short-crested waves. Since the energy is not particularly correlated from one frequency to another, we find a nearly Gaussian distribution of wave amplitudes with occasional very large waves—the MOHOLE barge criteria (for safety) were for 62-ft waves with 20-sec period and a maximum wave height of about 100 ft.

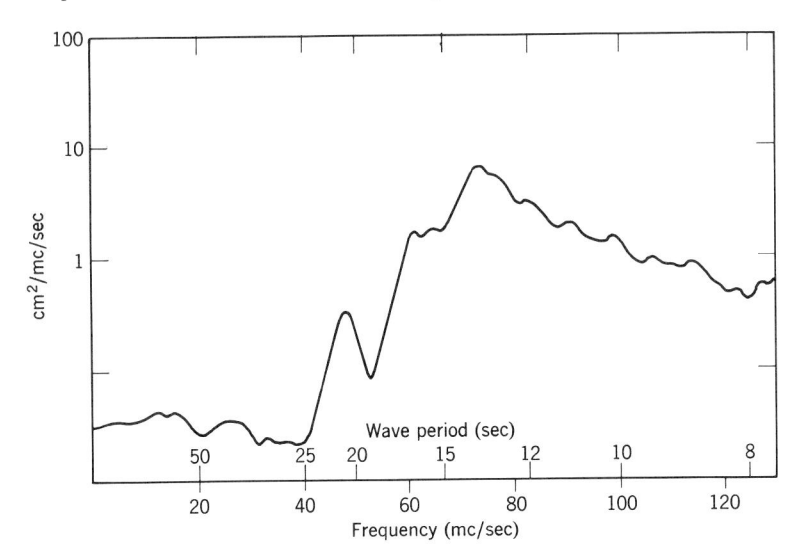

FIG. 15-3. Observed wave power spectrum.

Consideration of the seaway as a superposition of sine-wave components allows computation of several useful parameters and for an approach to the stability problems with a proper mathematical viewpoint. In particular, we are concerned with the manner in which these disturbances die out as we go deeper into the water. Analysis shows a relationship between wave period and wavelength and further reveals that the amplitude of motion and of related pressure fluctuation dies off exponentially, with longer-wavelength waves penetrating more deeply than short. Thus as we go down into the water we find that the peak of the wave spectrum shifts gradually to lower frequencies (longer period of oscillation) with over-all reduction in amplitude. The key relationships are, with T = wave period, L = wavelength, and g = acceleration due to gravity:

$$L = \frac{gT^2}{2\pi}$$

$$P = P_0 e^{-2\pi z/L}$$

where P is the pressure fluctuation amplitude at depth z and wavelength L. Approximate numerical relationships are

$$L = 5T^2 \ (L \text{ in feet, } T \text{ in seconds})$$

and wave amplitude drops by a factor of 2 for a depth difference

$$\Delta z = L \left(\tfrac{1}{9}\right)$$

The right hand portion of Fig. 15-4 shows the way in which the amplitude of the fluctuating pressure component and the displacement amplitude fall off with depth for waves of three different periodicities.

The horizontal and vertical displacement amplitudes associated with any particular frequency are essentially equal to one another at any given depth in deep water. In addition there is no "phase shift" as we go down in the water column—that is, the peaks and troughs of the pressure fluctuations at any point occur exactly at the same time as the peaks and troughs of the wave motion overhead.

Currents in the sea are generated by the wind, the tide, and by large-scale effects of the rotation of the earth and the shapes of ocean basins. (Sverdrup et al., 1942). Typically there will be a westerly flowing equatorial current, a concentrated poleward flowing western boundary current (e.g., Gulf Stream) and a broad, weak equatorially directed eastern boundary current (California current). Superposed on these are fluctuating tidal flows, large eddies (particularly where major currents interact with shallow sea bottom or islands) and local wind-driven currents. Whereas the latter (wind driven) move predominantly with the wind at the very surface, the effect of the

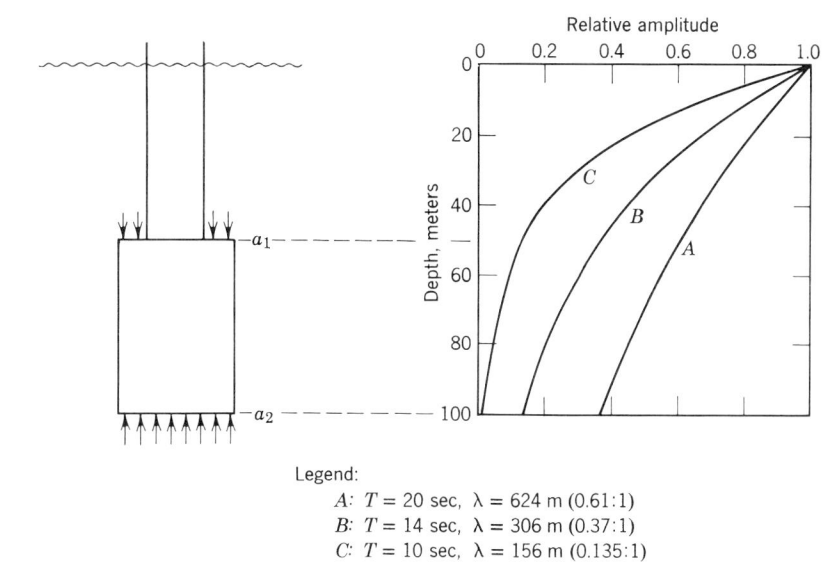

FIG. 15-4. Two-cylinder buoy and wave amplitude curves.

Legend:
A: $T = 20$ sec, $\lambda = 624$ m (0.61:1)
B: $T = 14$ sec, $\lambda = 306$ m (0.37:1)
C: $T = 10$ sec, $\lambda = 156$ m (0.135:1)

earth's rotation (Coriolis force) is such as to make the slightly deeper portion move to the right of the wind in the northern hemisphere. A local current of a knot or so can develop in a few hours of steady 30-knot wind.

Since most stable platforms tend to have rather significant extension down into the water compared with more conventional ships it is essential to take account of the variation (shear) in the current speed and direction as we go down from the surface. A rather extreme variation of this sort has been well documented in the case of the subsurface Cromwell current, (Knauss, 1960), which flows easterly exactly centered on the equator in the Pacific Ocean. A section through this is given in Fig. 15-5 which shows a core velocity of 125 cm/sec (about $2\frac{1}{2}$ knots) at a depth of 100 m with velocity about zero at the surface.

Actual sea experience with FLIP in three areas of the North Pacific shows the variability that we might expect. During a 27-day period on station, operating with 295- to 300-ft draft, near 40°N, and 150°W with winds continuously greater than 15 knots, the *total* drift experienced was 40 miles or an average current of 0.06 knots. (Northrop, 1964). Off the Hawaiian Islands steady drifts of $\frac{1}{4}$ to $\frac{1}{2}$ knot were experienced, to northwest in some areas and south in others. Finally in the continental borderland region off southern California it has occasionally been small, occasionally over half a knot and in several instances oscillating displacements of about tidal period have occurred.

In considering how to cope with the effects of waves and currents it is

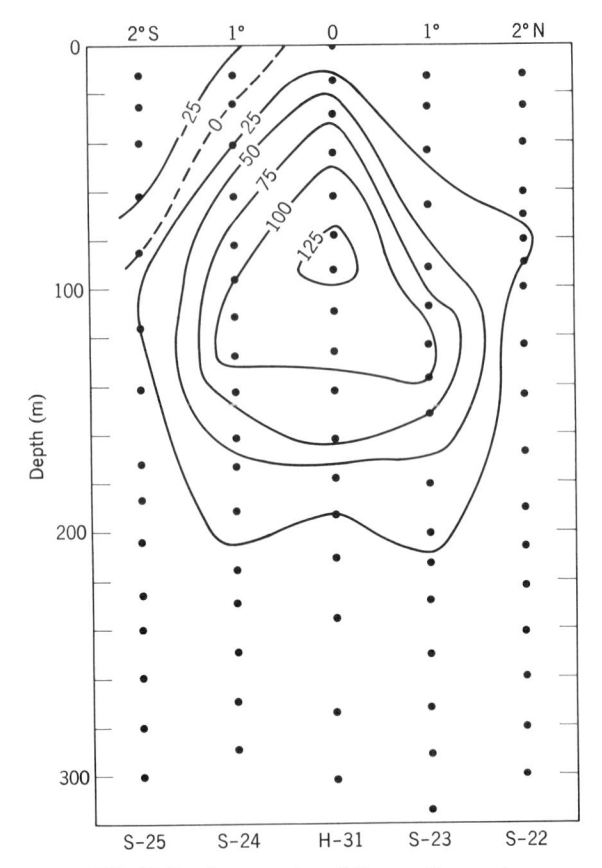

FIG. 15-5. Cross section of Cromwell current.

essential to remember that we cannot expect any agreement among the directions of sea, swell, and current at any given time and place—these are basically uncorrelated quantities.

15.2 GENERAL CONSIDERATIONS

When considering the design of a stable platform, it is important to keep in mind three basic classes of means for minimizing the motion of the ship. Although these three approaches are quite different, they are also not mutually exclusive and thus all three can be used, to some extent, simultaneously.

The first concept is that of reduction of coupling between the disturbing forces and the ship itself. This can be effected in many ways, principally by reducing drag in the more violently moving parts of the medium and by arranging to have such forces as cannot be eliminated work on the structure in a balanced way so that the net effect will be minimal. In using this approach (as in applying the others to be discussed below) it is essentail that one remember the entire problem, including the complexity of the ocean environment. For example, extreme fairing (or streamlining) might be considered as a way of reducing the oscillatory wave drag forces. The non-uniform direction of the waves, however, with swell propagating in several directions superposed on the locally generated sea in many instances leads to greater drag coupling for an extreme fairing than for a simple cylinder. The possibility of balancing the effects of the forces on various parts of the craft is one which leads to quite satisfying solutions for particular problems and is discussed at greater length below.

The second approach is through passive resistance to motion or to the forces which tend to produce it. This can be achieved by providing the craft with large inertia (mass or moment of inertia) and by enhancing drag coupling the structure to some quiet part of the surrounding environment (deeper water). In general, inertia provides the best means for achieving stability when a basically stable position exists; thus in combatting heaving or rolling motion, it is most appropriate. Increased drag in deep water is particularly useful for motions for which the craft has no intrinsically determined position or orientation. This latter situation exists in general with regard to horizontal displacements or rotations about the vertical axis. In either event it seems inevitable that larger dimensions in quiet parts of the medium lead to greater stability. It seems then to be dictated by the combination of these first two principles that for resistance to surface-wave induced motion, the structure should reach deep into the water and have large extent there with small dimensions near the surface.

The third means for minimizing motion is by providing dynamic resistance to the motion. In this case one must supply the necessary opposing energy and exert some sort of opposing thrust. This thrust can be derived from the reaction to forces exerted on the water or the air (as with propellers) or by inertial reaction (as with a flywheel mounted within the ship). Quite a variety of means exist for attacking the problem from this viewpoint, however it seems in most instances best to utilize the dynamic method primarily in conjunction with the two passive philosophies mentioned above. The principal class of problems in which the dynamic approach is essential and often must be the primary means is that of motion induced by currents having appreciable thickness. Here it may not be possible to provide drag in stationary water without making the structure very large, although consideration can be given even then to use of very deep "sea anchors" or to anchoring to the sea floor itself.

These generalities can be made somewhat clearer by considering specific classes of motions and examining the manner in which the principles might be brought to bear. Vertical motion (heave) provides the best initial target for discussion. Here the quiet deeper water is available and craft extending into it seem most appropriate. In addition, if we can make our structure very large in both horizontal dimensions, the forces exerted by the waves to lift one portion will be out of phase with those tending to lift another, thus (if sufficient rigidity is available) only the average elevation of the sea will contribute to produce vertical displacement. The horizontal extent required in this case is dictated by the nature of the surface-wave spectrum, since we must average over several wavelengths to find effective cancellation. In the open sea there will often be appreciable energy in 12- to 15-sec period waves, with corresponding wavelengths of 200 to 400 m. This would imply at least 500-m transverse dimensions for effective cancellation—a rather large structure for most applications although this might be quite a practical approach to oceanic airfield design.

Fortunately the wave field varies on a smaller dimensional scale in the vertical direction thus allowing the heave stability problem to be attacked quite effectively by providing vertical extent to the structure involved. The driving forces can best be visualized by considering the surface wave field as providing a fluctuating pressure in addition to the hydrostatic pressure at any point. For any sine-wave component of the wave field, the amplitude of this disturbance can be calculated as a function (exponential) of depth. The resulting driving force is then obtained by integrating this pressure field over the entire surface of the ship.

The simplest craft on which to demonstrate this effect is one having horizontal extent small compared to the relevant wavelength and having no variation of horizontal cross section with depth for its full length. Floating structures having this configuration (large ratio of vertical to horizontal dimensions) are usually called spar buoys and have been in use for many years as markers and as wave-measuring devices. Because of the lack of variation of cross section with depth and the small horizontal extent, the integral of the pressure over the surface of the ship vanishes except on the bottom surface. The response then becomes, to a good approximation (except close to resonance),

$$\frac{Z}{Z_0} = \frac{e^{-2\pi D/L}}{1 - (2\pi D/L)}.$$

Here Z is the amplitude of vertical motion of the spar, Z_0 the amplitude of the wave motion at the surface, D is the draft (depth to which the buoy extends down into the water), and L is the wavelength of the component under consideration. The actual total motion must be obtained by summing over all components of the spectrum. The motion will be quite large at the resonant wavelength ($2\pi D = L$) which corresponds to the period of free oscillation of the buoy Expressed in simple form this natural resonant period (T) is

$$T = 2\pi\sqrt{D/g}.$$

It thus appears that if we want the resonance of this sort of simple structure to be outside the range in which energy in the ocean is appreciable, we should pick T significantly greater than 15–20 sec. Using $T = 25$ sec gives a value D about 500 ft. Since this is a rather large number, it is obvious that we should ask what will happen if the cross section is not uniform with depth. First we note that a more general expression for the resonant period is

$$T = 2\pi\sqrt{V/ag},$$

where V is the submerged volume of the ship and a is the cross-sectional area at the waterline. We can thus, for any given draft, make the waterline area very small and make T as large as desired. In doing this, however, we must return to consideration of the driving force because now the spar has sloping side surfaces on which the pressure can generate downward or upward forces. This then opens an opportunity to investigate the interplay of these forces. The simplest situation to analyze is one in which uniform cross section is maintained except for a single stepwise transition at some depth. In this case there will be a downward force on the shoulder, opposing the upward one on the bottom. Because the fluctuating pressure amplitude is less at the bottom but its area of action is greater, there will be some particular wave period such that the two forces will be in exact balance and the resulting reponse curve will show a null. The left-hand portion of Fig. 15-4 shows a craft of this sort with 100-m draft and the transition at 50 m. If the step area (A_1) is 0.6 of the bottom area (A_2), then the null will occur for a wave period of 20 sec. A ratio of 0.37 to 1 puts the null at 14 sec and 0.14 puts it at about 10 sec. The smaller ratios correspond of course to larger waterplane areas ($A_2 = A_1 + a$), thus to higher-frequency resonances. The pertinent quantities are summarized below:

$\dfrac{A_1}{A_2}$	Null Period (sec)	Resonant Period (sec)
0.61	20	26
0.37	14	23
0.14	10	21
0	none	20.

The response curve for this situation is given by

$$\frac{Z}{Z_0} = A_2 e^{-2\pi D/L} - \frac{A_1 e^{-\pi D/L}}{a} - 2\pi \frac{V}{L}$$

or in terms of wave period rather than wavelength,

$$\frac{Z}{Z_0} = A_2 e^{-4\pi^2 D/gT^2} - \frac{A_1 e^{-2\pi^2 D/gT^2}}{a} - \frac{4\pi^2 V}{gT^2}.$$

At this point it is clear that we can trade structural complexity to some extent for deep draft and that this particular form of the force-canceling technique can be utilized with even more complicated cross-section variations designed to minimize the total response to the full spectrum of waves rather than having simply a null at a single point. Such approaches have not yet been implemented in any full-scale vehicle but can be considered as available for application.

Carrying these lines to their extreme leads to the general conclusion that the best way to minimize heaving motion is to have a large deep portion to the ship—the greater the depth and the horizontal extent the better, with minimal area at the surface and extending down as far as practicable. Constraints (usually in the form of construction cost and operating restrictions) will naturally arise to dictate compromises with this principle. Thus, in the case of the MOHOLE vehicle, a decision was made in favor of a large underwater body but a compromise was made limiting the depth for operating convenience. In FLIP the modest budget dictated a small structure and thus depth and hull form were used to meet the necessary requirements.

Stability against horizontal oscillatory motion (surge and sway) and against roll and pitch can chiefly be achieved by providing horizontal extent comparable to or greater than a wavelength. Although it was noted above that the required dimensions are quite large for longer wavelengths, some improvement can be had even with a horizontal spread of as little as 75 m corresponding to a wavelength at 7 sec period or two wavelengths at 5 sec. Such gains are important in the context of reduction of horizontal motions where they were not for vertical since the vertical motion at these higher frequencies (5 to 7 sec) can be very significantly decreased with quite modest vertical structural dimensions. Large vertical extent can play a major role in reducing angular motions (roll and pitch), but this is achieved without significant reduction of surge and sway. A simple spar will respond to the horizontal pressure gradient associated with the waves, with resulting motion at the surface about equal to the wave amplitude but with the bottom of the structure nearly motionless. The result is that, for a spar draft D

and wave amplitude Z_0 (crest to trough distance $= 2Z_0$) the angular deflection will be about $180Z_0/\pi D$ degrees. For a 100-m-deep craft and wave amplitude 2 m, the roll would be slightly over one degree and the most noticeable motion would be the horizontal displacement of $\pm$ 2m.

If minimization of either the angular or horizontal displacement motion is an essential requirement, it appears that the most effective means will be through use of multilegged structures. In this way the cancellation principle can be applied at least in a statistical manner since legs a half-wavelength apart will experience driving forces which are exactly out of phase. The difficulty here is not only that we must deal with many frequencies superposed (as in the vertical case), but that the waves will come from a variety of directions. Thus a two-legged structure would have to be oriented in some particular direction for maximum effectiveness, and even then, if the arriving energy were not angularly concentrated, it would not achieve great improvement. Structures having three or more legs would be much more effective.

Azimuthal stability presents a problem of a different sort than those discussed immediately above since, in absence of any driving forces from either air or sea a ship in general has no particular stable orientation in the horizontal plane. It may resist being rotated about a vertical axis but this is in the nature of a drag force and no tendency exists for it to move back to the position from which it has been turned. The wind and waves will in general tend to orient the ship in some particular direction depending on its shape. Such torques can be minimized if desired by making the structure cylindrically symmetrical.

If it is desired that some arbitrary orientation be maintained the principles of both passive and dynamic resistance can be used. The moment of inertia should be made large if feasible, which dictates a craft of considerable horizontal dimensions—a spar in general has a rather small radius of gyration about its vertical axis. Provision of drag in the deeper part of the water column should assist appreciably. Here we should remember that if the craft is successful the deep flow about the structure as it rotates will be very slow. We are thus led to consider drag provided by perforated plates or large numbers of thin wires. Cox, for example, in making free-drifting deep floats that he desired to have follow the varying rotation of the surrounding water in spite of their own rotational inertia, made a web of fine nylon monofilament line extending out to several times the diameter of the cylindrical float in order to achieve adequate coupling.

Having achieved what inertia and deep water drag we can, the next possibility is that of balancing out the applied torques. In the case of the wind, for example, we can alter the shape presented to it by use of adjustable panels or sails. It is more difficult to work in this fashion with the waves, however, without exposing the shape-altering components to rather severe transient loading, thus there is more incentive to provide cylindrical symmetry and derive the orienting torques from some other source.

Dynamic orientation systems of many sorts can be visualized and, in the case of FLIP at least, several have been tested. The basic choices are from among three: exert thrust on the water, on the air, or use some sort of inertial reaction. The three tried from FLIP were: motor-driven propellers in the air, propellers in the water (both were conventional air and water screw types), and reaction to a water jet discharged into the air. Of these three the most effective and easiest to implement with sufficient power was the screw in the water, used in an autopilot system to hold arbitrary orientation to a part of a degree in a 20-knot wind. Other systems considered have included jets in the water and large flywheels (not used because of the steady acceleration required to produce a counter to a steady wind-generated torque). With so many choices and so little experience, it does not appear possible at this time to make any generalizations about dynamic orientation systems.

Motion produced by current is a very obvious one which, in certain applications is not a severe problem. Most of the work conducted to date by FLIP, for example, has been concerned with the characteristics of a patch of water rather than sea floor and thus the only disturbing factors have been those due to variations of current with depth, such that one moves neither with the surface water nor with the slightly deeper current but rather with some average over the entire 300-ft depth of the structure. If the occasion demands, it should be possible to provide large drag at the depth of the water which one wants to follow. If the sea floor is the object of interest, then we can of course anchor directly to it although this still presents some formidable problems. Depending on the horizontal excursions allowed it may be necessary to provide a multiple-point mooring, which in the deep ocean is not a course of action for which we can find clear instruction. In any event the use of even a single line mooring must be considered with regard to the entire stability problem, particularly the means used to attach the ship to the mooring. For example Cousteau has moored his *Bouée Laboratoire* in the Mediterranean in a manner designed to prevent the craft from listing because of the couple between the surface current and the direction of pull of the line. This has been successful but the resulting point of attachment (effectively on the side of the ship) produces a moment which works against the wind-induced torque on the above-water portion producing constant fluctuation of orientation of the buoy.

If stability over a point in deep water is required, it seems best to utilize propeller or other drive mechanisms to provide thrust in the water. The need to provide this sort of control for drilling operations even in modest water depths had led to construction of a number of ships having highly versatile propulsion systems similar to that used on CUSS for the experimental MOHOLE drilling operations in 1960. Several of the oil companies as well as independent groups (e.g., Global Marine) have built these ships. Within the research field it appears that the first truly versatile ship of this sort will be the U.S. Navy's new class of oceanographic research ships (AGOR-14 class) which will be equipped with a pair (one forward, one aft) of Voigt–Schneider propulsion units. These techniques, however, are related more to propulsion plants than to ship form and thus will not be considered further in this chapter.

Generalizations about design philosophy can be conducted in a very abstract fashion. They are of little use, however, unless one is aware of the compromises which contact with reality must bring. Reality, too, can be a subject for generalization; thus it may be useful here to bring out a few broad points before proceeding to a discussion of real craft. The operation of a ship having unconventional form (either very deep or very broad or both) presents two major classes of problems—somewhat related to one another. These are the problems of use of port facilities and of transit from one area to another. These must be met or a conscious decision must be made to do without certain operational flexibility. This is particularly true of very deep draft platforms. For example Cousteau's group decided at the outset that their interest (with regard to their buoy laboratory) was confined to the western Mediterranean. They were thus willing, and consciously decided, to build a ship which would not be easy to move from place to place (1-knot towing speed) and which could only be serviced by approach to ports having adequate water depth nearby (e.g., Villefranche). On the other hand, with FLIP in the Pacific it was essential that the craft be reasonably mobile and able to come into San Diego Bay, thus the idea of laying the ship on her side for transit seemed an appropriate approach. The MOHOLE drilling barge study approached mobility and harbor entry by providing for variable draft by flooding or emptying ballast tanks. This produced some related restrictions on maximum feasible depth in the drilling condition and thus on the achievable heave stability. These points will be alluded to in the more specific treatment of these craft below.

One further practical matter has to do with Coast Guard and related inspection and licensing. At the time of construction of FLIP there was little precedent for action and she was thus treated as a simple barge. As of this writing, however, rules for research vessels in general are being formulated and it is expected that some notice of unconventional craft will be taken in them.

15.3 CRAFT IN BEING

True engineering lies not in generalizations nor in paper feasibility studies but rather in the production of some specific structure. Only in carrying through from concept to reality do we face up to the myriad detailed decisions that shape the eventual nature of the ship. The compromises that we must make to insure structural integrity, operational performance, and habitability in compliance with constraints arising from limitations on cost, time, or available materials and fabrication techniques are usually not faced realistically in the conduct of a feasibility study. The further steps involved in preparation of contract plans and specifications, review by regulatory bodies, drawing of detailed construction plans, and the actual act of construction itself all serve to shape the vehicle. For this reason it is useful to consider some craft that have actually survived these ordeals to become operating units—their performance can be certified as achievable and their shortcomings can be used to indicate the problems remaining as challenges to the engineer.

Four research platforms having basically spar-buoy shape have been built and are described below. Table 15-1 summarizes their basic characteristics.

TABLE 15-1
Basic Characteristics of Four Spar-Buoy Type Research Platforms

Name	Submerged Depth (ft)	Displacement (tons) (approx)	Vertical Resonance (sec)	Waterplane Area (sq. ft.)	Response at 8 sec (%)	Accommodations (men)
FLIP	300	2,200	27	125	5	12
Bouée Laboratoire	170	250	22	25	16	4
SPAR	300	1,700	19	200	0.1[a]	0
POP	180	1,115	21	7		2

[a]Calculated response.

FLIP (Fig. 15-6) built in 1961–1962 under the sponsorship of the Marine Physical Laboratory of the University of California Scripps Institution of Oceanography, was the first of these to be completed. Funding was from the U.S. Navy, engineering work was done by M. Rosenblatt and Sons (first phase) and L. Glosten and Associates (second phase), and construction was carried out by Gunderson Bros. in Portland, Oregon. This craft has variable underwater cross section, being composed of two cylinders (upper and lower, 12.5 ft and 20 ft diameter respectively) joined by a 90-ft long conical section. The operational problems of transit and harbor entry are solved by providing tankage such that the craft can be rotated (flipped) to place the

FIG. 15-6. FLIP in horizontal attitude.

long axis horizontally, giving in this condition a draft of about 12 ft. With provision of a conventional bow shape at one end and a set of skegs at the other, she is easily towed (at about 8 knots) from one location to another. The principal structural problem met in this design was that of high stresses due to bending moments near the center of the ship while horizontal in the seaway. This craft will be described in considerably more detail below.

The second craft of this type to be built was the *Bouée-Laboratoire* of the Musée Oceanographique of Monaco (Piccard, 1965). Funded by the French government, this craft was under construction at about the same time as FLIP and design information was exchanged occasionally between the two groups (Musée Oceanographique and Marine Physical Laboratory). Both craft are designed to be compatible with a wide variety of oceanographic tasks and both rely on deep draft and small waterplane area to provide stability. The two have grown from rather different operational philosophies, however, stemming from the different ocean characteristics faced by the two groups. The Mediterranean—operating area for *Bouée-Laboratoire*—being much smaller, does not have the long–period wave energy (15 to 18 sec) that is present in the Pacific. Development of significant energy in such long wavelengths requires both high winds and long distances over which they must blow. Absence of such long-period components makes it quite feasible to achieve reasonable stability without having such deep draft and without as low a frequency of natural resonance. In addition, stability performance criteria for *Bouée-Laboratoire* were not as stringent as those set for FLIP since the former did not include certain aspects of underwater acoustic research within its immediate research program.

The principal way in which *Bouée-Laboratoire* differs from FLIP (other than in gross size) is that the former is designed to remain permanently in the vertical condition. This would be quite a handicap in as large an ocean as the Pacific where there are many widely dispersed areas of oceanographic interest. The Mediterranean however, provides a single major basin immediately available off the French coast and this area can be covered nicely from shore stations (island and continental) with the addition of only one or two observation posts in midbasin. The *Bouée-Laboratoire* was designed to fill the need for such stations and thus does not impose undue operational hardships because of low towing speed, (about one knot) in her permanently vertical condition.

As a result of this situation the Musée group has been the principal one to devote heavy attention to moored, as opposed to drifting operations. Typically, the *Bouée* is moored between Corsica and Nice in about 2600 m of water. The usual compromises between positional uncertainty (related to scope of mooring line out) and mooring-line tension led the Musée group to use about 3500 m of line of which the lower 2000 m are positively buoyant polypropylene (density about 0.95) and the upper 1500-m section is negatively buoyant (about 1.1) nylon. With this configuration there is less tendency for the line to foul during calm weather by piling up either on the top or the bottom. Line of this sort is subject to degradation by abrasion; thus care is taken to isolate it at both ends. The anchor (a four-bladed propeller) is connected to the lower end of the polypropylene line by a length of chain. This is the conventional means for providing additional weight which can, as mooring tension fluctuates, be supported either by the sea floor or by the line tension. At the upper end of the mooring configuration, the strain is taken from the nylon by wire rope which is attached to the buoy by a yoke. One leg of this yoke is brought up through a sheave near the bottom of the buoy and the other through a second sheave well up the structure. Both lines are brought up to the house section at the buoy top and made available in such a way that the operators can alter the relative tension and thus adjust the effective depth of the attachment point.

As can be seen from Fig. 15-7 in comparison with Fig. 15-2 and 15-6, there is a significant advantage in habitability to be gained by building the ship to remain vertical. The need for a sturdy structure as viewed with the long axis horizontal makes it difficult to provide large deck areas (for example for helicopter landing) when the craft is vertical. In the case of the *Bouée-Laboratoire*, although it is basically a small structure, it does provide unbroken area topside of about 650 ft^2 and corresponding area immediately below very well fitted out to provide laboratory and living space for four men. The machinery space is immediately below the main living deck in a smaller transition section. Here are the two engine–generator combinations (25 kW each) and other auxiliary machinery.

The underwater portion gives the ship a draft of about 170 ft, which can be varied somewhat by changing the quantity of seawater in the variable ballast tanks. The cross section varies, making a gradual transition from

FIG. 15-7. *Bouée-Laboratoire.*

about 5.6-ft diam at the waterline to 6 ft at 26-ft draft. At this point the structure joins a simple cylinder about 10-ft (3 m) in diameter and 55 ft long in which are six accessible decks for laboratory and storage use reached by an internal elevator from the main deck. Below this there is another step in cross section to about 6.5-ft (2 m) diam from which the structure tapers to 5.6 ft at bottom. Outside this lower portion are ballast and fuel tanks.

The ship has been operated satisfactorily on station continuously for several years except for a period of rebuilding following a fire which caused extensive damage in the living quarters. As mentioned above, the only real disadvantage of this craft is that she must remain in deep draft condition (ballasted with concrete and scrap iron) and thus can only be towed at low speed and cannot easily be brought into harbor for conduct of underwater-body maintenance and installation of new equipment.

The third craft of this type (SPAR—Seagoing Platform for Acoustic Research) was built somewhat after FLIP and *Bouée-Laboratoire* under the guidance of the U.S. Naval Ordnance Laboratory, White Oak, Maryland. Engineering was done by M. Rosenblatt & Sons, New York, and she was built by Gibbs Shipyard in Jacksonville, Florida, and launched in 1964. Her purpose is to carry out experiments in underwater acoustics.

This platform, like the other two, has no propulsion and, like FLIP, is towed from place to place with its long axis horizontal (Fig. 15-8). Major dimensions are similar to those of FLIP—355 ft long, 300-ft draft in vertical. The craft has a simple cylindrical form, 16 ft in diameter for nearly its full length. As a result it has a shorter resonant oscillation period, about 19 sec. Although this should lead to a noticeably greater response to swell (wave periods in the 14–20 sec range), it should have much smaller response than either of the other two to wave periods shorter than 10 sec since they can only act on the bottom surface, at a depth such that the wave amplitude is greatly attenuated. To date no actual wave response data have been gathered, thus it is not yet clear whether other effects (e.g., drag-induced forces) will dominate in this range and prevent realization of the calculated response shown in Table 15-1. The amplitude of motion which will exist at the resonant period due to excitation from the usually very small component of energy in the wave spectrum at the same frequency is also not clear.

The bow configuration (Fig. 15-9) was chosen to reduce the control that the wind might have upon the azimuthal orientation when SPAR is vertical.

FIG. 15-8. SPAR in horizontal attitude.

FIG. 15-9. SPAR in vertical attitude.

In limited operation to date she has shown no tendency to respond to the wind. A propeller system is provided to position the craft dynamically in the desired orientation.

The major difference between SPAR and the others is that she has been designed to operate unmanned. This decision was made in order to avoid the problems associated with transfer of personnel at sea in heavy weather. The problems can be difficult ones as small craft usually must be used to make the transfer and they will respond in rapid fashion to the seaway whereas the spar ship will have negligible corresponding motion. In the case of offshore drilling platforms and even the *Bouée-Laboratoire* this problem is solved primarily by avoidance since they are usually close enough to their logistic bases that periods of manageable weather can be chosen to effect transfers of personnel and supplies. In the case of FLIP, although transfers at sea have been made in fairly heavy weather, the normal operating pattern is such as to allow all necessary people and equipment to be loaded in port and thus to avoid facing the problem at all.

Unmanned operation leads to certain operational difficulties. All ship operations and research observations must be carried out by remote control. This means that before even preliminary experiments can be carried out, all research equipment must have adequate reliability for at least a week of untended operation. In addition a specially equipped towing–tending ship must be used and must stand by, permanently attached to SPAR through a special combined towing and electrical telemetering cable arrangement. If sound-propagation experiments are to be conducted, this then requires use of a second ship to carry the sound source rather than allowing the towing ship to carry out this function. In addition we are restricted in the conduct of simple shakedown and equipment checkout operation since small, inexpensive tugs, obtainable on charter, cannot be used because of the unusual winch requirements. These conditions have led to the fact that as yet no research use has been made of SPAR.

The most recent addition to the family of spar buoy laboratories is the Perpendicular Ocean Platform (POP) built by the General Motors Defense Research Laboratory, Santa Barbara, California. Engineering was done by the laboratory staff and construction by Williams Welding Co., Long Beach, California, with launching in June 1966.

This is the smallest of the four craft in tonnage although its 180 ft draft is not much different from that of the *Bouée-Laboratoire*. It is a very slim spar with variable cross section—3 ft diameter from waterline to about 60 ft below at which point there is a transition to 6 ft diam, which is maintained nearly to the bottom. At the 75-ft level the hull is occupied by a decompression chamber which acts as an airlock so that divers can climb down the interior of the 3-ft cylinder and enter or leave the water submerged.

As with FLIP and SPAR, this ship also has ballast tanks which can be blown dry with compressed air to bring the lower section to the surface and allow for towing with shallow draft while the long axis is horizontal. Unlike FLIP and SPAR, however, the lower, rather than the upper, end becomes the bow when she is horizontal. The reasons for this are twofold—the bow-shaped section can then be mated to the thick rather than to the thin portion of the hull and the working platform at the top need not be given as much strength to withstand the sea. Disadvantages to this configuration are not serious in a small craft such as this which is not intended for long tows into the open sea. These are the submergence of the line-handling machinery on the bow and the separation of living quarters (which might best be above water in the vertical) from the work area involved in making up and dropping the tow (at the bow). In this case, however, no permanent living quarters are installed. Instead a trunion-mounted platform is provided which can be turned to be horizontal in either attitude of the main hull. Living and laboratory spaces are to be provided by placing vans on this deck, presumably at sea after the transition to the vertical has already been made. Installation of the vans will probably be feasible in modest weather but may give some problem if much of a sea is running.

POP has been involved in fitting-out activities since her launching and has made only one flip cycle for test purposes. It is anticipated that she will be used primarily off the southern California coast in the vicinity

Oceanographic and Experimental Platforms 577

of the Channel Islands. The research and development program which she is to support has not yet been formulated.

These four craft are the full extent of actually completed stable platforms for deep water use other than oil-well drilling ships. All four have relied on a simple, single, rigid, vertical structure. Many more complicated structures have been studied in which multiple legs (Fig. 5-10), hinged and telescoping structure have been visualized but not built (Pollock et al., 1960). Two examples are the MOHOLE drilling ship (McClure 1965) (Fig. 15-10) and Sea Legs (Glosten, 1966) (Fig. 15-11). Rigid multilegged structures, as in the MOHOLE design, have some awkward compromises to be made between maximum obtainable operating draft and ability to shift to shallow draft for harbor entry. Hinged multiple legs (Sea Legs concept of Glosten) which can be flipped between a horizontal and a vertical position are perhaps feasible but pose some problems in matching the transition times

FIG. 15-10. Model of MOHOLE drilling structure.

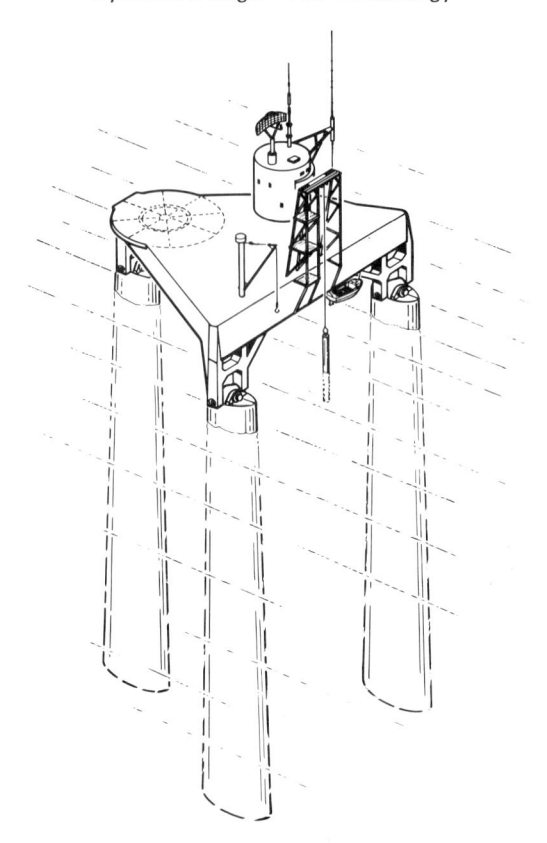

FIG. 15-11. Sketch of "Sea Legs" structure.

of all the legs. One concept which looks promising for larger craft in use of a rather large-diameter single cylindrical hull which can flip, but with the entire bow structure (living, machinery and laboratory space) trunion mounted to remain horizontal irrespective of the cylinder attitude. This would be an extension of the rotating after deck concept employed in POP.

15.4 FLIP

As an example for detailed discussion FLIP is an appropriate topic. Not only is its history particularly familiar to the author, it has in fact to date had the most extensive and diverse use of any of the stable platforms built for research. Its origin was rather complex, being primarily the result of a long series of acoustics research problems undertaken by Marine Physical Laboratory staff members. Invariably in these we had a requirement to hold a set of receiving hydrophones quietly in the water, hopefully knowing where they were with respect to one another and to the sound source. As time went on these location requirements became more and more stringent, leading us finally to consider possible vehicles for our work other than the submarines which had provided our support in earlier investigations. It appeared that a stable platform of some type was required and, if it could be simple enough, such a craft would not only solve the mechanical problems involved but would also give us freedom to schedule operations in scientifically interesting areas without the need to compete directly with the military tasks for which conventional submarines are primarily employed.

Initially we considered the possibilities of implementing an earlier idea of Allyn Vine to utilize a World War II submarine, rebuilt to allow it to be turned on end. It appeared, that this would be costly compared with the value of the raw materials involved, thus we began to consider possibilities starting from the keel up. Various concepts were examined, involving hinged, telescoping, inflatable, and other complicated structures but all were rejected (on the basis of possible acoustic noise and design uncertainties) in favor of a simple rigid sparlike structure which could be towed with its long axis horizontal and, by flooding tanks, swung into a deep draft vertical position to provide the desired stability. The long rigid underwater body would be quite appropriate for the mounting of hydrophones, temperature sensors and other instruments whose locations would be accurately known relative to the above water portion of the structure. It was decided that a draft of 300 ft would be adequate to allow investigations through the most complicated portion of the water column and it was felt that if deeper installations were required they could be hung with reasonable ease in the quiet water below. Knowing the nature of the wave spectrum in the Pacific we realized that a uniform cross section would give us a natural period of oscillation uncomfortably close to existing energetic wave periods. It was thus decided that the water plane area should be less than that at deeper points. At this juncture, in considering the effect of the resulting submerged shoulder, P. Rudnick (1960) produced the basic analysis of response which first made clear the possibility of a null at some wave frequency due to the opposition of the forces on the shoulder and the bottom of the structure. At this point it was decided that the shoulder should be at half draft and that the area ratio should be two to one. This gave a natural period about 2.5 sec and a null at about 18 sec (both these were subsequently modified).

To this point the studies of the craft had been carried out by physicists in Marine Physical Laboratory (principally Rudnick, F. H. Fisher, and F. N. Spiess). The next phase, however, had to do with construction feasibility, so a naval architectural firm, M. Rosenblatt & Sons, San

Francisco, was engaged to continue the study. Under their auspices, model towing tests were conducted at University of California, Berkeley, and wave response experiments (at 1/25 and 1/100 scale) were conducted by Lockheed, Sunnyvale, California (Vytlacil, 1961). Through these tests and structural calculations it was determined that a two-cylinder configuration with a shiplike bow (initially quite blunt) would be appropriate. Possible use of a D-shaped cross section for the slim part of the hull was ruled out in the wave response tests when invariably the flat face presented itself to the waves in a manner which would have required heavy structure to absorb the resulting dynamic loading.

While these investigations were proceeding, the group at Marine Physical Laboratory was tackling the problems associated with making the transition from horizontal to vertical. Calculations of the dynamical processes of tank flooding and varying conditions of loading did not seem convincing, thus F. H. Fisher and C. S. Mundy entered upon a model study program using 1/10 scale (35-ft long) craft which could actually carry out the flipping operation. These studies were quite valuable, particularly on two counts. First, in showing the need for control of free water surface and ballasting to prevent capsizing while returning to horizontal. Second, they demonstrated dramatically the problems to be encountered if the tanks were flooded too rapidly, leading to violent vertical surging motions upon arrival at the vertical. In some instances vertical oscillations of full-scale equivalent amplitude of 30 ft were observed. Restricting flooding rates of the main ballast tanks reduced this effect markedly and was thus employed in the actual craft as built.

Schematic plans and specifications were prepared, bids for construction received and Gunderson Bros. Engineering Co. chosen to build the craft. At this point, also, a decision was made to shift to another design agent and L. Glosten and Associates, Seattle, Washington, was chosen to carry out the detailed design. During this phase it was determined that the principal mechanical design problem lay in coping with the stresses encountered while the craft would be under tow in the horizontal position. Most severe were the calculated loadings resulting from bending moments in the hogging and sagging conditions (hull supported by wave crest amidships or crests at ends and trough amidships). As a result a decision was made to eliminate the abrupt step amidships, substituting a 90-ft long cone. In addition high-strength steel (ASTM A441) was specified by Glosten (1962) for all but the aftermost 70 ft of the shell. Plating thickness and frame spacing vary along the structure in accordance with expected loading. With these changes the calculated resonance period became 27 sec with null at 22 sec. In the sagging condition on a L/20 wave (wavelength 350 ft, period $8\frac{1}{2}$ sec) in the horizontal mode the deflection of the structure was calculated as a little over 1 ft.

Actual construction was begun in February of 1962 with launching in Portland, Oregon, in June of the same year, followed by initial trials in Dabob Bay (Puget Sound) in August, about two years after the initial studies were begun. Because of the unique nature of the craft the Marine Physical Laboratory, rather than the building yard, provided most of the trial crew, utilizing largely the staff members who had participated in the design and followed the construction closely (F. N. Spiess, F. H. Fisher, E. D. Bronson, and C. S. Mundy, plus William Gilmore of Gunderson Bros.). Total cost (exclusive of fitting out of laboratory and living spaces) was about $600,000 for the 600-ton structure (about 2000 tons displacement in vertical position).

The resulting craft is shown in Fig. 15-12. In the vertical, tanks 1, 2, 3, and 4 are filled, 5 is dry, 6, 7, and 8 have variable amounts of water and 9 is dry. To move from vertical to horizontal, tank 9B is flooded while compressed air (initially at 250 psi) is used to blow water from tank 3T. As the lower end moves up the air in 3T expands to expel water from 3B and (through a cross connection) from 1T. In horizontal towing condition, tanks 2 and 4 (free flooding) are about half full and the variable tanks (6, 7, 8, 9) are empty. The flipping operation from horizontal to vertical is made quite simply by adjusting the water in the variable tanks and opening the vent lines on tanks 1T, 3B, and 3T and waiting. Tank 3T is the last to fill as water access to it is through four, 1-in. holes in its bottom deck—in fact, the craft is vertical before flooding of this space is complete. Choice of this mode of operation was based on the $\frac{1}{10}$ scale-model studies referred to above. The time scale involved in the transition from one attitude to the other is shown in Figs. 15-13 and 15-14 which are based on actual tracking of transponders mounted on FLIP during its initial trial operations in Dabob Bay (Puget Sound), (Fisher and Spiess, 1963). The change of attitude is accomplished slowly enough that personnel (all on deck) can easily readjust their footing (as demonstrated by the Commanding Officer, E. D. Bronson, in Fig. 15-15) to maintain their own vertical positions.

Data on actual response to the seaway have been gathered on several occasions and analyzed both by Rudnick (1964) and by Snodgrass et al. (1966). A representative set of points is given by the dots in Fig. 15-16. The upper plot gives the log to base 10 of the ratio of FLIP heaving motion to wave amplitude with the solid curve showing the calculated response. The lower chart gives the log, base 10, of the ratio of the horizontal displacement to the wave amplitude as observed at the laboratory deck level. One sees that the vertical motion is less than 10% (log = -1.0) over most of the important part of the spectrum while the upper end of the structure moves

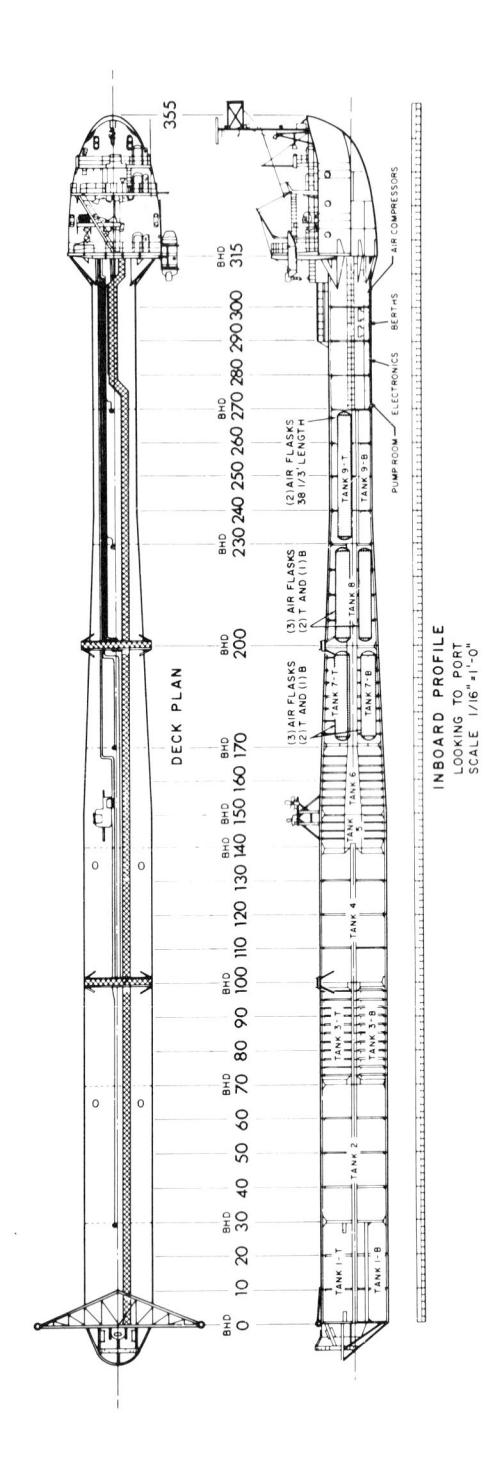

FIG. 15-12. FLIP—inboard profile.

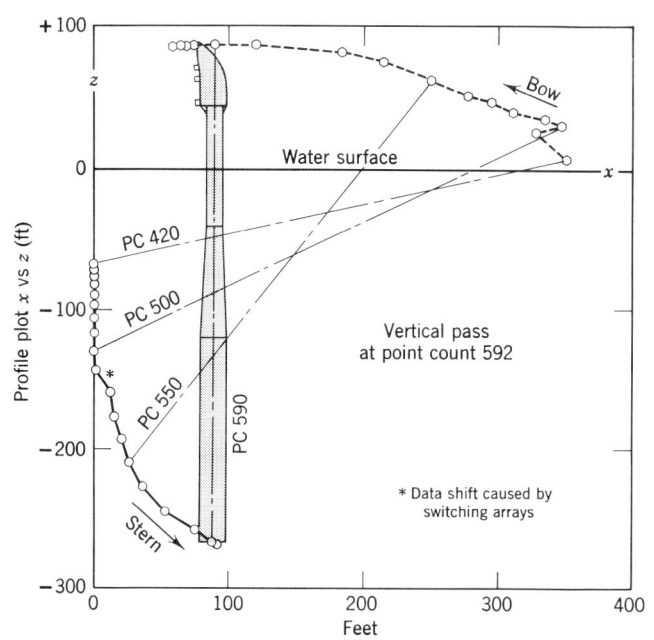

FIG. 15-13. Time required for transition from horizontal to vertical during initial trial operations.

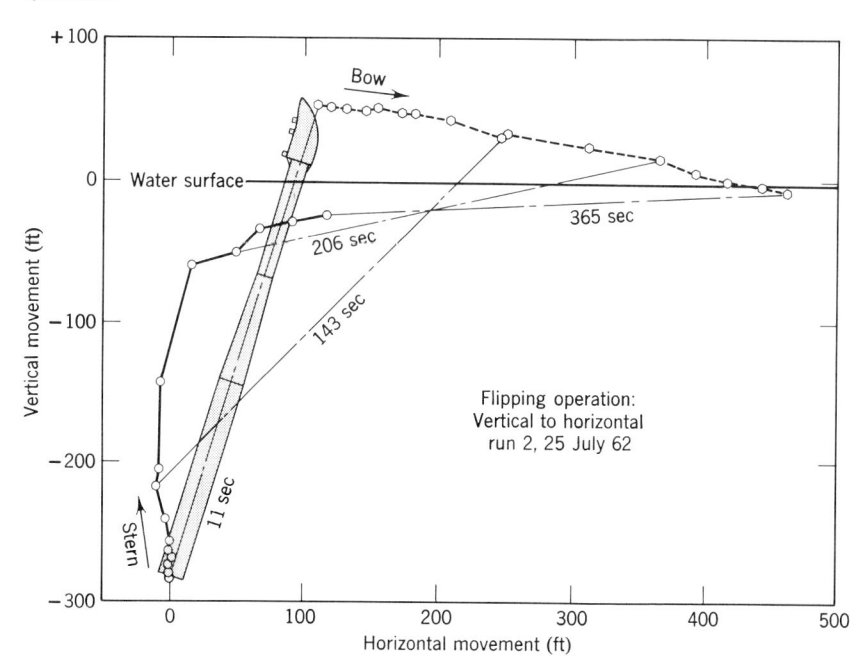

FIG. 15-14. Time required for transition from vertical to horizontal during initial trial operations.

FIG. 15-15. Personnel adjusting to attitude change of FLIP.

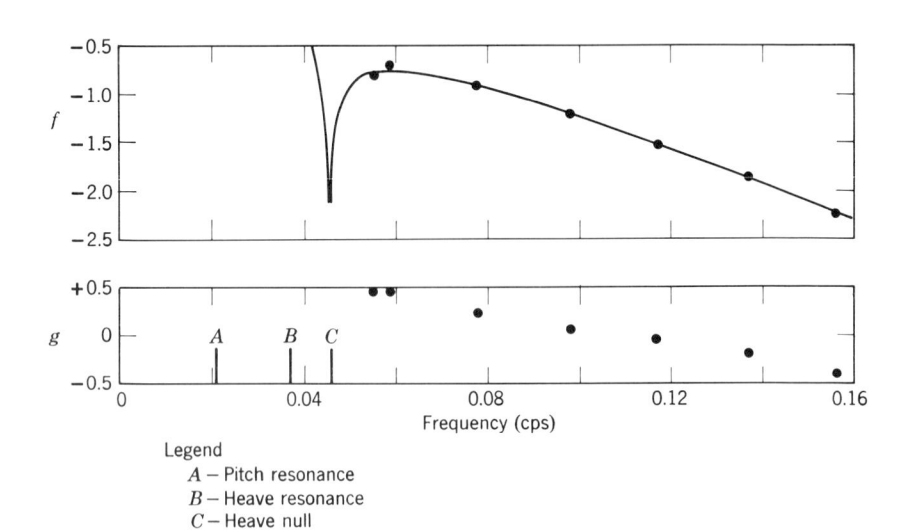

Legend
 A — Pitch resonance
 B — Heave resonance
 C — Heave null

FIG. 15-16. FLIP response curves.

horizontally approximately with wave amplitude (log = 0.0). The related angular motion (roll or pitch) in 10-ft waves (±5 ft from mean level) would thus be about ±1° since horizontal motion at the lower end is essentially zero.

Although FLIP has no propulsion plant capable of moving her from one station to another, she does have a set of propellers with electric-hydraulic drive to provide azimuthal orientation while in the vertical position. This system can be operated with manual control but also has an auto-pilot operating from a Sperry Mk XVIII gyro compass and capable of holding orientation to about one degree in a crosswind of 20 knots. (FLIP's normal heading is with "keel" into the wind.)

Towing has been done with a variety of commercial and research ship types with principal reliance on the Scripps Institution of Oceanography research ship *Horizon*, a converted tugboat (ATA). Speeds of 6 to 8 knots are typical, depending to a considerable extent upon the sea state. A 1 5/8-in. towing wire is used by *Horizon* to achieve the steady pull of 25,500 lb required for 7.5 knots in a calm sea. FLIP tows in good fashion with her two 10 × 12 ft skegs aft providing drag to assure her following the towline. One area in which good seamanship plays an essential part is in the close work required when hooking up the tow in the open sea. Here there is no substitute for a skilled towing-ship captain and a capable deck force.

Reference to Figs. 15-2 and 15-12 shows the occupied spaces, most of which are above the waterline in the vertical position. Easy access exists as far down the structure as the interior of compartment 9T. From there a trunk extends to the midships void (tank 5) which is available for conversion to an underwater observation space by replacing existing manhole plates with specially designed windows. No airfilled spaces exist below the 150-ft draft mark in order to minimize acoustic reflections. It was felt that the best viewing capability (when one is needed) would be obtained by providing an external, pressurized elevator compartment which could operate along the hull or below it at any level compatible with winch capability.

The principal spaces used are those in the bow section. The uppermost deck is used as a crew berthing space, the next as laboratory, third is galley, head, and scientific party berthing. The lowest level in the bow section is the principal machinery space, housing three diesel-generator sets (two 65 kW, one 10 kW), gyrocompass, and electrical distribution board. In the cylindrical section below the engine room are other machinery (air compressors, principally) and additional laboratory and living spaces. Bunks are provided for twelve total crew and scientific party. Normal operations at sea require a crew of five thus allowing a re-

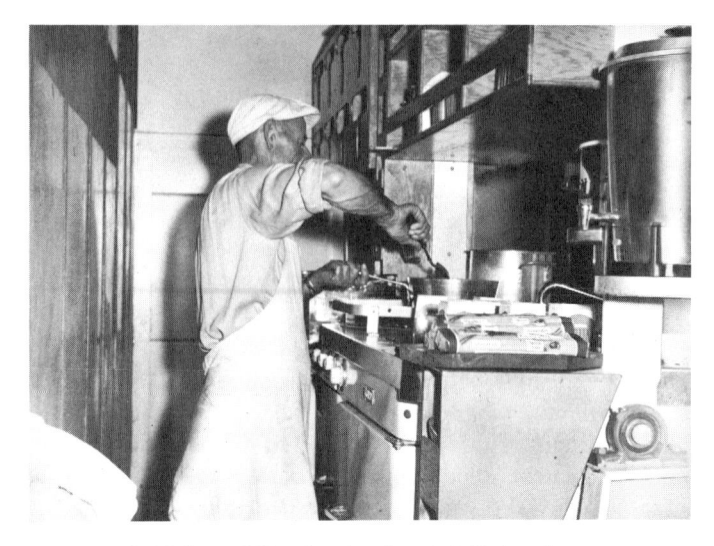

FIG. 15-17. FLIP galley showing gimballed equipment.

search group of up to seven. All living accommodations and machinery are useable in either the vertical or horizontal attitude, with most components [engines, generators, bunks, and galley equipment (Fig. 15-17)] swung in trunions with axes athwartships.

To date, FLIP has operated quite satisfactorily on a number of research problems, spending about 25% of her time at sea. The longest single operation was conducted during the summer of 1963 when she was used for wave and acoustic measurements 1500 miles northwest of San Diego. In this case she spent 45 days out of port, 27 of them drifting in the vertical condition (Northrop, 1964). During 1966 she operated in the Hawaiian area from February to September, involved in sound propagation and acoustics noise studies, internal wave observations, and seismic refraction operations. With over 100 flipping cycles behind her and a variety of experiments planned for the future, she appears to have demonstrated clearly both the usefulness and the effectiveness of ships of this type.

REFERENCES

Fisher, F. H., and F. N. Spiess, 1963, "FLIP—Floating Instrument Platform," *J. Acoust. Soc. Am.*, **35**, 1633 (1963).

Glosten, L. R., 1962, "FLIP—Some Remarks on Certain Design Considerations," Pacific Northwest Section, SNAME (October 6, 1962).

Glosten, L. R., 1966, Patent No. 3, 273, 526, "Stable Ocean Platform," issued September 20, 1966.

Knauss, J. A., "Measurements of the Cromwell Current," *Deep-Sea Res.* **6**, 265–286 (1960).

Korvin-Kroukovsky, B. V., 1961, *Theory of Seakeeping*, The Society of Naval Architects and Marine Engineers, New York.

McClure, A. C., 1965, "Development of the Project Mohole Drilling Platform," Annual Meeting SNAME, New York (November 11, 1965).

Northrop, J., 1964, "FLIP Returns from Mid-Pacific Research Cruise," *Trans. Am. Geophys. Union* **45**, 1, 165 (March 1964).

Piccard, Jacques, "*La Bouée-Laboratoire*," *Geo-Marine Tech.* **1**, 27 (September 1965).

Pollock, W. H., for M. Rosenblatt & Son, Inc., with F. H. Fisher and F. N. Spiess, 1960, "Report on Concept and Feasibility Discussions for a Manned Offshore Scientific Buoy," Report No. S-974-1, (May 1960).

Rudnick, P., 1964, "FLIP: An Oceanographic Buoy," *Science* **146**, 1268–1273 (1964).

Rudnick, P., 1960, "Motion of Spar Buoy in Swell", Part I, MPL-U-17/60 (June 14, 1960); Part II, MPL-U-21/60 (July 7, 1960).

Snodgrass, F. E., G. W. Grove, K. F. Hasselmann, G. R. Miller, W. H. Munk, and W. H. Powers, 1966, "Propagation of Ocean Swell Across the Pacific," *Phil. Trans. Roy. Soc. London* **A259**, 431–497 (1966).

Sverdrup, H. U., M. W. Johnson, and R. H. Fleming, 1942, *The Oceans*, Prentice Hall, New York.

Vytlacil, N., for M. Rosenblatt & Son, Inc., 1961, Lockheed Missles & Space Co., "Pitch and Heave Response in Waves of a Manned, Free-Floating Oceanographic Station," T. M. 81–73–6 (February 2, 1961).

CHAPTER 16

Materials Selection for Ocean Engineering

F. L. LaQUE

This review of the requirements of materials[1] for use in ocean engineering applications will concentrate on resistance to deterioration in ocean environments. This is justified by the fact that such requirements represent the principal distinction between ocean-engineering and other engineering applications. The usual properties related to the strength of structures and devices and how they may be fabricated are not peculiar to ocean-engineering applications and can be dealt with by reference to voluminous pertinent literature, much of it recently related to projected designs for pressure vessels, etc., proposed for undersea use.

16.1 STEELS

Except for limitations imposed by corrosion and fouling behavior, steels would represent the most economical choice for most ocean-engineering applications. They are available in all forms, are readily fabricated by conventional methods and can be provided at such high levels of strength as to qualify them for selection when a high strength-to-weight ratio is a critical factor.

This part of the discussion deals with the corrosion-resisting characteristics of steel that may indicate the circumstance under which either steel must be protected by coatings or cathodic currents or some other material that will not require protection must be substituted. Frequently the latter approach will be found to be more economical even when the substitute material costs more per pound. The behavoir of substituted materials is discussed later.

[1]Composition, trademarks, etc., of materials to be referred to in this Chapter are listed in Section 16.19.

Discussion of the behavior of steel can be handled most conveniently with reference to the different categories of marine environments, that is, salt atmospheres, tidal-range exposures, quiet immersion, and exposure to high velocities.

Atmospheric corrosion of steel

The resistance of steels to corrosion by salt atmospheres is influenced by the composition of the steel, the aggressiveness of the atmosphere as determined by its content of salt particles, temperature and humidity, and incidental conditions of exposure, particularly the degree of shelter from sun, wind, and rain.

Effects of composition of steel or iron

The resistance of steels to corrosion by salt atmospheres is improved by the presence of such alloying elements as copper, phosphorus, nickel and chromium alone or in various combinations. The most harmful element is sulfur, the effect of which is offset by that of copper, which should be present in greater amount than the sulfur. Fortunately, this is usually the case since ordinary carbon steel without any specification of copper content is likely to contain as much as 0.08% adventitious copper; this exceeds the usual sulfur content, which is under 0.05%.

The individual beneficial effects of copper, nickel, and chromium are shown by Figs. 16-1 to 16-4.

Most alloy steels likely to be used in ocean engineering contain combinations of elements that act together to improve resistance to atmospheric corrosion through the development of rust films that become quite protective, especially when they are used where there is frequent opportunity for drying. The protective value of such rust films is reflected in the shapes of weight-loss versus time curves that become quite flat with the better steels as illustrated by Fig 16-5.

Cast irons resist corrosion by marine atmospheres better than does unalloyed steel. As shown by Fig. 16-6, spheroidal graphite (ductile) cast iron is superior to flake graphite cast iron, whereas the highly alloyed austenitic cast irons (Ni-resist family) have exceptionally good resistance to salt atmospheres. Ductile cast iron is a favorite material for deck piping and valves on tankers, as illustrated by Fig. 16-7.

The advantage of the superior corrosion resistance of alloy steels carries over to the even better relative behavior when painted, as illustrated by Fig. 16-8, and discussed in greater detail in a paper by Larrabee and Copson (1959).

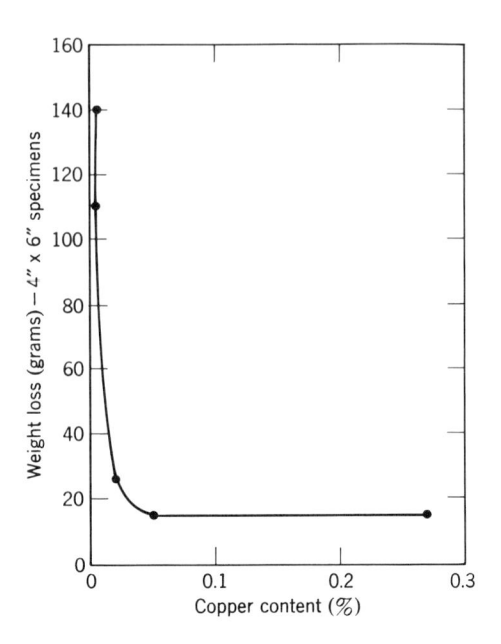

FIG. 16-1. Effect of copper content on corrosion of open-hearth iron in marine atmosphere at Kure Beach, N.C., 18 months exposure (*Corr. I-440*).

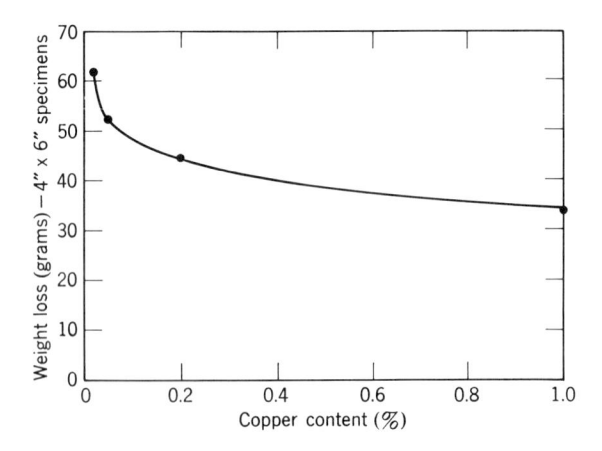

FIG. 16-2. Effect of copper content on corrosion of open-hearth steel in marine atmosphere at Kure Beach, N.C., 90 months exposure (*Corr. 1-438*).

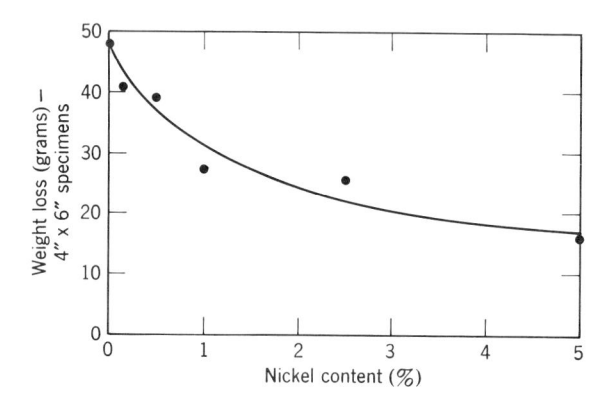

FIG. 16-3. Effect of nickel content on corrosion of steel in marine atmosphere at Kure Beach, N.C., 90 months (*Corr. I-407*).

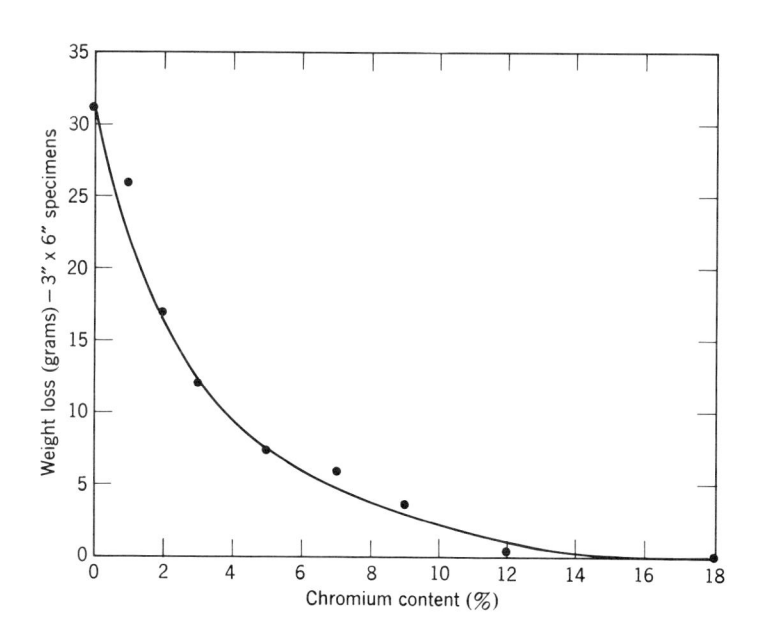

FIG. 16-4. Effect of chromium content on corrosion of steel in marine atmosphere at Kure Beach, N.C., 90 months (*Corr. I-408*).

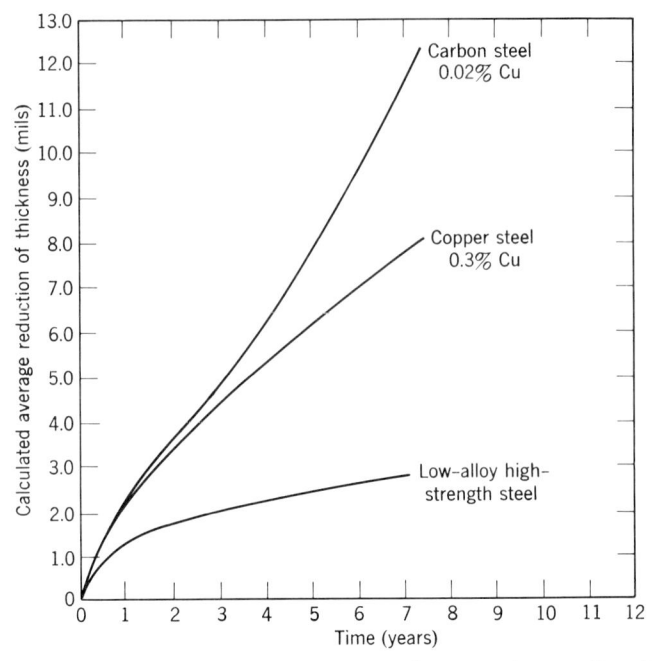

FIG. 16-5. Time-corrosion curves of steels in a marine atmosphere at Kure Beach, N.C. (*Corr. I-963*).

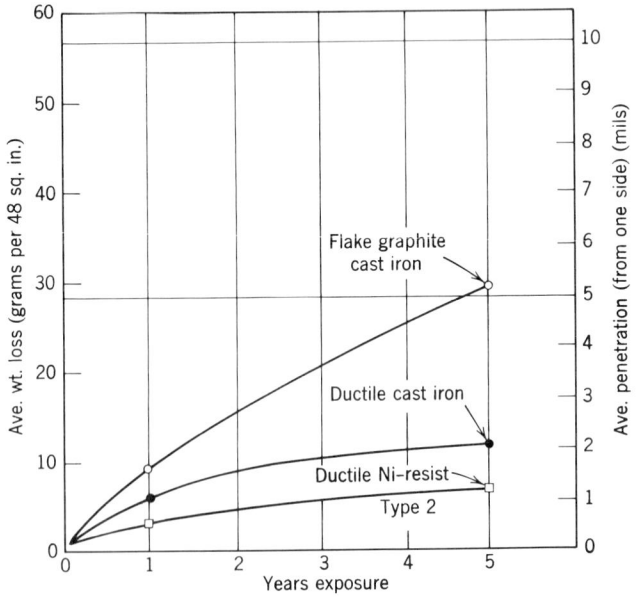

FIG. 16-6. Five Years' exposure 800 ft from the ocean at Kure Beach, N.C. (*Corr. I-926*).

FIG. 16-7. Ductile iron piping aboard a French tanker (*Corr. IV-620*).

	0.01 CU O.H. IRON	COPPER STEEL	LOW ALLOY HIGH STRENGTH STEEL
WT. LOSS IN GRAMS OF UNPROTECTED PANELS IN 18 MONTHS	130	14	7

FIG. 16-8. Effect of composition on corrosion resistance of steels and performance of protective coating (*Corr. IV-381*). (a) 0.01 Cu, O.H. iron; (b) copper steel; (c) low-alloy, high-strength steel.

Effects of content of salt particles

H. R. Ambler and A. A. J. Bain (1955) made a systematic study of how corrosion of metals in a tropical environment (Nigeria) was affected by distance from the ocean and the amount of salt suspended in the air as this varied with this distance. Their results are summarized in Fig. 16-9. The most significant results were the straight-line relationships between corrosion and salt content of the atmosphere when plotted on a logarithmic scale and the fact that there was a tenfold reduction of corrosion within the first 1000 yd from the ocean. Restated from the standpoint of an ocean engineer, there was a tenfold increase in corrosion of exposure on or close to the ocean as compared with a few hundred yards inland. The Mohurdi location covered in Fig. 16-9 was 200 miles from the ocean. The gross acceleration of corrosion in the immediate vicinity of the ocean will account for the substantially greater rates of deterioration of structures under such circumstances as compared with experience with similar structures located only a few hundred yards from the sea.

Temperature and humidity

The temperature and humidity at the Nigerian test location were higher than at Kure Beach, North Carolina, where a great deal of atmospheric

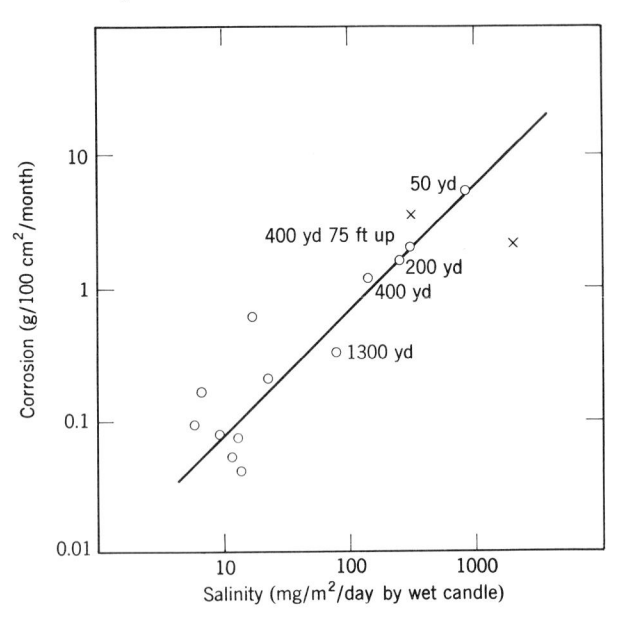

FIG. 16-9. Relation between corrosion of mild steel, salinity, and distance from ocean (*Corr. I-964*).

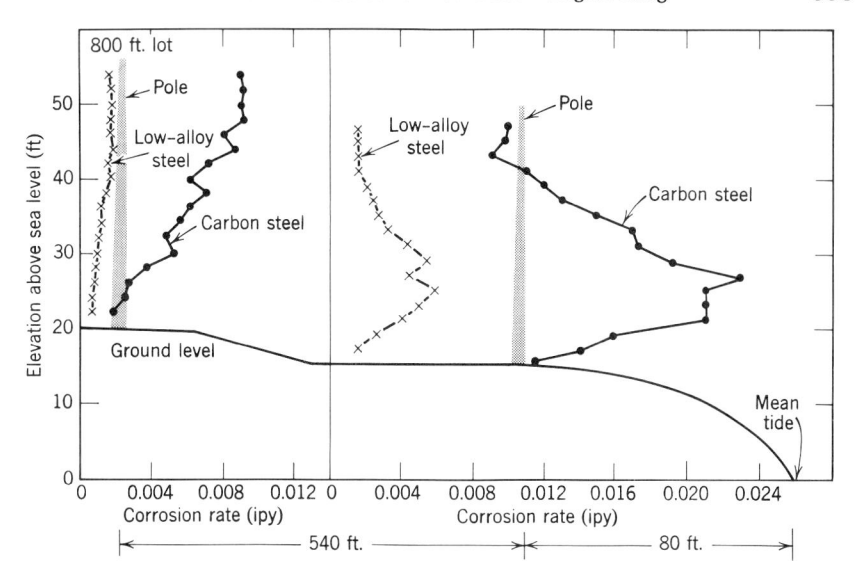

FIG. 16-10. Effect of height and location on the corrosion of steel exposed in the atmosphere at Kure Beach for 2.1 years (*Corr. 1-921*).

corrosion data have been accumulated. An indication of this effect is provided by comparing the rate of corrosion of unalloyed steel about 50 yd from the ocean at both locations. A typical rate for Kure Beach would be 15 mils/year and for Nigeria, 30 mils/year. The seasonal average temperature range at Kure Beach is from 18 to 92°F, whereas at Lagos it is from 76 to 92°F.

Effects of elevation, orientation, and shelter

Elevation, orientation, and degree of shelter can affect corrosion as they influence the amount of salt particles that reach metal surfaces and the beneficial washing effects of rain and drying effects of the sun. The effect of elevation along with distance from the ocean is illustrated by Fig. 16-10. Close to the ocean, corrosion reached a maximum at an elevation of about 25 ft whereas a short distance further inland corrosion increased directly with elevation, at least up to a height of 50 ft. It is significant, also, that the alloy steel was less sensitive to these factors and demonstrated maximum superiority under the most severe conditions.

The influence of orientation and shelter is illustrated by Fig. 16-11. based on exposure of specimens 800 ft from the ocean at Kure Beach, North Carolina, attached to four sides of a structure that provided for partial shelter of specimens exposed under an exaggerated, overhanging eave. The greatest corrosion occurred in the specimens on the ocean side (east) just

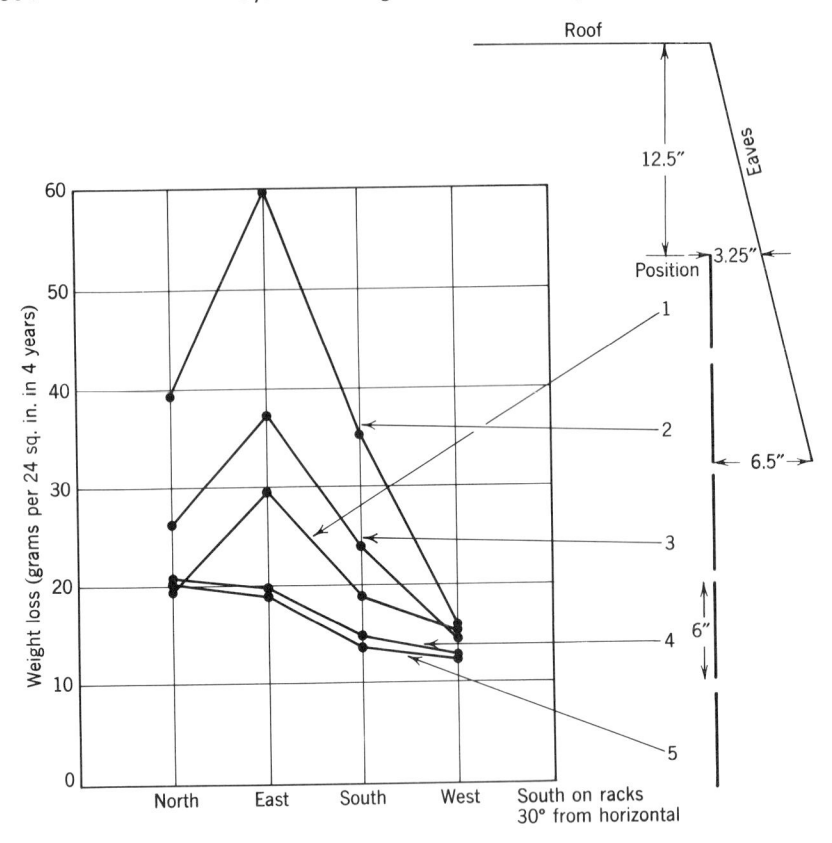

FIG. 16-11. Effect of degree of shelter and orientation on corrosion of carbon steel (0.04 Cu) specimens exposed at Kure Beach, N.C. (*Corr. I-406*).

under the eave where they encountered substantial amounts of wind-borne salt, but were shielded from the beneficial effects of rain and sun.

Distribution of corrosion of steel below, in, and above a tidal zone

Figure 16-12 illustrates a typical profile of the distribution of corrosive attack on a steel structure, such as piling, that extends from below the mud line to above the high-tide level. The actual rates of corrosion in the critical splash zone, and above it, will be influenced by the temperature and humidity of the atmosphere as previously discussed and may be higher or lower than the values shown, based on tests at Kure Beach, North Carolina. These values are useful as a reasonable general indication of what may reasonably be anticipated at other locations, with proper allowances for local conditions.

Splash zone

The maximum rates of corrosion in the splash zone are accounted for by the fact that rust films in this zone have little or no opportunity to dry so that they can develop the protective characteristics that account for lower rates of attack in the atmosphere above this zone. The high content of dissolved oxygen in the water that splashes above the high-tide level accounts for the higher rates of attack in this zone as compared with the continuously submerged surfaces below low tide where a combination of corrosion products and marine growths retards diffusion of oxygen to the corroding surfaces. In fact corrosion under these anaerobic conditions would be even less if sulfate-reducing bacteria were not usually present to take the place of oxygen in the cathodic half of the corrosion reaction.

The substantial reduction of corrosion in the tidal zone is accounted for by the action of a differential aeration or oxygen concentration cell. The tidal-zone surfaces become wetted by highly aerated seawater as the tide rises, whereas the surfaces below low tide remain in contact with practically oxygen-free water. The differential aeration cell established in this way makes the tidal-zone surfaces strongly cathodic to the submerged surfaces so that a current as high as 30 mA/ft² will flow to these surfaces from the submerged surfaces. This internally generated cathodic current is sufficient to protect the tidal-zone surfaces.

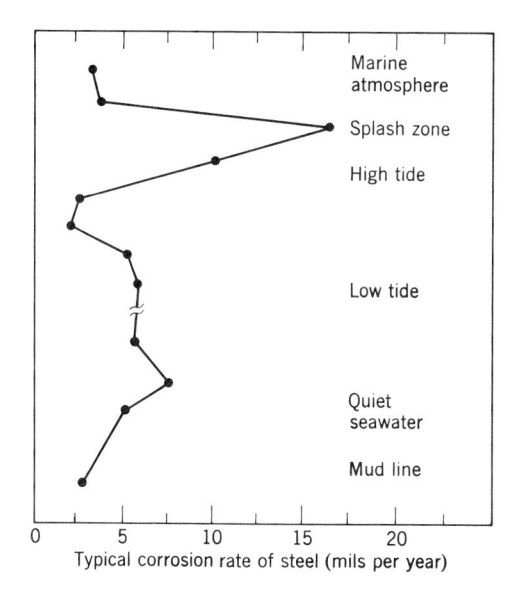

FIG. 16-12. Distribution of corrosion of steel piling partially immersed in sea water (*Corr. I-965*).

Corrosion by general wastage between low tide and the mud line will ordinarily be within the range of 3 to 6 mils/year in clean seawater at any geographic location. Rates outside this range may be associated with peculiar conditions of local pollution as disclosed by local experience. Accelerated attack in the form of generally rounded or flat-bottomed wide pits may be at several times the rate of general wastage. The ratio of pitting to general attack decreases with the duration of exposure. Over a period of several years, the pitting ratio will be about 3 to 1 for descaled steel and 5 to 1 for steel exposed with mill scale which accelerates pitting rather seriously, especially during the early period of exposure, for example, a factor as high as 20 to 1 for the first few months. It is very desirable to remove mill scale by pickling or sandblasting, especially if the thickness of the metal is small and if perforation by pitting must be avoided.

The presence of sulfate-reducing bacteria or other corrosion-accelerating constituents of sediments may bring about another zone of accelerated attack at or near the mud line. Where experience indicates that such conditions may exist and where it is practical to do so, the thickness of metal should be increased in this zone so as to provide for accelerated attack.

Behavior at extreme depths

Results of tests undertaken so far at depths up to 6000 ft (Reinhart, 1965) have shown that general wastage and pitting of steels at such depths are within the range shown by tests near the surface, as discussed already and plotted in Fig. 16-16 and 16-17. Present indications are that pitting may be less in very deep water except in the case of aluminum alloys. Corrosive muds and sediments may cause accelerated corrosion in some locations, as previously mentioned.

Protective measures

Measures that may be taken to protect steel in the several zones that have been discussed are listed below.

Above the splash zone. (a) Use alloy steels with superior resistance to atmospheric corrosion as previously discussed. (b) Apply protective coatings, such as paints, enamels, zinc or aluminum, including sprayed zinc or aluminum with appropriate sealers (American Welding Society, 1962).

In the splash zone. (a) Use an alloy steel especially formulated to achieve superior resistance to corrosion under splash-zone conditions. (Larrabee, 1958). A steel containing 0.5% nickel, 0.5% copper, and 0.12% phosphorus has been shown to be about 5 times better than carbon steel in tests in the

FIG. 16-13. Monel nickel copper alloy sheathing on offshore drilling platform after 18 years (*Corr. IV-845*).

critical splash-zone region. (b) Use protective coatings as for the upper zone, having in mind that it will be very difficult to renew such coatings under the unfavorable conditions that will prevail during attempts to recoat structures *in situ*. (c) Apply sheathing of a corrosion-resisting alloy such as the Monel[2] nickel-copper alloy first used successfully for this purpose nearly 20 years ago and still performing satisfactorily in this first installation, as shown by Fig. 16-13. (d) Cathodic protection will not be effective in the spalsh zone.

In the tidal zone. The protective measures suggested for the more difficult splash zone will be effective in the tidal zone. Cathodic protection from submerged anodes will also be helpful here.

Below low tide. Where attack at the rates indicated cannot be tolerated, cathodic protection at a current density of from 3 to 5 mA/ft^2 will be effective for uncoated steel depending on turbulence, etc. The current densities suggested are what will be required after the required polarization has

---

[2]Trademark, The International Nickel Company, Inc.

been established. To reach this state quickly will require substantially higher current at the start, for example, 3 to 5 times as much as will be needed to maintain the required polarization once it has been achieved.

The current requirement can be reduced substantially by the use of a supplementary protective coating. If zinc or aluminum coatings are used, it will be necessary to control the application of cathodic protection so that the polarized potential will not exceed about 0.9 V versus a saturated calomel reference electrode so as to avoid generation of alkali that would corrode the aluminum or zinc.

If an organic coating is used, the vehicle should be tolerant of cathodic alkaline conditions. Polarized potentials, for example, over 1.0 V, that would induce hydrogen evolution and blistering of relatively impermeable coatings should be avoided. A potential of 0.85 V is high enough for cathodic protection of steel in seawater. Another satisfactory criterion of protection would be a shift in potential of the order of 0.15 V in the less-noble direction.

Cathodic protection of steel ships or other vessels that move through the water (Graham et al., 1956) can be provided by current from galvanic anodes of magnesium, zinc, or aluminum of appropriate size, properly located. A more scientific method is the use of applied current systems with insoluble anodes and electronic control of the system to maintain polarization at the required level under different conditions of operation (Anderson and Anderson, 1965). To facilitate current distribution and reduce the amount of current required and to control fouling, it is necessary to use protective coatings as a supplement to cathodic protection. With the best coating systems in good condition, the current density required to protect ships will be about 2 mA/ft^2 of hull plus 20 mA/ft^2 of propeller when the propellor is in electrical contact with the hull, as is usually the case (Shreir, 1963).

As coatings deteriorate, or are removed accidentally, the current requirements will increase. Coatings should tolerate the alkali generated by cathodic reactions.

Soft organic coatings are penetrated by barnacles and will require supplementary cathodic protection to take care of bare spots exposed in this way. Antifouling coatings cannot be depended on to prevent fouling for more than about two years. Where more extended freedom from fouling is required, it will be necessary to apply an anitifouling metal sheathing, preferably a 90:10 cupronickel alloy, electrically insulated from adjacent uncoated steel so as to avoid galvanic effects in the form of accelerated corrosion of the steel and suppresion of the antifouling properties of the sheathing.

The special problem of maintaining a smooth surface on sonar domes

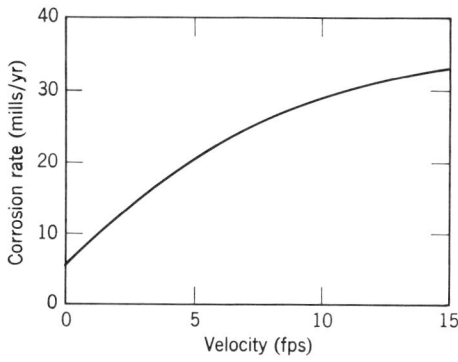

FIG. 16-14. Effect of sea water velocity on corrosion of steel at ambient temperature (*Corr. I-966*).

may be solved by sheathing with the 70:30 cupronickel alloy which can be expected to resist fouling with the help of the high-frequency vibrations.

Velocity effects on steel

Corrosion of bare steel increases as the velocity of movement of seawater increases as shown graphically by Fig. 16-14 based on tests using tubular specimens with water flowing through them (Field, 1945). Rates of attack continue to increase at much higher velocities. For example, tests designed to simulate velocities that might be encountered by hydrofoils, for example, 135 ft/sec, showed a rate of corrosion of steel of about 0.2 in./year or somewhat higher than could be calculated by extrapolating from Fig. 16-14.

These data refer to conditions under which the steel surfaces remain free from adherent fouling organisms or slimes which can be quite protective. A heavy growth of fouling organisms can reduce the velocity at the metal surface to practically zero, irrespective of how fast the water may be moving past the outer surfaces of the marine growth layer. Bacterial slimes can provide (Turner et al., 1948) similar protection against velocity effects at modest velocities. Chlorination or similar treatments to eliminate slimes and fouling can aggravate corrosion at high velocities by removing the protective effects of organisms (Volkenig, 1950; Anderson and Richards, 1965).

The corrosion of zinc coatings (galvanizing) commonly used to protect steel in the form of pipe and wire increases sharply as the seawater velocity increases, as shown by Fig. 16-15. This limits the extent to which such thin zinc coatings can extend the life of steel in moving seawater, for example, as encountered in piping systems (Field, 1945).

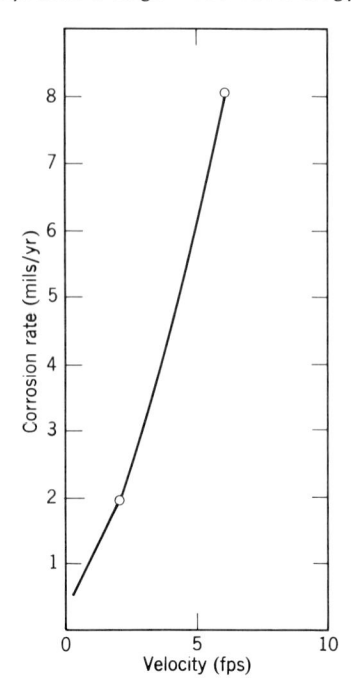

FIG. 16-15. Effect of sea water velocity on corrosion of zinc at ambient temperature (*Corr. I-967*).

16.2 GUIDES TO SELECTION OF MATERIALS OTHER THAN STEEL

The limitations imposed by the corrosion behavior of steels and irons in seawater and the cost or impracticality of protective measures, as discussed in the preceding section, frequently make it necessary to use more corrosion-resistant materials. Limitations of space and time will not permit detailed treatment of the corrosion-resisting characteristics of the many materials from which a choice might be made. What'll be provided will be some charts and tables reproduced, with some modification, from a paper by Tuthill and Schillmoller (1966) presented at the Ocean Science and Engineering Conference, Washington, D.C., June 1965, which summarize recorded data and experience. These serve as a framework or checklist so that all important aspects of a choice can be taken into account. The charts can be kept up to date by the user by insertion of new data as they appear, or revision of the data in the present charts on the basis of results of tests or experience more closely related to a particular location or material or a specific application of immediate concern to the user of the charts.

The charts and tables will permit what might be called a "first approximation" to an estimate of durability and performance of materials worthy of further consideration after they have been established as possessing the necessary mechanical properties, fabricability, and availability in the required forms. This "first approximation" can be refined by reference to pertinent experience and by consultation with producers and other users of materials indicated as being of particular interest by this "first approximation."

The charts show typical ranges of quantitative data for the specific materials listed. They may also be applied to "families" of the alloys represented, for example, the data for general wasting and pitting of carbon steel can be extended reasonably to cover wrought iron and low alloy steels. The data for Alloy 20 can be applied to Incoloy[3] Alloy 825.

16.3 GENERAL WASTING

Figure 16-16 shows ranges of rates of general wasting of materials when immersed in quiet seawater, as reported in the literature (Uhlig, 1948) or from tests undertaken at the International Nickel Company Marine Research Laboratory at Harbor Island, North Carolina. It will be noted that some of the materials that are practically immune to general wasting in quiet seawater, for example, stainless steels, are limited in their usefulness under such conditions by their susceptibility to local attack, pitting, for example, under marine organisms or other persistent deposits or in crevices. Conditions of high velocity flow or turbulence that keep surfaces clean are favorable to these materials. Data on pitting in quiet seawater are covered by Fig. 16-17.

Results of early tests of materials exposed to quiet seawater at great depths are shown in Figs. 16-16 and 16-17 by small circles. Present indications are that general wasting of most metals at great depths can be expected to fall within the ranges observed in tests near the surface. Results of visual observations of specimens exposed at a depth of 5640 ft at a particular location are summarized in Table 16-1 (Reinhart, 1964, 1965).

The mechanism of pitting in quiet seawater most commonly involves differential aeration cells or the even more aggressive cells in which the driving force is the potential difference between active surfaces at the base of a pit and usually much larger passive surfaces surrounding a pit.

Copper-base alloys are less susceptible to highly localized deep pitting because they do not develop such high potential differences between active and passive surfaces. In addition, metal ion concentration cells based on the soluble corrosion products of copper alloys oppose differential aeration cells so as to arrest pitting initiated by the latter mechanism.

[3]Trademark, The International Nickel Company, Inc.

Systems Design—The Technology

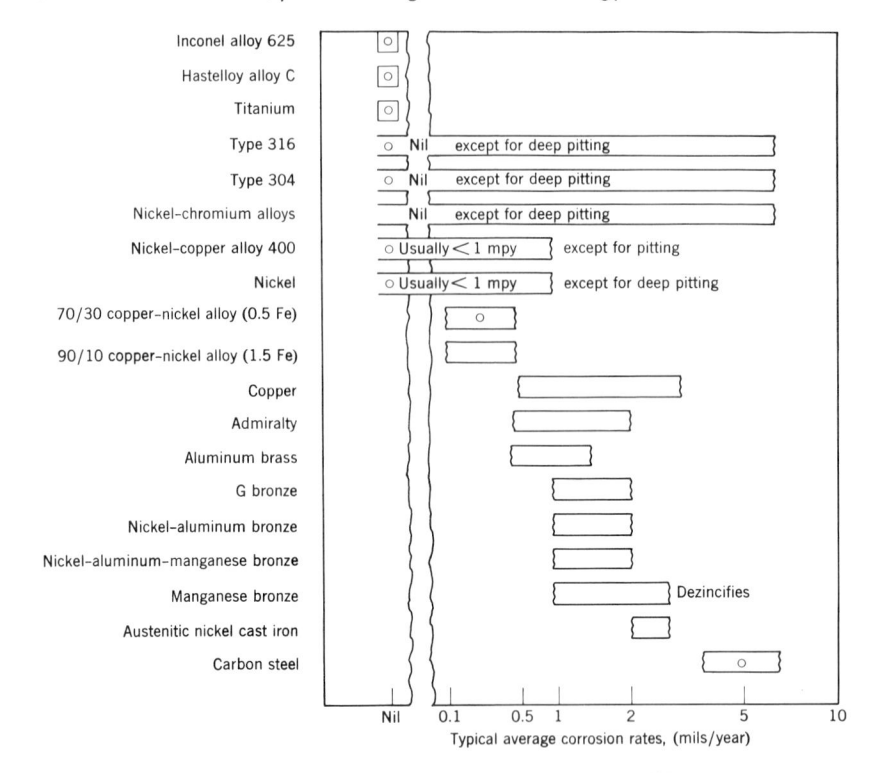

Material	Corrosion rate
Inconel alloy 625	○
Hastelloy alloy C	○
Titanium	○
Type 316	○ Nil — except for deep pitting
Type 304	○ Nil — except for deep pitting
Nickel-chromium alloys	Nil — except for deep pitting
Nickel-copper alloy 400	○ Usually < 1 mpy — except for pitting
Nickel	○ Usually < 1 mpy — except for deep pitting
70/30 copper-nickel alloy (0.5 Fe)	○
90/10 copper-nickel alloy (1.5 Fe)	
Copper	
Admiralty	
Aluminum brass	
G bronze	
Nickel-aluminum bronze	
Nickel-aluminum-manganese bronze	
Manganese bronze	Dezincifies
Austenitic nickel cast iron	
Carbon steel	○

Typical average corrosion rates, (mils/year): Nil, 0.1, 0.5, 1, 2, 5, 10

○ Data from results of early tests at depths of 2300 to 5600

FIG. 16-16. Rates of general wasting of metals immersed in quiet seawater—velocity less than 2 fps. (*Corr. I-968*).

Differences in the pattern and depth of pitting of the two classes of alloys in quiet seawater are illustrated by Fig. 16-18.

Data on pitting of specimens immersed at a great depth are shown by small circles in Fig. 16-17. The principal difference in behavior at a great depth was exhibited by the aluminum alloys which suffer relatively little pitting near the surface except when the water is contaminated by ions of heavy metals—particularly copper—which promote pitting. Pitting of aluminum appears to be accelerated at great depths, apparently as a result of a low concentration of the dissolved oxygen needed to maintain or restore the oxide films upon which aluminum depends for corrosion resistance. Perhaps for the same reason, the less noble aluminum alloys used as cladding to protect more noble aluminum compositions were not as effective at great depths as at the surface.

Extruded aluminum alloy 5086 in the as-extruded condition, as used for the test racks, suffered much less pitting than specimens of the same alloy in the form of flat-rolled plate and sheet. This can be explained by the

Materials Selection for Ocean Engineering 605

fact that in the as-extruded sections the Mg_2Al_2 compound in the alloy was dispersed through the grains whereas in the flat-rolled material, the heat treatment given the material to increase ductility resulted in concentration of the anodic magnesium aluminum compound along the grain boundaries so as to promote intergranular attack, associated with pitting. See Fig. 16-19.

16.4 CREVICE CORROSION

The mechanisms of corrosion within or around crevices are much the same as those involved in pitting. Consequently, susceptibility to crevice corrosion parallels susceptibility to pitting as indicated by Fig. 16-17 and

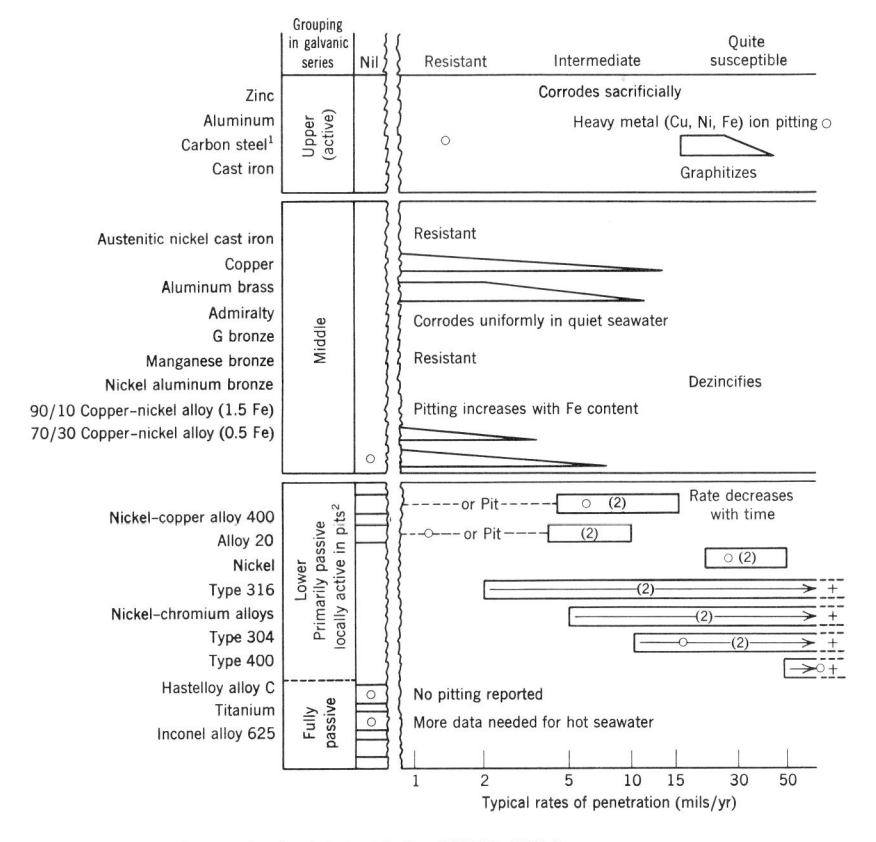

○ Data from results of early tests at depths of 2300 to 5600 ft
(1) Shallow round-bottom pits.
(2) As velocity increases above 3 fps pitting decreases. When continuously exposed to 5 ft per sec. and higher velocities these metals, except Type 400 series, tend to remain passive without any pitting over the full surface in the absence of crevices.

FIG. 16-17. Rates of pitting of metals immersed in quiet seawater—velocity less than 2 fps (*Corr. I-969*).

TABLE 16-1
Appearances of Specimens After Exposure for
123 Days at Depth of 5,640 Ft (Reinhart, 1964, 1965)

Temperature	37°F (3°C)
Oxygen	1.2 ppm
Velocity	1 ft/sec pH 7.3
Salinity	34.6 parts per thousand

Class of Material	Behavior
Aluminum alloys	Pitted and exfoliated to much greater extent than near surface
Copper alloys	Very shallow uniform attack under corrosion product films
Magnesium alloys	Scattered penetration by pits
Nickel–base alloys[a]	Apparently unaffected except for Monel nickel–copper alloy 400, which developed adherent corrosion product film
Ordinary and alloy steels	Uniform attack under thin layer of rust
Stainless steels—400 Series	Characteristic pitting and crevice corrosion with occasional tunneling from points of initial attack
Stainless steels—300 Series[a] and more highly alloyed	Generally free from attack
Titanium alloys	Apparently unaffected

[a] Local attack of nickel–base alloys of low molydenum content and the 300 Series stainless steels will probably be observed in more extensive tests in quiet water at great depths.

summarized in Table 16-2. The order of merit in this table is from left to right by classes and top to bottom in each class.

In using a material with low tolerance for crevices, as shown in Table 16-2, it is necessary to avoid shapes and joints that form crevices and opportunities for crevice effects to develop under persistent deposits or marine organisms. Crevices that exist under fastenings, washers, flanged joints, in cable sockets, in the interstices of stranded cable, etc., are common locations for crevice attack. When such crevices must be tolerated, materials resistant to such attack should be chosen. It is difficult to find caulking compounds or sealants that will not form equally damaging crevices. Greases containing zinc oxide have been found to be effective with stainless steels under circumtances where they can survive for the required length of time.

90/10 Copper-Nickel

Deepest Pit — 8 mils
Av. Pit Depth — 4 mils
(3 years)

Type 304 Stainless Steel

Deepest Pit — 250 mils
(perforated)
(3 years)

70/30 Copper-Nickel

Deepest Pit — 18 mils
Av. Pit Depth — 10 mils
(3 years)

Type 316 Stainless Steel

Deepest Pit — 113 mils
Av. Pit Depth — 72 mils
(3 years)

FIG. 16–18. Illustrative examples of localized attack of metals in quiet seawater under fouling conditions (*Corr. IV-846*).

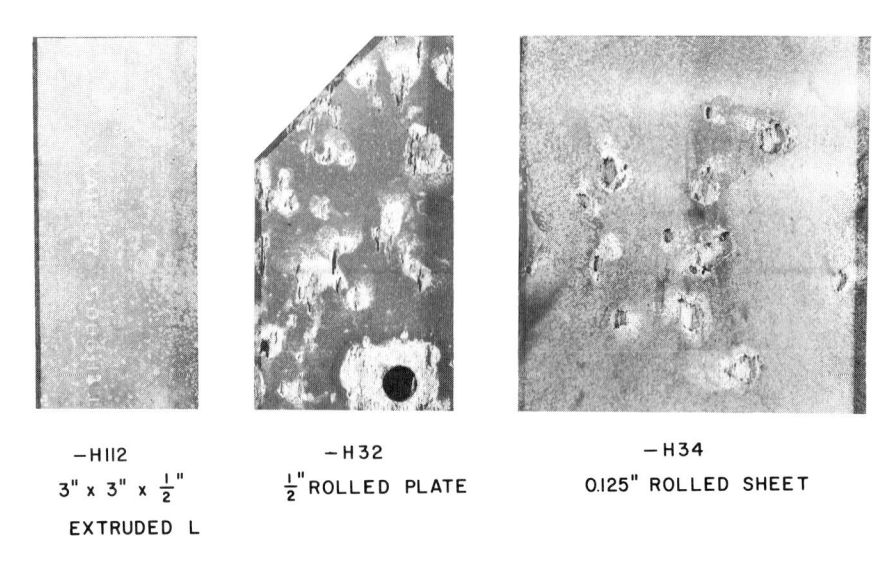

−H112　　　　　−H32　　　　　　−H34
3" x 3" x $\frac{1}{2}$"　　　$\frac{1}{2}$" ROLLED PLATE　　0.125" ROLLED SHEET
EXTRUDED L

FIG. 16–19. Effect of strain hardening on corrosion of 5086 aluminum alloy in deep ocean: depth, 5640 ft; 751 days (*Corr. IV–847*). (Photo courtesy F. M. Reinhart, U.S. Naval Civil Engineering Laboratory.)

16.5 FOULING

In some applications it is necessary or desirable that surfaces remain free of fouling without the benefit of an antifouling coating. Examples are the inside of piping systems, buoys or floats that should retain their initial buoyancy, or the sheathing of vessels.

It has been observed that the only alloys that remain free of fouling are copper alloys that corrode fast enough to release copper in corrosion products at a rate in excess of about 5 Mg/DM^2/DAY (La Que and Clapp, 1945). This is equivalent to general wastage at a rate of only about 1 Mil/year. When copper is alloyed with metals that cause corrosion under fouling conditions below this critical rate, fouling will occur. However, the firmness of attachment of organisms which are able to grow is also reduced by copper in corrosion products below the level required to prevent attachment. Galvonic effects on cathodic protection that can suppress corrosion of otherwise antifouling alloys below the level required to prevent fouling will allow fouling to occur. The iron modified 90:10 cupronickel alloy provides the best combination of antifouling properties and resistance to pitting and the effects of high velocities.

Corrosion products containing an optimum amount of copper may prove to be helpful in the growth and harvesting of shellfish (oysters). For example, observations of the attachment and growth of oysters on a series

TABLE 16-2
Tolerance for Crevices Immersed in Quiet Seawater

Inert Best	Good	Useful Resistance Neutral	Less	Crevices Tend to Initiate Deep Pitting
Hastelloy	90/10 Copper–nickel alloy (1.5 Fe)	Aus. nickel cast iron	Nickel–iron–chromium alloy 825	Type 316
Alloy C	70/30 Copper–nickel alloy (0.5 Fe)	Cast iron	Alloy 20	Nickel–chromium alloys
Titanium[a]	Bronze	Carbon steel	Nickel–copper alloy 400	Type 304
Nickel–chromium Alloy 625	Brass		Copper	Type 400 series

[a]Susceptible in hot seawater.

of copper–nickel alloys have indicated a possible advantage of the 45% nickel, 55% copper alloy. Oysters were able to grow on this alloy, were larger than on inert materials and were more easily removed.

16.6 VELOCITY EFFECTS

The most convenient way to discuss velocity effects is in terms of the services in which such effects present problems with respect to the choice and performance of materials.

Pertinent data related to piping systems and condensers or other heat exchangers are summarized in Fig. 16-20. This shows, for example, that copper can be expected to perform satisfactorily at flow velocities below 3 ft/sec but will suffer accelerated attack at higher velocities, whereas the stainless steels will behave well at high velocities but may suffer deep pitting at low velocities or in stagnant water. The more highly alloyed compositions of stainless steel, such as Incoloy Alloy 825, are preferred for the broad range of velocity. Even greater reliability is provided by the

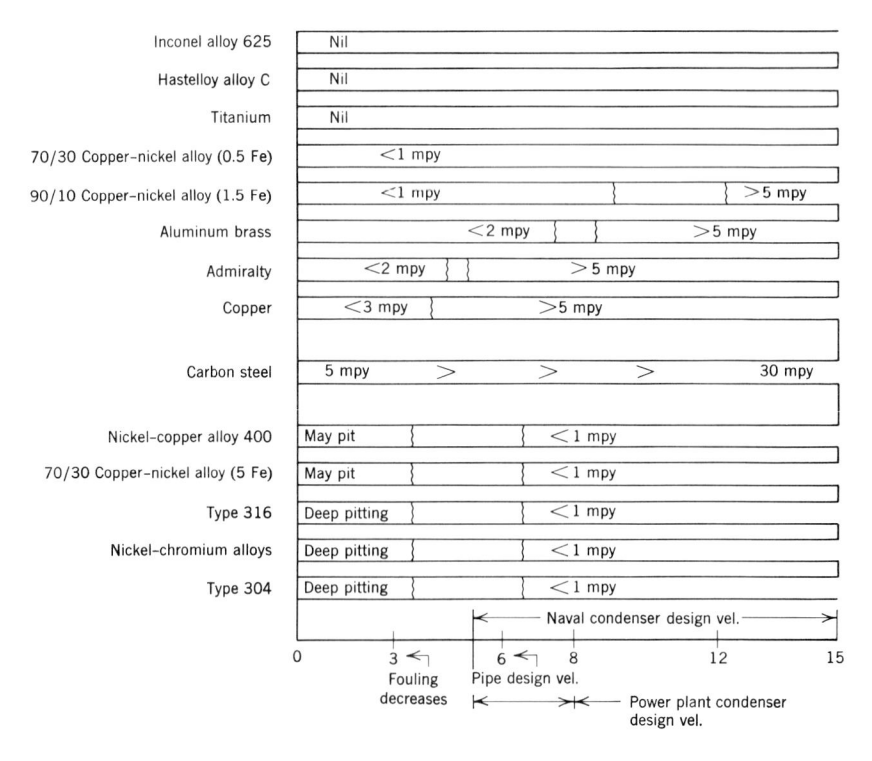

FIG. 16-20. Effects of velocity in pipe and tube service ranges on corrosion of metals in seawater (*Corr. I-970*).

nickel-base alloys with chromium and molybdenum represented by Hastelloy[4] Alloy C and Inconel[5] Alloy 625.

The values for design velocity used to define limits of performance in Fig. 16-20 take into account effects of local turbulence at the entrance of heat exchanger (condenser) tubes and downstream of short-radius elbows, the connections and throttled valves in pipe lines which give rise to velocity effects more intense than would be expected from the average velocity in the system. Anything that can be done in design or fabrication to avoid such sources of local turbulence will lengthen the life of any materials being used near the limits of tolerable velocity indicated by Fig. 16-20. By the same token, a choice of material capable of withstanding a wide range of velocities will reduce the chance of failure if considerable variation from the design velocity must be provided for or may be encountered accidentally.

Similar data for the higher ranges of velocity and greater turbulence encountered in valves and pumps and by hydrofoils are summarized in Fig. 16-21.

The materials listed in Fig. 16-21 fall into two general classes; those that will tolerate the highest velocity likely to be encountered and those where durability may be characterized as being velocity limited, as indicated by the data in this chart. Rates of attack may vary through wide limits in the case of velocity-limited materials near or above the threshold of the tolerable velocity limit, depending on the incidental conditions of use.

Velocities encountered by pump casings may range from 5 to 50 ft/sec. Higher velocities pertain to impellers and the tips of propellers and are modified in their corrosive effects somewhat by friction films which reduce the relative velocity between the metal surface and the liquid. With pump casings and other castings which have to be of substantial thickness, higher rates of corrosion can be tolerated than with the critical seating surfaces of valves or relatively thin vanes in pump impellers.

Stainless steels that will tolerate extremely high velocities are subject to pitting and crevice corrosion in stagnant water. Consequently, when both conditions of service must be povided for, the selection should be made by reference to Fig. 16-17 and Table 16-2, as well as Fig. 16-21.

16.7 CAVITATION EROSION

Cavitation erosion as encountered in pumps or propellers and on hydrofoils represents an extreme form of what might be called a velocity

---

[4]Trademark, Union Carbide Corporation.

[5]Trademark, The International Nickel Company, Inc.

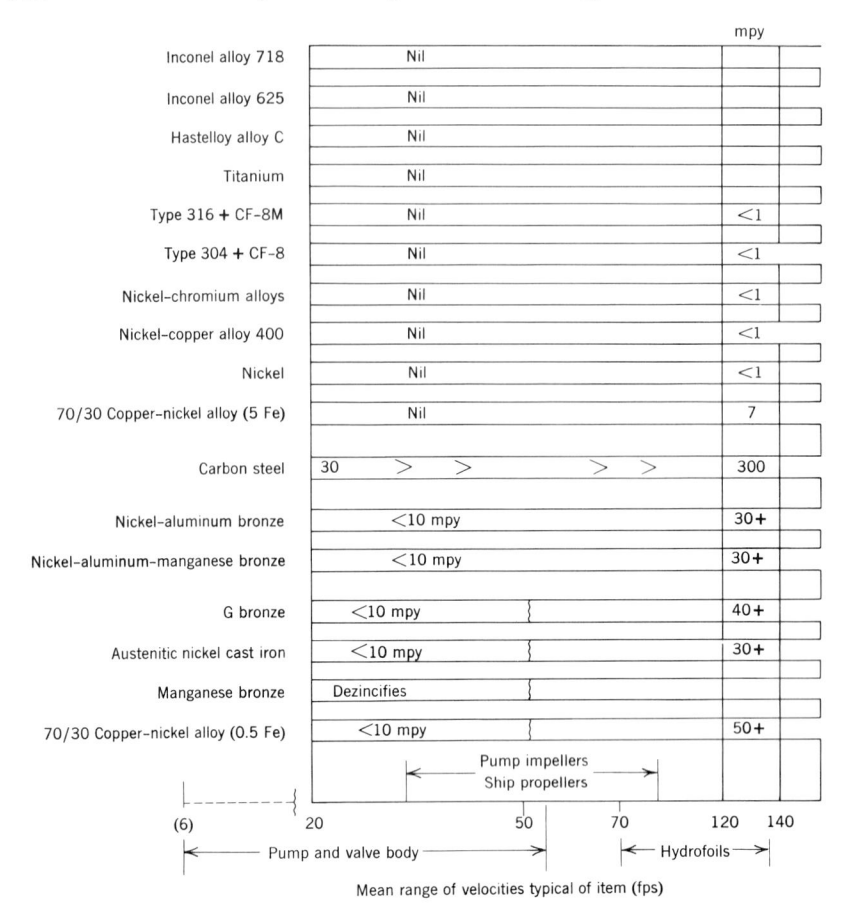

Material	Low velocity	High velocity (mpy)
Inconel alloy 718	Nil	
Inconel alloy 625	Nil	
Hastelloy alloy C	Nil	
Titanium	Nil	
Type 316 + CF-8M	Nil	<1
Type 304 + CF-8	Nil	<1
Nickel-chromium alloys	Nil	<1
Nickel-copper alloy 400	Nil	<1
Nickel	Nil	<1
70/30 Copper-nickel alloy (5 Fe)	Nil	7
Carbon steel	30 > > > >	300
Nickel-aluminum bronze	<10 mpy	30+
Nickel-aluminum-manganese bronze	<10 mpy	30+
G bronze	<10 mpy	40+
Austenitic nickel cast iron	<10 mpy	30+
Manganese bronze	Dezincifies	
70/30 Copper-nickel alloy (0.5 Fe)	<10 mpy	50+

Velocity ranges: Pump and valve body (to 70 fps); Pump impellers / Ship propellers (50–120 fps); Hydrofoils (70–140 fps). Mean range of velocities typical of item (fps).

FIG. 16-21. Effects of velocity in pump and hydrofoil ranges on corrosion of metals in seawater (*Corr. I-971*).

effect. There are continuing arguments as to the relative importance of mechanical effects and corrosion effects in the mechanism of damage by cavitation erosion. It is safe to conclude, however, that best peformance in resisting such damage is most likely to be provided by a material that combines a high level of resistance to corrosion and corrosion fatigue with at least a moderate level of hardness and the ability to develop greater hardness by the cold work induced by collapse of cavitation bubbles (Holl and Wood, 1964; Eisenberg et al., 1965).

If other requirements dictate the choice of a material having inadequate resistance to cavitation erosion, it will be necessary to apply some sort of protective coating. This can present problems of maintaining proper adherence of such coatings under cavitating conditions. Considerable research to find appropriate coatings has been undertaken (Lichtman et al., 1961). Indications are that organic coatings should possess a high degree of resilience to enable them to absorb the mechanical forces without cracking or spalling.

Most accelerated tests to rate materials with respect to their ability to resist cavitation damage in seawater tend to put a premium on hardness relative to corrosion resistance.

The order of merit shown in Table 16-3 reflects a combination of recorded experience and results of accelerated cavitation erosion tests. It differs slightly from a similar list provided by Rheingans (1957) and takes into account specific effects of seawater.

Reference to Table 16-3 will be useful as a guide to a choice of material where resistance to cavitation erosion is a prime requirement. It may be even more useful in indicating materials that could be substituted advantageously for one lower in the list that has suffered from unanticipated cavitation erosion in a particular service.

16.8 GALVANIC EFFECTS

The previous discussion of the behavior of metals and alloys in seawater has not taken into account disturbances in normal behavior as a result of galvanic action between dissimilar metals when they are in electrical contact in seawater. Electrical contact in this context requires that the galvanic circuit include an electronic, as well as an ionic, conductor. Simple proximity in the water or in water-soaked wood will not suffice. However, there can be corrosive effects resulting from migration of corrosion products of one metal that can affect corrosion of an adjacent one, for example, copper corrosion products accelerate corrosion of aluminum, zinc, and steel. These are not galvanic effects of the type being dealt with here.

In any galvanic couple the corrosion of one metal is accelerated and the corrosion of the other is reduced. This can lead to trouble if the effect is not anticipated and provided for or can be beneficial if the choice of metal, its thickness and the over-all arrangement is designed to take advantage of the corrosion of an expendable metal to provide built-in protection to another one where survival with minimum corrosion is a critical requirement; for example, the seating surfaces of a valve that can be protected by corrosion of a valve body designed to accommodate the amount of sacrificial corrosion that it must withstand.

The arrangement of metals in a galvanic series for seawater in Table 16-4 indicates which of the metals (the higher ones in the list) in a galvanic couple will suffer accelerated corrosion and which will have its corrosion

TABLE 16-3
Order of Resistance to Cavitation Erosion
in Seawater

Rating	Alloy or Class	Example of Class
1	Stellite	
2	Age-hardened nickel chromium Alloys	Inconel nickel chromium Alloy 718
2	Precipitation-hardened stainless steels	17-4 PH
2	Austenitic stainless steels	18 Cr–8 Ni
2	Matrix-stiffened nickel–Chromium alloys	Inconel nickel chromium Alloy 625
2	Titanium alloys	
3	Nickel–chromium alloys	Inconel nickel chromium Alloy 600
3	Age-hardenend nickel–copper alloy	Monel Alloy K 500
4	Nickel–copper alloy	Monel Alloy 400
5	Nickel–aluminum bronze	Ni–Br Al
6	Copper–nickel alloy	70% Cu 30% Ni
7	Manganese bronze	
8	Tin Bronze	Composition-G (10% Sn)
9	Valve bronze	Composition-M (6% Sn)
10	Austenitic flake graphite cast iron 3% Cr	Ni-Resist type 3
11	Austenitic spheroidal graphite cast iron	Ductile Ni-Resist
12	Heat-treated alloy steel	HY-80
13	Alloyed ductile cast iron	$1\frac{1}{2}$% Ni 0.5% Mo
14	Ductile cast iron	
15	Alloyed flake-graphite cast iron	$1\frac{1}{2}$% Ni 0.5% Mo
16	Cast carbon steel	
17	Flake-graphite cast iron	
18	Aluminum alloys	(6061-T-6 and 356-T-6)

reduced. The magnitude of the effects will be determined by several factors (Wesley, 1940) including, most importantly, the relative immersed areas of the dissimilar metals, their polarization characteristics, and the over-all resistance of the electrical circuit, including the metallic path, the water, and any films or deposits on the metal surfaces. The maximum accelerating effect occurs when the area of the more noble metal is relatively large as compared with that of the less noble metal; such combinations are most dangerous and must be avoided. It is for this reason that when insulating coatings are used to avoid galvanic corrosion, it is necessary to coat both metals in a galvanic couple. If only the less noble metal be coated, any break or imperfection in the coating will concentrate severe attack on the

TABLE 16-4
Galvanic Series of Metals in Flowing Seawater

Anodic or least noble
Magnesium and Magnesium alloys
CB75 aluminum anode alloy
Zinc
B605 aluminum anode alloy
Galvanized steel or galvanized wrought iron
Aluminum 7072 (cladding alloy)
Aluminum 5456
Aluminum 5086
Aluminum 5052
Aluminum 3003, 1100, 6061, 356
Cadmium
Aluminum 2117 rivet alloy
Mild steel
Wrought iron
Cast iron
13% Chromium steel type 410 (active)
17% Chromium steel type 430 (active)
18-8 Stainless steel type 304 (active)
18-12-3 Stainless steel type 316 (active)
Ni-resist
Lead
Tin
Muntz metal
Manganese bronze
Naval brass (60% copper 39% zinc 1% tin)
Yellow brass (65% copper 35% zinc)
Copper
Silicon bronze
Red brass (85% copper 15% zinc)
Aluminum brass
Composition G bronze
Composition M bronze
Admiralty brass
90% Copper 10% Nickel
70% Copper 30% Nickel
Nickel
Inconel (78% nickel, 13.5% chromium, 6% iron)
Nickel–aluminum bronze
Silver
Titanium
18-8 Stainless steel type 304 (passive)
Hastelloy alloy C
Monel nickel–copper alloy
Type 316 stainless steel (passive)
Graphite
Platinum
Cathodic or most noble

relatively small area of metal so exposed. A similar discontinuity in the coating on the more noble metal will have little effect.

The stainless steels are shown in two positions in this galvanic series. The more noble position represented by their passive state is the one that is hoped for. However, activity can develop within pits or in crevices with corrosion in such areas being accelerated by the galvanic action of the surrounding passive surfaces and by any metals in the galvanic series more noble than the active positions of the stainless steels. Metal less noble than the active positions of the stainless steels will give galvanic protection to the stainless steels and can prevent pitting or crevice corrosion, especially if the surfaces of the less noble metals are relatively large. In estimating possible effects of stainless steels in accelerating corrosion of other metals in galvanic couples, it should be anticipated that the stainless steels will occupy the positions of their passive states in the galvanic series.

Charts indicating the direction and extent of galvanic effects in seawater, with area effects being taken into account, may be found on pp. 420 to 426 of *The Corrosion Handbook* (Sorkin, 1965). In using such charts to predict probable behavior under conditions of atmospheric exposure, it can be assumed that the effective areas of each of the metals in a couple will be the same, irrespective of their actual dimensions.

The same principles have been used in the preparation of Figs. 16-22 and 16-23 as guides to appropriate selections of materials for fasteners and for pump and valve components, respectively.

16.9 SELECTIVE CORROSION

The internal structure of a metal or alloy sometimes influences corrosion and can lead to preferential or selective attack. There are many forms of selective corrosion including "Graphitic corrosion" or "graphitization," "dezincification," "dealuminumification,"—even "denickelification." These related phenomena are greatly intensified in seawater. Intergranular corrosion of stainless is another, but quite different, form of selective corrosion.

Cast iron generally contains about 3.5% carbon, along with silicon and other elements. Structurally it includes iron, iron carbide, and free carbon (graphite). As iron corrodes out, the graphite residue remains. The graphite retains the shape and size of the original casting, but because it consists largely of nonmetallic elements, it has little residual strength. It is soft and crumbles. Scraping the surface with a penknife is a practical field test, but care should be taken lest the knife go through the wall.

The 2 to 3% nickel gray cast irons and the ductile irons show slightly better resistance to graphitization, but cannot be justified on this basis alone. Graphitization is not a problem with the austenitic nickel cast irons.

BASE METAL	FASTENER								
	Aluminum[1]	Carbon steel	Silicon bronze	Nickel	Nickel-chromium alloys	Type 304	Nickel-copper Alloy 400	Type 316
Aluminum	Neutral	Comp.[2]	Unsatis-factory[2]	Comp.[2]	Comp.	Comp.	Comp.[2]	Comp.
Steel and cast iron	N.C.	Neutral	Comp.	Comp.	Comp.	Comp.	Comp.	Comp.
Austenitic nickel cast iron	N.C.	N.C.	Comp.	Comp.	Comp.	Comp.	Comp.	Comp.
Copper	N.C.	N.C.	Comp.	Comp.	Comp.[3]	Comp.[3]	Comp.	Comp.[3]
70/30 copper-nickel alloy	N.C.	N.C.	N.C.	Comp.	Comp.[3]	Comp.[3]	Comp.[3]	Comp.[3]
Nickel	N.C.	N.C.	N.C.	Neutral	Comp.[3]	Comp.[3]	Comp.	Comp.[3]
Type 304	N.C.	N.C.	N.C.	N.C.	May[4] vary	Neutral[3]	Comp.	Comp.[4]
Nickel-copper Alloy 400	N.C.	N.C.	N.C.	N.C.	May[4] vary	May[4] vary	Neutral	May[4] vary
Type 316	N.C.	N.C.	N.C.	N.C.	May[4] vary	May[4] vary	May[4] vary	Neutral[4]

[1] Anodizing would change ratings as fastener.
[2] Fasteners are compatible and protected but may lead to enlargement of bolt hole in aluminum plate.
[3] Cathodic protection afforded fastener by the base metal may not be enough to prevent crevice corrosion of fastener particularly under head of bolt fasteners.
[4] May suffer crevice corrosion, under head of bolt fasteners.

NOTE: Comp. = Compatible, Protected. N.C. = Not Compatible, Preferentially Corroded.

FIG. 16-22. Galvanic compatibility of fastener and base metal combinations in seawater (*Corr. 1-972*).

BODY MATERIAL	Brass or bronze	TRIM Nickel-Copper alloy 400	Type 316
Cast iron	Protected	Protected	Protected
Austenitic nickel cast iron	Protected	Protected	Protected
M or G bronze 70/30 copper-nickel alloy	May vary[1]	Protected	Protected
Nickel-copper alloy 400	Unsatisfactory	Neutral	May vary[2]
Alloy 20	Unsatisfactory	May vary	May vary

[1] Bronze trim commonly used. Trim may become anodic to body if velocity and turbulence keep stable protective film from forming on seat.

[2] Type 316 is so close to nickel-copper alloy 400 in potential that it does not receive enough cathodic protection to protect it from pitting under low velocity and crevice conditions.

FIG. 16-23. Galvanic compatibility of valve and pump trim with body materials in seawater (*Corr. 1-973*).

Copper-zinc alloys (brasses) that contain more than 15% zinc, for example, Muntz metal, naval brass, admiralty brass, and aluminum brass, are subject to a form of selective corrosion called "dezincification." The copper in the corrosion products redeposits on the corroding alloy surface as a spongy layer or plug so as to maintain the approximate original dimensions and shape but without any strength. With alloys containing less than 30% zinc, it is possible to inhibit dezincification by adding to the alloy a small amount (e.g., 0.03%) of arsenic, antimony, or phosphorus. These inhibitors are not effective with the higher-zinc-content alloys. They may also be ineffective at high temperatures, sometimes encountered in petroleum or chemical process heat exchangers.

An analogous form of attack occurs with aluminum bronzes. This can be avoided by having a minimum of 4% nickel in nickel–aluminum bronzes (Niederberger, 1964).

Denickelification of cupro–nickel alloys has been observed, in a few instances, in seawater in heat exchangers operating at high temperatures under low-velocity conditions favoring the development of local "hot pots." This can be avoided by maintaining high rates of flow, for example, over 5 ft/sec.

Stainless steels are subject to intergranular corrosion if they have been subjected to heat effects which enable carbon in excess of about 0.03% to precipitate as carbides along grain boundaries. Such adverse heating effects may result from welding, hot working, or improper annealing operations. (ASTM, 1950).

Intergranular corrosion can be avoided by

1. keeping the carbon content of the alloy below 0.03%, for example, in the "C" grades;
2. including in the alloy appropriate amounts of columbium (347 grade) or titanium (321 grade) which act as "carbide stabilizers" (Alloy 20 and Incoloy Alloy 825 are similarly stabilized);
3. quenching the alloys from temperatures high enough to keep the carbon in solution (e.g., 2050°F) after any heating operation that may have precipitated carbon.

Ways of dealing with selective corrosion are summarized in Table 16-5.

16.10 STRESS CORROSION CRACKING

Stress corrosion cracking presents a problem with high-strength steels, aluminum alloys and titanium alloys in seawater at ambient temperature and austenitic stainless steels in seawater at temperatures above about 150°F, especially if the circumstances of use permit the seawater to become concentrated by evaporation.

The behavior of the stainless steels was summarized in technical publication No. 264 of The American Society for Testing and Materials (1960).

Resistance of stainless steels to stress corrosion cracking in hot saltwater increases with nickel content, with immunity being achieved in alloys containing more than 40% nickel (Copson, 1959). Somewhat lower nickel contents (e.g., 30 to 40%) in alloy 20 and incoloy alloy 825 will suffice under many conditions of use.

Copper–nickel alloys, the nickel–copper alloy 400, and brasses containing less than 15% zinc, as ordinarily used in seawater, have not presented problems of stress corrosion cracking. There have been instances of cracking of brass tubes in heat exchangers resulting from the presence of ammonia in the vapors rather than any effect of seawater as such.

High-strength manganese bronzes are susceptible to stress corrosion cracking in salt environments, for example, when used for stressed fastenings.

The need for very high-strength alloys to cope with the stresses to which vessels may be subjected as a result of very high pressures at great depths has made it necessary to give particular attention to the resistance of such high-strength alloys to stress corrosion effects.

It has also been necessary to take into account the loss of strength of cold-worked or age-hardened alloys that results from annealing of zones

TABLE 16-5
Selective Corrosion

	Graphitization	Dezincification	Dealuminification	Denickelification	Intergranular Corrosion of Austenitic Stainless Steels
Susceptible	Cast iron Ductile iron	Copper alloys with more than 15% Zn. Examples: naval brass, admirality, aluminum brass, Muntz metal, manganese bronze	Aluminum bronzes with less than 4% Ni	70/30 copper nickel refinery condenser tubes at high temperature and low flow	Heat of welding or slow cooling of castings leads to selective attack of stainless steel in seawater but has little effect in marine atmospheres
Solutions	Use austenitic nickel cast iron	1. Use inhibited grade 2. Use alloys with less than 15% Zn. Examples: red brass, silicon bronze, tin bronze, copper-nickels	Use 4% Ni grade	Don't run condenser dry. Keep 3 fps minimum flow	1. Anneal 2. Use low carbon grades —304L, 316L, CF-4, CF-4M 3. Use stabilized grade—347 or 321 4. Avoid welding after annealing of susceptible grades

adjacent to welds. This limits the usable strength of alloys that are weakened by the heat of welding.

The limitations imposed by stress corrosion have been the principal factor in establishing the levels of stress at which it is considered to be safe to use many high-strength alloys in seawater.

Similar limitations are imposed by the susceptibility of some alloys, at certain strength levels, to embrittlement by hydrogen that may be generated by corrosion or released by the cathodic reactions in cathodic protection.

The present situation is summarized in Fig. 16-24 taken from the work of Brown and Birnbaum (1964), supplemented by recent observations on some otherwise desirable high-strength titanium alloys that indicate susceptibility to stress corrosion or hydrogen embrittlement effects not previously suspected.

It will be noted that the search must be continued for alloys usable for pressure vessels in seawater at tensile stresses above about 150,000 psi without risk of stress corrosion cracking or hydrogen embrittlement. This limitation does not apply to alloys for high-strength wires and ropes to be discussed later.

It may be possible to extend the stress limits for maraging steels by the application of cathodic protection controlled so as not to encounter hydrogen embrittlement by limiting the potential to which the alloy is polarized below that at which hydrogen evolution is the principal cathodic reaction.

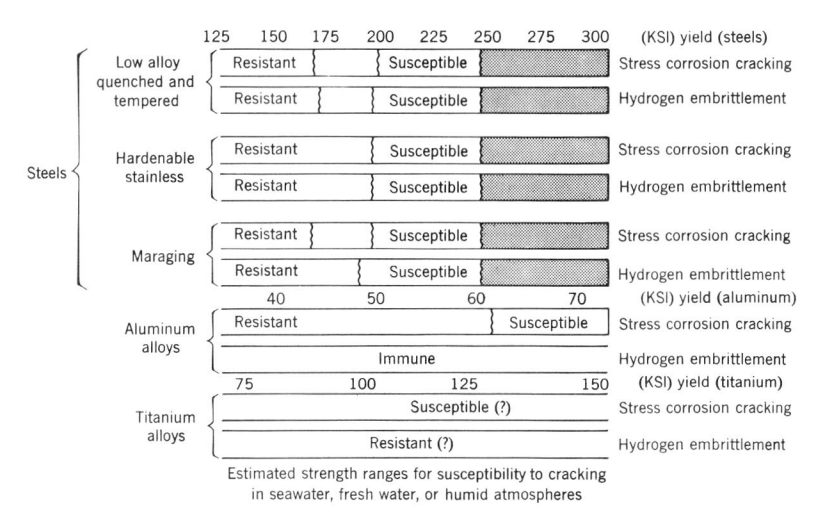

FIG. 16-24. Stress limits imposed by stress corrosion or hydrogen embrittlement in seawater (*Corr. I-974*).

The usefulness of these steels at high-strength levels may also be extended by protective coatings. Specific information on maraging steels was provided in a paper by Dean and Copson, (1965).

16.11 WIRES AND ROPES

Wires and ropes required for towing, lowering, and recovering instruments and devices must combine unusually high strength with adequate resistance to corrosion, kinking, and stress corrosion. Zinc coatings on the common high-strength "plough" steel wires have a limited life which is decreased further by effects of velocity of movement of or through the water. Aluminum coatings can be expected to be more durable than zinc but still will have a finite life.

High-strength stainless steels are subject to severe corrosion, pitting, or "tunneling" within the crevices of stranded cables. This has been prevented by coating stainless-steel wires with 90:10 or 70:30 cupronickel capable of providing galvanic protection to the base alloy and thereby preventing crevice corrosion, etc. The cupronickel alloy coatings, and particularly the 90:10 alloy, have the added advantage of suppressing fouling.

Because any weak metal or alloy coating reduces the strength of coated wires as compared with bare wires of the same dimensions there is a continuing search for a strong alloy wire capable of resisting corrosion effects without any coating. Experiments with wires made of the inconel 625 alloy have shown very promising results with respect to adequate strength and freedom from pitting, crevice corrosion and stress corrosion.

16.12 HOT SEAWATER

Most experience with metals in seawater has been at ambient temperature or the modest increase in temperature that results from the use of seawater as a cooling medium in condensers and other heat exchangers.

Current interest in conversion of seawater into fresh water by distillation processes is extending the range of temperature to much higher values.

To insure maximum recovery of heat, these distillation processes employ evaporation under high vacuum and with deaeration to reduce corrosion and improve heat transfer. At the higher temperatures, it is necessary to reduce the pH of the water to prevent carbonate scaling.

A number of observations have been made of the performance of tubes in service and in pilot-plant evaporators and of test coupons exposed in operating plants. More systematic studies of effects of temperature, pH, deaeration, and velocity are in progress. These will

Materials Selection for Ocean Engineering 623

add greatly to knowledge of how materials can be expected to perform in these plants.

Representative data already available from the tests in operating plants are provided by Fig. 16-25 for irons and steel and Fig. 16-26 for copper alloys. It is evident that at the higher temperatures, deaeration does not reduce corrosion of ordinary steel and cast iron as much as had been hoped. The austenitic Ni-resist cast irons performed well at all temperatures. Indications from these tests and a great deal of practical experience with the copper base alloys, and particularly the cupronickel alloys, support the conclusion that they will perform well in saltwater distillation equipment. Monel nickel–copper alloy 400 has also been used with excellent results in this service.

Behavior of aluminum is sensitive to the presence of corrosion products of heavy metals. In their absence, aluminum, and particularly the clad-aluminum alloys favored for such service, can be expected to demonstrate good resistance to corrosion with the exception of pitting, which was aggravated at temperatures above 290°F in tests reported by Bohlman and Posey (1965). The 5454 alloy was best. This investigation also disclosed that titanium is subject to rapid attack in crevices at temperatures over 212°F. Similar crevice attack of titanium has been observed in heat exchangers handling sea water at 250°F. Further research will be required

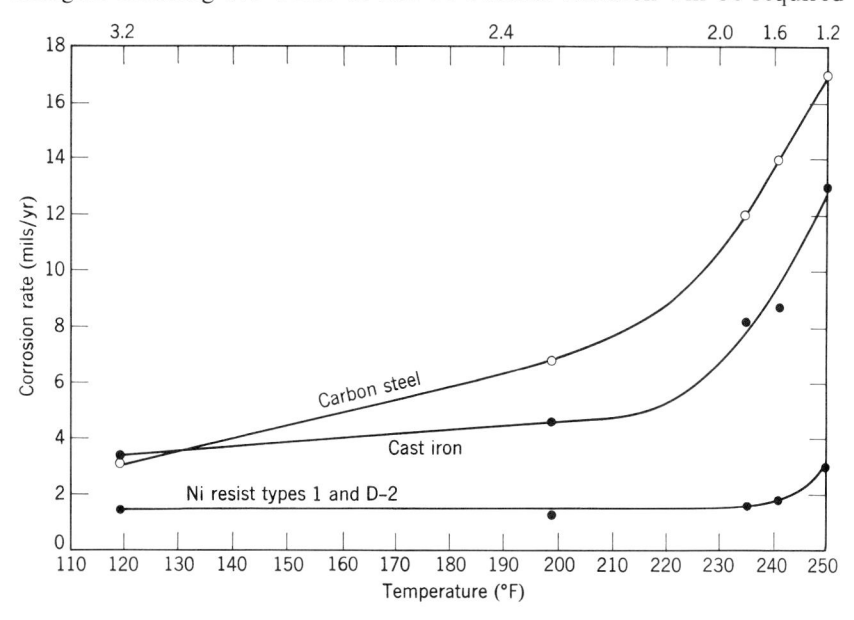

FIG. 16-25. Effect of temperature on corrosion rate of ferrous alloys in deaerated seawater brine (*Corr. I-961*).

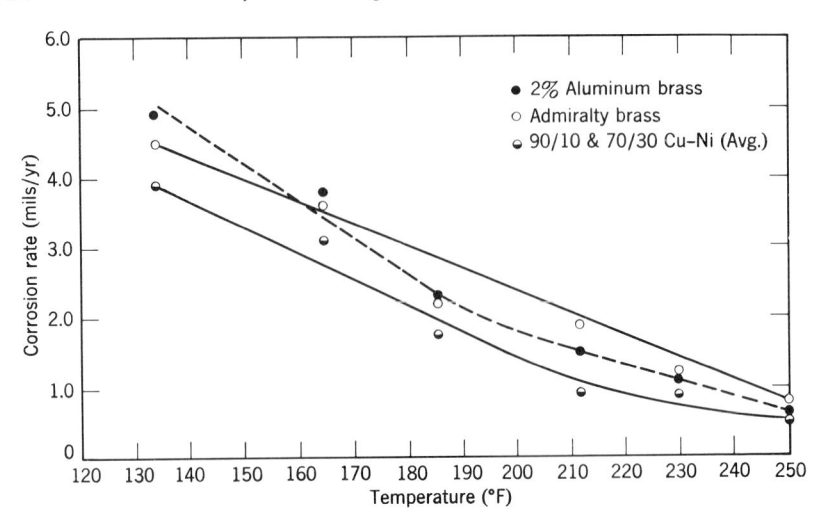

FIG. 16-26. Effect of temperature on corrosion rates of copper-base heat-exchanger tube materials in deaerated normal seawater (*Corr. 1-940*). Continuous exposure: 156 days: dissolved oxygen level: 40 to 600 ppb; pH: 6.2 to 7.8; brine conc. factor: 0.8 to 1.1; Turbulence: moderate – low at 250°F.

to establish the limiting temperatures above which pitting and crevice corrosion of aluminum and titanium may become troublesome.

In addition to corrosion by the water itself, attention will need to be paid to possible corrosive effects of vapors that may be released from polluted waters and which may aggravate corrosion on the steam side of tubes. The greater tolerance of cupronickel for ammonia in steam will give it an advantage over other copper base alloys under such circumstances.

16.13 ATMOSPHERIC CORROSION OF NONFERROUS METALS AND ALLOYS

With the exception of high-zinc-content brasses, which may suffer dezincification in heavily salt-laden, warm, humid atmospheres, none of the highly alloyed materials will suffer significant structural damage in marine atmospheres.

For example, atmospheric tests carried out by ASTM at La Jolla, California for 20 years (ASTM, 1956), showed the rates of attack and losses of strength and ductility for representative materials shown in Table 16-6.

The preferred aluminum alloys for use in marine atmospheres are identified by designations AA1150, AA3003, 3004, AA5082, 5083, AA4043 and AA6061, 6063. The alloys that contain over about 0.5% copper require protection by cladding with a less-noble aluminum composition.

Nickel–chromium alloys and austenitic stainless steels suffer no measurable loss of weight or strength properties after exposure for long periods in aggressive marine atmospheres (ASTM, 1956; 1946; Copson and Tice, 1964). The nickel-base chromium alloys retain their original appearance with little loss of luster and no significant loss of strength or ductility. The stainless steels acquire superficial rust tarnish films, least in the case of the type-316 alloy, which contains molybdenum in addition to chromium and nickel. The marginally austenitic 301 alloy in the sensitized condition is susceptible to stress-corrosion cracking. Spot-welded specimens of the 301 alloy in the heavily cold-worked condition have exhibited no evidence of stress corrosion in salt atmospheres.

16.14 BIOLOGICAL DETERIORATION

References have been made previously to indirect effects of microorganisms (slimes and sulfate-reducing bacteria) and microorganisms (barnacles, etc.,) on the corrosion of metals. Organisms can also have direct destructive effects.

The boring organisms, for example, Teredo, Pholads (Martesia), and Limnoria can be very destructive to substances that they are able to penetrate either for shelter or for sustenance. Some are able to penetrate and destroy manila, and cotton ropes, jute fibers, soft rocks, and weak grades of concrete. They can also enter certain plastics; for example, Martesia have been observed to bore into methyl-methacrylate rods in surface waters and borers identified as *Xylophaga Washingtonia*, Bartsch

TABLE 16-6
Results of Exposure of Tension Test Specimens to Marine Atmosphere at La Jolla, California for 20 Years

Metal	Average Penetration (mils/year)	Loss in Tensile Strength (%)
Copper	0.05	18.0
Silicon bronze	0.05	9.7
Tin bronze (8% Sn)	0.09	13.2
Aluminum bronze	0.006	5.3
Aluminum (1100)	0.025	30.5
Duralumin (2017)	0.098	55.8
Manganese bronze	0.35	100.0
Red brass (15% Zn)	0.01	3.5
Yellow brass (30% Zn)	0.006	11.8
Admiralty brass (29% Zn, 1%Sn)	0.08	19.8
Cupronickel (19% Ni, 1% Sn)	0.01	4.6
Monel nickel–copper alloy 400	0.006	0.1
Nickel	0.006	0.0

and *X Duplicata* were found to have penetrated woods and such plastics as teflon, nylon, and vinyl (unplasticized polyvinylchloride), and cellulose acetate after they had gotten a start in wood attached to the plastics. Perhaps the bases of fouling organisms can provide a stage from which boring organisms can enter plastics. Such attack was observed at a depth of 5300 ft off Port Hueneme, California for three years (Muraoka, 1965). Examples of borer attack in plastics and manila rope are shown in Fig. 16-27 and 16-28.

Nylon, saran, polypropylene, and polyethylene fiber ropes are greatly superior to manila ropes in resistance to attack by marine organisms. Manila ropes may lose practically all of their strength in a few weeks as a result of attack by marine organisms (Fig. 16-28). However, manila is superior to nylon in resisting deterioration above the water.

In the tests described the following materials were able to resist attack by marine organisms: rubber tubing, coral concrete, acrylic sheet, saran and polyethylene films, nylon, and ethylcellulose. The following cable-insulating materials: neoprene, butyl rubber, natural rubber, teflon, polyethylene, fep, bakelite, pvc, and nylon, were satisfactory. Silicone-rubber insulation and friction tape were susceptible to biological destruction. Insulating coatings lost a great deal of their insulating properties, for example, up to 90% loss after immersion for three years in the tests at 5300 ft.

Woods capable of receiving a heavy preservative creosote treatment will resist attack by borers with the noteable exception of Limnoria, which are able progressively to attack surface layers from which creosote has been leached. This problem can be dealt with by applying a protective plastic or metallic (e.g., 90:10 cupronickel) sheathing to creosoted piling. The sheathing retards leaching of the creosote and, in the case of cupronickel sheathing, creates an environment next to the wood made even less favorable to growth

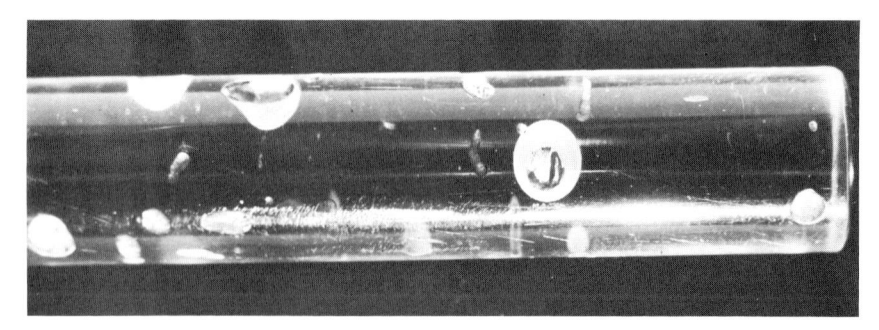

FIG. 16-27. Pholads in methylmethacrylate (*Corr. IV-848*) (Courtesy William F. Clapp Laboratories.)

of destructive organisms by the copper compounds leached from the cupronickel alloy.

16.15 FIBERGLASS REENFORCED PLASTICS

Fiber-glass re-enforced plastic (epoxy) vessels and structures appear to be immune to effects of organisms. However, there may be some degradation of cut filaments exposed at joints or vessel penetrations since the strength of glass fibers is reduced greatly by moisture. The principal problems in using glass re-enforced plastics will be in fabrication and the avoidance of voids (Kies and Wolock, 1964).

16.16 MASSIVE GLASSES

High-silicate glasses are attractive for applications in vessels for use at great depths (Perry, 1964) in which advantage can be taken of the fact that hollow glass structures develop improved resistance to damage by mechanical impact under the influence of the pressures encountered at great depths.

Such glasses are not significantly affected by the corrosive action of seawater nor by marine organisms. They are given treatments that leave the surface layers in compression so as to avoid surface tensile stresses under

FIG. 16-28. Destruction of manila rope by *Teredinidae* in three months at Guam (*Corr. IV-849*). (Courtesy William F. Clapp Laboratories.)

load which would lead to the customary brittle failure of glass when bent or under tension. The principal problems in the use of massive glass structures in ocean engineering will be in the areas of design and fabrication and the protection of very large vessels by treatments to achieve the surface layers in compression that are necessary.

16.17 CONCRETE

High-strength concrete (in excess of 5000 psi compressive strength) should be adequately durable in seawater (Klieger and Gustaferro, 1964) with maximum resistance to weakening by leaching of lime and attack by sulfates being achieved by limiting the tricalcium content of the cement. To avoid corrosion of conventional steel re-enforcement, the concrete cover should have a minimum thickness of $2\frac{1}{2}$ to 3 in. Prestressing with small-diameter wires is a questionable practice for seawater installations. The most aggressive deterioration influences are encountered in areas subject to alternate wetting and drying in the tidal and splash zones.

16.18 CONCLUSION

Although some problems remain and it will be necessary to pay close attention to definitions of performance requirements and the matching of material properties to these requirements, it seems likely that the availability of suitable materials will be able to keep pace with other advances in ocean engineering.

16.19 APPENDIX: IDENTIFICATION OF ALLOYS REFERRED TO IN TEXT

Irons and steels

Low-alloy high-strength steels. Steel alloyed with small percentages of nickel, chromium, copper, columbium, phosphorus in various combinations to provide yield strengths from 50,000 to 100,000 psi.

HY-80. Alloy steel with 2.75% Ni, $1\frac{1}{4}$% Cr and 0.30% Mo to provide yield strength of 80,000 psi.

Maraging steel. 17% Ni, 6% Co, 3% Mo.

Ductile cast iron. Cast irons treated with magnesium to achieve graphite in a spheriodal rather than a flake form.

Ni-resist. A family of cast irons containing enough nickel or nickel plus copper (in excess of 20%) to achieve an austenitic structure.

Ductile Ni-resist. Austenitic cast irons with spheroidal graphite.

Copper alloys

Muntz Metal. 60% Cu, 40% Zn.
Admiralty brass, 443, 444, 445. 70% Cu, 29% Zn, 1% Sn, plus As, Sb, or P.
Aluminum brass 687. 77% Cu, 21% Zn, 1% Sn plus 0.1% As.
Red Brass 230. 85% Cu, 15% Zn.
Silicon bronze. 2–3% Si, 1% Mn, balance Cu.
Composition G. bronze. 88% Cu, 10% Sn, 2% Zn.
Composition M bronze. 88% Cu, 6% Sn, 4.5% Zn, 1.5% Pb.
Manganese bronze. 58% Cu, 39% Zn, 1% Fe, 1% Al, 0.25% Mn.
Nickel–aluminum bronze (Ni Bral). 74% Cu, 10% Al, 4% Ni, 4% Fe, 3.5% Mn.
Nickel–aluminum manganese bronze. 74% Cu, 12.5% Mn, 8% Al, 3% Fe, 2% Ni.
90:10 copper nickel alloy 706. 88% Cu, 10% Ni, 1.5% Fe.
70:10 copper nickel alloy 715. 69% Cu, 30% Ni, 0.5% Fe.
70:30 copper nickel alloy 7152. 65% Cu, 30% Ni, 5% Fe.

Nickel–base alloys

Monel nickel–copper alloy 400. 66% Ni, 32% Cu.
Monel nickel–copper alloy K-500. 66% Ni, 3% Al, 0.5% Ti, balance Cu.
Incoloy nickel chromium alloy 825. 40% Ni, 21% Cr, 3% Mo, 2% Cu, balance Fe.
Inconel nickel chromium alloy 600. 76% Ni, 16% Cr, 8% Fe.
Inconel nickel chromium alloy 718. 53% Ni, 19% Cr, 5% Cb, 3% Mo.
Inconel nickel chromium alloy 625. 66% Ni, 21% Cr, 3.5% Cb, 9% Mo.
Hastelloy alloy C. 59% Ni, 16% Mo, 16% Cr, 4% W, 5% Fe.
Stellite [6]. 65% Co, 25% Cr, 10% W.

Stainless steels

Type 410. 13% Cr, balance Fe.
Type 430. 17% Cr, balance Fe.
Type 304. 18% Cr, 9% Ni, balance Fe.
Type 321. 18% Cr, 11% Ni, 0.5% Ti, balance Fe.
Type 347. 18% Cr, 12% Ni, 0.8% Cb, balance Fe.
Type 316. 18% Cr, 12% Ni, 3% Mo, balance Fe.
17–4 PH[7] 17% Cr, 4% Ni, 3.5% Cu.
CF-8. Cast equivalent of type 304.
CF-8M. Cast equivalent of type 316.
Alloy 20. 20% Cr, 29% Ni, 2% Mo, 3% Cu.

[6]Trademark, Union Carbide Corporation.
[7]Trademark, Armco Steel Corporation.

Aluminum Alloys

1100. 1% Si + Fe, 0.2% Cu.
2117. 2.5% Cu, 0.2% Mn, 0.3% Mg, 0.25% Zn, 0.8% Si, 1% Fe.
3003. 1.3% Mn, 0.6% Si, 0.7% Fe, 0.2% Cu.
3004. 1.3% Mn, 0.3% Si, 0.7% Fe, 0.25% Cu.
4043. 5% Si, 0.8% Fe, 0.3% Cu
5052. 2.5% Mg, 0.25% Cr, 0.45% Si + Fe, 0.10% Cu.
5083. 4.5% Mg, 0.15% Cr, 0.4% Si, 0.4% Fe, 0.10% Cu,
5086. 4.0% Mg, 0.15% Cr, 0.4% Si, 0.5% Fe, 0.10% Cu.
5454. 2.7% Mg, 0.15% Cr, 0.4% Si + Fe, 0.10% Cu.
5456. 5% Mg, 0.15% Cr, 0.4% Si + Fe, 0.2% Cu.
6061. 1% Mg, 0.25% Cr, 0.6% Si, 0.7% Fe, 0.3% Cu.
6063. 0.5% Mg, 0.1% Cr, 0.4% Si, 0.4% Fe, 0.1% Cu.
7072. 1% Zn, 0.7% Si + Fe, 0.10% Cu.
356. 7% Si, 0.3% Mg, 0.5% Fe, 0.20% Cu.
CB 75 (anode). 7% Zn.
B 605 (anode). 5% Zn.

REFERENCES

Ambler, H. R., and A. A. J. Bain, 1955, "Corrosion of Metals in Tropics," *J. Appl. Chem.* **5**, 437 (September 1955).

American Welding Society, 1962, "Corrosion Tests of Metallized Coated Steel 6-year Report," AWS C 2.8-62.

Anderson, D. B., and B. R. Richards, 1965, "Chlorination of Sea Water—Effects on Fouling and Corrosion," Paper before ASME Research Committee on Condenser Tubes, ASME Annual Winter Meeting, Chicago, Nov. 8, 1965.

Anderson, E. P., and D. B. Anderson, 1965, "Platinum Anode Systems for Cathodic Protection," Paper given during "Corrosion Week" at Free University of Brussels, June 11–15, 1965.

Bohlmann, E. G., and F. A. Posey, 1965, First International Symposium on Water Desalination Washington, D. C., October 3–9, 1965, Paper #SWD/53.

Brown, B. F., and L. S. Birnbaum, 1964, in W. S. Pellini, ed., U.S. Naval Res. Lab., Rep. No. 6167, Nov. 4, 1964, p. 222.

Copson, H. R., 1959, in T. N. Rhodin, ed., *Physical Metallurgy of Stress Corrosion Fracture*, Interscience Publishers, New York, p. 247.

Copson, H. R., and C. P. Larrabee, 1959, "Extra Durability of Paint on Low Alloy Steels," *ASTM Bull*, **68** (December 1959).

Copson, H. R., and E. A. Tice, 1964, "Investigation of Atmospheric Corrosion Behavior of Selected Nickel Alloys," *Werkstoffe Korrosion* **15**, 645 (1964).

Dean, S. W., and H. R. Copson, 1965, "Stress Corrosion Behavior of Managing Nickel Steels in Natural Environments," *Corrosion* **21**, 95 (Mar. 1965).

Eisenberg, P., H. S. Preiser, and A. Thiruvengadam, 1965, Soc. Naval Arch. and Marine Eng., Annual Meeting, November 1965, Paper No. 6.

Ffield, P., 1945, "Recommendations for Using Steel Piping in Salt Water Systems," *J. Am. Soc. Naval Eng.* **57**, 1, 1 (February 1945).

Graham, D. P., F. E. Cook, and H. S. Preiser, 1956, *Trans. Soc. Naval Arch. Marine Eng.* **64**, 241 (1956).

Holl, J. W., and G. M. Wood, eds., 1964, *Symposium on Cavitation Research Facilities and Techniques*, ASME, New York.

Kies, J. A., and I. Wolock, 1964, in W. S. Pellini, ed., U.S. Naval Res. Lab., Rep. No. 6167, November 4, 1964, p. 54.

Klieger, P., and A. H. Gustaferro, 1964, in W. S. Pellini, ed., U.S. Naval Res. Lab., Rep. No. 6167, November 4 1964, p. 120.

LaQue, F. L., 1950, *Drilling* **11**, 29 (June 1950).

LaQue, F. L., and W. F. Clapp, 1945, "Relationship Between Corrosion and Fouling of Copper-Nickel Alloys in Sea Water," *Trans Electrochem. Soc.* **87**, 103 (1945).

Larrabee, C. P., 1958, "Corrosion Resistant Experimental Steel for Marine Applications," *Corrosion* **14**, 501t (November 1958).

Litchman, J. Z., D. H. Kallas, C. K. Chatten, and E. P. Cochran, Jr., 1961, "Cavitation Erosion of Structural Materials and Castings," *Corrosian* **17**, 497t, (October 1961).

Muraoko, J. S., 1965, *Bioscience* **15**, 191, (1965). See also U. S. Naval Civil Eng. Lab., Techn. Rpt. R–428, February 1966.

Niederberger, R. B., 1964, *Trans. AFS* **75**, 115 (1964) Also in *Modern Castings* **45**, 115 (March 1964).

Perry, H. A., 1964, in W. S. Pellini, ed., U.S. Naval Res. Lab., Rpt. No. 6167, Nov. 4, 1964, p. 100.

Reinhart, F. M., 1954, U.S. Naval Civil Eng. Lab. Tech. Note N-605 (AD#601, 892) (June 1964), 13 pp; See also, *Geo-Marine Tech.* **1**, 19 (September 1965).

Rheingans, W. J., 1957, "Resistance of Various Materials to Cavitation Damage" Report of 1956 Cavitation Symposium, ASME, NewYork.

"Report on Stress Corrosion Cracking of Austenitic Chromium-Nickel Stainless Steels," 1960, ASTM Spec, Tech. Publ. No. 264.

Schreir, L. L., ed., 1963, *Corrosion*, 2 vols., George Newnes Ltd., London, p. 11.27.

Symposium on Atmospheric Corrosion of Non-Ferrous Metals, 1956, ASTM Spec. Tech. Publ. No. 175.

Symposium on Atmospheric Weathering of Corrosion Resistant Steels, 1946, *Proc. ASTM* **46**, 593–677 (1946).

Symposium on Evaluation Tests for Stainless Steels, 1950, ASTM Spec. Tech. Publ. No. 93.

Turner, H. J., Jr., D. M. Reynolds, and A. C. Redfield, 1948, *Ind. Eng. Chem.* **40**, 450 (March 1948).

Tuthill, A. H., and C. M. Schillmoller, 1966, "*Guidelines for Selection of Marine Materials*," Paper before Ocean Science and Ocean Engineering Conference Marine Technology Society, Washington, D. C., June 14–17, 1965; International Nickel Co., Inc., 1966, 37 pp.

Uhlig, H. H., ed., *Corrosion Handbook*, 1948, Sponsored by the Electrochemical Society, Wiley, New York, pp. 383–446.

Volkening, V. B., 1950, *Corrosion* **6**, 123 (Apr. 1950).

Wesley, W. A., 1940, *Proc. ASTM* **40**, 690 (1940).

SUPPLEMENTARY REFERENCES

In addition to the specific references mentioned in the text there are many other sources of pertinent information that will be helpful. These are listed here.

Probably the best source of additional information to supplement the first step provided by this paper in making a choice of materials for specific applications will be the producers of the materials indicated as being worthy of such further, more detailed, consideration.

Ailor, Jr., W. H., and F. M. Reinhart, 1963, "Ten Years Weathering of Aluminium Alloys," *Mat. Protection* **2**, 6, 30 (1963).

Copson, H. R., 1945, "A Theory of the Mechanism of Rusting of Low Alloy Steels in the Atmosphere," *Proc. ASTM* **45**, 554 (1945).

Craven, J. P., and H. Bernstein, 1965, "Materials for Deep Submergence," *Mat. Res. Std.* **551** (November 1965).

Drisko, R. W., and C. V. Brouiilette, 1966 "Comparing Coatings in Shallow and Deep Ocean Environment," *Mat. Protection* **5**, 4, 32 (1966).

Evans, U. R., 1948, *Metallic Corrosion, Passivity and Protection*, Longmens Green and Company, New York.

Harper, R., 1961 *Aluminium Alloys in Marine Environments*, Met. Ind., London, p. 454.

Krenzke, M. A., 1965 "Structural Aspects or Hydrospace Vehicles," *Naval Eng. J.* **77**, 4, 697 (1965).

LaQue, F. L., 1941 "The Behavior of Nickel Copper Alloys in Sea Water," *J. Am. Soc. Naval Eng.* **53**, 1, 30, (1941).

LaQue, F. L., and H. R. Copson, (eds.), 1963 *Corrosion Resistance of Metals and Alloys*, 2nd ed., Reinhold, New York, 712pp.

LaQue, F. L., and A. H. Tuthill, 1961, "Economic Considerations in the Selection of Materials for Marine Applications," *Trans. Soc. Naval Arch. Marine Eng.* **69**, 619–639 (1961).

May, T. P., E. G. Holmberg, and J. Hinde, 1962, "Sea Water Corrosion at Atmospheric and Elevated Temperature," *Sonderd. Dechema-Monographien* **47**, 253–274, (1962).

May, T. P., and B. A. Weldon, 1964, "Copper Nickel Alloys for Service in Sea Water," Int. Gongr. on Fouling and Marine Corrosion, Cannes, France, June 1964.

Muraoka, J. S., 1966, "Effect of Deep Sea Micro-Organisms on Rubber and Plastic Insulation," *Mat. Protection* **5**, 4, 35 (1966).

Reinhart, F. M., 1966, "Corrosion of Materials in Hydrospace," U.S. Naval Civil Engineering Laboratory, Port Hueneme, California, T.R. (R504) (December 1966).

"Report Sub-Committee VI on Atmospheric Corrosion," 1962, *Proc. ASTM* **62**, 216 (1962).

Rigo, J., 1966, "Corrosion Resistance of Stranded Steel Wire in Sea Water," *Mat. Protection* **5**, 4, 54, (1966).

Rogers, T. H., 1960, *The Marine Corrosion Handbook*, McGraw-Hill Company of Canada, Ltd., Toronto, 1960.

Schorsch, E., and C. Garland, 1965, "Maraging Steel and Other Competitive Materials for Deep Submersibles," *Soc. Naval Arch. and Marine Eng.*, Philadelphia Section, December 1965, Sun Shipbuilding and Dry Dock co., Document S 40–3.

Schreir, L. –L., ed., 1963, *Corrosion Handbook*, Vol. 2, George Newnes Ltd., London, p. 11.27.

Sorkin, G., 1965, "Materials for Submarine Hard Sea Water Systems," *Naval Eng. J.* **77**, 1, 93 (February 1965).

Uhlig, H. H., ed., 1948 *Corrosion Handbook*. Wiley, New York.

CHAPTER 17

Hydrospace-Environment Simulation

J. D. STACHIW

The success and reliability of any engineering structure in a given environment depends basically on three parameters. First, it depends on a knowledge of general engineering principles by the designer; second, the accurate prediction of the loading conditions that the mission of the engineering structure will have imposed on it; and third, the accumulated information available on the performance of materials in a particular structure under specific conditions in a given environment. The knowledge of the necessary engineering principles is readily acquired today by either the college-trained engineer or the self-educated person because, in addition to the enormous institutionalized education available, there exists a vast store of knowledge printed in engineering books and manuals which present, in easily digested form, the physical and chemical principles that govern the behavior of matter in an engineering structure. The accurate prediction of the loading conditions is predicated on the concise description of the mission that the engineering structure must fulfill, and on the detailed knowledge of parameters that govern the environment where the structure will operate. The concise description of the loading can be generally ascertained by a thorough analysis of the mission of the structure. The parameters that govern a given environment, however, can be understood only through scientific explorations of that particular environment. In terms of ocean engineering, this means that only through the actual exploration of hydrospace can we understand the forces that govern that space. Although some measure of understanding can be gained in the laboratory by the construction and operation of hydrospace models, the contribution of these models is directly proportional to the knowledge of hydrospace which the model's designer already possesses. Thus, before truly sophisticated models of hydrospace can be built, a thorough exploration of hydrospace must first take place.

The information on the behavior of materials in hydrospace can be acquired either through the observation of structures exposed to hydrospace, or by the subjection of materials to hydrospace conditions simulated on land inside a pressure vessel. The observation of material behavior in hydrospace presupposes a long time period during which numerous operational engineering structures penetrate the hydrospace and their performances are noted after successful return to land. Such an approach to the gathering of information on material behavior is time consuming and will not generate sufficient data rapidly enough to satisfy the requirements imposed on ocean engineering by the defense needs of our country. In the recent past the time-honored technique of build–fail–observe was used to increase knowledge in a particular area no matter how long it took. Now it is no longer feasible to use this technique as our times require that the engineer know, prior to building working vehicles, many of the complex parameters that contribute to failures so that the use of time may be optimized. It is specifically in the area of gathering data on the behavior of materials and structures under hydrospace conditions that environmental simulation can prove to be the golden key to the unlocking of hydrospace for man. If most of the failures of hydrospace structure materials and components take place in a hydrospace environment simulation facility and not in actual hydrospace, then the confidence of the hydronauts using such structures will amply reward all the hydrospace environment testing expenses by the rapid exploration and settling of the hydrospace, unhampered by fear of equipment failure.

17.1 HISTORY OF HYDROSPACE-ENVIRONMENT SIMULATION

Hydrospace-environment simulation, although a relatively new specialization, has its roots deep in classical mechanics and shares with the other fields of engineering the rich heritage handed down from the days of the Renaissance. The history of this field can be divided conveniently into two facets. One deals with the theoretical knowledge of the behavior of materials, with the unstated assumption that the environment is known and accounted for. The other facet deals with actual hydrospace environment simulation.

Da Vinci is usually considered to be the first modern man to be concerned with structural engineering (Timoshenko, 1963). His early attempts to apply statics in finding the forces which act in the various members of structures represented a radical departure for his day. Prior to his work, builders in all fields had been using the simple empirical tests of "Does it work" and

"Has it worked before" to guide them in their labors. Da Vinci, by experimentation, attempted to secure knowledge of probable results before actual construction. Although his results were often questionable, his approach laid a foundation upon which others built. Galileo was the next to apply experimentation and from his work dates the history of the mechanics of elastic bodies and dynamics as they are known today. Using and building on the work that had been done before, Robert Hooke devised his famous law which describes the linear relation between the force acting on a structural member and the subsequent deformation. Both da Vinci and Galileo had worked with experimentation as a tool for solving actual problems. Hooke, however, although interested in actual problems, was more interested in generalized theories and from this point on and for a number of years the greater emphasis in the field of mechanics was on theory and mathematical models.

Euler, Lagrange, Coulomb, and Navier all worked with mathematical models. Euler introduced calculus to the science of moving bodies, and thereby added a new tool to the study of mechanics. In the hands of Lagrange "mechanics became a branch of analysis which he called the 'geometry of four dimensions'." (Timoshenko, 1963, p. 37) Both Coulomb and Navier studied the behavior of elastic bodies, Navier devising the mathematical theory of elasticity. The theories these men devised in the eighteenth century were to provide the basis for the advances of the nineteenth century.

France had been the center for the superb mathematical advances in mechanics, especially in the eighteenth century. With the advent of the nineteenth century the work in mechanics began to disperse from the French fountainhead. However, many French scientists and engineers were still found in the ranks of the important. Cauchy introduced his theory of stress and strain to elasticity and, in addition, developed the equations necessary for solving problems of the elasticity of isotropic bodies. Lamé developed formulas for the stresses produced by uniform internal and external pressures and, extremely important to hydrospace-environment simulation, showed that the ultimate pressure a vessel can withstand cannot be any greater than the strength of the material that comprises the vessel's walls. Jourawski developed a method for determining shear of beams that is very useful today in certain cases related to thin-walled structures. Saint-Venant, unlike many of his fellow Frenchmen, began to argue that theoretical and experimental work must be combined in order for lasting advances to be made in the field of strength of materials. Because Saint-Venant followed his own advice, his principle regarding bending is still used by today's engineers when analyzing stresses in structures.

Wöhler, a German engineer of the period, was instrumental in establishing material testing laboratories which helped to bring about uniformity in the mechanical testing of metals. (Timoshenko, 1963, pp. 167–168.) Kirkaldy in England performed the same service for his country in addition to discovering that the shape of a specimen being tested determines the experimental results obtained.

In America the emphasis continued to be on the empirical test of "Does it work?" However, from the late 1700's on there was a growing interest in the theory of mechanics and many inventors and experimentors who wrote books on their work discussed these problems (Stachiw, 1963). These books were usually based on experimentation with generalized conclusions rather than on mathematical models as was popular with the French writers. Because most of the important American engineers of the day were self-taught men, they usually lacked the mathematical education necessary for an understanding of the elaborate equations devised by the European theoreticians. Therefore in America experimentation, preferably with full-size working models, became much more widespread than mathematical theories devoid of actual hardware.

In addition to the testing of materials and the growing use of mathematical models, some scientists were interested in the development of new tools for exploring strength of materials. Noteworthy in this area was James Maxwell's work with photoelastic strain gages. With the theory of frozen stresses and the use of polarized light, Maxwell made many contributions to the understanding of two-dimensional problems. Today these basic ideas have been expanded to the study of three-dimensional problems also.

Electric strain gages and sophisticated machines for the testing of creep also contributed to a more complete understanding of the strength of materials, so that by the twentieth century the basic tools and theories for stress analysis and strength of materials had already been achieved. On the bedrock of these tools and concepts, hydrospace-environment simulation was to be built.

Long before hydrospace-environment simulation for deep-ocean engineering became possible or necessary, other specializations had found the need to study the reaction of environment on models, and the parameters important to that environment. One of the oldest of these specializations is hydraulics which, through the work of Toricelli, and later Pascal, added much knowledge on pressure and vacuum. Both of these men, as well as others who became interested in this field, saw the necessity of building models, often full-scale models, to test the appropriateness of their theories and to gain new knowledge on the relationship between their models and the environment in which their models were to work. They found, for example, that a specific pump did one kind of work at sea level and quite another kind on a mountain top. Because of such experimentation hydraulics has a long history of environment simulation woven into its contributions to science.

Another very old specialization which has used models and environment simulation is the field of naval architecture. In the seventeenth century, it was decreed in France that before any ship was to be built for the royal navy that a scale model of that ship must first be built. Many of these beautifully executed models are today the prize of collectors.

Once a scale model had been built, it was but a short step to the testing of the behavior of that model in its planned environment—water. At first only a crude attempt was made to survey the reaction of model to water and wind. But as the materials for building the ships became better understood so did the behavior of the models. The first truly scientific attempt to control the variables occurred in England in 1872 when William Froude established his model basin. The basin was to be used to test the interaction of ships and waves in the hope that ship building could be improved. The American naval officer David Taylor became interested in this type of research and in 1889 won the support of the U.S. Congress in establishing a model testing basin in Washington, D.C. At this time, however, the models that were being tested were all surface crafts and the facilities were relatively primitive affairs in comparison to today's water tunnels. These tunnels or model basins are basically of two types. In one type, the true water tunnel, the model to be tested is held at rest by various means and the water is sent past the model at controlled velocities. In the other type, the towing tunnel, the water remains quiet and the model is towed through the water. Through the use of model basins, naval architects have gained much new and useful knowledge on the behavior of bodies moving through water and thus, by using this knowledge have greatly improved the design of modern ships.

Another group of scientists who have used model testing and environment simulation extensively are the aerodynamicists. Almost from its inception, aerodynamics has relied heavily on scale models tested in wind tunnels to ascertain the behavior of planned vehicles. Because of the interest in this facet of research, the aerodynamicist of today can more nearly optimize his design, construction method, and material selection and, therefore, his vehicle performs more predictably and reliably, thus making the actual testing of a prototype vehicle much safer.

With the background of what other fields had accomplished with environment simulation and models, it was only natural that those designers and builders of hydrospace vehicles should also turn to this type of research. The first submarine that could be described as successful was a submarine designed and built by Holland around the turn of this century. Holland, like those designing surface crafts, tested his models in a model basin. However, Holland's submarines were not expected to withstand large hydrostatic pressures and thus his main preoccupation was with the behavior of his model on the surface. When we consider that until very recently submarines were only expected to reach operational depths of 200–500 ft, which represent pressures of approximately 100–250 psi, and that these submarines were expected to remain submerged only a very short time, then we can appreciate the fact that Holland and the other early designers thought of their vessels as more surface crafts than submersibles. Only since the development of atomic power has the true submersible come into being.

However, by the 1920's hydrospace engineers were seeing the necessity of more extensive testing of submarine hulls, torpedoes, mines, hydrophones, transducers, and structural members of the submarines. These engineers found their basic tool to be the pressure vessel, now adapted, refined, and extremely complex. One man who helped advance the usefulness of the pressure vessel was Bridgman, a professor at Harvard. As early as 1905, he was working with pressures as high as 256,000 psi, and by 1940, he reached 6,000,000 psi (Bridgman, 1964). Although these pressures are much larger than those used by today's hydrospace engineers, the seals, pressure-vessel design, end closures and pumps used in vessels such as Bridgman's have made possible the advances in the basic tool for hydrospace-environment simulation.

The pressure vessel, however, is not the only component necessary to hydrospace environment simulation. Pressure is but one facet of that mystical area called the ocean. Temperature, salinity of the water, presence of marine organisms, movement of the water, and the presence of dissolved gases and minerals are all factors which the ocean engineer must contend with in actual experience and which the hydrospace environment simulation facility must attempt to duplicate in varying degrees. The successful coupling of all the components to simulate hydrospace forms an area of vital importance to the work done today with hydrospace.

17.2 HYDROSPACE-ENVIRONMENT SIMULATION

The simulation of the hydrospace environment can best be thought of in terms of the testing of hardware on land in order to assure the safe and reliable performance of that hardware in the depths of the ocean. However, since the land surface is not the ocean depths only simulation can take place. The accent is on *simulation*, as an attempt to create ocean conditions without having at one's disposal the ocean.

Simulation of hydrospace is really a misnomer, for the emphasis is not on simulation of hydrospace, as such, with its topography, currents, fauna and flora, and reactions to weather changes, but on simulation of the hydrospace forces that have an influence on the performance of hardware. Such forces as pressure, dynamic impulse, chemical attack of seawater, forces of underwater currents, to mention a few, are what the modern hydrospace-environment-simulation facility hopes to simulate within the confines of a pressure vessel with a limited volume and devoid of many characteristics exhibited by the ocean.

There are basically three concepts that provide the basis for hydrospace-environment simulation. The first concept is the response of material and equipment to hydrospace forces. In the simulated environment it is necessary to study the response of the structure to pressure, temperature, and reaction of materials and equipment to seawater. Pressure is important in discovering leaks, compressive strength, instability, stress distribution in hulls, and calibration of pressure transducers. Temperature has been found to be important to creep, fatigue, ultimate strength of the material, and reliability of many electrical components used in today's underwater structures. The reaction of materials and structures to the seawater environment is important in such things as corrosion of the material in saltwater, corrosion of working parts of a vehicle subjected to long-term submergence in the ocean depths, electrolytic processes induced by two or more metals emersed in saltwater, and in shorting out of electrical components forced to work in the seawater environment.

The second concept deals with the response of hydrospace to manmade forces. It has become necessary to study implosion and explosion shock-wave propagation, sound propagation, radiation decay, and containment, and behavior of sediments under propeller wash or ocean-floor traffic. Although these properties of ocean depths are of great importance to deep ocean research, the simulation of such has been difficult to attain and only recently has there been an attempt to provide facilities that could study one or more of these properties.

The third concept is that of hydrospace in equilibrium. In order to study equilibrium it is necessary to simulate a microcosm hydrospace by gathering marine organisms, subjecting them to the simulated hydrospace, and observing the biological and chemical processes involved. With this simulated hydrospace environment, it is possible to study symbiosis, life cycles, and chemical and organic changes in the hydrospace environment induced by the presence of such organisms, and reaction of such organisms to alien objects such as manmade materials. At the present time very few facilities are actually attempting to implement this concept to even a limited extent.

Over the years, however, hydrospace-environment simulation facilities have pushed back the frontiers of knowledge concerning hydrospace engineering by improving their capability in several areas. One such area has been the increase in the size of the pressure vessels used. The early high-pressure test vessels had diameters of a few inches and only very small models or structural members could be tested in them. With the employment of new alloys, designs and construction techniques, the present hydrospace engineer has at this command vessels up to 10 ft in diameter into which he can often place a full-size prototype structure. Thus with these large vessels it is often possible to prooftest underwater structures and effectively increase their safety margin.

Another area which has seen advancement is the increased hydrostatic pressure capability of modern pressure vessels. Until very recently, the major underwater structures which the United States employed were military submarines. These submarines were limited to depths of less than 1000 ft. To test models of these, the pressure vessel only required a pressure potential of 500–600 psi. With the present emphasis on vehicles and structures which can descend to the bottom of the ocean, there has arisen a need for pressure vessels which can test models of such structures. Therefore hydrospace simulation facilities are building or have built vessels with pressures in excess of 10,000 psi. Coupled with the increased size, the higher pressure capacity is adding new dimensions to the testing it is possible to undertake.

The development in the area of number of parameters simulated has also brought radical changes in the hydrospace-simulation facility. The early work in hydrospace simulation was limited to pressure testing. This parameter is not only the most obvious but is also the easiest to simulate. About 90% of all facilities today are still in this stage of development. The 10% of the facilities which are modernizing are attempting to simulate one or more of the other parameters exhibited by the ocean. These advanced facilities are simulating such things as ambient ocean temperature, the exact chemical parameters of seawater, various types of static and dynamic point-force loading of submerged structures, dynamic loading, sea-floor soils, and many others. The simulation of such parameters as these has allowed the engineer to more rigidly control the testing of models and structural members, and, at the same time, to provide a greater variety of conditions to be imposed on the specimens tested.

Another area which has seen great improvement has been the flow of information from the tests employed. The flow of information from the simulated hydrospace consists of data on the pressure, temperature, and chemical properties of the pressurizing media. From the specimens being tested, the information flow describes the loads imposed on the specimen, strain and strain rate within the specimen, corrosion of the specimen, permeability of the specimen, leakage into the specimen's interior if the specimen is a hollow structure, resistivity of the specimen, and many other responses of the specimen in the simulated hydrospace. With such information, the hydrospace engineer can be assured of better control of the simulated hydrospace and more realistic and reliable results from his tests.

Although truly tremendous strides have been made in the simulation of the hydrospace environment, it must be observed that since the ocean is very large in comparison to the simulated hydrospace, many parameters inherent to the ocean are either lost or mutilated in the transition to the simulated hydrospace. In general, almost all of the limitations imposed on simulated hydrospace stem from the fact that the ocean is, for all practical purposes, an infinite volume of seawater completely surrounding any object submerged at great depth, whereas the laboratory pressure vessel contains a relatively small volume of fluid surrounding a disproportionately large object. Therefore, the most obvious limitation on simulated hydrospace involves space. There are definite limitations on how large a pressure vessel it is possible to build and, too, how large a vessel a particular facility can afford to buy. The present state of the art imposes the limitation on the size it is possible to construct, and the funding limits the cost. These two factors have made a vessel of 10-ft diam and 10,000 psi a true rarity even at the present time.

Another limitation is the limited hydrostatic energy storage factor. The amount of potential energy available in a pressure vessel is the sum of that which is stored in a finite volume of compressed fluid and to the energy stored in the vessel's walls due to their elasticity. Although considerable energy can be stored in a compressed fluid, it is an infinitesimal amount when compared to the potential energy available at a corresponding depth in the ocean. Because of this limitation, the collapse of a structure being tested is severely affected. The collapse is frequently much slower than it would be in the ocean and, sometimes the specimen will not collapse at all, even though it would have at a corresponding depth. In order to reduce the effects of this factor, it is necessary to either have a very large volume of vessel versus the volume of test specimen, use very fast pumps to maintain the pressure as the model is collapsing or employ air accumulators. However, no matter what device is employed, the best that can be accomplished is a reduction in the effect of this limitation and not its elimination.

The limited heat-sink capability of the vessel is another limitation imposed on simulation. The problem of disposing of large quantities of heat generated by a test specimen, while maintaining realistic deep ocean temperature conditions around it, may become a serious limitation when the amount of heat involved is large or the volume of the vessel is small. The primary method employed to alleviate this factor is to use refrigeration. However, the capacity of the refrigeration unit itself is limited by the surface heat transfer coefficients of heat exchangers outside or inside the vessel, and although the effect of this factor may be reduced, it can not be eliminated in the way or to the extent that it is in the ocean.

Another limitation on simulating the hydrospace environment is the limited sound absorption of the pressure vessel. A large volume within the vessel compared to the volume of the specimen tends to reduce this factor somewhat. However, because the fabrication of huge vessels is expensive and frequently impossible, other areas are being investigated which hold promise of reducing this limitation to a minimum. One such area has been the use of anechoic lining for the vessel walls. Anechoic materials are being used in submarine research in an attempt to protect the submarine from sonar detection. Such materials, when used as a liner in a pressure vessel, increase the sound-absorption qualities of the vessel. At the present time, anechoic materials are only good for low pressures because at the high pressures they become too compressed to absorb the sound. When these materials or others of similar qualities become available for high pressures, the simulation of the hydrospace environment in terms of sound transmission pattern will have been accomplished.

Related to the limited sound absorption is the limited shock-wave absorption capabilty of pressure vessels. Results from the study of shock-wave phenomena in a pressure vessel would be exceedingly difficult to extrapolate to the ocean environment. The primary reasons for this are that, as in the case of sound studies, the reflection of the shock waves by the vessel's walls, the elastic response of the walls, and the changes in water density and viscosity which would result from pressure change all combined to give anomalous results.

Another limitation to hydrospace-environment simulation is the limited chemical-energy storage available. Seawater has much more than fresh water in which an appropriate amount of salts has been dissolved. Seawater contains live organisms, a suite of dissolved gases, various dissolved chemicals, and suspended organic and inorganic matter. As soon as a quantity of seawater is removed from the ocean, it starts to change both chemically and biologically. The native organisms die and decay, new organisms become predominant, the dissolved-gas system changes radically, the pH changes, and new chemical contaminants appear as a result of the new environment. Obviously, if the chemical composition of seawater is involved, as in a corrosion experiment, a continuous supply of fresh seawater of appropriate chemical properties must be supplied to the pressure vessel. However, even if this fresh seawater is supplied, there is no assurance that the water drawn from ocean surface is identical to seawater at particular ocean depth being simulated.

There is also an inherent limitation on the simulated hydrospace for generation of underwater currents existing in the ocean. Although water tunnels, similar to wind tunnels, exist for testing of underwater vehicle models to determine their hydrodynamic drag at various velocities, they are not capable of simulating depths over 100 to 200 ft. Also, the sizes of models that can be tested in them are rarely as much as 1 ft in diameter and 10 ft in length. Only one direction of water current flow can be generated in them, whereas in the ocean underwater currents and waves rarely are constrained to flow along only one directional vector. Because of it, water tunnels are popular mostly with hydrodynamicists that have as their sole objective the reduction of hydrodynamic drag of submarines, torpedoes, and towed sonar buoys, or the increase in efficiency of propellers used to drive above mentioned vehicles. For the investigation of ocean-bottom soil migration, erosion of soil under ocean-bottom structure foundations, and measurement of hydrodynamic forces on ocean-bottom structures, simulated hydrospace is required in which ocean currents have no artificial constraints for their movement. For this purpose, very large pressure vessels are required which not only serve to contain the pressurized hydrospace, but also serve as large underwater current-observation chambers. Needless to say, the major limitation on such a current chamber is its size and pressure rating. Too small a size of the vessel introduces unwanted boundary effects on the current flow and makes the test results less representative of actual hydrospace conditions.

In addition to these limitations imposed on research conducted in simulated hydrospace, there are problems that stem directly from these limitations which complicate the whole process. As stated before, the ocean is very large and the simulated hydrospace is very small in comparison and, therefore, in order to produce tangible results, most testing involves models of structures and structural components and not full-size structures.

In general there are two types of model: elastic and realistic (Rowe, 1964). The elastic model is used to establish, modify, and simplify theoretical solutions based on the classical theory of elasticity. With this model, usually one specific performance parameter is studied, such as how the strains are distributed in a structure (Fig. 17.1). These models are not necessarily fabricated from the same material or in the same manner as in the full-size structure. An example would be an acrylic or epoxy resin model of a steel oceanographic vehicle hull section built in order to examine the strain distribution around a hatch opening.

The other type of model is the realistic model whose performance is a scaled version of the full-size structure (Fig. 17.2). These models are constructed from the same material using the same fabrication and joining techniques as would be employed in a full-size structure. It must be

FIG. 17-1. Elastic model of a metallic underwater vehicle shell of helicoid-stiffener sandwich design. Model was fabricated from acrylic resin. (Courtesy of Pennsylvania State University.)

observed, however, that realistic models have several inherent problems. First, in order to scale a model properly, the designer must have an intimate knowledge of the nondimensional parameters involved. He must know how to scale the model for hydrostatic pressure, point forces, dynamic pressure, and many others if his model is to behave in the same manner as the full-size structure. Second, it is difficult to reproduce some material properties in a model as they exist in a full-size structure. In most materials, as thickness increases, the strength decreases, particularly in rolled metal sheets and plates. With brittle materials, like glass, this inherent weakness occurs as the volume of material increases because of the presence of many more flaws. Third, there is a size limitation on model fabrication and the reproducibility of surface roughness effects on the initiation and propagation of cracks. Fourth, there is the problem of reproducing in a model weld the qualities of a full-size weld and its effect on the over-all performance of the structure. Fifth, there is a limitation on the reproducibility of fillet radius effects on the stress concentrations in thick and thin structural shapes. If meaningful results are to be obtained, the designer of a model must solve the questions involved in these problems. If he does not, the test results will be anomalous and a great deal of time and money will have been wasted.

17.3 PRESSURE VESSELS FOR THE SIMULATION OF HYDROSPACE

In order to simulate hydrospace various tools are required. The most basic tool is the pressure vessel. Without the pressure vessel there is no pressurized space and hence simulation must begin here. As this is the case, it is necessary to look closely at the numerous aspects of the pressure vessel.

Shape

There are many shapes feasible for building pressure vessels. The reason that only two shapes, spherical and cylindrical, are popular is due to the fact that the spherical shape is the most economical in terms of material

FIG. 17-2. Realistic model of a concrete deep-submergence station for collection of oceanographic data. (Courtesy U.S. Naval Civil Engineering Laboratory.)

used to contain a given volume of pressurized liquid, and the cylindrical shape is the most economical in terms of thick-wall construction and space utilization. Thus we see spherical vessels for the containment of gases and fluids where only limited accessibility to the pressurized volume is required. In order to make complete accessibility to the interior of the sphere feasible (where we could test a specimen of D_o equal to D_i of sphere), a joint has to be located at the equator of the sphere, and the upper hemisphere must be removed every time we desire to place or remove an object from the sphere. In larger vessels this imposes a considerable loading requirement on hoisting equipment, as 50% of the vessel's weight is to be raised. Also every time the vessel is to be opened, 50% of the pressurizing fluid must be removed and, upon reclosing of the vessel, must be replaced, placing undue demands on the pumping system. The difficulty in fabricating thick-walled spheres will be discussed later when the various methods of constructing thick-walled vessels are discussed.

Cylindrical shapes are used almost exclusively in the creation of simulated hydrospace. Equipped with any array of head shapes, closure restraints, and types of seals, they are the mainstay of simulated hydrospace facilities. The reasons for it are first, cylindrical vessels are easier to fabricate than spherical. Second, there is a larger variety of end-closure restraints, seals, and head shapes which are applicable to cylindrical vessels than to spherical ones. These components can be selected to match the budget, test requirements, or already available hoisting and weight-handling equipment. Third, most structures, structural components, or mechanical devices requiring testing in simulated hydrospace are not spherical or cubical, but oblong. Thus a pressure vessel with one interior dimension larger than the other is more economical in terms of volume utilization. Fourth, when placed vertically, little or no fluid must be removed prior to opening of the end closure and removal of the test specimen, making rapid change of test specimens possible.

Types of vessel construction

Because the cylindrical vessel is the predominant type of pressure vessel for hydrospace simulation, the discussion will center on that type of vessel and all the remarks will be understood to pertain to it unless specifically mentioned otherwise.

There are several types of vessel-wall construction, each type representing an approach to circumvent some limitation in pressure vessel construction or operation.

1. *Solid-wall* construction (Fig. 17-3) is the oldest type of construction and is used almost exclusively with small-diameter vessels, and is also used with large-diameter vessels when the pressures contained are low.

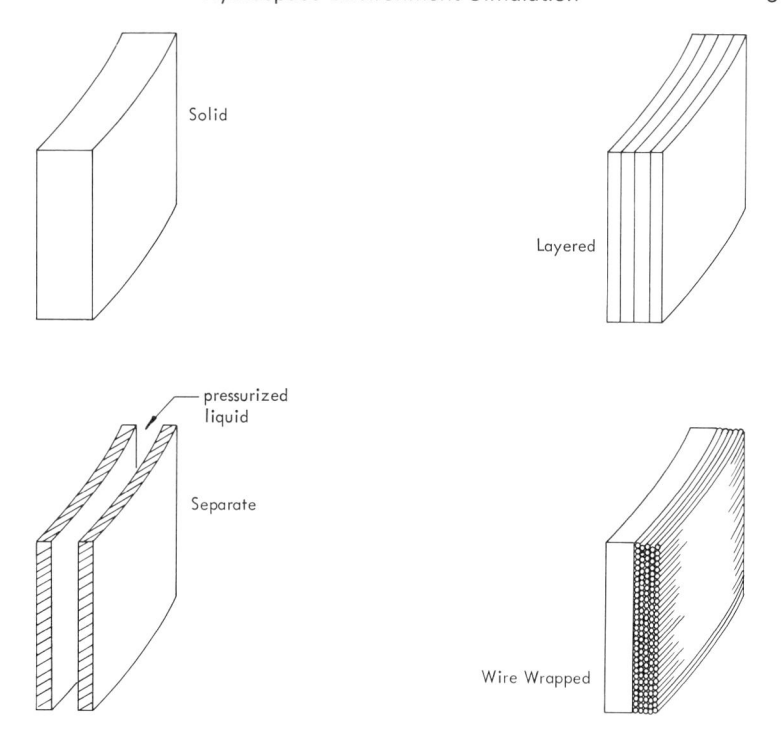

PRESSURE VESSEL WALL CONSTRUCTION

FIG. 17-3. Types of pressure-vessel wall construction; (a) solid wall; (b) layered wall; (c) separate wall; (d) wire-wrapped wall.

2. *Layered-wall* construction (Fig. 17-3) is used in the construction of large-diameter, high-pressure vessels. The layers are in intimate contact, and there is some pretensioning of layers, either by a special welding technique or by thermal shrinking. Basically, however, it has a thick-walled-vessel stress distribution in its wall.

3. *Separate-wall* construction (Fig. 17-3) has been proposed for constructing large-diameter high-pressure vessels in cylindrical or spherical shapes. The individual layers are not in contact, and the space between them is pressurized with fluid. The pressure in the annular space is a predetermined fraction of the vessel's internal pressure. In this manner each layer can be made to be stressed equally. A steel working model of such a vessel in the spherical shape has been build and it has been found to perform as predicted (De Hart and BasdeKas, 1963; Schmidt and Pickett, 1964). Intricate hydraulic pressure controls are utilized to maintain the pressures inside the annular spaces in a specified ratio to each other.

4. *Wire-wrapped* type of wall reinforcement construction (Fig. 17-3) has been used to a limited extent in the construction of air flasks. Wrapping high-strength wire around a vessel creates compressive stresses in the interior layer of the vessel wall that is advantageous to the distribution of stresses (Braun and Wesner, 1965).

5. *Filament-wound composite* construction (Fig. 17-2) in the form of glass-fiber epoxy laminates has been used widely in the fabrication of compressed-air bottles, missile fuel tanks and hydraulic accumulators. It has been proposed also for the construction of large pressure vessels. Instead of glass fibers, metallic fibers may be substituted to eliminate the effect of static fatigue always present with glass fibers.

6. *Composite reinforced with high-strength rods* (Figs. 17-4 and 17-5) is a typical steel-reinforced concrete structure. Its use has been proposed by many for containment of simulated hydrospace (Suarez and Kramarow, 1965), although at the present time it is primarily used in the nuclear-reactor structures where the pressures are relatively low but the volume of the contained gas large. Cylindrical structures with dimensions as large as 77 ft in inner diameter and 60 ft internal length are being built for 400 psi internal pressure service in Britain. Smaller ones have been in operation since 1960 in France (March and Rockenhouser, 1965).

7. *Stacked-ring* construction (Fig. 17-4 and 17-6) aims at overcoming the problem of fabricating large vessels from monolithic single wall tube forgings. By stacking many rings upon each other it is possible to achieve essentially the effect of a large monolithic tube without paying the economic penalty of preparing a single large tubular forging. The clamp bands tying the many individual rings together make it possible for the rings to carry also axial loads imposed by end closures attached to end rings. Some stacked-ring designs utilize external tie rods instead to carry axial loads imposed on the vessel by end closure (Fig. 17-4).

8. *Segmented-wall construction* relies on the assembly of many modular wall segments, pinned together with longitudinal shear pins, to achieve a pressure-resistant container. The longitudinal loads imposed on the pressure vessel are either carried by the shear pins, or by tie rods placed outside the vessel whose sole function is the retainment of end closures. Vessels of this type, except for working models (Fig. 17-7) have not yet been built, although several proposals to this effect are being evaluated by different fabricators.

Types of end-closure restraint

The previous section dealt with the many available and proposed methods for constructing vessel walls. This section deals with just as important a part of vessel structure—the end-closure restraint. The end-closure restraint

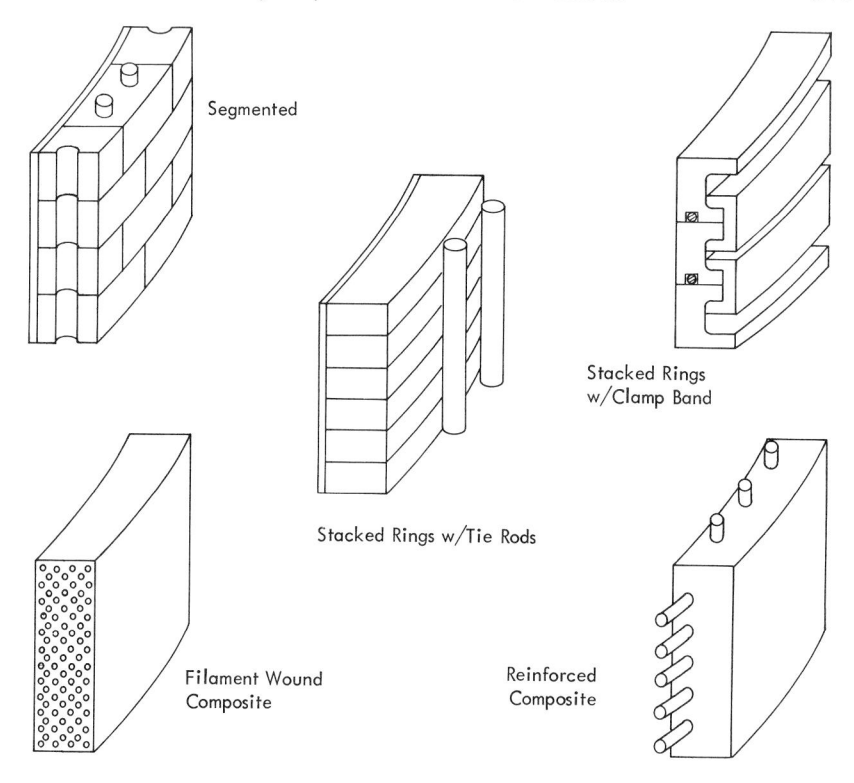

PRESSURE VESSEL WALL CONSTRUCTION

FIG. 17-4. Types of pressure-vessel wall construction: (a) filament-wound composite; (b) composite reinforced with high-strength rods; (c) stacked rings with tie rods; (d) stacked rings with clamp bands; (e) segmented wall.

is an important feature of any pressure vessel design because of the many stress concentrations that occur there. But for pressure vessels used to simulate hydrospace, the end-closure restraints are even more of a problem. This is because in this category of vessels, the end closures, or at least one of them, must be completely removable, to make the total diameter of the vessel accessible for test-specimen. Both this, and the fact that rapid removal of the end closure is desired, imposes difficult requirements on the vessel designer. However, once the approach to the end-closure requirements is formed, it must basically satisfy two requirements: (a) to restrain the end closure from separating from the cylindrical portion of the vessel under the action of hydrostatic pressure, and (b) to give speedy and unhindered access to the vessel's interior. Some of the end-closure restraints found today on simulated hydrospace vessels, or proposed for future ones,

FIG. 17-5. Prestressed-concrete pressure-vessel model of 99 in. I.D. and 75 in. internal length for 450-psi operational pressure. (Courtesy Gulf General Atomic.)

are discussed below. Although heads come in different shapes, for the sake of this discussion of end restraints they are all shown as flat.

1. *Continuously threaded head* (Fig. 17-8). The thread in the head is used to restrain the head from axial displacement under internal hydrostatic pressure. This is an economical design as it requires less of the vessel's length for threads than any other threaded end closure restraint. The disadvantages of this end-closure restraint are the rotation of the head during closing or opening of the pressure vessel, thus causing tangling and twisting of wires and hydraulic lines connecting the head to both the test specimen and the external pumping and instrumentation systems. Furthermore, it is very difficult and expensive to equip such an end-closure restraint with labor-saving power equipment that would speed up the locking and unlocking operations.

2. *Head with interrupted thread.* (Fig. 17-8). The disadvantages of this end-closure restraint are similar to the continuously threaded head design,

FIG. 17-6. Acrylic model of a stacked-ring wall construction pressure vessel with tie rod end closure restraint. (Courtesy U.S. Naval Civil Engineering Laboratory.)

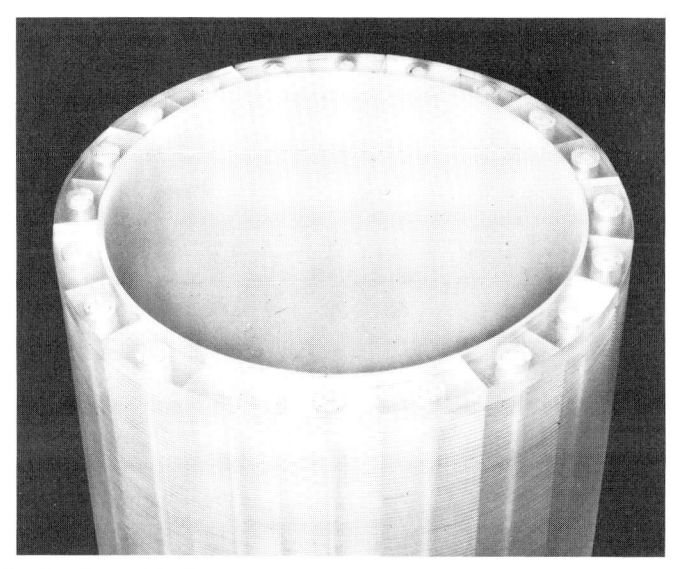

FIG. 17-7. Acrylic model of a segmented wall construction pressure vessel. (Courtesy U.S. Naval Civil Engineering Laboratory.)

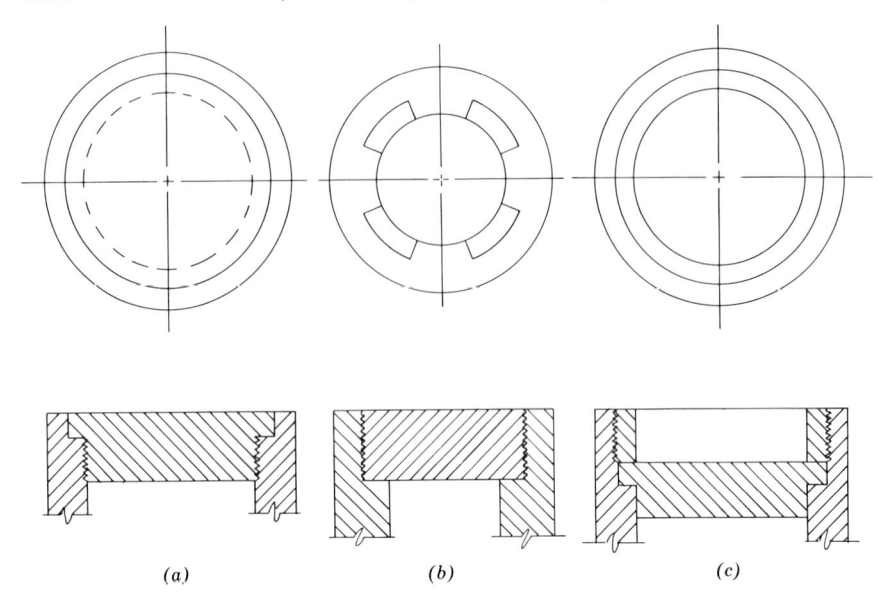

FIG. 17-8. Pressure-vessel end closure restraints: (*a*) threads in cover; (*b*) interrupted threads in cover; (*c*) threads in retaining ring.

except to a lesser measure, as only $\frac{1}{8}$ or a $\frac{1}{4}$ turn are required to open or close the vessel. This decreases the amount of twisting to which wires and hydraulic lines attached to the head are subjected. A new disadvantage is introduced, however, by the presence of interrupted thread in the form of high stress concentration factor at the threads. A very important advantage that is introduced by this design is the ease with which a hydraulically operated opening and closing mechanism can be incorporated into the end-closure restraint. Large vessels with interrupted thread in the head have been built for hydrospace-simulation facilities (DeHart, 1965) and have been proven to be economical in operation (Figs. 17-9 and 17-10).

3. *Continuously threaded retaining ring* (Fig. 17-8). The ring restrains the head from axial displacement under internal hydrostatic loading. The advantages are the nonrotating feature of the head, the breakdown of the end-closure mass into two more manageable units, the head (Fig. 17-11) and the retaining ring (Fig. 17-12), and alignment in azimuth feature of the head which makes coupling of hydraulic fittings in the head to external rigid piping easy. The major disadvantages are the same as for the continuously threaded head; it is difficult to motorize or automate this type of closure restraint mechanism.

4. *Retaining ring with interrupted thread* (Fig. 17-13). The interrupted

Hydrospace-Environment Simulation 653

FIG. 17-9. One of the largest vessels in the United States used exclusively for hydrospace simulation. Note the interrupted thread in the vessel flange. (Courtesy Southwest Research Institute.)

thread in the retaining ring restrains the head from axial displacement. The advantages are the nonrotational and azimuth alignment features of the head, the smaller individual masses of head and ring, and the rapidity of opening and closing the vessel since only 1/8 to 1/4 turn is required to close

FIG. 17-10. End closure with interrupted thread for vessel shown in fig. 17-9. Note the hydraulic cylinders mounted on the end closure for its locking and unlocking. (Courtesy Southwest Research Institute.)

FIG. 17-11. The head of an 18 in. I.D. × 36-in. long hydrospace simulation vessel for 20,000 psi hydrospace simulation service. Threads for the retainment of the head are located in a separate threaded retaining ring. (Courtesy U.S. Naval Civil Engineering Laboratory.)

FIG. 17-12. Threaded retaining ring for the head shown in Fig. 17-11. (Courtesy U.S. Naval Civil Engineering Laboratory.)

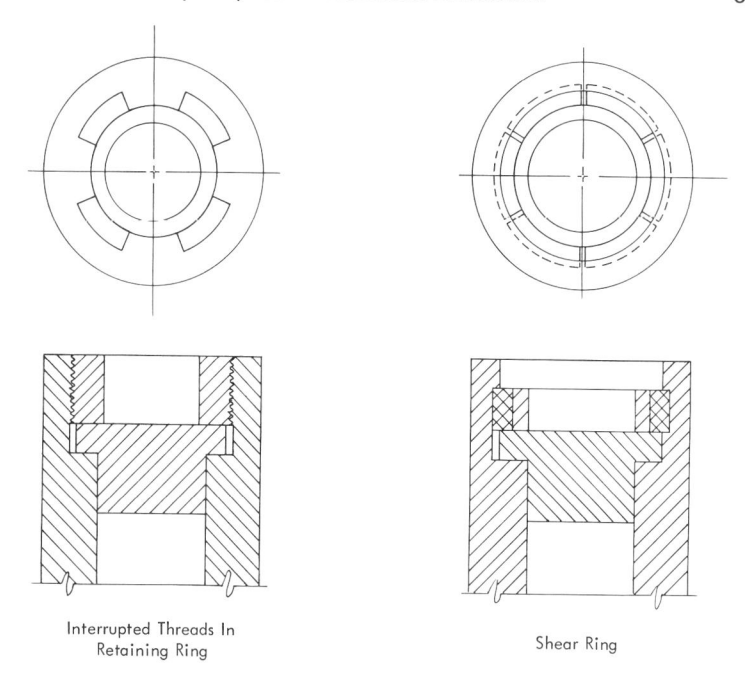

FIG. 17-13. Pressure-vessel end-closure restraints: (*a*) interrupted threads in retaining ring; (*b*) shear retaining ring.

or open the vessel as with the interrupted thread in the head type. The disadvantage lies mainly in increased length of threaded portion of ring and vessel in order to compensate for decreased cross-sectional area of individual threads. The interrupted character of individual threads also creates serious stress concentrations for each thread. Motorization and automation of this end-closure restraint is, however, relatively easy and inexpensive, as it was in the case of the head with the interrupted thread type.

5. *Sheer retaining ring* (Fig. 17-13). The restraint to the axial displacement of the end closure is provided by a segmented retaining ring subjected to shear. This type of end restraint, like the interrupted thread, provides a means for rapid opening and closing of the end closure besides providing the head with nonrotational alignment in azimuth for easy coupling of tubing. It is more economical to make than the threaded end-ring restraint, but it is much harder to be adapted to power equipment for its placement and removal.

6. *Bolts* (Fig. 17-14). Bolts are the oldest means of restraining the end closures from axial displacement. They are very inexpensive in terms of locking machanism, but expensive in terms of added head weight and flange weight. This type of end-closure restraint imposes serious bending stresses

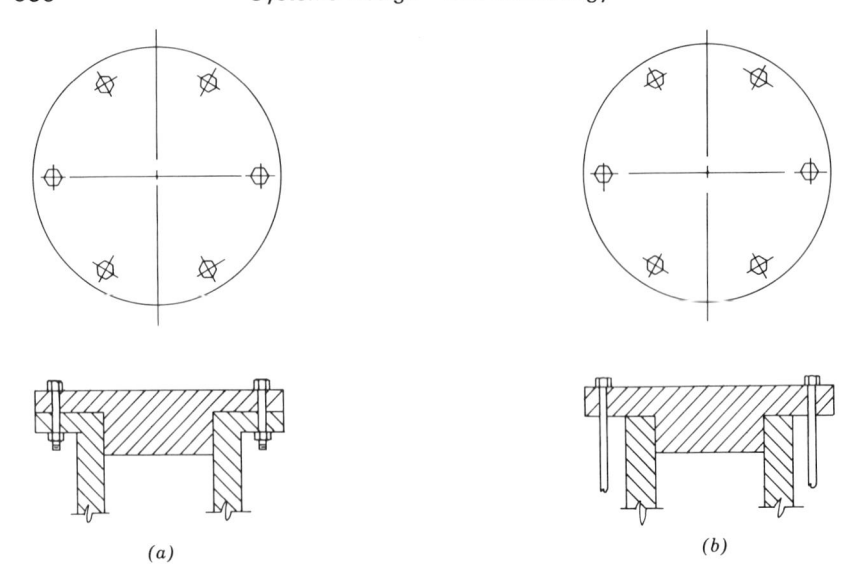

FIG. 17-14. Pressure-vessel end-closure restraints: (a) bolts; (b) tie rods.

on the end of the vessel. Another disadvantage is the very slow removal and closing of the vessel with this type of closure restraint, but it does permit alignment of head in azimuth for easy connection to piping. However, for small pressure vessel sizes, and long-term pressurization studies in simulated hydrospace where the head is removed only infrequently, the bolted end closure restraint is a very economical approach to vessel construction.

7. *Tie rods* (Fig. 17-15). With tie rods the same principle of endclosure restraint is used as with bolts, except that the axial load is carried here by the tie rods instead of by the vessel walls. This arrangement permits the elimination of vessel end flanges, as well as the assembly of the vessel from stacked rings or segment modules that by themselves cannot carry axial loads imposed by end closures. A penalty is generally paid in added material required for the tie rods, and large flange on the head.

8. *Shear pins* (Fig. 17-16). This end-closure restraint relies on the forces required to shear pins in retaining ring or head to resist the axial displacement of the end closure under the action of hydrostatic pressure. This type gives the head the nonrotational feature, and it permits rapid placement and removal of the shear pins and head. It is an extremely reliable, and particularly for smaller vessel sizes (Fig. 17-17), easy to engineer and fabricate end closure restraint.

9. *Clamp bands* (Fig. 17-16). Clamp-band end-closure restraint relies on the clamping action of the band to restrain the end closure from axial displacement. The advantages of this type of end-closure restraint are the

rapidity of placement and removal of the end closure, particularly where motorized clamp band locking devices are used, and the alignment of the head in azimuth (Fig. 17-18). Only small flanges are required on the vessel and on the cover. For applications where extremely rapid opening and closing operation of vessel is required, the interrupted clamp band (Fig. 17-19) and flanges with power assist are utilized.

10. *Straight laid-in thread* (Fig. 17-20). This type of restraint is very similar to the threads in retaining-ring restraint mentioned at the beginning of the end closure restraint series. The difference lies in the fact that in this case the threads are neither part of the retaining ring, or of the vessel, but are a separate part of the end-restraint mechanism. The laid in threads consist of a helical spring laid into concave grooves in the retaining ring and the vessel end (Fig. 17-21a). Because the helical spring gives under load application, it helps distribute the end load, thus decreasing the generally

FIG. 17-15. $8\frac{3}{4}$ in. I.D. × 11 in. long pressure vessel for 10,000 psi service. Tie rods are used for restraint of the end closures. (Courtesy U.S. Naval Civil Engineering Laboratory.)

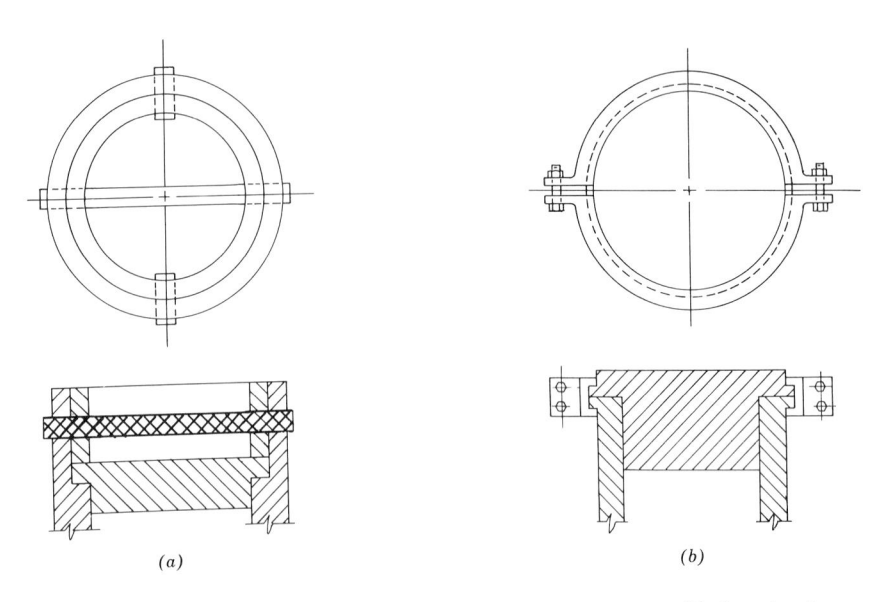

FIG. 17-16. Pressure-vessel end-closure restraints; (a) shear pins; (b) clamp band.

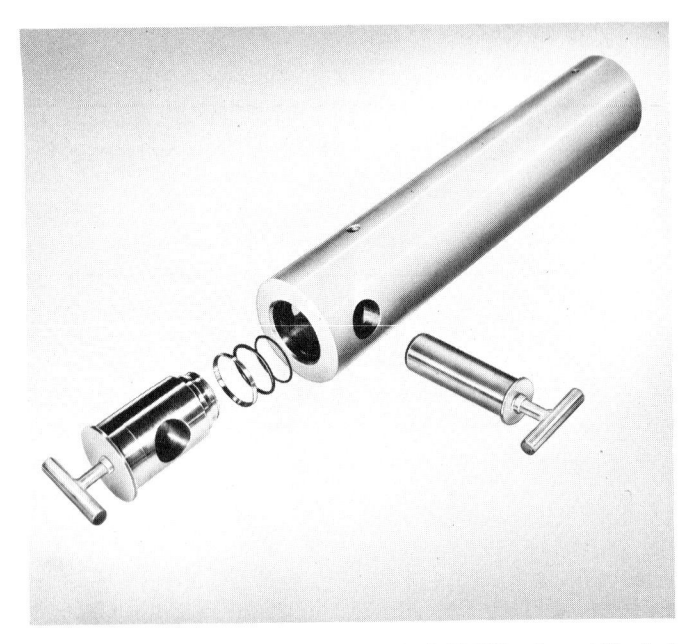

FIG. 17-17. A simple and reliable pressure vessel of 20,000 psi capability for hydrostatic testing of small components. Note the shear-pin restraint for locking of pressure vessel end closure. (Courtesy Autoclave Engineers.)

FIG. 17-18. Automatically inserted shear pins are used in this end-closure restraint to lock the two halves of the clamp band together. (Courtesy U.S. Navy Ordnance Laboratory.)

existing overload on the first thread and the stress concentrations associated with it. Also, since the hemispherical grooves have a gentle curvature, the usually present stress concentrations at the root of the threads are absent.

11. *Tapered laid-in thread* (Figs. 17-20 and 17-21b). This possesses the same advantages as the straight laid-in thread plus the further advantages that it requires fewer turns to open and close the vessel, and that the stress concentration at the lowest thread is further diminished below that of the straight laid in thread. The tapered thread is basically of the same type as in oil-field drill pipe.

12. *Yoke end-closure restraint* (Fig. 17-22). This type provides end-closure restraint without the creation of any stress concentrations in the cylindrical body of the vessel. In this respect it is similar to the tie-rod restraint. The biggest difference between the yoke restraint and the tie-rod restraint lies in the absence of flanges in both the cylinder and the head. Also the head may be much thinner than if the yoke were not present, as the yoke, through the bearing block, provides an even support for the external surface of the

FIG. 17-19. Typical vessel for industrial processes requiring rapid opening and closing of the end closure. Only a fraction of a turn is required to open or close the multiple cam-equipped clamp band (Courtesy Autoclave Engineers.)

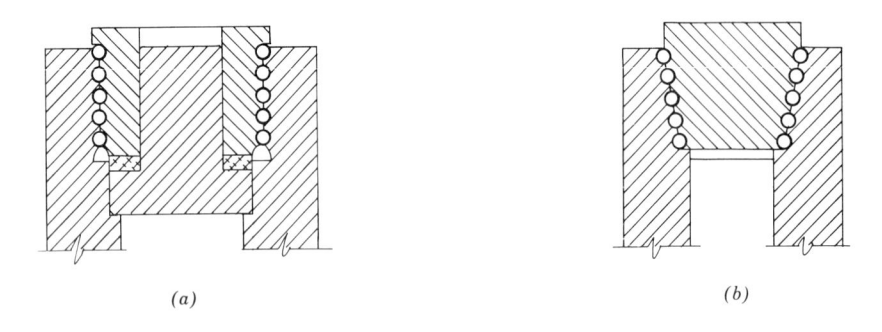

PRESSURE VESSEL END CLOSURE RESTRAINTS

FIG. 17-20. Pressure-vessel end closure restraints: (a) laid-in helical spring thread in a cylindrical retaining ring; (b) laid-in helical spring thread in a conical end plug.

FIG. 17-21. (a) Medium-size pressure vessel for service in the 50,000 to 100,000 psi range. Note the straight laid-in helical spring thread in the head retaining ring. (b) Laid-in helical spring thread in a conical end plug. Rapid opening and closing of the vessel is feasible with this type of end restraint, which in addition almost completely eliminates stress risers at the roots of the resilient spring thread. (Courtesy Autoclave Engineers.)

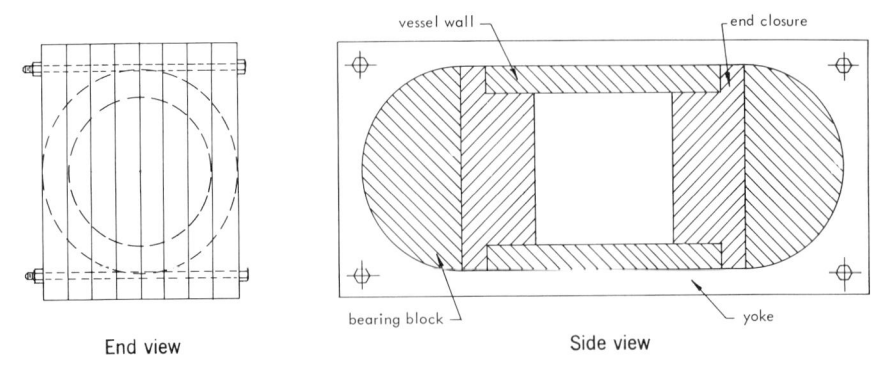

FIG. 17-22. Pressure-vessel end-closure restraint: yoke type.

cover. A further advantage of the yoke restraint is that it is generally fabricated from commercial plate stock bolted together to form the yoke structure. Because of the yoke's massiveness, stresses in it are low and nonpremium materials can be utilized. As the yoke is fabricated from steel plates, or sheets bolted together, *in situ* assembly from individual plate and sheets is feasible. The cylindrical portion of the vessel can, of course be assembled from rings or segments, as all of the axial loads acting on the vessel closures are carried by the yoke. The largest disadvantage of this design is that means must be provided for sliding or rolling out of the cylindrical portion of the vessel from inside the yoke so that the head may be removed and a test specimen placed inside the vessel.

End-closure shapes

The head shape is generally, for high-pressure internal vessels, either flat or hemispherical. Whether the selected shape is *flat, hemispherical convex,* or *hemispherical concave* (Fig. 17-23) depends on several factors. These factors are:

1. maximum permissible weight of end closure;
2. type of end closure restraint;
3. monolithic or polylythic head design.

Generally speaking, the flat head is the heaviest one for a given vessel diameter because of the high flexure stresses generated in a flat vessel head by hydrostatic pressure. Still, for small high-pressure vessels that have a diameter of less than 2 ft, the flat type of head is the most economical, as the saving in weight that accrues from using hemispherical heads is more than offset here by the high cost of forging hemispherical heads. High-pressure vessels with larger diameters cannot be equipped with flat heads, as in this case the thickness of the head assumes such magnitude that it either cannot be forged monolithically, or it becomes so heavy that it becomes not economically feasible, and the hemispherical type of construction becomes attractive although it requires expensive forging and machining operations.

The *hemispherical convex* type of head has two main advantages: (a) a favorable stress distribution and (b) additional internal volume that it contributes to the vessel's interior. Because of the favorable stress distribution, the thickness of the hemisphere is generally much less than if it were flat. Because of the additional volume that a hemispherical convex head provides for the vessel, the length of the cylindrical vessel can often be shortened, making its fabrication more economical.

The *hemispherical concave* type of head is generally used only in those vessels where, because of size, the hemispherical head must be fabricated from structural modules. Welding could be used to join the head modules into a monolithic head that could be employed in convex manner, but because of the many welds its strength would have to be somewhat derated. When instead of a convex hemisphere design a concave design is used, there is no need for welded joints with high structural efficiency, as now the end closure is loaded in compression rather than in tension. In concave hemisphere design, the head modules may be bolted, as the only loading requirement on the module joints is to keep the end closure from falling apart when it is lifted, or lowered onto its seat in the vessel. Using the modular head design in concave end closures, vessels can be built with extremely large end closures that otherwise could not be fabricated from a

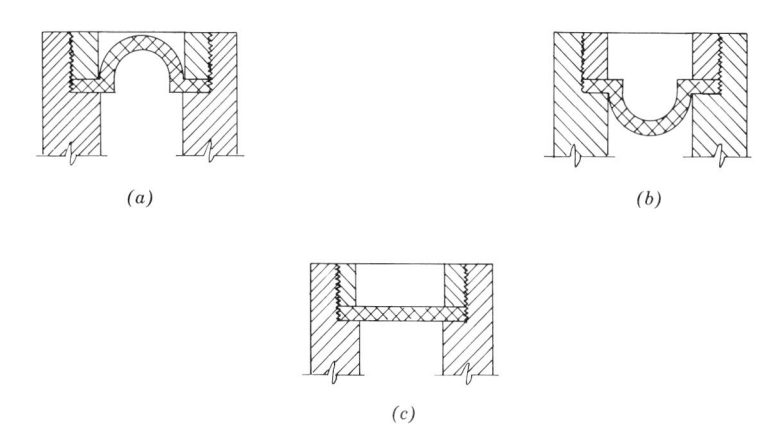

TYPICAL END CLOSURE SHAPES

FIG. 17-23. Typical end closure shapes: (*a*) hemispherical convex; (*b*) hemispherical concave; (*c*) flat.

single steel forging. There is only one severe disadvantage connected with the use of concave heads. When concave heads are employed, the internal volume of the vessel decreases considerably. To compensate for the decrease in internal volume of a vessel equipped with concave heads, the length of the cylindrical portion of the vessel must be increased. Thus a penalty in terms of vessel weight and cost must be paid for the use of such heads. Still, for vessels with very large diameters, the choice lies between yoke closure restraint with a flat head, and hemispherical end closure with concave head assembled by bolting of modules.

Sealing arrangements for pressure-vessel end closures

Seals, in addition to pressure-containing cylinder, end-closure restraint, and end closures, are a most important feature of pressure-vessel construction. Without seals containment of pressure is not feasible at all. With cumbersome or unreliable seals the containment of pressure becomes an unpredictable operation, prone to malfunctions which in turn will ruin the results of many experiments taking place in the simulated hydrospace environment. The loss of pressure, or inability to attain higher pressure than already achieved terminates experiments in which a mechanical or structural component of an underwater structure is being pressurized to implosion under short-term pressurization conditions. The interruption in pressurization, loss of pressure, or inability to reach higher pressure generally changes a short-term destructive, or for that matter nondestructive, hydrostatic pressurization experiment into an unforeseen pressure cycling experiment.

There are several methods now employed which provide adequate sealing under specific loading conditions. One of these is the use of an *elastomeric O-ring subjected to axial compression* (Fig. 17-24). This arrangement is used primarily in low-pressure system, or in inexpensive high-pressure system. Its advantages are the low cost of machining a single O-ring groove and the positive sealing under an external or internal pressure in the low-pressure ranges. Besides this, it is insensitive to changes in the vessel's internal diameter under pressure. Its disadvantages are that extrusion and ultimately the loss of the seal occur at high internal pressure, which causes the end closure to lift and thus creates a clearance between the head and its seat. Another disadvantage is the jamming of threaded end closures after the ring has extruded. With the use of backup rings or backup wedges the extrusion of the elastomeric O-ring under axial compression can be prevented to a great extent, or at least shifted to a higher pressure range. In general this seal is more applicable to external-pressure applications where the external pressure helps to keep together the metallic parts that contain the elastomeric O-ring. This feature makes the O-ring under

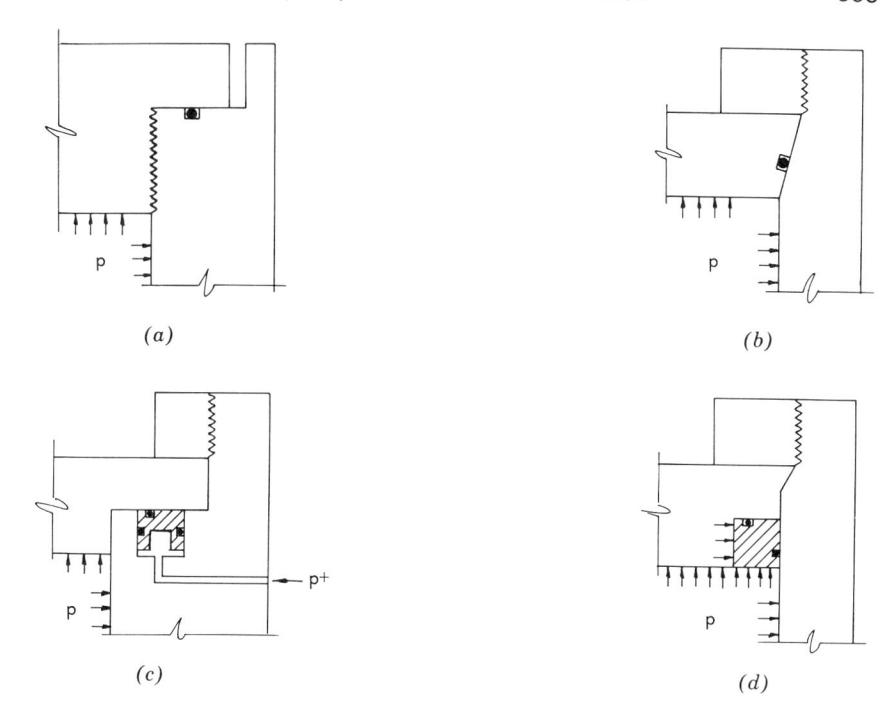

TYPICAL PRESSURE VESSEL END CLOSURE SEALS

FIG. 17-24. Typical pressure-vessel end-closure seals: (a) axial compression; (b) radial compression; (c) externally energized "floating" seal; (d) self-energizing "floating" seal.

axial compression a self-energizing seal under external pressure loading, whereas under internal pressure the separation of the metallic parts under increasing pressure makes it a self-de-energizing seal. Still, with a sufficient preload applied during bolting or threading-in of the end closure, and the use of O-ring extrusion preventors like plastic backup rings or wedges, this seal can safely contain internal pressures in the 0 to 20,000 psi range.

Another arrangement often used in an *O-ring under radial compression* (Fig. 17-24). This arrangement is used primarily in high-pressure systems to 20,000 psi. The advantage of this seal is the positive sealing at external pressure of any magnitude and internal pressure to around 20,000 psi. The disadvantages of this seal are similar to those of the axial seal, except that they are less pronounced than for the axial seal at the same pressure. The radial seal is less sensitive to axial displacement of the end closure because of the steep bevel angle on the mating steel sealing surfaces; however, it is quite sensitive to the vessel's radial dilation under pressure. Because of this lessened sensitivity to axial displacement of the head, but considerable sensitivity to the vessel's radial dilation, the radial seal is capable of sealing internal pressures only to 20,000 psi without the aid of backup rings or wedges.

A third type of seal arrangement is the *externally energized floating O-ring seal*, which utilizes an external source of hydraulic power to create, upon the operator's command, a positive seal for external or internal pressure in the 0 to 100,000 psi range. Since in this seal the lower metallic seal seat is movable and follows the axial displacement of the end closure, this seal is capable of maintaining the pressure inside the vessel even if the endclosure has moved 1/8 to 1/2 in axially (Fig. 17-24). The advantages of this type of seal are (a) activation of the seal upon command, (b) tolerance for very large clearances between mating steel surfaces of the head and vessel containing the seal (c) recessing of the seal when not in use. The disadvantages of this seal are (a) high cost of fabrication of the seal groove and metallic follower ring, (b) complexity of the system requiring additional pump and pressure controls, or a differential piston operated by the vessel's pressurization system.

In general, this seal represents a step above the nonenergized seals. First of all it does not require any preload upon the end closure prior to pressurization. This makes it possible to use end-closure restraints that do not generate preload, like shear rings, shear pins, interrupted thread with zero pitch, and others. Second, the radial clearance between the end closure and the closure seat can be very large and have a rough finish. Only the flat sealing surface on the head needs to be machined fine. This makes the machining of the head a relatively inexpensive process. Of course, some of the cost savings accrued from less machining on the head are offset by the machining of the groove in the vessel flange for the metallic ring follower.

The fourth type of seal assembly is the *self-energized floating O-ring seal*, which is considered to be the most reliable and sophisticated seal for internal pressure in the 0 to 100,000 psi range (Fig. 17-24). The popularity of the seal stems from (a) its self-energizing feature where the higher the pressure, the more intimate the contact between mating steel parts that contain the two O-rings making up the seal and (b) its ability to contain any pressure in the 0 to 100,000 psi range because of its "floating" feature, where the ring containing the seals follows both the axial displacement of the end closure and the radial expansion of the vessel. Because of the "floating" feature, the vessels equipped with this type of seal can utilize end closures whose restraint mechanism possesses a large amount of axial play. The major disadvantage of the self-energizing free-floating seal lies, primarily, in its high cost of fabrication. The O-ring retainer requires very small radial clearance when initially installed in the vessel's interior so that it may create a seal even when there is yet no internal pressure in the vessel. Because of this, the dimensional tolerances on the radial dimensions of the retainer and the vessel's diameter are very close requiring precise machining operations. As the retainer ring must be extremely limber in order to follow the radial expansion of the vessel, springing out of shape during machining operations complicates the matter further.

Typical seals

The most popular seal (Fig. 17-25) for internal pressure vessels is the elastomeric O-ring whose applications were discussed previously either by itself, or with different forms of backups that decrease considerably its extrusion through the clearance between mating metallic surfaces containing the O-ring seal. The inherent sealing capacity of the elastomeric O-ring is infinite, as long as a metal-to-metal contact is maintained between the mating surfaces of the seal. Once the metal-to-metal contact is lost, the sealing ability of the O-ring becomes a function of the O-ring's hardness and the width of the clearance between the separated metallic surfaces of the seal. To overcome the extrusion of the O-ring into the crack, plastic or metallic backup are used. The plastic backup rings are compressed together with the O-ring when the seal is initially preloaded, only to expand later with the O-ring when the metallic seal surfaces separate under hydrostatic pressure. The added stiffness of the plastic backup ring permits the O-ring to withstand higher pressures before extruding together with the plastic, through the crack. The wedge backup ring is generally of split construction, and made of metal. As mating surfaces of the seal separate, the wedge moves out

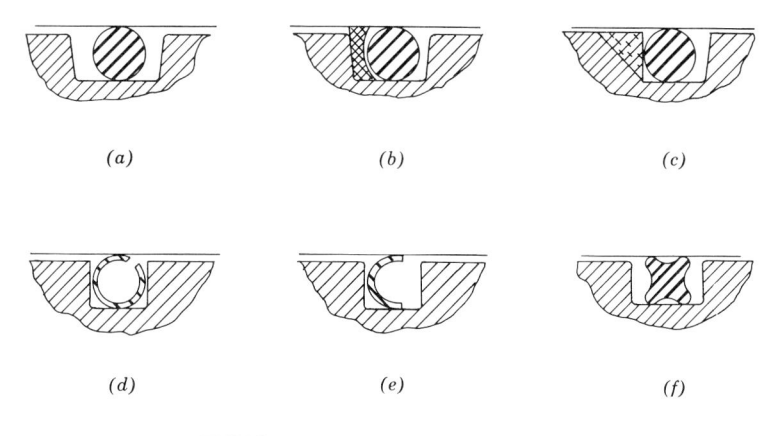

TYPICAL METHODS OF SEALING

FIG. 17-25. Typical methods of sealing (a) elastomeric O-ring; (b) elastomeric O-ring with plastic backup; (c) O-ring with metallic wedge backup; (d) self-energizing metallic O-ring; (e) self-energizing metallic O-ring; (f) elastomeric or metallic x-ring.

under the pressure of the O-ring completely filling the crack. Since the wedge is of split construction, it expands under very little hydrostatic pressure. When the mating seal surfaces separate further than the width of the wedge ring, a total loss of the pressure occurs.

A separate and distinct class of seals utilizes instead of elastomeric O-rings, metallic self-energizing O-rings, c-rings, or x-rings. In this group of seals, the high strength of metal is utilized to create a seal even where clearances between mating seal surfaces exist. Although steel seals are capable of sealing higher internal pressures than nonenergized and nonreinforced elastomeric O-ring seals, they are seldom employed in hydrospace simulation facilities because of some of the disadvantages that the metallic seals possess when used in an end closure that is often opened and closed. The major disadvantages of the metallic seals are (a) the large preload required to compress them properly so that they will seal at low pressures; (b) extremely fine finish required of the mating seal surfaces, as otherwise no sealing action can be generated; (c) very limited, if any, reusability of the seals. Because metallic seals are expensive, the limited reusability feature of such seals makes them an expensive method of sealing end closures that must be repeatedly opened and closed.

Stresses in vessels

When a cylindrical internal pressure vessel is pressurized, basically three kinds of stresses are generated inside the cylinder (Harvey, 1963). The stresses caused by the radial loading of the cylinder are membrane hoop stresses and radial stresses. The membrane hoop stress is the reaction of the vessel material to being stretched along the circumference of the vessel, whereas the radial stress is the reaction of the vessel material to being compressed radially. The axial loading of the cylindrical vessel caused by action of hydrostatic pressure on the end closures produces meridional membrane stress within the vessel wall. In a thin-walled pressure vessel (Fig. 17-26), that is, a vessel whose thickness to internal radius ratio is less than 0.1, the radial compression stresses are considered negligible, whereas the hoop and meriodional membrane stresses are considered to be constant throughout the thickness of the shell wall. Furthermore, the hoop-membrane stress is always twice as large in magnitude as the meridional-membrane stress. Thus, at first glance there would appear to be no particularly difficult problem in increasing the pressure capability of a vessel by increasing its shell thickness. Since in a thin-walled pressure vessel doubling the wall thickness doubles the pressure capability, no limit appears to be present to the magnitude of pressure that a vessel may contain if its wall is made sufficiently thick. However, this is a misleading concept based on the performance of thin-walled pressure vessels in which the magnitude of the hoop-membrane

Hydrospace-Environment Simulation

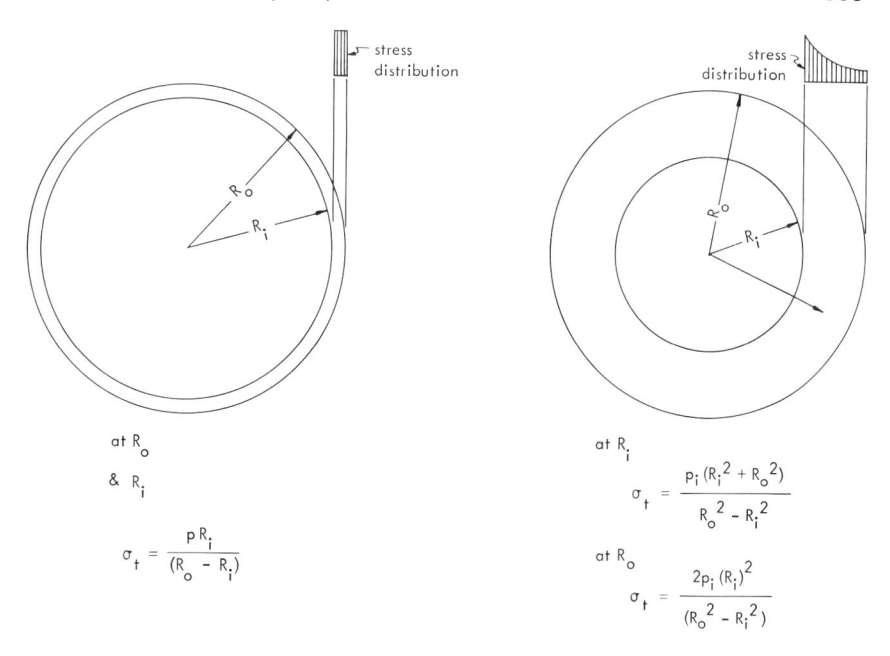

at R_o & R_i

$$\sigma_t = \frac{pR_i}{(R_o - R_i)}$$

at R_i

$$\sigma_t = \frac{p_i(R_i^2 + R_o^2)}{R_o^2 - R_i^2}$$

at R_o

$$\sigma_t = \frac{2p_i(R_i)^2}{(R_o^2 - R_i^2)}$$

STRESS DISTRIBUTION IN PRESSURE VESSEL WALLS

FIG. 17-26. Hoop-stress distribution in the pressure-vessel walls: (*a*) thin-wall cylinder under internal pressure, $1.0 < \frac{R_o}{R_i} < 1.1$; (*b*) thick-wall cylinder under internal pressure, $\frac{R_o}{R_i} > 1.1$.

stress is constant throughout the thickness of the wall. The actual magnitude of hoop stress in a thick pressure vessel is a function of its location in that wall. The highest hoop tensile stresses occur on the inside surface of the vessel, whereas the lowest ones occur on the outside surface of the vessel (Fig. 17-26). As the thickness of the shell is increased (Fig. 17-27), the tensile hoop stresses for a given internal pressure on the outside surface become smaller and smaller approaching zero, while those on the inside also become smaller, but instead of approaching zero they approach the magnitude of internal pressure. In other words, doubling the wall thickness of a shell does not double the pressure vessel's capability every time. It does so in the beginning when the shell is thin, but with increasing shell thickness, the gain in the pressure capability becomes less and less, until the maximum pressure capability of the vessel is reached, where the maximum shear stress on the interior of the vessel is equal to the yield point stress in shear of the material; that means in terms of stresses, σ_t(hoop stress) $- \sigma_r$(radial stress) $= \sigma_{yp}$(yield in simple tension). No further increase in the vessel's pressure capability beyond this pressure value is possible without

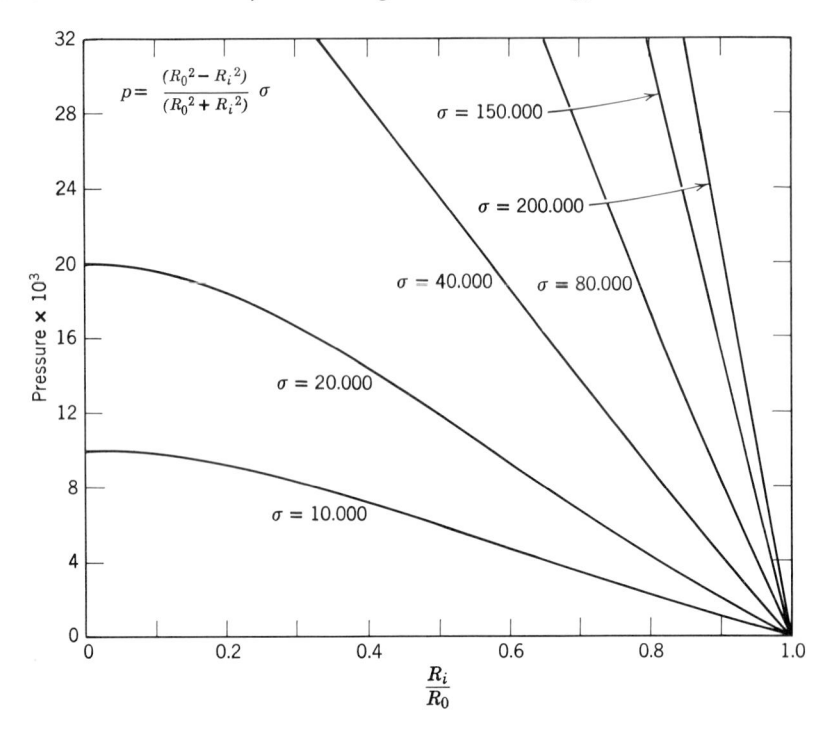

FIG. 17-27. Relationship between R_i/R_0 of the pressure vessel and the contained hydrostatic pressure when the magnitude of the hoop stress on the vessel's interior surface is held constant for a given material. Stress levels of 10,000, 20,000, 40,000, 80,000, 150,000, and 200,000 psi are plotted versus R_i/R_0 and internal pressure.

permanently deforming the walls of the vessel. The radial compressive stress varies also with its location inside the vessel wall. It is highest on the internal surface where its magnitude is equal to internal pressure, and it is lowest on the external surface where its magnitude is equal to zero.

Because of this unfavorable stress distribution, many approaches have been resorted to in order to make the magnitude of hoop stresses more uniform across the thickness of the wall. One such approach is *shrink fitting* of an outer shell upon the inner vessel shell (Fig. 17-28) as by this means compressive stresses can be generated on the interior surface of the cylinder prior to application of hydrostatic pressure to its interior. These compressive stresses on the interior of the thick vessel are of course generated at the expense of tensile hoop stresses created in the external shell being shrunk on the inner shell. When hydrostatic pressure is applied to the interior of the vessel, the typical thick-wall cylinder stresses are then superimposed on the stresses in the vessel created by previous preloading with the shrinking on of the external vessel layer. Since the preload hoop stress on the interior of the vessel was compressive, it tends to make the resulting tensile hoop stress smaller when the internal pressure is applied. By the same token, the tensile hoop stress preload on the exterior of the vessel tends to make the resulting tensile hoop stress larger upon application of internal pressure to the vessel. Thus, because of shrink-fit preload on the vessel, a more desirable stress distribution is achieved when the vessel is internally pressurized—the resultant tensile hoop stress on the interior of the vessel is less and the resultant hoop stress on the exterior of the vessel is larger than it would be without the super-imposition of stress created by shrink fitting of an external shell upon the pressure vessel.

Shrink fitting of shells upon large vessels is generally accomplished by welding of steel plates which are preformed into semicylindrical segments prior to assembly on the vessel's exterior. For this purpose, wide meridional gaps between segments of the assembled exterior shell are designed. During the welding process, these gaps are filled with weld material. Upon cooling of the weld material, the width of the gaps decreases considerably, pretensioning the outer shell. Small vessels, on the other hand, are provided with shrink-fitted shells by preheating an undersized shell to the

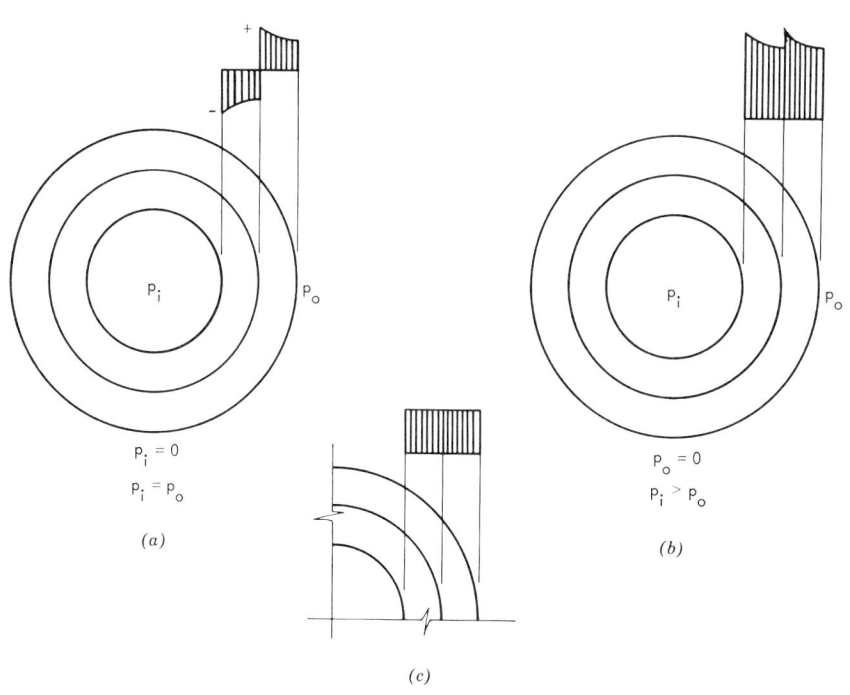

FIG. 17-28. Hoop-stress distribution in a thick vessel of shrink-fitted design. (*a*) Unpressurized; (*b*) pressurized; (*c*) ideal stress distribution.

temperature at which the shell will expand sufficiently to slip over the external diameter of the inner shell. Upon cooling of the outer shell, an interference fit results, causing tensile hoop stresses in the outer shell and compressive stresses in the inner shell.

Another method of creating compressive hoop stress preload on the interior of the vessel is by *autofrettage*. In this technique a sufficiently high internal overpressure is applied to the interior of the thick vessel to produce plastic flow on the inner surface of the cylinder. After removing this high internal hydrostatic overpressure, residual stresses remain both on the exterior and on the interior of the vessel (Fig. 17-29). The residual stress on the interior of the vessel is compressive, whereas the residual stress on the exterior of the vessel is tensile. When operational internal pressure, which is considerably lower than the overpressure used in the autofrettage treatment, is applied during hydrospace simulation experiments, tensile hoop stresses according to thick-wall vessel equation are superimposed upon the residual stresses caused by autofrettage. The result is a more desirable stress distribution, as the hoop stress on the interior of the vessel is less while the hoop stress on the exterior of the vessel is larger than without the effect of residual stress. The end result is a stress distribution similar to the one achieved by shrink fitting of an infinite number of thin shells upon each other till the desired vessel wall thickness is achieved.

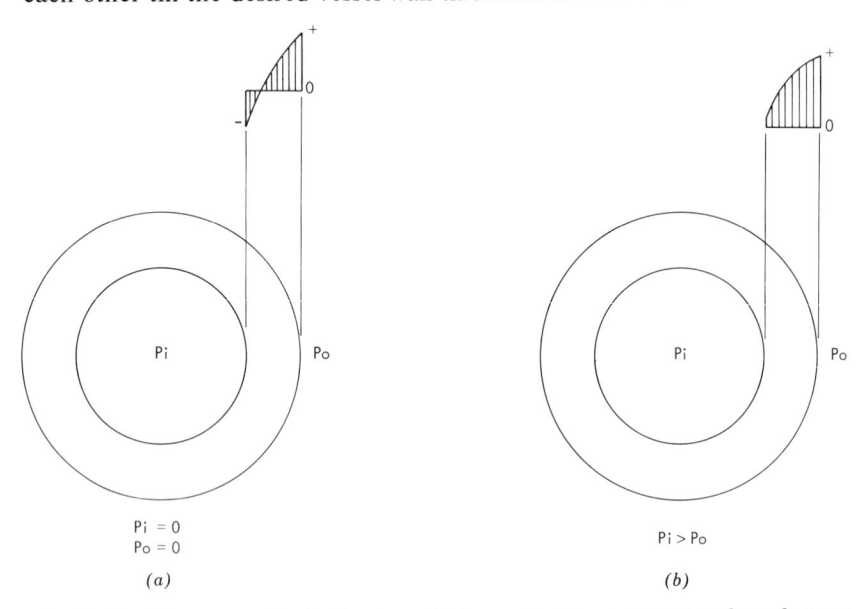

FIG. 17-29. Hoop-stress distribution in a thick vessel after performance of autofrettage operation (a) Residual stresses after operation; (b) stresses in an autofrettaged vessel after internal pressure is applied.

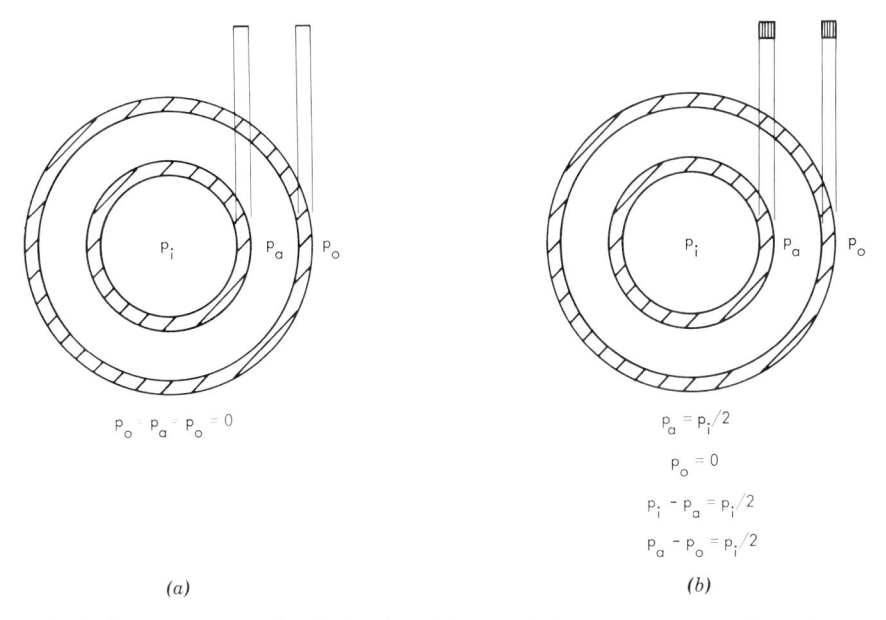

FIG. 17-30. Hoop-stress distribution in a thick vessel of concentric separated-wall design. (a) Unpressurized; (b) pressurized.

Besides shrink fitting and autofrettage, there is also the *wirewrap preloading technique*, which in a manner similar to shrink fitting, generates compressive stresses on the interior surface of the vessel by pretensioning the outer surface of the vessel made up of tightly wrapped high-tensile-stength wire. This technique was widely employed in the past for the fabrication of artillery guns and air storage flasks. Although at the present time it is not employed for fabrication of simulated hydrospace vessels, proposals have been made to this effect, as some economic advantages may accrue from this construction method.

A different approach to changing of the stress distribution present in thick-walled vessels is the *separated-wall construction technique* briefly discussed in a previous section on the types of pressure wall construction (Fig. 17-3). The attractiveness of this approach lies in its ability to completely transform the thick-walled vessel stress distribution to thin-walled vessel stress distribution where the magnitude of the stress is uniform across the whole thickness of the wall (Fig. 17-30). This is accomplished, as already previously mentioned, by pressurizing the annular spaces between individual shells to such a magnitude that the pressure drop across each shell is the same. Considerable problems are encountered, of course, in the design of such a vessel at the end closures and any other penetrations. Still the separated-wall construction technique is desirable for applications where shrink fitting or autofrettage techniques are not feasible as in a large diameter sphere with large end-closure opening (Fig. 17-31) for hydrospace simulation work.

The four above-mentioned techniques for changing the thick-walled vessel hoop-stress distribution to a thin-walled vessel membrane-stress

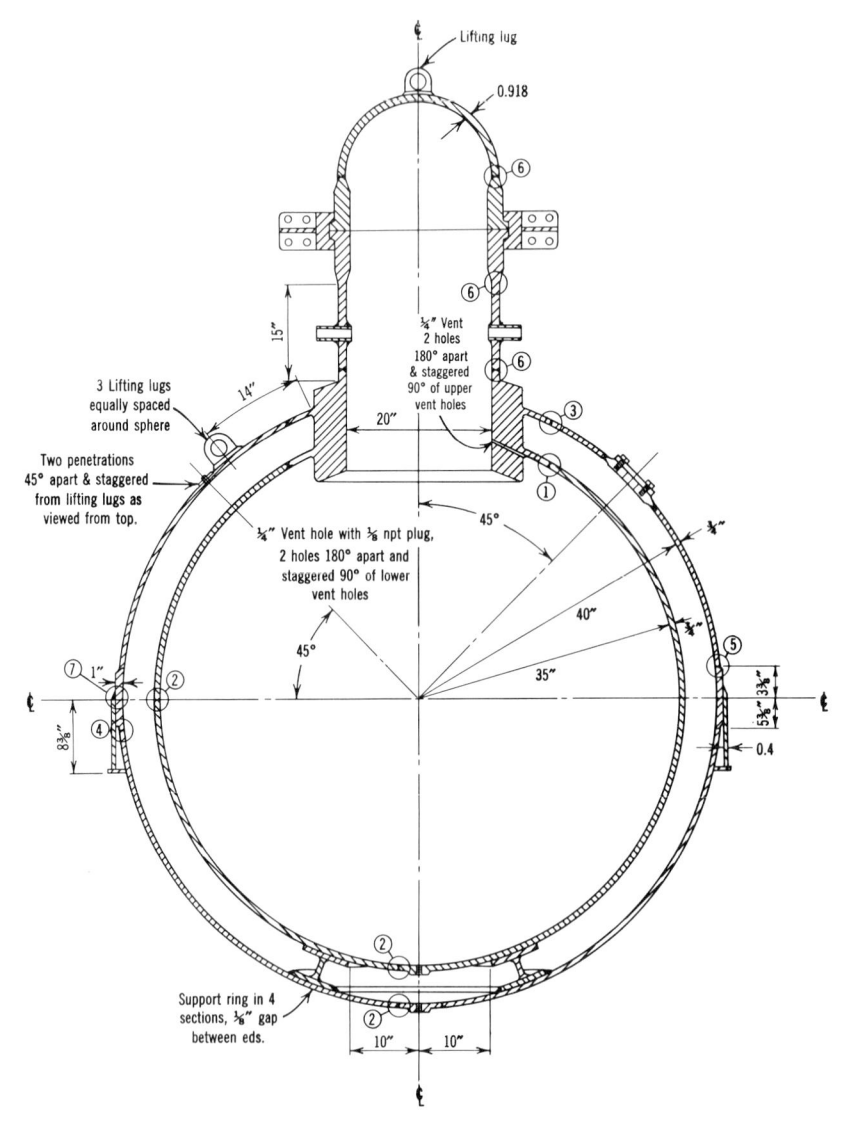

FIG. 17-31. 70-in I.D. steel model of a 420 in. I.D. spherical pressure vessel for 5000-psi service. (Courtesy Southwest Research Institute.)

distribution typify the design techniques utilized by vessel design engineers to economize the use of structural material in the fabrication of hydrospace-simulation vessels. No doubt, the future will see other techniques added to those that already exist, as requirements grow for larger vessels with pressure capability sufficiently high to simulate abyssal depths of the ocean.

Selection of safety parameters for pressure vessels

The selection of the proper safety parameters for a given pressure vessel to be used in a simulated hydrospace facility is one of the most difficult decisions that the facility manager must face before placing the order for the fabrication of the vessel. The reason for this lies in the two vital areas of management function on which the selection of the safety parameters impinges. These areas are safety and economics. Considerations of safety demand, on one hand, a vessel with as many safety aspects as possible, while economics, on the other hand, require the lowest possible investment in an operable vessel. Financial investment can be readily calculated, but it is hard to place a fixed price on safety, therefore, most of the debate revoles around the latter concept.

The problem can be, to a large extent, clarified by discussing briefly the possible financial losses that may be incurred by the catastrophic failure of any pressure vessel used for simulated hydrospace work. The losses are made up of three factors: (a) loss of life with entailing damage suits, (b) loss of the pressure vessel, and (c) damage to buildings. Assuming that the vessel is not located in a remote area or in a completely explosionproof barricade where no contact ever results between the operating personnel and the pressurized vessel, a reasonable assumption is that a catastrophic failure of the vessel will result either in the loss of life, or serious injury to the operating personnel or innocent bystanders. This assumption is valid, regardless of the size of the vessel used; the only modification to this assumption that is required is that the failure of a large vessel will probably kill or injure more than one person. Furthermore, as the size of the vessel increases, the damage to the buildings becomes more and more costly. The cost of the ruined vessel itself is self-explanatory and does not need to be discussed.

Because elimination of all contact between a pressurized vessel and the operating personnel is nearly impossible, and because containment of a vessel explosion is only feasible for very small vessel sizes, further discussion will proceed on the assumption that some contact does exist between operating personnel and the pressurized vessel and that failure of the vessel will result in injury and damage to buildings. Although the cost of an injury is hard to calculate, it can be conservatively estimated to be about $500,000, particularly since highly qualified scientific and technical personnel constitute the manpower complement of a hydrospace simulation facility. With vessels under 1 ft in diameter and pressure capability above 10,000 but less than 50,000 psi, the cost of the damage to the vessel and the building is not high, but still probably at least $10,000 to $20,000. Thus for vessels under 1 ft, the major item in damage cost estimates is the injury to scientific personnel.

For vessels in the 1- to -5-ft diam range with 10,000 to 20,000 psi capability, the cost of vessel and building damages easily reaches 1 million dollars, whereas for vessels in the 5- to -10-ft diam range with the same pressure capability, the cost begins to approach 10 million dollars. For vessels under 1ft in diameter, the assumption was made that only one person would be injured; the catastrophic failure of the larger vessels will most certainly injure more than one person, further adding to the cost of the explosion. Although the breakdown of pressure vessel sizes into three size categories is arbitrary, and the resulting damage costs approximate, the general damage estimates are of the right order of magnitude.

In view of these damage-cost estimates it is mandatory for the hydrospace-simulation facility management to take all steps at their disposal during the procurement of any vessel, except the small ones in a protective enclosure, that will result in a pressure vessel structure whose failure is not possible within the state of engineering knowledge present at that time, regardless of what cost it may entail.

The popular, although somewhat erroneous, concept of safety centers around the phrase "vessel safety factor," which to the casual reader would seem to denote a vessel quality, like weight, diameter, or pressure capability, only in this case denoting the amount of safety that the designer and the fabricator have incorporated in a given vessel. Because the safety factor only partially describes the safe aspects of a given vessel, other safety parameters will be mentioned which must be considered prior to setting down engineering specifications for a new vessel. However, because the safety-factor concept is the best-known parameter used for specifying a safe vessel, it will be discussed first.

Traditionally, the safety factor has meant the relationship between the vessel's working stress at operational pressure and the material's yield strength or ultimate strength. The safety factor as a relationship between the working stress and the material's yieldpoint has found support in Europe, whereas the relationship between the working stress and the ultimate strength is traditional in the United States. The minutely described relationship between the mechanical properties of materials and the allowable stresses permitted in a vessel design has been described in every country by a pressure-vessel code, which has the aim of safeguarding the public against accidents resulting from the faulty design or fabrication of pressure vessels. Wherever such codes are applicable, they should be considered not as the ultimate in safe and desirable design and fabrication practice, but as the minimum level of safe vessel engineering practice, which is open to further improvements in the safety of vessel design and fabrication.

As the purpose of this section is not to describe the existing state and federal pressure vessel codes, but to bring out some of the factors that must be remembered when specifying a vessel for a hydrospace simulation facility, some space will be devoted to a discussion of these factors. A major factor, other than the operational pressure, that must be considered in specifying a vessel is its pressure-cycling life. Because the hydrospace facility vessel is mainly employed for the short-term hydrostatic, hydrodynamic, or pressure cycling of underwater engineering hardware systems or subsystems, the service to which such a vessel is subjected is basically cyclical. Although for some specimen cyclical tests, the internal pressure of the specimen is cycled, rather than the pressure in the hydrospace vessel, for most specimens that is not applicable as they are either sealed mechanical or electronic components. Therefore, a safe assumption to make is that in any of the simulated hydrospace tests, the pressure will be cycled in the vessel, and thus the vessel itself. Most of the short-term hydrostatic or hydrodynamic tests conducted in a simulated hydrospace vessel will subject it to only a single pressure cycle per day, as the majority of time must be devoted to the placement and retrieval of the specimen from the vessel prior to, and after the test. During the pressure-cycling tests, which can be assumed will occur fairly often, the vessel may be subjected on the average to 1000 cycles a day. If we assume an average life of 10 years for the vessel with 25% of the time devoted to cyclical tests, the total number of working pressure cycles to which the vessel may be subjected lies in the million-cycles range. Such a service places the vessel design immediately into the realm of possible failure due to fatigue, rather than due to static pressure failure. Furthermore, implosion tests, although generally not considered as cycling tests beyond the fact that they impose a single pressure cycle on the vessel during each pressurization required for implosion, are in fact high-frequency short-duration pressure-cycling tests. In one of these tests many high-frequency cycles are generated imposing high-frequency stress cycles with fairly large amplitudes superimposed on the mean cycle stress that may be equal to the vessel's static working stress. Because of the possibility of fatigue, a reasonable approach to pressure vessel design would be based predominately on fatigue considerations rather than on static safety-factor aspects.

The fatigue life of a vessel depends basically on three major parameter areas; the fatigue properties of the material, the service to which the vessel will be subjected, and the design and fabrication techniques. The fatigue resistance of metals varies with the alloy, but for most of them it depends on the number of stress cycles, presence of notches and discontinuities, magnitude of the mean tensile stress and the amplitude of stress cycle, surface finish, and ambient atmosphere. Generally speaking, the fatigue failure occurs sooner if discontinuities are present, high mean tensile stress and large amplitude of stress variation exist during cycling, surface finish is rough, and the ambient atmosphere is corrosive. However, all ferrous alloys have a safe range of stress at the given ambient temperature, atmosphere, surface finish, and presence of discontinuities, below which failure will not occur in a given number of cycles.

The fatigue life of a pressure vessel is, in a large measure, dependent on the elimination of stress risers, or their recognition and the subsequent neutralization of their effect on the fatigue life by the use of lower design stresses in these areas. If the stress concentration factors, peak stresses and stress gradients in the vessel, notch sensitivity, variation of mechanical properties, and response of the material to the many stress combinations were known for each vessel design, no further requirements would be necessary in the specification for vessel design but that it successfully withstand the total number of pressure cycles projected for its life. Unfortunately with best presently available design theories, material properties control, and fabrication supervision, unforeseen stress risers will be introduced into the finished vessel that during cycling may cause the vessel to fracture at a much lower number of pressure cycles than those specified. For this reason it appears that the design stress based on 10^6 pressure cycles should be decreased to a lesser value by a minimum factor of 2, or preferably 3. The factor of 3 is based on the general consideration that with the best intents of the designer and manufacturer, stress risers will be included in the structure and that the presence of the stress risers will not be accounted for in the design. Because the magnitude of most stress concentration factors seldom surpasses a factor of 3, the decreasing of the design stress based on fatigue-life considerations by a factor of 3 would appear to be more than adequate to insure a safe vessel.

The presence of the above-mentioned safety factor is very useful in simulated hydrospace facility vessels where seawater is used continuously. Data available for fatigue properties of materials is generally based on materials in a noncorrosive atmosphere, but it is known that corrosion fatigue in the presence of saltwater is particularly severe and tends to shorten the fatigue life of the vessel. Use of corrosion-resistant liners for the vessel is certainly a cheap insurance against the major effects of corrosion fatigue, but even with such a liner, sufficient contact with saltwater will be made on the end-closure restraint mechanism to cause some deleterious effects. Besides the corrosion fatigue, there is also the thermal stress fatigue introduced into the hydrospace-simulation vessel by the refrigeration of the vessel's contents in simulation of temperatures present in the hydrospace.

Other safety aspects important in the specifying of metallic pressure-vessel construction are ductility, nil ductility transition temperature, notch toughness, and possibly yieldpoint and ultimate tensile strength.

Ductility is a measure of the material's plastic deformation at time of fracture in a tensile test. The larger the ductility, the more the material will plastically flow at a point of high stress concentration alleviating, to a large degree, the danger of crack initiation at that point and, furthermore, distributing the stress to lesser stressed areas. Materials with very little or no ductility are called "brittle" and are susceptible to brittle fracture at relatively low over-all stress levels at the presence of a flaw or notch in the structure. Because the elimination of all notches or flaws in a large welded and machined structure is impossible, brittle materials can not be used for pressure-vessel construction. Although materials are available with widely varying ductilities, a ductility range of 15 to 25% is generally considered adequate.

Nil ductility transition (NDT) temperature is the temperature at which a decrease of the material's ductility properties occurs to such an extent that the stress required for crack propagation is much lower than is required for crack initiation. Although NDT temperature occurs for most steels below freezing temperatures, only vessels operating at steady ambient temperatures above 100°F are outside the scope of danger that NDT imposes. Vessels that are either operated in climates with freezing temperatures or are equipped with refrigeration coils for simulation of near-freezing temperatures prevalent in ocean depths, are particularly subject to brittle fracture danger if fabricated from material with high NDT temperature. Because of this, a safety factor on NDT temperature of NDT + 60°F to + 100°F is recommended, so that to meet a + 32°F operational temperature requirement a material with − 30° to − 70°F, NDT temperature must be specified.

Notch toughness is the ability of the material to resist crack propagation, and it is usually measured by impact tests of the Charpy V-notch type. In practice, such impact tests are used on material coupons to determine its NDT temperature, as in that temperature range the magnitude of impact energy required to break a notched specimen radically decreases. Materials with low toughness are not suitable for pressure-vessel construction as they have no ability to distort under impact and furthermore, once a crack is initiated in such a material, it propagates explosively resulting in an instantaneous release of potential energy contained within the vessel. Materials with a notch toughness of 30 to 50 ft-lb at the lowest operational temperature of the vessel as determined by the Charpy V-notch impact tests, are generally considered satisfactory for pressure vessel construction.

The *tensile yield strength* and *ultimate tensile strength* of a material describe its yield and ultimate strength. Steels are available with yieldpoint as low as 15,000 psi, and as high as 350,000 psi. Their ultimate tensile strengths may be just slightly higher than their yield points, or they may be larger by a factor of 2. Generally speaking, the higher the yieldpoint of a material, the closer it will be to its ultimate tensile strength. The difference between the magnitudes of yield and ultimate tensile strength of a given material must receive consideration when the safety factor for static working stresses in the vessel is selected on the basis of the ultimate tensile strength. Otherwise the selection of an insufficiently high static working stress safety factor for material with large difference in their yieldpoint and ultimate strength may make the working stress to be equal to the yield strength of the material. From an economic consideration, the material with the lowest yieldpoint and ultimate strength should be specified that will meet the vessel's operational requirements without making the vessel so cumbersome that it cannot be transported. Premium materials with very high strengths become economically attractive only if the vessel cannot be otherwise transported by rail, or if by using premium materials the more advantageous thin-wall instead of thick-wall stress distribution can be generated in the vessel wall with the resultant economy in material. However, the higher the strength of the material, the more difficult it is to join it reliably be welding, and thus more costly to use on a per-pound basis. But without doubt, as the demands for higher pressure and larger diameter increase, more and more premium materials with strengths in excess of 150,000 psi will be utilized. Such a family of materials in the form of maraging steels already exists, but it requires a considerable amount of additional research on reliable welding techniques before they will find popular acceptance as safe pressure-vessel construction materials.

In summary, the selection of the safety aspects of a hydrospace-simulation pressure vessel is not a simple or easy undertaking. The vessel's projected utilization, pressurization fluid, fatigue life, ductility, nil ductility transition temperature, notch toughness, yield point and ultimate strength all must be included in the design and specification of the vessel. Only by including these numerous aspects can the facility management be assured of a safe operating vessel which will not fail, thus killing or maiming the scientific and technical personnel and causing extensive and costly damages to the vessel and buildings.

Vessel placement

There are several salient factors that must be considered in the placement of a pressure vessel (Fig. 17-32). One of these is the *operational factor*, which to some extent was discussed in the section on shape of the vessel. For use in short-term testing it is imperative that the vessel be easy to open and close and that the vessel be put quickly into full working order. If it is not, the number of tests will be drastically cut and the facility used only infrequently. From the viewpoint of operational simplicity and economy of operating time, the vertically oriented vessel installed in a pit is most desirable. With this orientation it is not necessary to pump out the fluid before opening the vessel and it is possible to prefill the vessel before inserting the specimen to be tested.

However, for some types of testing, such as large-diameter objects of considerable length, it is often necessary to use the horizontally oriented vessel. In this case the pressurization fluid must be pumped out before the vessel is opened and cannot be replaced prior to closing. In addition it is usually necessary to have special equipment which moves on rails to place the specimen inside the vessel, and other complex power-driven manipulators for inserting and removing the vessel end closure.

Another type of orientation is a pivoted system (Fig. 17-32 and 17-33). With this system it is possible to test in either the horizontal or vertical position or at some intermediate one. The size of the vessel, however, limits the use of this system because the vessel must be relatively small. The advantages of such a system are that both end closures can be removed for cleaning and repair and that the utility of the vessel is much greater than if it were placed either vertically or horizontally.

A second factor that is important to vessel placement is that of *safety*. Too many laymen believe that hydrostatic pressure is harmless. This is far from the truth. Anyone who has participated in hydrostatic pressure testing at high pressures knows the danger involved (Fig. 17-34). Therefore, when tests are in progress, particularly high-pressure tests, it is imperative that the operating personnel be located in a separate control room well protected against the force of possible vessel explosion by reinforced concrete walls and ceiling. If such an arrangement is unacceptable for a particular facility because of lack of space, portable shields must be employed. The portable shield (Fig. 17-35), although not capable of containing the whole force of a vessel explosion, will protect the personnel against high-pressure jets of water or shrapnel produced by a crack in the vessel wall, or failure of high-pressure plumbing components. When such shields are utilized, it is important that besides being sufficiently sturdy, they should also be anchored firmly to the building foundations to prevent them from becoming lethal projectiles themselves when struck by a large high-pressure jet of

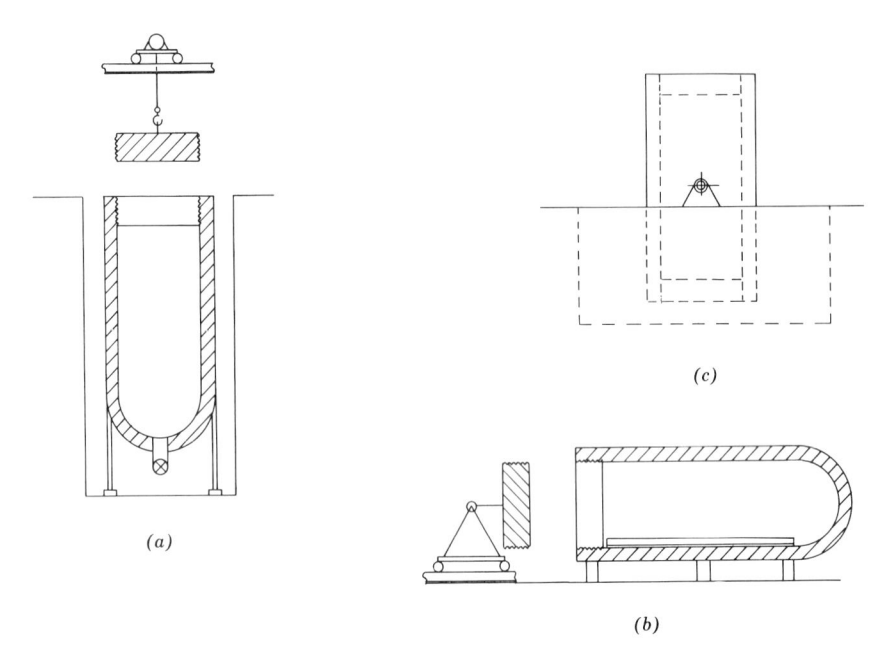

FIG. 17-32. Pressure-vessel orientation: (*a*) vertical; (*b*) horizontal; (*c*) pivoted.

FIG. 17-33. Pivoted 18-in. I.D. × 36-in. long vessel for 20,000-psi operation. (Courtesy U.S. Naval Civil Engineering Laboratory.)

FIG. 17-34. Results of 9.5-in. I.D. × 48-in. long pressure-vessel end-closure failure at 38,000 psi. Estimated kinetic energy of ejected head was 225,000 ft/lb. (Courtesy U.S. Naval Civil Engineering Laboratory.)

FIG. 17-35. Portable shields for protection of operating personnel against shrapnel and high-pressure jets of water accompanying pressure-vessel failures. (Courtesy U.S. Naval Civil Engineering Laboratory.)

684   Systems Design—The Technology

water. Small high-pressure pumping and vessel systems can be completely enclosed in steel shielding to contain all products of vessel or pumping system explosion (Fig. 17-36).

Besides protecting operating personnel, it is also necessary to give thought to the protection of people and buildings in the vicinity of the hydrospace-simulation facility. Because the calculated forces of a hydrostatic pressure-vessel explosion at pressures of 10,000 psi or higher are quite large, upward ejection of end closures upon vessel failure can be predicted and have been observed in some cases to be 100 to 500 ft high. To stop an end closure of several tons moving at sufficient velocity to reach such elevations is nearly impossible without locating the pressure vessel in mountain caves. It is therefore much more preferable to place a simulated hydrospace facility sufficiently distant from other buildings and traffic where an upward ejection of the end closure, a typical type of failure, would cause it to fall relatively harmless on unoccupied ground. The bursting of vessel walls is, of course, protected against by placing the vessel upright in a pit of sufficient depth where only upward ejection of end closure or wall shrapnel can take place. Personnel in all cases should be during tests in underground bunkers of sufficient strength to withstand the impact of an end closure falling upon it from elevation of 100 to 500 ft. Regardless of which approach is taken to the

FIG. 17-36. Enclosure for the hydraulic pressurization system and 1.5-in. I.D. × 10.5-in. long pressure vessel of 15,000-psi operational pressure. (Courtesy U.S. Naval Civil Engineering Laboratory.)

safety of personnel and equipment, it will be only crowned by success when the vessel and associated pressurized systems are considered to be made of an explosive substance only awaiting the spark of crack initiation. Although economic considerations are always uppermost in the layout and construction of a hydrospace-simulation facility, and safety is an extremely expensive feature of any design, it certainly is cheap in the long run where a single pressure-vessel failure may kill or mutilate the whole technical and scientific complement of a simulated hydrospace facility.

The third factor of importance to vessel placement is that of weather protection. Housing for the pressure facility must provide, at a minimum, protection from the elements and protection from windblown abrasive dust. The finely machined mating surfaces of the vessel and end-closure parts can be severely damaged by abrasive dust. This problem is aggravated by the fact that these parts must be lubricated and provided with anticorrosion coating which may also take the form of a grease or oil. If long-term testing is considered, the problem of air-temperature variations and the effect this has on the vessel temperature and, as a consequence, on the vessel's internal pressure must be considered. A temperature change of a few degrees in a vessel can change the internal pressure in the vessel by several hundred psi. For this reason, if high-quality simulation is to be the goal, then air conditioning must be considered as that is the most economical method available for assuring constant temperature.

The fourth aspect worthy of consideration is adequate lighting, drainage of floors and pits, convenient location of power, air and water outlets, and built in fire-fighting system where oil is used as pressurization fluid.

17.4 HOISTING EQUIPMENT

Another type of equipment that is necessary in a hydrospace-simulation facility is hoisting equipment. Such equipment must be an integral part of the facility and capable of being used at a moment's notice. This equipment is used every time the end closure of the vessel is placed on the vessel, or removed. In addition to the simple process of lifting and lowering the components of the end closure, the hoist must also support the end closure when it is being screwed and unscrewed, otherwise the weight of the end closure is so great on the seating surfaces of the threads that screwing is very difficult. The placement or removal of specimens, too, depends on the hoisting equipment. With very large or very heavy specimens, it would be impossible to place them on their supports inside the vessel unless the hoist was available. Hoisting equipment is also necessary for the maintenance of the pressure pumps and other heavy hydrospace-simulation-facility components. It is sometimes impractical or impossible to service these components *in situ* and the hoist must be available to lift, move, and replace the components.

There are several characteristics of good hoisting equipment for the hydrospace-simulation facility. The most desirable equipment has all of these characteristics. First, the equipment should consist of a bridge-crane-type hoist with sufficient travel and traverse to reach the vessel, the specimen preparation area, the equipment maintenance area, and the unloading area. Unless the hoist has this much travel and traverse, its use is limited and additional expenses will be incurred in purchasing other equipment, such as a fork lifts, which can travel to these areas.

Second, the hoist should be electrically powered. Electric power allows a one-hand operation which permits the other hand to guide the object being placed. Some facilities use manually-powered hoists but this type is certainly not desirable under the circumstances. Besides the hoist, the travel and traverse should also be electrically powered and provided wherever feasible with electric brakes. When we consider that often the end closure weighs many tons, it is obvious that the crane should be powered in all aspects or the operators will be spending excessive time and energy attempting to handle the equipment.

Third, it is best if the hoist is equipped with two speeds, a mating speed and a transport speed, whereas a single speed is sufficient for traverse and travel. The mating speed should be a very slow speed, about 6 in./min. This is imperative when the end closure is being placed on the vessel because if the closure is seated abruptly it can damage the seals, the mating surfaces, the threads and/or the feedthroughs. When a specimen is being placed or removed, the slow speed assures that there will be no damage to the specimen and that the specimen can be aligned with its supports inside the vessel. However, once the specimen or end closure is free of the vessel, the hoist should be able to lift its load quickly. A speed of 5 to 10 fpm is sufficient for this second speed.

There are bridge cranes available from commercial suppliers which have all of these features. However, many times the hydrospace-simulation facility cannot afford such a crane. It is possible, of course, to do without electrically powered equipment, but it is extremely unwise not to have some type of two-speed arrangement on the lift. Many of those facilities which have limited resources solve their speed problem by using a crane equipped with only a transport speed and attaching to the crane an air-cylinder lift attachment. The end closure or specimen being moved is then attached to the piston of the air cylinder. By slowly raising or lowering the pressure inside the cylinder, a second slower speed is obtained. In addition, the air cylinder acts as a shock absorber, thus providing protection to the equipment being moved and the structure supporting the crane.

17.5 HYDRAULIC SYSTEM FOR SIMULATED HYDROSPACE PRESSURIZATION

Besides the pressure vessel, the most important and expensive equipment of the hydrospace simulation system is the hydraulic system. The reason for its importance lies in the fact that it not only supplies the pressurized seawater to the pressure vessel, but also controls its pressure and rate of flow. The components of the hydraulic system can be readily classified into the following categories: *pumps, piping, fittings, pressure* and *rate-of-flow regulators*.

As mentioned previously in the discussion of the pressure-vessel construction, corrosion in the presence of seawater is a serious problem that must be guarded against. Not only can corrosion damage the equipment in the simulation system, it also presents a second problem of contamination of the fluid in the system. High-pressure needle valves, check valves, and pumps are easily damaged by particulate matter, such as rust. Dissolved corrosion products will contaminate the normal seawater system to be duplicated. For these reasons, corrosion of the system must be reduced to a minimum and the system should ideally be of the open-loop type so that if some of the fluid does become contaminated, it is not reintroduced into the system.

To insure that the hydraulic components of the facility do not contaminate the system by their corrosion products, they should be made of Hastelloy C, Monel, Type-316 stainless steel, or plastic. If for some reason it is impossible to use these materials, then the components must be at least nickel plated, or coated with plastic coatings. Only by using special materials or coatings can it be assured that the simulation facility is reasonably free from corrosion and its byproducts.

High-pressure pumps

A very important part of the hydrospace-simulation hydraulic system is the high-pressure pump or pumps which actually accomplish the pressurization of the hydrospace. Pumps for high-pressure work are generally of the positive-displacement-piston type, driven either by air (Fig. 17-37) or an electric motor. For small volumes, up to 100 in.^{3}/min at 10,000 psi or 50 in.^{3}/min at 20,000 psi, the air driven pumps provide the best and most economical solution. For higher-volume delivery requirements, particularly in the case of pressure cycling where relatively large volume must be pumped in a short time, the electric motor-driven pump is more advantageous.

There is a characteristic of air-driven pumps which makes them especially attractive to hydrospace simulation. These pumps operate on a differential

FIG. 17-37. Bank of individually controlled air operated positive displacement pumps with a total output of 50 in.^{3}/min of seawater at 20,000 psi (Courtesy U.S. Naval Civil Engineering Laboratory.)

area piston principle, with large piston area exposed to the driving air, and the small piston area exposed to the fluid to be pumped. The piston-area ratios vary from 1:1 for low-pressure pumps to 500:1 for high-pressure pumps. Because of the differential piston-area arrangement, the maximum output-pressure of an air-driven pump can be controlled by regulating the air input pressure. Furthermore, such pumps will automatically stall when the maximum hydraulic pressure for which the air pressure has been set is reached and will restart automatically if the hydraulic pressure on the output side drops. This in effect provides an inexpensive method of controlling the hydraulic pressure in a system, assuming, of course, that the sensitivity of the air-pressure regulator used with the particular pump is satisfactory.

The electric motor-driven pumps, on the other hand, have no such built-in feature. In order to have an automatic system for maintaining pressure these pumps require a separate pressure-sensing and control system which either shuts off and restarts the pump, or opens and closes a bypass valve as required. This type of pressure-control system is extremely expensive when large volumes, high pressures, and corrosive fluids like seawater are to be handled. However, whenever large volumes of fluid are to be pressurized rapidly, an electric motor-driven pump is usually essential (Fig. 17-38).

The life of air-driven or electric pumps for pressurization of seawater to high pressures is rather short due to corrosion fatigue of cylinders, and abrasion of reciprocating or rotating pump components caused by lack of surface lubrication. To alleviate this problem many large-volume and high-pressure pumping installations for simulated hydrospace facilities utilize a three-stage system for supply of pressurized seawater. One part of the system utilizes hydraulic oil as the pressurizing medium, and the other part of the system uses seawater. The pump's pressure regulators, pressure recorders, and flow regulators are located in the portion of the hydraulic system that uses oil in a closed-loop cycle, and the portion of the system utilizing saltwater encompasses only the pressure-vessel cavity. The exchange of hydraulic power between the two systems occurs in a single or multiple energy-transfer chambers where the two fluids are either separated by thin elastomeric membranes or reciprocating leaktight rigid pistons. Where an open-loop hydraulic system for the continuous supply of seawater is not required, the energy-transfer chamber may be a simple accumulator of sufficient volume to pressurize the simulated hydrospace vessel with a single expansion of elastomeric membrane, or one stroke of the accumulator piston. When continuous supply of seawater is required in an open-loop hydraulic system, the energy-transfer stage must incorporate a system of check valves and power operated valves to permit the energy-transfer chambers to perform similarly to positive displacement pumps with very large

FIG. 17-38. Portable self-contained pressurization system capable of pressurizing 5.5 gal of seawater/min at 10,000 psi. (Courtesy Kobe, Inc.)

slow-moving pistons. By the utilization of a two-fluid hydraulic system, great economies in initial investment and maintenance may be achieved, as pumps and pressure and volume regulators are readily available for hydraulic oil service in a large selection of materials, whereas for seawater service they must be in many cases custom made of seawater-corrosion-resistant materials.

Low-pressure pumps

Low-pressure pumps are utilized for supplying seawater from the ocean to the high-pressure pumps, and to evacuate it after it has seen its use in the vessel. The selection of commercially available pumps for low-pressure seawater service is large both as to type as well as material of construction. The most economical and corrosion resistant have proven to be, so far, centrifugal pumps of polyvinyl chloride plastic construction. They are commercially available in a large range of volume and pressure ratings. Generally speaking, the low-pressure transfer pumps can be considered a readily available and easily maintained part of the hydraulic system for simulated-hydrospace facilities.

Piping

Piping for the low-pressure part of the hydraulic system generally consists of polyvinyl chloride pipes equipped with PVC threaded or solvent-bonded fittings. The sizes of commercially available PVC pipes range from $\frac{1}{4}$-in. to 6-in. pipe with 40, 80, and 120 schedule walls. The PVC piping represents a very economical solution to containment of seawater in the -14.7-to-$+150$-psi pressure range, as it is readily available, can be cut to shape and bonded with ordinary tools, and is not attacked by seawater.

Piping for the high-pressure part of the hydraulic system employing seawater is commercially available as tubing in the Type-316 stainless-steel alloy. The tubing is available in a wide assortment of sizes and wall thicknesses. For the convenience of the user, the tubing stock is generally classified according to its working-pressure capability into 6000-, 15,000-, 30,000-, 60,000- and 100,000-psi pressure categories. In the 6000-psi range the tubing is available in outside diameter sizes to 1 in., in the 15,000-psi range to 1 in., in the 30,000-psi range to $\frac{9}{16}$ in., in the 60,000-psi to $\frac{9}{16}$ in., and in the 10,000-psi range to $\frac{5}{16}$ in. Besides tubing, Type-316 stainless-steel piping is also available for the 6000-psi pressure range in sizes to 3 in.

Fittings and valves (Fig. 17-39) in different connection styles are available for all the above mentioned pressure ranges and tubing and pipe sizes

FIG. 17-39. Typical high-pressure needle valve with NBS cone type connections. (Courtesy Autoclave Engineers.)

(Fig. 17-40). The connections with operational capability in all pressure ranges to 100,000 psi are the NBS cone, and lens types. The other types of connections are limited to lesser pressure ranges. Thus pipe-thread connections are limited to 3000 psi, AN-type connections to 6000 psi, and compression-type connections to 15,000 psi. The standard flanged pipe connections are generally used only with larger pipe sizes in the 6000 psi range. Although it is an accepted practice to make only those parts of the fittings from Type-316 stainless steel that are wetted by the pressurized seawater, it is highly recommended to use only fittings and connections in which all of the components are made from Type-316 stainless-steel, Hastelloy, Monel, or titanium. Otherwise, if due to leakage or spillage, the seawater comes in contact with the generally nonwetted components, it can cause extensive and unsightly rusting.

Pressure and rate-of-flow regulators

The pressure and rate of flow are very important parameters of pressurized seawater that must be continuously controlled in a hydraulic system feeding the simulated hydrospace chambers. In view of the many controllers available for that purpose only a few high points will be discussed. The major features of regulators that differentiate them from each other are their

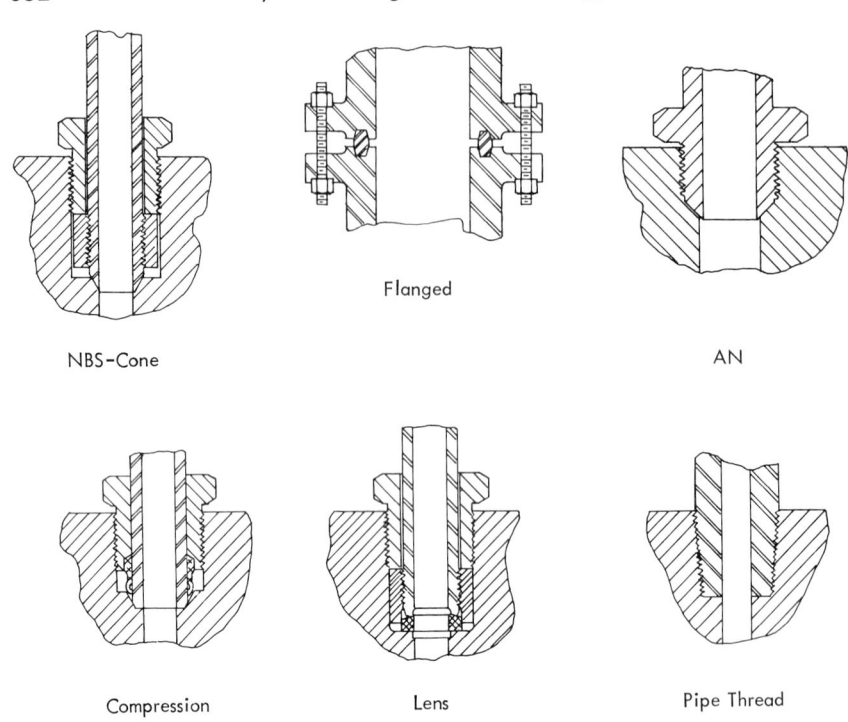

TYPICAL HIGH PRESSURE CONNECTIONS

FIG. 17-40. Typical high-pressure connections.

sensing circuits and their response circuits. The sensing circuits of the regulators may operate either on the feedback or the programmed principle. In one case the output of the regulator is continuously monitored and the information fed back to the control circuits, which activate the pneumatic or electric response circuits that physically move the proper needle valves according to information provided by the pressure or rate-of-flow feedback circuits. In the other case the regulator is mechanically or pneumatically programmed for a given pressure or rate-of-flow control setting. The sensing circuit in this case is generally a compressed spring, or compressed air, whereas the response circuit is a pneumatically, or hydraulically operated needle valve. In view of the large pressures, small flow rates, and corrosive fluid handled, the programmed regulators are not deemed as reliable as regulators with independent feedback sensing systems. With no independent feedback-sensing circuit, the regulator will attempt to compensate for wear or corrosion of the mechanical regulator components regulating the flow of seawater. On the other hand, the programmed regulators not provided with independent feedback-sensing circuits may have their programmed pressure or volume setting change considerably due to deterioration of the regulator components without the regulator responding to it, because the regulator has no engineered-in capacity of noticing the deterioration.

17.6 INSTRUMENTATION OF SIMULATED HYDROSPACE

In all of the previous sections it was inferred that instrumentation was used for ascertaining the results of the interaction of simulated hydrospace and the test specimen. Flow of information from tests conducted in simulated hydrospace actually involves information from two sources—the simulated hydrospace itself and the specimen being tested.

In instrumenting the hydrospace the objective is to monitor the various properties of the hydrospace. Instrumentation is necessary to know the static and dynamic pressure exhibited by the hydrospace, the temperature of the seawater, the pH factor, the salinity or change in salinity of the seawater, and other chemical and physical forces acting in the hydrospace (Fig. 17-41). The philosophical consideration behind this monitoring is that if the condition of the simulated hydrospace is unknown or uncertain, then the interaction of this hydrospace and the specimen will also be uncertain.

The instrumentation of the specimen too is of vital importance (Fig. 17-42). The objective in this type of instrumentation is to ascertain the changes in the specimen as a result of the action of the simulated hydrospace on the specimen. It is necessary to know the strains on the interior and exterior of the specimen if it is a structure, the strain rate, the reaction to dynamic forces such as impact, the deflection and translation of the specimen surface, the buoyancy or change in buoyancy, the change in volume of the specimen, the permeability of the specimen, the corrosion of its surface, creep, and other data.

To obtain data from both simulated hydrospace and the specimen, numerous instruments are employed. These can be grouped into hydraulic, mechanical, electronic, and photo-optical types of instrumentation. This classification is entirely arbitrary, as most of the instruments used to monitor properties of the simulated hydrospace and behavior of the specimen rely on a combination of hydraulic, mechanical, electronic, and optical principles to make them operational. Still, the classification of instruments into various categories is useful for the sake of better understanding their principle of operation.

Instrumentation relying on mechanical principles of operation is, generally speaking, extremely reliable, rugged, and simple to operate. Its greatest advantage is absence of drift under long-term test conditions. Thus

694   Systems Design—The Technology

FIG. 17-41. Typical instrumentation console for readout and recording of simulated hydrospace pressure, temperature, pH, dissolved oxygen, and salinity. (Courtesy U.S. Naval Civil Engineering Laboratory.)

its basic field of application is the instrumentation of hydrospace and specimens for long-term pressurization tests. Good examples of mechanical instrumentation are the Bourdon-type pressure gauges for measurement of simulated hydrospace pressure, and mechanical dial gauges for the measurement of specimen surface deflection and translation.

The major problem connected with mechanical instrumentation of specimens subjected to hydrostatic pressure inside a simulated hydrospace chamber is the readout of the mechanical gauges. For some of the mechanical gauges, it is possible to place the dials on the exterior of the vessel, while the mechanical transducer actuating the gauge by proper mechanical linkage is located in the interior of the vessel. An example is the measurement of surface deflection on the interior of a hemisphere mounted on the interior surface of the vessel closure. In this case the dial is located on the exterior of the vessel, and the mechanical linkage protrudes into the interior of the externally pressurized hemisphere through an unsealed opening in the end closure. In case of structure-model tests in which the mechanical gauges are located within a completely enclosed structure wetted on all sides by simulated hydrospace, the readout of the gauge dials is either accomplished optically via glass fibers, by photocells, by TV camera, by mechanical coupling to a potentiometer, or by mechanical coupling to a selsyn. The photocell, TV camera, potentiometer, or selsyn is then read externally by means of electronic instruments.

Instrumentation relying on hydraulic principles of operation is, in general, fairly reliable, rugged, and relatively simple to operate. Its greatest advantage is the ability to measure directly forces, changes of volume, and pressures with little or no drift. Good examples are the manometer for measurement of relatively low hydrospace pressures, and graduated for the measurement of liquid volume displaced from the interior of a liquid-filled structure under external hydrostatic pressure. The measurement of the displaced liquid volume from the structure interior represents the change of the structure's displacement, and thus buoyancy. This is the most direct way known to integrate all the local deflections and translations of the structure's

FIG. 17-42. Hollow concrete sphere instrumented with electric resistance strain gauges equipped with water proof wiring plugs. (Courtesy U.S. Naval Civil Engineering Laboratory.)

hull. Miniature hydraulic cylinder and piston assemblies mounted on the interior of the structure can be also used as motion transducers to measure local deflections and translations of structural components in a structure. By having the output of the hydraulic motion transducer connected via tubing to a set of hydraulic actuator and mechanical dial-indicator assemblies on the outside of the pressure vessel, the deflections of the structure undergoing hydrostatic testing can be read directly. If the area of the transducer piston is large in comparison to the area of the actuator piston, magnification by a factor of 100 of the structure deflection can be readily achieved. A great advantage of hydraulic instrumentation lies in its ability to transmit the signals from the transducers located inside the structure undergoing testing to readout instruments in the laboratory via hydraulic piping and tubing.

Electronic instrumentation is, in general, versatile, capable of large signal magnification, and adaptable to many already existing readout and recording systems. Its greatest advantages are the miniature sizes of electronic transducers, use of inexpensive and readily sealed wire conductors for transmission of signals from the transducers to remote recording equipment, and the relative low cost of transducers for the measurement of a variety of mechanical, physical, and chemical parameters describing the properties of simulated hydrospace (Fig. 17-41), and the behavior of the test specimens. The shortcomings of electronic instrumentation are primarily its tendency to drift with time and its susceptibility to malfunction in the presence of seawater. Typical examples of electronic instrumentation are the piezo electric transducer for measurement of dynamic pressures in simulated hydrospace, and the electric resistance strain gauge for measurement of strains on the exterior and interior of test specimens.

Instrumentation employing optical principles for its operation is, generally speaking, very reliable, rugged, and does not require wires, tubes or mechanical linkages for transmission of signals from the specimen to the readout instruments. Its greatest advantage lies in its fine resolution of details under observation. Typical applications of optical instrumentation are the observation of biological specimens in hydrospace, changes of specimen shape under pressure, deterioration of specimen surfaces, and measurement of strains on specimen surfaces by photoelastic techniques. Optical instrumentation has not been employed extensively in the past due to lack of reliable data for the design of windows in hydrospace-simulation vessels, and nonavailability of pressure-resistant miniature light housings containing intense light sources for illumination of simulated hydrospace. Such hardware is becoming available (Fig. 17-43) and modern hydrospace simulation facilities are incorporating them into their simulated hydrospace vessels (Fig. 17-44).

FIG. 17-43. Miniature (625 W, 120 V) electric-light assembly for illuminating the interior of simulated hydrospace vessels at 10,000 psi operational service. (Courtesy U.S. Naval Civil Engineering Laboratory.)

FIG. 17-44. Acrylic window for 10,000-psi service located in the head of an 18-in. diameter vessel used to observe behavior of models under pressurization. (Courtesy U.S. Naval Civil Engineering Laboratory.)

From the discussion of the four general types of instrumentation, a conclusion can be drawn that there exists a large selection of instruments with which the properties of simulated hydrospace and the behavior of specimens undergoing testing can be accurately ascertained. In the past, our knowledge of electronic, mechanical, hydraulic, and optical principles has not been utilized to its full advantage for the generation and transmission of information from the interior of simulated hydrospace vessels. As the requirements for more complete understanding of the interaction between the simulated hydrospace and test specimens increase, the development of instrumentation for such purposes will follow suit.

17.7 TESTING IN SIMULATED HYDROSPACE

At the present time in most hydrospace-environment simulation facilities, "to simulate the ocean depths" simply means to simulate the pressure and temperature parameters adequately. The seawater chemistry system can be only roughly approximated by using fresh seawater in an open-loop water system. There is nothing that can be done about the divergence between pressure-vessel size in comparison with ocean size or about the other limitations imposed on simulation. However, as pressure, saltwater environment, and temperature are three of the most important parameters, simulation expenses are justified on that basis alone.

There are several types of tests being conducted in simulated hydrospace—materials testing, structure testing, structural component testing, soils testing, bio-organism testing, and construction-processes testing. Each of these special problems often requires highly specialized equipment.

Materials testing

Materials for use in hydrospace can be tested (Fig. 17-45) to determine the effect of the hydrospace environment on their corrosion rates at great depths, as well as effects of such an environment on their mechanical properties. The problems involved in establishing a suitable environment for valid corrosion studies are considerable. The best that is attempted at present is to use the open-loop water system to provide a constant supply of fresh seawater. By using this type of system, it is possible to study both ordinary corrosion of the materials and stress corrosion under static or cyclical load. However, constant pressure and constant flow of water are required and at present there are few, if any, pumps that can provide such conditions over a long period or numerous cycles. The pump itself is seriously attacked by seawater if made from anything except Hastelloy C, Monel, or Type-316 stainless steel; in addition, most pumps just are

FIG. 17-45. Placement of a plastic container with material test specimens for corrosion testing into a simulated hydrospace vessel. (Courtesy U.S. Naval Civil Engineering Laboratory.)

not designed to operate at high pressures over extended periods of time when the fluid pumped is seawater.

Much could be accomplished in the determination of mechanical properties of materials in the hydrospace environment, however, such tests require equipment that will operate while submerged in pressurized simulated hydrospace. It is possible for example, to test tensile strength, compressive strength, shear strength, hardness, and impact strength when the equipment becomes available. In all of these tests, the equipment for the specific test would be placed, with the material sample, inside the pressure vessel and, after pressurizing the vessel, the test performed. The only reason that such tests have not already been performed is that such test machines have not yet been designed and built.

Structure testing

Another type of testing in simulated hydrospace is structure testing. These tests usually involve either a section of a full-size vehicle (Fig. 17-46) or structure, or a scaled model (Fig. 17-47) of a vehicle or structure. This particular area of testing is probably the most highly developed and widely used of all types. There are many reasons for this, the most basic being that structures for the hydrospace environment form the basis for

FIG. 17-46. Section of an experimental underwater vehicle shell of circumferentially stiffened-sandwich construction. (Courtesy Pennsylvania State University.)

FIG. 17-47. $\frac{1}{8}$ scale model of an acrylic hull for underwater habitat for placement in a simulated hydrospace vessel. (Courtesy U.S. Naval Civil Engineering Laboratory.)

hydrospace exploration. Without these vehicles and structures, there is only superficial knowledge of the underwater biology, geography, soils, oceanography, and acoustics. With tested and proven underwater structures and vehicles, the hydronauts can descend to the bottom of the ocean and add knowledge and power to man's conquest of his universe.

There are several types of tests employed in structure testing. One such test is the *static test*, which may be either destructive or nondestructive.

In the destructive test the structure is tested to implosion (Fig. 17-48). The object of the test is to find the ultimate strength of the structure. By careful instrumentation, it is possibe to learn the nature of the failure, the weaknesses of the design, properties of the material, and the shortcomings of joining techniques used in fabrication. With such information, the designer can improve or discard the particular structure design and thus

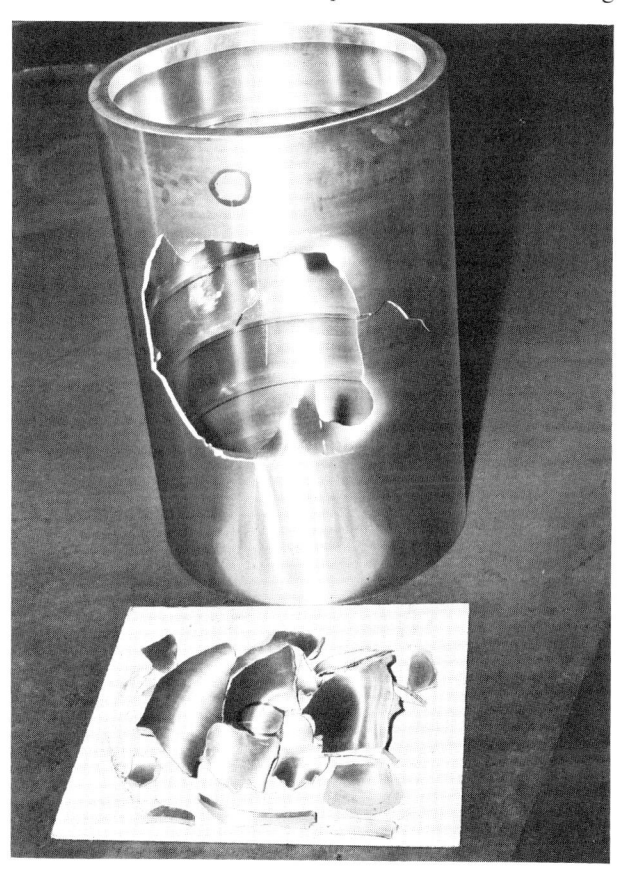

FIG. 17-48. Section of an experimental deep-submergence vehicle of ring-stiffened construction after destructive hydrostatic testing. (Courtesy Southwest Research Institute.)

create a backlog of tested design parameters, materials, and joining techniques.

In the nondestructive test the structure is either exhaustively strain gauged or prefilled with water (Fig. 17-49) and during the hydrostatic pressurization either the volume displacement, the generated internal pressure or strain increments minutely analyzed. When the volume increase becomes disproportionate to the rise in the external pressure, or when the internal pressure rise begins to equal the external pressure rise, or when abnormal strains appear, failure is emminent. In any such test the object of the test is to stop just short of implosion so that the structure can be reused. If failure has not occurred, the specimen may be subjected to other tests, such as long-term pressurization.

A special case of nondestructive structure testing is the proof test. The objective of this type of test is to overload the structure by some percentage of its operational depth so that the structure can be certified safe at its

FIG. 17-49. $\frac{1}{8}$ scale model of an acrylic hull for underwater habitat prefilled with water for measurement of displacement change during the model's pressurization in a pressure vessel with external hydrostatic pressure. (Courtesy U.S. Naval Civil Engineering Laboratory.)

FIG. 17-50. Prooftesting of 10 in. diam glass buoys for 20,000-ft submergence service. (Courtesy U.S. Naval Civil Engineering Laboratory.)

operational depth (Fig. 17-50). Prooftesting is always considered to be of the nondestructive type, even though occasionally implosion does occur in an improperly fabricated structure. Because prooftesting constitutes only a final step in the certification of any structure, destructive tests of static, cyclical, and dynamic pressure type must have been already performed either on realistic models of the structure, or in case of small structures, on duplicate full-size specimens of that structure. Prooftesting without prior destructive tests to models, prototypes, or duplicates of that structure are not conclusive and only develop a false sense of security.

Another type of structure testing is *cyclical testing*. This test, again, may be either destructive or nondestructive. In either case the structure may be pressure cycled from outside or from inside. Outside pressurization of the structure for cycling is more desirable but it is very hard on the pressure vessel, and therefore wherever possible is avoided, although test results obtained by inside pressurization are somewhat questionable. The object of cyclical testing is to ascertain the number of cycles a particular structure can withstand. In the destructive test the cycles continue until the structure implodes. In the nondestructive test the cycling ceases after the appearance of cracks in the structure. It is necessary in the nondestructive test to remove the structure periodically from the vessel and inspect the structure with magnetic flux, X-rays, crack penetrating dyes, or sonic rays so that cracks can be detected in the structure prior to its implosion.

704 Systems Design—The Technology

Still another type of structure testing is *dynamic testing*. The object of this type of testing is to determine the ability of the vehicle or structure to withstand dynamic loading. In the impact-type test the structure, while pressurized, is struck by impact hammers. Generally these hammers are hydrostatically actuated when the burst diaphragm ruptures, however, there are also explosive-actuated hammers now available. Numerous experiments of this kind have been performed on glass models to determine impact resistance (Fig. 17-51).

In addition to impact dynamic tests, there are the explosion tests in which explosives are used to generate shock waves inside the pressure vessel while the structure is under pressure. However, it is not known definitively yet whether the energy released by the explosive is the same regardless of pressure. These tests are generally performed to simulate depth-charge explosions but much research must be done in this area before reliable results can be obtained in simulated hydrospace. Yet another type of dynamic test is the implosion test in which the object of the test is to determine the resistance of the structure to sympathetic implosion. So far the only experiments in

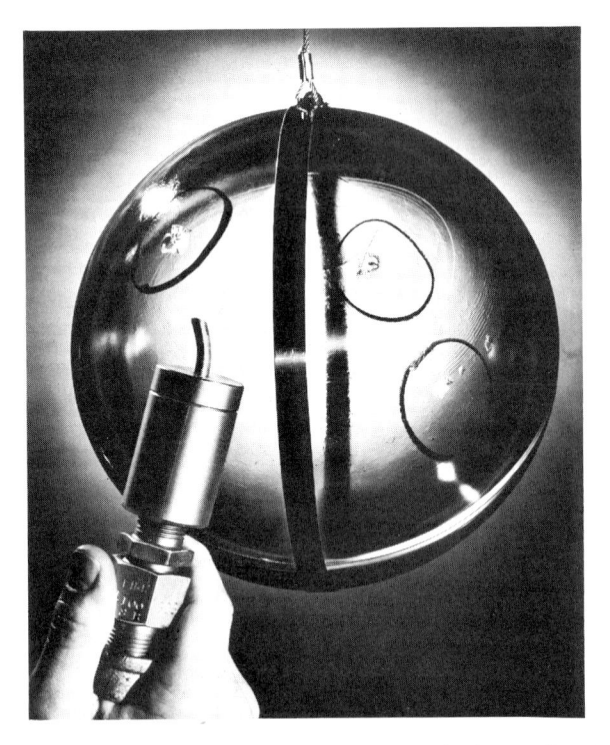

FIG. 17-51. Hollow glass sphere after being subjected to point impacts by a hydrostatically impulsed steel pin at simulated hydrospace depths of 20,000 ft. (Courtesy Benthos Co.)

this area in simulated hydrospace have been conducted to determine the forces released by implosions. These tests are very hard on the pressure vessel and utmost care must be taken to protect the experimenters in case of failure of the pressure vessel. Plans are being made to follow up single-specimen implosion tests with twin-specimen sympathetic implosion tests, where the implosion of one hollow specimen triggers the sympathetic implosion in another hollow specimen located in close proximity. The crucial factor in such experiments is the ratio between the volumes of the test specimens and of the vessel. If the specimen/vessel volume ratio is too high, the pressure of the simulated hydrospace will drop instantaneously upon implosion of the first specimen, making it impossible for the second specimen to fail by the sympathetic-implosion mechanism, in which case, the sympathetic implosion test performed in simulated hydrospace will not be representative of the sympathetic-implosion machanism found in actual hydrospace. What constitutes a desirable specimen/vessel volume ratio is unknown, but generally speaking it is considered acceptable when upon implosion of both sympathetic-implosion-test specimens, the resultant pressure in simulated hydrospace has dropped off less than 5% of the initial pressure existing before the implosion of the test specimens.

Of the several types of structure tests conducted, the dynamic tests are the most undeveloped. They impose the most strain on the hydrospace-simulation facility and are very unsafe unless carefully guarded. However, these dynamic tests must be refined and improved if the simulation of hydrospace is to advance.

Structural-component testing

In addition to testing structures and vehicles it is often necessary to test specific structural components. Many of these components can be mounted on the end closure of the pressure vessel instead of having to have an extra bulkhead. Some components tested in this manner are feedthroughs, manipulators, rods and shafts, and windows (Fig. 17-52). When the windows are tested to destruction, extreme care must be taken to protect the operators from flying fragments and the escaping pressurized media (Fig. 17-53). Tests on the closure-mounted components can, of course, be either static, dynamic, cyclical, and destructive or nondestructive. However, little has been done at present to test these components dynamically.

Another series of components tested in simulated hydrospace are the self-contained components such as lights (Fig. 17-54), motors, batteries, and storage flasks. For example, much vital information has accrued from the testing of motors (Fig. 17-55) subjected to simulated hydrospace where power output, seal life, bearing life, integrity of housing, and insulation

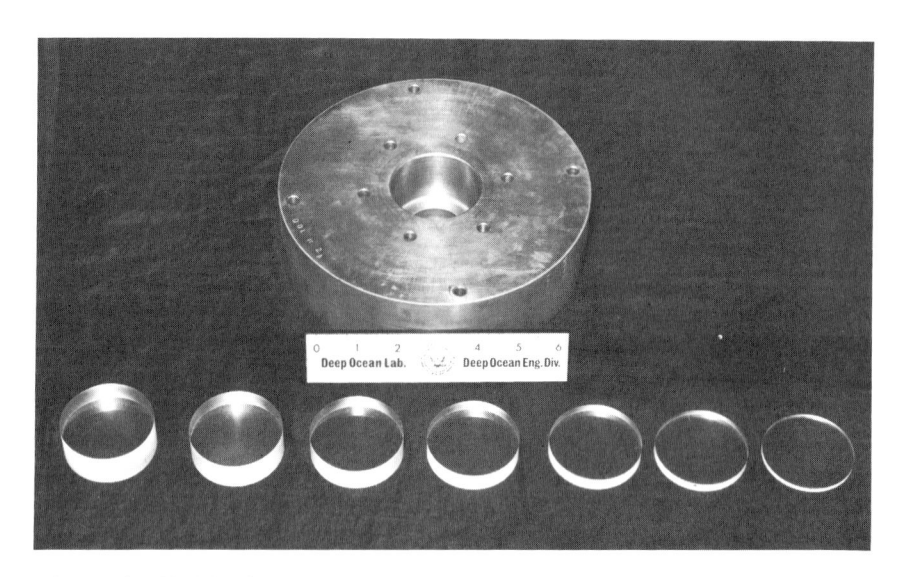

FIG. 17-52. Models of acrylic windows and the retaining flange for deep-submergence applications prior to testing in simulated hydrospace vessel. (Courtesy U.S. Naval Civil Engineering Laboratory.)

FIG. 17-53. Ejection of window fragments upon window failure from end closure of a hydrospace vessel. (Courtesy U.S. Naval Civil Engineering Laboratory.)

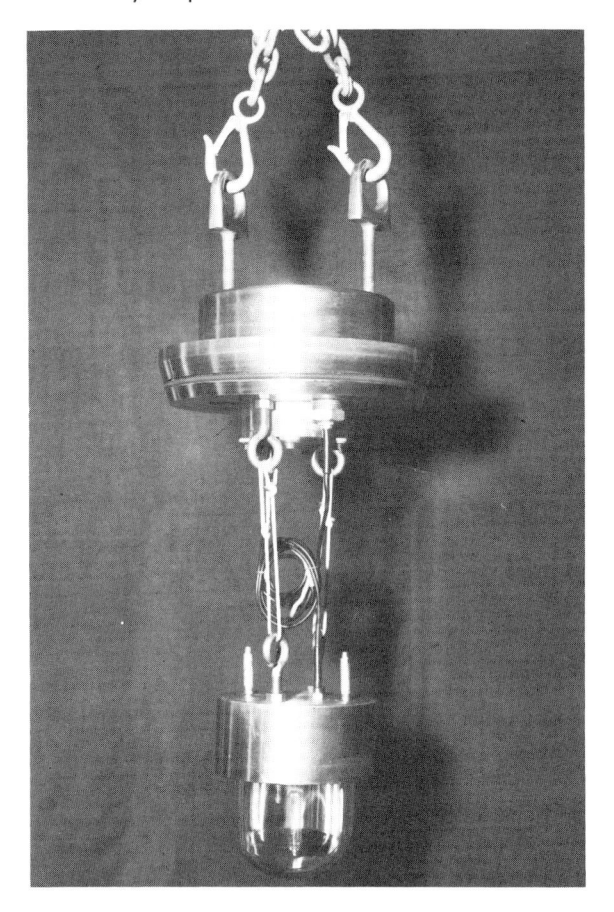

FIG. 17-54. Deep-submergence electric light assembly prior to testing in a simulated hydrospace vessel. (Courtesy U.S. Naval Civil Engineering Laboratory.)

deterioration have been monitored. Such information allows the designer of the component to know where his design needs improvement.

Soil testing

Investigation of physical, chemical, and mechanical properties of ocean-bottom soils in the simulated hydrospace environment has only recently begun to be considered a valuable tool in the understanding of ocean-bottom soils. The investigation of soils in a simulated hydrospace environment has among its objectives the validation of ocean-soil properties measured under atmospheric pressure, the evaluation of methods for the stabilization of ocean-floor soils, the study of forces acting on a buried object subjected

FIG. 17-55. Electric motor for deep-submergence service in free-flooded condition prior to operational tests at 15,000 psi in a refrigerated simulated hydrospace vessel. (Courtesy U.S. Naval Civil Engineering Laboratory.)

to salvage operations, and the calibration of devices for *in situ* measurement of mechanical properties of soils.

Bio-organism testing

Bio-organism testing is also possible in the simulated hydrospace environment. At the present time, however, this testing is in its infancy. It would be possible, for example, to test marine life for the effects of temperature changes, pressure changes, changes in pH, alkalinity, etc., but equipment is not available yet which will allow this. Only in a well-regulated aquarium is anything like a real sea environment preserved, and even in such an artificially maintained environment only the surface or near-surface conditions vital to certain life forms are simulated.

In addition the real interest in this area is in the micro-organisms and not in such things as the fish life in the ocean depths. These microorganisms are the ones that eat the wooden structures, form colonies on the underwater cables, and advance corrosion of vehicles and components. To capture such minute organisms alive is the first problem; to keep them alive long enough to transport them to the hydrospace-simulation laboratory is a second problem; not to kill them by subjecting them to the simulated hydrospace, is the third. Even if these problems are overcome, the studying of the organisms presents yet another problem. At present, the study of these micro organisms would presuppose the availability of special microscopes which would be mounted partially inside the pressure vessel, and partially outside. To keep such microscopes in working order and to keep lenses clean in seawater environment would require the services of a host of additional devices not yet conceived. For these reasons, it will probably be some time before bio-organism testing reaches the level attained by the other types of testing.

Construction-processes testing

The construction processes testing differs from the other types of tests in one important aspect. Whereas the other types of tests conducted in simulated hydrospace deal with the study, evaluation, or prooftesting of materials, structures, equipment, soils, or organisms, the construction-processes testing deals mostly with processes. Equipment utilized to perform the construction processes is also studied, but only peripherally, for the attention is focused in these tests on the construction processes themselves rather than on the equipment. Among processes to be studied in simulated hydrospace will be welding and cutting, in-place mixing of concrete, and in-place casting of concrete, bonding of structural elements.

17.8 SUMMARY

The simulation of the hydrospace environment is an undertaking (Fig. 17-56) that is difficult, expensive, and attended by a certain degree of danger. At the present, it is really only possible to simulate adequately the pressure and temperature of the ocean depths, with some limited simulation of the seawater chemistry by using fresh seawater in an open-loop hydraulic system. The other limitations on simulating hydrospace, such as the heat-sink capability and shock-wave absorption, are not to be solved easily, or soon.

However, as vessels are constructed that are larger and more pressure resistant, and other components become available that are not only reliable, durable, and safe under the severe service encountered in operating a high-pressure and large-volume hydrospace-simulation facility, but are also more refined and ingenious, additional hydrospace parameters will be incorporated in the land-based simulated hydrospace system. The scientific

FIG. 17-56. Interior of a typical simulated hydrospace facility for testing of small models, or small equipment components under 20,000 psi hydrostatic pressure in temperature controlled seawater environment. (Courtesy U.S. Naval Civil Engineering Laboratory.)

principles describing the operation of simulated hydrospace are known, what is required is a high-level engineering effort to translate those scientific principles into an operational hardware system that will permit the creation of a simulated hydrospace indistinguishable from the hydrospace of the oceans. Certainly the technology capable of launching and retrieving man from outer space can also be directed to create simulated hydrospace of sufficient capacity and complexity to be capable of pretesting any engineering system or subsystem designed for the exploration and exploitation of hydrospace for man's welfare. There are indications that the dawn of such technological effort is on the horizon.

REFERENCES

Braun, H. E, and J. W. Wesner, Jr., 1965, "Study of Cable Wrapped Pressure Vessel for Nuclear Reactors," ASME-65-Wa/NE-4.

Bridgman, P. W., 1964, *Studies in Large Plastic Flow and Fracture, with Special Emphasis on the Effects of Hydrostatic Pressure*, Harvard University Press, Cambridge, Mass.

DeHart, R. E., 1965, "Environment Test Chambers for Submersible Vehicles," *Mater. Res. Stds.* 5, 11 (November 1965), ASTM.

DeHart, R. E., and N. L. Basdekas, 1963, "Feasibility Study for a Large High Pressure Anechoic Tank," SWRI Report 1178-3 (January 1963).

Harvey, J. F., 1963, *Pressure Vessel Design, Nuclear and Chemical Applications*, Van Nostrand, Princeton, N. J.

March, R. O., and W. Rockenhauser, 1965, "Prestressed Concrete Structure for Large Power Reactors," ASME-65-WA/NE-9.

Rowe, R. E., 1964, *Lecture 1. Proceedings: Model Testing*, Cement and Concrete Association, London, (March 17, 1964).

Schmidt, W. R., and A. G. Pickett, 1964, "Experimental Stress Analysis of a 1/6 Scale Model of an Anechoic Pressure Vessel," SWRI Report of 03-1178 (May 1964).

Stachiw, J. D., 1963, "The Birth of a Profession: the Transition from Mechanic to Engineer" doctoral dissertation, Pennsylvania State University.

Suarez, M. G., and N. M. Kramarow, 1965, "Large Prestressed Concrete Vessels for Deep Submergence Testing," ASME-65-WA/UNT-8.

Timoshenko, S.,P., 1963, *History of Strength of Materials*, McGraw-Hill, New York, (a detailed discussion).

INDEX

Acoustic noise, 449
Acoustic telemetry, 419
Acoustic waves, 253
Adiabatic temperature changes in seawater, 267
Aesthetics, in ocean engineering, 199
AGOR (-3), research ship, 358; (-14), (-15), 363; (-16), 366
Air lift, dredging, 497
Albacore, exploration submarine, 376
Alkalinity, seawater, 274
Alloys, identification by chemical composition, 628
Aluminaut, deep diving vehicle, 49, 374, 384
Alvin, 49, 374, 386, 540
Ambient pressure: aquanaut and astronaut exposure, 479; factor in engineering development for Man-in-the-Sea project, 482
Amphidromic point, 242
Amplitude, vertical motion in floating platforms, 564
Anaerobic corrosion, 272
Anchorage systems, 327
Angular momentum in water column, function of latitude, 219
Aquaculture, 17
Aquanaut, 485; environmental operations compared to astronaut, 478; team selection (Man-in-the-Sea), 486, training, 487
Archimède, 48
Asherah, 49
Atlantis II, research ship, 359, 400
Atmospheric corrosion, 624; steel, 588
Atolls, coral, 167
Auguste Piccard, 49
AUTEC, work vehicle, 546
Autofrettage for stress distribution in pressure vessels, 672
Azimuthal stability in floating platforms, 567

Bathythermograph (BT), 413
Batteries, instrumentation, 405
Beaches, source of commerical products, 300
Bearing thrust loads, current meter, 442
Biological cycles, carbon, nitrogen, phosphorus, silicon, oxygen, 288
Biological deterioration, materials, 625
Bio-organism testing, simulated hydrospace, 708
Biotelemetry, engineering problems, 485
Blood enzyme levels, physical and emotional stress indicator in divers, 485
Body-heat, loss in aquanaut, 488
Boiling point of seawater, 265
Bond, George F., 51
Boring, organisms, 296
Bottom, ocean: signalling device, 427; vehicle (crawling), 529
Bouée Laboratoire (Cousteau), 568, 571
Breathing mixture control, 479
Buoyancy-weight, relationship to deep submergence hull design, 379
Buoy, Giant, 471
Buoys, oceanographic data, 462

Calibration, current meter, 448
Capillary waves, 228
Carbon cycle, 279
Carbon dioxide, relation to pH in seawater, 272
Carpenter, Scott, 52
Cathodic effect: protection against, 600; steel in splash zone, 597
Cavitation erosion, 611

Centrifugal reaction, tidal motion, 238
Chassis, electronic: current velocity meter, 442; temperature gradient recorder, 432
Chemical content of the oceans, relationship to sediments, 168
Circulation, ocean, 169
Climate: intervention by man, 178; relation to changes in the ocean, 186
Coastline, element of sea power, 42, 43
Colligative properties, seawater, 265
Commercial products from the sea, 299
Communications, aquanaut, 488
Component approach to plans implementation, 91
Component testing in simulated hydrospace, 705
Composite reinforced structure, pressure vessel, 648
Compressibility: filling fluids in oceanographic instruments, 399; pressure hulls, 383; seawater in relation to deep-diving submarines, 350
Computer programming, plotting signal-noise ratios, 469
Concentration: distribution of seawater variables, 283, trace elements in marine organisms, 280
Concept evaluation, system design, 308
Concrete deterioration, 628
Concrete structures, 344
Conservation, living resources of the high seas, 147
Conservative properties of seawater, defined, 283
CONSHELF Two, Three, undersea habitat, 376
Construction, marine, long range technological forecast, 347
Construction processes, testing in simulated hydrospace, 709
Coring operations, soil mechanics investigations, 320
Coriolis deflection of water particles in wave motion, 241, 249
Coriolis effect of force on ocean circulation, 215
Corrosion: deep diving submarines, 350; effects of composition of steel, 589; special problem for ocean engineers, 197; steel affected by air-suspended salt particles, 594; steel in the tidal zone, 596
Coupling between disturbing forces, floating platform, 562
Cousteau, Jacques-Yves, 49
Crevice corrosion, 606
Criteria, selecting aquanaut team, 486
Cromwell current, 226
Current: effect on floating platforms, 248; meter, telemetering, 439; pattern relationship to wind systems, 214
CURV, underwater recovery vehicle, 533
Customs zone, 117
Cutting and welding, underwater, 513

Data channel proposals, 464
Data collection, 462; world-wide network, 463
Dealuminification, 616
Deep-diving capabilities forecast, 490
Deep-diving craft, 378
Deep Quest, 49
Deepstar, 49
Deep Submergence Systems Project, (DSSP), rescue submarine, 366
Defense, Department, effect on system planning environment, 58
Denickelification, 616
Denitrification in seawater, 271
Density, seawater, 209, 262
Density flows, a special problem for ocean engineers, 197
Density gradient, relation to ocean circulation, 169
Depths, relation to corrosion of steel, 598
Desalination, seawater, 200
Design considerations for marine installations, 313
Design philosophy for stable platforms, 569
Dezincification, 616
Diamond mining, 496
Diffusion coefficients, distribution of variables in seawater, 284
Disposable instruments, 410
Distribution, biologically altered constituents, 291
Diver work systems, 511
DOTIPOS, Deep Ocean Test Instrumentation Placement and Observation System, 324

714   Index

Dredging equipment, 493, 497
Drilling ship, mobile, 368
DSRV, rescue vehicle, 50, 390, 549
Dual radiance meters, 454
Ductility, pressure vessels, 679
Dynamic orientation systems, floating platforms, 568
Dynamic resistance to motion, forces producing motion, floating platforms, 563
Dynamic testing, simulated hydrospace, 704

Earthquake shock, special problems for ocean engineers, 194
Ekman wind-drift layer, 217
Electrical conductivity: measure of salinity, chlorinity, or density, 262; sea water, 264
Electrical noise, 408
Electromagnetic energy, attenuation in the ocean, 426
Electronics, early application to oceanographic instrumentation, 397
Elevation, effect on corrosion of steel, 595
E.M.F. sources, inside and outside the oceans, 408
Emplacements, seafloor structures, 337
End-closure, pressure vessels, 662
Energy spectrum, surface waves, 230
Equipment, aquanaut, 488
Evaporation, sea water, 176; effect on salinity, 205; latent heat, seawater, 267
Evolutionary cycle, in maritime society, 44
Exchanges across ocean boundaries: gases, solids, solutes, 282, 283; water, 281
Expendable bathythermograph, 418; costs, logistics, 421
Exploitation of the ocean, industry approach, 63
Explosion, shock wave propagation, 639
External environment, saturation diver, 481
Extra vehicular activity (EVA), astronaut compared with aquanaut, 478

Factors, needs analysis, 61
Falcon nozzle, diver operated, dredging, 514
Fatigue life, pressure vessels, 677
Feasibility study, sequence of steps, 308
Fiber glass reinforced plastics, 627
Filament-wound composite pressure vessel, 648
Fish, farming, 14
Fisheries: growth of harvest, potential harvest, kinds of fish, 10; maximum sustainable yield, 122; production by nations, continents, 14
Flare, hull design, research ships, 353
FLIP, floating instrument platform, 554, 570, 578
Float submerged, measurement of deep water flows, 223
Floating-rig, 506
Floor, deep-sea, source of commercial products, 301
Flotation materials, 382
Flows, deep water, measurement, fluctuation, 224
Fluid-breathing, long range future diving operations, 491
Fluid friction, effect on circulation, 219
Fluids, filling oceanographic instruments, 399
Foam, syntactic, 47
Foam-in-salvage, 521
Food, animal protein, current fisheries harvest, potential harvest, trophic levels, 8–10
Fouling: marine organisms, 296; materials, 608
Franklin, Benjamin, oceanographic investigator, 393
Freeboard, research ships, 353
Free-diving aquanaut, 478
Freedom of the seas, 114
Free-swimming manned vehicle, 535
Freezing point, sea water, 265
Friction, bearing operation in current meter, 442
Functional flow diagrams, 81

Galvanic effects, 613; series, 615
Gas: dissolved in seawater, 267; diving mixtures, 31; effect on properties of seawater when dissolved, 274; solubili-

Index   715

ties at air-sea interface, 273
Genesis Project, 52
Geostrophic motion, 215
Glaciation, 177
Glass: hull material, 51; massive structures, 627; properties under hydrostatic pressure, 429; reinforcement for plastics, 382
Glauconite mining, 501
Gold mining, 500
Government, element of national sea power, 42, 43
Graphitic corrosion, graphitization, 616
Group velocity, surface waves, 232
Gulf Stream, 220

Habitability, offshore fixed towers, 317
Habitats: manned, 341; saturation diver, 484
Hardrock ore mining, 502
Hardware: defined for system planning, 81; performance in Man-in-the-Sea program, 481
Hard-wire telemetry, 419
Hazard: bone damage in divers, 486; decompression sickness (bends), 479
Heat: flux through bottom sediments, 430; factor in engineering development for Man-in-the-Sea program, 482
Heave, ship motion, 564
Helium gas, pervasive qualities in diver equipment, engineering problems, 483
Hemispherical (convex, concave), heads for pressure vessel end closure, 663
High-pressure pumps, 587
History, hydrospace-environment simulation, 634
Hoisting equipment, hydrospace environment simulation laboratory, 685
Holm, Carl, 405
Howland, John, 47
Hull fraction, buoyancy-weight relationship, 380
Humidity, effect on corrosion of steel, 594
Hunley-Wischhoefer recovery system, 520
Hydrocarbons, power sources for oceanographic instrumentation, 403
Hydrocycle, great, 176
Hydrodynamic winch, 526

Hydrogen, use in future deep diving operations, 491
Hydrographic data, global distribution, 207
Hydrojet lift, dredging, 497

Ice formation, effect on salinity, 210
Implementation, system plans, 89
Implosion, 51; shockwave propagation, 639
Industrial wastes, 298
Industry, defined for systems planning, 58
Inertial forces, effect on tidal motion, 240
in situ: processes affecting seawater constituents, 288; recording radiance levels, 454; soil mechanics testing, 321
Instrumentation, simulated hydrospace, 693
Instrument damage, internal condensation, 435
Interdisciplinary functions, system planning, marine installations, 311
Intergranular corrosion, 619
Internal environment, saturation diver, 481
Internal waves, 244, 249
International Frequency Registration Board, (IFRB), 464
Investment, return (ROI), a planning factor, 77
Iron mining, 500
Irradiance meter, 452

Keach, Donald, 47
Kelvin edge wave, 242
Kinematic viscosity, breathing gases, effect on diver voice communications, 484

Law of the Sea: air transport, 124; contiguous zone, 145; fishing, 121; freedom of navigation, 119; Hague Conference (1930), 117; high seas, 146; innocent passage, 153; internal waters, 142; land under sea, 139; neutrality, 120; new weapons systems, 125; petroleum minerals, 125; Roman Law, 113; superadjacent atmosphere, 138; territorial sea, 143, 150; 12 mile fishing rule, 151
Layered-wall pressure vessel, 647
Licensing (Coast Guard), stable platforms, 569

Index

Lift attachments, heavy, 526
Lift pad, diver attached, 516
Limitations on simulation of hydrospace-environment, 641
Link, Edwin, 52
Luminescent organisms, 454
Luminous flux, measurement, 452
Lunar gravity, relation to tidal motion, 238

McCann rescue chamber, 366, 374
Magnetic field, 166
Mahan, Alfred Thayer, 38, 41
Maintenance, offshore fixed towers, 317
Manganese nodules, mining, 501; future prospects, 501
Man-in-the-Sea, 482, 53; aquanaut team selection, 486; critical engineering problems, 482
Manned habitats, 341
Marine environment, design of stable platforms, 558
Marine vehicles, competition from aircraft in ocean transportation, 352
Marketing, ocean exploitation, 64
Martin, George, 47
Mass movements, submarine slope failures, 325
Materials engineering: effect on instrument design, 440; testing in simulated hydrospace, 698
Mathematical models, relation to hydrospace-environment simulation, 635
Mazzone, Walter, 51
Microelectronics, 426
Micronutrients in seawater, 275
Military use of the ocean, 32
Mineral resources, 17, 19, 299
Mining systems, 493; platforms, 496
Missions, military, 69; deep submergence, 92; selection for system planning, 79
Mixing, relation to δ_t, 210
Mobile drilling platform, 506
MOBOT, manipulator-operated system, 507
Models, types for hydrospace-environment simulation, 643
MOHOLE Project, 373
Moisture damage to instruments, 435
Mooney, Bradford, 47

Moray, 49

Naval requirements, influence on development of ships and submarines, 352
Navigation requirements, mining operations, 496
Needs: analysis in system design, 308; military, 67; non-military, 6, 66, 69; ocean exploitation, 72; ocean system, 70
Nil ductility, transition temperature, 679
Nitrogen, cycle, 278; fixation in the ocean, 271
Nodules, manganese, 20, 501
Noise, acoustic, 449; electrical, 408
Non-ferrous materials: atmospheric corrosion, 624; guides to selection for ocean engineering, 602
Notch toughness, 679
NR-1, nuclear powered research submarine, 49
Nuclear power, oceanographic instrumentation, 407
Nuclear-powered submarine, support vehicle for diver operations, 485
Nutrient cycles in the sea, 276

Occupation of the sea bed, 33, 55
Ocean engineering, 3, 194; compared to terrestrial engineering, 197; design, managing, planning, 3; education, 3; problems identified, 199
Ocean-floor, area distribution as function of depth, 351
Oceanographic research ships, requirements, 352
Off-bottom wellheads, 509
Oil-well drilling, 370; completion, 505
Optical telemetry, 419
Organisms, diurnal migration, interaction with internal waves, 182
Orientation effect, corrosion of steel, 595
Oscillatory motions, floating platforms, 557
Osmotic pressure, in sea water, 266
Oxygen: apparent utilization (A.O.U.), 289; monitoring and flow control for saturation divers, 490

Packaging, electronic instrumentation, 397

Index 717

Parameters, hydrospace simulation, effect on testing facilities, 640
Partial pressure, breathing gases, relation to inert gas in blood solution, 479
Passive resistance, motion or forces producing motion in floating platforms, 563
Perry, H. L., 51
Petroleum deposits, exploration, 19
pH, seawater, 274
Phosphorite mining, 500, 502
Phosphorus cycle, 277
Photo cells, 402
Photosynthesis (and respiration), effect on oxygen and carbon dioxide concentrations, 269
Physical qualities, required in aquanauts, 486
Physiology, of man: application to sea bed strategy, 47; as a free swimmer, 51
Phytoplankton, effect on depth of light penetration, 269
Piccard, Auguste, 47
Piccard, Jacques, 49
Piping, hydrospace-environment simulation facilities, 690
Piracy, 118
Pitch damping, 355
Pitting, materials exposed to ocean environment, 603
Planning, systems approach, 1, 4; advance planning for industry, 65; factors, 74; marine installations, 311, 314; methodology, 73; planning defined, 59
Plate bearing tests, soil mechanics investigations, 322
Platforms, offshore, 41; stable, 368; superstable, 200
Platinum mining, 500
Policy, national: Marine Science Affairs (Public Law 89-454), 1; goals, 2
Political situation: Argentina, 129; Chile, 129; Costa Rica, 129; Ecuador, 130; El Salvador, 129; Iceland, 132; international, 1945-58, 126; Mexico, 128; Peru, 129; United States, 127
Polyethylene, material for fabrication of oceanographic instruments, 441
POP, perpendicular ocean platform, 576
Population, requirements, 6

Potential energy, power source for oceanographic instrumentation, 404
Power: atomic fission, fusion, 26; from the sea, 25; thermal, 26
Power sources for diver habitat, 484
Power-stud-attached lift pad, 516
Pressure, effects on instrument components, 400
Pressure-equalized instruments, 399
Pressure-hull, materials, 381
Pressure-protected instruments, 398
Pressure vessel, hydrospace simulation, 638, 639, 645, 646, 681
Pressurization, pressure vessels for hydrospace-environment simulation, 687
Problem definition, system design, 308
Processes in natural ocean systems: affecting distribution of chemical properties of seawater, 281; continental drift, 163; convective transport, 163; geothermal energy, 160; key to the origin of the planet, 161; volcanic activity, 164
Production facilities, deep-water wells, 509
Productivity of the sea, devices for increasing, 173
Protective measures, corrosion of steel, 598
Psychological qualities, required in aquanauts, 486
Pyrheliometer, measurement of radiant energy, 451

Radiant-energy recorder, 451
Radioactive wastes, 180, 181, 298
Radio frequencies for telemetering oceanographic data, 461
Radio telemetry, 419
Reactor, deep-ocean emplacement, 341
Readout unit, current meter, 445
Recording, temperature gradient recorder, 436
Recreation, 31
Refraction, wave, 251
Refractive index, seawater, 264
Regulators, pressure and rate of flow, 691
Reliability, oceanographic data communications, 466
Remote-controlled vehicles, 529, 533
Requirements: allocation sheets, 81, 96–

98; design, 96; large-scale system design and interdisciplinary aspects of ocean engineering, 308
Resonant period, free oscillation in floating platforms, 565
Resources, marine: conservation, 8, 36; defined, 7; food, 8; ownership (jurisdiction), 34; utilization, 33
Retrieval, seafloor structures, 337
Rigging, emplantment and retrieval of bottom resting structures, 331
ROBOT, manipulator-operated system, 507
Roll stabilization, platforms, 356
Ropes, deterioration, 622
Rotor, ocean current sensor, 439
RUDAC, remote drilling and completion system, 507

Safety parameters, pressure vessels, 675
Salinity, 204, 260; measurement, 206
Salt fountain, 176
Salvage: diver, 512; heavy lifts, 524; pontoon, 519
Satellites, oceanographic data, 462
Saturation diving, 480
Scale model testing, 637
Scattering, acoustic energy, 254
Sea, use, 39; constraints, Law of the Sea, speed of transit, 39; geopolitical and geologistic relationships, 39
SEA LAB, (I), (II), 52, 480; (III), 519
Sealing techniques, pressure vessel end closure, 664
Sea power: introduction of nuclear power, steam, 47; lake and river societies, 45
Seas, Convention on High Seas, 35
Sea state, power spectra, 353
Seawater, nature, 259; chemical constituents, 261; at high temperature and effect on materials, 622; properties dependent on salinity, 262
Sea-worthiness, oceangoing research ships, 353
SEDCO-135, oil drilling rig, 370
Sediments, 168; geological record of climate changes, 189; measurement of temperature gradients, 430; relation to chemistry of sea, 168
Segmented-wall pressure vessel, 648

Seismic disturbances, source of acoustic background noise in the sea, 450
Selective corrosion, 616
Separate-wall pressure vessel, 647
Sewage, domestic, 297
Shallow-diving submarines, 375
Shear strength tests, soil mechanics investigations, 321
Shelf, Continental, source of commercial products, 300
Shelf, Convention on the Continental Shelf, 34
Shelter, effect on corrosion of steel, 595
Ship motion, 353
Shrimp, source of acoustic background noise in the sea, 450
Shrink fit, technique for distributing stresses in design of pressure vessels, 670
Shumaker, Lawrence, 47
Simulation of hydrospace-environment, reaction on models, 636, 638
Soil mechanics: sea-floor investigations, 317; testing in simulated hydrospace-environment, 705
Solar energy flux at sea surface, 170
Solar power for oceanographic instrumentation, 404
Solar tides, 240
Solid-wall pressure vessel, 646
Solubility, gases in seawater, 268
Solutes, seawater, 260
Sonnenberg, Robert, 52
Sound: fixing and ranging (SOFAR), 254; speed of transmission in seawater, 264
Souscoupe, exploration submarine, 49, 376
Space diversity, communications reliability, 466
SPAR, sea-going platform for acoustic research, 573
Specific heat, seawater, 266
Speed, phase, gravity waves, 229, 232
Spillhaus, Athelstan F., 416
Splash zone, effect on corrosion of steel, 597; protective measures, 598
Stabilization, ships and platforms, 554
Stable platforms, needs, 554
Stacked-ring pressure vessel, 648
Stagnation, seawater, 294
State-of-the-Art (SOTA), planning factor, 77

Steels, ocean engineering materials, 588
Strategy, business, planning, 85
Stratification, density, effect on wave motion, 245
Strength of materials, relation to development of hydrospace-environment simulation, 636
Strength-to-weight ratio, pressure hulls, 382
Stress, diver, psychological and physiological, 485
Stress corrosion, 619
Stresses, pressure vessels for hydrospace-environment simulation, 668
Structural characteristics, offshore fixed towers, 315
Structural testing, simulated hydrospace, 699
Stud gun, diver tool, 515
Submarine rescue ship, (ASR), 366
Submerged installations, future applications, 338
Submersible Test Unit (STU), system components, 336
Surface-tethered vehicles, 533
Surge, storm, 234
Surge, and sway, floating platforms, 566
Swimmer propulsion, 490
Systems approach: design, 307; plans implementation, 91
Systems development, background, planning, 60

Technology: factor in system planning, 82; maritime, element of sea power, 43
Telecommunication, high frequency, oceanographic data, 462
Telemetry, 419
Temperature, ocean: effect on corrosion of steel, 594; function of latitude, 207; function of maximum seawater density, 263; relation to climate changes off California coast, 186; variation with depth, 206
Temperature-gradient recorder, 430
Territorial sea, cannon-shot rule (Law of the Sea), 115
Test, submersible unit (STU), 328
Testing, simulated hydrospace environment, 698

Tethered manned vehicles, 530
Thermal conductivity, effect in Sea Lab II experiments, 483
Thermal probe, 431
Thermocline, 206; relation to geostrophic flows, 218
Thermohaline flow, 212
Three-mile limit, 115
Thresher (USS), 47, 390
Tidal crest, progress, 243
Tidal zone, protective measures, steel corrosion, 599
Tides, 235
Tin mining, 499
Tools, diver, 489, 513
Topology, land and water masses, determinant of sea power, 42
Towers, offshore fixed, 315
Trace elements, seawater, 280
Training aquanauts, 487
Transportation, ocean-borne, 26; ship design, construction, 27; system optimization, 28
Treaty of the Continental Shelf, 54
Trichodesmium, algae effective in the fixation of nitrogen, 271
Trieste, bathyscaphe, 48; (*I*), 384; (*II*), 384, 537
Tsunami, 233

Undersea habitations, effect of helium gas properties, 482
Undersea mining systems, future, 501
Underwater vehicles, 374
United Nations, Convention on Law of the Sea, 39; fishery conservation, 133; International Law Commission (1947), 130
Unmanned operation of floating platforms, 575
Upwelling, near coastlines, 225

Vaisala frequency, relation to internal waves, 245
Vapor pressure, seawater, 265
Variables, distribution in the sea, 283
Vehicle-manipulator systems, 528
Vehicles, manned, sea bottom strategy, 47
Velocity, gradients and distribution of

variables in the sea, 284
Velocity effects: and corrosion of steel, 601; and materials, 610
Vertical oscillation, floating platform response to wave period, 564
Vine, Allyn, 400
Viscosity of seawater, 267
Visibility, aquanaut, 488
Voice distortion, related to breathing mixtures for divers, 483

Waste disposal: effects on environment, radioactive isotopes, pesticides, heat, 29; heat, solids, 30; pollution at sea, 297
Wasting of materials, ocean environment, 603
Water, desalination of seawater, 24; economics, 23; fresh, 23
Water mass analysis, 212
Water-transparency measurements, 411

Water transport, Gulf Stream, 221
Wave, motion, 226; effect on floating platforms, 559; factor in design of surface ships, 351; source of electrical noise, 408; source of power for oceanographic instrumentation, 404
Weaponry, uniqueness of technology, 64
Welding and cutting underwater, 513
Wellhead completion, 507
Wenk, Edward, 49
Wind, distribution, 213; effect on floating platforms, 558; relation of wind system to current system, 214; stress, 216
Wires, corrosion, 622
Wirewrap, technique for distributing stresses in pressure vessels, 648, 673
Workman, Robert, 51
Work manipulators, vehicle-mounted, 542
Work systems, deep-ocean, 493
World Administrative Radio Conference (WARC), 462

Date Due

WITHDRAWN